OXYGEN TRANSPORT
TO TISSUE – V

ADVANCES IN EXPERIMENTAL MEDICINE AND BIOLOGY

Recent Volumes in this Series

A Continuation Order Plan is available for this series. A continuation order will bring delivery of each new volume immediately upon publication. Volumes are billed only upon actual shipment. For further information please contact the publisher.

OXYGEN TRANSPORT TO TISSUE—V

Edited by

D. W. Lübbers
H. Acker
E. Leniger-Follert

Max Planck Institute for System Physiology
Dortmund, Federal Republic of Germany

and

T. K. Goldstick

Northwestern University
Evanston, Illinois

PLENUM PRESS • NEW YORK AND LONDON

Library of Congress Cataloging in Publication Data

Main entry under title:

Oxygen transport to tissue — V.

   (Advances in experimental medicine and biology; v. 169)
   "Proceedings of the meeting of the International Society on Oxygen Transport to
Tissue, held September 15-17, 1982, in Dortmund, Federal Republic of Germany" — P.
   Includes bibliographical references and index.
   1. Oxygen transport (Physiology) — Congresses. 2. Oxygen in the body — Congresses.
I. Lübbers, D. W. II. International Society on Oxygen Transport to Tissue. III. Title:
Oxygen transport to tissue — five. IV. Series.
[DNLM: 1. Biological transport — Congresses. 2. Oxygen — Blood — Congresses. 3.
Oxygen consumption — Congresses. W1 AD559 v.169/QV 312 I6120 1982]
QP99.3.O90936   1984   599′.011   83-27049
ISBN-13: 978-1-4684-1190-4   e-ISBN-13: 978-1-4684-1188-1
DOI: 10.1007/978-1-4684-1188-1

Proceedings of the meeting of the International Society on
Oxygen Transport to Tissue, held September 15-17, 1982, in
Dortmund, Federal Republic of Germany

©1984 Plenum Press, New York
Softcover reprint of the hardcover 1st edition 1984

A Division of Plenum Publishing Corporation
233 Spring Street, New York, N.Y. 10013

PREFACE

On the understanding that few people ever read the preface to any book and also on the understanding that even those few people who do read the preface realize that virtually nothing of any substance is ever said, I shall write at such length as will be proportional to my expected readership.

The meetings of the International Society on Oxygen Transport to Tissue provide a forum for discussion amongst scientists who, although being from very diverse and specialized backgrounds, have tissue oxygenation as a unifying theme of interest.

The wide variety of research material presented in this volume and the multiplicity of the experimental techniques described, should serve as an adequate gauge to the range of expertise and knowledge of the society's members. Such diversity should also stress the importance of the need for multidisciplinary approaches to complex biological problems.

In attempting a fundamental characterization of a biological process such as tissue oxygenation, the application of very many separate research skills are necessary, such as mathematics, engineering, biophysics, biochemistry, physiology, histology and clinical medicine. The success of the ISOTT has - and we hope - will continue to be causing a combination of individuals to direct their specialized knowledge to the many facets of a single process - tissue oxygenation.

All of the papers in this volume have received some editorial revision and, where queries have arisen, questions have been put directly to the authors in the hope that their subsequent replies will help to clear up uncertainties. Where uncertainties remain, the reader is of course recommended to take issue with the authors directly.

Grateful thanks go to Mr. A.J. BAKER and Mr. J.F. O'RIORDAN for valuable editorial assistance. Gratitude is also due to our secretarial staff Frau E. MENNE, Frau D. SÄNGER-KRAUSE and Frl. D. MÄGDEFESSEL for the most taxing of duties, the preparation of the manuscripts.

The society is indebted to the MAX-PLANCK-GESELLSCHAFT for the generous support of the meeting and the assistance in editing this book.

In addition, our meeting was generously supported by the following firms:

Dr. THIEMANN GmbH, 4760 Lünen

H.C. BÖHRINGER, 6507 Ingelheim

SANDOZ AG, Ch-4002 Basel

BÖHRINGER MANNHEIM AG, 6800 Mannheim

BAYER AG, 5600 Wuppertal

PICKER INTERN. GmbH, 4992 Espelkamp

CARL ZEISS, 7082 Oberkochen

SCIENCE TRADING, 6000 Frankfurt

H.P. HORN, 7080 Aalen

ERNST LEITZ GmbH, 5000 Köln 1

WALDECK, 4400 Münster

For the editors

D.W. Lübbers

CONTENTS

BLOOD AND OXYGEN TRANSPORT

BRAIN

CONTENTS

OXYGEN RADICALS

MUSCLE

ABDOMINAL ORGANS

TUMOR

MAIN LECTURES

FACILITATED DIFFUSION OF OXYGEN:

POSSIBLE SIGNIFICANCE IN BLOOD AND MUSCLE

F. Kreuzer, and L. Hoofd

Department of Physiology, University of Nijmegen
Nijmegen, The Netherlands

Facilitated diffusion of a permeant is transport with a rate faster than that according to the diffusive conditions of the permeant. This enhancement may be effected by a molecular carrier reversibly reacting with the permeant. The kind of facilitated transport where the permeant is transported both in free form and combined with a carrier (as pertains to oxygen) is called "carrier-mediated transport" (Schultz et al., 1974). Such a carrier shows saturation behavior due to the finite concentration of carrier and can be inhibited by molecules structurally similar to and thus competing with the carried species. The total transport is the sum of the two components, free diffusion and the permeant and the diffusion of the carrier loaded with the permeant. These two movements obey their respective concentration gradients which are not independent but related by the rates of chemical reactions between permeant and carrier. Carrier-mediated transport therefore belongs to the class of reaction-enhanced transport and is analyzed as a problem of diffusion coupled with reversible chemical reactions. After the establishment of facilitation of O_2 diffusion in the presence of hemoglobin or myoglobin by Klug et al. (1956), Wittenberg (1959) and Scholander (1960), much experimental and theoretical work during the past two decades has clarified the mechanism and the conditions of this phenomenon. It is the purpose of this brief review to give an impression of the present state of the art and to discuss some physiological implications. The emphasis will be on the diffusion of O_2 facilitated by hemoglobin or myoglobin.

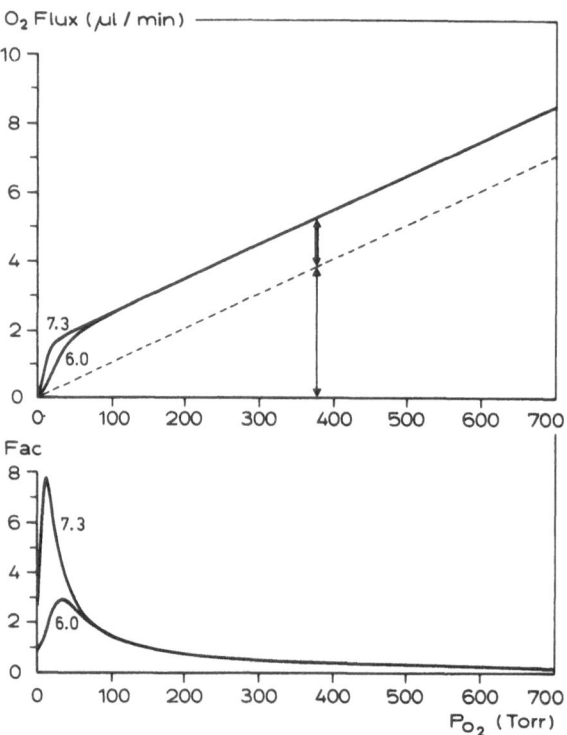

Fig. 1. Basic characteristics of facilitated diffusion of oxygen.
Upper panel: Steady-state oxygen fluxes (in ul/min) through
a layer of hemoglobin solution vs. O_2 pressure Po_2 at
chemical equilibrium in the absence of back pressure at the
low-pressure side. Millipore membrane 150 um, Hb concentration 19 g%, temperature 25°C. The plain diffusion component
(Hb inactivated as metHb, broken line) of the total O_2 flux
(solid line) is indicated by the light arrow, the hemoglo-
bin-augmented component by the heavy arrow. In the ascending
part of the facilitated flux the two curves reflect two
HbO_2 dissociation curves according to pH 7.3 and 6.0 respectively. Lower panel: Dimensionless relative facilitation
(total flux - plain flux)/plain flux = facilitated flux/
plain flux vs. O_2 pressure Po_2. Maximum facilitation is
about 8 at pH 7.3 and about 3 at pH 6.0. The maximum at pH
7.3 occurs at a lower O_2 pressure than at pH 6.0 due to
the respective position of the HbO_2 dissociation curve.
The two curves do not pass through the origin where the
quotient mentioned above is 0/0 or undefined. From Kreuzer
and Hoofd (1982).

BASIC MECHANISM

The basic features of facilitated O_2 transport are visualized
in Fig. 1 for the system Hb + O_2 (Kreuzer and Hoofd, 1982). The
plain diffusive O_2 flux in the solution with Hb inactivated as
metHb (upper panel, broken line) is linearly related to Po_2 as
expected. The line for total (facilitated + plain) O_2 flux (upper
panel, solid line) rises from zero with increasing Po_2, but its
slope decreases as the point of saturation of the carrier with O_2
is reached; then the line runs parallel to that of plain diffusion
with further increase of Po_2 (at chemical equilibrium), i.e., the
enhancement becomes constant (heavy arrow). The ascending part of
the facilitated O_2 flux is a result of the prevailing O_2 dissociation curve (ODC) and thus depends on the pH, whereas above the
point of saturation and the facilitated O_2 flux becomes independent
of the ODC. The dimensionless relative facilitation, usually expressed as (total flux-plain flux)/plain flux = facilitated flux/
plain flux, plotted against Po_2 (lower panel) starts with an undefined value (due to 0/0), reaches a maximum depending on the position
of the ODC (i.e., on the pH here), and then gradually declines. The
maximum relative enhancement is found with low Po_2 values, and the
O_2 flux due to HbO_2 will depend not only on the Po_2 difference across
the layer but also on the absolute pressures at the layer boundaries.
Below the point of saturation the facilitated O_2 flux will depend
on the position and slope of the operating range of the ODC. Higher
affinity of the carrier for the permeant will promote facilitation
at low pressures, but also renders the system more sensitive to back
pressure. Above the point of saturation, however, the facilitated
flux does not increase further with rising Po_2 even though the
HbO_2 profile steepens and moves toward the low-pressure side; this
apparent paradox is resolved by the facilitation being independent
of the ODC in this range. In the fully oxygenated part of the membrane the enhanced transport is carried by an increased Po_2 gradient due to the flux being augmented by the HbO_2 gradient near the
low-pressure boundary. Thus the two gradients contribute to the
total O_2 transport (equal in all planes of the layer) very differently in the upper and lower portions of the layer, plain diffusive
flux being dominant near the high-pressure side and carrier-mediated
flux dominating near the low-pressure side (Kreuzer, 1970; Wittenberg, 1970).

QUANTITATIVE INTERPRETATION OF FACILITATED O_2 DIFFUSION

Carrier-mediated transport due to translational carrier diffusion and reversible permeant-carrier reaction throughout the membrane is a classical diffusion-reaction problem with a bimolecular
chemical reaction between the carrier C and the permeant S to form
a carrier-permeant complex CS. The differential equations coupling
diffusion and reversible chemical reaction are based on local

material balance or conservation of species at steady state. The
assumption that the diffusion coefficients D of free and combined
carrier are equal and independent of substrate concentration leads
to the constraint that the total carrier concentration is constant
everywhere in the layer. A partial solution provides an exact relationship between the flux and the boundary concentrations of
permeant and permeant-carrier complex:

$$D_S \ (C_{S_{x=0}} - C_{S_{x=L}}) + D_{CS} \ (C_{CS_{x=0}} - C_{CS_{x=L}}) = J \ L \qquad (1)$$

where D = diffusion coefficient, C = concentration, x = 0 at high-
pressure boundary, x = L at low-pressure boundary, J = flux, L =
layer thickness.

This equation clearly shows the total flux J to be the sum of
the fluxes of the permeant and the permeant-carrier complex. A final
solution requires a definition of the concentrations of both species
(S and CS) at the upper and lower boundaries. The boundary conditions for both boundaries include Henry's law (possibly corrected
for interfacial effects) and the fact that the boundaries are imper-
vious to carrier C and carrier complex CS, i.e., the species C and
CS are confined to the layer, whereas species S can enter and leave
the layer. The boundary concentrations of S are under direct experimental control, whereas those of C and CS are not. A direct solution
of equation (1) can be obtained only at chemical equilibrium due to
infinite reaction rates, using the equilibrium relationship between
CS and S (ODC). Whenever the reaction rates are finite, however,
specific numerical solutions or analytic approximations must be
developed to solve this system of nonlinear (due to the bimolecular
kinetics with saturation behavior) second-order differential equations which cannot be solved analytically.

Numerical solutions have the advantage of being applicable
also to nonlinearities in the equations and of providing solutions
of any desirable accuracy in principle. However, they are not generally applicable and do not provide any information about mechanisms,
require a priori specification of parameters and may involve excessive computer time (Goddard, 1977).

When trying to account for reaction rate limitations one is
confronted with problems involving "stiff" differential equations
(for situations including steep local gradients) of the "boundary-
layer" or "singular-perturbation" variety (Goddard, 1977). In order
to appreciate the possibilities and limitations of approximate
analytic solutions it may be useful to introduce the Damköhler
number which permits one to evaluate the relative contributions of
diffusion and chemical reaction to a mixed process. This number is
defined as the ratio of characteristic or relaxation times of
diffusion and reaction, τ_D and τ_R respectively:

$$\gamma = \frac{\tau_D}{\tau_R} = \frac{\Theta L^2}{CD} = \left(\frac{L}{\lambda}\right)^2 \tag{2}$$

where Θ = typical reaction rate and $\lambda = \sqrt{CD/\Theta}$ is a characteristic reaction-diffusion (reaction-layer) length scale (according to Friedlander and Keller, 1965).

A large Damköhler number characterizes the near-equilibrium regime where the chemical reaction is so predominant that it determines the course of events and the facilitated flux approaches its maximum corresponding to chemical equilibrium. A small Damköhler number characterizes the near-diffusion regime. Here the residence time of reacting species in the layer may be so short as to preclude an appreciable chemical reaction, so that diffusion prevails over chemical reaction. In this regime, facilitation is much smaller than the maximum possible, although it may be substantial compared to with free diffusion. In the limit of $\gamma \rightarrow 0$ the near-diffusion regime provides the nonreactive (passive) physical permeability for the permeant.

Approximate analytic solutions are possible and may be useful for the two extreme asymptotic regimes, the near-diffusion regime and the near-equilibrium regime. In the limiting case of chemical equilibrium an analytic solution can be obtained. Over the entire range of Damköhler numbers, an analytic treatment with linearized kinetics is possible when the concentration gradient becomes very small (Friedlander and Keller, 1965).

Departure from chemical equilibrium due to finite reaction rates must imply nonequilibrium near the boundaries (experimentally confirmed by Weigelt, 1975). For, because S can cross the boundaries whereas C and CS cannot, there must always be an association of S with C at the high-pressure side and a dissociation of S from CS at the low-pressure side. Thus three parts of the layer can be distinguished: two boundary zones with thicknesses determined by the penetration depth λ and a middle core as shown in Fig. 2 for carrier Mb and permeant O_2. At the boundaries the gradient of S or P_{O_2} carries the total flux (no gradient in CS or saturation) and is equal on both sides at steady state (solid lines in Fig. 2). The boundary zones can be interpreted as boundary layer resistances (Gijsbers and van Ouwerkerk, 1976). Two different approaches have been discerned and identified as weak and strong boundary layer methods (Schultz et al., 1974) which, however, cannot be discussed in detail here. The earlier work of Kreuzer and Hoofd (1970, 1972), which was recognized as a useful tool for application to numerous situations (Goddard, 1977), led to a very generally applicable solution by Hoofd and Kreuzer (1979) (see also Fig. 3 below). To complete the discussion of Fig. 2, in the strong boundary layer method there is equilibrium in the middle core, but this equilibrium is uncoupled from the boundaries and is coupled to the nonequi-

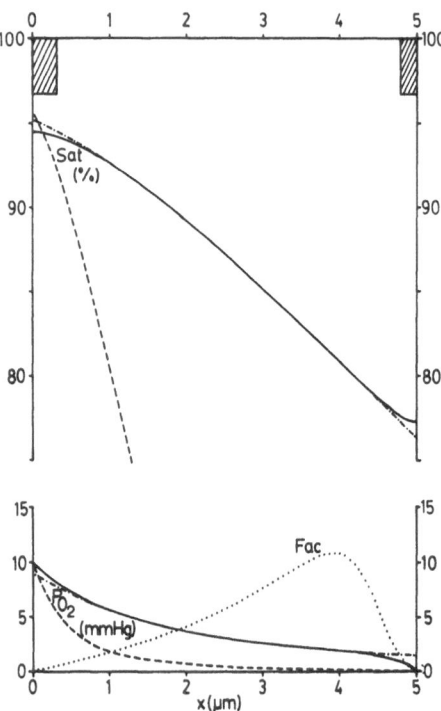

Fig. 2.   Profiles of O_2 saturation, (Sat, top), O_2 pressure (Po$_2$,
          bottom), and nonequilibrium facilitation (Fac, dotted line)
          plotted against distance (x) in a 5 um thick layer of a
          15 g% Mb solution at steady state, exposed to a Po$_2$ of 10
          (left) and 0 (right) Torr respectively. The curves are
          scaled for the same flux contribution. The saturation scale
          only covers the range from 75 to 100 %. Solid lines are
          calculated profiles for nonequilibrium; at the boundaries
          the slopes of Po$_2$ are equal (steady state) and the slopes
          of MbO$_2$ saturation are zero (no flux). Broken-dotted lines
          are equilibrium core extensions whereas broken lines show
          profiles for equilibrium presumed throughout the layer;
          the broken line for saturation starts at an equilibrium
          value of less than 100% because a Po$_2$ of 10 Torr is not
          sufficient for complete saturation, and would continue to
          zero on the right side. Hatched areas at the top indicate
          the penetration depth λ for non-equilibrium; although this
          penetration depth is smaller at the low-pressure boundary,
          its effect of raising the MbO$_2$ saturation is enormous
          (from 0% at equilibrium to 77% at nonequilibrium) due to
          the steep course of the MbO$_2$ dissociation curve in this
          range. From Kreuzer and Hoofd (1982).

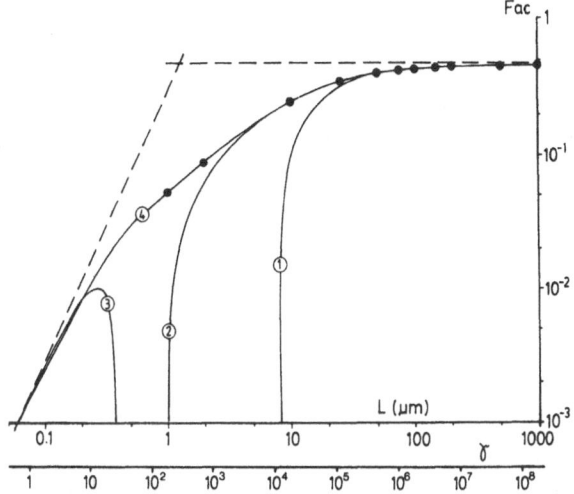

Fig. 3. Plot of facilitation Fac against layer thickness L and
         Damköhler number γ for a Hb solution of 15 g% and a Po_2
         difference of 200 vs. 2 Torr. The two broken lines are the
         asymptotes; the near-diffusion asymptote (left) coincides
         with a thin-layer first-order solution and the near-equi-
         librium asymptote (top) with thick-layer zero-order solu-
         tions (maximum facilitation). The individual lines indicate:
         1. Weak boundary layer first-order (Goddard et al., 1970;
            Mitchell and Murray, 1973).
         2. Strong-layer first-order (Smith et al., 1973).
         3. Strong-layer second-order (Smith et al., 1973).
         4. Wide-range, improved for flat layers (Hoofd and Kreuzer,
            1979). Dots are numerical results of Kutchai et al.
            (1970). Simplified from Kreuzer and Hoofd (1982).

librium boundary zone solutions; extension of the core solution
provides virtual values of Po_2 and saturation (broken-dotted lines).
Broken lines show profiles for equilibrium presumed throughout the
layer; the broken line for saturation starts at an equilibrium value
of less than 100% because a Po_2 of 10 Torr is not sufficient for
complete saturation and would continue to zero on the right side;
note the enormous difference in saturation on the low-pressure side
between nonequilibrium and equilibrium. Hatched areas at the top
indicate the penetration depth λ for non-equilibrium. The dotted
line "Fac" demonstrates the nonequilibrium facilitation along the
distance in the layer.

    Fig. 3 presents a comparison of the performance of various
quantitative interpretations of facilitated O_2 diffusion for the
entire range of Damköhler numbers. There is no space here to enter
into details. What is important here, however, is to point out the
approach of the various solutions to the limiting regimes of near-

diffusion (small Damköhler numbers, left) and near-equilibrium (large Damköhler numbers, top), and the problems occurring in the middle range of intermediate Damköhler numbers, where the various solutions can be seen to be of very different validity.

All the experimental and theoretical studies referred to so far were conducted on models under conditions often widely different from physiological circumstances. They were indispensable to clarify the concepts for both biological and chemical engineering problems, but for the biologist the question of a possible physiological importance of facilitation is imperative.

POSSIBLE SIGNIFICANCE OF FACILITATED O_2 DIFFUSION IN THE BLOOD

The conditions in the interior of the red blood cells (RBC) containing the oxygen carrier Hb differ markedly from those in most model arrangements. The Hb is surrounded by a membrane, it occurs in high concentration (about 35 g%) and in a thin layer corresponding to the small size of the RBC. Besides, the Hb in the RBC is exposed to the influence of effectors (H^+, CO_2, organic phosphates, small anions), ionic fluxes and a potential across the membrane. Early experiments on RBC showed marked enhancement (review by Kreuzer, 1970, and Kreuzer and Hoofd, 1982) but in some cases the results may have been impaired by the presence of hemolysis.

Since there is much evidence that the diffusive resistance of the RBC membrane for O_2 is negligible (Kreuzer and Yahr, 1960), a heterogeneous two-media theory can be applied to a suspension of RBC in plasma. The application of heterogeneous media theories showed substantial facilitation, though well below its value at equilibrium. Under physiological conditions (hematocrit 45, driving force from 95 to 40 Torr Po_2, 37°C), however, equilibrium may be presumed with a maximum facilitation of 15% (Stroeve et al., 1976); facilitation would be considerably larger at hypoxia.

In nonsteady-state systems which are more important physiologically, deviations from chemical equilibrium may be more marked. Kutchai (1970; 1971) calculated some facilitation in the initial rate of O_2 uptake by a Hb solution simulating the conditions in the RBC. Gonzalez-Fernandez and Atta (1981) computed that for diffusion path lengths of 0.75 to 1 um the hemoglobin-facilitated O_2 flux in capillaries contributes some 30% of the total O_2 flux at normoxia and ca. 60% at hypoxia.

It is well known that in flowing blood the RBC are deformed more or less according to the velocity of flow. Therefore there must be considerable mixing of the contents of the RBC under certain conditions. This mixing would promote the transport of O_2 and may be more effective than facilitation. A quantitative approach to the mixing problem has been tried only recently (Diller and Mikic, 1980).

POSSIBLE SIGNIFICANCE OF FACILITATED O_2 DIFFUSION IN MUSCLE

   The situation in muscle is even more complicated than that in
the RBC. For a possible carrier function of Mb not only the concentration of Mb is important but also the fact that muscle is highly
structured and contains a high concentration of other proteins
(about 20 g%). Distribution and mobility of Mb in the cell remain a
controversial issue (Kreuzer, 1970; Wittenberg, 1970). The Mb diffusion coefficient has been measured in solution and in muscle homogenates but not in intact muscle so far.

   The requirements for physiological facilitation of O_2 transport by Mb are: 1) Mb must be present in sufficiently high concentration; 2) Mb must be mobile, at least partially; 3) there must be
a partial (preferably marked) desaturation along a Po_2 gradient in
muscle at low Po_2 in view of the high Mb-O_2 affinity. The transport
of O_2 in respiring muscle includes three basic processes, free diffusion of O_2, diffusion of MbO_2 coupled with reversible reaction
between Mb and O_2 (facilitated O_2 diffusion), and the irreversible
reaction of O_2 consumption in the mitochondria.

   Wittenberg et al. (1975) measured the steady-state O_2 uptake
of fiber bundles from resting pigeon breast muscle in a chamber
perfused with a succinate-containing Krebs-Ringer solution at $37^{\circ}C$.
The O_2 consumption with Po_2 values below 150 Torr was reduced by
inactivation of Mb to about half the value found with active Mb,
leading to the conclusion that Mb in intact muscle facilitates the
O_2 transport to the mitochondria. However, a mathematical treatment
was impossible and therefore no quantitative data in terms of diffusional parameters were presented. Later the same group (Cole et al.,
1978) failed to demonstrate any difference in O_2 uptake with and
without nitrite inactivation of Mb below a Po_2 of 400 Torr, when
perfusing an isolated dog heart preparation with intact microcirculation with a fluorocarbon suspension at $28^{\circ}C$. The authors suspected
that a heterogeneous distribution of the Po_2 gradients in the myocardium with predominant regions of high and near-zero Po_2 might
have precluded any facilitation. In view of these difficulties the
same group (Wittenberg and Wittenberg, 1981) isolated fresh cells
from adult rat ventricles providing better controlled experimental
conditions. Steady-state O_2 uptake of this preparation was O_2 limited
only at very low Po_2. Cole (1982) perfused the gastrocnemius-plan-
taris muscle of the dog with Ringer lactate solution at $37^{\circ}C$ and
measured its O_2 consumption and tension generation during contraction before and after adding H_2O_2 to the perfusion solution, H_2O_2
inactivating the Mb without interfering with the other functions of
the muscle. He found that H_2O_2 inactivation of Mb reduced the O_2
consumption to 65% and the twitch tension amplitude to 54% of the
control values. He concluded that Mb is important for a proper functioning of the working muscle although this muscle did not become
hypoxic in the absence of active Mb.

Gayeski and Honig (1978) found a uniformly high MbO_2 saturation of more than 70% in resting dog gracilis muscle at 37°C and concluded that this was not compatible with facilitation. However, a relatively high mean MbO_2 saturation need not exclude local desaturation with steep MbO_2 gradients around the mitochondria, where therefore the conditions might be favorable for facilitation, as shown by Fletcher (1980).

De Koning et al. (1981) studied the possible significance of facilitated O_2 diffusion in thin layers (300-700 um) of respiring chicken gizzard smooth muscle at steady state and 37°C. O_2 flux measurements at high Po_2 yielded maximum O_2 consumption and inert O_2 permeability, whereas a mathematical analysis of O_2 uptake from measurements at lower (subcritical) Po_2 provided values of the Po_2 for half maximum O_2 consumption and facilitation. It appeared that the effects of Po_2-dependent O_2 consumption and facilitation counteract one another. Zero facilitation was defined as that obtained in the presence of 1% CO to inactivate Mb, maximum possible facilitation was calculated from the prevailing Mb concentration assuming full mobility of Mb. Facilitation without CO was 50-100% of the theoretical maximum as shown in Fig. 4 where $J_0^2/2\rho M$ (dimension of pressure) is plotted against P_O (J_O = O_2 flux entering the layer, ρ = O_2 permeability, M = maximum O_2 consumption, P_O = upstream O_2 pressure). The identity line (broken line) holds for zero-order O_2 consumption and zero facilitation. The shaded area below the identity line holds for the theoretical range of O_2 consumption according to Michaelis-Menten kinetics with a Po_2 for half maximum O_2 consumption between 0.75 and 3 Torr in the presence of CO, that above the identity line shows the theoretical range of O_2 consumption for the same Po_2 at half maximum O_2 consumption in the absence of CO for a maximum facilitation (equivalent facilitation pressure P_F = 30 Torr). These two shaded areas show that an O_2 consumption of nonzero order decreased the flux term and that any facilitatory effect starts from a lower nonfacilitatory level and thus produces a facilitatory effect lower than it would be with zero-order O_2 consumption. The majority of the experimental results with CO (circles) fall within the range of the theoretical shaded area of Po_2 for half maximum O_2 consumption between 0.75 and 3 Torr. The experimental results for facilitation (no CO, squares) lie within the range of the theoretical shaded area or somewhat below. This means that the data correspond to facilitations between about 50 and 100% of the maximum possible. The mean Po_2 for half maximum O_2 consumption from the experiments is about 2.2 Torr. Microcryophotometric measurements of the MbO_2 gradients in this muscle suggested that at most 40% of the O_2 transport could be mediated by Mb (Schwarzmann and Grunewald, 1978). Application of Maxwell's heterogeneous media theory to a muscle with cells of radius 3.5 um showed that facilitation is about the same as inert O_2 permeability, and that it is 85% of its maximum equilibrium value (Stroeve and Eagle, 1980). This agrees with the results of de Koning et al. (1981) and justifies their assumption

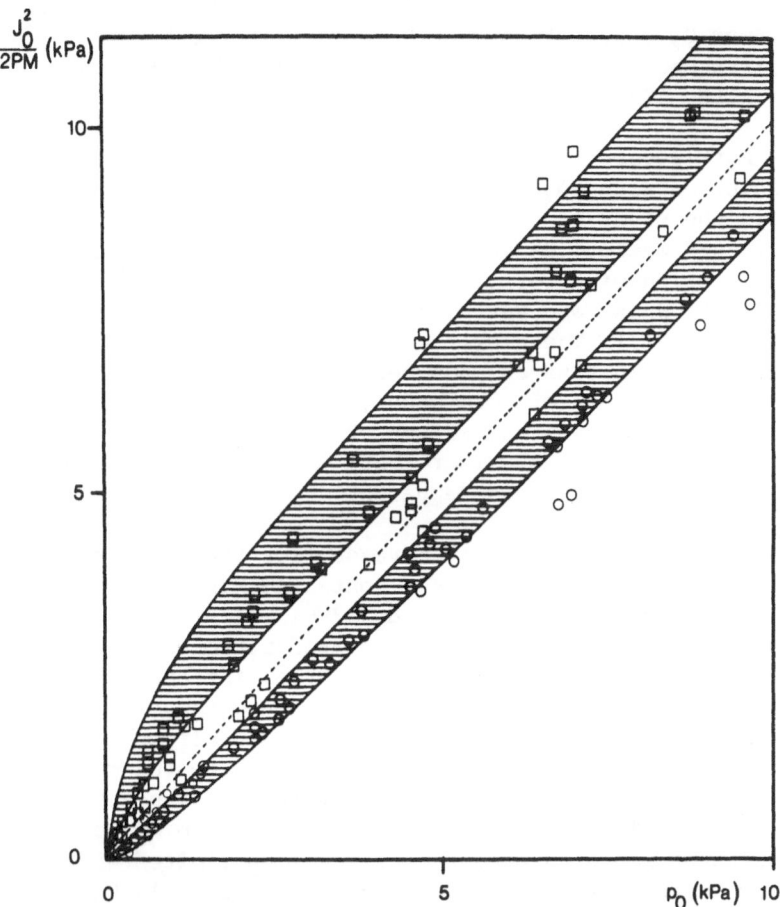

Fig. 4.   Flux parameter $J_O^2/2\rho M$ as a function of O_2 pressure P_O at
          high-pressure side for a layer of chicken gizzard smooth
          muscle with Michaelis-Menten kinetics of O_2 consumption.
          The two hatched areas indicate the theoretical areas for
          no facilitation (below) and for maximum facilitation
          (above) with a range of apparent K_m between 0.75 and 3
          Torr. The circles and squares refer to measurements with
          or without CO, respectively. Pressure on ordinate and
          abscissa in kPa (1 kPa = 7.5 Torr). For further details
          see text. From de Koning et al. (1981).

of chemical equilibrium between Mb and O_2. The reason for the
facilitation being lower than its maximum possible is not clear.
Four factors may intervene:
    1) An erroneously high value for D_{Mb} in the calculation of
       maximum facilitation;
    2) Only partial mobility of Mb in the intact muscle;

3) Nonzero-order O_2 consumption (increasing Po_2 for half maximum O_2 consumption decreases facilitation); or
4) Presence of membranes in the muscle (Gonzales-Fernandez and
   Atta, 1982).

We must revert to the problem of Michaelis-Menten kinetics of
O_2 consumption with Po_2 for half maximum O_2 consumption around
2 Torr in the study of de Koning et al. (1981). Mitochondrial K_m
(Po_2 for half maximum O_2 consumption) is known to be very low, i.e.,
a small fraction of a Torr, as found by several authors (e.g.,
Chance, 1965; Bârzu and Satre, 1970; Oshino et al., 1972; Starlinger
and Lübbers, 1973). These extremely low values come close to zero-
order kinetics. In tissues, however, other authors have also found
an O_2 consumption not of zero order, e.g., Buerk and Longmuir (1977),
Buerk and Saidel (1978) and Wilson et al. (1979), reporting K_m
values in the order of one Torr. Thus one has to distinguish between
"true" mitochondrial K_m and "apparent" K_m in tissues.

We must ask now what is the reason for the difference between
mitochondrial K_m and apparent K_m. Jones and Mason (1978) found that
the K_m in hepatocytes was 1.9 uM at 37^OC, which is an order of
magnitude greater than the K_m of isolated mitochondria. The authors
attributed this difference to a substantial intracellular O_2 gradient between the outer cellular membrane and the mitochondrial
inner membrane. Tamura et al. (1978) provided direct experimental
evidence for possible sharp O_2 gradients between the cytosol and the
mitochondria in the beating isolated perfused rat heart. Wittenberg
and Wittenberg (1981) found that the steady-state O_2 uptake of their
preparation of isolated cardiac cells was O_2 limited with a K_m near
0.1 Torr, whereas isolated mitochondria, measured in the same way,
had a very much lower K_m. Thus there is reliable evidence now that
the apparent K_m in tissues is larger than the mitochondrial K_m due
to diffusion resistances of the mitochondria and their cellular
environment.

The notion of an apparent K_m is well established in enzyme
kinetics (e.g., Engasser, 1978; Müller and Zwing, 1982), particularly in the important field of immobilized enzymes which are analogous to the situation of the enzymes in the mitochondrion. When
the vicinity of a bound enzyme is depleted of substrate, the actual
enzyme activity is usually lower than the intrinsic activity in the
absence of any diffusion limitation, i.e., for a uniform substrate
concentration equal to the bulk concentration. The relative decrease
of enzyme activity due to diffusion limitations may be expressed by
an effectiveness factor defined as the ratio of actual activity and
intrinsic activity. The effectiveness factor approaches unity in
the absence of diffusion effects and is smaller than one as diffusion effects decrease the actual activity. When O_2 diffuses from
outside across a cellular medium to and into the mitochondria, the
lowered enzyme activity is due to two diffusion resistances in

series, resulting in an external (cytosol) and internal (mitochon-drion) O_2 pressure gradient. This does not affect the O_2 consumption in the presence of zero-order kinetics, but with nonzero-order kinetics, e.g., Michaelis-Menten kinetics, it will increase K_m to values above the intrinsic mitochondrial value which would prevail in isolated mitochondria suspended in a well-stirred medium without external diffusion limitation. A remaining internal diffusion limitation would be expected to depend on the size of the mitochondrion or on the size of possible mitochondrial aggregations.

We have calculated the drop of Po_2 and the course of O_2 consumption as a function of the penetration depth in a layer of tissue for four situations, i.e. zero-order kinetics ($K_m = 0$) without and with facilitation and Michaelis-Menten kinetics (apparent $k_m > 0$) without and with facilitation. The results are shown in Fig. 5. In the plot of Po_2 (above) it may be seen that both with zero-order and Michaelis-Menten kinetics the Po_2 gradient (slope) is first increased by facilitation but further along the diffusion path it is decreased, i.e., it extends further into the layer. This means that the O_2 reaches further with facilitation in the presence of low values of Po_2, particularly with Michaelis-Menten kinetics. The O_2 consumption (below) shows a sharp front with zero-order kinetics, which reaches further with facilitation. Beyond the front Po_2 and O_2 consumption are zero. In the presence Michaelis-Menten kinetics the O_2 consumption falls more and more below the maximal value with increasing depth of penetration; at larger distances with low Po_2 and O_2 consumption is higher with facilitation.

Fig. 6 shows the dependence of O_2 consumption on tissue Po_2 for Michaelis-Menten kinetics as a function of the ratio of the permeabilities inside and outside a mitochondrion of radius 1 um. An increase in the permeability ratio implies an increase in the external diffusion limitation relative to a constant internal permeability. The graph shows that an increased external diffusion resistance (decreased external permeability) reflects in an increased apparent K_m. There is a linear relationship between apparent K_m and permeability ratio or external diffusion limitation. The effect of an increased diffusion limitation is also seen in Fig. 7 where the apparent K_m is plotted against the mitochondrial radius for three values of the permeability ratio. Apparent K_m increases with increasing mitochondrial radius and with increasing permeability ratio or increasing diffusion limitation. These calculated results show that apparent K_m increases with larger diffusion limitation and depends on geometric factors such as mitochondrial radius. Numerical calculation from our simple model provided a K_m of 0.11 times the Po_2 just sufficient for maximal O_2 consumption (critical Po_2) in the absence of external diffusion limitation. When the permeabilites are equal inside and outside the mitochondrion (permeability ratio = 1) K_m is 10 times larger and increases further as the external diffusion limitation becomes larger (or the internal diffusion

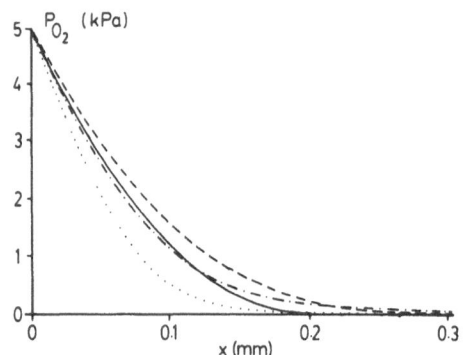

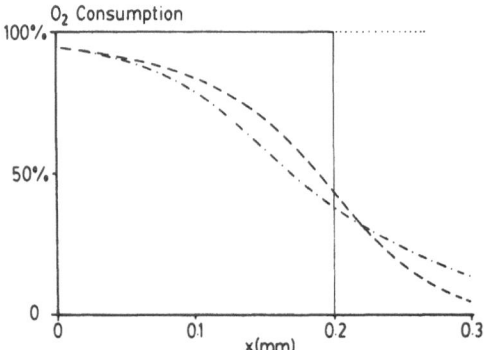

Fig. 5.   Above: Drop of Po_2 in a tissue layer against depth of
          penetration x for four situations: _____ zero-order kinet-
          ics without facilitation, ... zero-order kinetics with
          facilitation, ---- Michaelis-Menten kinetics without faci-
          litation, -·-·- Michaelis-Menten kinetics with facilitation.
          Values assumed: apparent Michaelis-Menten constant = 2.2
          Torr, P_{50} of MbO_2 dissociation curve = 1.5 Torr, equivalent
          facilitation pressure = 30 Torr. The initial slope of the
          curves is a measure of the flux. Below: Course of O_2 con-
          sumption against depth of penetration x for the same four
          situations and values assumed as above.

resistance becomes smaller, e.g., by a possible transport facilitation within the mitochondrion).

PHYSIOLOGICAL SIGNIFICANCE OF FACILITATED O_2 DIFFUSION IN TISSUES
OTHER THAN MUSCLE

     Longmuir and McCabe (1964) found that the respiration rate of
various tissues (particularly liver and kidney) behaved according to
Michaelis-Menten kinetics rather than to zero-order O_2 consumption
assumed in the classical diffusion model. They suspected a possible
facilitation by a carrier which, however, remains unidentified.

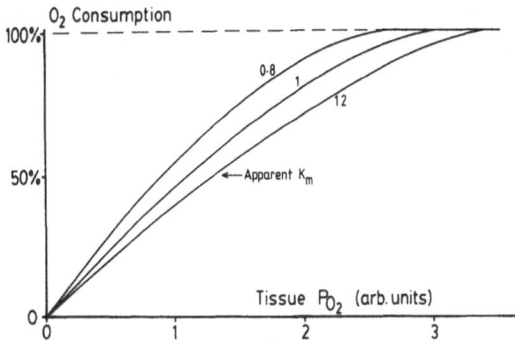

Fig. 6. Plot of relative O_2 consumption against tissue Po_2 in
         arbitrary units for three values of the ratio of the permea-
         bilities inside and outside a mitochondrion of radius 1 um.
         Increasing permeability ratio means increasing external
         diffusion resistance and implies an increase in apparent K_m.

        Facilitated O_2 and CO transport has been suggested for the
sheep placenta (Gurtner and Burns, 1972) and for the lungs of sheep
and dogs (Burns and Gurtner, 1973), invoking cytochrome P-450 as a
carrier. The data for the placenta were found to not be good enough
for a conclusive proof (Longo, 1978), but recently Gurtner et al.
(1982) once more obtained results on the sheep placenta consistent
with their hypothesis of a placental O_2 transport being carrier-
mediated partially. As far as the lung is concerned, Burns and
Shepard (1981) assembled a number of arguments against an O_2 and

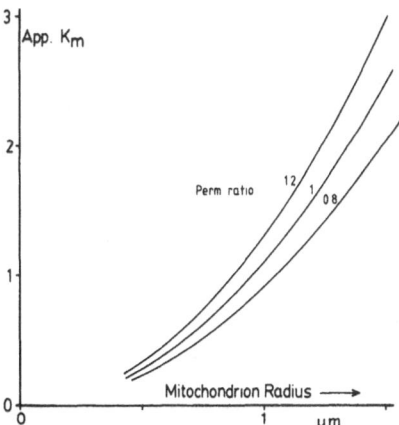

Fig. 7. Apparent K_m as a function of the mitochondrial radius for
         three values of the permeability ratio as defined in Fig. 6.
         K_m increases with larger mitochondrial radius and with
         increasing permeability ratio or increasing external diffu-
         sion resistance.

CO carrier playing a role in pulmonary gas diffusion. Gurtner and Peavy (1981) as well as Knoblauch et al. (1981) maintained their carrier hypothesis and explained the discrepancy with the results of Burns and Shepard (1981) by systematic differences in experimental design.

Recent experimental evidence by other groups of workers (Kawashiro et al., 1978; Meyer et al., 1981; Rubin et al., 1981; Jones et al., 1981) does not support the contentions of Longmuir or Gurtner and their coworkers.

REFERENCES

Bârzu, O., and Satre, M., 1970, Determination of oxygen affinity of respiratory systems using oxyhemoglobin as oxygen donor, Anal. Biochem., 36:428-433.

Buerk, D.G., and Longmuir, I.S., 1977, Evidence for nonclassical respiratory activity from oxygen gradient measurements in tissue slices, Microvasc. Res., 13:345-353.

Buerk, D.G., and Saidel, G.M., 1978, A comparison of two nonclassical models for oxygen consumption in brain and liver tissue, in: "Oxygen Transport to Tissue - III", I.A. Silver, M. Erecińska, H.I. Bicher, eds., Adv. Exper. Med. Biol., Vol. 94, Plenum Press, New York - London, pp. 225-232.

Burns, B., and Gurtner, G.H., 1973, A specific carrier for oxygen and carbon monoxide in the lung and placenta, Drug Metab. Dispos., 1:374-379.

Burns, B., and Shephard, R.H., 1981, Membrane diffusion: Comparison between dithionite Do_2 and DLCO, in: "Progress in Respiration Research", Vol. 16, J. Piiper, P. Scheid, eds., Karger, Basel-München-Paris-London-New York-Sidney, pp. 130-141.

Chance, B., 1965, Reaction of oxygen with the respiratory chain in cells and tissues, J. Gen. Physiol., 49:163-188.

Cole, R.P., 1982, Myoglobin function in exercising skeletal muscle, Science, 216:523-525.

Cole, R.P., Wittenberg, B.A., and Caldwell, P.R.B., 1978, Myoglobin function in the isolated fluorocarbon-perfused dog heart, Am. J. Physiol., 234:H567-H572.

Diller, T.E., and Mikic, B.B., 1980, Modeling the oxygen diffusion effects of red cell motions in flowing blood, in: "Advances in Bioengineering", C.V. Mow, ed., Am. Soc. Mech. Eng., New York, pp. 177-180.

Engasser, J.M., 1978, A fast evaluation of diffusion effects on bound enzyme activity, Biochim. Biophys. Acta, 526:301-310.

Fletcher, J.E., 1980, On facilitated diffusion in muscle tissues, Biophys. J., 29:437-458.

Friedlander, S.K., and Keller, K.H., 1965, Mass transfer in reacting systems near equilibrium. Use of the affinity function, Chem. Eng. Sci., 20:121-129.

Gayeski, T.E.J., and Honig, C.R., 1978, Myoglobin saturation and calculated Po_2 in single cells of resting gracilis muscle, in: "Oxygen Transport to Tissue - III", I.A. Silver, M. Erecińska, H.I. Bicher, eds., Adv. Exper. Med. Biol., Vol. 94, Plenum Press, New York - London, pp. 77-84.

Gijsbers, G.H., and van Ouwerkerk, H.J., 1976, Boundary layer resistance of steady-state oxygen diffusion facilitated by a four-step chemical reaction with hemoglobin in solution, Pflügers Arch., 365:231-241.

Goddard, J.D., 1977, Further applications of carrier-mediated transport theory - a survey, Chem. Eng. Sci., 32:795-809.

Goddard, J.D., Schultz, J.S., and Bassett, R.J., 1970, On membrane diffusion with near-equilibrium reaction, Chem. Eng. Sci., 25:665-683.

Gonzales-Fernandez, J.M., and Atta, S.E., 1981, Transport of oxygen in solutions of hemoglobin and myoglobin, Math. Biosci., 54: 265-290.

Gonzales-Fernandez, J.M., and Atta, S.E., 1982, Facilitated transport of oxygen in the presence of membranes in the diffusion path, Biophys. J., 38:133-141.

Gurtner, G.H., and Burns, B., 1972, Possible facilitated transport of oxygen across the placenta, Nature, 240:473-475.

Gurtner, G.H., and Peavy, H.H., 1981, Evidence for facilitated transport of O_2 and CO in the lungs, in: "Progress in Respiration Research", Vo. 16, J. Piiper, P. Scheid, eds., Karger, Basel-München-Paris-London-New York-Sidney, pp. 161-165.

Gurtner, G.H., Traystman, R.J., and Burns, B., 1982, Interactions between placental O_2 and CO transfer. J. Appl. Physiol.: Respirat. Environ. Exercise Physiol., 52:479-487.

Hoofd, L., and Kreuzer, F., 1979, A new mathematical approach for solving carrier-facilitated steady-state diffusion problems, J. Math. Biol., 8:1-13.

Jones, D.P., and Mason, H.S., 1978, Gradients of O_2 concentration in hepatocytes, J. Biol. Chem., 253:4874-4880.

Jones, H.A., Buckingham, P.D., Clark, J.C., Forster, R.E., Heather, J.D., Hughes, J.M.B., and Rhodes, C.G., 1981, Constant rate of CO uptake with variable inspired CO concentration, in: "Progress in Respiration Research", Vol. 16, J. Piiper, P. Scheid, eds., Karger, Basel-München-Paris-London-New York-Sidney, pp. 69-171.

Kawashiro, T., Piiper, J., and Scheid, P., 1978, Dependence of O_2 uptake on surface Po_2 in intact, excised skeletal muscle of the rat: validity of the Warburg model, J. Physiol., 284: 45P-46P.

Klug, A., Kreuzer, F., and Roughton, F.J.W., 1956, The diffusion of oxygen in concentrated hemoglobin solutions, Helv. Physiol. Pharm. Acta, 14:121-128.

Knoblauch, A., Sybert, A., Brennan, N.L., Sylvester, J.T., and Gurtner, G.H., 1981, Effect of hypoxia and CO on a cytochrome P-450-mediated reaction in rabbit lungs, J. Appl. Physiol.: Respirat. Environ. Exercise Physiol., 51:1635-1642.

Koning, de, J., Hoofd, L.J.C., and Kreuzer, F., 1981, Oxygen
    transport and the function of myoglobin. Theoretical model and
    experiments in chicken gizzard smooth muscle, Pflügers Arch.,
    389:211-217.
Kreuzer, F., 1970, Facilitated diffusion of oxygen and its possible
    significance; a review, Respir. Physiol., 9:1-30.
Kreuzer, F., and Hoofd, L.J.C., 1970, Facilitated diffusion of
    oxygen in the presence of hemoglobin, Respir. Physiol.,
    8:280-302.
Kreuzer, F., and Hoofd, L.J.C., 1972, Factors influencing facili-
    tated diffusion of oxygen in the presence of hemoglobin and
    myoglobin, Respir. Physiol., 15:104-124.
Kreuzer, F., and Hoofd, L., 1982, Facilitated diffusion of O_2 and
    CO_2, in: "Handbook of Physiology: Respiration", L.E. Farhi,
    S.M. Tenney, eds., American Physiological Society, Bethesda,
    Md., in press.
Kreuzer, F., and Yahr, W.Z., 1960, Influence of red cell membrane
    on diffusion of oxygen, J. Appl. Physiol., 15:1117-1122.
Kutchai, H., 1970, Numerical study of oxygen uptake by layers of
    hemoglobin solution, Respir. Physiol., 10:273-284.
Kutchai, H., 1971, O_2 uptake by 100 u layers of hemoglobin
    solution, Respir. Physiol., 11:378-383.
Kutchai, H., Jacquez, J.A., and Mather, F.J., 1970, Nonequilibrium
    facilitated oxygen transport in hemoglobin solution, Biophys.
    J., 10:38-54.
Longmuir, I.S., and McCabe, M.G.P., 1964, Evidence for an oxygen
    carrier in tissue, J. Polarogr. Soc., 10:45-48.
Longo, L.D., 1978, Placental diffusing capacity for carobon monox-
    ide. Letter to the Editor, J. Appl. Physiol.: Respirat.
    Environ. Exercise Physiol., 45:155.
Meyer, M., Lessner, W., Scheid, P., and Piiper, J., 1981, Pulmonary
    diffusing capacity for CO independent of alveolar CO concen-
    tration, J. Appl. Physiol: Respirat. Environ. Exercise
    Physiol., 51:571-576.
Mitchell, P.J., and Murray, J.D., 1973, Facilitated diffusion: The
    problem of boundary conditions, Biophysik, 9:177-190.
Müller, J., and Zwing, T., 1982, An experimental verification of
    the theory of diffusion limitation of immobilized enzymes,
    Biochim. Biophys. Acta, 705:117.123.
Oshino, R., Oshino, N., Tamura, M., Kobilinsky, L., and Chance, B.,
    1972, A sensitive bacterial luminiscence probe for O_2 in
    biochemical systems, Biochim. Biophys. Acta, 273:5-17.
Rubin, D.Z., Fujino, D., Mittman, C., and Lewis, S.M., 1981,
    Competitive inhibition of carbon monoxide transport: evidence
    against a carrier, J. Appl. Physiol.: Respirat. Environ.
    Exercise Physiol., 50:1061-1064.
Scholander, P.F., 1960, Oxygen transport through hemoglobin solu-
    tions, Science, 131:585-590.

Schultz, J.S., Goddard, J.D., and Suchdeo, S.R., 1974, Facilitated transport via carrier-mediated diffusion in membranes. Part I. Mechanistic aspects, experimental systems and characteristic regimes, AIChE J., 20:417-445.

Schwarzmann, V., and Grunewald, W.A., 1978, Myoglobin-O_2-saturation profiles in muscle sections of chicken gizzard and the facilitated O_2-transport by Mb, in: "Oxygen Transport to Tissue - III", I.A. Silver, M. Erecińska, H.I. Bicher, eds., Adv. Exper. Med. Biol., Vol. 94, Plenum Press, New York - London, pp. 301-310.

Smith, K.A., Meldon, J.H., and Colton, C.K., 1973, An analysis of carrier-facilitated transport, AIChE J., 19:102-111.

Starlinger, H., and Lübbers, D.W., 1973, Polarographic measurements of the oxygen pressure performed simultaneously with optical measurements of the redox state of the respiratory chain in suspensions of mitochondria under steady-state conditions at low oxygen tensions, Pflügers Arch., 341:15-22.

Stroeve, P., Colton, C.K., and Smith, K.A., 1976, Steady state diffusion of oxygen in red blood cell and model suspensions, AIChE J., 22:1133-1142.

Stroeve, P., and Eagle, K., 1980, Myoglobin-facilitated oxygen transport in heterogeneous red muscle tissue, in: "Advances in Bioengineering", C.V. Mow, ed., Am. Soc. Mech. Eng., New York, pp. 341-344.

Tamura, M. Oshino, N., Chance, B., and Silver, I.A., 1978, Optical measurements of intracellular oxygen concentration of rat heart in vitro, Arch. Biochem. Biophys., 191:8-22.

Weigelt, H., 1975, Mikrophotometrische Messungen zur Untersuchung des erleichterten Sauerstofftransports in Gegenwart von Hämoglobin, Dissertation, Ruhr-Universität Bochum.

Wilson, D.F., Erecińska, M., Brown, S., and Silver, I.A., 1979, The oxygen dependence of cellular energy metabolism, Arch. Biochem. Biophys., 195:485-493.

Wittenberg, J.B., 1959, Oxygen transport - a new function proposed for myoglobin, Biol. Bull., 117:402-403.

Wittenberg, J.B., 1970, Myoglobin-facilitated oxygen diffusion: Role of myoglobin in oxygen entry into muscle, Physiol. Rev., 50:559-636.

Wittenberg, B.A., Wittenberg, J.B., and Caldwell, P.R.B., 1975, Role of myoglobin in the oxygen supply to red skeletal muscle, J. Biol. Chem., 250: 9038-9043.

Wittenberg, J.B., and Wittenberg, B.A., 1981, Facilitated oxygen diffusion by oxygen carriers, in: "Oxygen and Living Processes", D.L. Gilbert, ed., Springer, New York.

MUSCLE O_2 GRADIENTS FROM HEMOGLOBIN TO CYTOCHROME:

NEW CONCEPTS, NEW COMPLEXITIES

C.R. Honig[1], T.E.J. Gayeski[1], W. Federspiel[2],
A. Clark, Jr.[2], and P. Clark[3]

[1]The University of Rochester, School of Medicine and
Dentistry
[2]College of Engineering and Applied Science
601 Elmwood Avenue, Rochester, NY 14642, USA
[3]Rochester Institute of Technology

INTRODUCTION

The purpose of this paper is to reappraise the sequential
barriers to O_2 transport from red cells to mitochondria in light of
a more accurate model of intracapillary O_2 transport, and measurements of myoglobin (Mb) saturation in subcellular volumes. We find
that transcapillary gradients are larger and tissue gradients smaller
than predicted by existing models of O_2 diffusion. During exercise
the principal "resistance" to O_2 transport resides in the capillary
and extracellular space; the limiting variables are rate of O_2
release and red cell transit time. Mb plays a major role in over-
coming "resistance" at the capillary by buffering Po_2 well below
capillary Po_2 ($P_{cap}o_2$). These findings are incompatible with classical concepts of O_2 delivery to red muscle. Alternatives to Kroghian
thinking are proposed.

METHODS

Intracapillary O_2 gradients are evaluated mathematically by
use of techniques described below. Transcapillary and intracellular
O_2 gradients are evaluated from measurements of Mb saturation in
dog gracilis muscles frozen in situ (Gayeski, 1981). Freezing "stops
the clock", so we can make as many measurements as required and
consider all to be simultaneous. Covariates include BP_a, flow/g,
$P_a o_2$, $S_a o_2$, pH_a, $P_v o_2$, $S_v o_2$, hemoglobin (Hb) concentration,

functional capillary density (FCD), and tissue concentrations of
various metabolites.

Our conclusions depend on our ability to "trap" O_2 gradients.
We therefore modelled the interrelation between the rate of prop-
agation of the freezing front and chemical reactions involving O_2.
Results indicate that the dominant factors are the high diffusivity
of heat and the kinetics of O_2 binding to Mb and Hb. The rate of O_2
consumption ($\dot{V}_{O_2}$) during cooling has no significant effect on saturation in mean cell volume. The possible effects of $\dot{V}_{O_2}$ on small-
scale O_2 gradients are under study (Clark, A. and Clark, P., 1983).
The calculated error in Mb saturation attributable to freezing is
approximately 0.1% in dog gracilis. These calculations are supported
by the following methodologic results. 1) The initial rate of
cooling determined experimentally is about 10 um/ms, in good agreement with calculations. 2) Measured saturation is not critically
dependent on the temperature of the heat sink, as predicted by the
mathematical model of the freezing process. 3) Probability distributions of Mb saturation measured 50 and 500 um from the surface of
the same muscle are identical. 4) Small-scale saturation gradients
can be "trapped"; see Fig. 4. Though the Mb saturation measurements
are surprising they cannot be explained as freezing artifact.

Saturation is determined by use of a 4 wavelength method described in detail previously (Gayeski, 1981). Depth of penetration
of light is on the order of 2 - 3 ice crystals, or < 2 um; spatial
resolution in the plane of observation is presently about 4 um (< 2
mitochondria in dog gracilis). The overall error of the saturation
measurement is < 5%. P_{O_2} is calculated from saturation using a Mb
P_{50} of 5.3 Torr at $37^{\circ}C$ (Gayeski, 1981).

RESULTS AND DISCUSSION

Intracapillary O_2 Transport

The model for the intracapillary transport of O_2 includes the
free and Hb-facilitated diffusion of O_2 inside the red cell (Federspiel et al., 1982). Non-linear reaction kinetics between Hb and
O_2 are taken into account. The dimensionless transport equations
for the red cell region are

$$\frac{\partial \Psi}{\partial t} = R\nabla^2\Psi - \beta R^2 \left\{ (1-S)\Psi^n - S \right\} \tag{1}$$

$$\frac{\partial S}{\partial t} = \rho\nabla^2 S + \beta R \left\{ (1-S)\Psi^n - S \right\} \tag{2}$$

where Ψ is the free O_2 concentration normalized by N_{50} (the O_2

concentration that saturates 50% of the Hb), S is the Hb saturation, t is the dimensionless time, n is the exponent for the Hill
equation, ρ , ß and R are dimensionless parameters. The nonlinear
reaction term is chosen so that equations (1) and (2) predict the
proper Hb-O₂ equilibrium behavior. Dimensionless parameter values
are consistent with those used in previous model studies (Moll,
1969; Kutchai, 1970; Sheth and Hellums, 1980).

The transport equations for the red cell region are analyzed
using boundary layer theory. Hb and O₂ are in equilibrium at the
center of the red cell but depart from equilibrium near the red cell
membrane. The extent of the non-equilibrium region depends on the
Po_2 at the red cell membrane and is greatest (approximately 0.9 um)
when the surface Po_2 equals zero (Federspiel et al., 1982). It is
also shown that the assumption of Hb-O₂ equilibrium everywhere in
the red cell can lead to a significant error, and that it is important to use a reaction mechanism that reduces to the actual Hb-O₂
equilibrium behavior (consistent with the findings of Sheth and
Hellums, 1980). The nonlinear transport equations are solved nume-

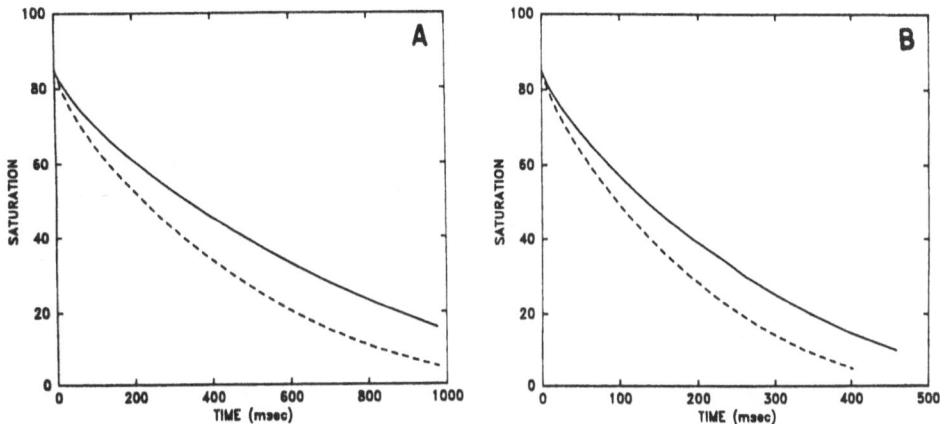

Fig. 1 A and B.   The change in mean saturation of a red cell
                  (modelled as a cylinder ignoring end effects) of
                  radius 2 um (A) or 1 um (B). The red cell is sur-
                  rounded by a "resistance layer" of 1 um (dashed
                  line) or 2 um (solid line). The resistance layer
                  accounts for the diffusional resistance of the
                  plasma, endothelium, and extracellular fluid layers
                  that separate the red cell from a muscle cell.
                  Since the Po_2 is set at 0 Torr outside the resist-
                  ance layer, the curves correspond to maximum release
                  rate or minimum unloading time. Initial saturation
                  determined from models of precapillary O₂ loss
                  (Popel and Gross, 1979).

rically using the method of double orthogonal collocation (Villadsen and Sorensen, 1969).

In capillaries of skeletal muscle working near $\dot{V}o_{2max}$, the flux of O_2 from the red cells is largely radially outward, i.e., an observer on the capillary wall would see a "burst" of O_2 as a red cell passes (Federspiel and Sarelius, 1983). To a first approximation then we can ignore O_2 diffusing from the ends of the red cell. The red cell membrane per se is not an O_2 barrier (Kreuzer and Yahr, 1960; Huxley and Kutchai, 1981). However, O_2 must diffuse through a narrow gap of plasma between the red cell and the endothelium, the endothelium itself, and the extracellular space before reaching the muscle cell membrane. For this reason the model includes a "resistance layer" outside the red cell where free O_2 moves by simple diffusion. The diffusion coefficient and solubility for O_2 in the resistance layer is taken to be the same as in plasma. The resistance layer is a functional (diffusional) barrier rather than an anatomical barrier because the rate of release of O_2 from red cells covered by endothelium is about the same as the rate when the red cells are covered by a plasma layer of equal thickness (Sinha, 1969).

A useful quantity to calculate from the transport model is the minimum release time for unloading O_2 from a red cell surrounded by a resistance layer. The red cell is in a maximum demand environment when the Po_2 outside the surrounding resistance layer is maintained at zero. The resulting unloading time can be considered as a lower bound estimate for physiological release time. The minimum unloading time for O_2 is shown in Fig. 1 A. The red cell is modelled as a cylinder of Hb solution (ignoring transport from the ends) of radius 2 um inside a capillary of 2.5 um radius. The saturation at the capillary inlet is obtained from estimates of precapillary O_2 losses during exercise (Popel and Gross, 1979). The time course is shown for a 1 um resistance layer (dashed line) and a 2 um resistance layer (solid line), corresponding to 1/2 um and 1 1/2 um of endothelium and extracellular fluid space, respectively. The same information is shown for a 1 um cell in Fig. 1 B. Since the conclusion we will draw is qualitatively the same for 1 and 2 um cells, exact treatment of red cell size and shape is not crucial. For both cases, the resistance layer around the red cell acts to "throttle" the release of O_2.

Some Evidence Against Kroghian Thinking

Modellers have focused on O_2 transport within the muscle cell, adopting the Kroghian view of a capillary supplying its own cylindrical section of surrounding tissue. In this view the tissue "resistance" is paramount, and the critical variable is diffusion distance. Often the capillary, endothelium, and extracellular fluid space are regarded as of no consequence for O_2 transport, and the

tissue Po_2 adjacent to the capillary is equated with mean capillary
Po_2. In view of the complexity of the problem this simplification
can be appreciated but should be well understood.

Consider a fluid flowing through a capillary of radius R_C with
a uniform axial velocity $\bar{U}$; the capillary is surrounded by a Krogh
cylinder of radius R_T. The fluid contains a diffusible species that
is consumed in the annular ("tissue") region at a rate of M moles
volume^{-1} time^{-1}. The concentration in the capillary, $C_B(rZ)$, and
tissue, $C_T(r,Z)$ depend on radial (r) and axial (Z) position. At
steady state, ignoring axial diffusion, an overall mass balance
leads to the following equation for the capillary mean concentration
$\bar{C}_B(Z)$:

$$\frac{d\bar{C}_B(Z)}{dz} = \frac{-M}{\bar{U}} \, (\gamma^2 - 1) \tag{3}$$

where γ is the ratio of the Krogh cylinder radius to capillary
radius. Equation (3) depends only on the foregoing assumptions, and
holds regardless of the relative transport resistances in the capillary and tissue.

Suppose more information is desired and one proceeds to study
the tissue (annular) region. Assuming simple radial diffusion and
uniform consumption M, the concentration profile in this region is:

$$C_T(r,Z) = C_T^*(R_C,Z) - \frac{M}{4D_T} \left\{ R_C^2 (2\gamma^2 \ln \frac{r}{R_C} + 1) - r \right\} \tag{4}$$

where D_T is the diffusion coefficient in the tissue region. $C_T^*(R_C,Z)$
is the concentration in the tissue at the capillary-tissue interface
and cannot be specified without either an assumption about the relative transport resistance in the capillary or the solution of some
model for the capillary transport.

The simplest way to proceed is to assume that the transcapillary O_2 gradient is essentially flat:

$$C_T^*(R_C,Z) = \bar{C}_B(Z) \tag{5}$$

The only other choice is to model the capillary transport and
obtain a value for $C_B^*(R_C,Z)$. Then

$$C_T^*(R_C,Z) = C_B^*(R_C,Z) \tag{6}$$

The latter choice allows an estimate of the relative transport resistances in the capillary and tissue. Oxygen transport modellers
face the decision either to ignore capillary resistance, or to model
capillary transport.

Hellums (1977) attempted to evaluate the effect of modelling
the capillary as a continuum rather than a particulate fluid. He
estimated that placing Hb in "packets" makes capillary "resistance"
about 1/2 of the total "resistance" to O_2 transport. Despite major
shortcomings, his model directed attention to the possibility that
capillary "resistance" to O_2 mass transfer is not negligible. To
test this possibility, consider our simple mass transfer example
again. We follow a volume of fluid in the capillary and determine
the time t* for the fluid to go from some initial value of mean
concentration $\bar{C}_{Bi}$ to some final value $\bar{C}_{Bf}$. We compare t* to the
minimum release time obtained by setting the concentration of diffusing species at zero outside the capillary. If the minimum release
time is comparable to t*, capillary resistance is large and the
transcapillary gradient is substantial for the chosen conditions.

The foregoing test can be applied to the Kroghian representation of O_2 transport. Equation (3) can be written for blood where
$\bar{C}_B$ would be the mean concentration of total O_2, bound and unbound.
To a first order approximation, hemoglobin carries almost all the
O_2 so that

$$\bar{C}_B \approx N_{Hb} \; \bar{S} \; H, \tag{7}$$

where N_{Hb} is the total heme concentration in the red cell, $\bar{S}$ is the
average saturation of the red cell, and H is the capillary hematocrit. Using equation (7) and integrating equation (3), we can calculate the distance required to go from some value $\bar{S}_i$ to $\bar{S}_f$. Since
we assumed a uniform axial velocity, this distance can be converted

Table 1. Comparison of minimum release times, t_m for unloading O_2
with the Krogh release time, t*

EXERCISE CASE:

Saturation change		t	$t_m^{(1)}$	$t_m^{(2)}$	$t_m^{(3)}$	$t_m^{(4)}$ (msec)
in	out					
85%	50%	322	223	334	97	135
85%	40%	427	327	481	139	192
85%	30%	522	451	657	190	261

$t_m^{(1)}$ and $t_m^{(2)}$ correspond to a red cell radius of 2 um with a 1 um
and 2 um resistance layer, respectively. $t_m^{(3)}$ and $t_m^{(4)}$ correspond
to a red cell radius of 1 um with a 1 um and 2 um resistance layer,
respectively. Parameters for t*: $\dot{V}_{O_2}$ = 15 ml O_2/(100g·min), γ = 7.5,
H = .30. N_{Hb} = 1.97 x 10^{-5} mole Hb/cm^3.

to the time, $t*$, it takes a control volume of fluid to go from $\bar{s}_i$ to $\bar{s}_f$:

$$t* = \frac{HN_{Hb}}{M(\gamma^2 - 1)} \left\{ \bar{s}_i - \bar{s}_f \right\} \tag{8}$$

Table 1 compares $t*$ to the minimum release time, t_m, for unloading O_2 from a red cell through a resistance layer, for several red cell radii and resistance layer thicknesses. The parameters for equation (8) are chosen to be representative of a heavily working skeletal muscle. The values for t_m are on the same order of magnitude as those for $t*$. Thus, Table 1 suggests that under these conditions O_2 is unloading near its maximum rate, transcapillary gradients must be large, and capillary resistance is substantial. All three suggestions are inconsistent with classical Kroghian thinking.

Flux Densities

We can make an order-of-magnitude analysis of the ratio of flux density at the capillary to that at the mitochondria for skeletal muscle working at 70% of $\dot{V}_{O_2 max}$. The O_2 flux density at a mitochondrion is given by the following calculation:

$$4.7 \times 10^{-9} \frac{\text{moles } O_2}{mm^3 \; min} \div 1.9 \times 10^7 \frac{\text{mito.}}{mm^3} \div 14 \frac{um^2}{\text{mito.}} =$$

$$1.8 \times 10^{-17} \frac{\text{moles } O_2}{um^2 \; min}$$

To calculate the flux density at the capillary, we first calculate approximate capillary area per volume of tissue:

$$1000 \frac{cap}{mm^2 \; tiss} \times \frac{2\pi(2.5 \; um)}{cap} \times \frac{1000 \; um}{1 \; mm} = 1.5 \times 10^7 \frac{um^2}{mm^3 \; tiss}$$

However, at high $\dot{V}_{O_2}$ the O_2 leaving a red cell travels, to a first approximation, radially outward; since the red cell spacing is about 1 cell length, only half the capillary wall is functional for O_2 transport. Therefore, the flux density at the capillary is:

$$4.7 \times 10^{-9} \frac{\text{moles } O_2}{mm^3 \; min} \div \frac{1}{2} \div 1.5 \times 10^7 \frac{um^2}{mm^3} = 6.3 \times 10^{-16} \frac{\text{moles } O_2}{um^2 \; min}$$

The flux density ratio (functional capillary to mitochondria) is about 40. This indicates that the capillary could be the "bottleneck" for O_2 transport.

Release Time vs. Transit Time

    The red cell cannot be thought of as equilibrating with tissue; it simply releases an amount of O_2 determined by the particular conditions which happen to exist in and around the capillary it traverses. The longitudinal intracapillary gradient is governed by the relation between rate of O_2 release and red cell transit time. Even in resting muscle Mb "buffers" cell P_{O_2} around 20 Torr, whereas $P_v O_2$ is 40 - 50 Torr - far from "equilibrium" with tissue (Gayeski and Honig, 1978; Gayeski, 1981). Since the transcapillary gradient is at least 20 Torr and $\dot{V}O_2$ is small at rest, the minimum release time for O_2 (t_m) is on the order of 50 ms whereas measured red cell transit time in muscle capillaries is on the order of 1000 ms (Sarelius and Duling, 1983). Thus, the rate of release of O_2 should never limit O_2 delivery at rest, in accord with earlier calculations (Honig and Odoroff, 1981).

    The following discussion focuses on exercising skeletal muscle and myocardium where high demand stresses the O_2 delivery system. The relation between capillary transit times and rate of O_2 release is complicated by the variances of both parameters. However, variability in the capillary population should approach a minimum during exercise because red cell velocity (Honig et al., 1981), capillary hematocrit (Klitzman and Duling, 1979), and the transcapillary gradient (Gayeski and Honig, 1982) are all more uniform. Consequently, the minimum times for release of O_2 furnished by the intracapillary transport model can be used to obtain a conservative estimate of the fraction of the capillary population in which t_m exceeds red cell transit time.

    Transit time heterogeneity in dog gracilis was evaluated by use of a probability model. Probability densities for capillary path length and red cell velocity were well-fitted by two-parameter gamma distributions (Honig and Odoroff, 1981). The empirical distribution of capillary transit times was computed as the ratio of the two gamma distributions by use of an F distribution. The result was remarkably insensitive to variability of lengths and velocities, and was largely determined by the means. Predicted capillary transit times for dog gracilis working at 4/s and 8/s (70 and 100% of $\dot{V}O_{2max}$) are shown in Fig. 2 A. The median transit time at 4/s is 210 ms; approximately 10% are < 100 ms. Minimum times for release of O_2 are shown in Table 1 for the same conditions of work. Even for the thinnest "resistance layer" outside the red cell transit times in the lower tail of the distribution are shorter than the minimum release time. If this were not the case, O_2 extraction would be larger and $\dot{V}O_2$ would have been satisfied at lower flow. Consequently, $S_v O_2$ in exercise reflects a convective O_2 shunt attributable to a functional O_2 barrier at the capillary. $S_v O_2$ is also influenced by effluent from capillaries in the upper tail of the transit time distribution; note that 20% are > 400 ms in Fig. 2 A. Table 1 suggests that such

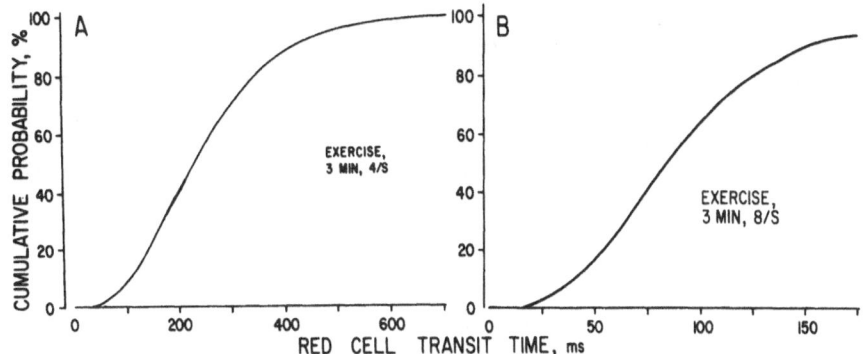

Fig. 2.   Red cell transit times in dog gracilis muscles calculated
          as described in Honig and Odoroff (1981). A: 4 twitches/s,
          B: 8 twitches/s.

capillaries should be capable of releasing almost all their O_2
provided P_TO_2 approaches zero.

      Proof that the capillary is indeed a functional O_2 barrier is
shown in Fig. 3. Measured S_vO_2 for dog gracilis <u>rose</u> significantly
when work rate was increased from 4/s to 8/s, despite higher $\dot{V}O_2$ and
lower Mb saturation and cell PO_2. Calculated transit times for work
at 8/s are shown in Fig. 2 B. All calculated transit times during
work at 8/s are less than t_m in Table 1. The total amount of O_2
shunted, estimated from bulk parameters, is about 1/3 the total consumed. This is consistent with the discrepancy between transit and
release times because the Hb oxydissociation curve insures that even
the fastest red cells will release a substantial amount of O_2.

      As $P_{cap}O_2$ approaches tissue-cell PO_2 (P_TO_2), O_2 delivery becomes
small relative to local demand, because conditions for release of O_2

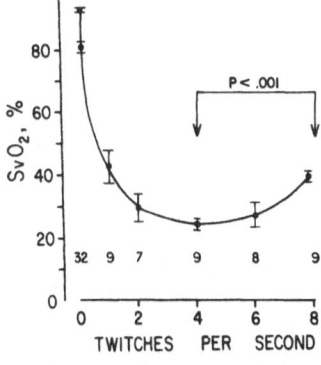

Fig. 3. S_vO_2 for dog gracilis muscles during twitch contraction.
          n = number of muscles; bars represent 1 S.E.M.

are unfavorable. In effect, part of the capillary is deleted from
the supply network. If median transit time is short enough to prevent such maladaptive behavior for capillaries at the uper end of
the probability distribution the amount of O_2 released from the
capillaries at the lower end will be limited by the functional
capillary barrier. Thus, high flow and a modest convective O_2 shunt
are essential to achieve high $\dot{V}o_2$.

Conditions comparable to those during twitch contraction at 8/s
seldom occur during natural exercise in skeletal muscle, but appear
to be the rule in myocardium. The calculated median capillary
transit time for unstressed dog heart is about 150 ms, if capillary
path length is about the same as in dog gracilis. The expansion
factors for canine coronary flow and functional capillary density
are about 4 and 2, respectively (Honig, 1981). Consequently, linear
velocity could double and mean transit time could be halved. Even
if capillary path length was 2 mm, a substantial convective shunt
would be expected during heavy cardiac work. The predicted shunt
has recently been observed in dog heart by Rose and Goresky (1982).
They report a non-exchanging, or "throughout" component in washout
curves for O^{18} and Cr^{51}-labelled erythrocytes, indicative of a
capillary barrier of the sort predicted by our mathematical models.

Myoglobin and the Transcapillary O_2 Gradient

The Mb P_{50} is 1/5 that of Hb (Tamura et al, 1978; Gayeski,
1981) but approximately 5 times the critical Po_2 for maximum turnover of cytochrome a, a_3 (Chance et al., 1973). The Mb dissociation
curve is hyperbolic, so large changes in saturation below the P_{50}
are associated with small changes in Po_2. Mb is, therefore, ideally
suited to buffering P_To_2 well below $P_{cap}o_2$, without compromising
$\dot{V}o_2$. By creating a low, uniform cell Po_2, Mb maximizes the transcapillary O_2 gradient. This essential function is shown in Fig. 4.
Profiles of 6 gracilis muscle cells are surrounded by 44 irregularly spaced capillaries (filled circles). The muscle was stimulated
at 40/s for 15 s and frozen during the tetanic contraction. P_vo_2
was 27 Torr and P_To_2 was < 4 Torr at all loci. Similar P_vo_2 and P_To_2
were observed during steady twitch contraction at $\dot{V}o_{2max}$. Since the
cells and capillaries were sectioned at arbitrary points in their
longitudinal gradients, Po_2 in many capillaries should have been
greater than 27 Torr. The percent saturation of Mb is shown for each
measurement site; numbers in parenthesis denote the corresponding
Po_2. The latter are about as low as can be attained without compromising respiration. This is true even 5 um from capillaries, which
is as close as we can measure with the present system optics. The
near-maximum transcapillary gradient should promote $\dot{V}o_2$ by acceler-
ating release of O_2 from Hb. It is, therefore, important that $\dot{V}o_2$
and active tension are severely curtailed in dog gracilis if Mb is
oxidized to its ferric form (Cole, 1982). The Mb saturation measurements again confirm the principal conclusion of the intracapillary

modelling and transit time analysis: The capillary is indeed a major
O_2 barrier at high $\dot{V}o_2$.

Intracellular and Intercellular O₂ Gradients

We previously reported that the O_2 distribution within a working muscle cell is remarkably uniform (Gayeski and Honig, 1982).
Fig. 4 was chosen to illustrate the largest intracellular gradients
we have observed. They amount to only 14% in Mb saturation or
1.2 Torr! This implies that the tissue "resistance" to O_2 transport
is small relative to that of the capillary as predicted by our calculations. Presumably the Mb-facilitated O_2 flux plays an important
role. Notice that the shapes of measured gradients are incompatible
with existing models of O_2 diffusion in muscle, since Mb saturation
is not necessarily lowest most remote from capillaries. For example,
the highest saturation in the cell cluster (38%) is observed at the
center of the large cell at upper right, whereas saturation is only
17% in the pericapillary region of cells at lower left.

Judging from the behavior of red cells (Kreuzer and Yahr, 1960;
Huxley and Kutchai, 1981) the sarcolemma per se is unlikely to be a

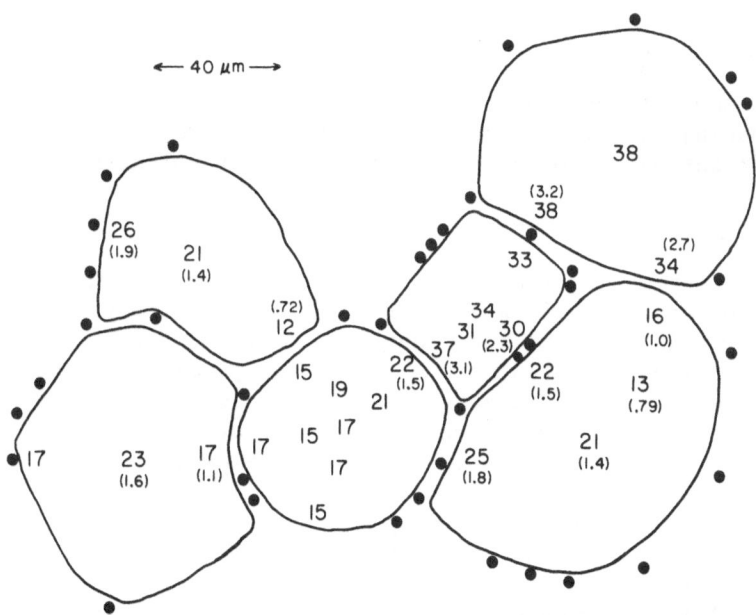

Fig. 4.   Profiles of muscle cells and capillaries in dog gracilis
           muscle freeze-clamped during tetanic contraction. The Mb
           saturations are shown for various measurement sites. Values
           in parentheses are corresponding Po_2 values if the Mb
           $P_{50} = 5.3$ Torr.

significant diffusion barrier. It is, in fact, ignored in mathematical models. Our measurements indicate, however, that the tissue
cannot be regarded as a homogeneous medium since large intercellular
gradients exist over very short distances. For example, saturations
in contiguous regions of the two cells at far right are 34% and 16%.
Such intercellular gradients could be accounted for if the extracellular fluid offers a resistance to transport greater than that found
in myoplasm. The mechanism of such resistance could be a resistance
layer homologous with that already discussed for plasma and Hb.

Interaction of Diffusion Fields

Almost all modellers deal with muscle capillaries in arrays of
crystalline regularity. Regularity is necessary to fix the domain
of each capillary in order to set boundary conditions. Examination
of the actual distribution of red cell-containing capillaries indicates, however, that there is no repetitive pattern. Pairs or
clusters of capillaries are common as in Fig. 4. Obviously, the
diffusion fields around capillaries separated by only a few microns
cannot be independent. Diffusion interaction can also take place
between Hb and Mb, because the muscle cells can be O_2 sources as
well as O_2 sinks. This is possible even though intercellular concentration gradients may be smaller than transcapillary gradients,
because the surface areas of adjacent muscle cells are large relative to the surface area of contiguous capillaries. Since the flux
density is low, O_2 could be redistributed by a small driving force.
In Fig. 4 for example, we can visualize O_2 moving from the large
cell at upper right and the adjacent small rectangular one into two
neighbouring cells. The following section suggests that MB may be
as important an O_2 source as Hb in support of voluntary movement.

Comparison of Old and New Concepts of Tissue O_2 Transport

Existing mathematical descriptions of O_2 distribution in tissue
regard the capillary as a virtual point source supplying a large
spatially uniform sink; see right hand portion of Fig. 5. Since the
capillary is not treated as a barrier, the predicted gradient between
plasma and sarcolemma is very small, as indicated schematically on
the graph below. The principal site of resistance to O_2 mass transfer is classically thought to be the tissue cell because the calculated Po_2 falls very steeply with distance from the capillary. This
drop reflects the Kroghian assumption that O_2 is consumed at the
same rate at every point along the diffusion path, so distance from
the capillary source is the dominant variable. The assumed small
transcapillary O_2 gradient is incompatible with: 1) the ΔPo_2 required
to release O_2 from red cell monolayers (Thews, 1959), 2) with the
experimentally observed behavior of the capillary as O_2 barrier
(Honig and Odoroff, 1981; Rose and Goresky, 1982), 3) with our
intracapillary model and transit time analyses (Honig and Odoroff,
1981; Federspiel et al., 1982). Most importantly, a small trans-

capillary gradient is directly refuted by our Mb measurements. These
latter also prove that intracellular gradients in O_2 content and
Po_2 are much less steep than predicted by current models of free
and facilitated diffusion (Grunewald and Sowa, 1977; Fletcher, 1980).
Finally, the Mb measurements document extensive interaction of
diffusion fields. Clearly, the conventional model must be abandoned.

The left-hand portion of Fig. 5 summarizes our current conception of O_2 delivery during a normal voluntary movement of red muscle.
A centrally located working fiber is shown surrounded by resting
cells, as expected for normal motor unit function (Buchtal and
Schmalbuch, 1980). The resting cells should contain well-saturated
Mb at O_2 tensions near 20 Torr (Gayeski and Honig, 1978; Gayeski,
1981). In contrast, Mb in working fibers buffers Po_2 below 5 Torr
(Gayeski and Honig, 1982). Since O_2 gradients from capillaries to
the working fibers are much larger than to resting fibers the former
receives the bulk of the transcapillary O_2 flux. The effect of the
motor unit organization is to make the volume of tissue supplied by
each capillary smaller than in the conventional model at the right.
The Po_2 gradient from resting to working fibers is substantial, and
the surface area of the muscle cell is greater than capillary surface
area. Consequently, the amounts of O_2 delivered to the active

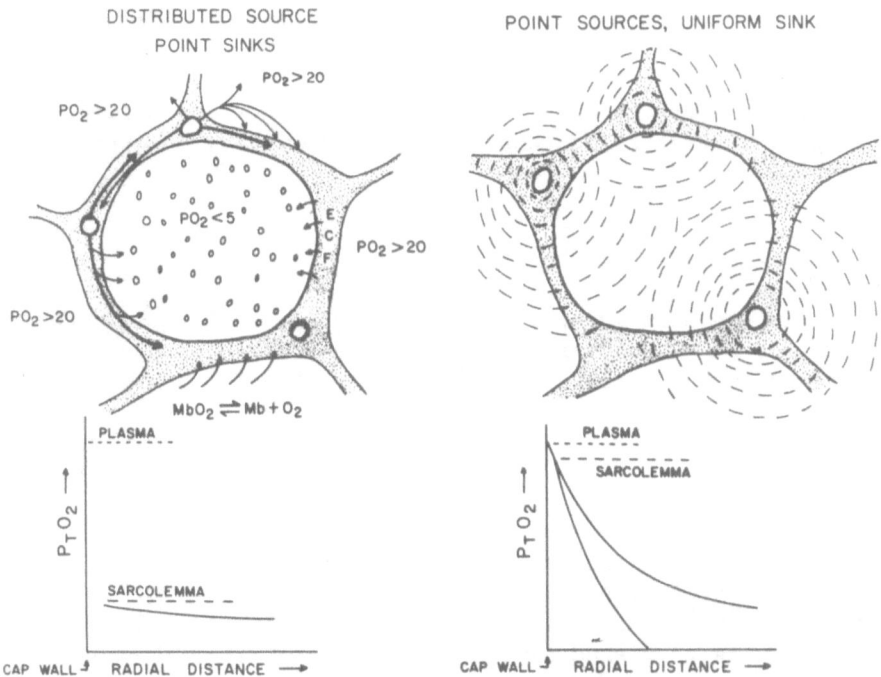

Fig. 5.   Schematic comparison of Kroghian viewpoint (right) with
          new concepts suggested by data (left).

fiber from contiguous muscle cells and capillaries could be comparable. In effect, the working fiber is supplied by the extracellular fluid acting as a distributed O_2 source much as though the fiber were suspended in a solution. The Po_2 of the solution (ECF) would be buffered by Hb and Mb.

O_2 disappears into mitochondria, which are virtual point sinks on the scale of a whole cell. The mitochondria (open dots in Fig. 5), occupy only 10% of cell volume in dog gracilis. The ΔPo_2 between mitochondria is very small because such tissue does not consume O_2. Note that the geometry we propose (distributed, interactive O_2 sources and point sinks) is precisely the reverse of the conventional model of the capillary as virtual point source delivering O_2 at high flux density into a uniform sink. The same geometric principles apply to simulated exercise in the laboratory, and offer an explanation for the shallow Mb gradients we observe. Mathematical modelling of Mb saturation for the geometry shown in Fig. 5 is in progress.

SUMMARY

1. The capillary is the principal barrier at high $\dot{V}o_2$.

2. Mathematical modelling indicates that at high flow and $\dot{V}o_2$ the time required for release of O_2 is greater than red cell transit time in some capillaries. This convective shunting appears to be particularly important in the myocardium.

3. The Mb acts to buffer Po_2 below 5 Torr during muscle contraction. This greatly increases the transcapillary O_2 gradient and promotes O_2 delivery.

4. During voluntary movements, Mb should act as a major O_2 source in parallel with capillaries.

5. The Kroghian model of the capillary as a "low-resistance" point source of O_2 supplying a spatially uniform sink appears to be the reverse of the actual geometry of O_2 supply to working red muscle.

REFERENCES

Buchtal, F., and Schmalbuch, H., 1980, Motor unit of mammalian muscle, Physiol. Rev., 60:90.
Chance, B., Oshino, N., Sugano, T., and Mayevsky, A., 1973, Basic principles of tissue oxygen determination from mitochondrial signals, Adv. Exp. Med. Biol., 37a:277.
Clark, A., and Clark, P., The capture of chemical reations in tissue by freezing, Math. Biosci., in preparation.

Cole, R.P., 1982, Myoglobin function in exercising skeletal muscle, <u>Science</u>, 216:523.

Federspiel, W.J., Clark, A. Jr., and Cokelet, G.R., 1982, Oxygen delivery: the view from the red cell, <u>Microvasc. Res.</u>, 23:251.

Federspiel, W.J., and Sarelius, I.H., Role of red cell spacing on the intracapillary transport of oxygen, <u>Microvasc. Res.</u>, in preparation.

Fletcher, J.E., 1980, On facilitated oxygen diffusion in muscle tissues, <u>Biophys. J.</u>, 29:437.

Gayeski, T., 1981, A cryogenic microspectrophotometric method for measuring myoglobin saturation in subcellular volumes; Application to resting dog gracilis muscle. Ph.D. Dissertation, University of Rochester, Rochester, N.Y.

Gayeski, T.E.J., and Honig, C.R., 1978, Myoglobin saturation and calculated Po$_2$ in single cells of resting gracilis muscles, <u>Adv. Exp. Med. Biol.</u>, 94:77.

Gayeski, T.E.J., and Honig, C.R., 1982, Direct measurement of intracellular O$_2$ gradients: role of convection and myoglobin, <u>Adv. Exp. Med. Physiol.</u>, in press.

Grunewald, W.A., and Sowa, W., 1977, Capillary structures and O$_2$ supply to tissue, <u>Rev. Physiol. Biochem. Pharmacol.</u>, 77:149.

Hellums, J.D., 1977, The resistance to oxygen transport in the capillaries relative to that in the surrounding tissue, <u>Microvasc. Res.</u>, 13:131.

Honig, C.R., 1981, "Modern Cardiovascular Physiology", Little Brown Co., Boston, pp. 223-227.

Honig, C.R., and Odoroff, C.L., 1981, Calculated dispersion of capillary transit times: significance for oxygen exchange, <u>Am. J. Physiol.</u>, 240:H199.

Honig, C.R., Odoroff, C.L., and Frierson, J.L., 1982, Active and passive capillary control in a red muscle at rest and in exercise, <u>Am. J. Physiol.</u>, 243:H196.

Huxley, V.H., and Kutchai, H., 1981, The effect of the red cell membrane and a diffusion boundary layer on the rate of oxygen uptake by human erythrocytes, <u>J. Physiol.</u>, 316:75.

Klitzman, B., and Duling, B.R., 1979, Microvascular hematocrit and red cell flow in resting and contracting striated muscle, <u>Am. J. Physiol.</u>, 237:H481.

Kreuzer, F., and Yahr, W.Z., 1960, Influence of red cell membrane on diffusion of oxygen, <u>J. Appl. Physiol.</u>, 15:1117.

Kutchai, H., 1970, Numerical study of oxygen uptake by layers of hemoglobin solution, <u>Respir. Physiol.</u>, 10:273.

Moll, W., 1969, The influence of hemoglobin diffusion on oxygen uptake and release by red cells, <u>Respir. Physiol.</u>, 6:1.

Popel, A.S., and Gross, J.F., 1979, Analysis of oxygen diffusion from arteriolar networks, <u>Am. J. Physiol.</u>, 237:H681-H689.

Rose, C.P., and Goresky, C.A., 1982, Barrier-limited transport of oxygen in the coronary circulation, <u>Fed. Proc.</u>, 41:1252.

Sarelius, I.H., and Duling, B.R., Direct measurement of microvascular hematocrit, red cell flux and transit time, <u>Am. J. Physiol.</u>, in press.

Sheth, B.V., and Hellums, J.D., 1980, Transient oxygen transport in
    hemoglobin layers under conditions of the microcirculation,
    Ann. Biomed. Eng., 8:183.
Sinha, A.K., 1969, Oxygen uptake and release by red cells through
    capillary wall and plasma layer (thesis), Univ. of California,
    San Francisco, CA.
Tamura, M., Oshino, N., Chance, B., and Silver, I.A., 1978, Optical
    measurements of intracellular oxygen concentration of rat heart
    in vitro, Arch. Biochem. Biophys., 191:8.
Thews, G., 1959, Untersuchung der Sauerstoffaufnahme und -abgabe
    sehr dünner Blutlamellen, Pflügers Arch. Ges. Physiol., 268:308.
Villadsen, J., and Sorensen, J.P., 1969, Solution of parabolic
    partial differential equations by a double collection method,
    Chem. Eng. Sci., 24:1337.

ACKNOWLEDGEMENT

    This research was supported by grants HLB 03290, GM-07136, and
HL-18208 from the U.S. Public Health Service.

METABOLIC RATE AND MICROCIRCULATION[*]

W. Kuschinksky

Department of Physiology, University of Munich
Pettenkoferstr. 12, 8000 München 2, FRG

THE CONCEPT OF METABOLIC CONTROL OF MICROCIRCULATION

The first hypothesis about the dependency of microcirculation on the metabolic rate of the corresponding tissue was put forward more than a hundred years ago (Roy and Brown, 1879; Gaskell, 1880). "It is surely worth while to see whether it is not possible that the chemical changes going on in the organ itself may not directly bring about a dilation of the blood vessels of that organ, and so, without the intervention of the nervous system, regulate its own blood supply according to its own needs" (Gaskell, 1880). This hypothesis seemed reasonable, since this mechanism allows an adjustment of the blood flow to the actual demand of the tissue. This means an economical distribution of blood flow resulting in a low heart work and energy consumption. It also means a linkage between the blood flow and that parameter, which should be its final determinant, i.e. the metabolic rate of the tissue.

Numerous attempts have been made to test the validity of this concept, as described by Barcroft (1963). The techniques available in the past allowed mainly an approach which tries to define the nature of the signal which is released by the tissue and which influences the microcirculation. This approach was chosen, since it does not necessitate a direct measurement of the local metabolic rate of the tissue. In two unifying hypotheses, an attempt was made to specify the signal: adenosine was proposed as the metabolic regulator of coronary blood flow (Berne, 1963) and the extracellular pH as the main factor controlling cerebral blood flow (Skinhoj, 1966; Lassen, 1968).

[*]Supported by the Deutsche Forschungsgemeinschaft

According to the concept of a metabolic control of microcircul-
lation these hypotheses implied the following events: with an increase in the metabolic rate of an organ, the metabolically dependent factor, either adenosine or H^+, should be released at a higher
rate thus resulting in a raised tissue concentration of this factor.
If the higher concentration of this factor induces a dilation of the
resistance vessels in the respective organ, it could be responsible
for the increased blood flow in this organ which is needed to match
the higher demand of oxygen and nutrients. This enhanced blood flow
could take place exactly at the point of increased metabolism in the
tissue thus allowing a local control of blood flow. Concerning the
signal which triggers the increase in blood flow, a unifying hypothesis could not be sustained: the gathered data indicated that a
variety of metabolic factors participate in the adjustment of blood
flow to the local metabolic demands; the importance of these factors
seemed to vary from organ to organ (Haddy and Scott, 1968). Even in
the same organ, the significance of the different metabolic factors
varied with time, as shown for skeletal muscle blood flow during
longer lasting functional hyperemia (Haddy and Scott, 1975).

These investigations have yielded information about the participation of metabolic factors in the control of microcirculation by
focussing the attention on measurements of tissue or blood concentrations on the one hand and local blood flows or vessel reactions
on the other hand. One problem of these investigations is that conclusions were drawn about the interdependency of metabolic rate and
microcirculation, which were based on measurements of only one of
these parameters. The other parameter, the metabolic rate, could
not be quantified locally. This is due to the fact, that in most
organs, the total metabolism is the result of the oxidation of a
number of substrates, and that even the utilization of the different
substrates varies. This makes the measurements of the actual turnover rate of the different substrates nearly impossible.

One organ shows a unique feature in this respect, the brain:
The brain uses only one substrate for oxidation, at least under
normal conditions. The fact that glucose is the only relevant fuel
for the brain has recently been utilized for the development of a
method to quantify the metabolic rate in the brain on a local level,
as will be described later. The brain is especially suited for the
quantification of the relationship between local metabolic rate and
microcirculation since there is, even under normal conditions, a
wide range of local blood flows and metabolic rates, which help to
establish the relationship on a broad basis.

Therefore, the brain was selected to gather information on the
following topics:

1. Metabolic control: is the concept of a metabolic control
applicable to the regulation of the cerebral circulation?

2. Relationship between metabolic rate and microcirculation:
when local metabolic rate and microcirculation are quantified for
the different brain structures, does a correlation exist between
these parameters? Can this relationship be influenced by altered
metabolic or circulatory conditions?

METABOLIC CONTROL OF THE CEREBRAL MICROCIRCULATION

It is generally accepted that at least two criteria must be
fulfilled for a substance to be accepted as a metabolic regulator
of the vascular resistance of an organ: 1. the concentration of the
substance should change, when the metabolism is altered, and 2. the
substance should have a vasoactive effect on the resistance vessels
of that organ, which should occur at concentrations comparable to
those measured during different metabolic activities. The most
likely candidates for a metabolic control in the brain will now be
discussed.

Hydrogen Ions

The vasodilating effect of increased concentrations of H^+ and
the constrictory effect of decreased H^+ concentrations have been
verified in pial arteries using local perivascular microapplication
(Kuschinsky et al., 1972; Haller and Kuschinsky, 1981) or flushing
of the brain surface under a pial window (Kontos et al., 1977). The
intraparenchymal resistance vessels in the brain apparently show
the same behavior, although their reactivity can only be inferred
from more indirect approaches (Pannier et al., 1972; Cameron and
Caronna, 1976; Britton et al., 1979). Whereas the vasoactive action
of H^+ has been generally accepted, the release of H^+ during functional activation has been a matter of debate for several years.
The initial data obtained from pH microelectrodes inserted into the
cortical tissue showed an increased H^+ activity starting 7 seconds
(Urbanics et al., 1978) or 25 seconds (Astrup et al., 1976) after
the onset of cortical seizure activity. The latency between the
immediate increase in blood flow parallel to the cortical activation
and the delayed increase in cortical H^+ acitivity argued against a
role of H^+ in the mediation of the early phase of functional hyperemia (Urbanics et al., 1978) or was even taken as evidence against
H^+ as a main factor controlling cerebral blood flow (Astrup et al.,
1976; Purves, 1978). Meanwhile, electrode measurements have been
published which show an immediate acidosis in the cortical tissue
starting with seizure acitivity without any time delay (Heuser,
1978; Kuschinsky and Wahl, 1979). An example is given in Fig. 2.
Although seizure activity represents extreme neuronal activation
and may induce changes which do not occur under more physiological
conditions of activation, these data are in favor of the hypothesis,
that H^+ can be considered as one of the main metabolic factors
which control cerebrovascular resistance. A more detailed analysis
of the experiments is given elsewhere (Kuschinsky, 1982).

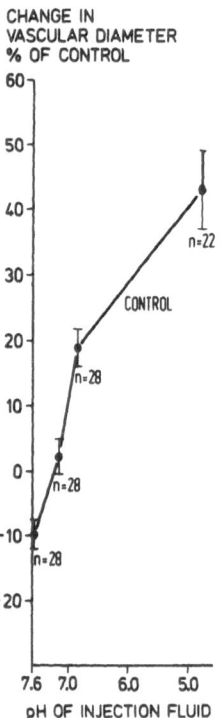

Fig. 1. Concentration response curve for H^+. Local micro-application to the outside of pial arteries and arterioles. From
Haller and Kuschinsky (1981).

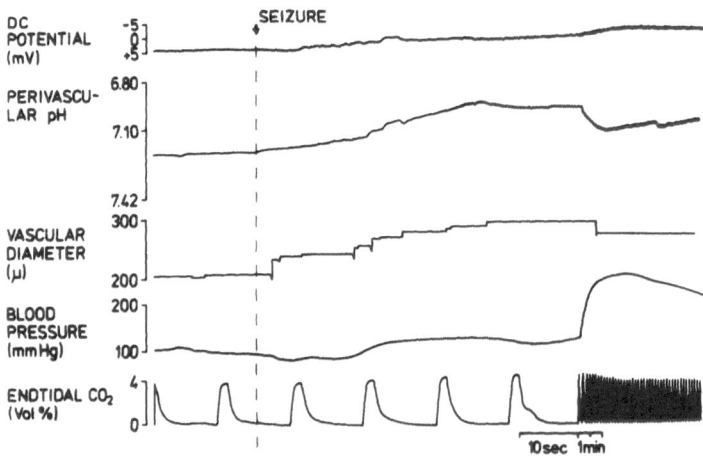

Fig. 2. Recording of an experiment, which demonstrates the effect
of seizure activity on the reference signal, which represents the DC potential, on the pial arterial diameter, on
the arterial blood pressure, and on the endtidal CO_2
concentration. From Kuschinsky and Wahl (1979).

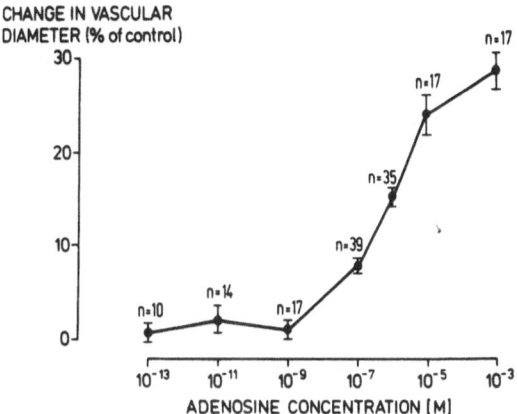

Fig. 3.   Concentration response curve for adenosine. Local micro-
          application to the outside of pial arteries and arterioles.
          From Wahl and Kuschinsky (1976).

Adenosine

The role of adenosine in the regulation of cerebral blood flow
has been reviewed recently (Winn et al., 1981a). Similar to the data
about H^+, a vasoactive action of adenosine has been accepted earlier
than the release from brain tissue during functional activation.

The dilating effect of extravascular adenosine on pial arteries
has been verified congruently by several groups (Berne et al., 1974;
Wahl and Kuschinsky, 1976; Gregory et al., 1980). The dilations
start at concentrations of adenosine (Wahl and Kuschinsky, 1976,
Fig. 3) which are comparable to those measured in the cerebrospinal
fluid (Berne et al., 1974) or calculated for the extracellular space
of the brain from measurements of tissue concentrations under the
assumption of a predominant extracellular localization of adenosine
(Schrader et al., 1980; Winn et al., 1981a). Calculations of the
perivascular concentrations of adenosine under physiological conditions are necessary to assign adenosine a physiological role in the
regulation of cerebrovascular resistance, since a concentration of
about 10^{-8} M has to be passed over to obtain a vasodilation (see
Fig. 3). Below this concentration there exists no vascular effect
of adenosine, in contrast to H^+ which induce constrictions, when
lowered and dilations, when increased, independent of the starting
concentration of H^+ (Kuschinsky et al., 1972; Haller and Kuschinsky,
1981, Fig. 1).

As to the release of adenosine during enhanced neuronal
activity doubts came up concerning its functional significance,
after Rehncrona et al. (1978) were unable to find any change in
brain adenosine content during epileptic seizures induced by i.v.

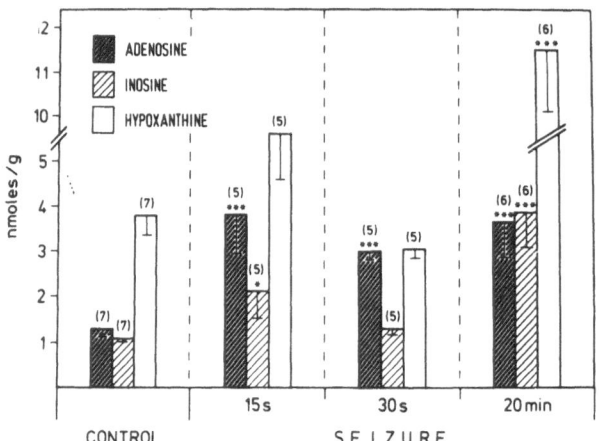

Fig. 4. Changes in the brain cortical content of adenosine, inosine
         and hypoxanthine following seizures induced by bicuculline.
         Asterisks indicate significant differences from control
         values. * p<0.025 *** p<0.005. From Schrader et al.
         (1980).

bicuculline, a GABA antagonist. These authors could also not find
an increase in brain adenosine content during hypoxia, which was at
variance with earlier results of Rubio et al. (1975). This finding
could also be matched with earlier data on brain activation, which
showed an increased brain adenosine content during eletrical stimulation in vivo (Rubio et al., 1975) and in brain slices (Pull and
McIlwain, 1972). Therefore, it was not too surprising, that two
other groups could indeed demonstrate a significant increase in
brain adenosine concentration during bicuculline induced seizures
(Schrader et al., 1980; Winn et al., 1980; Fig. 4). The reason for
the diverging results of the Rehncrona et al. (1978) is not evident.
It may be suspected that the speed of brain freezing was too low in
these experiments (Winn et al., 1981b). Taking all data together,
there is considerable evidence for a release of adenosine from
brain tissue activated by bicuculline or electrical stimulation.
The discussions and investigations now concentrate more on the
origin and localization of the measured adenosine: the original
concept implied a cellular breakdown of ATP to adenosine via AMP;
adenosine thus formed would diffuse into the extracellular space;
an additional site of origin may be extracellular ATP release from
purinergic nerves (Burnstock et al., 1978); this ATP could be broken
down to adenosine in the extracellular space. For these two pathways
the membrane-bound ectoenzyme 5'-nucleotidase is needed. Recently,
the possibility is being considered, that considerable amounts of
adenosine, in brain and heart, may be formed intracellularly via a
well known pathway from S-adenosylhomocysteine. S-adenosylhomocysteine is formed from S-adenosylmethionine by the transfer of a

methyl group. This transfer is necessary to yield a variety of compounds, like epinephrine, choline, creatine and others. Thus, the intracellular production of adenosine from S-adenosylhomocysteine would be linked to a number of essential transmethylation reactions.

Potassium Ions

The listing of K^+ in the context of a metabolic control of the cerebral microcirculation needs some explanation, but first, the experimental data will be presented. K^+ ions are released from neuronal cells into the extracellular space of the central nervous system during increased neuronal activity. This fact has been demonstrated by a number of experimental groups and is, therefore, a well-known phenomenon. Increases in K^+ activity in the extracellular fluid of the brain cortex have been detected during seizure activity as well as during increased sensory inputs using K^+-sensitive ion exchanger microelectrodes (for references see Kuschinsky and Wahl, 1978; Somjen, 1979). After K^+ release, increased pump activity is necessary to pump the K^+ back into the cells from where it has been released (Cordingley and Somjen, 1978), since a transport of K^+ to the blood stream is not a significant factor in the rapid clearing of K^+ from the extracellular space of the brain (Mutsuga et al., 1976). Increase pump activity means increased cell metabolism. A higher metabolic rate necessitates a higher blood flow in the corresponding brain area, when a metabolic control of cerebral blood flow exists. This higher blood flow can be met by a direct effect of extracellular K^+ on the resistance vessels of the brain. This effect is shown in Fig. 5. This figure shows the effect of increasing the extracellular K^+ concentration on the vascular diameter of pial arteries of anesthetized cats and confirms the original data (Kuschinsky et al., 1972). The higher sensitivity to K^+, when compared to the former data (Kuschinsky et al., 1972) is due to the use of a mock cerebrospinal fluid whith a composition, which is closer to the physiological conditions. An increased K^+ concentration, as occurs during enhanced neuronal activity, is the trigger for a vasodilatation which starts within seconds. The effect of a lowered K^+ concentration, vasoconstriction, may be of less physiological significance, but undershoots of K^+ activity have been measured following stimulus - induced rises in K^+ activity (Heinemann and Lux, 1975).

The action of K^+ in the control of blood flow takes place in a slightly different manner, when compared to the action of K^+ and adenosine. Whereas the release of K^+ and adenosine takes place after metabolism has increased, K^+ ions are released in the moment of electrical activity, their release being a stimulus for increased pump activity, increased metabolism and increased blood flow. This means, that the order of events is different: H^+ and adenosine are released after an increase in metabolism and can exert their dilating effects only secondary to the increased metabolism; K^+ ions are released parallel to or even before an increase in metabolism takes

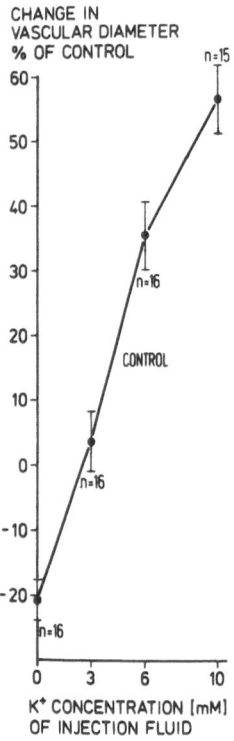

Fig. 5.   Concentration response curve for K^+. Local micro-applica-
          tion to the outside of pial arteries and arterioles. From
          Haller and Kuschinsky (1981).

place; thus K^+ ions are responsible for a quick adjustment of the
microcirculation to the metabolic needs of the tissue.

Calcium Ions

        The contribution of Ca^{++} to the metabolic control of the micro-
circulation can be described in analogy to the contribution of K^+.
The movements of Ca^{++} take place during the course of the action
potential. The pump activity, necessary to reestablish the normal
ionic gradients for Ca^{++}, represents the metabolic event which
necessitates a higher blood flow. The difference between K^+ and Ca^{++}
lies in the direction of the concentration changes: whereas the
action potential is accompanied by an increase of extracellular
K^+, which is released by the neurons, extracellular Ca^{++} is varied
in the opposite direction: during the action potential, Ca^{++} permeability of the neurons is increased, thus allowing a movement of Ca^{++}
along its concentration gradients, which is from the extracellular
to the intracellular compartment (Heinemann et al., 1977; Nicholson,
1980). This decreased extracellular Ca^{++} concentration during
enhanced neuronal activity induces a dilation of pial arteries (Betz

and Csornai, 1978). However, the quantitative contribution of Ca^{++}
to the functional hyperemia in the brain seems extremely limited:
the decrease of extracellular Ca^{++} activity in the somatosensory
cortex during repetitive stimulation of the contralateral forepaw
is about 0.1 $mmol \cdot l^{-1}$. For comparison, extracellular K^{+} activity is
increased simultaneously by more than 2 $mmol \cdot l^{-1}$ (Heinemann et al.,
1977). Whereas the decrease in Ca^{++} is too small to induce a
detectable change in pial arterial diameter (Betz and Csornai, 1978),
the simultaneous increase in extracellular K^{+} can induce a dilation
of pial arteries of about 25% (Haller and Kuschinsky, 1981). This
comparison demonstrates a minor importance of Ca^{++} in the regulation
of the cerebral microcirculation.

Metabolic Control under Pathophysiological Conditions:

The data presented here are strong evidence for the existence
of a metabolic control of the cerebral microcirculation under conditions of an acute rise in the energy demand of the tissue. The basis
of this metabolic control is the increased release of metabolic
factors during enhanced metabolic activity and the vascular action
of these metabolic factors. On the basis of these two components,
concentration change and vasoactivity, the metabolic control mechanism is effective in the normal control of the cerebrovascular
resistance. One interesting question is, how these components of
the metabolic control mechanism are affected under pathophysiological conditions, especially ischemia.

As to the first component, the concentration change of metabolic factors, there is evidence that the concentration of all
mentioned factors are changed in such a way that vasodilation would
be expected to occur under physiological conditions. During ischemia, extracellular activities of H^{+}, K^{+} (Astrup et al., 1977; Hossmann et al., 1977) and adenosine (Berne et al., 1974) are increased
and Ca^{++} activity is decreased (Harris et al., 1981). Although
interactions between these factors might limit the extent of vascular reaction (Wahl and Kuschinsky, 1977), there is no doubt, that
these changes in activity represent a strong stimulus for a vasodilation. The second component of the metabolic control mechanism,
the vascular reactivity, has been less convincingly analyzed during
ischemic and postischemic conditions. Studies have been concentrated
on the reactivity to changes in CO_2, probably because of the easy
experimental approach to this question. These studies showed con-
sistently a reduction or abolishment of the cerebrovascular responsiveness to CO_2 in the ischemic area. This reduced reactivity was
found in experimental studies in baboons (Symon et al., 1974), dogs
(Nemoto et al., 1975) and cats (Hossmann et al., 1973) as well as
in clinical investigations in man (Hoedt-Rasmussen et al., 1967;
Paulson, 1970; Paulson et al., 1970). Although there was improvement
in responsiveness to CO_2 within weeks after the ischemic impact
(Waltz, 1970), an impaired reactivity to CO_2 remained for as long

as 3 years after an experimental infarction (Symon et al., 1975).
In the same model of a chronic stable stroke, there was no evidence
of autoregulation in the ischemic center (Symon et al., 1976). The
general conclusion from these experiments was the existence of vasoparalysis in cerebral vessels during and after an ischemic impact.
However, from our knowledge about metabolic control factors in the
regulation of cerebral blood flow it is evident, that experiments
in which the CO_2 reactivity is tested, can yield information only
about the sensitivity to one metabolic factor, i.e. H^+, since it is
well known that the changes in cerebrovascular resistance induced
by CO_2 are mediated by H^+ (for references see Kuschinsky, 1982).

It seemed, therefore, to be appropriate to evaluate the concept
of postischemic vasoparalysis by more extensive investigation of
the reactivity of cerebral vessels after ischemia. As a model of
ischemia, air embolism was chosen as described by Fritz and Hossmann
(1979). This model has been extensively investigated by these
authors in connection with cerebral blood flow, electroencephalography, and cerebral metabolic rate for oxygen. This form of
ischemic insult affects the portion of the cerebral arteries located
proximally to the capillaries and distally to the large segments.
It is transient, and has been observed clinically as a hazard of
deep sea diving and a complication of open heart surgery (for
references see Haller and Kuschinsky, 1981). Using this model, the
microapplication technique was chosen to assess the vascular reactivity. This technique enables one to test the reactions of individual
pial arteries to local changes in the perivascular environment,
both under control conditions and after an ischemic insult. Thus,
each vessel can serve as its own control. In addition, using this
method, the occlusion of the vessel under investigation can be
verified by direct observation, and the onset of re-perfusion is
directly visible.

Under these experimental conditions, the reactivity of the
pial vessels to changes in the perivascular pH was tested during
the phase of postischemic re-perfusion. The results are shown in
Fig. 6. It is evident from this figure, that the reactivity to
changes in the perivascular pH is abolished after air embolism. The
slope of the concentration response curve after air embolism is not
significantly different from zero. These data show that the model
of air embolism, as chosen for the experimental approach to the
question of postischemic vasoparalysis, is acceptable for investigating this question, since the result of an abolished reactivity
of pial vessels to changes in perivascular pH in this model is in
accordance with the result obtained from experimental animals and
patients with ischemic brain diseases which has revealed an abolished reactivity of cerebral vessels to changes in arterial Pco_2.
The micro-application technique in combination with the ischemic
model of air embolism allows one to evaluate the concept of vasoparalysis by testing the postischemic vascular reactivity to factors
other than H^+.

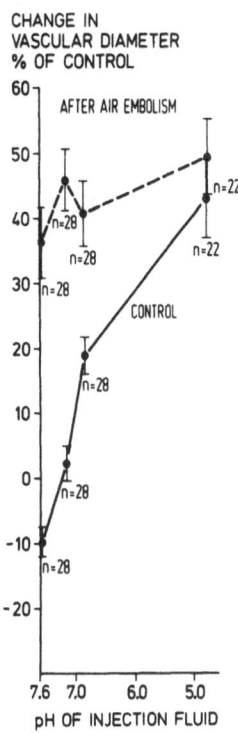

CHANGE IN
VASCULAR DIAMETER
% OF CONTROL

Fig. 6.   Concentration response curves for K^+. The control curve is
          represented by the solid line, and the curve after air
          embolism by the broken line. From Haller and Kuschinsky
          (1981).

        Therefore, the reactivity to K^+ was tested under identical
conditions, since K^+ is certainly one of the most relevant local
factors in the regulation of the cerebral microcirculation (see
above). Fig. 7 shows the result of these experiments. In contrast
to the results obtained for H^+ (Fig. 6), these data show that the
vascular reactivity to K^+ is mainly preserved after the ischemic
impact. There is only a slight reduction in the vascular reactivity
to K^+ after air embolism. The differential impairment of vascular
reactivity after an air embolism does not support the concept of
postischemic vasoparalysis.

        This graded impairment of sensitivity is the consequence of
different specific vascular effector mechanisms to H^+ and K^+, which
vary in their vulnerability to the ischemia. The vascular response
to K^+ has been attributed to an alteration in membrane conductivity
of K^+ (Wahlström, 1971) and/or to the effect of an electrogenic Na^+
pump (Chen et al., 1972). As far as pH is concerned, hydrogen ions
have been shown to modify Ca^{++} fluxes in the vascular smooth muscle
membrane (van Breemen et al., 1972). Another possibility of ex-

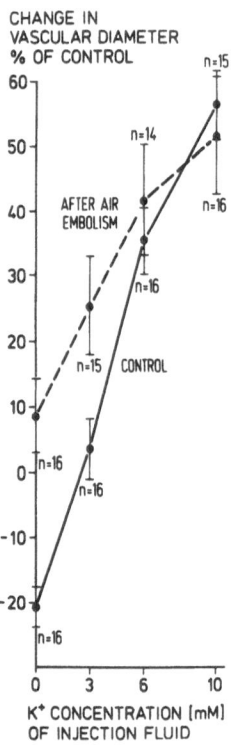

Fig. 7.   Concentration response curve for K^+. The control curve
          is represented in the solid line, and the curve after air
          embolism by the broken line. From Haller and Kuschinsky
          (1981).

plaining the vascular effect of pH is, that H^+ interfere with the
effect of Ca^{++} on the actomyosin-system thus modifying the activity of the myofibrillar ATP'ase by competitive antagonism; there is
also evidence for a direct inhibitory effect of H^+ on the activity
of this enzyme (Schädler, 1967). The fundamental mechanism of vascular smooth muscle mechanical action does not seem to be severely
damaged by ischemia, since the reaction to one of the factors tested,
i.e., K^+, was only slightly reduced after ischemia. From this one
may conclude that the damage of vascular reactivity to H^+ occurs
somewhere in the sequence of reactions which regulate the activity
of the myofibrillar ATP'ase; this would have to be at an earlier
stage where the H^+ and K^+ mechanisms are still dissociated. This
hypothesis would imply that the activation of this final common
pathway, i.e. the chemo-mechanical process of the actomyosin systems,
takes place via two separate chains of reactions, one for H^+ (and
therefore CO_2) and the other for K^+. Because of their differential
vulnerability to ischemia, the concept of total vasoparalysis after
ischemia does not seem justifiable. It appears more likely, that
the ischemic impact leads to an accumulation of metabolic factors,

some of which mediate the postischemic hyperemia, while others have lost their vascular acitivity.

QUANTIFICATION OF BLOOD FLOW IN EACH BRAIN STRUCTURE

As explained earlier in this article, it is now possible to quantify the relationship between local metabolic rate and micro-circulation in the brain. The prerequisite for such a quantification is the availability of methods, which allow a reliable measurement of both of these parameters, preferably in the conscious animal. Therefore, the methods will be described which were applied for this quantification, before the results of this approach are presented.

Global Blood Flow

The basis of a quantitative measurement of cerebral blood flow was laid by Kety's idea to apply Fick's principle to the measurement of cerebral blood flow, as documented by his mentor, C.F. Schmidt (1982). This principle "postulates in its simplest form that the quantity of any substance taken up in a given time by an organ from the blood which perfuses it is equal to the total amount of the substance carried to the organ by the arterial inflow less the amount removed by the venous drainage during the same period" (Kety and Schmidt, 1948). The nitrous oxide method developed on this basis (Kety and Schmidt, 1948) became the classical method to measure cerebral blood flow in experimental animals and man. This method allows the quantification of the global cerebral blood flow, when the course of the arterial and cerebral venous concentration of nitrous oxide are known. The other parameter necessary for a quanti-fication is the tissue concentration of the indicator substance. This cannot be determined directly, but is calculated from the cerebral venous concentration by multiplying the cerebral venous concentration of nitrous oxide with the partition coefficient for nitrous oxide between brain and blood, under the assumption of an equilibrium between brain and cerebrovenous blood with respect to nitrous oxide tension. This results in the following equation:

$$\frac{F}{W} = \frac{c_v \cdot \lambda}{\int_0^T (c_a - c_v)\,dt} \tag{1}$$

where $\frac{F}{W}$ equals the amount of blood flow per unit mass of tissue and minute, c_a and c_v equal the arterial and cerebrovenous concentration of the indicator substance, e.g. nitrous oxide: λ is the brain-to-blood partition coefficient of the indicator substance, e.g. nitrous oxide; and T represents the duration of the experiment.

Local Blood Flow, Inert Gases

Application of these methods for the measurement of global blood flow in man revealed results, which were a bit disappointing to the investigators in Kety's lab: when they subjected the brains of the experimental persons to a state which they considered to be maximal activation (mental arithmetics) or inactivation (sleep) they found only small or no changes in global brain blood flow (Sokoloff et al., 1955; Mangold et al., 1955). Therefore, they soon tried to apply the same principles of measurement to the development of a method which allows one to measure local cerebral blood flow in the brain. It became apparent, that local concentrations of the indicator gas could be detected in the brain, when radioactive gases were used as indicators and autoradiographic methods were applied after cryosectioning of the brain. Determination of local tissue concentrations from the optical densities in the autoradiographs had also the advantage, that sampling of venous blood from the brain could be avoided since

$$c_v = \frac{c_i}{\lambda} \tag{2}$$

where C_i equals the tissue concentration of the radioactive tracer. Insertion of equation (2) into equation (1), expression in differential form and integration yields

$$C_i(T) = \lambda \cdot K \int_0^T c_a(t) \cdot e^{-K(T-t)} dt \tag{3}$$

K is defined as follows:

$$K = \frac{F}{W \cdot \lambda} \tag{4}$$

This autoradiographic method was introduced by Landau et al. (1955) and its theoretical basis was given by Freygang and Sokoloff (1958). The advantages of this method had to be bought with some disadvantages: since each set of measurements requires the sacrifice of the animal, repeated determinations of local blood flow in the same brain struture of one animal are impossible, and the method cannot be applied in man.

Local Blood Flow, Nonvolatile Tracers

The method of Landau et al. (1955) and Freygang and Sokoloff (1958) could have opened a new era in the measurement of cerebral blood flow, because of its theoretical soundness in the presentation

of values of local flow in each brain structure of the awake animal.
However, this kind of study with an inert radioactive gas indicator
has never been repeated, neither in the laboratory of origin nor
elsewhere. This is due to the technical disadvantage of using a
radioactive inert gas like (131J) CF_3J. This gas is not commercially available, is rather difficult to synthesize and purify, and the
processing of the tissue has to be performed under extreme precautions to avoid a loss of the highly volatile tracer.

It was attempted to circumvent these difficulties by using
nonvolatile radioactive tracers. A substance, which revealed a
number of advantages, when compared to (131J) CF_3J, was (14C)-
antipyrine: in addition to the easier processing of the brain sections, it yields a better autoradiographic resolution and, because
of the long half-life of (14C), permanent radioactive standards can
be made (Reivich et al., 1969). It was later recognized by Eckman
et al., (1975), that the application of (14C)-antipyrine causes a
serious problem: these authors could demonstrate in simulation
studies and by a direct measurement of cerebral blood flow in cats
using (14C)-antipyrine, that the brain tissue is apparently not in
diffusion equilibrium with the venous blood. This was in accordance
with an earlier study of Eklöf et al. (1974) which pointed into the
same direction. The uptake of (14C)-antipyrine by the brain tissue
is not entirely flow limited, although this is essential for a blood
flow marker, but it is partly also permeability limited. This means
that equation (2) does not hold for conditions of higher blood flow.
The quantitation of blood flow is still possible, if a value is
introduced into equation (4), which quantifies the degree of dis-
equilibrium between brain tissue and cerebral venous blood:

$$K = \frac{m \cdot F}{W \cdot \lambda} \tag{5}$$

A value of m = 1 exists with flow limitation (= inert gases), whereas a value of m between 0 and 1 indicates diffusion limitation. For
accepting a substance as a blood flow marker, it should have a value
of m = 1. The problem is then that the actual value of m should be
determined for each brain structure, because it must vary with blood
flow. Since such a local determination of m is impossible, the
search for nonvolatile markers of cerebral blood flow was continued.

Since Sakurada et al. (1978) have introduced (14 C)-iodoanti-
pyrine as a blood flow marker, this substance is accepted as the
best available indicator for the measurement of local cerebral blood
flow by tissue sampling (e.g. Ohno et al., 1979). Therefore, this
indicator was also chosen in the present study for the quantification of the relationship between local cerebral glucose utilization
and local cerebral blood flow. Local cerebral blood flow was determined in the awake rat by using the method, as described by Sakurada
et al. (1978). The only methodological difference was that, in the

present study, (14 C)-iodoantipyrine was infused with increasing
speed during the 60 seconds' measuring period. This was done to
ensure a high arteriovenous difference for the blood flow marker
over the whole measuring period.

The results of these measurements are shown, for selected brain
structures, in the right part of Table 1. The heterogeneity of blood
flow in the brain, depending on the brain structure, is evident.
Table 1 shows high values of local blood flow, when compared to
values for the same brain structures using other methods of measurement of brain flow (Lacombe et al., 1980). This argues against a
diffusion limitation of iodoantipyrine. However, a direct evidence
that there exists no diffusion limitation for iodoantipyrine in the
structures of the rat brain, which have high flow rates, has not
been given. Sakurada et al. (1978) could only compare local cerebral
blood flow values, as obtained using (131 J) CF_3J and (14 C)-iodo-
antipyrine in the cat. Unfortunately, studies using (131 J) CF_3J
in the rat brain have not been performed. Therefore, the suitability
of (14 C)-iodoantipyrine for the measurement of high flow rates has
been subjected to some doubts (Goldman et al., 1980; Tomita and
Gotoh, 1981). Goldman et al. (1980) have proposed 1-butyl-3-phenyl-
thiourea as a blood flow marker, since it exerted a higher single
pass extraction value than iodoantypyrine. However, even the proposed marker, 1-butyl-3-phenylthiourea, did not seem to be freely
diffusible in the brain. Apparently an ideal marker of local cerebral blood flow using autoradiographic techniques has not yet been
found. The alternative, to choose other methods for quantifying
local cerebral blood flow, is also not satisfying, because other
methods have different inherent problems (Edvinsson and MacKenzie,
1977; Kuschinsky and Wahl, 1978; Lacombe et al., 1980).

QUANTIFICATION OF THE METABOLIC RATE IN EACH BRAIN STRUCTURE

The idea of also using a quantitative autoradiographic technique for the measurement of the local metabolic rate in the brain
tissue was difficult to pursue. The normal substrates of cerebral
energy metabolism are oxygen and glucose. Both of these substrates
are so rapidly converted to CO_2 in the brain, that a quantitative
trapping of one of them for autoradiographic purposes is impossible. Sokoloff has recently introduced a method which uses a labeled
analogue of glucose, 2-deoxy-D-(14 C) glucose ((14 C)DG) (Sokoloff
et al., 1977). (14 C)DG is metabolized through the main pathway of
glucose metabolism. The product of phosphorylation, however, (14 C)
DG-6-phosphate, is trapped in the tissue, which is necessary for
the application of the quantitative autoradiographic technique. In
order to achieve a state, in which the concentration of radioactivity in the tissue could be correlated with the rate of glucose
utilization, Sokoloff has developed a model, which is based on the
biochemical properties of deoxyglucose and glucose (Sokoloff et al.,

Table 1. Glucose utilization and blood flow in selected brain
         structures of the awake rat.

	local cerebral glucose utilization (u moles/(100 g.min))	local cerebral blood flow (ml/(100 g.min))
Sensory motor system		
Somatosensory cortex	109	167
Inferior olivary nucleus	74	139
Cerebellar cortex	61	102
Vestibular nucleus	111	190
Extrapyramidal system		
Caudate nucleus	103	147
Globus pallidus	56	57
Substantia nigra	65	93
Limbic system		
Hippocampus	82	114
Dentate gyrus	70	111
Amygdala	53	104
Septal nucleus	59	114
Nucleus accumbens	81	150
Hypothalamus	55	94
Mamillary body	113	172
Visual system		
Visual cortex	112	165
Lateral geniculate body	95	150
Superior colliculus	92	147
Thalamus: posterior lateral nucleus	106	179
Auditory system		
Auditory cortex	149	307
Medical geniculate body	117	255
Inferior colliculus	181	326
Lateral lemniscus	106	193
Superior olivary nucleus	140	228
Cochlear nucleus	125	202
White matter		
Corpus callosum	44	44
Internal capsule	37	46

1977). The diagrammatic representation of this model is shown in
Fig. 8. The model takes advantage of the fact that deoxyglucose and
glucose are competitive substrates for both blood-brain transport
and hexokinase-catalyzed phosphorylation. The quantity of (14 C)DG-
6-phosphate accumulated in the tissue at any time after introduction
of (14 C)DG into the circulation is equal to the integral of the rate
of (14 C)DG phosphorylation by hexokinase during that interval of
time. This integral is related to the amount of glucose that has
been phosphorylated over the same interval, and this relationship
can be quantified and expressed as glucose utilization of the brain
tissue. The operational equation derived by Sokoloff is shown in
Fig. 9 (Sokoloff, 1978). The numerous applications of the deoxyglucose method have been reviewed recently (Sokoloff, 1981). Our
results of the local quantification of glucose utilization in the
brain of normal conscious rats are shown in Table 1 for selected
brain structures.

CORRELATION BETWEEN LOCAL CEREBRAL GLUCOSE UTILIZATION AND LOCAL
CEREBRAL BLOOD FLOW

The availability of methods which allow the quantification of
local glucose utilization and local blood flow in each brain structure makes it now possible to define the quantitative relationship
between these parameters. Because of the wide variability of
measured values of both local cerebral blood flow and local cerebral
glucose utilization, it was of interest to investigate this relationship in normal conscious rats. Therefore, in one experimental group
of rats, local gluose utilization was measured in 39 structures of
grey and white matter, and in the other group local blood flow was
determined in the same structures. The results are shown in Fig. 10.
The excellent correlation between local cerebral glucose utilization
and local cerebral blood flow is apparent. These data can be taken
as evidence for a coupling between local metabolism and local blood
flow in the brain, which takes place on a long-term basis. The data
are in accordance with a study of the relationship between the local
metabolic rate for oxygen and local blood flow in the brain cortex
of man (Raichle et al., 1976). These authors found a correlation
between the local metabolic rate of oxygen and local blood flow.

It would be of interest to compare the values found for local
glucose utilization and local blood flow with the values determined
for global metabolism and blood flow in many earlier studies. An
additional measurement of global glucose utilization and blood flow
became possible by the aid of computer-assisted densiomètry (Goochee
et al., 1980). The films with the autoradiograms were scanned, and
the optical densities of all sections were measured. These values
were then converted to either blood flows or glucose utilizations.
The average of all these values allows one to compare the results
of the autoradiographic studies with the results of other studies,
in which global methods had been employed, whereas taking just the

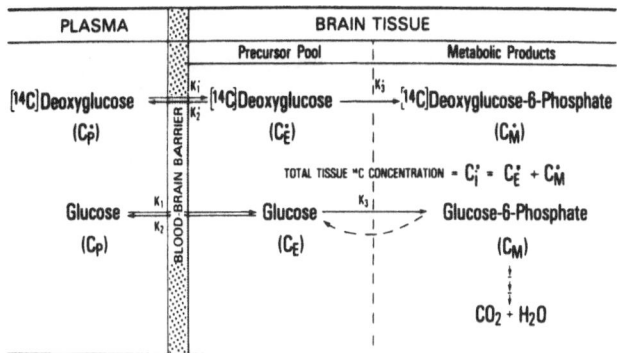

Fig. 8.   Theoretical model of the deoxyglucose method. C_i^* represents
the total (14 C) concentration in brain tissue. C^* and C_p
represent the concentrations of (14 C)deoxyglucose ((14 C)DG)
and glucose in the arterial plasma, respectively; C^* and
C_E represent their respective concentrations in the
tissue pools that serve as substrates for hexokinase. C_M
represents the concentration of (14 C)deoxyglucose-6-phos-.
phate in the tissue. The constants k_1^*, k_2^*, and k_3^*, represent the rate constants for carrier-mediated transport of
(14 C)DG from plasma to tissue, for carrier-mediated transport back from tissue to plasma, and for phosphorylation
by hexokinase, respectively. The constants k_1, k_2, and k_3
are the equivalent rate constants for glucose. (14 C)DG
glucose share and compete for the carrier that transports
both between plasma and tissue and for hexokinase which
phosphorylates them to their respective hexose-6-phosphates.
The dashed arrow represents the possibility of glucose-6-
phosphate hydrolysis by glucose-6-phosphate activity, if
any. From Sokoloff et al. (1977).

average of the 39 measurements could give a different value, since
it would only average the values of some selected structures. The
average in this study, obtained by scanning the whole films, was
for gluose utilization 68.6 u moles/(100 g·min) and for blood flow
106.7 ml/(100 g·min). These values are in the range expected for
the awake rat brain from other, global methods (Siesjö, 1978).

The question, which mechanism mediates the tight relationship
between local glucose utilization and local blood flow, as shown in
Fig. 10, can be investigated by testing the influence of changing
experimental conditions on the relationship. A change in systemic
acid base status was chosen as the experimental model because
metabolic acidosis is a rather common clinical feature and the H^+
concentration is one of the factors which mediate the acute adjustment of local flow to changing metabolic needs, as discussed
earlier in this paper. The experimental animals obtained 0.35 M
NH_4Cl in their drinking water for 5-6 days. By this procedure,

General Equation for Measurement of Reaction Rates with Tracers:

$$\text{Rate of Reaction} = \frac{\text{Labeled Product Formed in Interval of Time, 0 to T}}{\left[\begin{array}{c}\text{Isotope Effect}\\ \text{Correction Factor}\end{array}\right]\left[\begin{array}{c}\text{Integrated Specific Activity}\\ \text{of Precursor}\end{array}\right]}$$

Operational Equation of $[^{14}C]$ Deoxyglucose Method:

Labeled Product Formed in Interval of Time, 0 to T

$$R_i = \frac{\overbrace{C_i^*(T)}^{\substack{\text{Total } ^{14}\text{C in Tissue}\\ \text{at Time, T}}} - \overbrace{k_1^* e^{-(k_2^*+k_3^*)T} \int_0^T C_p^* e^{(k_2^*+k_3^*)t}\, dt}^{^{14}\text{C in Precursor Remaining in Tissue at Time, T}}}{\underbrace{\left[\frac{\lambda\cdot V_m^*\cdot K_m}{\phi\cdot V_m\cdot K_m^*}\right]}_{\substack{\text{"Isotope Effect"}\\ \text{Correction}\\ \text{Factor}}} \underbrace{\left[\underbrace{\int_0^T \left(\frac{C_p^*}{C_p}\right) dt}_{\substack{\text{Integrated Plasma}\\ \text{Specific Activity}}} - \underbrace{e^{-(k_2^*+k_3^*)T} \int_0^T \left(\frac{C_p^*}{C_p}\right) e^{(k_2^*+k_3^*)t}\, dt}_{\substack{\text{Correction for Lag in Tissue}\\ \text{Equilibration with Plasma}}}\right]}_{\text{Integrated Precursor Specific Activity in Tissue}}}$$

Fig. 9. Operational equation of deoxyglucose method. T represents the time at the termination of the experimental period; λ equals the ratio of the distribution space of deoxyglucose in the tissue to that of glucose; ϕ equals the fraction of glucose which, once phosphorylated, continues down the glycolytic pathway; and K_m^* and V_m^* and K_m and V_m represent the familiar Michaelis-Menten kinetic constants of hexokinase for deoxyglucose and glucose, respectively. The other symbols are the same as those defined in Fig. 8. From Sokoloff (1978).

arterial base excess was decreased by 16.7 mmol·l^{-1} (local glucose utilization group) and 15.9 mmol·l^{-1} (local blood flow group). The acidotic animals were alert and noncomatose. Metabolic acidosis induced a reduction of local cerebral glucose utilization in all brain structures tested. The average glucose utilization, obtained by scanning the whole films, was reduced by 29%. In contrast to this, blood flow was only insignificantly reduced by 8%. This dissociation between a decrease in glucose utilization and blood flow does not indicate a major impairment of the relationship between local glucose utilization and local blood flow. This is demonstrated in Fig. 11. In this figure the values for local glucose utilization are plotted against the local blood flows of the same structures as in Fig. 10, but for the two groups of acidotic rats. It is evident that the degree of acidosis induced in these experiments, although strong enough to reduce average glucose utilization by 29%, did not affect the correlation between local glucose utilization and local

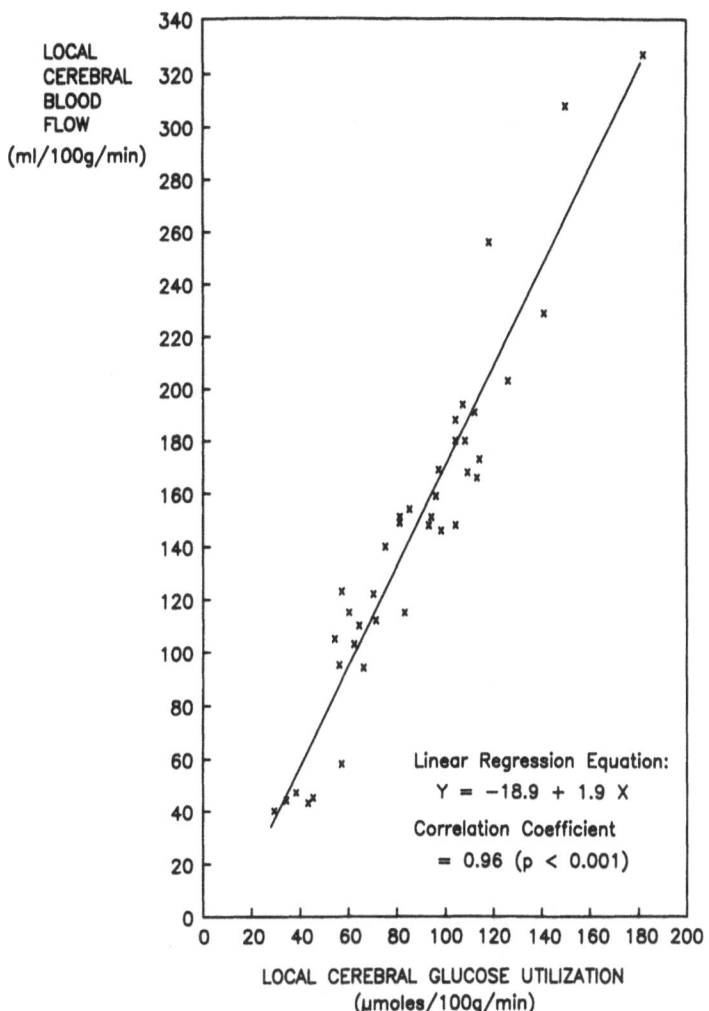

Fig. 10. Relationship between local cerebral glucose utilization
          and local cerebral blood flow in 39 anatomical structures
          of normal conscious rats. Fom Kuschinsky et al. (1981b).

blood flow, as indicated by the nearly unchanged correlation coefficient (0.95), compared with control conditions (0.96). On the other
hand, there existed a difference between both groups, as shown in
Fig. 12. This figure shows, for comparison, the regression lines of
Fig. 10 and 11. The values for the individual structures during
control conditions have been connected with the values during acidosis. Although there was excellent coupling under both conditions,
the amount of change in blood flow per amount of change in glucose
utilization was altered by acidosis, as indicated by the different
slopes of the two regression lines. The difference between the
slopes was statistically significant (p<0.01).

W. KUSCHINKSKY

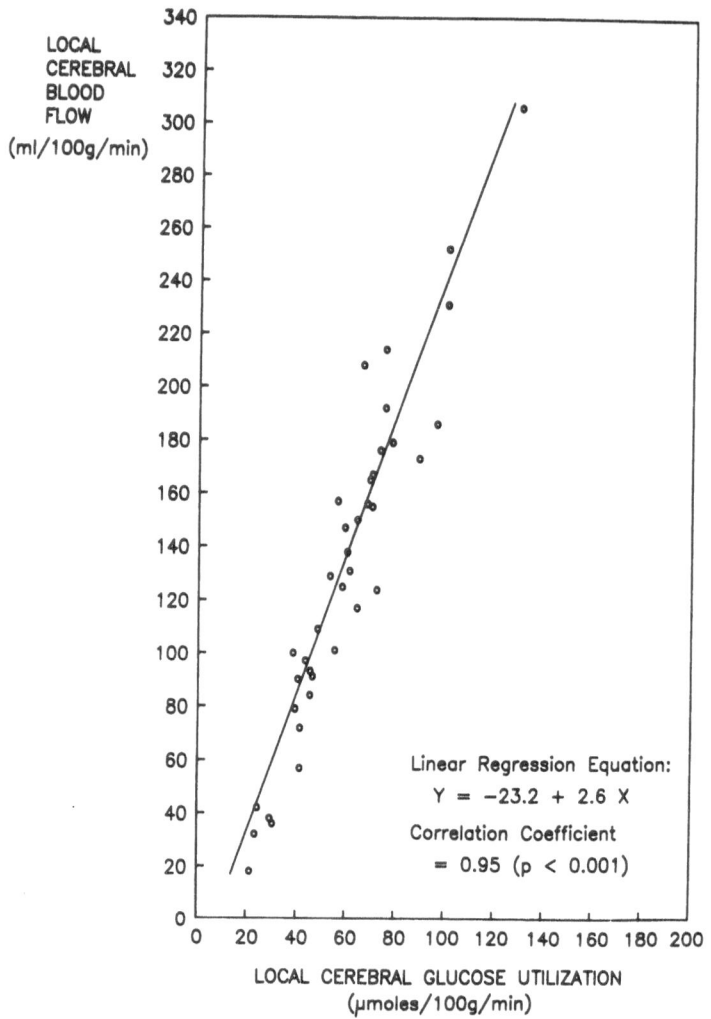

Fig. 11.   Relationship between local cerebral glucose utilization
          and local cerebral blood flow in 39 anatomical structures
          of acidotic rats. From Kuschinsky et al. (1981b).

        Corresponding experiments were performed to test the relation-
ship between local cerebral glucose utilization and local cerebral
blood flow during norepinephrine infusion (Kuschinsky et al., 1981a)
and during the action of gamma-hydroxybutyrate (Kuschinsky et al.,
1982). Under both conditions, we could again measure an increased
steepness of the slope of the regression line. This shows that the
amount of blood flow needed to perfuse a brain structure is not
fixed to its metabolic rate, but can vary even at the same metabolic
rate, dependent on the experimental conditions. This variability of
local blood flow is observed although a tight correlation between
local metabolic rate and local blood flow still exists.

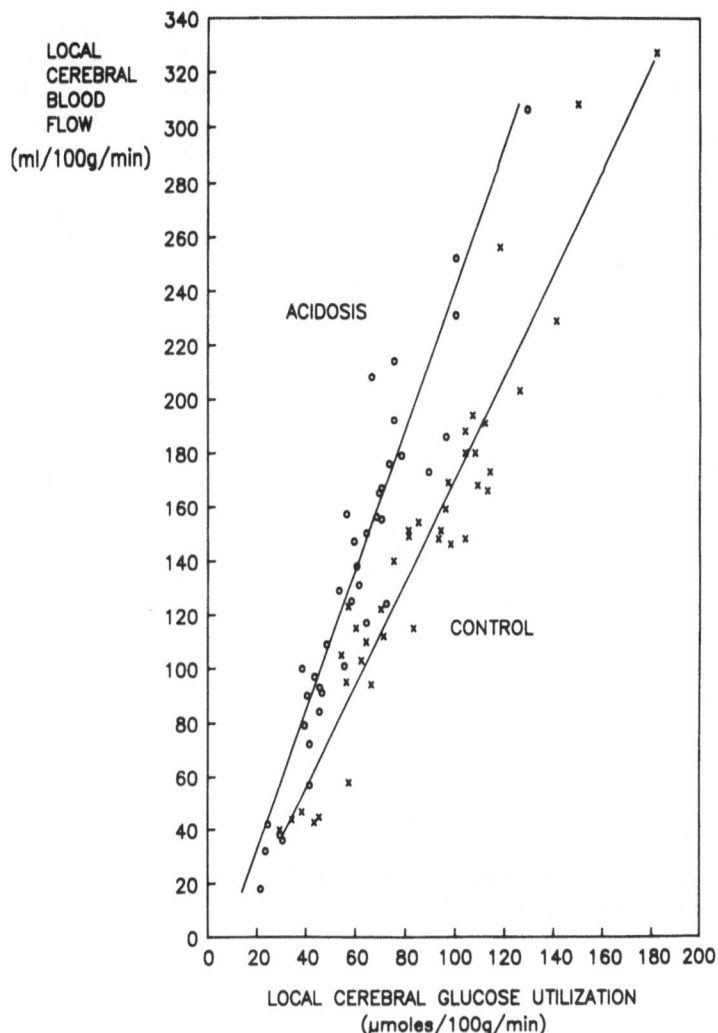

Fig. 12. Comparison and relationship between local cerebral
          glucose utilization and local cerebral blood flow during
          control acidosis. From Kuschinsky et al. (1981b).

POSSIBLE MECHANISMS MEDIATING THE CORRELATION BETWEEN LOCAL CEREBRAL
GLUCOSE UTILIZATION AND LOCAL CEREBRAL BLOOD FLOW

Since different correlations exist between local cerebral glucose utilization and local cerebral blood flow under different conditions, it is evident that the amount of local perfusion is not a
simple function of the local metabolic rate. This raises the question as to the mechanisms which mediate the correlation between
local cerebral glucose utilization and local cerebral blood flow.
The fact that a metabolic depression, as induced by gamma-hydroxy-

butyrate, does not fundamentally disturb the correlation, could be taken as an argument against metabolic factors as mediators of the correlation between local cerebral glucose utilization and blood flow. Parallel with a metabolic depression a decrease of metabolic factors can be expected to occur. This should be followed by a corresponding reduction of blood flow, when the local metabolic control mechanisms are the only determinants of microcirculation. Therefore, one interpretation of the present results is, that local metabolic factors are of minor importance for the normal correlation between local cerebral glucose utilization and local cerebral blood flow. In this case, the concept of local metabolic factors controlling cerebral microcirculation would be confined to functional and reactive hyperemia. The normal distribution of blood flow in the brain could then be dependent on the amount of capillary density, which may have developed according to the metabolic demands. Differences in capillary density in the brain are well known (Zeman and Innes, 1963; Lierse and Horstmann, 1965). An alternative interpretation can sustain the concept of local metabolic control: one can assume that the altered concentration of metabolic factors under the different experimental conditions tested may induce a changed reactivity of the cerebral resistance vessels, since blood flow was not depressed proportional to metabolism. Preliminary experiments using the micro-application technique support this concept: when the reactivity of pial arteries to changes in perivascular pH was tested before and after systemic application of gamma-hydroxybutyrate an increased sensitivity to changes in pH was detected during the action of gamma-hydroxybutyrate (Haller and Kuschinsky, unpublished results). However, more experiments are necessary to completely elucidate the mechanisms which mediate the tight correlation between local metabolic rate and microcirculation in the brain under different conditions.

REFERENCES

Astrup, J., Heuser, D., Lassen, N.A., Nilsson, B., Norberg, K., and Siesjö, B.K., 1976, Evidence against H^+ and K^+ as the main factors in the regulation of cerebral blood flow during epileptic discharges, acute hypoxemia, amphetamine intoxication, and hypoglycemia. A microelectrode study, in: "Ionic Actions on Vascular Smooth Muscle", D. Betz, ed., Springer, Berlin-Heidelberg-New York.

Astrup, J., Symon, L., Branston, N.M., and Lassen, N.A., 1977, Cortical evoked potential and extracellular K^+ and H^+ at critical levels of brain ischemia, Stroke, 8:51.

Barcroft, H., 1963, Circulation in skeletal muscle, in: "Handbook of Physiology", Section 2, Vol. 2, Chapter 40, American Physiological Society, Williams & Wilkins, Baltimore.

Berne, R.M., 1963, Cardiac nucleotides in hypoxia: possbile role in regulation of coronary blood flow, Am. J. Physiol., 204:317.

Berne, R.M., Rubio, R., and Curnish, R.R., 1974, Release of adeno-
    sine from ischemic brain. Effect on cerebral vascular resis-
    tance and incorporation into cerebral adenine nucleotides,
    Circ. Res., 35:262.
Betz, E., and Csornai, M., 1978, Action and interaction of perivas-
    cular H^+, K^+ and Ca^{++} on pial arteries, Pflügers Arch., 374:67.
Breemen, van, C., Farinas, B.R., Gerba, P., and McNaughton, E.D.,
    1972, Excitation-contraction coupling in rabbit aorta. Studied
    by the lanthanum method for measuring cellular calcium influx,
    Circ. Res., 30:44.
Britton, S.L., Lutherer, L.O., and Davies, D.G., 1979, Effect of
    cerebral extracellular fluid acidity on total and regional
    cerebral blood flow, J. Appl. Physiol., 47:818.
Burnstock, G., Cocks, T., Kasakov, L., and Wong, H.K., 1978, Direct
    evidence of ATP release from non-adrenergic, non-cholinergic
    ("purinergic") nerves in the guinea-pig taenia coli and bladder,
    Eur. J. Pharmacol., 49:145.
Cameron, I.R., and Caronna, J., 1976, The effect of local changes in
    potassium and bicarbonate concentration on hypothalamic blood
    flow in the rabbit, J. Physiol., 262:415.
Chen, W.T., Brace, R.A., Scott, J.B., Anderson, D.K., and Haddy,
    F.J., 1972, The mechanism of the vasodilator action of potas-
    sium, Proc. Soc. Exp. Biol. Med., 140:820.
Cordingley, G.E., and Somjen, G.G., 1978, The clearing of excess
    potassium from extracellular space in spinal cord and cerebral
    cortex, Brain Res., 151:291.
Eckman, W.W., Phair, R.D., Fenstermacher, J.D., Patkak, C.S.,
    Kennedy, C., and Sokoloff, L., 1975, Permeability limitation
    in estimation of local brain blood flow with (^{14}C)antipyrine,
    Am. J. Physiol., 229:215.
Edvinsson, L., and MacKenzie, E.T., 1977, Amine mechanisms in the
    cerebral circulation, Pharmacol. Rev., 28:275.
Eklöf, B., Lassen, N.A., Nilsson, L., Norberg, S., Siesjö, B.K.,
    and Torlöf, P., 1974, Regional cerebral blood flow in the rat
    measured by the tissue sampling technique; a critical evalua-
    tion using four indicators C^{14}-antipyrine, C^{14}-ethanol, H^3-
    water and Xenon 133, Acta Physiol. Scand., 91:1.
Freygang, W.H., Jr., and Sokoloff, L., 1958, Quantitative measure-
    ment of regional circulation in the central nervous system by
    the use of radioactive inert gas, Adv. Biol. Med. Phys., 6:263.
Fritz, H., and Hossmann, K.-A., 1979, Arterial air embolism in the
    cat brain, Stroke, 10:581.
Gaskell, W.H., 1880, On the tonicity of the heart and blood vessel,
    J. Physiol. (London), 3:48.
Goldmann, S.S., Hass, W.K., and Ransohoff, J., 1980, Unsymmetrical
    alkyl aryl thiourea compounds for use as cerebral flow tracers,
    Am. J. Physiol., 238:H776.
Goochee, C., Rasband, W., and Sokoloff, L., 1980, Computerized
    densitometry and color coding of (^{14}C)deoxyglucose autoradio-
    graphs, Ann. Neurol., 7:359.

Gregory, P.C., Boisvert, D.P.J., and Harper, A.M., 1980, Adenosine response on pial arteries, influence of CO_2 and blood pressure, Pflügers Arch., 368:187.

Haddy, F.J., and Scott, J.B., 1968, Metabolically linked vasoactive chemicals in local regulation of blood flow, Physiol. Rev., 48:688.

Haddy, F.J., and Scott, J.B., 1975, Metabolic factors in peripheral circulatory regulation, Fed. Proc., 34:2006.

Haller, C., and Kuschinsky, W., 1981, Reactivity of pial arteries to K^+ and H^+ before and after ischemia induced by air embolism, Microcirculation, 1:141.

Harris, R.J., Symon, L., Branston, N.M., and Bayhan, M., 1981, Changes in extracellular calcium activity in cerebral ischemia, J. Cereb. Blood Flow Metab., 1:103.

Heinemann, U., and Lux, H.D., 1975, Undershoots following stimulus induced rises of extracellular potassium concentration in cerebral cortex of cat, Brain Res., 93:63.

Heinemann, U., Lux, H.D., and Gutnick, M.J., 1977, Extracellular free calcium and potassium during paroxysmal activity in the cerebral cortex of the cat, Exp. Brain Res., 27:237.

Heuser, D., 1978, The significance of cortical extracellular H^+, K^+, and Ca^{++} activities for regulation of local cerebral blood flow under conditions of enhanced neuronal activity, in: "Cerebral Vascular Smooth Muscle and its Control", Ciba Foundation Symposium 56, Elsevier, Amsterdam.

Hoedt-Rasmussen, K., Skinhoj, E., Paulson, O., Ewald, J., Bjerrum, J.K., Fahrenkrug, A., and Lassen, N.A., 1967, Regional cerebral blood flow in acute apoplexy, Arch. Neurol., 17:271.

Hossmann, K.-A., Lechtape-Grüter, H., and Hossmann, V., 1973, The role of cerebral blood flow for the recovery of the brain after prolonged ischemia, Z. Neurol., 204:281.

Hossmann, K.-A., Sakaki, S., and Zimmermann, V., 1977, Cation activities in reversible ischemia of the cat brain, Stroke, 8:77.

Kety, S.S., and Schmidt, C.F., 1948, The effects of altered arterial tensions of carbon dioxide and oxygen on cerebral blood flow and cerebral oxygen consumption of normal young men, J. Clin. Invest., 27:484.

Kontos, H.A., Raper, A.J., and Patterson, J.L., Jr., 1977, Analysis of vasoactivity of local pH, Pco_2 and bicarbonate on pial vessels, Stroke, 8:358.

Kuschinsky, W., 1982, Role of hydrogen ions in regulation of cerebral blood flow and other regional flows, Adv. Microcirc., 11:1.

Kuschinsky, W., Suda, S., Bünger, R., and Sokoloff, L., 1981a, The effect of norepinephrine on the local coupling between brain metabolism and blood flow, Pflügers Arch., Suppl., 391:R31.

Kuschinsky, W., Suda, and Sokoloff, L., 1981b, Local cerebral glucose utilization and blood flow during metabolic acidosis, Am. J. Physiol., 241:H772.

Kuschinsky, W., Suda, and Sokoloff, L., 1982, The relationship
     between local cerebral glucose utilization and local cerebral
     blood flow during the action of gamma-hydroxybutyrate, Pflügers
     Arch., 392:R10.
Kuschinsky, W., and Wahl, M., 1978, Local chemical and neurogenic
     regulation of cerebral vascular resistance, Physiol. Rev.,
     58:656.
Kuschinsky, W., and Wahl, M., 1979, Perivascular pH and pial arte-
     rial diameter during bicuculline induced seizures in cats,
     Pflügers Arch., 382:81.
Kuschinsky, W., Wahl, M., Bosse, O., and Thurau, K., 1972, Perivas-
     cular potassium and pH as determinants of local pial arterial
     diameter in cats. A microapplication study, Circ. Res., 31:240.
Lacombe, P., Meric, P., and Seylaz, J., 1980, Validity of cerebral
     blood flow measurements obtained with quantitative tracer
     techniques, Brain Res. Rev., 2:105.
Landau, W.M., Freygang, W.H., Jr., Rowland, L.P., Sokoloff, L., and
     Kety, S.S., 1955, The local circulation in the living brain;
     values in the unanesthetized and anesthetized cat, Trans. Am.
     Neurol. Assoc., 80:125.
Lassen, N.A., 1968, Brain extracellular pH: the main factor con-
     trolling cerebral blood flow, Scand. J. Clin. Lab. Invest.,
     22:247.
Lierse, W., and Horstmann, E., 1965, Quantitative anatomy of the
     cerebral vascular bed with especial emphasis on homogeneity
     and inhomogeneity in small parts of the grey and white matter,
     Acta Neurol. Scand., Suppl., 14:15.
Mangold, R., Sokoloff, L., Conner, E., Kleinermann, J., Thermann,
     P.-O.G., and Kety, S.S., 1955, The effects of sleep and lack
     of sleep on the cerebral circulation and metabolism of normal
     young men, J. Clin. Invest., 34:1092.
Mutsuga, N., Schuette, W.H., and Lewis, D.L., 1976, The contribu-
     tion of local blood flow to the rapid clearance of potassium
     from the cortical extracellular space, Brain Res., 116:431.
Nemoto, E.M., Snyder, J.V., Carroll, R.G., and Morita, H., 1975,
     Global ischemia in dogs: cerebrovascular CO_2 reactivity and
     autoregulation, Stroke, 6:425.
Nicholson, C., 1980, Modulation of extracellular calcium and its
     functional implications, Fed. Proc., 39:1519.
Ohno, K., Pettigrew, K.D., and Rapoport, S.I., 1979, Local cerebral
     blood flow in the conscious rat as measured with ^{14}C-anti-
     pyrine, ^{14}C-iodoantipyrine and ^{3}H-nicotine, Stroke,
     10:62.
Pannier, J.L., Weyne, J., Demeester, G., and Leusen, I., 1972,
     Influence of changes in the acid base composition of the ven-
     tricular system on cerebral blood flow in cats, Pflügers Arch.,
     333:337.
Paulson, O.B., 1970, Regional cerebral blood flow in apoplexy due
     to occlusion of the middle cerebral artery, Neurology, 20:63.

Paulson, O.B., Lassen, N.A., and Skinhoj, E., 1970, Regional cere-
    bral blood flow in apoplexy without arterial occlusion, Neuro-
    logy, 20:125.
Pull, I., and McIlwain, H., 1972, Metabolism of (^{14}C)adenine and
    derivates by cerebral tissues, superfused and electrically
    stimulated, Biochem. J., 126:965.
Purves, M.J., 1978, Control of cerebral blood vessels: present state
    of the art, Ann. Neurol., 3:377.
Raichle, M.E., Grubb, R.L., Gado, M.H., Eichling, J.O., and Ter-
    Pogossian, M.M., 1976, Correlation between regional cerebral
    blood flow and oxidative metabolism, Arch. Neurol., 33:523.
Rehncrona, S., Siesjö, B.K., and Westerburg, E., 1978, Adenosine
    and cyclic AMP in cerebral cortex of rats in hypoxia, status
    epilepticus and hypercapnia, Acta Physiol. Scand., 104:453.
Reivich, M., Jehle, J., Sokoloff, L., and Kety, S.S., 1969, Meas-
    urement of regional cerebral blood flow with antipyrine-14 C
    in awake cats, J. Appl. Physiol., 27:296.
Roy, C., and Brown, J.G., 1879, the blood-pressure and its varia-
    tions in the arterioles, capillaries and smaller veins, J.
    Physiol. (London), 2:323.
Rubio, R., Berne, R., Bockman, E.L., and Curnish, R., 1975, Rela-
    tionship between adenosine concentration and oxygen supply in
    rat brain, Am. J. Physiol., 228-1896.
Sakurada, O., Kennedy, C., Jehle, J., Brown, J.D., Carbin, G.L.,
    and Sokoloff, L., 1978, Measurement of local cerebral blood
    flow with iodo(^{14}C)antipyrine, Am. J. Physiol., 234:H59.
Schädler, M., 1967, Proportionale Aktivierung von ATP-ase Aktivität
    und Kontraktionsspannung durch Ca^{++} Ionen in isolierten kon-
    traktilen Strukturen verschiedener Muskelarten, Pflügers Arch.
    Ges. Physiol., 296:70.
Schmidt, C.F., 1982, The early days of the indifferent gas method
    for measuring cerebral blood flow, J. Cereb. Blood Flow
    Metabol., 2:1.
Schrader, J., Wahl, M., Kuschinsky, W., and Kreuzberg, G.W., 1980,
    Increase of adenosine content in cerebral cortex of the cat
    during bicuculline-induced seizure, Pflügers Arch., 387:245.
Siesjö, B.K., 1978, "Brain Energy Metabolism", John Wiley and Sons,
    Chichester.
Skinhoj, E., 1966, Regulation of cerebral blood flow as a single
    function of the interstitial pH in the brain. A hypothesis,
    Acta Neurol. Scand., 42:604.
Sokoloff, L., 1978, Mapping cerebral functional activity with radio-
    active deoxyglucose, Trends Neurosci., 1:75.
Sokoloff, L., 1981, Localization of functional activity in the
    central nervous system by measurement of glucose utilization
    with radioactive deoxyglucose, J. Cereb. Blood Flow Metab.,
    1:7.
Sokoloff, L., Mangold, R., Wechsler, R.L., Kennedy, C., and Kety,
    S.S., 1955, The effect of mental arithmetic on cerebral circu-
    lation and metabolism, J. Clin. Invest., 34:1101.

Sokoloff, L., Reivich, M., Kennedy, C., DesRosiers, M.H., Patlak, C.S., Pettigrew, K.D., Sakurada, O., and Shinohara, M., 1977, The (^{14}C)deoxyglucose method for the measurement of local cerebral glucose utilization: theory, procedure, and normal values in the conscious and anesthetized albino rat, J. Neurochem., 28:897.

Somjen, G.G., 1979, Extracellular potassium in the mammalian central nervous system, Ann. Rev. Physiol., 41:159.

Symon, L., Branston, N.M., and Strong, A.J., 1976, Autoregulation in acute focal ischemia: an experimental study, Stroke, 7:547.

Symon, L., Crockard, H.A., Dorsch, N.W.C., Branston, N.M., and Juhasz, J., 1975, Local cerebral blood flow and vascular reactivity in a chronic stable stroke in baboons, Stroke, 6:482.

Symon, L., Pasztor, E., and Branston, N.M., 1974, The distribution and density of reduced cerebral blood flow following acute middle cerebral artery occlusion: an experimental study by the technique of hydrogen clearance in baboons, Stroke, 5:355.

Tomita, M., and Gotoh, F., 1981, Local cerebral blood flow values as estimated with diffusible tracers: validity of assumptions in normal and ischemic tissue, J. Cereb. Blood Flow Metab., 1:403.

Urbanics, R., Leniger-Follert, E., and Lübbers, D.W., 1978, Time course of changes of extracellular H^+ and K^+ activities during and after direct electrical stimulation of the brain cortex, Pflügers Arch., 378:47.

Wahl, M., and Kuschinsky, W., 1976, The dilatory action of adenosine on pial arteries of cats and its inhibition by theophylline, Pflügers Arch., 362:55.

Wahl, M., and Kuschinsky, W., 1977, Influence of H^+ and K^+ on adenosine-induced dilation at pial arteries of cats, Blood Vessels, 14:285.

Wahlström, B., 1971, The effects of changes in the ionic environment on venous smooth muscle distribution of sodium and potassium, Acta Physiol. Scand, 82:382.

Waltz, A.G., 1970, Effect of P_aCO_2 on blood flow and microvasculature of ischemic and nonischemic cerebral cortex, Stroke, 1:27.

Winn, H.R., Rubio, R., and Berne, R.M., 1981a, The role of adenosine in the regulation of cerebral blood flow, J. Cereb. Blood Flow Metab., 1:239.

Winn, H.R., Rubio, R., and Berne, R.M., 1981b, Brain adenosine concentration during hypoxia in rats, Am. J. Physiol., 241:H235.

Winn, H.R., Welsh, J.E., Rubio, R., and Berne, R.M., 1980, Changes in brain adenosine during bicuculline induced seizures in rats. Effects of hypoxia and altered systemic blood pressure, Circ. Res., 47:568.

Zeman, W., and Innes, J.R.M., 1963, "Craigie's Neuroanatomy of the Rat", Academic Press, New York.

TISSUE O_2 SUPPLY UNDER NORMAL AND PATHOLOGICAL CONDITIONS

M. Kessler[1], J. Höper[1], D.K. Harrison[1], K. Skolasinska[1],
W.P. Klövekorn[2], F. Sebening[2], H.J. Volkholz[1], I. Beier[3],
C. Kernbach[3], V. Rettig[3], and H. Richter[3]

[1]Institut für Physiologie und Kardiologie der Universität Erlangen-Nürnberg, Waldstr. 6, 8520 Erlangen, FRG
[2]Deutsches Herzzentrum, 8000 München, FRG
[3]Institut für Anästhesiologie der Universität Erlangen-
Nürnberg, Waldstr. 6, 8520 Erlangen, FRG

INTRODUCTION

Systematic investigations of a variety of local parameters which
form part of the complex functional chain responsible for delivery
of oxygen to tissues revealed that most relevant information can be
obtained by direct measurements of the oxygen tension field (Po_2
histogram) by means of Clark-type and Po_2 needle electrodes.

The special importance of local Po_2 results from the fact that
we are measuring at the end of a long transport chain, linked to the
energy producing system of the cell by the intracellular flux of
oxygen molecules. This diffusion flux is determined by the local Po_2
gradients between the capillaries and the oxygen consuming enzymes
in the cells.

Investigations of spatial Po_2 fields in tissues of various organs (see Kessler et al., 1973) have demonstrated that, under physiological conditions, two types of Po_2 histograms exist in the
tissues. The first type of Po_2 distribution curve found in brain
(Skolasinska et al., this volume), liver (Görnandt et al., 1973),
and skeletal muscle (Harrison et al., this volume) (Fig. 1) is characterized by the fact that the lowest Po_2 values approach 1 to 2
Torr and that the modes lie within a Po_2 range of 15 - 30 Torr. The
second type of Po_2 histogram which is found in the lung (Volkholz et
al., this volume), the heart (Kessler et al., this volume) and the

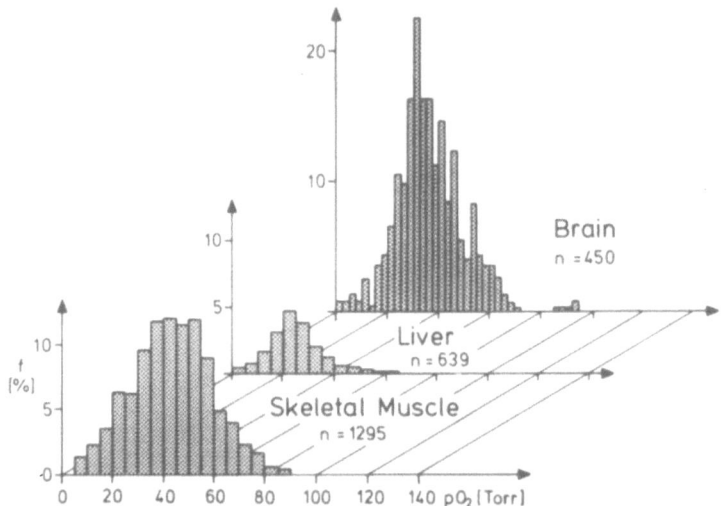

Fig. 1.   Po$_2$ histograms of skeletal muscle, liver and brain.

outer cortex of the kidney (Sinagowitz, 1977) shows a distinct shift
of the distribution curves to the right with the lowest Po$_2$ values
much higher than 1 - 2 Torr (Fig. 2).

    The fact that the Po$_2$ histograms have a very similar configu-
ration in all organs with the exception of lung, heart and outer

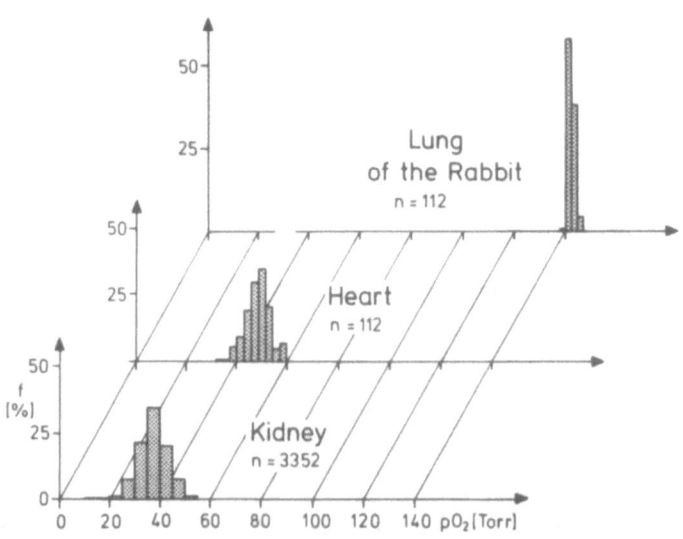

Fig. 2.   Po$_2$ histograms of kidney, heart and lung.

cortex of kidney, indicates that during the course of evolution, the biological systems of man and other mammals have reached an optimal level of functional organization that provides a secure oxygen supply.

Possible Mechanisms Involved in the Local Regulation of Microcirculation

Within the past decades it was assumed that the local factors which play a decisive role in the systemic regulation of macro- and microcirculation are metabolic products (ADP, phosphate, adenosine, lactate, CO_2) and/or ions (H^+, K^+).

In addition to this classical concept, our experimental work reveals increasing evidence that the tissue Po_2 seems to be a predominant integrating factor which determines the turnover rate of signal oxidases and thus activates intra- and intercellular signal chains responsible for the local regulation of macro- and microcirculation.

In recent years we have performed investigations in skeletal muscle, lung, heart and liver in order to analyze the regulation of local oxygen supply at rest, stimulation of energy metabolism, arterial hyperoxia and hypoxia, critical arterial stenosis and hypovolemic shock.

In experiments performed by Harrison et al. (this volume) in the skeletal muscle of the anesthetized dog energy metabolism was increased by means of electrical stimulation of the femoral nerve (Fig. 3).

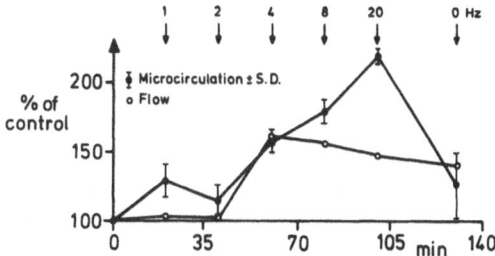

Fig. 3.   Changes in microcirculation of skeletal muscle and total flow in femoral artery during stimulation of the femoral nerve. The numbers at the top of the figure indicate the stimulation frequency. All values are given in percent of the control. Total flow was measured by a magnetic flow meter, microcirculation by hydrogen washout curves after inhalation of a hydrogen gas mixture. (n = 5)

   Blood flow in the femoral artery and microflow in the sartorius
muscle were measured simultaneously during the experiments. After
an early and transient change in microcirculation a concomitant increase of blood flow and microcirculation is observed at a stimulation frequency of 4 Hz. At higher frequencies a further increase in
microcirculation is induced whereas the arterial blood flow shows a
tendency to fall.

   This experiment demonstrates two interesting phenomena: when
microflow rate increases its standard deviation decreases, indicating that the microcirculation becomes more homogeneous. At rest a
bimodal distribution of micocirculation is found in the skeletal
muscle (Fig. 4) which disappears when a redistribution of microcirculation is induced by muscle work with low stimulation frequencies.

   This early redistribution seems to develop within a local microcirculatory volume which may be supplied by a few arterioles. Fig. 5
shows the Po_2 field of such a volume from the skeletal muscle of the
rat as measured by Beier and Rettig (Beier and Rettig, in preparation). The systematic measurements of surface Po_2 were performed in
anesthetized animals at rest by moving a surface electrode in steps
of 50 um.

   Another interesting finding is the further 50% increase in microcirculation during stimulation at 20 Hz with no concurrent increase
in arterial blood flow. This pronounced change in microcirculation
may be caused by regional redistribution of blood flow between small
arterioles that supply adjacent fascicles (Kessler et al., 1976).
Our recent experimental work resulted in the finding that the tissues contain oxygen regulating systems, capable of monitoring cellular oxygen supply and directly modifying microcirculation when
local oxygen demand changes or when cells are threatened by lack of
oxygen (Höper et al., 1981b).

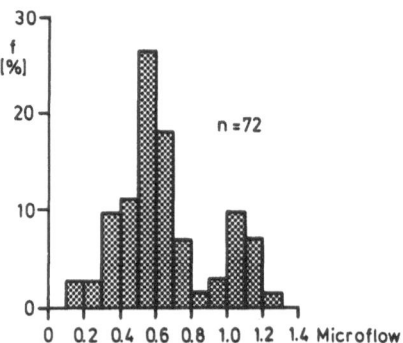

Fig. 4. Histogram of relative microflow values as measured by hydro-
         gen washout clearance in skeletal muscle of the dog.

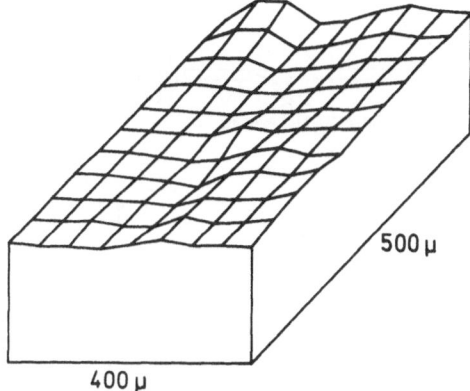

Fig. 5.   Po₂ profile as measured in the skeletal muscle of an anes-
          thetized resting animal (rat).

Furthermore, it was shown that not only the microcirculation
but also the cellular oxygen uptake rate can be modulated when
local oxygen supply deteriorates (Kessler et al., 1981).

Our investigations of local factors affecting regulation of
microflow and oxygen uptake rate opened the insight into a system
of possible cellular signal chains which have "signal" oxidases at
their sensing side and seem to be able to transfer information from
the "lethal corner" to the vascular smooth muscles as well as to the
mitochondria of adjacent parenchymal cells via the endothelial tubes
of the capillaries. It may be that two independent signal chains may
exist, one responsible for regulation of microcirculation whereas
the second may effect mitochondrial oxygen turnover rate and energy
consumption when cellular oxygenation reaches critical thresholds
(Kessler et al., 1981).

Our investigations performed in the isolated perfused rat liver
give strong hints that the "signal" oxidases might be mitochondrial
enzymes (Fig. 6). The studies of Höper performed in the perfused rat
liver indicate that the flow regulating signal oxidases might be
monoaminooxydases of type B (Höper et al., 1981).

Recently, Höper (1983) was able to show that tissue hypoxia
induces a rapid depolarization of hepatocytes in the range of 4 mV
and a concomitant change in intercellular coupling.

It seems to be likely that in the liver the Po₂ dependent
signal chains transfer the information via interaction between
adjacent hepatocytes. In the capillaries of the liver (the so-called
sinusoids) endothelial cells are not found.

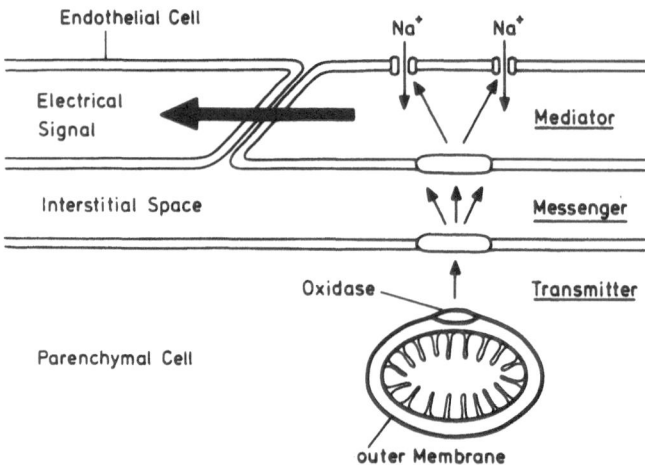

Fig. 6.   Schematic drawing of the hypothetical signal chain carrying
          information from the parenchymal cell to the endothelial
          cells.

        This raises the question whether such an intercellular mode of
transmission, modulated by signal oxidases, may also exist in other
organs and that the mode of transfer of oxygen specific signals
might be the interaction between adjacent endothelial cells (Fig. 6).

        Investigations performed in the beating hearts in dogs have
shown that a critical coronary stenosis which causes partial anoxia
only within small local areas, induces a decrease in regional contractility of the myocardium.

        The time constants of these contractility changes as induced
by the stenosis and the reopening of the coronary artery lie within
the range of several hundred milliseconds to a few seconds (Kernbach,
thesis in preparation).

        This observation excludes the existence of long diffusion dis-
tances in the presumable signal chains and emphasizes the possibility of a fast information transfer between parenchymal and endothelial cells (Fig. 6).

Oxygen Supply under Pathological Conditions

        Under physiological conditions an optimal autoregulation of
local oxygen supply is achieved which produces rather uniform non
organ-specific Po_2 histograms.

        Under pathological conditions an early decompensation of tissue
oxygenation can be prevented by the autoregulation of oxygen supply.

   Such compensatory autoregulation can be quantified when either
O$_2$ supply or demand are changed in order to test the capacity of
functional reserve.

   The high sensitivity of the Po$_2$ monitoring technique enables
one to detect early disturbances at a time when global hemodynamic
parameters or other diagnostic techniques for the control of tissue
homeostasis do not indicate any signs of local disturbances (Kessler
et al., 1981; 1982).

Local Oxygen Supply in Skeletal Muscle

   In patients under intensive care Po$_2$ histograms were measured
in the musculus sartorius femoris of patients ventilated mechanically with various oxygen concentrations in the inspired air (Fig. 7).

   In A a situation of a patient ventilated with an FiO$_2$ of 0.5 is
shown. Even though the arterial Po$_2$ is only 65 Torr a normal Po$_2$
histogram is obtained.

   Ventilation with hyperoxic mixtures can result in an almost bell-
shaped distribution curve, shifted to the right as shown in B.

   However, hyperoxia can also give rise to a disintegration of
the Po$_2$ histogram, either without local hypoxia or combined with
anoxic values.

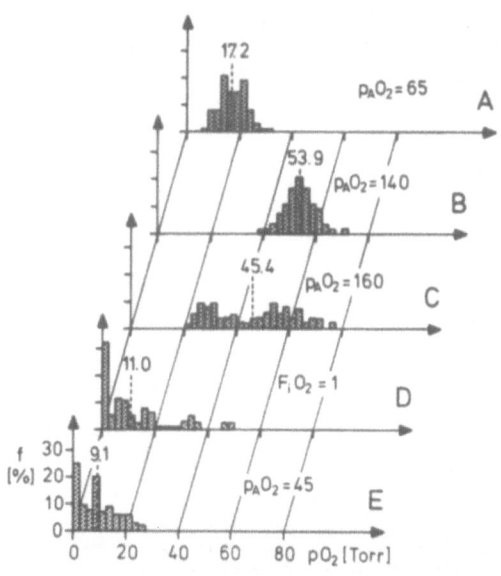

Fig. 7. Po$_2$ histograms obtained in skeletal muscle of critically
         ill patients under various conditions. For details see text.

In C ventilation with the same FiO_2 as in B gave rise to a distribution curve with two peaks. We interpret this type of curve as a "disintegrated normoxic Po_2 histogram".

An extreme situation during ventilation with an FiO_2 of 1.0 is shown in D. The Po_2 distribution presents a very disintegrated appearance and reveals a high percentage of hypoxic values. Finally, E represents a typical situation caused by arterial hypoxia.

Local Oxygen Supply in the Lung

Investigations of local oxygen supply of the lung performed by Volkholz et al. (this volume), revealed Po_2 histograms with values that correspond to the alveolar Po_2. This indicates that both the blood in the capillaries and the lung tissue are very quickly equilibrated to alveolar Po_2 values (Fig. 8).

Ventilation with hyperoxic gas mixtures causes a broadening of the Po_2 histogram from a ΔPo_2 of 100 to 360 Torr. The broadening of the histogram might be the result of disturbances of ventilation induced by hyperoxic gas mixtures. Experimental observations (Volkholz et al., this volume) indicate that the tonus of the smooth muscle of branchioli might be affected by high oxygen pressure.

Local Oxygen Supply in the Heart

In collaboration with Klövekorn and Sebening from the German Heart Center in Munich we tried to find the factors that cause hypokinetic zones in the myocardium when a critical stenosis develops in the coronary artery (Kessler et al., this volume). We therefore studied the local oxygen supply of the dog heart before and during critical stenosis. In order to be able to adjust the critical stenosis more precisely with our mechanical occluding device we decreased the oxygen transport capacity by isovolemic hemodilution.

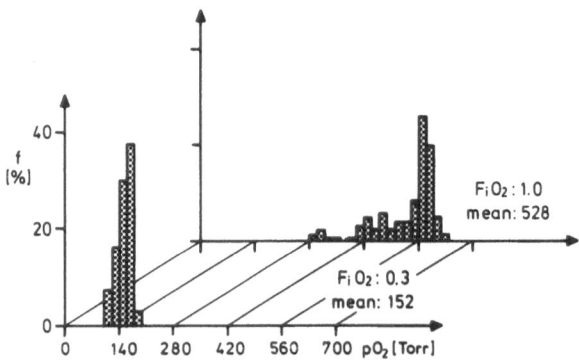

Fig. 8. Po_2 histograms obtained at the lung surface at FiO_2 values of 0.3 (lower histogram) and 0.1 (upper histogram).

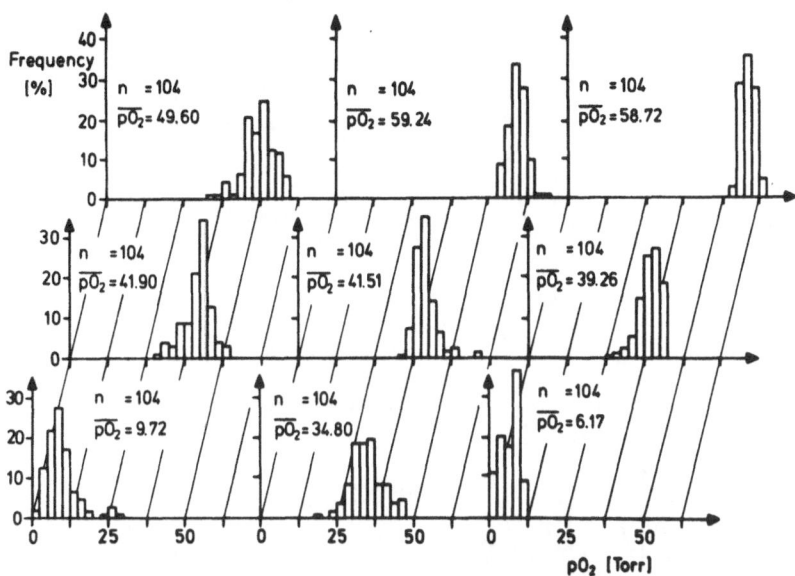

Fig. 9.   Po₂ histograms obtained in three different areas of the
          beating heart. Top: Po₂ histograms measured under control
          conditions. Middle: Po₂ histograms measured after hemo-
          dilution with HES (hydroxy ethyl starch). Bottom: Po₂
          histogram obtained during critical stenosis.

The Po₂ histograms measured in three different areas of the
myocardium supplied by the LAD (left anterior descendent artery)
demonstrate a rather pronounced heterogeneity of local oxygen
supply (Fig. 9). Hemodilution down to a hematocrit of 10% causes a
moderate shift of the distribution curves towards lower Po₂ values.
When critical stenosis is induced it becomes evident that the
deterioration of local oxygen supply does not show uniform changes
and cannot be predicted from the initial control histograms.

Concomitant measurements of myocardial wall motion, by use of
the technique of recording ultrasonic transient time, show that contractility starts to decrease when the first Po₂ values approach
zero Torr (Kessler et al., this volume).

This result means that in the heart muscle local Po₂ is a key
parameter responsible for controlling myocardial function when tissue is threatened by anoxia.

Local Oxygen Supply in the Liver

Some years ago we investigated local oxygen supply of dog liver
during hemorrhagic shock (Fig. 10). To our great surprise the observed changes in the Po₂ histogram were much less distinct than we had

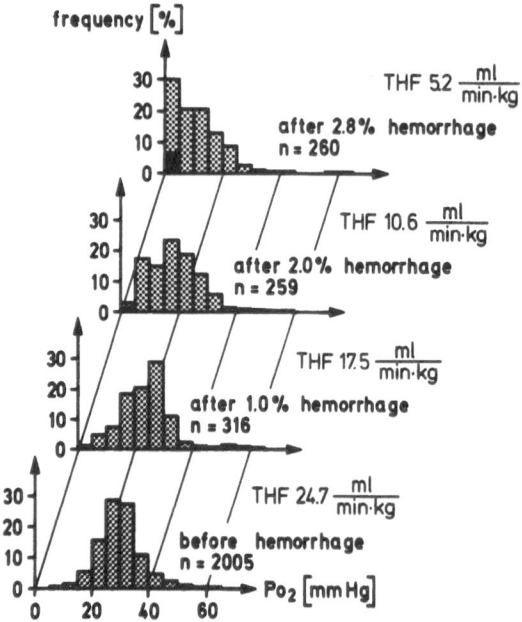

Fig. 10. Local oxygen supply of dog liver during hemorrhage (unpub-
          lished data).

expected. Even though the decrease in total hepatic blood flow (THF)
was very pronounced, only moderate changes were found in the histogram.

   A severe pathological alteration of the histogram was not found
before the blood flow had fallen to 20% of its initial value.

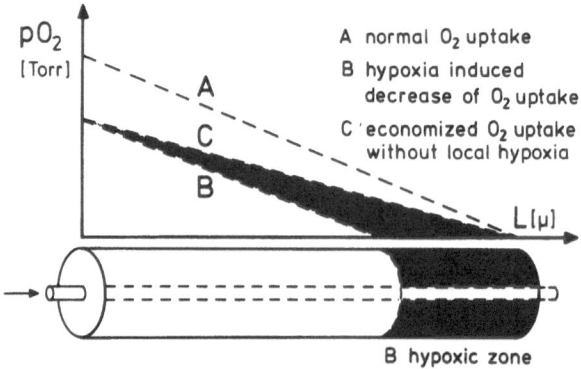

Fig. 11. Schematic drawing of changes of Po_2 gradient along a
          capillary.

Further investigations of these unexpected results were performed in the isolated perfused rat liver. They revealed clear evidence that critical oxygen supply can induce a decrease in oxygen uptake rate and energy consumption thus preventing local tissue anoxia within a certain and of course limited range of regulation (Kessler et al., 1981). Our investigations in the perfused liver showed that the Po_2 gradient along the capillary (Fig. 11) found after a decrease of arterial Po_2, did not shift from A to B as would be expected in a non regulated cellular system but underwent a transition from A to C (Höper et al., 1981a).

The observed change in the slope of the gradient can be explained by a compensatory decrease of oxygen uptake rate in all cells lining the capillary.

CONCLUSION

Summarizing our experimental results we can conclude that within certain limits organs have extremely efficient protective mechanisms able to prevent the deleterious consequences of cellular anoxia.

ACKNOWLEDGEMENT

We thank A. Brehm, G. Kerl and G. Schuster for skillful technical assistance.

Dedicated to Prof. Dr. D.W. Lübbers

REFERENCES

Beier, I., and Rettig, V., Dissertation, Erlangen, in preparation.

Görnandt, L., and Kessler, M., 1973, Po_2 histograms in regenerating liver tissue, in: "Oxygen Supply. Theoretical and Practical Aspects of Oxygen Supply and Microcirculation of Tissue", M. Kessler, D.F. Bruley, L.C. Clark Jr., D.W. Lübbers, I.A. Silver, J. Strauss, eds., Urban & Schwarzenberg, München-Berlin-Wien, pp. 288-289.

Harrison, D.K., Höper, J., Günther, H., Vogel, H., Brunner, M., Ellermann, R., and Kessler, M., 1983, Microcirculation and Po_2 in skeletal muscle during rest and nerval stimulation, this volume.

Höper, J., 1983, Dissertation, Erlangen, in preparation.

Höper, J., and Kessler, M., 1981a, Influence of buflomedil on oxygen uptake rate of liver tissue, in: "Microcirculation and Ischemic Vascular Diseases", K. Messmer, ed., Academy Professional Information Services, pp. 243-254.

Höper, J., and Kessler, M., 1981b, Po$_2$ and sodium dependent mechanism regulating liver blood flow, in: "Oxygen Transport to Tissue", Adv. Physiol. Sci. Vol. 25, A.G.B. Kovách, E. Dóra, M. Kessler, I.A. Silver, eds., Pergamon Press, Akadémiai Kiadó, Budapest, pp. 163-164.

Kessler, M., Bruley, D.F., Clark, L.C. jr., Lübbers, D.W., Silver, I.A., and Strauss, J., 1973, "Oxygen Supply - Theoretical and Practical Aspects of Oxygen Supply and Microcirculation of Tissue". Urban & Schwarzberg, München-Berlin-Wien.

Kessler, M., Höper, J., and Krumme, B.A., 1976, Monitoring of tissue perfusion and cellular function, Anesthesiology, 45:184-197.

Kessler, M., Höper, J., Lübbers, D.W., and Ji, S., 1981, Local factors affecting regulation of microflow, O$_2$ uptake and energy metabolism, in: "Oxygen Transport to Tissue", Adv. Physiol. Sci. Vol. 25, A.G.B. Kovách, E. Dóra, M. Kessler, I.A. Silver, eds., Pergamon Press, Akadémiai Kiadó, Budapest, pp. 155-162.

Kessler, M., Klövekorn, W.P., Höper, J., Sebening, F., Brunner, M., Frank, K.H., Harrison, D.K., Richter, H., Kernbach, C., and Ellermann, R., 1983, Local oxygen supply and regional contractility during critical stenosis of the LAD, this volume.

Sinagowitz, E., 1977, Die lokale Sauerstoffversorgung der Nierenrinde bei Hydronephrose und Nierenischämie; ihre klinische Bedeutung in der Urologie, Habilitations-Schrift, Freiburg.

Skolasinska, K., Günther, H., and Höper, J., 1983, Oxygen supply to the brain cortex in SHR and normotensive rats, this volume.

Volkholz, H.J., Höper, J., Brunner, M., Frank, K.H., Harrison, D.K., Ellermann, R., and Kessler, M., 1983, Measurement of local Po$_2$ and intracapillary hemoglobin oxygenation in lung tissue of rabbits, this volume.

RELATIONSHIP BETWEEN STEADY REDOX STATE AND BRAIN ACTIVATION-

INDUCED NAD/NADH REDOX RESPONSES

A.G.B. Kovách, E. Dorá, and L. Gyulai

Experimental Research Department and 2nd Institute of
Physiology, Semmelweis University Medical School
Budapest, Hungary

INTRODUCTION AND METHODS

It is well recognized that activation of the brain leads to
an overwhelming increase in cerebral blood flow (CBF), oxygen, and
glucose consumption (Siesjö, 1978; Sokoloff, 1981). Because ATP
usage is augmented, the ratio of ATP/ADP decreases, and according
to the in vitro data of Chance and Williams (1955), the rate of
mitochondrial electron transport and ADP phosphorylation will be
accelerated. Consequently, mitochondrial NADH should be oxidized
(Chance and Williams, 1955). The increased need of mitochondrial
electron transport for reducing equivalents is matched by the increased production of pyruvate via stimulation of glycogenolysis
and glycolysis (Siesjö, 1978; Sokoloff, 1981). Though it is unlikely that brain suffers from hypoxia under augmented electrical activity (Siesjö, 1978), a considerable amount of pyruvate is converted into lactate, and NADH accumulates in the cytosol (Howse and
Duffy, 1975; Siesjö, 1978). This NADH reduction is explained as
being due to the restricted capability of the so-called "H-shuttle"
mechanisms to transfer H^+ from cytosolic NADH to mitochondrial NAD
(Howse and Duffy, 1975; Siesjö, 1978). Interestingly, when the
mitochondrial NAD/NADH ratio has been determined with the oxidized-
reduced substrate ratio method during epileptic seizures, discernible NADH oxidation was not obtained (Siesjö, 1978). Using surface
fluororeflectometry in vivo and spectrophotometry in vitro, some of
the investigators found only NADH oxidation (Jöbsis et al., 1971;
Rosenthal and Somjen, 1973; Hempel et al., 1980), while others
revealed pronounced NAD reduction (Cummins, 1971; Lipton, 1973;
Dorá and Kovách, 1979; Gyulai et al., 1982) during excessive and
prolonged activations of the brain cortex. Since our preliminary

studies (Dorá and Kovách, 1978) indicated that activation can lead
both to NAD reduction and NADH oxidation, probably depending on the
prestimulatory steady redox state of the brain cortex, we decided
to explore this possibility more precisely.

To alter steady redox state and to see the relative importance
of oxygen availability in the activation-elicited cortical vascular
and metabolic responses, several experimental models (arterial
hypotension-reinfusion, arterial hyper- and hypoxia, spreading cortical depression) were used. Direct electrical stimulation of the
cerebral cortex, as a test, was superimposed on these interventions.

The experiments were carried out on 40 cats weighing 2.5 -
3.5 kg, anaesthetized with chloralose, immobilized with flaxedil,
and respired artificially. The artificial respiration was adjusted
so as to have the arterial Po_2 and Pco_2 around 100 and 32 mm Hg,
respectively. The trachea, femoral arteries and veins, and one of
the lingual arteries were cannulated. For optical monitoring of
cerebrocortical vascular volume and NAD/NADH redox state a cranial
window was implanted into the right parietal bone. In the wall of
the cranial window metal tubes were glued for superfusion of the
brain cortex, and for measuring intracranial pressure. ECoG of the
exposed brain cortex was monitored by silver electrodes built also
into the wall of the cranial window. The same electrodes were used
to stimulate the brain cortex electrically. Cerebrocortical NADH
fluorescence (450 nm) and vascular volume (366 nm) were measured by
a microscope fluororeflectometer using appropriate optical filters
and EMI photomultipliers for light detection (Dorá and Kovách, 1982).

The reflected light, measured at 366 nm, was used to follow the
changes in cerebrocortical vascular volume. The increase in cerebral
blood content, i.e. vascular volume, decreases the intensity of
reflected light and the intensity increases when the blood content
is diminished (Eke et al., 1979; Dorá and Kovách, 1982). Changes in
the oxygenation of hemoglobin do not discernibly alter the intensity
of the reflected light measured in vivo in the intact brain cortex
(Dora and Kovach, 1982). The normal values of cerebrocortical
vascular volume (CVV) determined by reflectometry and other methods
are comparable (Eke et al., 1979).

The virtual changes in NADH fluorescence, caused by the so-
called "hemodynamic artifact", were eliminated by the correction
method based on artificial hemodilution (Harbig et al., 1976). The
corrected NADH fluorescence shows the true alterations of NAD/NADH
redox state. Blood gases and hemoglobin concentration of arterial
blood samples were determined in every experiment. The rectal temperature of the animals was maintained at 37°C. The following parameters were recorded continuously: cerebrocortical uncorrected and
corrected NADH fluorescence, reflectance, ECoG of the exposed brain
cortex and the other hemisphere, arterial blood pressure and intracranial pressure.

The experiments were devided into 4 groups. In the first series of experiments arterial blood pressure was decreased by bleeding from the control value to 80, 60, and 40 mm Hg. Each hypotensive period was maintained for 30 min by a buffer-reservoire system connected to the femoral artery. After bleeding the shed blood was reinfused to restore arterial blood pressure to its reference level. Electrical stimulation of the brain cortex, lasting for 30 sec, was applied during the control period, at the 20th min of each hypotensive period, and 30 min after reinfusion. In all the experiments the intensity of the electrical stimulation was adjusted to be 1.5 - 2 times the threshold voltage (threshold: 2 - 5 V). Other parameters of the electrical stimuli were as follows: 0.5 msec, 15 Hz.

In the 2nd and 3rd series of experiments electrical stimulation was superimposed on arterial hyperoxia (respiration with oxygen and low CO_2 gas mixture, or pure oxygen), or arterial hypoxia (respiration with a gas mixture containing 6% oxygen balanced in nitrogen and low CO_2, or only 6% O_2 + 94% N_2). Hyperoxic and hypoxic episodes were maintained approximately for 5 - 8 min. The brain cortex was activated electrically when cortical NAD/NADH redox state reached a quasi-steady state.

In the 4th series of experiments electrical stimulation was superimposed on the NAD reductive phase of spreading cortical depression (SD) appearing spontaneously, or evoked by topical administration of 0.1 M KCl containing artificial CSF. Mock CSF of Wahl and Kuschinsky (1976) was kept at 38°C and equilibrated with a gas mixture of 5% CO_2 balanced in air. The composition of mock CSF was the following: Na^+ 156 mmol, K^+ 3 mmol, Ca^{2+} 1.5 mmol, Cl^- 151 mmol, HCO_3^- 11 mmol, pH 7.15 - 7.17. For superfusion of the brain cortex a 2-channel Harvard infusion pump with a perfusion rate of approximately 1 ml/min was used. The perfusion pressure was the same as the intracranial pressure.

RESULTS

Under normal conditions 30 - 50 sec electrical stimulation of the brain cortex resulted in mostly an initial slight and transient NADH oxidation (phase I) which was always followed by a more pronounced NAD reduction (phase II, Fig. 1). When stimulation was stopped NAD reduction overshot (phase III), and it took approximately 10 min until the redox state recovered to its reference level (phase IV). Due to excessive activation of the brain cortex, cerebrocortical vascular volume (CVV) increased considerably (R decreased), and because cerebral perfusion pressure (CPP) remained unaltered, CBF should have also been increased overwhelmingly. Following the cessation of stimulation uncorrected NADH fluorescence (F) overshot and the restoration of CVV was much faster than the recovery of NAD/NADH redox state (CF).

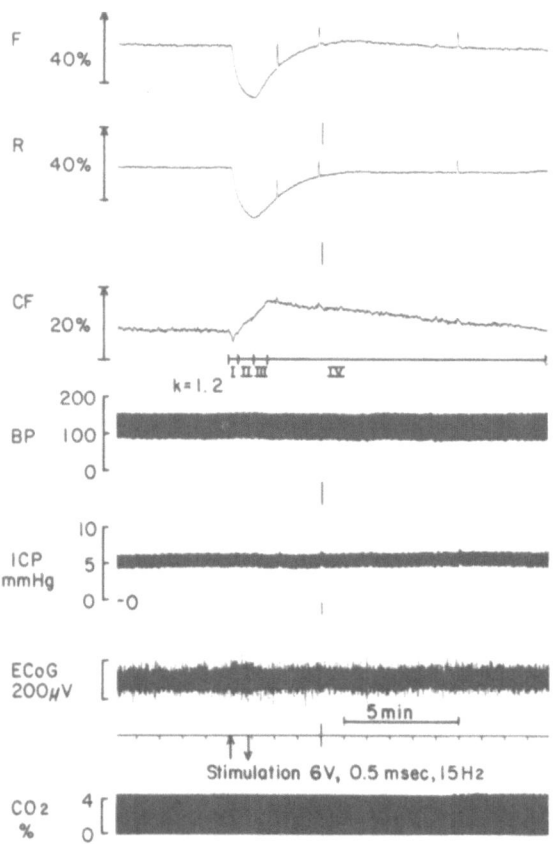

Fig. 1.   Effect of direct electrical stimulation of the brain cortex
          on the cortical uncorrected (F) and corrected (CF) NADH
          fluorescence, reflectance (R) as well as other routinely
          recorded parameters in a typical experiment. The duration
          of stimulation is marked by arrows on the time scale. Para-
          meters of stimulation and calibration of time are shown
          above the N-tidal CO_2 trace. Calibrations and directions
          of increase are marked before each parameter. Other abbre-
          viations: BP arterial blood pressure, ICP intracranial
          pressure, CO_2 % = exspired CO_2 percentage. Spikes on F and
          R traces were evoked by injecting 0.1 ml oxygenated dextran
          solution into the lingual artery for determining the cor-
          rection factor, k that is used to calculate true changes in
          cerebrocortical NAD/NADH redox state. The correction is
          made electronically. Note that the changes in reflectance
          (R) are inversely related to alterations in vascular volume.
          Decrease in R and CF means increase in cerebrocortical
          vascular volume and oxidation of NADH, respectively. The
          four phases of activation-evoked complex redox state
          responses are marked by Roman numbers under the CF trace.

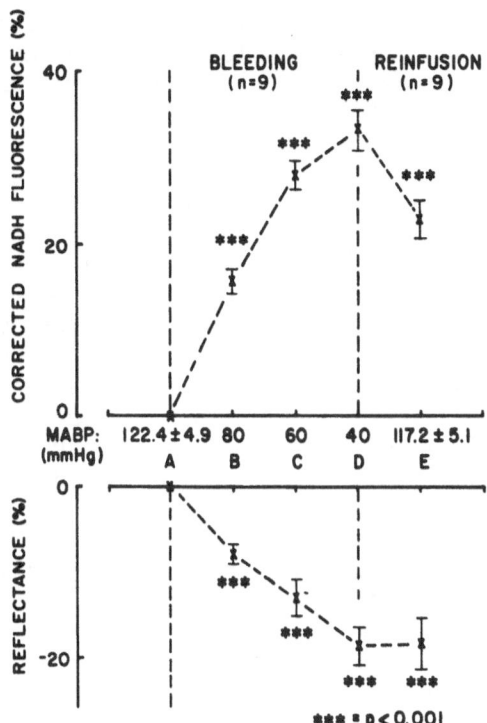

Fig. 2.   Effect of arterial hypotension and reinfusion on cerebrocortical corrected NADH fluorescence and reflectance.
Standard deviation of the mean values is represented by
vertical lines closed at both ends. Asterisks show the
degree of significant changes taking the reference values
to zero percent. Abbreviations: MABP = mean arterial blood
pressure, n = number of experiments averaged. The various
periods of the bleeding-reinfusion experimental model are
marked as follows: A: initial control period, MABP = 122.4
$\pm$ 4.9 mm Hg; B: MABP = 80 mm Hg; C: MABP = 60 mm Hg; D:
$\overline{\text{MABP}}$ = 40 mm Hg; E: period after reinfusion of the shed
blood, MABP = 117.2 $\pm$ 5.1 mm Hg. MABP before bleeding and
after reinfusion did not differ significantly from each
other. Note the marked NAD reduction within the autoregulatory range of cerebral blood flow (MABP until 60 mm Hg).

        Stepwise arterial hypotension resulted in a gradual and pro-
nounced increase in cortical NAD reduction and CVV (Fig. 2). Within
the autoregulatory range of CPP (lower threshold of CBF autoregulation being about 60 mm Hg mean arterial blood pressure) corrected
NADH fluorescence increased approximately by 28%. Albeit after
reinfusion of the shed blood, mean arterial blood pressure (MABP)
did not differ significantly from its prebleeding value, and the
cerebrocortical vessels remained dilatated (R was decreased, NAD/

NADH redox state was only partially reoxidized (Fig. 2E). This
result indicates that hemorrhage led to loss of CBF autoregulation
and perturbed irreversibly the intracellular NAD/NADH ratio. In
contrast to the control NADH oxidation-reduction cycle (Fig. 3A),
electrical stimulation of the brain cortex resulted in only NADH
oxidation at 80 and 60 mm Hg MABP (Figs. 3B and 3C), and because
CVV was increased by the autoregulatory mechanism of CBF, the
stimulation-induced CVV reactions became depressed. At 40 mm Hg
MABP, when the pial vessels are maximally dilatated, electrical
activation of the brain cortex did not bring about vasodilatation
but in fact had an inverse effect, a decrease of CVV was obtained
(Fig. 3D). In spite of the probable decrease in CBF, electrical
stimulation elicited NADH oxidation at 40 mm Hg MABP. After reinfusion of the shed blood electrical activation of the brain cortex
led to a markedly depressed CVV response and shifted the NAD/NADH
redox state monotonously toward a more oxidized state (Fig. 3E).

During arterial hyperoxia Po_2 of arterial blood samples increased to 431.5 + 14.9 mm Hg. When CO_2 was mixed into the hyperoxic gas mixture pH and Pco_2 in the arterial blood were not altered,
without additional CO_2 pH increased, Pco_2 decreased significantly.
Intracranial pressure remained unaltered, MABP decreased slightly
during arterial hyperoxia. Under respiration of 6% oxygen balanced
in nitrogen gas with or without CO_2 addition, Po_2 of arterial blood
samples decreased to 22.5 + 0.7 mm Hg, pH increased, Pco_2 decreased.
Arterial hypoxia did not cause significant change in MABP but increased intracranial pressure by around 20 mm Hg above its normoxic
reference value. Arterial hyperoxia resulted in a slight decrease
in cerebrocortical vascular volume (reflectance increased) and
shifted the NAD/NADH redox state toward oxidation (Fig. 4A). During
arterial hypoxia both cerebrocortical vascular volume and NAD reduction increased approximately by 30% (Fig. 4A). Arterial hyperoxia (constant arterial pH and Pco_2) had no influence on activation-
elicited cortical vascular and metabolic responses (Fig. 4B, see
Fig. 5). The effect of additional arterial hypocapnia and alkalosis
is shown in Fig. 6. When electrical stimulation of the brain cortex
was superimposed on arterial hypoxia the stimulation-elicited vascular responses were depressed considerably but the redox reactions
did not differ significantly from the normoxic ones.

In the fourth series of experiments electrical stimulation of
the brain cortex was superimposed on cortical spreading depression
evoked by topical application of 0.1 M KCl containing mock CSF, or
occurring spontaneously. Topical administration of 0.1 KCl to the
surface of the brain cortex resulted in a biphasic change in NAD/
NADH redox state, namely NADH oxidation (phase I) and consecutive
NAD reduction (phase II). Cerebrocortical vascular volume was always
increased during SD (Fig. 7). When brain activation was superimposed
on the second phase of SD, instead of the control NADH oxidation-
reduction cycle, only NADH oxidation was obtained, and the stimulation-induced increase in CVV became markedly depressed.

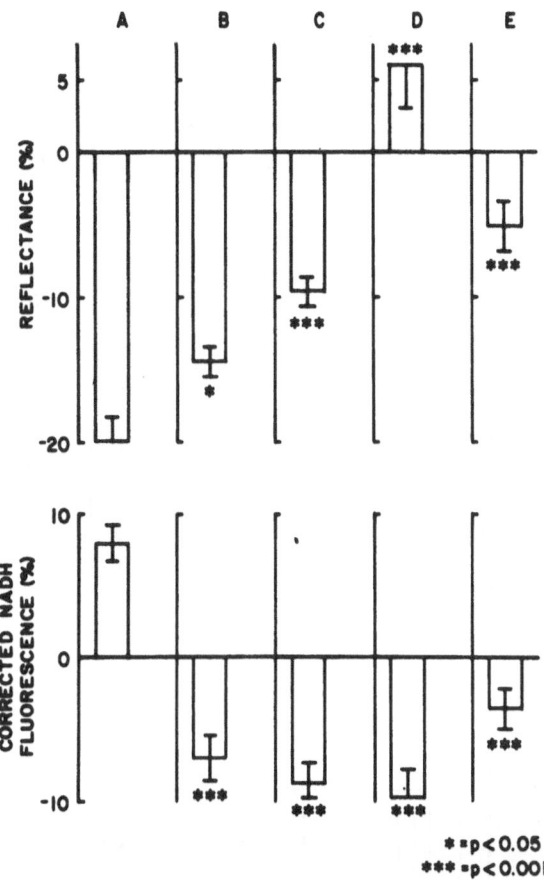

Fig. 3. Effect of direct electrical stimulation of the brain cortex
on cerebrocortical vascular volume (inversely related to
reflectance) and corrected NADH fluorescence in the same
experiments as shown in Fig. 2. Capital letters show the
various phases of the bleeding-reinfusion experimental
model where electrical stimulation with constant parameters
was applied. The standard deviation of the mean values is
represented on the columns by vertical lines closed at both
ends. Asterisks show the significant changes as compared to
control (Fig. 3A) responses. Note that though mean arterial
blood pressure before bleeding and after reinfusion did not
differ significantly, electrical stimulation of the brain
cortex before bleeding resulted in NAD reduction, but after
reinfusion in NADH oxidation. The stimulation-induced
vascular responses were also markedly depressed following
reinfusion (Fig. 3E).

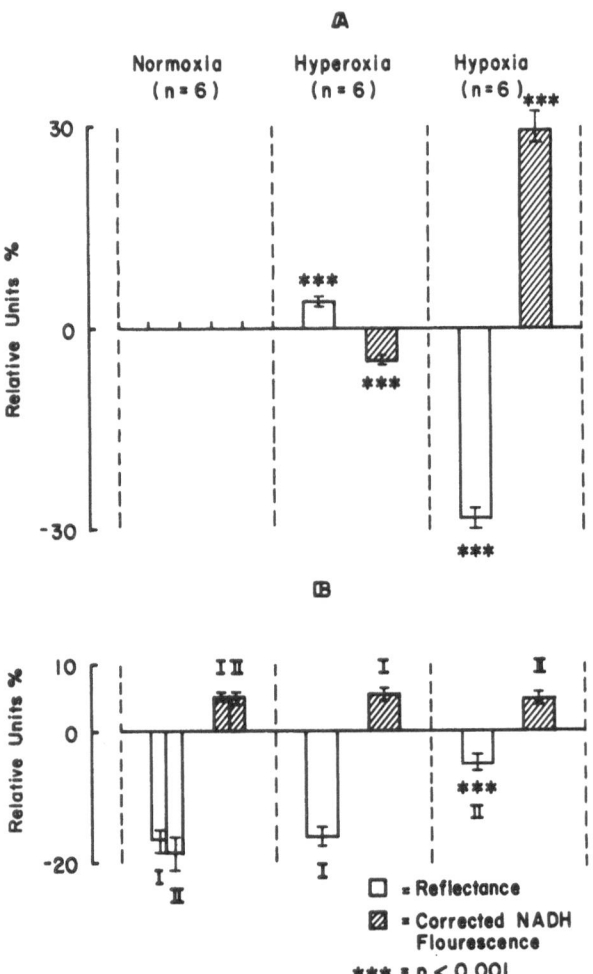

Fig. 4. Effect of arterial hyperoxia and hypoxia on the steady
         redox state and vascular volume of the brain cortex (Fig.
         4A), and the superpositioned activation-induced cerebro-
         cortical vascular and redox responses (Fig. 4B). Arterial
         hyperoxia and hypoxia were applied in separate series of
         experiments which are marked by Roman numerals (hyperoxia:
         I; hypoxia: II). Standard error of the mean values is
         represented on the columns by vertical lines closed at both
         ends. Asterisks show the significant changes. Note that
         when arterial pH and Po_2 are maintained as normoxic re-
         ference values during arterial hyperoxia, the activation-
         elicited vascular and metabolic reactions are not altered.

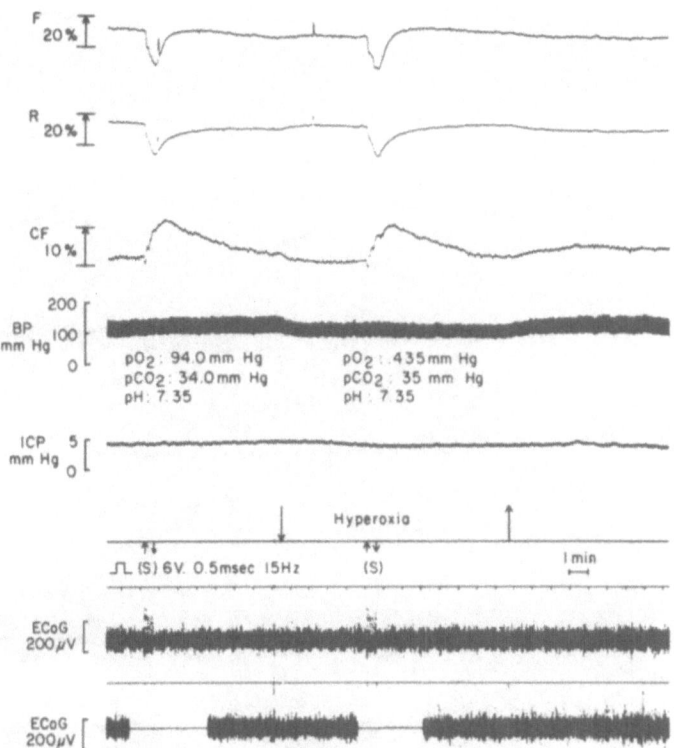

Fig. 5.   Effect of arterial hyperoxia on activation-induced cerebrocortical vascular volume and NAD/NADH redox state responses
with constant arterial pH and Pco_2 in a typical experiment.
The duration of electrical stimulation (S) and arterial
hyperoxia is marked by arrows above the time scale. Spikes
on F and R traces were evoked by artificial hemodilution
for determining the correction factor. Other abbreviations
and marks are the same as in Fig. 1.

Second administration of 0.1 M KCl evoked similar changes in
cerebrocortical vascular volume and redox state to the first one,
and electrical stimulation also resulted in NADH oxidation in the
second phase of SD. After the first two SD cycles, SD appeared
spontaneously, however both phases of SD-related redox alterations
became smaller.

     In many of our previous studies, when the cerebrocortical NAD/
NADH redox state was irreversibly perturbed, i.e., shifted steadily
toward a more reduced state (arterial hypotension, arterial hypoxia,
harsh brain surgery and brain trauma, etc.) electrical stimulation
of the brain cortex resulted in either no discernible change in
intracellular NAD/NADH redox state or NADH oxidation. In such

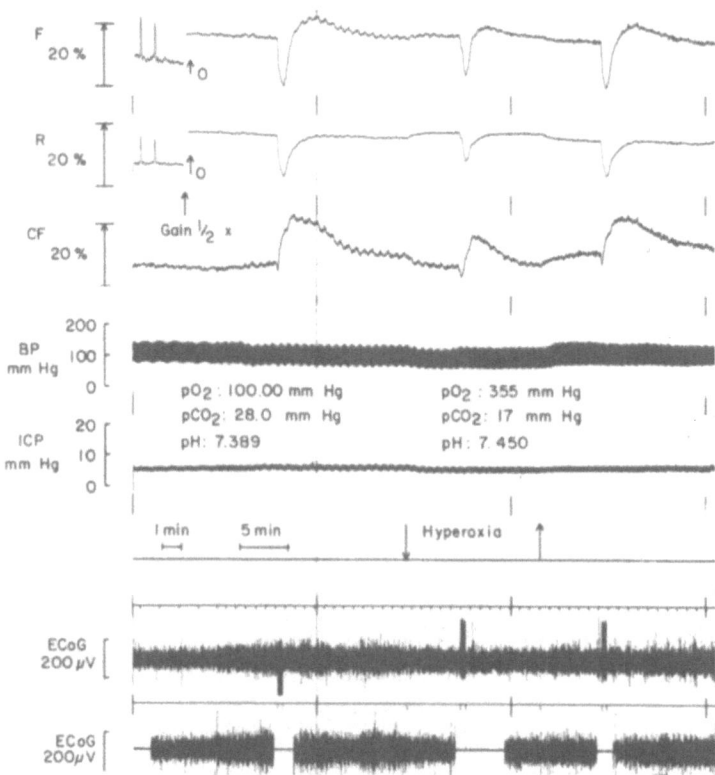

Fig. 6.   Effect of arterial hyperoxia on activation-induced cerebro-
          cortical vascular volume and NAD/NADH redox state responses
          with additional arterial hypocapnia and alkalosis in a
          typical experiment. The duration of electrical stimulation
          (S) and arterial hyperoxia is marked by arrows above the
          time scale. Spikes on F and R traces were evoked by arti-
          ficial hemodilution for determining the correction factor.
          Other abbreviations and marks are the same as in Fig. 1.

experiments SD appeared spontaneously often. Since in these experiments the local electrical activity of the exposed brain cortex was
not yet measured, these experiments were not evaluated until now.
As Fig. 8 shows, the vascular and redox responses with spontaneously
appearing SD were practically the same as with 0.1 M KCl-evoked SD.
The NADH became initially transiently oxidized and consecutively
reduced. The cerebrocortical vascular volume decreased during the
first phase of SD and it became markedly augmented during its NAD
reductive phase. Electrical stimulation of the brain cortex before
SD led to no discernible change in the NAD/NADH redox state but when
it was superimposed on the second phase of SD, the same stimulus
resulted in a considerable NADH oxidation.

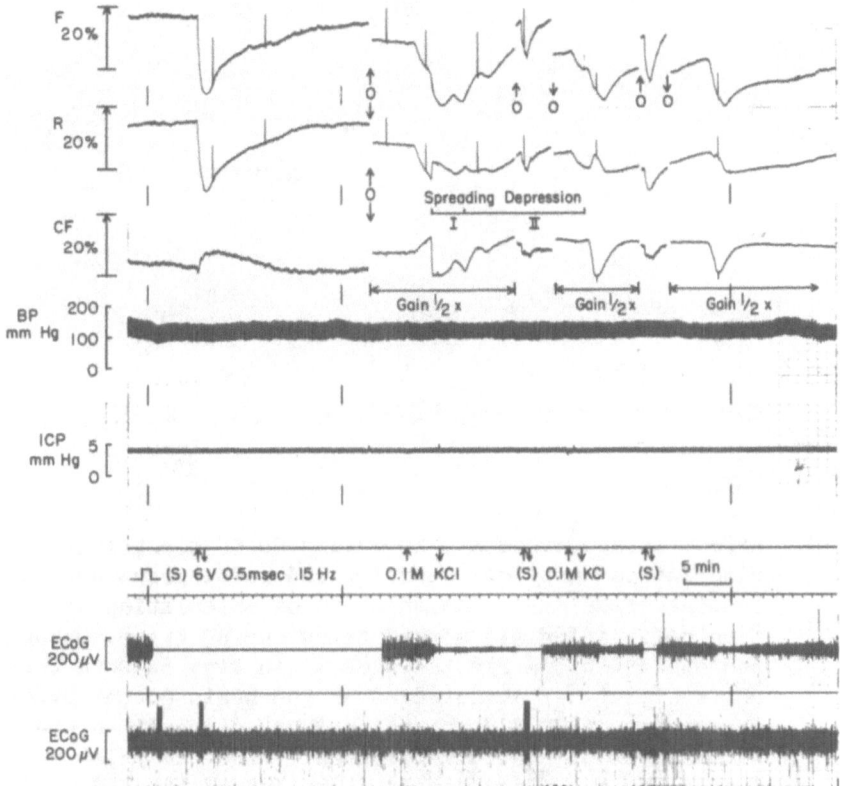

Fig. 7.   Effect of topically administered KCl on cerebrocortical vas-
          cular volume and NAD/NADH redox state, and the superposi-
          tioned activation-induced cortical vascular and redox re-
          sponses in a typical experiment. The duration of electrical
          stimulation (S), and superfusion of the brain cortex with
          0.1 M KCl containing CSF is marked by arrows above time
          scale. Spikes on F and R traces were evoked by artificial
          hemodilution for determining the correction factor. The two
          phases of spreading cortical depression (SD) are marked by
          Roman numerals above the CF trace (phase I: NADH oxidation;
          phase II: NAD reduction). Note the zero shifts and the
          change in amplification on F, R, and CF traces. Other marks
          and abbreviations are the same as in Fig. 5. When SD appear-
          ed during 0.1 M KCl superfusion the amplitude of the ECoG
          of the exposed brain cortex became abruptly and markedly
          depressed.

     Fig. 9 also demonstrates the importance of the steady redox
state in determining whether NAD reduction or NADH oxidation develop
during brain activation. As the local ECoG of the optically
monitored brain cortex shows, electrical stimulation superimposed
on arterial hypoxia resulted in spreading depression. Beside the

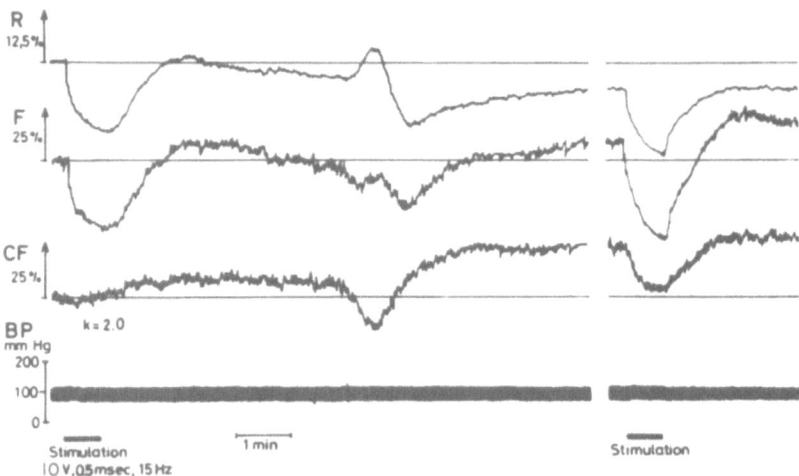

Fig. 8.   Effect of spontaneously appearing SD on electrical stimula-
          tion-induced cortical vascular and redox responses in a
          typical experiment. The duration of stimulation and the
          time calibration are marked below the BP trace. Other marks
          and abbreviations are the same as in Fig. 5. Note that
          while electrical stimulation of the brain cortex before the
          occurrence of SD did not bring about discernible change in
          NAD/NADH redox state, considerable NADH oxidation was ob-
          tained for the same stimulus, applied in the NAD reductive
          phase of SD.

changes of ECoG, the appearance of SD was characterized by an
initial NADH oxidation and a consecutive NAD reduction. During the
second phase of SD cerebrocortical vascular volume remained considerably augmented and NAD is reduced. When electrical stimulation
of the brain cortex was superimposed on this reduced state, instead
of the control NAD reduction, only NADH oxidation was obtained.

DISCUSSION

        Corresponding to our previously published data (Dorá and
Kovách, 1978, 1979; Gyulai et al., 1982), sustained direct electrical stimulation of the brain cortex leads to a pronounced NAD reduction and an increase in vascular volume. In the present study
the various phases of brain activation-induced intracellular NAD/
NADH redox state responses were explored more precisely. Accordingly, brain activation shifted the NAD/NADH redox state initially
toward a slightly more oxidized state and subsequently toward NAD
reduction. When electrical stimulation was terminated NAD became
transiently further reduced. The kinetics of these redox state re-

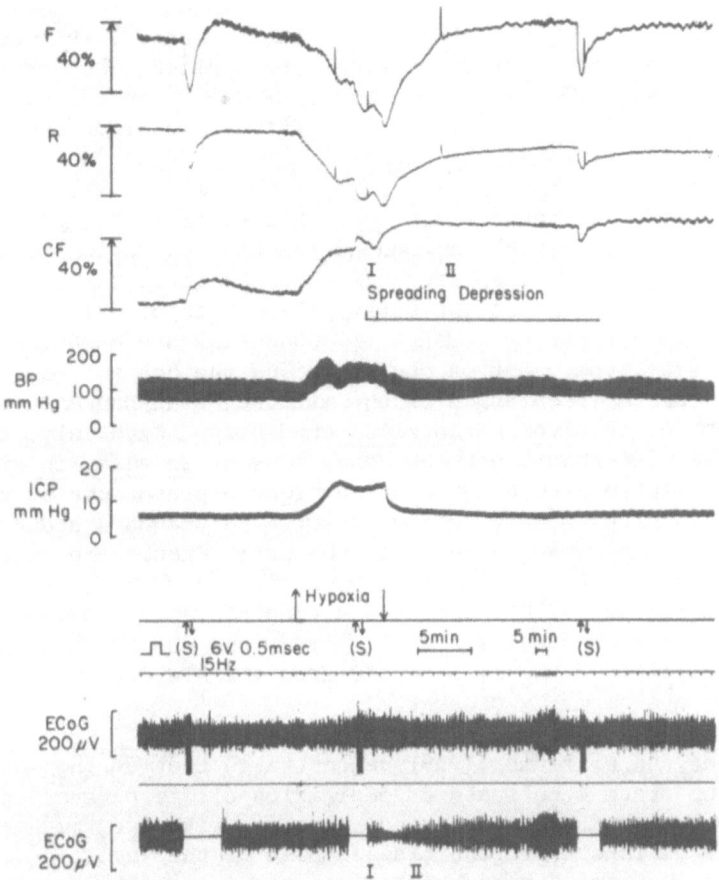

Fig. 9. Effect of the lack of recovery of cerebrocortical steady
NAD/NADH redox state from arterial hypoxia and additional
SD on the activation-induced cortical vascular and redox
responses in a single experiment. The duration of electrical stimulation (S) and arterial hypoxia is marked by
arrows. Roman numerals indicate the two phases of SD.
Spikes on F and R traces were evoked by artificial hemodilution for determining the correction factor. Time calibrations are shown above the time scale. Other abbreviations
and marks are the same as in Fig. 5. The ECoG at the bottom
of the figure was recorded from the exposed brain cortex.
Note the stimulation-induced opposite direction changes in
NAD/NADH redox state before and after arterial hypoxia.

sponses are fully comparable to those obtained in the isolated rat
cervical ganglion and brain slices (Brauser et al., 1970; Cummins,
1971; Lipton, 1973) but are in contrast to other data gained on in
vivo brain cortex (Jöbsis et al., 1971; Rosenthal et al., 1973;
Hempel et al., 1980), though in both cases excessive activation and
principally the same instrumentation were applied. The latter investigators revealed only monophasic NADH oxidation during epileptic seizures and prolonged electrical stimulation upon the brain
cortex. They interpreted the observed NADH oxidation as being in
agreement with the in vitro findings of Chance and Williams (1955),
as if mitochondrial respiration were shifted from state 4 to state 3.
However, since discernible mitochondrial NADH oxidation could not be
detected with the oxidized-reduced substrate ratio method during
epileptic seizures (Howse and Duffy, 1975; Siesjö, 1978), if this
method is appropriate to measure mitochondrial NAD/NADH ratio
(Siesjö, 1978), the question can be raised whether the same rules
hold true for the regulation of the mitochondrial NAD/NADH redox
state both for in vivo and in vitro conditions. Certainly, there are
substantial differences between conditions of in vivo, in situ and
in vitro isolated mitochondria which might explain some of these
disparate results. Namely, while in vitro in states 3 and 4 mitochondria have an excess amount of substrate (Chance and Williams,
1955), the substrate supply of in vivo mitochondria is strictly
regulated (Siesjö, 1978). Isolated mitochondria are suspended in a
highly optimized but artificial medium, and due to this they lack
of any control from the cytosol and from the ATP consumer cell
where they are normally present.

Chance and Williams (1955) demonstrated that during the transition from state 4 to state 3 respiration all respiratory chain-
linked cytochromes became reduced. However, with spectrophotometric
techniques on intact nervous tissues some of the investigators
revealed oxidation of cytochrome b and c and also a (Brauser et al.,
1970; Hempel et al., 1980), others found reduction of all cytochromes (Cummins, 1971) during excessive activations. Although
organ-specific properties, and methodological difficulties to
measure the oxido-reduction state of cytochrome a in blood perfused
brain (Bashford et al., 1982) might explain some of these contro-
versies, the inherent cause of the disparity of the above cited
data is not obvious.

We presumed, if brain activations were superimposed on a previously shifted steady redox state, a better understanding of the
regulation of intracellular NAD/NADH redox state can be achieved.

This approach seemed to be reasonable because our earlier preliminary studies (Dorá and Kovách, 1978) suggested that the activation-induced cerebrocortical NADH oxidation, reported by a group
of investigators (Jöbsis et al., 1971; Rosenthal et al., 1973;
Hempel et al., 1980), might be mainly attributed to the marked re-

duction of the cerebrocortical steady redox state before stimulation. This assumption was based mainly on our observations obtained in the past 10 years, namely when the brain cortex was traumatized by surgery or its redox state was steadily shifted to a more reduced state (falling arterial blood pressure, hemorrhagic shock, severe and long lasting arterial hypoxia, etc.) electrical stimulation of the brain cortex never resulted in NAD reduction but NADH oxidation (Dorá and Kovách, 1979; Gyulai et al., 1982). As we have already discussed (Dorá and Kovách, 1979; Gyulai et al., 1982), for the explanation of excessive brain activation-induced cortical NAD reduction the following should be considered: relative brain hypoxia, cytosolic versus mitochondrial NADH fluorescence changes, enhanced substrate influx and the mitochondrial NAD/NADH redox state.

The question whether the brain cortex becomes hypoxic during massive activation has been disputed many times because the rate of "anaerobic glycolysis" and lactate production increase considerably in such a condition (Siesjö, 1978). Our present data give further evidence against relative brain hypoxia, and intracellular NAD reduction for these reasons: At controlled arterial hyperoxia the stimulation-induced vascular and redox state responses were not altered; electrical stimulation of the brain cortex at 80 mm Hg MABP, when CBF, cerebral oxygen consumption and adenylate charge are maintained at normal level (Siesjö, 1978; Dorá and Kovách, 1982), resulted in NADH oxidation; after reinfusion, when MABP was not significantly different from its prebleeding level but redox state remained partially reduced, electrical activation of the brain cortex led to NADH oxidation; when electrical stimulation was superimposed on the NADH reductive phase of SD NADH oxidation was obtained. On other hand our data indicate indirectly that the NAD reduction which occurred within the autoregulatory range of CBF and during the second phase of SD cannot be attributed to tissue hypoxia and mitochondrial NAD reduction from this reason either. This conclusion disagrees with the one of Kontos et al. (1978) who claimed that relative brain hypoxia has a crucial role in the autoregulatory dilatation and functional hyperaemic responses of cerebrocortical (pial) vessels: However, since Kontos et al. (1978) did not measure pH and Pco_2 in the oxygenated fluorocarbon superfused brain cortex, and the dilatated pial vessels might react differently than the normal ones to abundant supply of oxygen, their findings do not necessarily prove the existence of tissue hypoxia during CBF autoregulation and brain activation.

The NAD reductive shift in the steady redox state of the brain, obtained during arterial hypotension and hypoxia, is consistent with other data from literature obtained by surface fluororeflectometry in vivo (Mayevsky and Chance, 1975; Hempel and Jöbsis, 1979). Spreading cortical depression led to initial NADH oxidation and consecutive NAD reduction in our experiments. Such an oxidation-reduction cycle during SD can also be observed in some of the re-

cordings of Mayevsky and Chance (1975) but these investigators did
not emphasize the secondary NAD reductive phase.

Corresponding to our findings considerable increase in cytoplasmic NAD reduction was calculated with the lactate/pyruvate ratio
method during epileptic seizures, arterial hypotension and hypoxia,
and during the secondary phase of SD (Howse and Duffy, 1975; Siesjö,
1978; Gjedde et al., 1981). This close correlation suggests that the
NAD reduction obtained by surface fluororeflectometry in vivo might
be mainly attributed to cytoplasmic NAD reduction due to the stimulation of "anaerobic glycolysis" under these experimental conditions.
In this case the reversal of the activation-induced NAD reduction to
NADH oxidation might be explained so that the masking effect of
cytoplasmic NADH fluorescence increase on mitochondrial NADH fluorescence decrease was less pronounced or became negligible when
electrical stimulation was applied during arterial hypotension and
reinfusion, and the secondary phase of SD. However, if surface
fluororeflectometry measures almost exclusively mitochondrial NADH
fluorescence, as some of the investigators stated (Jöbsis et al.,
1971; Rosenthal et al., 1973; Mayevsky and Chance, 1975; Hempel et
al., 1980), the interpretation of our results becomes rather
complex. Unfortunately very little is known about the exact changes
of mitochondrial NAD/NADH redox state and the redox state of mitochondrial cytochromes under the above mentioned experimental conditions. To calculate the mitochondrial NAD/NADH ratio the oxidized-
reduced substrate ratio method cannot be used for the brain (Siesjö,
1978), and also controversial results were obtained applying
reflectance spectrophotometry, surface fluororeflectometry and the
Fp/Fp+PN (Fp = flavine dinucleotides, PN = pyridine dinucleotides)
method to reveal changes in the redox state of mitochondrial NAD/
NADH and cytochrome a during arterial hypoxia. With reflectance
spectrophotometry in vivo Rosenthal et al. (1976) showed marked
cytochrome a reduction already when the percent oxygen of the
inspired gas mixture was decreased from 50% to 20%, contrast to
this Bashford et al. (1982) did not observe any reduction of the
redox state of cytochrome a until the percent oxygen of the inspired
gas mixture was not diminished below 6%. The same controversy holds
true if one compares the changes in intracellular NAD/NADH redox
state during arterial hypoxia; determined by surface fluororeflectometry in vivo 'and Fp/Fp+PN ratio method on freeze-trapped brains.
Mayevsky and Chance (1975) and Dora et al. (1979) reported that
cerebrocortical NADH fluorescence was increased by as slight hypoxia
as 10% oxygen in the inspired gas mixture, contrast to this Bashford
et al. (1982) who found no change in Fp/Fp+PN ratio at this severety
of arterial hypoxia. The inherent causes of these basically different results are hardly understandable if one presumes that the
hemodynamic artifact is properly corrected for in both measurements,
and if the freeze-trapping procedure and handling of frozen brain
samples do not induce such changes in NADH fluorescence that would
mask the early changes of NADH fluorescence detected by surface
fluorometry in vivo at moderate arterial hypoxia.

As the correction of the hemodynamic artifact is concerned in the in vivo NADH fluorescence measurements, the appropriateness of the correction procedure applied in the present studies has been proved by several laboratories (Jöbsis et al., 1971; Mayevski and Chance, 1975; Harbig et al., 1976; Dorá et al., 1979. Dorá and Kovach, 1982). Consequently, it is very likely that the early changes in cerebrocortical NADH fluorescence that develop during slight arterial hypoxia and arterial hypotension are real. These changes probably cannot be observed by the Fp/Fp+PN ratio method because either it has a lower sensitivity to changes in NADH fluorescence than surface fluororeflectometry, or the freezing procedure and handling of frozen brain samples mask the early changes in NADH fluorescence.

Finally, the question whether surface fluororeflectometry in vivo measures only mitochondrial NADH fluorescence, or both mitochondrial and cytoplasmic, has been long disputed but not yet clarified (Brauser et al., 1970; Jöbsis et al., 1971; Lipton, 1973; Mayevsky and Chance, 1975; Dorá et al., 1979; Hempel et al., 1980). Trying to solve this problem we reported previously (Dorá et al., 1980) that 2 - 3 min nitrogen gas respiration in normotensive animals resulted in only approximately 50% of the NAD reduction as compared to that which was obtained when arterial hypotension (MABP: 40 mm Hg) and superimposed anoxia-induced NAD reductions were additive. Since according to all available data it can be presumed that mitochondrial electron transport ceases during 2 - 3 min anoxia and hence respiratory chain linked NAD becomes fully reduced, it was concluded that surface fluororeflectometry measures the sum of mitochondrial and cytoplasmic NADH fluorescence. However, if a considerable amount of mitochondrial NAD remains in an oxidized state during 2 - 3 min nitrogen gas respiration this conclusion might be wrong. In this respect our studies, which are in progress, gave surprising results. Namely, when we killed the animals with nitrogen anoxia and waited at least 20 min from the beginning of nitrogen gas respiration, cerebrocortical NADH fluorescence increased approximately twice as much during this period of time than during 2 - 3 min nitrogen gas respiration. Since when we superfused the brain cortex of dead animals with oxygenated mock CSF, the NAD/ NADH redox state returned mostly to the level that was characteristic for the living normoxic animals, it might be concluded that this method measures mostly mitochondrial NADH fluorescence. However, if this is true, we have to presume that 2 - 3 min nitrogen anoxia reduces only about 50% of the total mitochondrial NAD, i.e., considerable part of the mitochondrial NAD is not very sensitive to the dramatic reduction of cerebrocortical oxygen tension.

Taking these arguments into consideration the most reasonable explanation for the NAD reducing effect of brain activation, and the reversal of stimulation-induced NAD/NADH responses during arterial hypotension and reinfusion, and during the second phase of spreading cortical depression, would be as follows:

In control conditions sustained electrical stimulation of the brain cortex resulted in NAD reduction because the supply of reducing equivalents into the mitochondria exceeded the rate of mitochondrial NADH oxidation.

When electrical stimulation of the brain cortex was superimposed on arterial hypotension and reinfusion, and on the secondary phase of SD, NADH oxidation was obtained because mitochondria have already been supplied with such a great amount of reducing equivalents that is characteristic for in vitro mitochondria under state 4 and state 3 respirations.

Relative brain hypoxia does not occur during electrical activation of the brain cortex, and hence it is not responsible for the NAD reducing and vasodilatatory effects of sustained electrical stimulation. Since under physiological conditions electrical activation resulted in NAD reduction, this response seems to be more characteristic for the in vivo intact brain than the NADH oxidation that was obtained only when the steady redox state of the brain cortex has already been shifted to a more reduced state.

REFERENCES

Bashford, C.L., Barlow, C.H., Chance, B., Haselgrove, J., and Sorge, J., 1982, Optical measurements of oxygen delivery and consumption in gerbil cerebral cortex, Am. J. Physiol., 242:C365.
Brauser, B., Bücher, T., and Dolivo, M., 1970, Redox transitions of cytochromes and pyridine nucleotides upon stimulation of an isolated rat ganglion, FEBS Lett., 8:297.
Chance, B., and Williams, R.G., 1955, Respiratory enzymes in oxidative phosphorylation. III. The steady state, J. Biol. Chem., 217:409.
Cummins, J.B., 1971, Spectral changes in respiratory intermediates of brain cortex in response to depolarizing pulses, Biochim. Biophys. Acta , 253:39.
Dorá, E., and Kovách, A.G.B., 1978, Electrically evoked cerebrocortical NADH fluorescence changes influenced by the steady state redox level, Fed. Proc., 37:498.
Dorá, E., and Kovách, A.G.B., 1979, Reactivity of the cerebrocortical vasculature and energy metabolism to direct cortical stimulation in hemorrhagic shock, Acta Physiol. Acad. Sci. Hung., 54:347.
Dorá, E., and Kovách, A.G.B., 1982, Effect of acute arterial hypo- and hypertension on cerebrocortical NAD/NADH redox state and vascular volume, J. Cereb. Blood Flow Metab., 2:209.
Dorá, E., Satori, O., Szabo, L., and Kovách, A.G.B., 1980, Shock-induced cytoplasmic NADH fluorescence changes in living cat brain cortex, Acta Physiol. Acad. Sci. Hung., 56:219.

Dorá, E., Zeuthen, T., Silver, I., Chance, B., and Kovách, A.G.B., 1979, Effect of various severity of arterial hypoxia on cerebrocortical redox state, vascular volume, oxygen tension, electrical activity and potassium ion concentration, Acta Physiol. Acad. Sci. Hung., 54:319.

Eke, A., Hutiray, G., and Kovách A.G.B., 1979, Induced hemodilution detected by reflectometry for measuring microregional blood flow and blood volume in cat brain cortex, Am. J. Physiol., 236:H759.

Gjedde, A., Hansen, A.J., and Quistorff, B., 1981, Blood-brain glucose transfer in spreading depression, J. Neurochem., 37:807.

Gyulai, L., Dorá, E., and Kovách, A.G.B., 1982, NAD/NADH redox state changes on cat brain cortex during stimulation and hypercapnia, Am. J. Physiol., in press.

Harbig, K., Chance, B., Kovách, A.G.B., and Reivich, M., 1976, In vivo measurement of pyridine nucleotide fluorescence from cat brain cortex, J. Appl. Physiol., 41:480.

Hempel, F.G., and Jöbsis, F.F., 1979, Comparison of cerebral NADH and cytochrome aa_3 redox shifts during anoxia and hemorrhagic hypotension, Life Sci., 25:1145.

Hempel, F.G., Kariman, K., and Saltzman, H., 1980, Redox transitions in mitochondria of cat cerebral cortex with seizures and hemorrhagic hypotension, Am. J. Physiol., 238:H249.

Howse, D.C., and Duffy, T.E., 1975, Control of the redox state of the pyridine nucleotide in rat cerebral cortex. Effect of electro-shock-induced seizures, J. Neurochem., 24:935.

Jöbsis, F.F., O'Connor, M., Vitale, A., and Vreman, H., 1971, Intracellular redox changes in functioning cerebral cortex. I. Metabolic effects of epileptiform activity, J. Neurophysiol., 34:735.

Kontos, H.A., Wei, E.P., Raper, A.J., Rosenblum, W.I., Navari, R.M., and Patterson, J.L., 1978, Role of tissue hypoxia in local regulation of cerebral microcirculation, Am. J. Physiol., 234:H582.

Lipton, P., 1973, Effects of membrane depolarization on nicotinamide nucleotide fluorescence in brain slices, Biochem. J., 136:999.

Mayevsky, A., and Chance, B., 1975, Metabolic responses of awake cerebral cortex to anoxia, hypoxia, spreading depression and epileptiform activity, Brain Res., 98:149.

Rosenthal, M., LaManna, J.C., Jöbsis, F.F., Levasseur, H.A., Kontos, H.A., and Patterson, J.L., 1976, Effects of respiratory gases on cytochrome a in intact cerebral cortex: is there a critical Po_2, Brain Res., 108:143.

Rosenthal, M., and Somjen, G., 1973, Spreading depression, sustained potential shifts, and metabolic activity of cerebral cortex of cats, J. Neurophysiol., 36:739.

Siesjö, B.K., 1978, "Brain Energy Metabolism", John Wiley & Sons, Chichester-New York-Brisbane-Toronto.

Sokoloff, L., 1981, Relationships among local functional activity, energy metabolism, and blood flow in the central nervous system, Fed. Proc., 40:2311.
Wahl, M., and Kuschinsky, W., 1976, The dilatatory action of adenosine on pial arteries of cats and its inhibition by theophylline, Pflügers Arch., 362:55.

PHYSIOLOGY OF OXYGEN
TRANSPORT TO TISSUE

THEORETICAL ASPECTS

MATHEMATICAL ANALYSIS OF TRANSPORT AND CONSUMPTION OF MOLECULES IN

HETEROGENEOUS BRAIN TISSUE (METHODOLOGY)

R.H. Kufahl[1], D.F. Bruley[1], N.A. Busch[1], and
J.H. Halsey, Jr.[2]

[1]Louisiana Tech. University, Ruston, Louisiana 71272
USA
[2]University of Alabama at Birmingham, Birmingham
Alabama 35294, USA

ABSTRACT

A computer model of metabolite transport and consumption in
heterogeneous brain tissue, using a combination of probabilistic
and deterministic techniques is being developed. The metabolites
are put into two separate classes: (I) those that have reached a
membrane for the first time during a small time step, Δt, and (II)
those that have not yet reached a cell membrane for the first time
during that time step. The time dependent spatial distribution of
class (I) molecules is determined using random walk theory, which
takes into account the actual paths of the molecules. The variation
of the spatial distribution of class (II) molecules with time is
determined using the time dependent diffusion equation with a
boundary condition of zero concentration on the enclosing membrane
boundaries.

INTRODUCTION

The understanding of the cause and effect of damage to the
central nervous system and the mechanism of subsequent nerve degen-
eration will eventually lead to answers concerning the repair and
regeneration of nerves. This paper presents the results of efforts
involved in developing techniques which will lead to better understanding of the functioning of single and multiple neuron systems.
The future of such efforts will be the understanding of the functioning of entire regions of the central nervous system.

The need to understand transport and metabolism of vital molecules in capillary-neuron systems has been demonstrated by many
researchers. At the University of Alabama, Birmingham, they have
shown that for two neurons, which are nearly the same distance from
a capillary, one is apparently healthy while the other is dead. The
mechanism for determining the fate of neurons in similar capillary-
neuron systems is the application of mathematical analysis of the
transport of molecules in the heterogeneous system. Classical
deterministic analysis of physiological systems, in particular brain
tissue, leaves many questions unanswered. Brain tissue is known to
be a heterogeneous media which cannot be modeled adequately with
deterministic methods. When the behaviour of single or multiple
neuron systems is to be investigated, the classical deterministic
methods prove to be inadequate. The method presented in this paper
is better suited to these heterogeneous problems than are the deterministic modeling techniques. This technique uses combined random
walk theory and analytical solution methods to handle the heterogeneities. The regions which, on the scale of a single neuron-
capillary system, may be assumed homogeneous are treated using
analytical solutions to the deterministic equations. The heterogeneities are handled using random walk theory. This technique is
able to handle multicomponent problems as well as normal and pathological systems.

METHODOLOGY

One of the most important events in a molecule's life is when
it enters a cell. For this reason, during any time interval Δt, the
molecules are put into two classifications:

(I) Those molecules which have reached a membrane boundary for
the FIRST TIME during the time step Δt.

(II) Those molecules which have NOT YET reached a membrane
boundary for the FIRST TIME during the same time step Δt.

During each time step the molecules are reclassified. A class (II)
molecule in the previous time step could become a class (I) during
the next time step and vice versa. The total concentration of any
molecular species at any point and time is the sum of its members
in class (I) and class (II).

The techniques used to calculate the spatial distribution, at
any instant, of the molecules in each of the two classes, (I) and
(II), are very different.

CLASS (I) - RANDOM WALK theory (Chandrasekhar, 1943; Karlin and
Taylor, 1975; Spitzer, 1976) is used to calculate the probability
of a molecule reaching a membrane for the first time step Δt.

Random walk theory is also used to determine the spatial distribution at any time, of molecules which have bounced off a membrane which is saturated with respect to that species. The DIFFUSION EQUATION ALONE CANNOT BE USED to determine the probability of hitting a membrane for the first time or the spatial distribution of the molecules bouncing off a saturated membrane because a solution to the diffusion equation alone CANNOT GIVE the PATH a molecule takes in diffusing from one point to another. A finite difference solution to the diffusion equation gives the concentration at any point after diffusion has occurred for a time step Δt. This solution does not yield the paths molecules take to reach a point, it just gives the final destinations of these paths after a time interval Δt. Because a random walk model of diffusion includes the molecular paths, it provides a more accurate description of diffusion in a heterogeneous media than does the diffusion equation alone. For example, for O_2 molecules near a neuron, the diffusion equation cannot determine the percent of time they would have spent in the neuron, and therefore their chance of being consumed, but the random walk model of diffusion can. Although the actual random walk paths are not computed, the number of molecules which reach a membrane for the first time during time Δt and the spatial distribution of molecules, at the end of a time step, which have bounced off a saturated membrane are found using random walk theory, which takes into account the molecular paths.

CLASS (II) - For molecules which have NOT YET reached a membrane for the FIRST TIME during Δt, the path taken to reach a neighbouring point is not important. Therefore, the DIFFUSION EQUATION can be used to calculate the concentration of CLASS (II) molecules at any point in space and time. Since the molecules which belong to class II have not yet reached a membrane boundary during Δt, the boundary condition on the diffusion equation will be one of ZERO CONCENTRATION ON THE MEMBRANE BOUNDARY for class (II) molecules. The diffusion equation for class (II) molecules is solved using the GREEN'S FUNCTION technique, an advantage of which is that the Green's function depends only on the diffusion coefficient and the shape of the boundaries. Therefore, the flow rate, metabolic rates, etc. can be changed without affecting the Green's function. The Green's function will be calculated both analytically, using the method of images approximation, and numerically, using the finite element technique.

Some of the advantages, both physiological and computational, of separating the molecules into two classifications are:

1) A membrane can be modeled by a two-state membrane which is EITHER SATURATED OR UNSATURATED with respect to the passage of a particular molecular species. When incorporating a two-state membrane, the number of molecules of a particular species which reach a membrane boundary for the FIRST TIME during Δt is determined. If the membrane is saturated with respect to this species, the maximum

number of this species is passed during the time interval Δt while the remainder bounce off.

2) Because random walk theory takes into account the path a molecule takes in diffusion from one point to another, it can more accurately determine the concentration of a particular species near a neuron. Although the diffusion equation can determine the concentration of molecules at a point near a neuron, after diffusion has occurred for a time Δt (i.e., normal distribution), it cannot determine the percent of time they spent in the neuron and therefore, an accurate estimate of the percent which are consumed.

3) Because CLASS (II) molecules must have ZERO CONCENTRATION ON THE MEMBRANE BOUNDARIES, THE DIFFUSION EQUATION CAN BE SOLVED INDEPENDENTLY IN EACH REGION ENCLOSED BY MEMBRANES for the concentration of class (II) molecules. Therefore, the solving of the diffusion equation is greatly simplified, since the separate membrane enclosed regions are not connected by flux boundary conditions (i.e., regions are uncoupled).

Probability Distributions for Class (I) Molecules

1) Fraction of molecules reaching an infinite wall for the FIRST TIME, $P_{a\infty}$, during time interval Δt (Chandrasekhar, 1943; Karlin and Taylor, 1975; Spitzer, 1976).

$$P_{a\infty}(x,\Delta t;a) = \frac{(a-x)}{\Delta t}\ \frac{1}{2(\pi D\Delta t)^{1/2}}\ \exp(-(a-x)^2/(4D\Delta t)) \qquad (1)$$

where a is the distance to the wall (see Fig. 1).

To illustrate the general stochastic-physical technique (Groom et al., 1976), the diffusion and possible consumption of O_2 near a neuron will be modeled for the simple geometry shown in Fig. 1. The number of O_2 molecules which reach the mitochondria membrane for the first time (i.e., class (I)), during the time interval from t to $t+\Delta t$, is found using eq. (1); and they are then normally distributed on the membrane boundary. During the next time step Δt (i.e., $t+\Delta t<t'<t+2\Delta t$), these O_2 molecules random walk freely back and forth across the membrane. Also during the time interval, $t<t'<t+\Delta t$, those O_2 molecules which have not reached the membrane during time Δt (i.e., class (II)) are allowed to undergo diffusion governed by the diffusion equation with the boundary condition of zero concentration on the membrane boundary. The number of the O_2 molecules, which random walked back and forth across the membrane during the time interval from $t+\Delta t$ to $t+2\Delta t$, which are consumed by the mitochondria is proportional to the time they spent in the mitochondria as well as a function of the concentrations of the rate controlling (i.e., irreversible reactions) species.

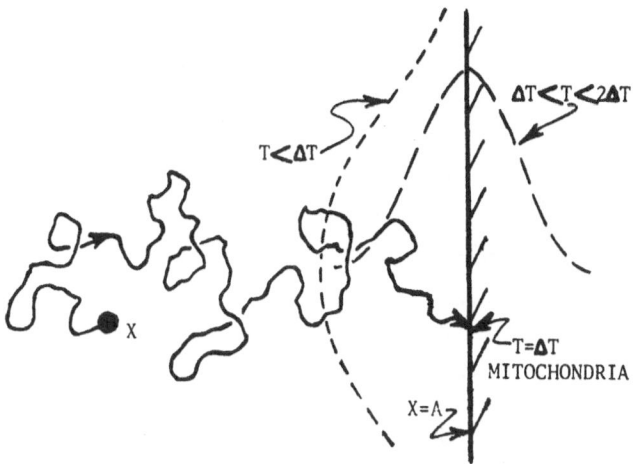

Fig. 1. Molecule reaches the membrane at t = Δt, for the first time, and then random walks back and forth across the membrane during Δt<t<2Δt. (T=t)

The percent of time an O_2 molecule spends in a mitochondria, after being placed on the membrane and random walking back and forth across it for time Δt, depends on the path taken by the random walk. Fig. 2 shows all possible combinations of a six step random walk in which the molecule's final destination is two space steps (i.e., 2Δx) inside the mitochondria. Each of these possible random walks has the same probability of occurring. The percent of time an O_2 molecule spends to the right of the membrane can be determined graphically from Fig. 2, to determine the spatial distribution of class (I) O_2 molecules, the molecules, after reaching mitochondria for the first time, random walk about the membrane, without consumption, and their final destination calculated using a normal distribution. The number of these molecules at each location is then reduced in proportion to the percent of time spent in the mitochondria, using a random walk analysis like that shown in Fig. 2. If the diffusion coefficient is smaller within a mitochondria or a red blood cell, the random walk step in Fig. 2 is made smaller within these cells than the surrounding media. A difference in diffusion coefficient across a membrane will show up as a drift velocity in the normal distribution which governs the random walk back and forth across the membrane.

2) Transport of glucose through a SATURATED membrane. The number of glucose molecules which reach the neuron membrane for the first time (i.e., class (I)) during Δt is calculated and compared with the maximum number the membrane can pass during Δt. If the membrane is saturated, the maximum number of these glucose molecules is allowed inside the neuron, while the remainder of these glucose

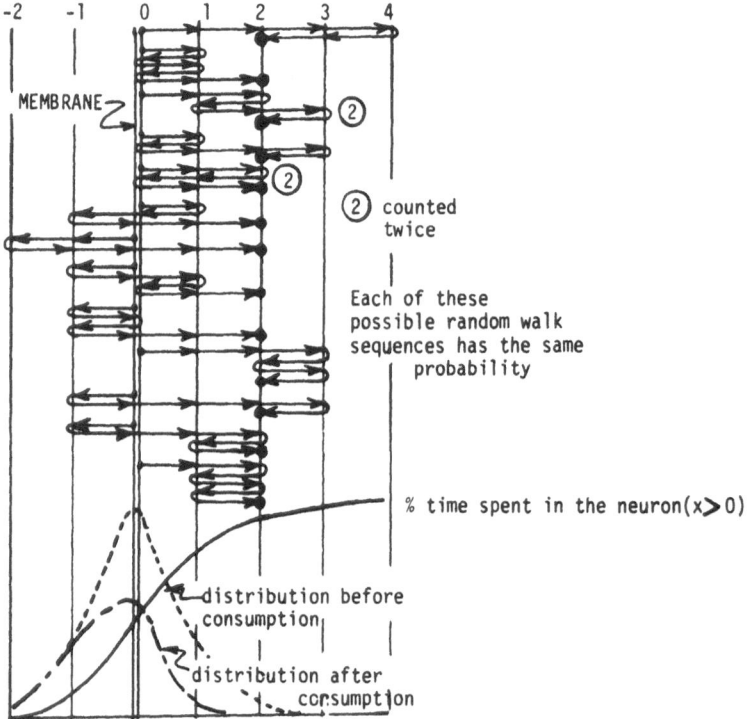

All possible sequences of 6 random walk steps for a molecule starting
on the membrane(X=0) and ending up 2 steps to the right.

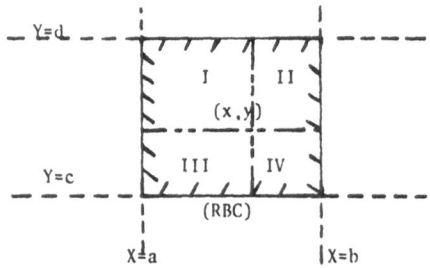

Fig. 2. The four sectors of a rectangle used in calculating the
probability of a molecule leaving the box (RBC) for the 1st
time and all possible sequences of 6 random walk steps for
a molecule starting on the membrane (X=0) and ending up
2 steps to the right.

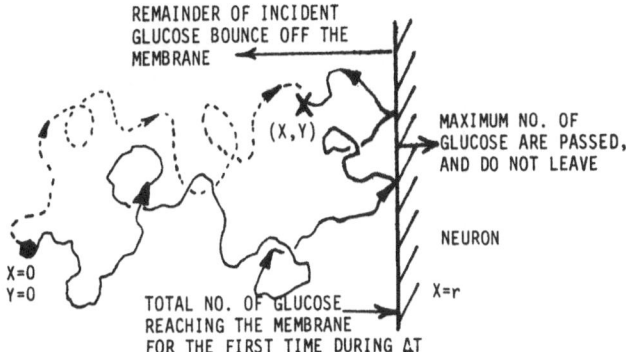

Fig. 3. Position of a glucose molecule after a time Δt after it has left x = 0, y = 0, and has either bounced off a saturated membrane or random walked to (x,y), without reaching the membrane.

molecules, which reached the membrane for the first time during Δt, bounce off the membrane, with their spatial distribution, together with those which have not reached the membrane, given by eq. (2).

$$P_{r\infty} = \frac{1}{2(\pi D \Delta t)^{1/2}} \left(\exp(-x^2/(4D\Delta t)) - \exp(-(2r-x)^2/(4D\Delta t)) \right)$$
$$\cdot \exp(-y^2/(4D\Delta t)) \qquad (2)$$

where r is the location of the reflecting boundary and $P_{r\infty}$ is the fraction of molecules between x and x+Δx. Because once a glucose molecule enters a neuron it cannot leave, the spatial distribution of glucose within the neuron cytoplasm will be assumed uniform.

The only important features of the paths shown in Fig. 3 are that the molecules bounced off the membrane and the locations of their final destinations after a time interval Δt are known. The number of times they have bounced off the membrane is not important.

3) The fraction of molecules which reach a corner, from the INSIDE, for the first time is given by:

$$P_{ab}(x,y,\Delta t;a,b) = P_{a\infty}(x,\Delta t;a) + P_{b\infty}(y,\Delta t;b)$$
$$-2P_{a\infty}(x,\Delta t;a) P_{b\infty}(y,\Delta t;b) \qquad (3)$$

The third term on the right represents the fraction of molecules that will cross both (infinite) lines x = a and y = b at least once and prevents the first two terms from counting the crossing of x = a or y = b twice. Since the random walks in the x and y directions occur independently, the third term can be written as a product.

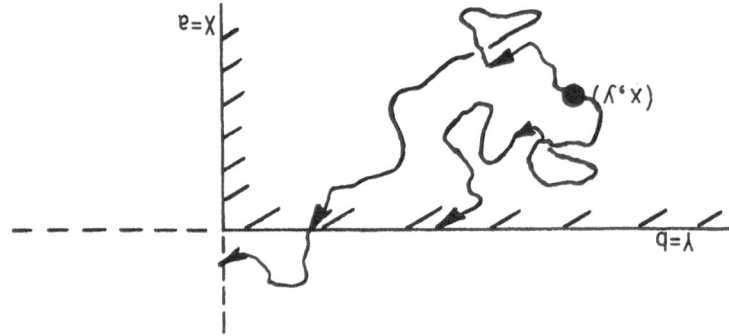

Fig. 4. A molecule reaching the inside of an (infinite) corner for
the first time during the time interval Δt.

4) The fraction of molecules LEAVING a box (red blood cell)
for the first time. The probability that a molecule will reach x =
b for the first time before it reaches x = a is given by:

$$P(x(T) = b|x(0) = x) = (x-a)/(b-a) \qquad (4)$$

where T is the first time the molecule reaches a or b. The box is
divided into 4 corners, and the fraction of molecules leaving any
corner, from point (x,y) is calculated using eq. (4a) (see Fig. 2).
For example, the fraction of molecules at point (x,y) which reach
corner II for the first time during Δt is given by:

$$P_{box}(x,y,\Delta t;a,b) = \frac{(x-a)}{(b-a)}\frac{(y-c)}{(d-c)} P_{ab}(x,y,\Delta t;a,b) \qquad (4a)$$

where the product is used in the first term since the random walks
in the x and y directions are independent.

5) The fraction of molecules which ENTER a corner, from its
surroundings, for the first time (see Fig. 5). This fraction is
calculated by dividing the entrance of molecules into the corner,
starting at point (x,y), into disjoint, first-time events.

a) The number of molecules leaving the corner x<a and y<b for
the first time is found from eq. (3):

$$P_{ab}(x,y,\tau_1;a,b) = P_a(x,\tau_1;a) + P_b(y,\tau_1;b)$$

$$-2P_a(x,\tau_1;a) P_b(y,\tau_1;b) \qquad (5)$$

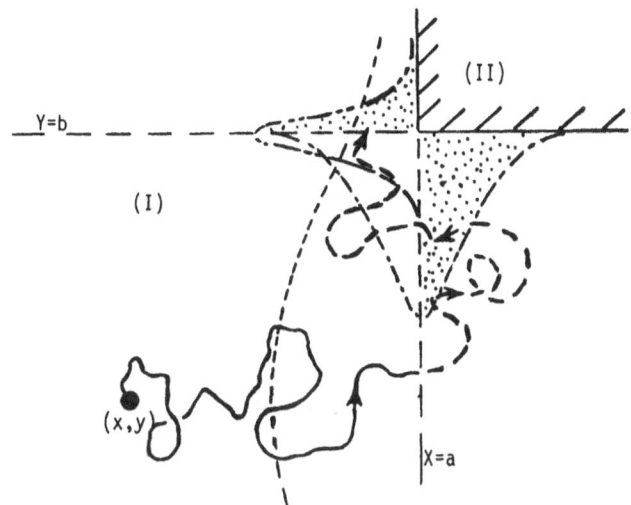

Fig. 5. The fraction of molecules entering a corner for the first
time from (x,y) (outside the corner) during the time Δt.

where τ_1 is the time to reach x = a for the first time. Because the
first two terms on the right of eq. (5) do not represent disjoint
events, they cannot be added without correcting for the possibility
of a molecule reaching BOTH x = a and y = b for the first time
(i.e., the third term in eq. (5)). The fraction of the molecules
which reach the side x = a of the corner I for the first time is
given by:

$$P_{ao}(x,y,\tau_1;a,b) = \frac{(b-y)}{(a-x)+(b-y)} P_{ab}(x,y,\tau_1;a,b)$$

These molecules which reach x = a for the first time are normally
distributed since there are no boundaries for x a to distort the
distribution. The spread of this normal distribution increases with
the time τ_1. Before a molecule can enter the corner II for the first
time, it must have first reached the lines x = a AND y = b for the
first time. Therefore, hitting corner II for the first time can be
divided into two steps: a) reaching x = a (or y = b) for the first
time from point (x,y) and b) then random walking (normal distribution) with these molecules initially on x = a, y<b, and determining
how many of these enter corner II for the first time during Δt.

$$\left.\begin{array}{l}\text{fraction of}\\ \text{molecules}\\ \text{reaching}\\ \text{y=b, x>a}\\ \text{from x,y}\end{array}\right\} = 1/2 \int_{-\infty}^{b} dy' \int_{\tau_1=0}^{t} d\tau_1 \int_{\tau_2=\tau_1}^{t} d\tau_2 \; P_{ao}(x,y,\tau_1;a,b)$$

$$\cdot \exp(-(y-y')^2/(4D\tau_1))P_{b\infty}(y',(\tau_2-\tau_1);b) \qquad (6)$$

where

$P_{ao}(x,y\tau_1;a,b)d\tau_1$ = fraction of molecules which arrive at $x = a$, $y<b$ for the first time between τ_1 and $\tau_1+d\tau_1$ from point x,y.

$\exp(-(y-y')^2/(4D\tau_1))dy'$ = fraction of these first time molecules on $x = a$ that lie between y' and $y'+dy'$.

$P_{b\infty}(y',(\tau_2-\tau_1);b)d\tau_2$ = fraction of molecules reaching $y = b$ for the first time from y' between τ_2 and $\tau_2+d\tau_2$ where $\tau_1<\tau_2<\Delta t$.

Equation (6) is multiplied by 1/2 because only 1/2 the molecules on $x = a$ enter the corner ($y=b$, $x>a$) for the first time at time τ_2. Of the 1/2 that missed the corner, and are therefore on $y = b$, $x<a$ at time τ_2, 1/2 can still enter the corner through the face $x = a$, $y> b$ at time τ_3; and these are included in eq. (7).

fraction of molecules reaching $y=b$, $x>a$ at τ_2 and $x=a$, $y>b$ at τ_3 for the first time during Δt } =

$$\frac{1}{2^2} \int_{-\infty}^{a}dx' \int_{-\infty}^{b}dy' \int_{\tau_1=0}^{\Delta t}d\tau_1 \int_{\tau_2=\tau_1}^{\Delta t}d\tau_2 \int_{\tau_3=\tau_2}^{\Delta t}d\tau_3\ P_{ao}(x,y,\tau_1;a,b)$$

$$\cdot \exp(-(y-y')^2/(4D\tau_1))P_{b\infty}(y',(\tau_2-\tau_1);b)$$

$$\cdot \exp(-(a-x')^2/(4D(\tau_2-\tau_1)))P_{ao}(x',(\tau_3-\tau_2);a)$$

$$(7)$$

The total fraction entering the corner is eq. (6) + eq. (7). This process may be continued until the number of molecules which enter the corner during Δt is small.

Those molecules which have not reached a membrane boundary for the first time during Δt (i.e., class (II)), are diffused for a time Δt using the time dependent diffusion equation, as explained in the next section.

The Diffusion and Consumption of Molecules which have not Reached a Membrane Boundary for the First Time During Δt (Class II)

Because only the final destination of the random walks, during Δt, of class II molecules is important, the diffusion equation is used to determine the concentrations of class II molecules. Since these class II molecules have not yet reached a membrane boundary, during Δt, the diffusion equation will have only the simple boundary condition of zero concentration on the boundary. The solution to the diffusion equation is expressed in terms of the Green's function, $G(r,\phi,t|r',\phi',t')$ and is given by:

$$c_i(r,\phi,t) = \pi \int_0^{2\pi} \int_a^b r'G(r,\phi,t|r',\phi',0)F_i(r',\phi')dr'd\phi'$$

$$-\pi \int_{t'=0}^{t} dt' \int_0^{2\pi} \int_a^b r' \, G(r,\phi,t|r',\phi',t')$$

$$\cdot g_i(c_i(r',\phi',t'),c_j,\ldots,c_m)dr'd\phi' \quad (8)$$

where c_i is the concentration of species i, F_i is the initial concentration of i, and g_i is the consumption of species i. The Green's function satisfies:

$$D \left(\frac{\partial^2 G}{\partial r^2} + \frac{1}{r}\frac{\partial G}{\partial r} + \frac{1}{r^2}\frac{\partial^2 G}{\partial \phi^2}\right) + \delta(r-r')\delta(\phi-\phi')\delta(t-t') = \frac{\partial G}{\partial t} \quad (9)$$

where $G=0$ on all boundaries, $G(r,\phi,t|r',\phi',t') = 0$ if $t<t'$, and (r',ϕ') is the location of an instantaneous unit source introduced at time t'. Equation (8) does not contain a boundary term, which would cause the solutions in each membrane enclosed region to become dependent on each other, and would cause eq. (8) to be very difficult to solve. This is one advantage of separating the metabolites into class (I) and class (II). Applying the mean value theorem to the integration with respect to time, for a time step Δt, eq. (8) becomes:

$$c_i(r,\phi,t+\Delta t) = \pi \int_0^2 \int_a^b r'G(r,\phi,t+\Delta t|r',\phi',t)c_i(r',\phi',t)dr'd\phi'$$

$$-\pi\Delta t \int_0^2 d\phi' \int_a^b r'G(r,\phi,t+h\Delta t|r',\phi',t)$$

$$\cdot \; g_i(c_i(r',\phi',t+h\Delta t), \; c_j,\ldots,c_m)\,dr' \tag{10}$$

where $0 \leq h \leq 1$. Because $G(r,\phi,t|r',\phi',t')$ is not defined when $h=0$ (see eq. (9)), only the approximation with $h=1$ will be used. With the GREEN's FUNCTION KNOWN on a set of grid points, the integrals are approximated by a composite Simpson's rule. Equation (10) then becomes a system of linear or nonlinear, depending on the kinetics, equations, which are solved for the unknown concentrations of class II molecules on the grid. As shown in eq. (10), the Green's function does not depend on the kinetics; it is only a function of geometry.

The Green's functions for simple geometries are found both analytically and numerically, using the finite element technique.

1) Analytical (approximate) solution for the Green's function. a) Exterior of a circle (neuron). The Green's function for a region bounded internally by a circle of zero concentration of radius, $r = b$, is given by:

$$G(r,\phi,t|r',\phi',t') = \sum_{n=-\infty}^{n=\infty} \cos n(\phi-\phi')$$

$$\cdot \int_0^{\infty} \beta \exp(-\beta^2 tD) \; \frac{U_n(\beta r)U_n(\beta r')}{J_n^2(\beta b)+Y_n^2(\beta b)} \, d\beta \tag{11}$$

where $U_n(\beta r) = J_n(\beta r)Y_n(\beta b)-J_n(\beta b)Y_n(\beta r)$. The above integral is being evaluated numerically using a composite Simpson's rule with increment halving. This integral is also being evaluated approximately. Because the integrand is near zero for β near 0, the Bessel functions may be replaced by their asymptotic approximations for large arguments and small orders, n.

$$J_n(\beta r) = (\frac{2}{\pi\beta r})^{1/2} \; \cos \, (\beta r - \pi/4 - n\pi/2)$$

$$Y_n(\beta r) = (\frac{2}{\pi\beta r})^{1/2} \; \sin \, (\beta r - \pi/4 - n\pi/2) \tag{12}$$

Substituting these asymptotic approximations of J_n and Y_n into the integral and evaluating the integral, an approximate Green's function is obtained.

$$G(r,\phi,t|r',\phi',0) = \sum_{n=0}^{m} \frac{\cos n(\phi-\phi')}{2\pi^2 tD(rr')^{1/2}} \left(\exp\left(\frac{-(r-r')^2}{4tD}\right)\right.$$

$$\left.-\exp\left(\frac{-(r+r'-2b)^2}{4tD}\right)\right) \tag{13}$$

where m must be a small integer in keeping with the approximation of the Bessel functions. Since only a few terms can be used in the sum, the variation of the Green's function with ϕ' will be gradual, which corresponds to the case of large times. This approximate Green's function has a simple physical interpretation. The first exponential term represents diffusion from a source at r' and the second exponential term denotes diffusion toward a sink located at the image point of r'. Preliminary results from the numerical integration of the exact expression for the Green's function, eq. (11), indicate that the integrals in the sum approach zero rapidly with increasing values of the angular eigenvalue, n. This rate depends on the parameters t and D. Therefore, t and D can be chosen such that only the term for n=0,1 in the approximate Green's function eq. (13) will be needed. The Green's function for the case of more than one neuron is approximated using the method of images. The exact expression for the Green's function, using the nearest neuron, is added to the sum of the images (i.e., second exponential in eq. (13)) of all the surrounding neurons, but the nearest one (see Fig. 6).

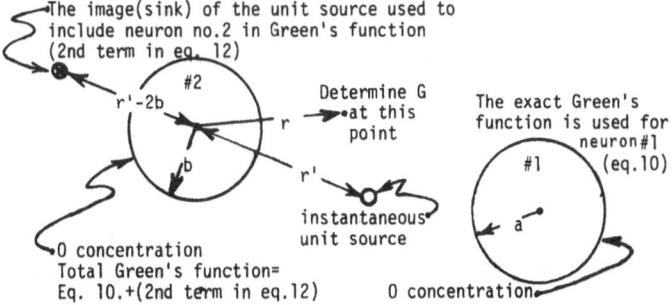

Fig. 6. Explanation of the Green's function for 2 neurons in an infinite media.

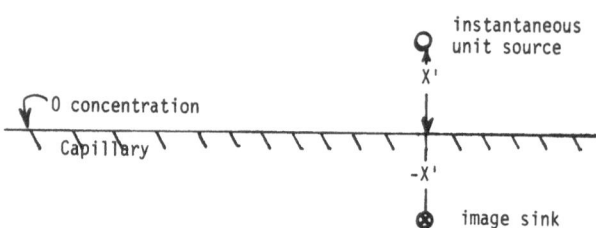

Fig. 7. Construction of a Green's Function for a straight line using the method of images (i.e., sources and sinks).

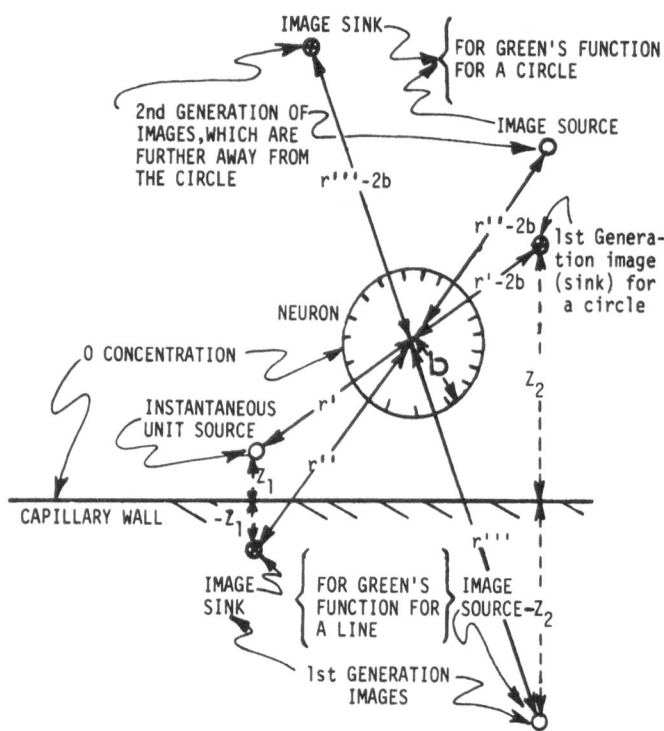

Fig. 8. Explanation of the construction of an approximate Green's function for a neuron near a capillary wall, including first and second generation images.

If the second generation of images (i.e., images of images) is included, their effect on the Green's function will be less than the first generation because they are located at farther distances away.

2) The Green's function for a line, with zero concentration on the line x = 0 given by:

$$G(x,y,\Delta t|x',y',0) = \frac{1}{2(\pi D\Delta t)^{1/2}} \; (\exp(-(x-x')^2/(4D\Delta t)).$$

$$-\exp(-(x+x')^2/(4D\Delta t))) \cdot \exp(-(y-y')^2/(4D\Delta t))$$

$$(14)$$

3) Construction of an approximate Green's function (2-D) of a neuron near a capillary. Eq. (12) is used to approximate the Green's function of the neuron using a source and a sink. Images of this source and sink about the capillary wall is then taken using eq. (13).

Here again second generation images can be included, but these will be located farther away than the first generation of images, and therefore, be of less importance (see Fig. 8).

Numerical Solution for the Green's Function Using the Finite Element Technique.

The finite element approximation is formed by applying Galerkin's method to eq. (9). The usual finite element approximation for the solution is made.

$$G^*(x,y,t) = \sum_{e}^{\substack{\text{no.}\\ \text{elements}}} \; \sum_{j\varepsilon e} \; G_j^e(t) \; N_j^e(x,y) \tag{15}$$

Where j is a node belonging to element e, G_j^e is the unknown concentration of class (II) at that node at time t, and $N_j^e(x,y)$ are the nodal basis functions which have the property that N_j^e = at node j of element e, N_j^e = 0 at all other nodes of element e, and N_j^e = 0 over all other elements other than e. The approximate solution, eq. (15), is substituted into eq. (9) to yield the error (i.e., residual) in the approximation. The integral of this error, times a weighting function, is then set equal to zero. This integral is over the entire domain.

$$\iint_D (\nabla^2 G^* - \frac{1}{D}\frac{\delta G^*}{\delta t} + \frac{1}{D}\;\delta(R-R')\delta(t-t'))W_L(x,y)dA = 0 \tag{16}$$

Where L = 1,2,...,NNODE, and NNODE are the total number of nodes.
Integrating the unit impulse (i.e., Dirac delta function), eq. (16)
becomes:

$$
\iint\limits_{D} (\nabla^2 G^* - \frac{1}{D} \frac{\delta G^*}{\delta t})\, W_L(x,y)\, dA = \begin{cases} 1, & R=R' \text{ and } T=T' \\ 0, & T>T' \end{cases} \tag{17}
$$

Where the term on the right of the equal sign is an instantaneous
source of unit strength applied at the point R' at time t'. The
presence of a Dirac delta function (i.e., an infinite spike) is
easily handled using the finite element technique, whereas the
presence of a delta function makes eq. (9) very difficult to solve
using the finite difference technique. Applying Green's theorem to
the first integral in eq. (17) and making use of eq. (15), eq. (17)
becomes:

$$
\sum_{e}^{\substack{\text{no.}\\\text{elements}}} \iint\limits_{D^e} \sum_{j \in e} (\frac{\partial N_j^e}{\partial x} \frac{\partial W_L}{\partial x} + \frac{\partial N_j^e}{\partial y} \frac{\partial W_L}{\partial y})\, G_j^e(t)\, dxdy
$$

$$
-\frac{1}{D} \iint\limits_{D^e} \sum_{j \in e} N_j^e\, W_L\, dxdy\, \frac{dG_j^e}{dt} + \int_{\Gamma} \sum_{i \in s} (\frac{\partial N_i^e}{\partial x} n_x + \frac{\partial N_i^e}{\partial y} n_y)\, G_i^e(t)\, W_L\, ds
$$

$$
= \begin{cases} 1, & R=R' \\ & \text{and } T=T' \text{ where } L = 1,2,\ldots \text{ NNODE} \\ 0, & T>T' \end{cases} \tag{18}
$$

Where Γ is the boundary of the entire domain, j is a node on this
boundary, and n_x and n_y are the direction cosines along the boundary. Because there is a W_L for each node, there will be as many
equations as there are unknowns. On the boundary points of the
domain where $G_j^e(t) = 0$, W_L is set equal to zero on these boundary
nodes, in order to keep the number of equations and unknowns equal.
Although $G_j^e(t) = 0$ along the boundary of the domain, the presence
of lines of symmetry may be used to lessen the number of unknowns
(see Fig. 9). Therefore, along the domain boundary where $G_j^e(t) = 0$,
$W_L = 0$; and along the lines of symmetry, grad $N_i^e = 0$. Thus, the
boundary integral in eq. (18) is zero along the boundary of the
domain and lines of symmetry; and eq. (18) becomes:

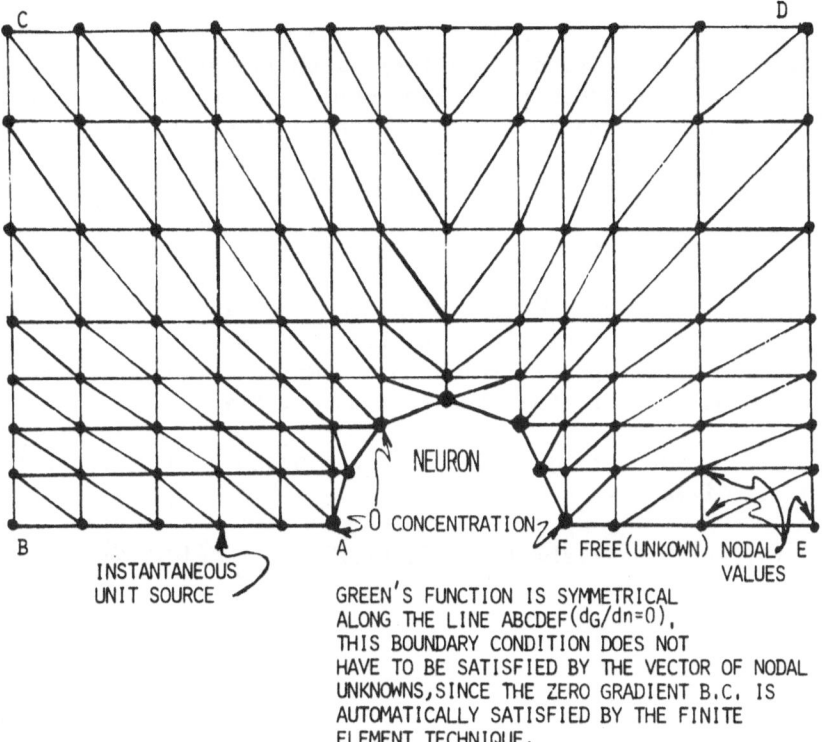

Fig. 9. Finite element mesh used to compute the Green's function
for class (II) molecules outside a neuron.

$$\sum_{e}^{\substack{\text{no.}\\ \text{elements}}} \iint_{D^e} \sum_{j \in e} \left(\frac{\partial N_j^e}{\partial x} \frac{\partial W_L}{\partial x} + \frac{\partial N_j^e}{\partial y} \frac{\partial W_L}{\partial y} \right) G_j^e(t) \, dxdy$$

$$-\frac{1}{D} \iint_{D^e} \sum_{j \in e} N_j^e(x,y) \, W_L \, dxdy \, \frac{dG_j^e}{dt} = \begin{cases} 1, & R=R' \text{ and } T=T' \\ 0, & T>T' \end{cases} \quad (19)$$

Where $L = 1,2,\ldots$ NINTERIOR (number of interior nodes). Along the
domain boundary, $G_j^e(t) = 0$; and no equation need be written for
these nodes (i.e., $W_L = 0$). Along the lines of symmetry, $G_j^e(t)$ in
equation are unknowns.

Triangular elements are used since they can more easily follow
the outline of a circle (see Fig. 9). The Green's function throughout the triangular element is interpolated linearly between nodal
values, $G_j^e(t)$. The Green's function over a triangular element is
therefore approximated by a space triangle above it. The approximate
Green's function over the entire domain will consist of a surface
composed of triangles attached along their edges. Although the
surface is necessarily continuous, the slope of the Green's function
will be discontinuous along the edges of the triangular elements.
According to eq. (9), the slope of the Green's function is necessarily discontinuous at the time t' at the location R' of the instantaneous unit source. Therefore, at least at the initial instant,
the finite element approximation to the Green's function with a
discontinuous slope at the source is acceptable.

For a linear triangular element the Green's function in that
element is given by:

$$G^e(x,y,t) = N_i^e(x,y)G_i^e(t) + N_j^e(x,y)G_j^e(t) + N_k^e(x,y)G_k^e(t) \qquad (20)$$

where

$$N_i^e(x,y) = \frac{1}{2A^e} (x_j y_k - x_k y_i + (y_j-y_k)x + (x_k-x_j)y)$$

$$N_j^e(x,y) = \frac{1}{2A^e} (x_k y_i - x_i y_k + (y_k-y_i)x + (x_i-x_k)y)$$

$$N_k^e(x,y) = \frac{1}{2A^e} (x_i y_i - x_j y_i + (y_i-y_j)x + (x_j-x_i)y)$$

and A^e is the area of the element

$$2A^e = (x_i y_j - x_j y_i) + (x_k y_i - x_i y_k) + (x_j y_k - x_k y_j)$$

and x_i and y_i are the coordinates of the vertices of the triangles.

In the Galerkin method, the weighting functions are set equal
to nodal functions (i.e., $W_L = N_j^e$). Because the Green's function
varies linearly over an element, the gradient of the nodal function
is a constant.

For any fixed node L, excluding those whose value of G_j^e is
known to be zero, the integrals in eq. (19) will be nonzero only
over the region where N_L is not zero (see Fig. 10). In eq. (19), the
integral is written as the sum of the integrals over each triangular

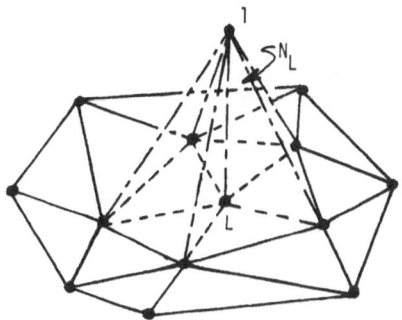

Fig. 10. Plot of a single linear basis function $N_L(x,y)$.

element; and therefore the integral is evaluated over a single triangular element. The first integral for a linear triangle becomes:

$$\iint_e \left(\frac{\partial N^e}{\partial x}\frac{\partial N_L}{\partial x} + \frac{\partial N^e}{\partial y}\frac{\partial N_L}{\partial y}\right) G^e(t)\,dxdy =$$

$$\iint_e \left(\frac{\partial N_i^e}{\partial x}G_i^e(t) + \frac{\partial N_j^e}{\partial x}G_j^e(t) + \frac{\partial N_k^e}{\partial x}G_k^e(t)\right)\frac{\partial N_L}{\partial x}$$

$$+ \left(\frac{\partial N_i^e}{\partial y}G_i^e(t) + \frac{\partial N_j^e}{\partial y}G_j^e(t) + \frac{\partial N_k^e}{\partial y}G_k^e(t)\right)\frac{\partial N_L}{\partial y}\,dxdy \qquad (21)$$

Where L can equal i, j, or k (i.e., vertices of triangle e) for eq. (21) to be nonzero (see Fig. 10). The second integral in eq. (19), for a triangular region, becomes:

$$\iint_e N^e(x,y)N_L(x,y)\frac{dG^e}{dt}\,dxdy =$$

$$\iint_e \left(N_i^e \frac{dG_i^e}{dt} + N_j^e \frac{dG_j^e}{dt} + N_k^e \frac{dG_k}{dt}\right) dxdy\, N_L \qquad (22)$$

where again L can equal i, j, or k for eq. (22) to be nonzero. Eq. (19) can be written in matrix notation.

$$(k) \quad G(t) + (M) \; \frac{dG}{dt} = \begin{cases} 1, & R=R' \text{ and } T=T' \\ \\ 0, & T>T' \end{cases} \tag{23}$$

Eq. (21), in matrix notation can be expressed as a 3 x 3 matrix, whose location in (k) is i^{th} row and column, the jth row and column, and the kth row and column. A single element of this matrix is equal to:

$$k_{ij} = (\frac{\partial N_i^e}{\partial x} \frac{\partial N_j}{\partial x} + \frac{\partial N_i^e}{\partial y} \frac{\partial N_j}{\partial y}) \iint_e dxdy$$

$$= \frac{1}{4A^e} (b_i b_j + c_i c_j) \tag{24}$$

where $b_i = y_j - y_k$ and $c_i = x_k - x_j$.

Eq. (22) is also expressed as a 3 x 3 matrix:

$$(M) = \iint_e \begin{bmatrix} N_i^2 & N_i N_j & N_i N_k \\ N_i N_j & N_j^2 & N_j N_k \\ N_i N_k & N_j N_k & N_k^2 \end{bmatrix} dxdy = \frac{A^e}{12} \begin{bmatrix} 2 & 1 & 1 \\ 1 & 2 & 1 \\ 1 & 1 & 2 \end{bmatrix} \tag{25}$$

From Fig. 10, it can be seen that the Lth row of (K) or (M) will have nonzero elements in the columns corresponding to nodes in the patch surrounding node L.

CONCLUSION

A new technique is being developed to analyze the transport of molecules in 3-dimensional, time dependent heterogeneous brain tissue. The model employs a combination of analytical solution techniques and random walk theory. The analytical techniques are used in the regions which are considered to be homogeneous (interstitial tissue) and the random walk theory is used to handle the heterogeneities. Preliminary results indicate that the technique is a powerful method of analyzing the transport of molecules in brain tissue and can be used to investigate the behavior of single or multiple capillary-neuron micro systems.

ACKNOWLEDGEMENTS

The authors wish to express their appreciation for support
from the National Institute of Health and the Louisiana Division of
Vocational Rehabilitation - Grant Numbers NIH P50-NS08802 and
83-006.

REFERENCES

Chandrasekhar, S., 1943, Stochastic Problems in Physics and Astrono-
    my, Rev. Mod. Phys., Vol. 15, No.1:1.
Groome, L.J., Bruley, D.F., and Kniseley, M.H., 1976, A stochastic
    model for the transport of oxygen to the brain, Adv. Exp. Med.
    Biol., 75:267-277.
Karlin, S., and Taylor, H.M., 1975, "A First Course in Stochastic
    Processes", Academic Press, New York.
Spitzer, F., 1976, "Principles of Random Walk", Springer, New
    York.

COUPLED TRANSPORT OF O_2 AND CO_2 WITHIN THE UPPER SKIN SIMULATED BY THE CAPILLARY LOOP MODEL

U. Großmann, P. Winkler, and D.W. Lübbers

Max-Planck-Institut für Systemphysiologie
Rheinlanddamm 201, 4600 Dortmund 1, FRG

INTRODUCTION

The transport of respiratory gases as oxygen and carbon dioxide through tissues and blood and their functional relations has been investigated in the past by numerous authors (for example: Roughton, 1964; Severinghaus, 1966; Thews, 1968; Lübbers, 1979). Due to the introduction of transcutaneous Po_2 and Pco_2 measuring techniques in clinical medicine (Huch et al., 1981) the transport of oxygen and carbon dioxide through the skin has become of considerable interest. To study those processes in detail we developed a so-called capillary loop model, which simulates the gas and heat transport through the skin including the loop-shaped capillary structure of the upper skin and the nonlinear effects due to the binding of O_2 by hemoglobin and temperature shifts (Großmann et al., 1980; Großmann, 1982).

The purpose of this paper is to study the simultaneous transport of oxygen and carbon dioxide through the upper skin under various conditions using the capillary loop model.

The Model

That part of the skin we are interested in is shown on the schematical drawing of the left side of Fig. 1. It consists of the epidermis (E) and the neighbouring stratum papillare containing the characteristic capillary loops.

To analyze the combined transport of oxygen and carbon dioxide through this part of the skin we used the microcirculatory unit shown on the right side of Fig. 1 as a model. Its position is marked

125

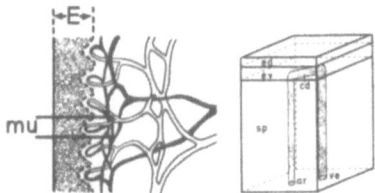

Fig. 1. Left side - schematical drawing of the skin with epidermis
        (E) and neighbouring stratum papillare containing the
        capillary loops; right side - microcirculatory unit (mu)
        containing one capillary loop (position is marked at the
        left side by mu).

in the schematical drawing on the left side by mu. The unit consists
of three spatial compartments, the dead part of the epidermis (ed),
the viable part of the epidermis (ev), and the stratum papillare
(sp) containing a single capillary loop. (ar) indicates the arterial
inflow, (ve) the venous outflow of the capillary, and (cd) the
capillary dome.

Assuming that a defined region of the upper skin is built
symmetrically by several, equal microcirculatory units and assuming
that diffusion conditions, oxygen consumption, carbon dioxide production, blood flow etc. are homogeneous, the lateral boundaries of
each microcirculatory unit are surfaces of symmetry for Po_2 and
Pco_2. Thus, the concentration gradients perpendicular to the lateral
boundaries vanish. With Fick's law of diffusion (Crank, 1955) it
follows that no diffusion flux occurs across these boundaries. This
means that each mu is isolated from its neighbours and can be analyzed as a representative unit of this defined skin region.

In the tissue compartments, O_2 and CO_2 are transported by
diffusion. The relationship of O_2 consumption and CO_2 production
is given by the respiration quotient. In the capillary, O_2 and CO_2
are transported by diffusion and convection. "Blood" is assumed to
be a homogeneous plasma-hemoglobin solution. Oxygen is physically
dissolved in the plasma and chemically reversibly bound by hemoglobin. CO_2 is physically dissolved in the plasma, stored as bicarbonate and reversibly bound by hemoglobin carbamino compounds.
Equilibrium is assumed for all reactions of O_2 and CO_2.

Transforming these assumptions into mathematical language
yields the following steady-state equations for Po_2 and Pco_2:

$$0 = \nabla \left\{ \alpha o_2 \cdot Do_2 \, \nabla Po_2 - \vec{v} \cdot Co_2 \, (Po_2, \, Pco_2) \right\} \tag{1}$$

$$0 = \nabla \left\{ \alpha co_2 \cdot Do_2 \, \nabla Pco_2 - \vec{v} \cdot Cco_2 \, (Po_2, \, Pco_2) \right\} \tag{2}$$

for the capillary compartment, and

$$0 = \nabla \left\{ \alpha_{O_2} \cdot D_{O_2} \nabla P_{O_2} \right\} - \dot{V}_{O_2} (P_{O_2}) \tag{3}$$

$$0 = \nabla \left\{ \alpha_{CO_2} \cdot D_{CO_2} \nabla P_{O_2} \right\} + RQ \cdot \dot{V}_{O_2} (P_{O_2}) \tag{4}$$

for the tissue compartments. Where:

α_{O_2} (ml O_2/(g·atm)) indicates the oxygen solubility coefficient of plasma, α_{CO_2} (ml CO_2/(g·atm)) that for CO_2, D_{O_2}, D_{CO_2} (um^2/sec) the diffusion coefficients for oxygen and carbon dioxide, $\vec{V}$ the velocity field of the flowing blood. C_{O_2} (P_{O_2}, P_{CO_2}) (ml O_2/ml) indicates the total oxygen concentration of the blood depending on P_{O_2} and P_{CO_2}. C_{CO_2} (p_{O_2}, P_{CO_2}) (ml CO_2/ml) the total carbon dioxide concentration of the blood depending on P_{O_2} and P_{CO_2} as given by Roughton (1964). $\dot{V}_{O_2}$ indicates the oxygen consumption (ml O_2/(g·min)), and RQ the respiration quotient.

Equations (1) – (4) together with suitable interface- and boundary conditions yield a coupled nonlinear, elliptic boundary value problem, which has at least one solution (Großmann, 1980). A numerical approximation of that solution is obtained with slightly modified techniques as given by the authors (Großmann, 1982). The anatomical data and the data concerning the transport of oxygen are described in detail in the above mentioned paper. In addition, we used data concerning the transport of carbon dioxide as given by Kawashiro et al. (1975).

The total CO_2 concentration, depending on P_{CO_2} for oxygenated and deoxygenated hemoglobin was given by Roughton (1964). We approximated both curves by straight lines in the range between 20 and 60 Torr. Intermediate states are interpolated according to the saturation of hemoglobin with oxygen.

Since heat transport obeys similar laws as the transport of oxygen and carbon dioxide, it is possible to use the capillary loop model to calculate temperature fields within the upper skin (Großmann, 1982). The temperature fields calculated are almost homogeneous. Thus, in the following calculations we assume a homogeneous temperature in the whole microcirculatory unit.

RESULTS

Using the capillary loop model we simulated four typical situations of respiratory gas transfer through the skin:
a) the skin being in contact with air, a resting value of blood flow and a temperature of 37°C,
b) the skin being impermeable to oxygen and carbon dioxide, a resting value of blood flow and a temperature of 37°C,
c) a situation of measuring P_{CO_2} at the skin surface and
d) a situation of measuring P_{O_2} at the skin surface.

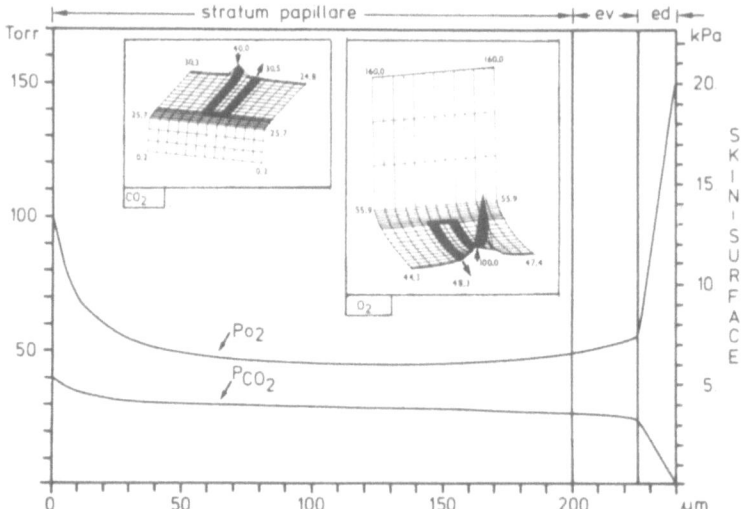

Fig. 2. Po_2 and Pco_2 profiles along the arterial limb and across
the epidermis (as inserts Po_2 and Pco_2 fields over a cross
section of the microcirculatory unit drawn as a surface);
blood flow 0.01 ml/(g·min), temperature 37°C, skin surface
in contact with air.

Fig. 2 shows the result of the calculation concerning case a).
Blood flow corresponds to a value of 0.01 ml/(g·min), the skin
surface is in contact with air and the whole mu has a homogeneous
temperature of 37°C. Fig. 2 shows Po_2 and Pco_2 profiles along the
arterial limb of the mu and across the epidermis drawn from the
arterial inflow to the skin surface. As inserts the Po_2 and Pco_2
fields over a cross section of the mu are drawn perspectively as
surfaces. Blood enters the arterial limb with a Po_2 of 100 Torr
(13.3 kPa) and decreases to 46 Torr (6 kPa) at the middle of the
arterial limb. Then it reincreases due to an additional O_2 supply
through the skin surface. 57% of the consumed oxygen is transported
from the surrounding air through the skin surface into the mu.

The arterial Pco_2 is 40 Torr (5.3 kPa) and decreases to 37.5
Torr (5 kPa) at the capillary dome. There is a steep decrease of
Pco_2 across the dead epidermis. An amount of 63% of the CO_2 diffusing out of the mu across the skin surface is not produced in the
mu, but is transported into the mu by the arterial blood.

Fig. 3 shows the result of case b). In this case, the skin
surface is impermeably covered for oxygen and carbon dioxide as
during presence of an electrode at the skin surface. Again a resting
value of blood flow and a temperature of 37°C are assumed. The Po_2
and Pco_2 fields change dramatically.

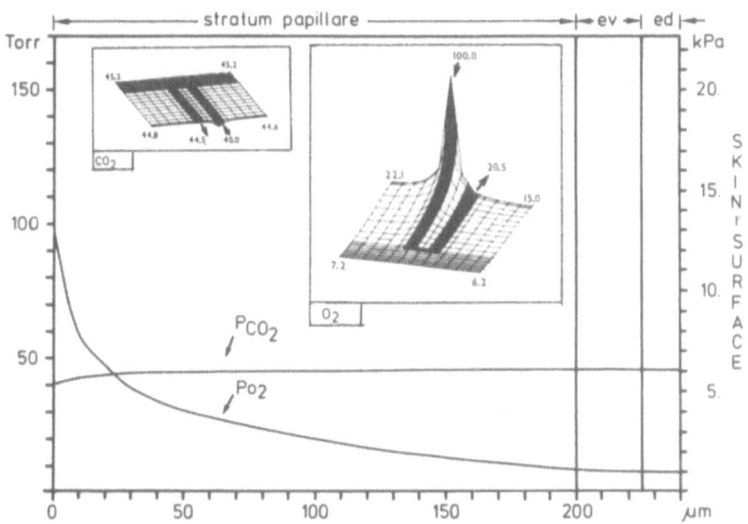

Fig. 3. P_{O_2} and P_{CO_2} profiles along the arterial limb and across
the epidermis (as inserts P_{O_2} and P_{CO_2} fields over a cross
section of the microcirculatory unit drawn as a surface);
blood flow = 0.01 ml/(g·min), temperature = 37°C, skin
surface impermeably covered.

The P_{O_2} decreases from 100 Torr (13.3 kPa) at the arterial
inflow to 7 Torr (0.9 kPa) at the skin surface. Moreover, a considerable amount of oxygen diffuses from the arterial limb into the
venous limb. 16% of the total O_2 extraction is due to shunt diffusion. The P_{CO_2} field increases from 40 Torr (5.3 kPa) at the arterial inflow to 45.2.Torr (6.4 kPa) at the skin surface. 11% of
the total amount of CO_2 entering the capillary is due to shunt
diffusion. This amount diffuses out of the venous limb into the
arterial limb.

Fig. 4 shows the result of a calculation simulating a situation of tcP_{CO_2} measurement. It is similar to the situation simulated
in the preceding case. The difference is that the skin is heated up
to 43°C and thus hyperemia is induced, by which the resting blood
flow is increased by a factor of 100.

Due to the temperature shift of the oxygen binding curve of
hemoglobin the P_{O_2} at the arterial inflow is elevated to 139 Torr
(18.5 kPa) and decreases to 111 Torr (11.8 kPa) at the skin surface. P_{CO_2} at the arterial inflow is elevated to 52.4 Torr (7 kPa)
and increases to 53.3 Torr (7.1 kPa) at the skin surface. We observe
a ΔP_{CO_2} of 0.9 Torr from the arterial inflow to the skin surface.

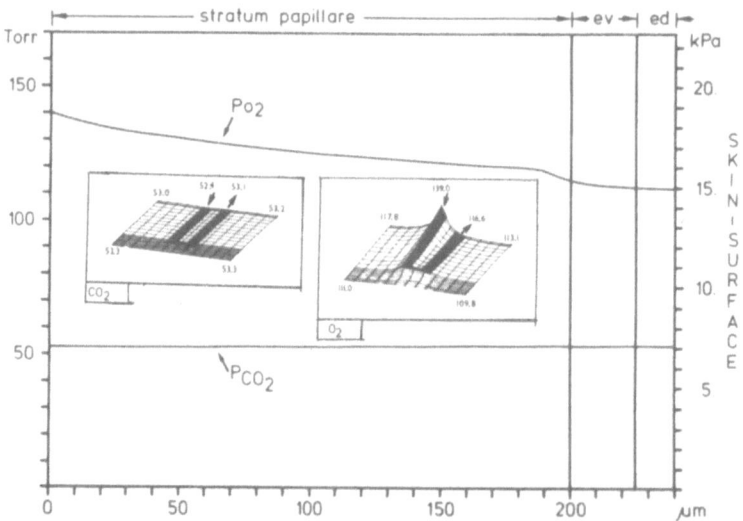

Fig. 4. Po_2 and Pco_2 profiles along the arterial limb and across
         the epidermis (as inserts Po_2 and Pco_2 fields over a cross
         section of the microcirculatory unit drawn as a surface);
         blood flow = 1.0 ml/(g·min), temperature = 43°C, skin
         surface impermeably covered (tcPco$_2$ measurement).

Fig. 5 shows the result of a calculation simulating a situation of measuring Po_2 on the skin surface with a heated electrode
but without a membrane. The Po_2 decreases from 139 Torr (18.5 kPa)
to 93 Torr (12.4 kPa) at the interface between the dead and the
viable epidermis. If we consider the dead epidermis as the membrane
of the electrode the figure shows that heating makes it possible to
obtain a Po_2 at the interface between membrane and skin, which
sufficiently approximates arterial Po_2 at 37°C. With a normally
membranized electrode the interface Po_2 would correspond to the
tcPo$_2$. The tcPco$_2$ field looks similar as in the preceding figure.
We have a Pco_2 difference of 0.8 Torr (0.11 kPa) between the
arterial inflow and the skin surface.

DISCUSSION AND CONCLUSION

The conception of a microcirculatory unit allows the use of a
single capillary loop with its surrounding tissue as a representative unit of a defined region of the skin.

Our model calculations simulate different oxygen and carbon
dioxide exchange situations in the upper skin, namely:
1. the situation at resting blood flow and
2. the situation with conditions similar to those found during Po_2
and Pco_2 measurements at the skin surface.

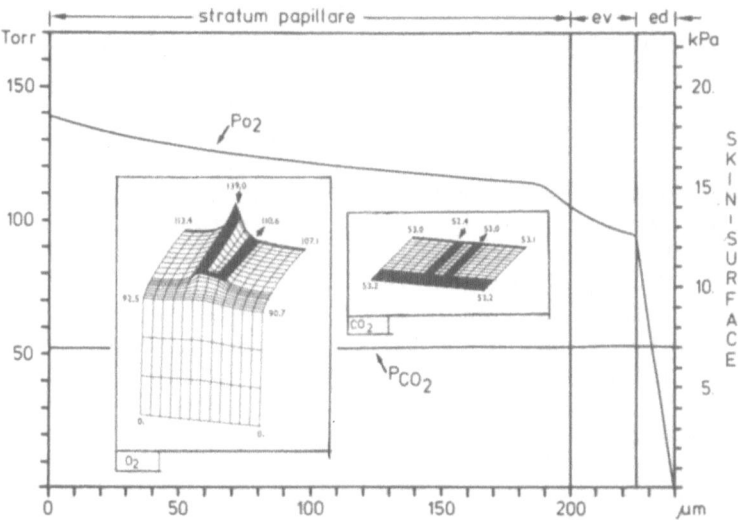

Fig. 5. Po_2 and Pco_2 profiles along the arterial limb and across
          the epidermis (as inserts Po_2 and Pco_2 fields over a cross
          section of the microcirculatory unit drawn as a surface);
          blood flow 1.0 ml/(g·min), temperature 43°C, skin surface
          impermeably covered for carbon dioxide (tcPo₂ measurement).

        In the case of oxygen our model calculations are in good
agreement with experimental findings.

        In the case of carbon dioxide there is a discrepancy between
the hyperemic Pco_2 differences between arterial inflow and skin
surface as calculated by our model (0.9 Torr) and the experimentally determined value given for example by Severinghaus (1981)
(4 Torr). The difference may be due to the fact that the transit
time of flowing blood in our model (approximately 1 sec) is in the
range of the reaction time of the carbonate process, i.e., we have
used an overall binding curve of Pco_2, which is too steep and, thus
obtained smaller Pco_2 differences.

        Further work should investigate the relation of transit and
reaction times under hyperemic conditions.

REFERENCES

Crank, J., 1955, "The Mathematics of Diffusion", at the Clarendon
     Press, Oxford.
Grossmann, U., 1980, Existence and uniqueness of solutions of
     quasilinear transmission problems of both elliptic and pseudo-
     parabolic type simulating oxygen transport in capillary and
     tissue, Math. Meth. Appl. Sci., 2:34-47.

Grossmann, U., 1982, Simulation of combined transfer of oxygen and heat through the skin using a capillary loop model, Math. Biosci., 61:205-236.

Grossmann, U., Huber, J., Fricke, K., and Lübbers, D.W., 1981, A new model for simulating the oxygen pressure field of skin, in: Adv. Physiol. Sci., Vol. 25, "Oxygen Transport to Tissue", A.G.B. Kovách, E. Dora, M. Kessler, I.A. Silver, eds., Pergamon Press, Akademiai Kiadó, Budapest, pp. 319-320.

Huch, R., Huch, A., and Lübbers, D.W., 1981, "Transcutaneous Po_2", Thieme, Stuttgart and New York.

Kawashiro, T., Nüsse, W., and Scheid, P., 1975, Determination of diffusivity of oxygen and carbon dioxide in respiring tissue: Results in rat skeletal muscle, Pflügers Arch., 35:231-251.

Lübbers, D.W., 1979, Cutaneous and transcutaneous Po_2 and Pco_2 and their measuring conditions, in: "Continuous Transcutaneous Blood Gas Monitoring", A. Huch, R. Huch, J.F. Lucey, eds., Original Article Series - Birth Defects, The National Foundation March of Dimes, Vol. 15.4, A.R. Liss, New York, pp. 13-21.

Roughton, F.J., 1964, Transport of oxygen and carbon dioxide, in: "Handbook of Physiology. Respiration", Section 3, Vol. I, W.O. Fenn, H. Rahn, eds., Am. Physiol. Soc., Washington D.C., pp. 767-825.

Severinghaus, J.W., 1966, Blood gas calculator, J. Appl. Physiol., 21:1108-1116.

Severinghaus, J.W., 1981, A combined transcutaneous Po_2-Pco_2 electrode with electrochemical HCO_3-stabilization, J. Appl. Physiol., 51 (N4):1027-1032.

Thews, G., 1968, The theory of oxygen transport and its application to gaseous exchange in the lung, in: "Oxygen Transport in Blood and Tissue", D.W. Lübbers, U.C. Luft, G. Thews, E. Witzleb, eds., Thieme, Stuttgart, pp. 1-20.

FACILITATED DIFFUSION AND ELECTRICAL POTENTIALS IN PROTEIN

SOLUTIONS WITH IONIC SPECIES

L. Hoofd, P. Breepoel, and F. Kreuzer

Department of Physiology, University of Nijmegen
Geert Grooteplein Noord 21a, 6525 EZ Nijmegen
The Netherlands

INTRODUCTION

The diffusion of oxygen or carbon monoxide is facilitated, i.e., the diffusion is greater than that expected from its concentration gradient in solutions of hemoglobin or myoglobin. Since the early discovery of this phenomenon (Wittenberg, 1959; Scholander, 1960) much theoretical work has been done that recognizes the facilitation as a cocurrent diffusion of the species bound to the respective protein (Kreuzer and Hoofd, 1982).

However, little or no attention was paid to the possible role of electrical potentials in the process. In binding oxygen, hemoglobin changes its charge and a gradient of H^+ (and/or OH^-) ensues across the diffusion layer. Early studies (Fox and Landahl, 1965; Bright, 1967) estimated this effect as negligible in buffered solutions, but recently a clear-cut effect was shown when measuring the electrical potential difference across a layer of hemoglobin solution (Breepoel et al., 1981). Diffusion of oxygen through salt-free hemoglobin solutions elicited a potential difference of up to 4 mV. Although this is not large as compared with potentials from CO_2 diffusion (Koning et al., 1978; Meldon et al., 1978) it may affect the facilitation of oxygen diffusion in these solutions. Previous measurements (de Koning et al., 1981a) gave an indication of such an effect.

A theoretical treatment can be given in terms of differential equations coupling chemical reaction with diffusion of charged species. Similar treatments were able to successfully explain facilitated CO_2 diffusion through protein solutions as well as O_2 diffusion in the absence of electrical potentials (Kreuzer and Hoofd, 1982).

THEORY

Mathematical Treatment

One-dimensional steady-state diffusion of charged species is described by the Nernst-Planck equation:

$$J_Y = -D_Y \left(\frac{d[Y]}{dx} + z_Y[Y] \frac{F}{RT} \frac{dU}{dx} \right) \tag{1}$$

where J_Y, D_Y, z_Y, $[Y]$ are flux, diffusion coefficient, (mean) charge and concentration of species Y, respectively, F Faraday constant, R universal gas constant, T absolute temperature and U electrical potential. The location in the layer is denoted by x.

Most of the species are involved in reactions, including H^+ and OH^- which are coupled by the water dissociation equilibrium $[H^+][OH^-] = K_W$ so that both can be expressed in terms of pH:

$$[H^+] = 10^{-pH} \qquad [OH^-] = K_W \cdot 10^{+pH} \tag{2}$$

If a salt is added, denoted by MA and dissociation into M^+ and A^-, the cation M^+ is assumed to be the only inert (non-reacting) ion in the solution and thus its flux $J_{M^+} = 0$. The ion is accordingly redistributed by the potential field:

$$[M^+] = M_0 \cdot e^{-FU/RT} \tag{3}$$

where M_0 depends on the total mean salt concentration.

The same is not true for the anion A^-; it is allowed to bind to the protein Pr so that $J_{A^-} \neq 0$ but the sum of free and bound A^- flux is zero. If different amounts of A^- are bound at either side of the layer, a 'facilitated A^- flux' will counteract and offset the 'free A^- flux' J_{A^-}:

$$-D_A - \left(\frac{d[A^-]}{dx} - [A^-]\frac{F}{RT}\frac{dU}{dx} \right) - D_{Pr} \left(\frac{d[PrA]}{dx} + z_{PrA}[PrA]\frac{F}{RT}\frac{dU}{dx} \right) = 0 \tag{4}$$

where $[PrA]$ denotes the amount of A^- bound to the protein, and z_{PrA} is the mean charge of this complex; $[PrA] = C_{PrA}S_A$ where C_{PrA} is binding capacity for A^- and S_A saturation of the protein with A^-.

The concentration of total protein C_{Pr} also is influenced by the potential field. Total Pr flux is zero so that:

$$\frac{dC_{Pr}}{dx} = -z_{Pr}C_{Pr} \frac{F}{RT} \frac{dU}{dx} \tag{5}$$

C_{Pr} is redistributed if the mean (total) protein charge z_{Pr} is nonzero.

The electrical potential in turn arises from the fact that there is no net charge flux (no current) in the layer:

$$J_{H^+} - J_{OH^-} + J_{M^+} - J_{A^-} + J_{zPr} = 0 \tag{6}$$

where J_{zPr} symbolically denotes the summed charge flux of all protein subspecies. The equation of zero current can be worked out in several ways, leading to an equation for the potential field dU/dx. One form is:

$$\frac{F}{RT}\frac{dU}{dx} = \frac{\ln(10)\left\{(D_{H^+} - D_{Pr})[H^+] + (D_{OH^-} - D_{Pr})[OH^-]\right\}\frac{dpH}{dx} + (D_A - D_{Pr})\frac{d[A^-]}{dx}}{D_{H^+}[H^+] + D_{OH^-}[OH^-] + D_{A^-}[A^-] + D_{Pr}([M^+] + \overline{z^2}C_{Pr})} \tag{7}$$

where $\overline{z^2}$ is the mean squared protein charge. This equation establishes the fact that the potential field arises from the difference in mobilities of the various ions in the solution, i.e., $D_y - D_{Pr}$. Together with the gradients of pH and free A^- concentration they constitute the potential field. As pointed out above an $[A^-]$ gradient is counter to a gradient in bound A^- so that also:

$$\frac{F}{RT}\frac{dU}{dx} = \frac{\ln(10)\left\{(D_{H^+} - D_{Pr})[H^+] + (D_{OH^-} - D_{Pr})[OH^-]\right\}\frac{dpH}{dx} - fD_{Pr}C_{PrA}\frac{dS_A}{dx}}{D_{H^+}[H^+] + D_{OH^-}[OH^-] + D_{Pr}\left\{[M^+] + [A^-] + \overline{z^2}C_{Pr} + fC_{PrA}S_A(z_{PrA} - z_{Pr})\right\}} \tag{8}$$

where $f = 1 - D_{Pr}/D_{A^-}$.

Although the volatile species G (oxygen or carbon monoxide) itself is uncharged, a gradient in saturation with G, S_G, will be accompanied by gradients in pH and S_A and thus by a gradient in potenial U. The potential field in turn affects the total flux of G, J, being the sum of free and bound (facilitated) fluxes:

$$J = -DG\frac{d[G]}{dx} - D_{Pr}C_{PrG}\left\{\frac{dS_G}{dx} + (z_{PrG} - z_{Pr})S_G\frac{F}{RT}\frac{dU}{dx}\right\} \tag{9}$$

where again C_{PrG} = binding capacity for G and z_{PrG} mean charge of the protein-gas complex.

Elaboration

For thick layers the set of equations found from mass conservation as derived above can be completed using the equilibrium
relation of the reactions between A^-, Pr, H^+ and G. This is possible
for thick layers (thicker than roughly some 100 um) (Wittenberg,
1966) in situations without a potential but also with a potential
(Hoofd et al., 1981).

If these equilibrium relations are used, some early conclusions
can be drawn from the above set of equations. In salt-free solutions
of hemoglobin, the oxy form is more 'acidic' than the deoxy form.
This implies a H^+ gradient cocurrent with the O_2 gradient and a pH
gradient counter to the O_2 gradient. According to eq. (7) there is
a potential gradient along the pH gradient. In eq. (9) the charge
of oxyHb, z_{PrG}, is less than the mean total Hb charge z_{Pr} so that
$(z_{PrG}-z_{Pr})$ is negative and the term $(z_{PrG}-z_{Pr})dU/dx$ again has the
same sign as dS_G/dx. This means that facilitated flux is enhanced
(Fig. 1a).

However, the enhancement must be small. $(F/RT)dU/dx$ at most is
equal to $\ln(10)dpH/dx$, leading to a maximum potential difference of
about 7 mV for a pH difference of 0.13 (as measured), but it is
expected to be very much smaller because of the small values of
$[H^+]$ and $[OH^-]$ in eq. (7). The term between braces in eq. (9),
accounting for facilitation, then becomes less than or equal to:

$$\frac{dS_G}{dx} + (z_{PrG}-z_{Pr})S_G\ln(10)\frac{dpH}{dx} \qquad (10)$$

If this expression is integrated from $S_G = 0\%$ to $S_G = 100\%$, its
value at most is 104% (Fig. 1a).

When a salt such as KCl is added the situation is completely
different. Deoxyhemoglobin binds Cl^- ions much more strongly than
oxyhemoglobin does (van Beek et al., 1979). The effects are best
seen from eq. (8). A gradient in S_A develops, again counter to the
gradient in O_2 (Fig. 1b; see also Fig. 3 below). This implies a
potential gradient now cocurrent with the O_2 gradient, i.e., the
potential changes sign on addition of KCl. Also when KCl is added
the oxyhemoglobin is more 'acidic' than deoxyhemoglobin (Rollema et
al., 1975) so that the reversed potential field will depress
facilitation.

Adding large amounts of KCl will have two effects:
- both oxy- and deoxyHb will become saturated with Cl^- and the
saturation gradient in S_A disappears;
- the terms $[M^+]$ and $[A^-]$ in the denominator of eq. (8) become
large.

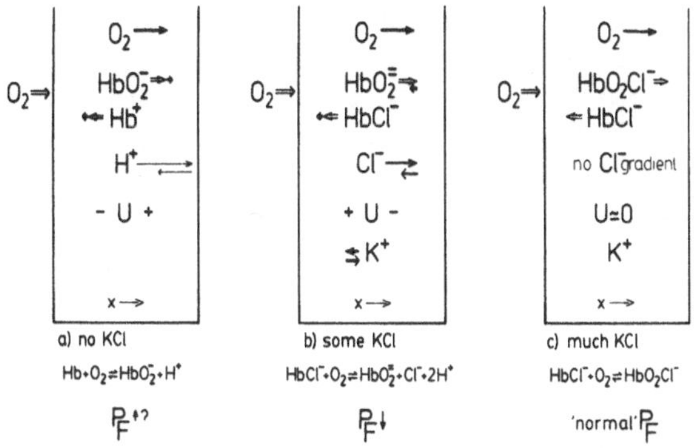

Fig. 1. Scheme of the influence of diffusion potentials on the
fluxes of species in the system hemoglobin + oxygen. Oxygen
is applied to the left side of the layer and diffuses to
the right side. Arrows indicate diffusion due to concentration (first or upper arrow) and potential U (second or
lower arrow), respectively. Below the schematic layer, the
main reaction and the influence on P_F are shown (see text).
Three situations are shown: a) No KCl present in the
solution; b) Intermediate amounts of KCl added; c) Large
amounts of KCl added.

Both effects cause the potential gradient to disappear and
remove any possible influence on facilitation (Fig. 1c).

RESULTS

It is quite difficult to do calculations in the system Hb +
O_2 + Cl^-. First, a method must be developed to model binding of the
different species (including H^+) to hemoglobin. Binding of H^+, O_2
and Cl^- to Hb are all interrelated. A computer subroutine (PROZ)
was set up to calculate charges and saturations in all possible
situations. The subroutine, which will not be detailed further here,
is based on the theory of linked functions and allows calculations
for any set of variables (pH, $[O_2]$, $[Cl^-]$) using fixed binding
parameters (pK's) and shifts in pK. The subroutine produces theoretically exact results but is time-consuming.

An equally large problem concerns the values of these binding
parameters. Our own measurements of titration curves of bovine
hemoglobin seemed to be very similar to the data of Rollema et al.
(1975) for human Hb. Therefore, as a first attempt, we decided to
use the binding data elaborated by the same group (van Beek et al.,
1979), and to superimpose them on a straight buffer line to match
charge data as measured by our own group for salt-free bovine Hb.
O_2 binding data (P_{50}, n_{Hill}) were chosen to match O_2 saturation
curves as measured for salt-free bovine Hb.

Another difficulty is the value of the diffusion parameters. These parameters depend on protein concentration (Kreuzer, 1970), and recently a (weak) dependence of oxygen permeability Po_2 on KCl concentration was found also (Breepoel et al., in prep.). With 'intermediate' KCl concentrations, gradients in all ionic species, including total protein concentration, are developed (Fig. 1b; Fig. 3 below), accompanied by gradients in the diffusion parameters. This makes a simple quantitative interpretation of the theory, as given for [KCl] = 0, impossible.

A major difficulty of the salt-free situation, however, is the diffusion coefficient of H^+ (and, to a lesser extent, OH^-). No measurements are available, and because of the structured ('ice-like') water in Hb solutions the value of this diffusion coefficient is uncertain (Kunst and Warman, 1980). We adopted 10^{-4} cm^2/sec for H^+ and 10^{-5} cm^2/sec for OH^- as guesses.

Measurements of facilitated O_2 flux were performed in a diffusion chamber described by de Koning et al. (1981b). Earlier data (de Koning et al., 1981a) were completed and all data were recalculated to a mean Hb concentration of 2.5 mmol·l^{-1} tetramer. The data were adapted to the 'overall' describing equation:

$$J = -Po_2 \left(\frac{\Delta P}{L} + P_F \frac{\Delta S}{L} \right) \tag{11}$$

where P and S are O_2 partial pressure and O_2 saturation difference across the layer. P_F is called 'facilitation pressure' (de Koning et al., 1981a, b) and quantifies facilitation including electrical effects. In diffusion without potential comparison with eq. (9) shows:

$$P_F = D_{pr} C_{prO_2}/Po_2 \qquad \text{if } U \equiv 0 \tag{12}$$

where C_{prO_2} is binding capacity for oxygen. The value of $P_F = 29.1$ kPa as found for large [KCl] where a potential is measured to be absent (see Fig. 2) implies a D_{Hb} from $4.7 \cdot 10^{-7}$ to $3.5 \cdot 10^{-7}$ cm^2/sec for the range of [KCl] from 0.1 to 0.5 mol·l^{-1} at 25°C, in agreement with former data (Spaan et al., 1980; Kreuzer and Hoofd, 1982).

Results of measurements and calculations are shown in Fig. 2. Values of U were taken from Breepoel et al. (1981). Although the qualitative behaviour of P_F is as predicted, the calculations show a much smaller effect than the measurements. Also the agreement between measured and calculated potential difference is poor though both are of the same order in at least the intermediate salt concentration range.

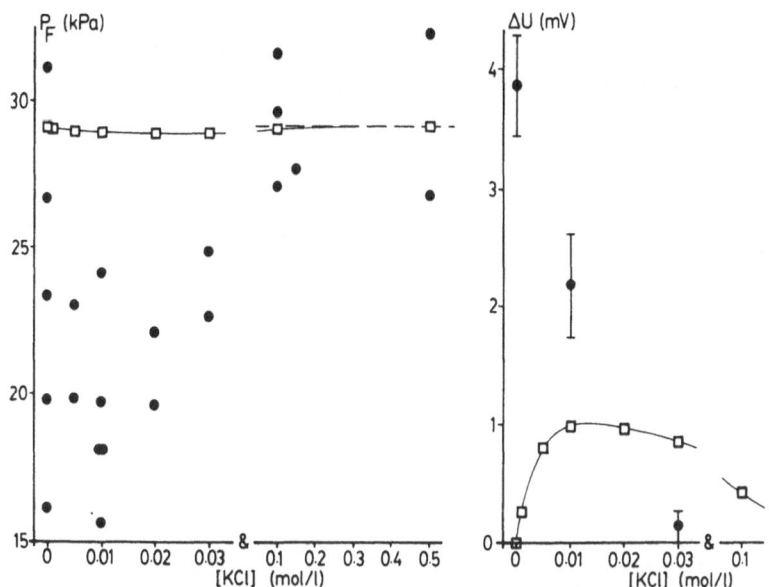

Fig. 2.   Plot of facilitation pressure (P_F, left panel) and poten-
          tial difference across the layer (ΔU, right panel) due to
          oxygen diffusion through a layer containing 2.5 mmol·l^{-1}
          of bovine Hb with various amounts of KCl added at 25°C.
          Black dots are measurements; error bars for ΔU (absolute
          value was measured) denote SD. Open squares are calculated
          values. Above [KCl] = 0.1 mol·l^{-1} no potential difference
          could be detected; the mean P_F (broken line in left panel)
          of the corresponding range is 29.1 kPa.

    To continue the calculation, a graph of the profiles in the
layer is shown in Fig. 3 for a salt concentration of 20 mmol·l^{-1}.
Some interesting features can be seen from this figure. First, the
potential U more or less follows the oxygen saturation profile; in
particular there is hardly any change in U as soon as the point of
saturation is reached. This is explained by eq. (8): once the hemoglobin is saturated with oxygen, no more Cl$^-$ is released and S_A no
longer changes so that $dS_A/dx = 0$ and dU/dx becomes small. Consequently, all profiles flatten off in this region (except the oxygen
pressure, of course).

    Clearly K$^+$, though being an inert ion, is seen to be redistri-
buted across the layer according to the potential (eq. (3)). A
similar redistribution seems to occur with Cl$^-$ but the (free) Cl$^-$
concentration is less because some of the ion is bound to hemoglobin. The resulting difference in charge allows the hemoglobin to
become negatively charged, but the amount of charge is not large
because of the small amount of Cl$^-$ that can maximally be bound (2
ions per tetramer). Consequently, the redistribution of protein
along the potential (eq. (5)) is only minor.

O_2 saturation decreases with decreasing oxygen partial pressure and in turn Cl^- saturation increases according to the shift in binding strength upon oxygenation.

DISCUSSION

The main question is: why is there so little quantitative agreement between measurements and calculations?

First of all the measured value of ΔU in salt-free solutions is (in absolute terms) 4 mV whereas -0.01 mV is calculated. Among the variables used to calculate ΔU, the value of D_{H^+} is probably the most questionable. To reach a value of ΔU as high as the measured value of 4 mV, a value of D_{H^+} of about 0.1 cm^2/sec would be required, which seems impossibly high.

Another possibility is that the solutions were not completely salt-free. However, with the parameter set used, this still could not explain such a high potential difference, or else Cl^- binding to Hb would have to be much stronger.

There is a possibility of other reactions (tetramer dissociation?) occurring in the layer, but this could not be investigated here.

For nonzero salt concentrations the disagreement is less and it seems that a reasonable adaptation of the parameter values (e.g., stronger Cl^- binding) would lead to results compatibel with the measurements.

Another, and larger, problem is P_F. Measurements may be compatible with the finding that P_F is hardly affected at zero salt concentration, but the decrease in the range of 5 to 30 mmol·l^{-1} KCl is at least 10 times larger than calculated.

It seems questionable whether other values of the parameters might explain the discrepancy. The problem of Cl^- (and H^+) binding to hemoglobin in particular could not be resolved as yet due to lack of conclusive results from preliminary calculations.

The same argument holds for the diffusion parameters; they seem unable to cause large differences in calculated P_F, except D_{Hb}. A change in D_{Hb} would reflect directly in P_F (see eq. (12)). However, it is unclear why D_{Hb} should show a 'dip-like' behaviour as P_F in Fig. 2 does.

There remains the question of the model. In CO_2 diffusion through protein and buffer solution, a theoretical model of the same type led to satisfactory agreement between calculations and

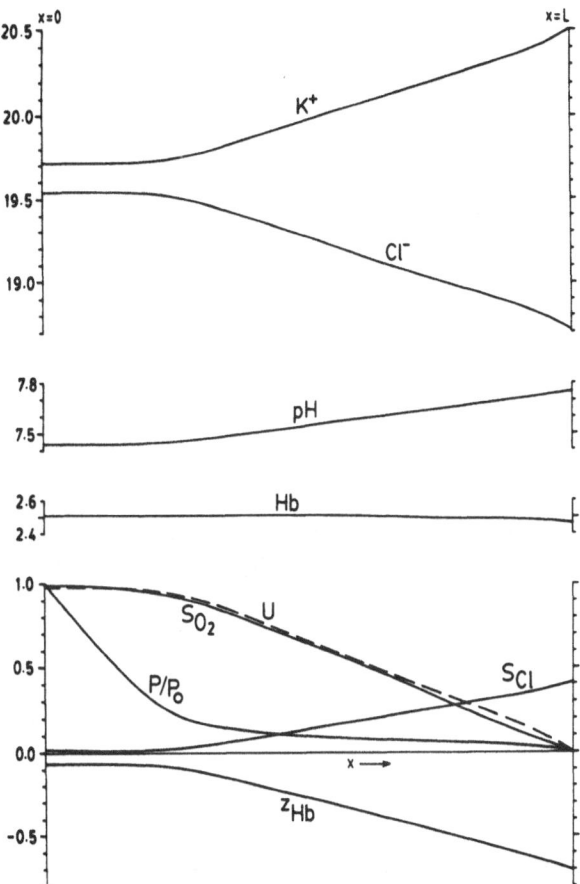

Fig. 3. Calculated profiles in a layer of Hb solution containing
20 mmol$\cdot$l^{-1} KCl exposed to an oxygen partial pressure P_o =
10 kPa against 0. U (broken line) is potential in mV, z_{Hb}
is mean hemoglobin charge, S_{O2} and S_{Cl} are saturation with
O_2 and Cl$^-$ respectively; P is oxygen partial pressure.
Concentrations are in mmol$\cdot$l^{-1}.

measurements for both potential and flux (Meldon, 1975; de Koning
et al., 1978; Meldon et al., 1978; Stroeve and Ziegler, 1980). The
reaction model of hemoglobin is much more uncertain. Without going
into detail, we found some profound disparities between binding
parameters and binding measurements, apart from the question whether
hitherto unknown phenomena might play a role. However, because of
the very time consuming computer program, a systematical survey of
parameter values (including diffusion parameters) was impossible.
This remains a purpose for further investigation. Also the effect
of nonequilibrium may be larger than estimated; calculations again
are difficult because of essentially unknown reaction parameters.
However, no indication of nonequilibrium was found in comparing
measurements in layers of different thicknesses (between 476 and
667 um).

One final remark must be added. The measured values of P_F were
evaluated by fitting eq. (11) to flux measurements in Hb layers
confined between Teflon membranes and exposed to a series of different oxygen pressures, where a correction had to be made for the
resistance of the Teflon membranes. Theoretical values were calculated only for a situation of 10 kPa against 0 oxygen driving force,
as shown in Fig. 3. In our opinion, however, there should not be
any discrepancy.

REFERENCES

Beek, G.G.M. van, Zuiderweg, E.R.P., and de Bruin, S.H., 1979, The
     binding of chloride ions to ligated and unligated human hemo-
     globin and its influence on the Bohr effect, Eur. J. Biochem.,
     99:379.
Breepoel, P.M., de Koning, J., and Hoofd, L.J.C., 1981, Facilita-
     tion of oxygen diffusion in hemoglobin solutions: measurement
     of electrical effects, in: "Oxygen Transport to Tissue", Adv.
     Physiol. Sci. Vol. 25, A.G.B. Kovách, E. Dóra, M. Kessler,
     I.A. Silver, eds., Pergamon Press, Akadémiai Kiadó, Budapest,
     p. 315.
Breepoel, P.M., de Koning, J., and Hoofd, L.J.C., in prep., Diffu-
     sion of oxygen in methemoglobin solutions. Dependence on salt
     concentration.
Bright, P.B., 1967, The basic flow equations of electrophysiology
     in the presence of chemical reactions; II. A practical appli-
     cation concerning the pH voltage effects accompanying the
     diffusion of O_2 through hemoglobin solution, Bull. Math.
     Biophys., 29:123.
Fox, M.A., and Landahl, H.D., 1965, Theory of hemoglobin facili-
     tated oxygen transport, Bull. Math. Biophys., 27:183.
Hoofd, L., Breepoel, P.M., and de Koning, J., 1981, Facilitation
     of oxygen diffusion in hemoglobin solutions: new theoretical
     aspects, in: "Oxygen Transport to Tissue", Adv. Physiol. Sci.
     Vol. 25, A.G.B. Kovách, E. Dóra, M. Kessler, I.A. Silver,
     eds., Pergamon Press, Akadémiai Kiadó, Budapest, p. 321.
Koning, J. de, Hoofd, L.J.C., and Breepoel, P.M., 1981a, Facili-
     tation of oxygen diffusion in hemoglobin solutions: influence
     of various salt concentrations, in: "Oxygen Transport to
     Tissue", Adv. Physiol. Sci. Vol. 25, A.G.B. Kovách, E. Dóra,
     M. Kessler, I.A. Silver, eds., Pergamon Press, Akadémiai
     Kiadó, Budapest, p. 323.
Koning, J. de, Hoofd, L.J.C., and Kreuzer, F., 1981b, Oxygen trans-
     port and the function of myoglobin, Pflügers Arch., 389:211.
Koning, J. de, Stroeve, P., and Meldon, J.H., 1978, Electrical
     potentials during carbon dioxide transport in hemoglobin
     solutions, in: "Oxygen Transport to Tissue - III", Adv. Exp.
     Med. Biol. Vol. 94, I.A. Silver, M. Erecínska, H.I. Bicher,
     eds., Plenum Press, New York - London, p. 181.

Kreuzer, F., 1970, Facilitated diffusion of oxygen and its possible
    significance; a review, Respir. Physiol., 9:1
Kreuzer, F., and Hoofd, L.J.C., 1982, Facilitated diffusion of O_2
    and CO_2, in: "Handbook of Physiology: Respiration", American
    Physiological Society, Washington, D.C., in press.
Kunst, M., and Warman, J.M., 1980, Proton mobility in ice, Nature,
    288:465.
Meldon, J.H., 1975, Theory of the effect of diffusion potentials
    on the transport of carbon dioxide in protein solutions,
    5th Intern. Biophysics Congr., Copenhagen, Abstract 413.
Meldon, J.H., de Koning, J., and Stroeve, P., 1978, Electrical
    potentials induced by CO_2 gradients in protein solutions and
    their role in CO_2 transport, Bioelectrochem. Bioenerg.,
    5:77.
Rollema, H.S., de Bruin, S.H., Janssen, L.H.M., and van Os, G.A.J.,
    1975, The effect of potassium chloride on the Bohr effect of
    human hemoglobin, J. Biol. Chem., 250:1333.
Scholander, P.F., 1960, Oxygen transport through hemoglobin solu-
    tions, Science, 131:585.
Spaan, J.A.E., Kreuzer,,F., and van Welny, F.K., 1980, Diffusion
    coefficients of oxygen and hemoglobin as obtained simultaneously
    from photometric determination of the oxygenation of
    layers of hemoglobin solutions, Pflügers Arch., 384:241.
Stroeve, P., and Ziegler, E., 1980, The transport of carbon dioxide
    in high molecular weight buffer solutions, Chem. Eng. Commun.,
    6:81.
Wittenberg, J.B., 1959, Oxygen transport - a new function proposed
    for myoglobin, Biol. Bull., 117:402.
Wittenberg, J.B., 1966, The molecular mechanism of hemoglobin-
    facilitated oxygen diffusion, J. Biol. Chem., 241:104.

DIFFUSIONAL COUPLING IN A HEMOGLOBIN-FREE PERFUSED CAPILLARY-TISSUE

STRUCTURE

J.E. Fletcher[1], and R.W. Schubert[2]

[1]Laboratory of Applied Studies, Division of Computer
Research and Technology, N.I.H., Bethesda, Md., 20205
USA
[2]Department of Biomedical Engineering, Louisiana Tech.
University, Ruston, La., 71272, USA

INTRODUCTION

The theoretical prediction of substrate levels in tissue from
a mathematical model has been an intensely investigated topic since
Krogh initiated the concept in the period 1918 to 1929 (Krogh, 1919).
Oxygen has been the most widely investigated substrate because of
its obvious necessity for cell viability. However, attempts to study
this viability, both experimentally and by means of a mathematical
model, have been frustrated by the presence of the blood hemoglobins.
The hemoglobin has the effect of a nonlinear buffer which complicates
the description by mathematical model and introduces experimental
difficulties in measuring actual oxygen levels in perfused tissues.
Schubert (1976) had attempted to circumvent these difficulties by
using an oxygen saturated modified Krebs-Henseleit perfusate.

The experimental designs for the study of isolated perfused cat
heart and rabbit brain being used by Schubert (Schubert and Whalen,
1976; Schubert et al., 1978) (a modified Langendorff preparation)
include the use of a mathematical model to aid in the interpretation
of data from these experiments. The need to understand the histograms
of measurements from these isolated-perfused organ studies has required that mathematical models and the boundary conditions that
describe such experiments be reexamined in the context of these
measurements. These mathematical models envision an ideal tissue-
capillary arrangement as parallel cylinders with cocurrent concentric capillaries. With this arrangement, a single cylinder, sometimes called a Krogh cylinder, is representative of average tissue

145

conditions and this cylinder can be modeled mathematically. The experiment and idealized capillary-tissue geometry are illustrated in
Fig. 1.

Schubert has discovered that Blum's original paper (Blum, 1960)
on Krogh models of this type did not obtain the correct mathematical
solution for the axial diffusion effects. A subsequent review of the
existing literature on parametric studies and mathematical analyses
for the models of Blum and others (Fletcher, 1978; Fletcher and
Schubert, 1982a) prompted us to question the presumed understanding
of the role of axial diffusion in a parallel flow capillary-tissue
structure. The absence of the non-linear effects of blood hemoglobins
and the presumed absence of Michaelis-Menten tissue metabolic kinetics suggests that analytic (closed form) solutions for these mathematical models should be available. The model that is considered
here is an extension of the Blum type model. It includes the often
neglected capillary axial diffusion, removes the singular behavior
at the venous end of the capillary, and allows a parametric study
of the parameters of the model. The model will use venous boundary
conditions representative of the perfused organ experiment, and will
permit a comparison of analytically determined histograms with histograms of experimentally obtained results.

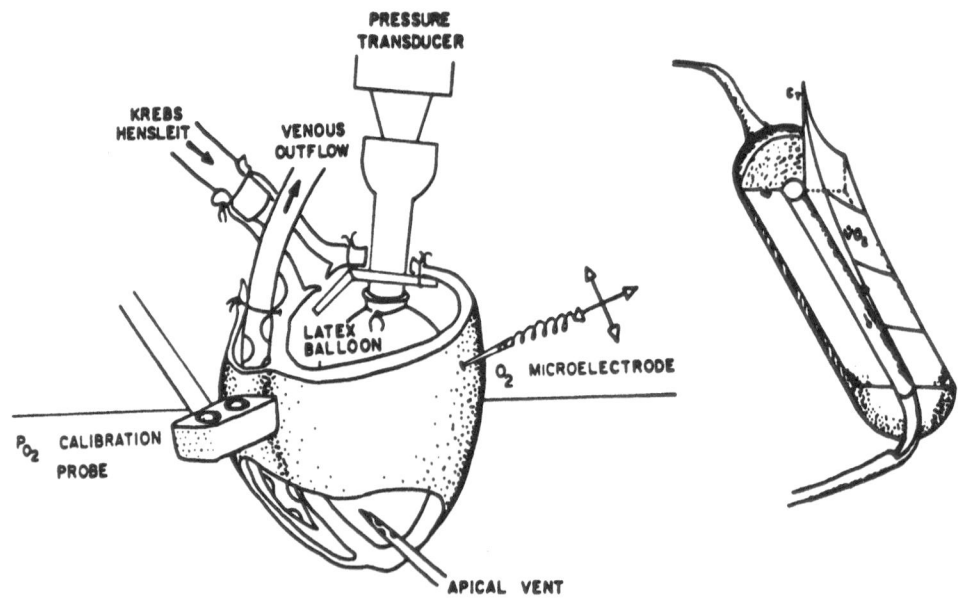

Fig. 1.   The perfused heart experiment and idealized Krogh cylinder
          geometry.

strate axial diffusion coefficient in the perfusate. Flow transport
is the dominant transport process in the capillary, therefore equation (1.3) is divided by

$$\frac{s_b v_o}{z_c}$$

in order to study the relative effects of the respective terms. The
local substrate partial pressure $P_b(y)$ is referenced to the nor-
malizing value P_s by defining $W(y) = P_b(y)/P_s$. The normalized form
of equation (1.3) becomes

$$- D_1 \frac{\partial^2 W}{\partial y^2} (y) + \frac{\partial W}{\partial y} (y) = \delta \frac{\partial U}{\partial x} (x_c, y). \qquad (1.4)$$

The characteristic capillary "diffusion" and radial "flux" parameters
are

$$D_1 = \frac{D_c}{z_c v_o} \quad \text{and} \quad \delta = \frac{2 D_r z_c}{r_c r_t v_o} \frac{s_t}{s_b}.$$

A customary procedure in studying model equations has been to
normalize the dependent variables to the entering concentration (or
pressure), and to reference all computations to unit value at the
capillary entrance. For organ perfusion studies, a typical entrance
condition is not well known. An autoregulating organ appears to adjust its flow to approximately uniform venous conditions. Furthermore, meaningful comparisons of parametric studies and experiments
can be made only if the parametric results have common conditions
of reference. Both these features are incorporated into the model
by assigning the following boundary conditions:

(1) $P_b(1) = P^o$ a fixed constant,

$$\qquad (1.5)$$

(2) $\dfrac{\partial P_b}{\partial y} (1) = 0$

All computed solutions will have the same venous Po_2, i.e., P^o and
there will be no substrate flux out of the capillary at the venous
end, by diffusion, i.e.

$$D_c \frac{\partial P_b}{\partial y} (1) = 0.$$

The second of conditions (1.5) has the additional advantage that it
dampens the troublesome singular behavior, present in the Blum model,
at the venous end of the capillary (Fletcher and Schubert, 1982a).

AN IDEALIZED MODEL OF CELL-FREE TISSUE PERFUSION

The Krogh geometry and notation is illustrated in Fig. 1. The local tissue substrate concentration (partial pressure) P_t is represented by the dimensional equation in cylindrical coordinates

$$\frac{1}{r}\frac{\partial}{\partial r}\left(r D_r S_t \frac{\partial P_t}{\partial r}\right) + \frac{\partial}{\partial z}\left(D_z S_t \frac{\partial P_t}{\partial z}\right) = \kappa, \qquad (1.0)$$

where S_t is the tissue substrate solubility, κ is the (constant) substrate metabolic rate in tissue, and D_r and D_z are the respective (constant) substrate tissue diffusion coefficients in the radial and axial direction. This equation is transformed to nondimensional coordinates by the natural transformations $x = r/r_t$, $y = z/z_c$, and $x_c = r_c/r_t$. The characteristic tissue "diffusion" and "metabolic" parameters are defined as

$$D_2 = \frac{D_z\, r_t^2}{D_r\, z_c^2}\quad,\quad K = \frac{\kappa\, r_t^2}{D_r S_t P_s}\quad;$$

and with $U(x,y) = P_t(x,y)/P_s$, where P_s is a reference pressure to be defined later, the nondimensional, normalized equation for the tissue substrate distribution is

$$\frac{1}{x}\frac{\partial}{\partial x}\left(x\frac{\partial U}{\partial x}\right) + D_2 \frac{\partial^2 U}{\partial y^2} = K. \qquad (1.1)$$

The no flux boundary conditions at the cylinder ends and outer tissue surface in nondimensional normalized form are

$$\frac{\partial U}{\partial y}(x,0) = \frac{\partial U}{\partial y}(x,1) = 0 \quad\text{on}\quad x_c < x < 1,$$
and $\qquad\qquad\qquad\qquad\qquad\qquad\qquad\qquad\qquad\qquad (1.2)$
$$\frac{\partial U}{\partial x}(1,y) = 0, \quad\text{on}\quad 0 \le y \le 1.$$

Because the perfusate is cell-free, the transport of capillary substrate (or partial pressure) P_b, involves only flow transport and axial diffusion, with radial flux at the capillary-tissue interface. The equation representing the average substrate pressure in a transverse capillary section is, in nondimensional coordinates,

$$-\frac{S_b D_c}{z_c^2}\frac{\partial^2 P_b}{\partial y^2}(y) + \frac{S_b v_0}{z_c}\frac{\partial P_b}{\partial y}(y) = \frac{2D_r}{r_t r_c}S_t \frac{\partial P_t}{\partial x}(x_c,y), \qquad (1.3)$$

where S_b is the solubility of substrate in the perfusing fluid, v_0 is the (average) steady capillary flow velocity, and D_c is the sub-

The capillary-tissue interface condition is either continuity of normalized pressure

$$W(y) = U(x_c, y) \quad ;$$

or if there is a restistance to substrate passage through the capillary wall,

$$\frac{\partial U}{\partial x}(x_c, y) = -\mu \left[W(y) - U(x_c, y) \right] ,$$

where

$$\mu = \frac{v r_t}{S_t D_r}$$

is the characteristic "permeability coefficient", and v is the physical permeability coefficient. Normalized conditions for relating $W(y)$ and $U(x,y)$ are given by

(i) $W(1) = P_o/P_s,$

(ii) $\frac{\partial W}{\partial y}(1) = 0$,   and either   (1.6)

(iii) $W(y) = U(x_c, y)$ or

$$\frac{\partial U}{\partial x}(x_c, y) = -\mu \left[W(y) - U(x_c, y) \right] .$$

EXACT SOLUTIONS OF THE MODEL EQUATIONS

It is instructive to consider solutions of equations (1.1), (1.2), (1.4) and (1.6) in a linear component form. In this way the significance and the coupling of each component in the complete model can be elucidated.

Consider first the simplest case, $D_1 = 0$, $D_2 = 0$, i.e., no axial diffusion. The solution components are the Krogh-Erlang terms

$$U(x,y) = W_o(y) + U_o(x) - \frac{\Omega}{\mu}$$

$$W_o(y) = \theta + \frac{P_o}{P_s} - \theta y,$$ (2.1)

where

$$U_0(x) = -\frac{K}{2}\left[\ln\left(\frac{x}{x_c}\right) - \frac{1}{2}(x^2 - x_c^2)\right], \quad \Omega = \frac{K}{2}\left(\frac{1}{x_c} - x_c\right),$$

and $\theta = \delta\Omega$.

It will be convenient in the following analyses to define the reference value, P_s, from the Krogh-Erlang model. P_s is chosen such that

$$W_0(0) = 1.0 = \theta + P_0/P_s$$

or

$$P_s = P_0 + \frac{\delta}{2}\frac{\kappa r_t^2}{D_r S_t}\frac{1}{\left(\frac{1}{x_c} - x_c\right)}.$$

This choice of P_s gives equations (2.1) the more familiar forms

$$W_0(y) = 1 - \theta y$$

$$U(x,y) = W_0(y) + U_0(x) - \frac{\Omega}{\mu}. \tag{2.2}$$

With this choice of P_s, the more complex solutions can be referenced to the normalized Krogh-Erlang results.

We shall present complete solutions of equations (1.1), (1.2), (1.4) and (1.6) in the following form:

$$W(y) = W_0(y) + W_2(y), \quad \text{and}$$

$$U(x,y) = W(y) + U_0(x) + U_2(x,y) - U_3(x,y). \tag{2.3}$$

The components of the solutions can be identified as representing the radial, axial, and wall permeability effects. This is the form of the solutions that Blum and others have sought, but failed to correctly or completely develop. The details of our development cannot be given in the limited space, but are described elsewhere (Fletcher and Schubert, 1982b). The component identified with capillary axial diffusion is

$$W_2(y) = -\theta D_1\left[1 - \exp\left(-\frac{1-y}{D_1}\right)\right]$$

$$\tag{2.4}$$

$$+ \frac{\pi}{2}\exp\left(-\frac{1-y}{D_1}\right)\sum_{n=1}^{\infty} c_n a_n (-1)^n \frac{D_1 n\pi}{1 + (D_1 n\pi)^2}$$

$$- \frac{\pi}{2}\sum_{n=1}^{\infty} c_n a_n \left[\frac{\sin(n\pi y) + D_1 n\pi \cos(n\pi y)}{1 + (D_1 n\pi)^2}\right].$$

The determining equations for A_o, and for the coefficients $\{a_n\}$ are

$$A_o = 1 - \frac{\theta}{2} - \theta D_1 \left[1 - D1(1 - \exp(-1/D_1)) \right]$$

$$+ \frac{\pi}{2} D_1 \left[1 - \exp(-1/D_1) \right] \sum_{n=1}^{\infty} c_n a_n \, (-1)^n \frac{D_1 n \pi}{1 + (D_1 n \pi)^2}$$

$$- \sum_{k=1}^{\infty} c_{2k-1} a_{2k-1} \frac{1}{(2k-1)} \frac{1}{1 + \left[D_1 (2k-1) \pi \right]^2} \quad ;$$

for $m = 2k$, (m even) $\hspace{6cm}$ (2.5)

$$a_{2k} \left[1 + \frac{\pi}{4} c_{2k} \frac{D_1 2k \pi}{1 + (D_1 2k \pi)^2} \right]$$

$$- \frac{\pi}{2} D_1 \frac{1 - \exp(-1/D_1)}{1 + (D_1 2k \pi)^2} \sum_{n=1}^{\infty} c_n a_n \, (-1)^n \frac{D_1 n \pi}{1 + (D_1 n \pi)^2}$$

$$+ \sum_{l=1}^{\infty} c_{2l-1} a_{2l-1} \frac{1}{1 + \left[D_1 (2l-1) \pi \right]^2} \frac{(2l-1)}{(2l-1)^2 - (2k)^2}$$

$$= \frac{\theta D_1^2 \left[1 - \exp(-1/D_1) \right]}{1 + (D_1 2k \pi)^2} \quad ;$$

and for $m = 2k-1$, (m odd)

$$a_{2k-1} \left[1 + \frac{\pi}{4} c_{2k-1} \frac{D_1 (2k-1) \pi}{1 + \left[D_1 (2k-1) \pi \right]^2} \right]$$

$$+ \frac{\pi}{2} D_1 \frac{1 + \exp(-1/D_1)}{1 + \left[D_1 (2k-1) \pi \right]^2} \sum_{n=1}^{\infty} c_n a_n \, (-1)^n \frac{D_1 n \pi}{1 + (D_1 n \pi)^2}$$

$$+ \sum_{l=1}^{\infty} c_{2l} a_{2l} \frac{1}{1 + (D_1 2l \pi)^2} \frac{2l}{(2l)^2 - (2k-1)^2}$$

$$= \frac{2\theta}{(2k-1)^2 \pi^2} - \theta D_1^2 \frac{\left[1 + \exp(-1/D_1) \right]}{1 + \left[D_1 (2k-1) \pi \right]^2} \quad .$$

The solution of the system (eqs. 2.5) is approximated by truncating the infinite sums at a finite number of terms and inverting the finite system. The results are demonstrated to asymptotically converge to the solution for the infinite system. The axial diffusion component in tissue is

$$U_2(x,y) = 2 \sum_{n=1}^{\infty} a_n \left[\frac{R_n^o(x)}{R_n^o(x_c)} - 1 \right] \cos(n\pi y) \quad ,$$

and the effect of a finitely permeable wall is   (2.6)

$$U_3(x,y) = \frac{\Omega}{\mu} + 2 \sum_{n=1}^{\infty} a_n (1-b_n) \frac{R_n^o(x)}{R_n^o(x_c)} \cos(n\pi y)$$

where the coefficients are $b_m = \left[1 - \frac{n\pi\sqrt{D_2}}{\mu} \frac{R_m^1(x_c)}{R_m^o(x_c)} \right] - 1$.

The cylindrical functions involved in the solutions are

$$R_n^o(x) = I_0(\rho_n x) K_1(\rho_n) + K_0(\rho_n x) I_1(\rho_n)$$

$$R_n^1(x) = I_1(\rho_n x) K_1(\rho_n) - K_1(\rho_n x) I_1(\rho_n) \quad .$$

where for $n > 0$, $\rho_n = n\pi\sqrt{D_2}$, and I_0, I_1, K_0, and K_1 are modified Bessel functions of the first and second kinds respectively, (see Olver, 1964).

RESULTS AND DISCUSSION

    Attempts at modeling microcirculatory structure and substrate supply are numerous in the biological literature. Many model results, though from similar geometric structures, are difficult to compare because of different initial conditions, different metabolic rates, and may or may not consider the effect of a finitely permeable wall. We have attempted here to unify the studies of radial and axial diffusion in the tissue space, of axial diffusion in the capillary, and of a finitely permeable wall, within the concepts of a single model with consistent boundary conditions. It is our hope that a more coherent picture of these various effects and their interactions will emerge in this setting. We illustrate our results with the parameters of the perfused cat heart experiment (Schubert and Whalen, 1976; Schubert et al., 1978).

$$D_r = 9.9 \times 10^{-4} \text{ cm}^2/\text{sec} \qquad\qquad r_t = 10.49 \text{ um}$$

$$D_z = 9.9 \times 10^{-4} \text{ cm}^2/\text{sec} \qquad\qquad r_c = 2.50 \text{ um}$$

$$D_c = 2.5 \times 10^{-5} \text{ cm}^2/\text{sec} \qquad\qquad \text{Temp.} = 32.5 \text{ deg. C.}$$

$$v_o = 538.00 \qquad \text{um/sec} \qquad\qquad z_c = 500 \text{ um}$$

$$S_t = 1.30 \times 10^{-9} \text{ moles/(g·mm Hg)} \quad S_b = 1.30 \times 10^{-9} \text{ moles/(g·mm Hg)}$$

$$\kappa = 3.2 \times 10^{-6} \text{ moles/(g·min)} \quad V/S_t = 10, \ldots, 1000 .$$

The technical aspects of making computations with the series solutions derived in section two cannot be discussed in the limited space of this paper. We refer instead to detailed studies to be published elsewhere (Fletcher and Schubert, 1982a; Fletcher and Schubert, 1982b).

The coupling effects of axial diffusion in both the capillary and in tissue in a parallel capillary-tissue structure are exhibited in the left half of Fig. 2. The solid straight line represents the solution with no axial diffusion. The dashed curve just under it is the effect of capillary diffusion with no axial diffusion in the tissue. The axial diffusion effect in the capillary does not become significant until it is coupled with axial diffusion in the tissue. The curve labeled 1 is the case tissue axial diffusion coefficient equal to the radial diffusion coefficient. The remaining successive curved profiles show the effect of increasing the tissue axial diffusion coefficient by 3, 5, and 10 times the radial diffusion value.

The overall effect of axial diffusion on the capillary profile and the tissue distribution is to lower the capillary entrance value and the entire substrate distribution in tissue below the corresponding values when there is no axial diffusion. This could account for the observation by many experimentalists that Po_2 values, measured in in vivo tissues, tend to be lower than those predicted by mathematical models, particularly in the capillary entrance regions. Findings here that do not appear to have been clearly understood are that axial diffusion lowers, not raises, substrate (Po_2) levels in most of the tissue region and that the role of capillary axial diffusion becomes important in that it permits the tissue region to influence the substrate levels in the capillary through this diffusional coupling.

A second important observation is that, for all practical purposes, the radial direction is "well-stirred" with only a very small negative gradient from capillary interface to outer tissue boundary. In the last ten per cent of the tissue region the model suggests a plateau effect with the venous side being maintained at substrate

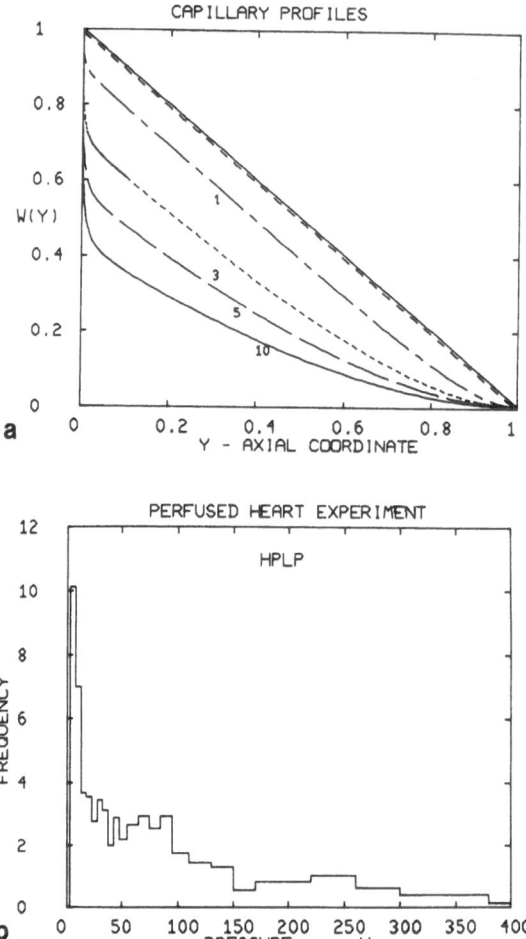

Fig. 2.   The effect of increasing axial diffusion in tissue on the
          capillary substrate profile (a), and a histogram of experi-
          mentally measured Po$_2$ values from the perfused heart ex-
          periment (b).

levels near the adjacent capillary concentration. Our computed results
suggest that the lowest Po$_2$ in the tissue differs from that of end
capillary blood by less than 1 mm Hg.

    It is desirable to represent the model predictions of substrate
levels in the Krogh cylinder in a form that can be compared to experimental measurements. In this experiment, a histogram of Po$_2$
values, shown in the right half of Fig. 2, is the experimental data.
A histogram for the model is constructed by dividing the cylinder

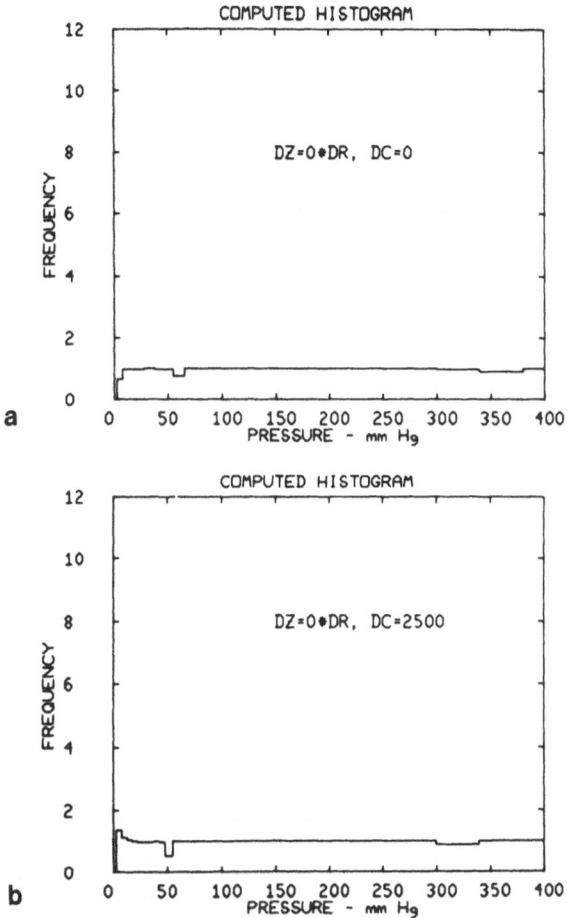

Fig. 3.   Theoretical histograms of tissue substrate distribution
          when there is no axial diffusion (a), and with axial
          diffusion only in the capillary fluid (b).

into a mesh of curvilinear cells comparable to the volume an oxygen
electrode might measure if it were inserted into the tissue cylinder.
An average theoretical pressure in each of these cells is computed
and the cell volume is recorded. This table of pressures and volumes
is cumulated in histogram bins and are normalized according to the
procedure described elsewhere (Schubert, 1976). The model histogram
thus becomes a relative volume versus pressure representation of the
theoretical substrate predictions. Trends in the model histograms
produced by parametric variation can then be directly compared to
experimental histograms determined from measurements under analogous
experimental conditions.

Space does not permit detailed parametric studies to be presented here, therefore we illustrate the histogram trends only for
changes in the axial diffusion coefficient. When there is no axial
diffusion and no wall resistance to substrate passage, the computed
histogram is featureless, as shown in the left half of Fig. 3. The
right half of Fig. 3 shows that permitting axial diffusion in the
capillary adds very few modifications to the first, featureless
histogram. With axial diffusion in the tissue space and coupling
with the capillary diffusion, more features begin to emerge as shown
in Fig. 4. As the axial diffusion coefficient is increased, the tissue

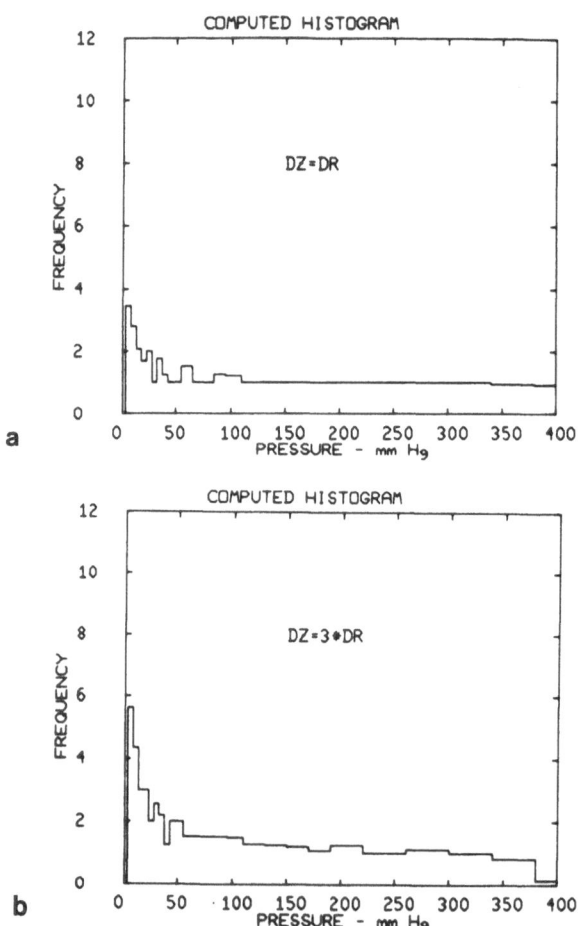

Fig. 4.   Theoretical histograms of tissue substrate distribution for
          the cases axial diffusion coefficient equal to the radial
          value (a), and three times the radial value (b).

substrate distribution moves downward which is reflected in the histogram in the left half of Fig. 4 by the leftward shift and the peaking
in the lower pressure ranges. As the axial diffusion coefficient is
increased still further, the substrate distribution moves lower and
the histograms in Fig. 5 exhibit more features, particularly in the
lower pressure ranges.

Fig. 6 repeats the theoretical histogram for $D_z = 10\ D_r$ and
the experimental histogram obtained by Schubert under conditions
analogous to the model conditions that we have used. Note that this

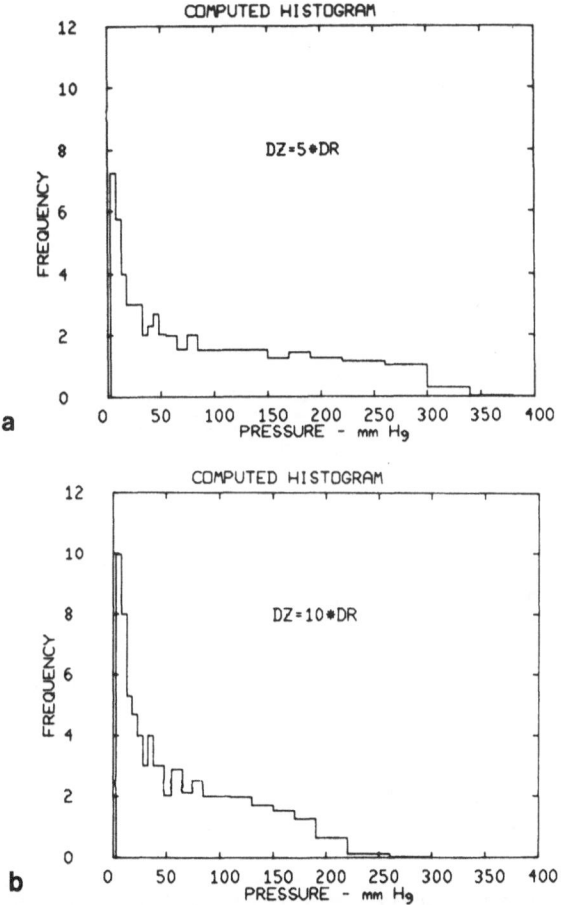

Fig. 5.   Theoretical histograms of tissue substrate distribution for
          the cases axial diffusion coefficient five times (a) and ten
          times the radial value (b).

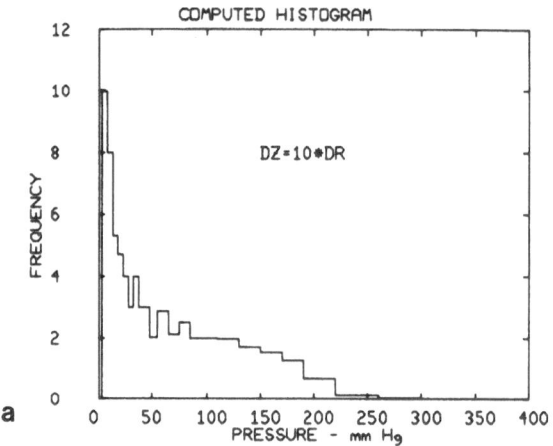

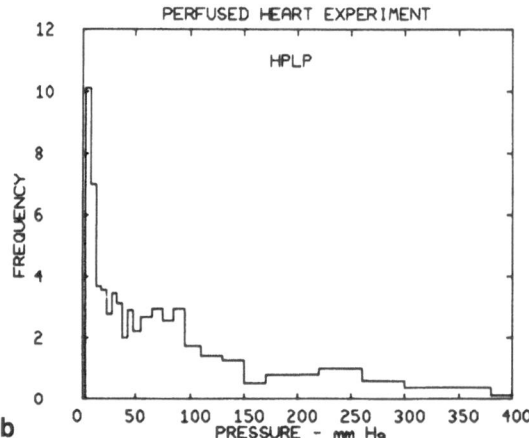

Fig. 6. A theoretical histogram of tissue substrate distribution
         with axial coefficient ten times the radial value, and a
         histogram of experimentally measured values from the per-
         fused heart experiment.

histogram exhibits features similar to the experimental data. However,
the fact that the model results require an axial diffusion coefficient
ten times greater than that measured in in vitro tissue preparations
is difficult to explain physiologically. The model results suggest
a preferential axial transport, possibly by an active mechanism, in
the direction parallel to the muscle striations.

    By introducing a finite wall permeability, we are able to ob-
tain similar results with a lower tissue axial diffusion coefficient,
but a relatively high wall resistance is required. An alternative
explanation could be that the network geometry we have used is in-

appropriate. However, any known alternative geometry would suggest
higher not lower average Po_2 values. Thus, other geometries would
tend to shift the histograms to the right and would tend not to
agree with the peaks in the measured Po_2 histograms. We currently
have no working physiological hypothesis that would explain our
results that is more convincing than a facilitated or heterogenous
diffusion property of the muscle tissues. The role of wall permeability is clearly important and demonstrates the need for a better
understanding of these effects. Such an understanding is particularly important in perfusion dependent treatments such as for shock or
drug therapies and chemotherapies.

REFERENCES

Blum, J., 1960, Concentration profiles in and around capillaries,
    Am. J. Physiol., 198:991-998.
Fletcher, J., 1978, Mathematical modelling of the microcirculation,
    Math. Biosci., 38:159-202.
Fletcher, J., and Schubert, R., 1982a, On the computation of substrate
    levels in perfused tissues, Math. Biosci., in press.
Fletcher, J., and Schubert, R., 1982b, Diffusional coupling in per-
    fused capillary-tissue structures, in press.
Krogh, A., 1919, The number and distribution of capillaries in
    muscles with calculations of the oxygen pressure head necessary
    for supplying the tissue, J. Physiol., 52:409-415.
Olver, F.W.J., 1964, Bessel functions of integer order, in: "Hand-
    book of Mathematical Functions", M. Abramowitz, I. Stegun, eds.,
    AMS 55, NBS, U.S. Dept. of Commerce, Washington D.C., pp. 374-429.
Schubert, R., 1976, A physiological and mathematical study of oxygen
    distribution in the autoregulating isolated heart, PhD thesis,
    Case Western University, Cleveland, Ohio.
Schubert, R., and Whalen, W., 1976, A mass transport model for pre-
    dicting O_2 distribution in the autoregulating heart, Micro-
    vasc. Res., 11:127.
Schubert, R., Whalen, W., and Nair, P., 1978, Myocardial Po_2 distri-
    bution: relationship to coronary autoregulation, Am. J. Physiol.,
    234(4):H361-H370

DISCUSSION

Question:
In the model-generated histograms, how many cells were used? Is the
FREQUENCY on the graphs really given as a percent?

Answer of the authors:
The histograms were constructed from a 16 (radial) x 285 (axial) array
of cells that would correspond roughly to a 0.5 by 2.0 um volume of
tissue. Average pressure values and exact volume are calculated in

each of these regions and are cumulated in 35 histogram bins according to the statistical procedure described, as stated, in our reference 2 (Schubert, 1976). Space restrictions have prevented giving
more of the specific details, which are available elsewhere.
The FREQUENCY is per cent volume per 5 mm Hg. Again, the histogram
construction is discussed fully in our reference 2 (Schubert,1976),
3 and 4 (Schubert and Whalen, 1976; Schubert et al., 1978).

Question:
What is the basis for the parameters used in the model?
What is the value used for γS_t in generating the figures?

Answer of the authors:
The origin of our parameters was given in the paper as reference 3
and 4 (Schubert and Whalen, 1976; Schubert et al., 1978), and they
are representative of the conditions of the perfused cat heart
preparation, as stated.
The value of γS_t for all figures in the paper was $+\infty$, i.e., the
first of equations (1.6, iii). We stated general conclusions for
values other than $+\infty$, but the space restrictions prohibited their
inclusion as figures.

Question:
What is the justification for assuming zero radial gradients in the
capillary?

Answer of the authors:
The radial gradients are very small relative to capillary transport
and axial gradients, thus steady state radial gradients within the
capillary are neglected. The more important point is that capillary
axial diffusion can modify the tissue substrate profile through a
tissue interface coupling effect, whereas the capillary radial diffusion is negligible.

BLOOD AND OXYGEN
TRANSPORT

ON THE SEEMINGLY DIMINISHED CO_2-BOHR EFFECT IN HYPOXIC CHEMODENERVATED RABBITS

H. Kiwull-Schöne, B. Gärtner, and P. Kiwull

Dept. of Physiology, Ruhr-University, 4630 Bochum, FRG

INTRODUCTION

The O_2-Hb dissociation curve (ODC) in mammalian blood
representing the O_2-Hb saturation as function of the O_2 partial
pressure has multiple underlying influences. The position of the
ODC as well as the magnitude of the Bohr effects show considerable
species differences. This may be partly due to intrinsic properties
of the different hemoglobins and partly due to different concentrations of organic phosphates (Bartels and Baumann, 1977). Furthermore, the Bohr shift is numerically greater if a pH change is
caused by CO_2 than if it is caused by fixed acids (Wranne et al.,
1972; Bauer, 1974; Duhm, 1976). This implies that the position of
the ODC can only be determined, if accompanying acid-base changes
of respiratory or metabolic origin are thoroughly distinguished.
The general question arises whether this requirement can be sufficiently fulfilled under in vivo conditions. During hypoxia in vivo,
respiratory (high altitude, asphyxia) or metabolic (anaerobic
glycolysis) acid-base disturbances of various degrees may occur at
the same time, far from representing well defined boundary conditions.

Earlier observations pointed to a particular significance of
the carotid chemoreflexes in determining the O_2-Hb affinity of
rabbit blood under comparable blood gas conditions in vivo: during
constant arterial hypoxia and various levels of hypercapnia, higher
O_2-Hb affinities were observed than were to be expected from the
CO_2-Bohr shift, as long as the animals were chemodenervated
(Kiwull-Schöne et al., 1976). Three questions therefore arise:
1. Is this seemingly diminished CO_2-Bohr effect in hypoxic
chemodenervated rabbits entirely to be explained by the actual
acid-base status? 2. Is it caused by an altered concentration of

163

organic phosphates (e.g. 2,3-DPG) during the course of the experiments? 3. Is it due to additional factors?

In the present study, CO_2- and lactic acid-Bohr factors for rabbit blood were determined in vitro under defined boundary conditions, in order to predict the position of the ODC in vivo, whereby the actual respiratory and metabolic pH changes were considered separately. The predicted results were compared to direct measurements of the P_{50} values.

METHODS

Two series of experiments were carried out. In series A, blood samples from air-breathing, anaesthetized rabbits were equilibrated in vitro with two O_2 partial pressures slightly above and below the half-saturation pressure (P_{50}) as well as with three different CO_2 partial pressures in the range of 20, 40, and 60 Torr. The O_2-Hb saturation and the plasma pH were measured and the corresponding values for P_{50} or pH_{50} were interpolated from a Hill-transformation of the data. The same procedure was repeated after adding constant amounts of lactic acid (5, 10, 14 $mmol \cdot l^{-1}$). Series B consisted of combined in vivo and in vitro experiments. Six spontaneously breathing rabbits (2.7 ± 0.1 kg) were anaesthetized and heparinized (sodium pentobarbital; Vetren; Promonta). Initially, the animals inspired a hyperoxic gas mixture (F_IO_2 = 0.3–0.35) for about twenty minutes. During the subsequent hypoxic period of about one hour, the inspiratory N_2/O_2 mixture was varied in such a way that the arterial Po_2 remained constant (46.4 ± 4.2 Torr) but the Pco_2 was enhanced stepwise, each step for ten minutes (see Table 1). During about one hour of hyperoxic recovery, both carotid sinus nerves were cut. After chemodenervation, the hypoxic period at different levels of Pco_2 was repeated.

Measurements

Tidal volume, respiratory rate, tracheal Pco_2 and blood pressure were continuously recorded. Under steady state blood gas conditions, a blood sample of about 1 ml was taken anaerobically. Concentrations of 2,3-diphosphoglycerate (2,3-DPG), lactate (LA) and hemoglobin (Hb) were determined photometrically (Hitachi) after treatment with the appropriate test solutions (Biochemica Test, Boehringer; Merckotest, Merck). Plasma pH (pH_a), CO_2 partial pressure (P_aco_2) and standard bicarbonate concentration ($HCO_3^-_{st}$) were measured by the Astrup technique (Radiometer), whereby the pH value was corrected for hypoxia according to v. Mengden et al. (1969) using computer assisted numerical iteration. The actual O_2-Hb saturation (So_2) was calculated from the O_2 content (Lex O_2 Con Apparatus, Lexington), considering the physically dissolved amounts. Another portion of the blood sample was equilibrated with a Pco_2,

corresponding to the value in vivo, as well as with two O_2 partial
pressures slightly above and below the P_{50} value to be determined.

Calculations

The slopes (Hill's number n) as well as the Po_2 and pH
values at half-saturation (P_{50}, pH_{50}) of the ODC were determined
from the Hill-transformation of measured pairs of values for So_2
and Po_2 or So_2 and pH, respectively (range $20\% \leqslant So_2 \leqslant 80\%$).

The data of series A enabled a quantitative determination of
the Bohr factors from the slope of lgP_{50} as a function of pH_{50}.
Since the CO_2-Bohr factor (φ_{CO_2}) and the lactic acid-Bohr factor
(φ_{LA}) in rabbit blood are numerically different, it is necessary
to distinguish between respiratory and metabolic pH deviations from
the standard condition (ΔpH_{CO_2}, ΔpH_{LA}). Consequently, the deviation
of P_{50} from its value under standard conditions may be calculated
as

$$\Delta P_{50} = \varphi_{CO_2} \cdot \Delta pH_{CO_2} + \varphi_{LA} \cdot \Delta pH_{LA} \tag{1}$$

The CO_2 induced deviation of pH from the standard value can be
estimated by

$$\Delta pH_{CO_2} = \frac{lgPco_2 - lg40}{\beta} \tag{2}$$

whereby the slope β of the buffer line ($lgPco_2$ vs. pH) was determined by the Astrup method.

Finally, the fixed acid induced deviation of pH from the
standard value is given by

$$\Delta pH_{LA} = pH_a - 7.4 - \Delta pH_{CO_2} \tag{3}$$

For statistical evaluation, mean values $\pm$ S.E.M. of different
variables were calculated for the whole animal population and
linear regression analysis of the means was carried out. Alternatively, single regression lines were calculated and their slopes
and intercepts averaged for the whole population. Significance of
differences was accepted for $P_d \leqslant 0.05$, if calculated by paired or
unpaired t-tests.

RESULTS

1. CO_2- and Lactic Acid-Bohr Effects in vitro

Under standard conditions (pH = 7.4, Pco_2 = 40 Torr, HCO_{3st}^- =
24 mmol·1^{-1}, 38^oC), the average Hill's n was 2.94 ± 0.06 (N = 10)

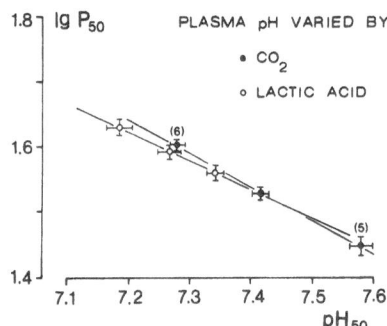

Fig. 1.   Comparison of the CO_2-Bohr effect and the lactic acid-Bohr
          effect on the half-saturation O_2 pressure (P_{50}) as a func-
          tion of plasma pH_{50} for rabbit blood <u>in vitro</u>. Linear
          regression lines of the mean values $\pm$ S.E.M., n = 7, or
          numbers in brackets.

and the P_{50} value 35.5 $\pm$ 0.4 Torr (4.73 $\pm$ 0.05 kPa). For a given
change in plasma pH, the deviation from the standard position was
greater if this pH change was caused by CO_2 than if it was caused
by lactic acid. This is quantitatively shown by the slopes of the
Bohr lines in Fig. 1, where pH changes are either induced by CO_2
without addition of lactic acid or by lactic acid at a constant
Pco_2 of about 40 Torr.

    The slopes of CO_2-Bohr lines do not change significantly if
constant amounts of lactic acid are added (even up to 14 $mmol \cdot l^{-1}$).
Thus, the average slope of 22 single Bohr lines, irrespective of
the concomitant lactic acid concentration, may be regarded as a
satisfactory numerical approach to the CO_2-Bohr factor of rabbit
blood φ_{CO_2} = -0.507 $\pm$ 0.014. Vice versa, the slopes of lactic acid-
Bohr lines do not change either for different but constant levels
of Pco_2 (between 20 and 60 Torr). The average slope of 18 of this
type of single Bohr lines, representing the lactic acid-Bohr factor
for rabbit blood, was φ_{LA} = -0.428 $\pm$ 0.011.

2. CO_2-Bohr Effect in vivo

    From Fig. 1 can be derived that an increase of Pco_2 <u>in vitro</u>,
which would decrease the plasma pH from 7.4 to 7.3, should lead to
an increase of P_{50} from 35.5 to 39.9 Torr. The question arises
whether this prediction is also valid for <u>in vivo</u>, for hypoxic
animals subjected to different levels of Pco_2 within the physiological range.

    Fig. 2 shows the <u>in vitro</u> CO_2-Bohr line in comparison to the
actually measured data <u>in vivo</u>. Whereas for the hypo- and normocapnic range, the <u>in vivo</u> values fit the <u>in vitro</u> Bohr line rather

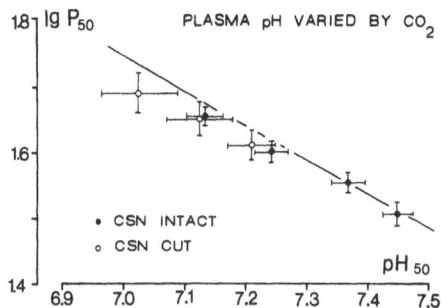

Fig. 2. Comparison of the CO_2-Bohr effect _in vivo_ and _in vitro_ on
the half-saturation O_2 partial pressure (P_{50}) as a function of plasma pH_{50}.
Mean values $\pm$ S.E.M. of hypoxic rabbits _in vivo_. N = 6
before and N = 5 after carotid chemodenervation. The
straight line represents the CO_2-Bohr line of Fig. 1 _in
vitro_.

well, they deviate progressively during hypercapnia, particularly
after carotid chemodenervation. A straight line representing the
course of the _in vivo_ mean values would appear to be distinctly
flatter than the plotted line. Numerically, the averaged slopes
of the individual regression lines for each rabbit were -0.456 $\pm$
0.026 with intact and -0.400 $\pm$ 0.050 with eliminated chemoreflexes.
These values agree neither with the CO_2- nor with the lactic acid-
Bohr factor determined for this species.

Supplementary to Fig. 2, it can be seen from Table 1 that the
pH change is not exclusively caused by CO_2 but also by metabolic
acid-base disturbances of various degrees. For instance, under
condition IV with intact carotid chemoreflexes, the increase in

Table 1. Acid-base conditions during constant arterial hypoxia
(P_aO_2 = 46 Torr) at different levels of P_aCO_2 with intact
(I-IV) or cut (V-VII) carotid sinus nerves.

	PCO_2 (Torr)	HCO_3^-st (mmol$\cdot$l^{-1})	Lactate$^-$ (mmol$\cdot$l^{-1})	pH_{50}
I	18.3 $\pm$ 1.5	19.8 $\pm$ 1.0	9.9 $\pm$ 1.8	7.450 $\pm$ 0.026
II	26.3 $\pm$ 1.2	19.0 $\pm$ 1.1	9.8 $\pm$ 1.4	7.370 $\pm$ 0.028
III	40.1 $\pm$ 1.2	17.7 $\pm$ 1.1	10.4 $\pm$ 1.5	7.244 $\pm$ 0.028
IV	55.0 $\pm$ 2.6	16.3 $\pm$ 1.1	11.2 $\pm$ 1.6	7.134 $\pm$ 0.032
V	27.4 $\pm$ 2.2	13.4 $\pm$ 1.4	14.1 $\pm$ 2.0	7.211 $\pm$ 0.040
VI	37.1 $\pm$ 2.7	12.5 $\pm$ 1.5	14.4 $\pm$ 2.2	7.126 $\pm$ 0.054
VII	52.2 $\pm$ 3.5	12.0 $\pm$ 1.8	15.9 $\pm$ 2.5	7.025 $\pm$ 0.063

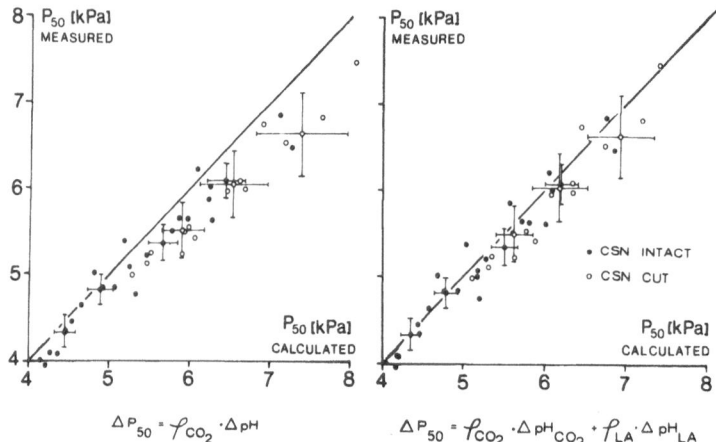

Fig. 3. Identity test of calculated and measured half-saturation
         O_2 partial pressures (P_{50}).
         Mean values $\pm$ S.E.M. and single values of N = 6 rabbits
         with intact and N = 5 rabbits with eliminated carotid
         chemoreflexes.
         Left-hand diagram: Calculation considering the CO_2-Bohr
         factor and the actual pH change, regardless of origin.
         Right-hand diagram: Calculation considering both the CO_2-
         and the lactic acid-Bohr factor together with the correspond-
         ing respiratory or metabolic pH changes.

Pco_2 above 40 Torr contributed to the decrease in pH_{50} below 7.4 by
0.064 ± 0.009 units ($\beta = 2.13 \pm 0.21$), according to equations (2)
and (3). In other words, the acidosis is 24% of respiratory and 76%
of metabolic origin. Correspondingly, under condition VII with
eliminated chemoreflexes, the CO_2 induced decrease in pH is 0.049
$\pm$ 0.014 units ($\beta = 2.26 \pm 0.20$), i.e. only 13% of the total change,
whereas now 87% is induced by fixed acids.

      Since it is thus possible to differentiate between pH changes
either due to CO_2 or due to lactic acid, the presumable position
of the ODC in terms of the P_{50} value can be estimated by equation
(1). Fig. 3 shows the calculated and the directly measured P_{50}
values, when correlated with each other (right-hand diagram). The
data refer to the different acid-base conditions occurring during
hypoxia and at different levels of Pco_2 in vivo. For the best
fitting linear regression line of the 39 single values one obtains
a slope of 0.95 ± 0.03 and an intercept with the ordinate of $1.3 \pm$
1.4 Torr, indicating no significant difference from the identity
line.

      For comparison, if only the actual pH change, regardless of
whether origin, is considered together with the most commonly used
CO_2-Bohr factor (Fig. 3, left-hand diagram), the correlation

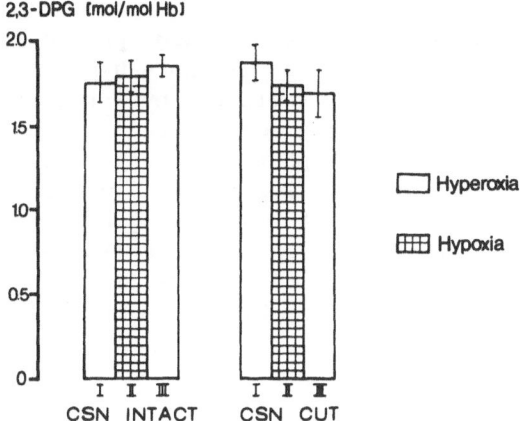

Fig. 4.   The molar ratio of 2,3-diphosphoglycerate (2,3-DPG) in
          rabbit blood before (I), during (II) and after (III) a
          hypoxic period of approximately one hour. Mean values
          + S.E.M. of N = 5 rabbits with intact and eliminated
          carotid chemoreflexes.

between calculated and measured P_{50} values distinctly deviates
from the identity line. In this case, the slope of the best fitting
linear regression line is 0.82 + 0.03, i.e. significantly smaller
than one, and the intercept is 5.2 + 1.3 Torr, i.e. significantly
different from zero.

There is no doubt that the measured values are better approached by considering respiratory and metabolic pH changes or
corresponding Bohr factors separately, as proposed by equation
(1).

3. The Possible Influence of 2,3-DPG on the CO_2-Bohr Effect in vivo

The average molar ratios of 2,3-DPG before, during and after a
hypoxic period of about one hour are shown by Fig. 4. The initial
control value during hyperoxia was 1.76 + 0.11 mol/mol Hb or 27.2
+ 1.7 umol/g Hb. The concomitant spontaneous values of P_{aCO_2}, pH
and $HCO_3^-{}_{st}$ were 34.4 + 2.9 Torr, 7.379 + 0.029 and 20.6 + 0.9
mmol·1^{-1}, respectively. With intact carotid chemoreflexes, a slight
but not significant increase in the 2,3-DPG concentration towards
the end of the hypoxic period could be observed, a tendency which
was maintained throughout the ensuing hyperoxic recovery period. On
the other hand, after carotid chemodenervation, a significant
decrease in the 2,3-DPG concentration occurred at the end of the
hypoxic period (-2.1 + 0.7 umol/g Hb, P_d < 0.05). This tendency
again was continued during the subsequent hyperoxic period.

In this context, it has to be taken into account that hypoxia with intact carotid chemoreflexes was accompanied by hypocapnia (P_aCO_2 = 20.8 $\pm$ 1.6 Torr) due to pulmonary hyperventilation. Therefore, the metabolic acidosis, indicated by a fall in HCO_{3st}^- to about 16 mmol$\cdot$l^{-1}, was entirely compensated by the respiratory alkalosis, so that the actual pH was close to 7.4, namely 7.398 $\pm$ 0.047. Without carotid chemoreflexes, the P_aCO_2 during hypoxia was rather higher (36.7 $\pm$ 8.1 Torr) than during hyperoxia. Together with the low HCO_{3st}^- concentration of about 12 mmol$\cdot$l^{-1}, this led to a distinct fall in pH to 7.141 $\pm$ 0.089. In summary, there is a small decrease of 2,3-DPG during hypoxia, but only if accompanied by acidosis.

DISCUSSION

Hypoxic chemodenervated rabbits subjected to increasing levels of hypercapnia in vivo, show a smaller decrease in the O_2-Hb saturation (Kiwull-Schöne et al., 1976) and hence a smaller increase in the P_{50} values (present results) than to be expected from the common CO_2-Bohr effect.

Among various possible factors to explain this phenomenon, we first tried to get information about the possible role of 2,3-DPG, which is known to interfere with the Bohr effect (Wranne et al., 1972; Bauer, 1974; Duhm, 1976). During hypoxia, the 2,3-DPG concentration should be changed depending on the concomitant acid-base situation (Lenfant et al., 1971). Accordingly we found a small but significant decrease of the 2,3-DPG concentration after carotid chemodenervation, whereby a severe metabolic acidosis prevailed. However, this decrease of 2,3-DPG should rather lead to an increase than decrease of the CO_2-Bohr effect, due to allosteric competition (Bauer, 1974) and/or due to an influence on the intracellular/extra-cellular pH difference (Duhm, 1976). Furthermore, the absolute changes in 2,3-DPG (less than 10% of the resting level) seem to be much too small to be responsible for any measurable interference with the CO_2-Bohr effect (Duhm, 1976). The reason for this may be that relatively short hypoxic periods have been applied in the present experiments, if compared to longer lasting high-altitude conditions (Lenfant, 1971).

Since it was thus possible to exclude influences of 2,3-DPG, we focussed our interest on the acid-base status under the present experimental conditions. In this context, it is well known, particularly for human blood, that pH changes due to CO_2 cause a greater change in P_{50} than those due to fixed acids (Sigaard-Andersen, 1972; Wranne et al., 1972; Bauer, 1974; Duhm, 1976). Accordingly, we also found for rabbit blood the CO_2-Bohr factor to be numerically greater than the lactic acid-Bohr factor (Kiwull-Schöne, 1981). However, compared to the great variation of the CO_2-Bohr factors among different mammalian species, the difference between the CO_2- and the lactic acid-Bohr factor appears to be small. Hilpert et al. (1963) even tend to neglect these species differences, as long as the pH changes are not too great, and Bauer (1974) argues that pH changes occurring in vivo were mainly due to CO_2 (except under extreme conditions like placental gas transfer or severe muscular exercise).

In contrast to many other laboratory animals, the rabbit exhibits a special ability to produce high concentrations of lactic acid during hypoxia even under resting conditions, up to about 20 $mmol \cdot l^{-1}$. On the one hand it can be derived from the present in vitro studies that high but constant concentrations of lactic acid do not change the numerical value of the CO_2-Bohr factor. On the other hand, we have to deal with a permanent lactic acid accumulation of varying speed and extent, under the present experimental in vivo conditions. Therefore one must thoroughly distinguish between respiratory and metabolic pH changes, in order to calculate the CO_2- and the fixed acid-Bohr effect separately for any actual acid-base disturbance. If we did not differentiate between respiratory or metabolic Bohr effects and pH changes, the estimated P_{50} values were distinctly overestimated, e.g. the corresponding So_2 values would be substantially underestimated.

The small discrepancy between estimated and measured P_{50} values in the range of severe metabolic acidosis remains to be explained. Although not significant, there was a small tendency for the CO_2-Bohr factor to decrease at high lactic concentrations, as well as a small tendency for the lactic acid-Bohr factor to decrease during severe hypercapnia. These tendencies in principle agree with the non-linear titration curves of Siggaard-Andersen (1974) for humans. Possibly, in the severe acidotic range, we used a value slightly too high for φ_{LA}.

In much the same way the assumption of linear Bohr lines may cause a resting difference between estimated and measured values, also the assumption of linear buffer lines may limit accuracy.

Finally, the question arises whether the chemoreflexes are beneficial for the oxygen supply of the tissues. Turek et al. (1978) pointed out that although during moderate hypoxia, a right-ward shift of the ODC is required for better unloading of O_2, during severe hypoxia, a left-ward shift appears to be advantageous. This implies that acute severe hypoxia might be better tolerated with intact carotid chemoreflexes. Thereby, the effectiveness of pulmonary ventilation for improving the arterial Po_2 decreases rather than increases at low inspiratory O_2 levels, so that the P_aO_2 values of intact and chemodenervated animals are quite similar (Schöne et al., 1973). Thus, the physiological significance of the ventilatory chemoreflex drive during severe hypoxia is predominantly to cause respiratory alkalosis and a subsequent left-ward shift of the ODC.

SUMMARY

In hypoxic rabbits, different levels of Pco_2 before and after carotid chemodenervation were applied in order to get information about the acid-base status and the position of the O_2-Hb dissociation curve (ODC). A CO_2-induced change in pH caused a smaller change in the half-saturation pressure (P_{50}) than was to be expected from the CO_2-Bohr effect alone. Considering both, the numerically different CO_2- and fixed acid-Bohr factors as well as the corresponding respiratory or metabolic pH changes, a method is presented to calculate the position of the ODC with high accuracy. From these considerations it can be derived that the seemingly diminished CO_2-Bohr effect in hypoxic rabbits in vivo, especially after chemodenervation, is due more or less to accumulation of lactid acid. This leads to an increasing error if only the CO_2-Bohr factor and the actual pH change are taken into account.

ACKNOWLEDGEMENT

We would like to thank Ms. E. Hoffmann for her help with the computations, Mrs. R. Richter for the drawings and Mrs. M. Duparc for typing the manuscript.

REFERENCES

Bartels, H., and Baumann, R., 1977, Respiratory function of hemo-
    globin, in: "International Review of Physiology", Respiratory
    Physiology II, Vol. 14, J.G. Widdicombe, ed., University Park
    Press, Baltimore, p. 107.
Bauer, Ch., 1974, On the respiratory function of hemoglobin, Rev.
    Physiol. Biochem. Pharmacol., 70:1.
Duhm, J., 1976, Dual effects of 2,3-diphosphoglycerate on the Bohr
    effects of human blood, Pflügers Arch., 363:55.
Hilpert, P., Fleischmann, R.G., Kempe, D., and Bartels, H., 1963,
    The Bohr effect related to blood and erythrocyte pH, Am. J.
    Physiol., 205:337.
Kiwull-Schöne, H., Gärtner, B., Mückenhoff, K., and Kiwull, P.,
    1981, Interaction of CO_2 and lactic acid upon the O_2-Hb
    affinity of mammalian blood and the theoretical role of red
    cell pH, in: "Oxygen Transport to Tissue", Adv. Physiol. Sci.,
    Vol. 25, A.G.B. Kovách, E. Dorá, M. Kessler, I.A. Silver,
    eds., Pergamon Press, Akadémiai Kiadó, Budapest, p. 311.
Kiwull-Schöne, H., Kiwull, P., Mückenhoff, K., and Both, W., 1976,
    The role of carotid chemoreceptors in the regulation of
    arterial oxygen transport under hypoxia with and without
    hypercapnia, in: "Oxygen Transport to Tissue II", Adv. Exp.
    Med. Biol., Vol. 75, J. Grote, D. Reneau, G. Thews, eds.,
    Plenum Publishing Corporation, New York, p. 469.

Lenfant, C., Torrance, J.D., and Reynafarje, C., 1971, Shift of the
    O$_2$ Hb dissociation curve at altitude: mechanism and effect,
    J. Appl. Physiol., 30:625.
v. Mengden, H.-J., Schultehinrichs, D., and Thews, G., 1969,
    Dependence of plasma pH on oxygen saturation, Respir. Physiol.,
    6:151.
Schöne, H., Wiemer, W., and Kiwull, P., 1973, The role of the
    carotid chemoreflexes in the regulation of arterial oxygen
    pressure, in: "Oxygen Transport to Tissue"., Adv. Exp. Med.
    Biol., 37A, H.I. Bicher, D.F. Bruley, eds., Plenum Publishing
    Corporation, New York, p. 603.
Siggaard-Andersen, O., 1974, "The acid-base status of the blood",
    4th Edition, Munksgaard, Copenhagen.
Turek, Z., Kreuzer, F., and Ringnalda, B.E.M., 1978, Blood gases at
    several levels of oxygenation in rats with a left shifted
    blood oxygen dissociation curve, Pflügers Arch., 376:7.
Wranne, B., Woodson, R.D., and Detter, J.C., 1972, Bohr effect:
    interaction between H$^+$, CO$_2$ and 2,3-DPG in fresh and stored
    blood, J. Appl. Physiol., 32:749.

DISCUSSION

Questions: F. Kreuzer
1) Did you also determine the titration curves for ΔZ vs. pH?

2) What is the possible effect of 2,3-DPG in respiratory alkalosis?

Answers of the authors:
ad 1) We did not determine complete titration curves of the lactic
acid-Bohr coefficients (φ_{LA}) vs. plasma pH. However, we could
derive some information about the dependency of φ_{LA} upon pH by
pooling the slopes of our Bohr lines at 0,1 unit intervals within
the pH range between 7.6 and 7.1. Compared to the titration curves
of Siggaard-Andersen for human blood, the maximum φ_{LA} values of
rabbit blood were shifted towards alkalinity. At three constant
levels of Pco$_2$ of 2.8, 5.6 and 8.3 kPa, the average numerical
values of φ_{LA} varied slightly with pH, between -0.46 and -0.43,
between -0.47 and -0.40, and between -0.44 and -0.36, respectively.

ad 2) First of all, it has to be questioned whether or not during
hypoxia of about one hour duration any change in the 2,3-DPG concentration would occur. If so, we would expect an increase in 2,3-
DPG with intact carotid chemoreflexes, due to the accompanying
respiratory alkalosis. This was in fact the case, as shown by
Fig. 4. However, the absolute changes in 2,3-DPG were by far too
small as being responsible for any change in the O$_2$-Hb affinity.

OXYGEN UPTAKE INTO THE SHEARED FLOWING BLOOD: EFFECTS OF RED CELL

MEMBRANES AND HAEMATOCRIT

K. Motthaghy, C.W.M. Haest, J. Cremer, and W. Derissen

Dept. of Physiology, Technical University
5100 Aachen, FRG

INTRODUCTION

The non-Newtonian behaviour of blood, due to its fluid droplet like behaviour and deformability of red blood cells (RBC), is responsible for its variable rheological properties under different flow conditions. This has been an object of intensive investigations such as blood viscometry and filtration as well as single cell deformation measurements etc. Inspite of this, only a few reports deal with oxygen transport of blood under shearing. It is to be expected that fluid-drop-like behaviour of RBC's in analogy to deformed suspended liquid droplets (Schmid-Schönbein et al., 1981; Torza et al., 1972) influences mass transport operation. Moreover, secondary flow (in addition to main flow) is known to enhance the mass transport in a homogenous liquid. Therefore, two different phenomena are considered here to improve gas exchange of flowing blood:

1. The intracellular convection within RBC e.g. due to their deformability properties and rotation of red cell membrane around the cell interior as proposed (Fischer and Schmid-Schönbein, 1977).

2. Secondary flow within flowing whole blood, occurring e.g. in vessel branches (Karino et al., 1979).

To investigate the effect of the above mentioned phenomena on gas transport a suitable model is necessary. In previous reports (Motthagy and Hanse, 1982; Motthagy et al., 1982) such a model has been introduced, in which the blood is sheared and concomitantly a well defined secondary flow is produced. This model is now used to investigate the effect of changes of RBC volume fraction in blood (haematocrit) and to discuss effects of membrane stiffness on oxygen uptake under different shear flow conditions.

THE MODEL SYSTEM AND THEORETICAL CONSIDERATIONS

The shear flow (couette-flow) is produced by means of two co-axial cylinders, where the annular gap of both is filled with blood. Secondary flows can be obtained in addition to shear flow, if the inner cylinder is rotating (see Fig. 1 and 2) as described first by Taylor (1923) (Taylor vortices). The model consists of a couette arrangement, with gas permeable membranes. The blood is led to the base of the device and is pumped out at the top. During its passage blood is sheared and additionally describes a secondary flow. The geometry of the couette arrangement and the rotational speed of the inner cylinder determines the extent of shearing and the intensity of vortices.

The circuit is arranged as follows: Blood circulates at constant volume flow from the top of the apparatus to a reservoir, from here to a blood-gas equilibrium system (a rotating disc oxygenator) where it is again pumped (roller pump) back to the couette apparatus. In order to obtain the desired "venous" blood inlet data, blood is gased in the rotating disc oxygenator with gas mixtures of O_2, N_2 and CO_2. The actual oxygen uptake of the circulating blood occurs in the couette apparatus with pure oxygen.

By analysing O_2 content of blood samples at inlet and outlet of the device (see Fig. 1) the amount of the oxygen uptake can be measured since the blood flow rate is kept constant.

Two dimensionless numbers are used to evaluate the data (for details see Motthagy et al., 1982; Motthagy and Hanse, 1982).

1. Taylor-number:

$$Ta = (\omega \cdot Rm \cdot 0.5\ d \cdot 1.5\ \rho)/(\eta \cdot Fg)$$

where
Rm is the mean radius of cylinders: $(R_1+R_2)/2$, d the gap width, Fg a geometrical factor, ω the angular velocity, ρ the density and η the dynamic viscosity of blood. An increase of Taylor number indicates an increase of vorticex-"velocity" (intensity).

2. Axial Reynolds-number:

$$Re = \frac{Dh \cdot w \cdot \rho}{\eta}$$

where

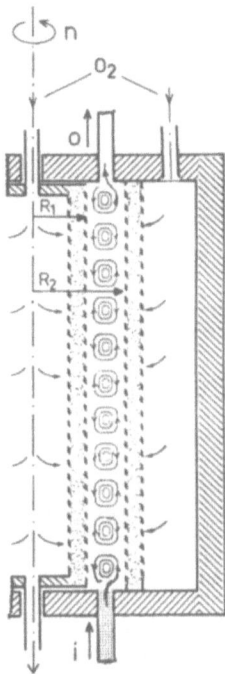

Fig. 1. Axial section of membrane-couette arrangement: R_1 and R_2
        inner and outer cylinder; i and o inlet and outlet of blood;
        O_2-inlet at the top and O_2-outlet at base of the device
        (the outlet for the outer cylinder is not drawn). Dotted
        areas: membranes.

Dh the "hydraulic diameter" of geometrical arrangement and w is the
axial velocity of blood through the gap of the angular section. This
number is proportional to the blood flow rate if η is assumed to be
constant. In other words Re-number represents the contact time of
blood within the membranes. In this report we inserted for η the
plasma viscosity of 2.7 mPa·s, which is a constant value.

Rigidification of RBC Membranes:

    Recently it was reported on a method (Haest et al., 1980) which
deals with a controlled artificial rigidification of erythrocytes.
This can be obtained by a chemical oxidation of membrane SH-groups
to disulfide bonds. The most suitable agent for this purpose is
diamide (diazine dicarboxylic acid bis-dimethylamide), which
oxidizes membrane SH-groups in a concentration dependent way and
cross-links preferably the extrinsic protein, spectrin. This
protein among others, located at the cytoplasmatic membrane surface
is supposed to be arranged as cytoskeleton and responsible for
membrane shear elasticity. After a pretreatment of the cells with

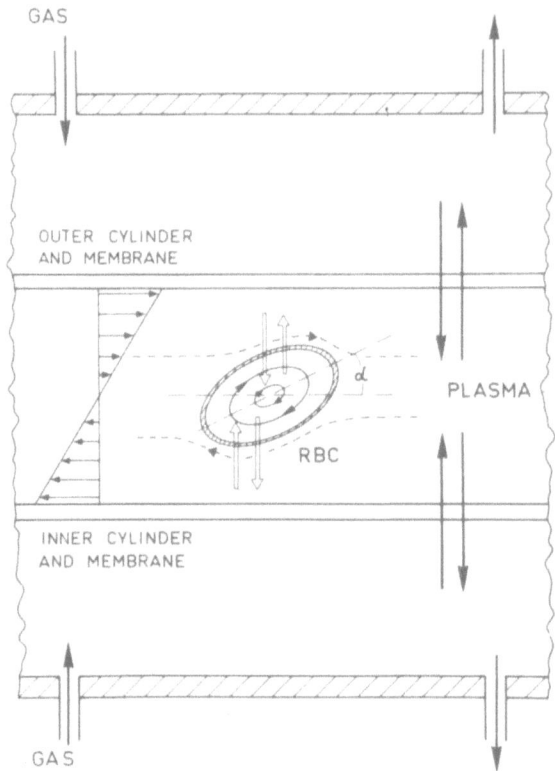

Fig. 2. Schematic horizontal view of the couette arrangement of
        Fig. 1. A single elongated RBC in analogy to droplet fluids
        in shear feld with inner circulation. The arrows (middle)
        indicate O_2 and CO_2 exchange, the upper and lower (gas)
        oxygen perfusion of the device.

iodoacetate, which blocks SH-groups of intracellular glutathion, a
short exposure of cells to diamide is sufficient to obtain an increase of shear stiffness.

METHOD

      Freshly heparinized pig blood is used for investigations on
haematocrit changes. Different haematocrit values of blood are
adjusted by centrifugation and resuspending of the cells in the
native plasma. For studies concerning cell rigidity freshly drawn
heparinized human blood is used. Rigidification of the membranes is
achieved by treating RBC's with 5 $mmol \cdot l^{-1}$ diamide (Calbiochem).
This treatment is preceded by a pretreatment with 5 $mmol \cdot l^{-1}$ iodoacetate.

Maximal shear rigidification without complete solidification meaning, the cells can still be haemolyzed osmotically, is obtained by a treatment with 0.03% glutardialdehyde (for details see Fischer et al., 1978; Haest et al., 1980; Motthagy et al., 1982).

Blood gas data are measured under STPD-conditions. Po_2, Pco_2 and other acid-base parameters are measured with an IL 312/214 pH-blood-gas analyser and calculator, blood oxygen saturation with IL 182-CO-Oxymeter (all from Instrumentation Laboratory Inc., Lexington, Mass.), and oxygen content with Lex-O_2-Con. TL (Lexington Instr., Woltham, Mass.). Haemoglobin concentration is quantified spectralphotometrically. Blood viscosity is measured by a DEER-Rheometer at constant shear stress (Integrated Petronic Systems Ltd., London).

The circulation system is primed at room temperature with the chosen blood type (with rigidified cells or different haematocrits) and the equilibration of blood starts. Afterwards the blood circulation through the couette system is started and reaching a steady state, blood-gas data, and the oxygen uptake, $\dot{V}o_2$, are measured by analysing the samples of the inlet and outlet blood. The blood flow rate is measured by calibrating the roller pumps and is kept constant during the whole experiment at 5 ml/min.

RESULTS

In a series of introductory investigations the axial blood flow rate was varied systematically, in order to find the optimal contact time of blood. Based on these results a flow rate of 5 ml/min which corresponds to an axial Reynolds-number of Re = 0.78 has been evaluated to deliver the most efficient contact time between blood and the gas compartment for the present study.

Fig. 3 demonstrates the oxygen uptake of blood with different haematocrits as a function of shear rate. For every haematocrit value there is a defined shear rate from which on no significant enhancement of O_2 uptake is obtained. For shear rates below 40 s^{-1} the curves are crossing. Thus, a defined relationship between O_2 uptake, shear rate and hct is not directly evident. This point is discussed below.

An interesting observation is that a decrease of haematocrit (hct) values from 45 to 30 induces only a slight decrease in the O_2 uptake. The rheological effect of hct is best visualized by calculation of oxygen uptake per gram of haemoglobin. In Fig. 4 the O_2 uptake of blood is plotted for red blood cell (RBC) volume fractions of hct = 45 and 80%. Actually a blood volume with 80% hct, as it can be easily calculated is capable of binding more oxygen than a sample with 40% hct. However, under rheological conditions the oxygen

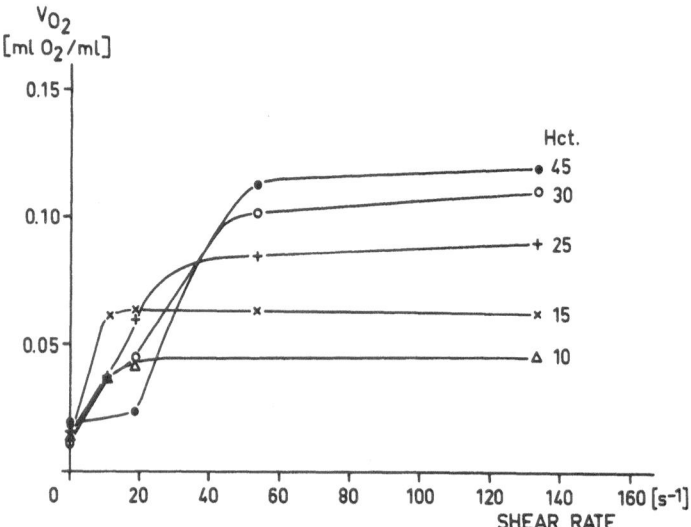

Fig. 3. Oxygen uptake (in- and outlet O_2-content-difference, $\dot{V}O_2$
of blood with different haematocrit values. Each point represents the mean value of 5 measurements.

transport rate "efficiency" of bood with a hct = 45 is higher. This
is especially pronounced at higher shear rates. Results obtained
from experiments with rigidified cells (hct = 41%) are depicted in
Fig. 5. It is evident that the O_2 uptake by diamide cells is markedly lower than that of the control cells and the difference is more
pronounced by increasing the shear rate. Glutardialdehyde treated
cells show even lower values (the meaning of arrows in Fig. 5 are

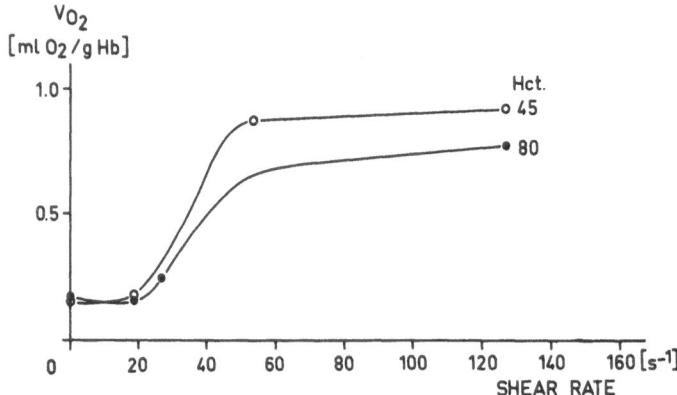

Fig. 4. Oxygen uptake per mg haemoglobin as a function of shear rate
for haematocrits of 45 and 80%. Each point represents the
mean value of 5 measurements.

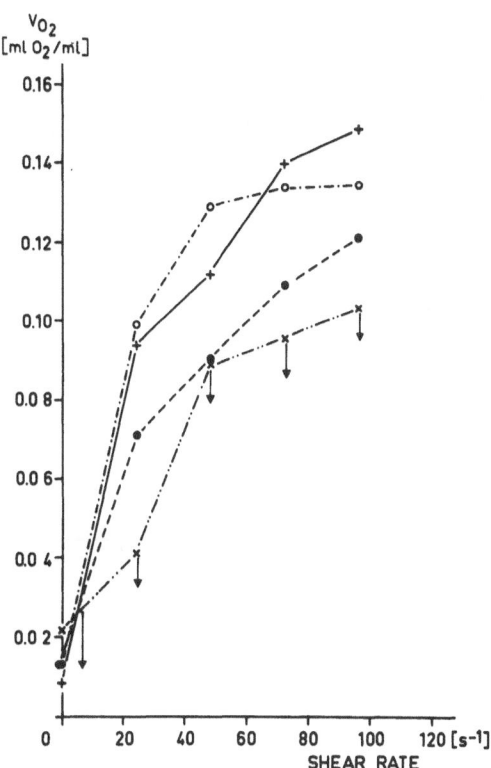

Fig. 5. Oxygen uptake of blood suspensions of normal (+), iodoacetate
        (o), diamide (●), and glutardialdehyde (x) treated cells
        with a haematocrit of (41%). Each point represents the mean
        value of 3 measurements.

discussed later). As already mentioned the hardened cells are pretreated with iodoacetate. Thus, as an additional control the oxygen
uptake by iodoacetate treated cells is measured. It can be seen
from Fig. 5 that there is no significant difference between iodoacetate treated and control cells.

    The effect of secondary flow in combination with shear flow upon
oxygen transport as a function of Taylor-number is depicted in Fig. 6.
The haematocrit is varied from 80% down to 10%. It is important to
note, that the viscosity of blood, which is necessary for the calculation of the Taylor-number, is a function of shear rate and not
a constant value. Especially at lower shear ranges the viscosity of
blood increases with increasing haematocrit values as seen in
Fig. 7. Therefore, we substituted the measured viscosity for the
corresponding shear rate and haematocrit value in the Taylor-number.
As demonstrated in Fig. 6 the slope of the curves at Taylor-numbers
below 100 is decreased by a decrease of haematocrit. On the other
hand higher Taylor-numbers and necessarily higher shear rates, see

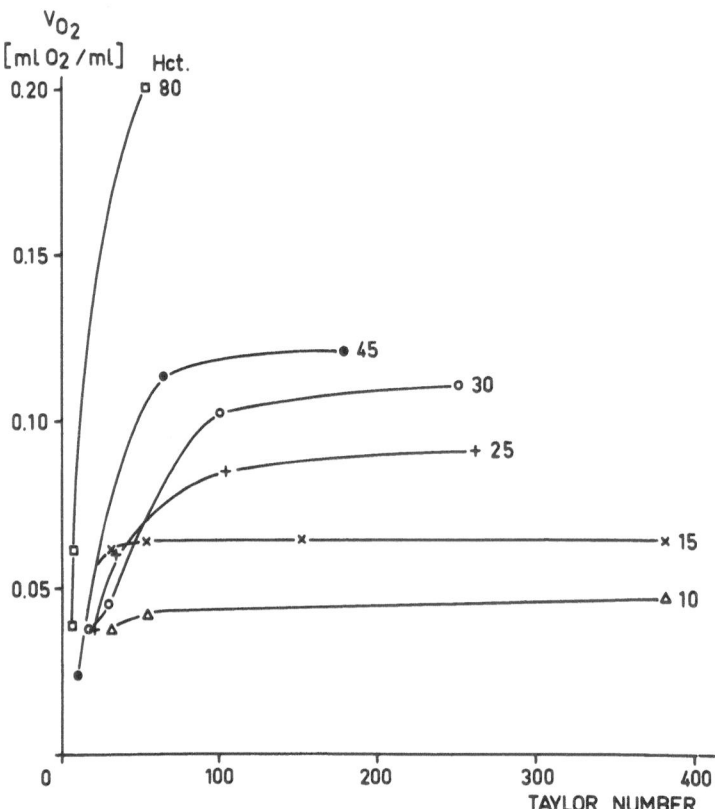

Fig. 6. Oxygen uptake of blood with different haematocrit values
         as a function of Taylor-number. Each point represents the
         mean value of 5 measurements.

also Fig. 3, are necessary to reach the asymmetrical range of $\dot{V}_{O_2}$
values. E.g. for the physiological haematocrit value of 45% it can
be seen from Fig. 6, that the increase of Taylor-number above 200
does not increase the oxygen uptake of blood, which according to
these results, means that the optimum of secondary flow "utilization" is reached.

DISCUSSION

     It was already mentioned that a clear interpretation of the
data hampered at low shear rate ranges and at low Taylor-numbers
below approx. 50. For this particular case an extensive theoretical
consideration for blood with physiological hct value and some other
fluids is discussed in previous reports (Motthagy et al., 1982;
Motthagy and Hanse, 1982). It should be only pointed out here, that
this is mainly the result of instabilities of vortices at low

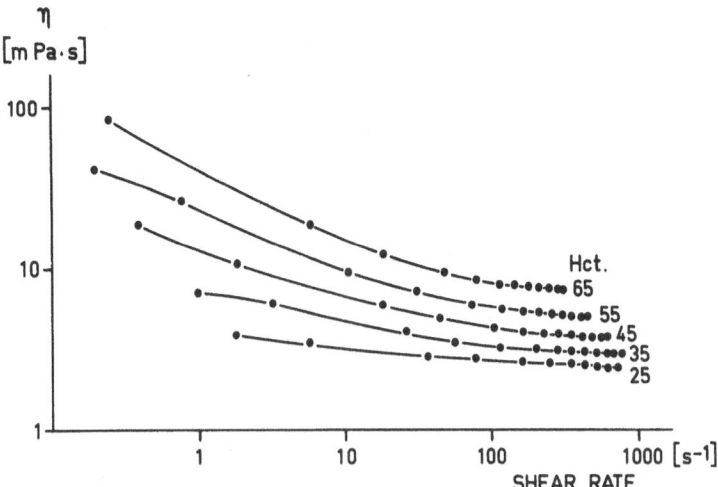

Fig. 7.   Dynamic blood viscosity with different haematocrits as a
          function of shear rate. Each point represents the mean
          value of 3 measurements.

Taylor-numbers (critical Taylor-numbers). Furthermore, for various
degree of haemodilution at low shear rates, in contrast to high
shear rates, because of sedimentation and aggregation tendencies it
is difficult to define a homogenous haematocrit value. The results
indicate, that if it is intended to study only the effect of low
shear rates the outer cylinder should rotate, while the inner one
is stationary. However, effect of red cell deformation upon oxygen
uptake becomes of interest at higher shear rates, which have been
shown (Lipowsky et al., 1978; Lipowsky et al., 1980) to occur in
microcirculation.

      There is no doubt that the present model, due to the relative
wide anular gap (4 mm) does not simulate microvessels geometrically. Nevertheless, since physiological shear rates can be simulated,
conclusions from experiments on O_2 uptake or release by single cells
can still be deduced. This can be discussed with the help of Fig. 5.
Fischer and Schmid-Schönbein (1977) reported that the degree of
elongation of RBC's (it is defined as an ellipsoidal shape in analogy
to fluid droplets derived from long and short axis, see Fig. 2) by
increasing of shear rate, is systematically reduced if the cell membrane is rigidified with diamide. This tendency is closely paralleled
by a reduction of O_2 uptake (Fig. 5). The data show not only an O_2
uptake reduction because of "macro" convection deficiency but in
addition indicate a reduction of an intracellular convection. This
could be explained by changes of elongation and concomitant variation of the intracellular convection under shear forces. It is also
shown (Haest et al., 1980) that iodoacetate does not diminish the
elongation of the cells, which is again in accordance with the essentially unchanged O_2 uptake data.

A rigidification of cells by glutardialdehyde causes the complete loss of elongation and further reduces oxygen uptake (Fig. 5). Possible effects of the chemical treatments on O_2 saturation curves of blood could be excluded. We measured the O_2 saturation curves with a method described by Sick and Gersonde (1980) for every type of blood and noted that only for glutardialdehyde treated cells a left shift of O_2 saturation curve has occurred. Since this increase of affinity facilitates the O_2 uptake so that the actual data in Fig. 5 for glutardialdehyde should be lower, as indicated with arrows. (A comparison of Fig. 3 and 5 shows, that the values for the control series are slightly higher than in Fig. 3 for hct = 41%. This is due to slight differences of inlet blood data and haematocrit value in both series.)

Concerning the effect of haemodilution, the model can deliver for any given arbitrary haematocrit value, the lowest Taylor-number of shear rate for a maximum of oxygen uptake. This fact can find its application in design of artificial organs especially oxygenators (Gaylor and Smeby, 1976) where secondary flows are utilized and blood trauma due to high shear forces is to be minimized. It is interesting to note, that a moderate haemodilution as shown in Fig. 3 and 6 for haematocrits of 30% still provides an "arterio/venous" oxygen content value of 10 Vol %. That means rheological consequences of haemodilution which is clinically also applied (Schmid-Schönbein et al., 1981) could be connected to studies by the present model system.

As a summary it can be concluded that this model allows direct measurements of oxygen uptake of flowing sheared blood with or without effect of a secondary flow. By perfusion of the device with N_2 and CO_2 gas mixtures the oxygen delivery can be investigated at desired flow conditions. Further studies aim to simulate the physiological conditions more effectively by reducing blood film thickness (Fig. 1), and to investigate the effects of 2.3 DPG and red cell rigidity upon oxygen release from blood using this modified model system.

ACKNOWLEDGEMENTS

The authors wish to thank Mr. H. Malotta for viscosimetry measurements. We also appreciate the excellent technical help of Mrs. J. Freund and Mr. B. Oedekoven.

REFERENCES

Fischer, T., Haest, C.W.M., Stöhr, M., Kamp, D., and Deuticke, B., 1978, Selective alteration of erythrocyte deformability by SH-reagents-evidence for an involvement of spectrin in membrane shear elasticity, Biochim. Biophys. Acta, 510:270.

Fischer, T., and Schmid-Schönbein, H., 1977, Tank tread motion of
    red cell membranes in viscometric flow: Behavior of intracel-
    lular and extracellular markers, Blood Cells, 3:351.
Gaylor, J.D.S., and Smeby, L.C., 1976, The Taylor vortex membrane
    oxygenator design analysis based on a predictive correlation
    for oxygen transfer, in: "Physiological and Clinical Aspects
    of Oxygenation Design", S.G. Dawids, H.C. Engell, eds., Else-
    vier, Amsterdam-Oxford-New York.
Heast, C.W.M., Driessen, G.K., Kamp, D., Heidtmann, H., Fischer,
    T.M., and Stöhr-Liesen, M., 1980, Is "deformability" a para-
    meter for the rate of elimination of erythrocytes from the
    circulation?, Pflügers Arch., 388:69.
Haest, C.W.M., Fischer, T.M., Plasa, G., and Deuticke, B., 1980,
    Stabilization of erythrocyte shape by a chemical increase in
    membrane shear stiffness, Blood Cells, 6:539.
Karino, T., Kwong, H.H.M., and Goldsmith, H.L., 1979, Particle flow
    behaviour in models of branching vessels - I. vortices in
    90° T-junctions, Biorheology, 16, No. 3:231.
Lipowsky, H.H., Kovalcheck, S., and Zweifach, B.W., 1978, The
    distribution of blood rheological parameters in the microvas-
    culator of cat mesentery, Circ. Res., 43, No. 5:738.
Lipowsky, H.H., Usami, S., and Chien, S., 1980, In vivo measure-
    ments of "apparent viscosity" and microvessel haematocrit in
    the mesentery of the cat, Microvasc. Res., 19:297.
Mottaghy, K., Haest, C.W.M., and Schleuter, H.J., 1982, Effect of
    red cell rigidity on gas transport by sheared flowing blood,
    Chem. Eng. Commun., 15:157.
Mottaghy, K., and Hanse, H.J., 1982, Oxygen uptake of red blood
    cells in flowing blood, submitted to Blood Cells.
Schmid-Schönbein, H., Messmer, K., and Rieger, H., eds., 1981,
    "Hemodilution and Flow Improvement", Bibl. Haematol., No. 47,
    Karger, Basel.
Sick, H., and Gersonde, K., 1980, Rapid measurements and computer
    analysis of complete oxygen dissociation curves of red blood
    cells, J. Clin. Chem. Clin. Biochem., 18, No. 10:686.
Taylor, G.I., 1923, VIII. Stability of a viscous liquid contained
    between two rotating cylinders, Philos. Trans. R. Soc. Lond.
    (A), 223:289.
Torza, S., Cox, R.G., and Mason, S.G., 1972, Particle motions in
    sheared suspensions. XXVII. Transient and steady state defor-
    mation and hurst of liquid drops, J. Colloid Sci., 38:395.

DIABETIC OXYGEN-HEMOGLOBIN EQUILIBRIUM CURVES EVALUATED BY NONLINEAR

REGRESSION OF THE HILL EQUATION

J.F. O'Riordan[1], T.K. Goldstick[1], J. Ditzel[2],
and J.T. Ernest[3]

[1]Dept. of Chem. Eng., Northwestern Univ., Evanston
IL 60201, USA
[2]Dept. of Medicine, Aalborg Regional Hospital
DK-9000 Aalborg, Denmark
[3]Dept. of Ophthalmology, Univ. of Illinois Eye and
Ear Infirmary, Chicago, IL 60612, USA

SUMMARY

The oxygen-hemoglobin equilibrium curves (OHECs) were measured
on whole blood samples from 131 individuals (33 normal and 26 diabetic adults and 30 normal and 42 diabetic juveniles) using a Radiometer Dissociation Curve Analyzer (DCA-1). All measurements were
made in the morning following an overnight fast and without exogenous insulin. The saturation versus Po_2 data were fitted to the
Hill equation using a previously described nonlinear regression algorithm to yield the parameters describing the position (P_{50}) and
shape (n) of each OHEC. It was found that the Hill model could be
used to describe OHECs of both normal and diabetic subjects. A
small (approximately 10%) but significant decrease in P_{50} was found
for the diabetic juveniles compared to normal juveniles. There appeared to be no change in P_{50} with diabetes in adults. However, in
these diabetic subjects, the P_{50} had been increased by the somewhat
elevated levels of 2,3-DPG. No difference in n was found between
either group of diabetics and their corresponding group of normals
but n was approximately 5% lower in juveniles than in adults. The
ability of blood to release oxygen to tissue may be transiently impaired in diabetic juveniles because of the left shift of their
OHECs.

INTRODUCTION

The oxygen-hemoglobin equilibrium curve (OHEC) gives the percent oxygen saturation of hemoglobin versus oxygen tension, usually
at the standard conditions of a blood pH of 7.4 and a blood Pco_2 of
40 Torr. It is well known that the position and shape of the OHEC
determine, to a considerable extent, the delivery of oxygen to
tissue. For example, a left shift of the OHEC, indicating an increased hemoglobin (Hb) affinity for oxygen, could cause hypoxia if
not compensated for by other changes (e.g., increased Hb content,
increased blood flow).

One proposed pathogenic mechanism in the etiology of the vascular complications of diabetes is so-called affinity hypoxia
(Ditzel, 1980). Diabetics have chronically high levels of glycosylated hemoglobin (HbA_{Ic}). Although HbA_{Ic} has been shown to have an
increased oxygen affinity in vitro (Bunn and Briehl, 1970), the
pathological significance of this increase has yet to be firmly
established. If the elevated level of HbA_{Ic} caused the overall OHEC
to be shifted left, tissue hypoxia could result, assuming conditions
of constant O_2 consumption, blood flow, pH, etc. Tissues with sensitive oxygen supply systems, such as the retina, or tissues with a
reduced ability to autoregulate oxygen delivery would be most subject
to this problem (Woodson, 1979). In addition, diabetics receiving
exogenous insulin may undergo acute fluctuations in the position of
their OHECs related to phosphate shifts and changes in intraerythrocytic 2,3-diphosphoglycerate (2,3-DPG) (Ditzel et al., 1981).

We have studied the chronic component of the affinity hypoxia
hypothesis by examining a number of OHECs from normal (NA) and diabetic (DA) adults and normal (NJ) and diabetic (DJ) juveniles in
the fasting state with no exogenous insulin. To simplify description of the OHEC we have employed a previously described method
(O'Riordan et al., 1982) to characterize the data using the Hill
equation (Hill, 1910). This method fits the data between 20% and
97% saturation obtained with a Dissociation Curve Analyzer (Model
DCA-1, Radiometer, Copenhagen) to the equation:

$$Y = \frac{(Po_2/P_{50})^n}{1 + (Po_2/P_{50})^n} \tag{1}$$

The two calculated parameters, P_{50} (Po_2 at half-saturation) which
indicates position of the curve and n (the Hill coefficient) which
describes the shape, have been previously found to accurately characterize normal OHECs (O'Riordan et al., 1982). The average values
of the parameters found by this method for our group of normal
adults, $P_{50} = 26.2 \pm 0.8$ Torr ($\pm$ SD) and n = 2.50 ± 0.07, were in
excellent agreement with normal values found by other investigators
(Perutz, 1970; Roughton et al., 1972; Winslow et al., 1978).

Table 1. Characteristics of subjects

		Group			
PARAMETER		NA	DA	NJ	DJ
SEX	M	24	20	14	14
	F	9	6	16	28
AGE, yr.	M	22.5 + 3.5*	43.2 +13.8	11.3 + 1.6	11.1 + 2.9
	F	26.2 + 3.9	50.0 + 4.4	11.4 + 1.4	11.8 + 3.7
Hb, g/dl		14.5 + 1.2	14.2 + 1.4	13.4 + 0.7	13.7 + 0.6
Hct, %		41.0 + 3.0	41.9 + 8.7	39.8 + 2.3	38.6 + 1.8
HbA_{Ic}, %		4.7 + 1.3	9.4 + 1.8	4.3 + 0.7	10.7 + 2.1
2,3-DPG, uM		139 + 25	152 + 19	134 + 14	150 + 13
Pi, mM		1.17 + 0.26	1.14 + 0.20	1.35 + 0.18	1.40 + 0.20
pH (art.)		7.43 + 0.02	7.40 + 0.04	7.43 + 0.02	7.39 + 0.02

NA = normal adults; DA = diabetic adults; NJ = normal juveniles;
DJ = diabetic juveniles; M = male; F = female; * mean $\pm$ SD

METHODS

All subjects for this study were ambulatory and all were non-smokers. Their blood chemistry parameters and ages are shown in Table 1. Analytical methods for these results were reported by Ditzel (1980).

One of us (J.D.) collected all of the experimental data analyzed here using the DCA-1 which is based on the method of Duvelleroy et al. (1970) as we have previously discussed (O'Riordan et al., 1982). Briefly, it oxygenated (in ca. 40 min) a stirred, 6.6 ml constant volume sample of previously deoxygenated whole blood by having it in contact with a 5.8 ml constant volume of oxygen-rich gas (Po_2 approximately 670 Torr initially). The Pco_2 and total pressure were maintained at 40 Torr and atmospheric pressure, respectively, and the temperature was kept at 37 degrees Celsius throughout the determination. While the blood was being oxygenated, its Po_2 and pH were continuously recorded as was the Po_2 of the gas phase. The device produced a graphical record of the Po_2 of the gas and the pH of the blood versus blood Po_2. Data were recorded from

an initial blood Po_2 of zero to nominally 300 Torr. The output was
digitized every two Torr of blood Po_2 from 0 to 100 Torr. For each
point we then computed the oxygen saturation as described in detail
elsewhere (O'Riordan et al., 1982). Blood Po_2 data were corrected
to a pH of 7.40 using a mean Bohr coefficient, $d(\log (Po_2))/d(pH)$,
of −0.38 (Garby and Meldon, 1977). The saturation versus Po_2 data,
above 20% saturation (as suggested by Roughton, 1964), were then
fitted using a locally adapted nonlinear regression software package
(Robinson, 1981) to give the Hill parameters P_{50} and n. The saturation, for a measured Po_2 of 100 Torr, was 96-97%. All computations
were made using the Cyber 170/730 computer at the Vogelback Com-
puting Center of Northwestern University. The data and the fitted
curve were graphed (Model 1136, California Computer Products, Anaheim,
Calif.) as percent saturation versus Po_2. A Hill plot was also made
to verify that the Hill equation was applicable over the region of
the curve fitted.

To evaluate the sensitivity of the computed Hill parameters to
correction for the Bohr effect, we also used a Bohr coefficient of
zero and the Pco_2 dependent Bohr coefficient of −0.48 (Severinghaus,
1966) in the analysis of eight, randomly-selected subjects from
each of the four subject groups. The first of these values, which
effectively ignores the Bohr effect, is used in a standard commercial device (HemoScan, Aminco, Silver Spring, MD 20910). The second
has been reported by other investigators (Hellegers and Schruefer,
1961; Hilpert et al., 1963) and accounts for both the direct effect
of molecular CO_2 as well as that of pH. Our value of −0.38 reflects
pH changes only and is appropriate for this study because Pco_2 was
maintained constant at 40 Torr.

Table 2.   The effect of varying the Bohr coefficient beta

GROUP (#)	beta = 0		beta = −0.48	
	$P_{50} \pm SD$	$n \pm SD$	$P_{50} \pm SD$	$n \pm SD$
NA (8)	−4.9 ± 2.5*	−5.8 ± 0.9	+1.3 ± 0.7	+1.7 ± 0.3
DA (8)	−6.2 ± 1.2	−5.9 ± 0.3	+1.7 ± 0.4	+1.7 ± 0.1
NJ (8)	−6.0 ± 1.5	−5.9 ± 0.1	+1.7 ± 0.4	+1.7 ± 0.3
DJ (8)	+3.0 ± 2.6	−7.9 ± 0.8	−0.6 ± 0.6	+2.3 ± 0.2

*values represent the percentage changes from parameter values
 calculated using beta = −0.38

Table 2 shows that when no correction is made for sample pH (i.e., a Bohr coefficient of zero), the P_{50} and n values decrease by approximately six percent each except for P_{50} in the diabetic juveniles. This exception results from the average blood pH values during the measurement being less than 7.40 in the eight diabetic juveniles selected and greater than 7.40 in all other cases. The effect of using a Bohr coefficient of -0.48 is a slight increase in P_{50} and n (Table 2), again with the one exception of diabetic juveniles. This suggests to us the importance of using the correct Bohr coefficient especially when looking for small differences in P_{50} and n.

Whole blood consumes oxygen at a normal rate but was not accounted for in the analysis. Since a significant amount of oxygen may have been consumed in the approximately 40 minutes required for the generation of each OHEC, we examined the effect of including oxygen consumption in the analysis. Correction of our data for oxygen consumption would give a negligible (0.5%) decrease in P_{50} and a larger (1.6%) increase in n as shown in Table 3 for the same eight, randomly-selected subjects evaluated previously. The details of our calculations are presented elsewhere (O'Riordan et al., 1982). Clearly, neglecting oxygen consumption is of little consequence.

RESULTS

Fig. 1 shows a typical set of experimental data and the fitted curve for a diabetic adult. Data outside the region of fit (less than 20% saturation), shown as triangles, deviate markedly from the model in most cases. The final sum of squared residuals was more sensitive (usually about twice) to relative changes in P_{50} than in n (i.e., convergence was achieved faster on P_{50} than on n)

Table 3. The effect of including O_2 consumption in the determination of P_{50} and n

GROUP (#)	% CHANGE	
	$P_{50} \pm SD$	$n \pm SD$
NA (8)	-0.3 ± 0.1	+0.7 ± 0.4
DA (8)	-0.5 ± 0.1	+1.5 ± 0.3
NJ (8)	-0.8 ± 1.1	+2.6 ± 3.9
DJ (8)	-0.5 ± 0.6	+1.6 ± 2.2

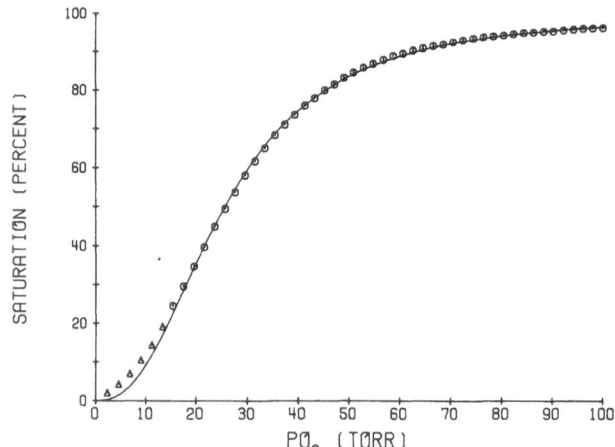

Fig. 1. Typical OHEC for a diabetic adult (A-8). The circles are
         points between 20% and 97% saturation which were included
         in the nonlinear regression, while the triangles are
         points outside the fitted region. The solid line is the
         fit. The parameters found were P_{50} = 25.7 $\pm$ 0.4 ($\pm$ SD)
         Torr, Hill coefficient n = 2.44 $\pm$ 0.09 and the RMS error =
         0.52% saturation.

suggesting that P_{50} was more accurately determined than n. The
overall quality of the fit was judged by the root-mean-squared
(RMS) error, which was 0.52% saturation in the case of Fig. 1 and
averaged approximately 0.5% saturation (range of all 131 subjects
0.23% to 0.83%) for each of the four groups studied. The data always deviated from the model in the same way as in Fig. 1. For
another view of the data, each data set was transformed and a Hill
plot made using the transformed data and a line calculated from the
nonlinearly fitted parameters (Fig. 2). As expected there is
greater deviation of the data below 20% saturation.

     Table 4 lists the means of P_{50} and n with their standard de-
viations for each of the four experimental groups studied. The actual distributions of our fitted values of P_{50} and n for the four
experimental groups studied are seen in Figs. 3 and 4, respectively. In the juveniles we see a significant left shift of the P_{50} in
the diabetic group (p < 0.001). Also, the normal juveniles have P_{50}
values higher than normal adults (p < 0.001) and both groups of
juveniles exhibit a lower n than found in the adults (p < 0.03). In
these groups there were no significant differences between the normal and diabetic adults.

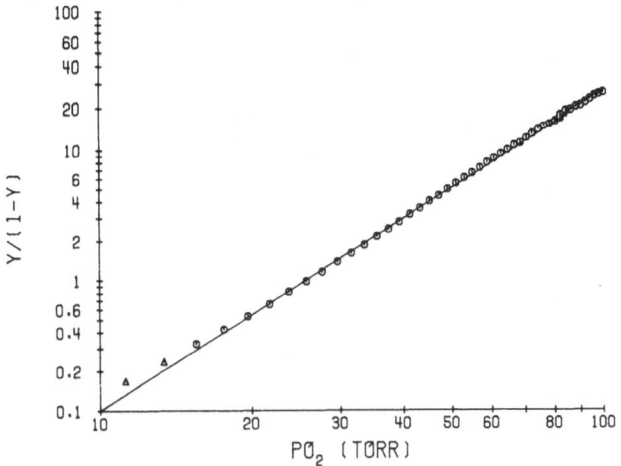

Fig. 2. Hill plot of data from diabetic adult of Fig. 1.

DISCUSSION

The nonlinear regression of the diabetic data of Fig. 1 is remarkably similar in its fit (RMS error, regions of deviation, etc.) to that of the normal adults fitted here and to fits of the Hill equation to other published OHEC data as discussed by us previously (O'Riordan et al., 1982). The Hill equation also fits the juvenile data in a similar way. Therefore, nonlinear regression of the Hill equation can be used to characterize the OHECs of normal and diabetic adults and juveniles. Accordingly, the average values found for the two parameters P_{50} and n, for each of the four groups, can be compared.

Table 4. Hill parameters for normal and diabetic adults and juveniles

GROUP	NUMBER	$P_{50} \pm$ SD	$n \pm$ SD
NA	33	26.2 ± 0.8	2.50 ± 0.07
DA	26	26.4 ± 0.9	2.45 ± 0.11
NJ	30	28.2 ± 1.2	2.38 ± 0.07
DJ	42	25.4 ± 1.1	2.35 ± 0.08

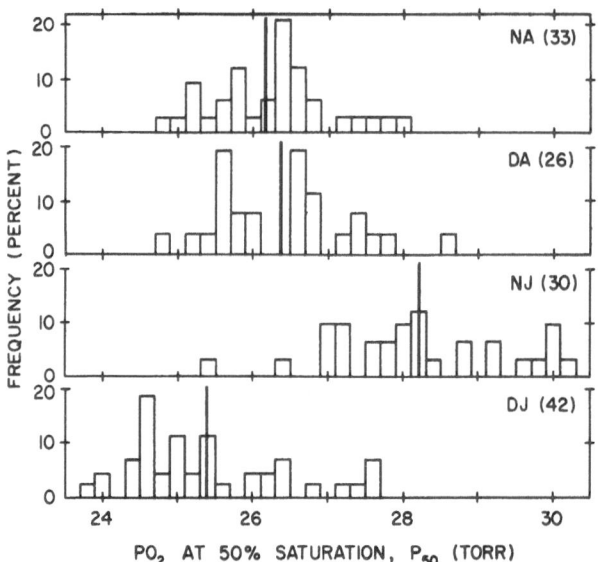

Fig. 3. Histograms of P_{50} for the four subject groups. NA, normal
         adults; DA, diabetic adults; NJ, normal juveniles; DJ,
         diabetic juveniles. The numbers in parentheses are the
         number of individuals in each group.

     Fig. 3 shows a significant ($p < 0.001$) shift in the P_{50} of the
juveniles with diabetes but this may have been partially compensated
for in vivo by the Bohr effect resulting from their lower arterial
pH (Table 1) and the right shift resulting from their elevated 2,3-
DPG (Table 1), as previously suggested by Ditzel (1979). Diabetes
does not seem to affect the P_{50} of adults but, again, the elevated
2,3-DPG may have compensated for the diabetic shift. Moreover, it
may not be appropriate to compare these two adult groups because of
the difference in age (Table 1). Also, unlike the juveniles, the
diabetic adults had a long history of diabetes with both micro- and
macroangiopathy. Possibly, the diabetic adults would have had a significantly different P_{50} if compared to normal adults of the same
age and with the same vascular complications.

     Fig. 4 shows no significant effect of diabetes on the Hill
coefficient n but a moderately significant ($p < 0.03$) difference in
n between adults and juveniles. Our values, however, were 12% and
8% lower, respectively, than the value ($n = 2.7$) for the standard
data of Severinghaus (1966), as noted previously for normal adults
(O'Riordan et al., 1982). Other investigators have also found a
lower value for n. Rossi-Bernardi et al. (1972) report a value of
2.6 for the data of Severinghaus (1966).

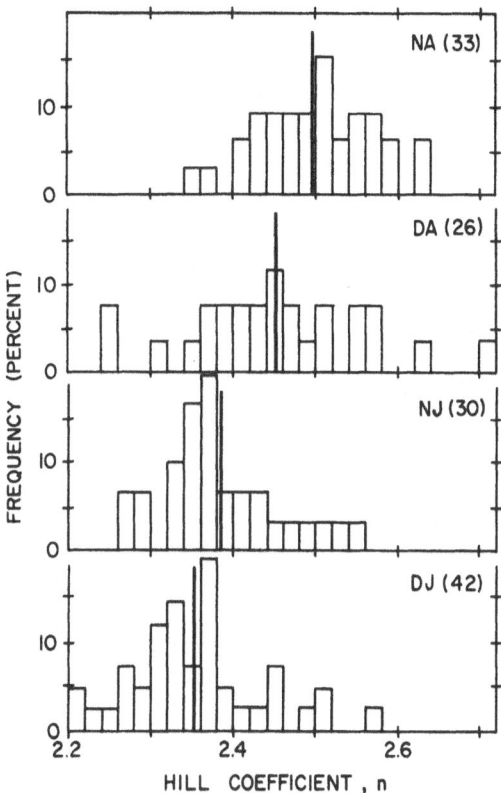

Fig. 4. Histograms of n for the four subject groups. NA, normal adults; DA, diabetic adults; NJ, normal juveniles; DJ, diabetic juveniles.

The OHECs computed using the average values of the parameters P_{50} and n, for the two juvenile groups, are shown in Fig. 5. The average effect of the left shift in the diabetic juveniles is small but statistically significant. One might expect that this shift would cause some affinity hypoxia in the diabetic juveniles unless compensated for by adaptive phenomena such as polycythemia or increased blood flow. No blood flow data for these individuals was available but polycythemia was not observed. Furthermore, experiments with dogs suggest that diabetics may have impaired oxygen autoregulation (Goldstick et al., 1981) making compensation appear less likely. Any hypoxic insult may therefore play an important role in the disease process. This hypoxia would tend to be a more severe insult because it would be transient and fluctuating and therefore uncompensated. Chronic hypoxia would probably play a less important role in the disease process. In fact diabetics undergo a shift in pH which partially compensates for any leftward shift of the OHEC. If not fully compensated, transient tissue hypoxia in diabetics could lead to vasodilatation and, possibly, to vasodilator-induced macro-

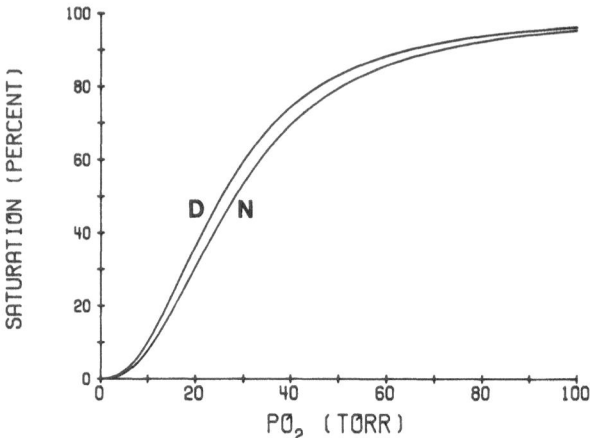

Fig. 5. Superimposed OHECs generated with the Hill equation using
         the average P_{50} and n for the two juvenile groups,
         diabetic (D) and normal (N).

molecular leakage in oxygen sensitive tissues such as the retina,
and this could lead to capillary occlusion and so on. These diabetic
changes in oxygen release may partly explain the observed microvascular dilatation in the preretinopathy stage of juvenile diabetes.

     In conclusion, nonlinear regression of the Hill equation
provides a means of characterizing the OHECs of normal and diabetic
adults and juveniles. Comparing the diabetic and normal juveniles,
there is a mean 2.7 Torr ($p < 0.001$) left shift in the fasting
(chronic) P_{50}s of diabetics.

ACKNOWLEDGEMENTS

     Supported in part by USPHS NIH grants EY 04085 and HL-17517,
Evanston Hospital, the Danish Heart Association, Novo Fund, and
Nordsk Insulin Foundation. We are also grateful for the early work
of Thomas L. Vaughan on this project.

REFERENCES

Bunn, H.F., and Briehl, R.W., 1970, The interaction of 2,3-Diphos-
     phoglycerate with various human hemoglobins, J. Clin. Invest.,
     49:1088-1095.
Ditzel, J., 1979, Changes in red cell release capacity in diabetes
     mellitus, Fed. Proc., 38:2484.

Ditzel, J., 1980, Affinity hypoxia as a pathogenic factor of micro-
    angiopathy with particular reference to diabetic retinopathy,
    Acta Endocrinol., 94 Suppl. 238:39-55.
Ditzel, J., Kawahara, R., Mourits-Andersen, T., Ostergaard, G.Z.,
    and Kjaergaard, J.J., 1981, Changes in blood glucose, glyco-
    sylated hemoglobin and hemoglobin-oxygen affinity following
    meals in diabetic children, Eur. J. Pediatr., 137:171-174.
Duvelleroy, M.A., Buckles, R.G., Rosenkaimer, S., Tung, C., and
    Laver, M.B., 1970, An oxyhemoglobin dissociation analyzer,
    J. Appl. Physiol., 28:227-233.
Garby, L., and Meldon, J., 1977, "The Respiration Functions of
    Blood", Plenum Medical Book Co., New York, p. 38.
Goldstick, T.K., Ernest, J.T., and Engerman, R.L., 1981, Impaired
    retinal vascular reactivity in diabetic dogs, Invest. Ophthal-
    mol. Vis. Sci., 20 ARVO Suppl.:92.
Hellegers, A.E., and Schruefer, J.J.P., 1961, Nomograms and empiri-
    cal equations relating oxygen tensions, percentage saturation
    and pH in maternal and fetal blood, Am. J. Obstet. Gynecol.,
    81:377-388.
Hill, A.V., 1910, The possible effects of the aggregation of the
    molecules of hemoglobin on its dissociation curves, J. Physiol.
    (Lond.), 40: (Proceedings) iv.
Hilpert, P., Fleischmann, R.G., Kempe, D., and Bartels, H., 1963,
    The Bohr effect related to blood and erythrocyte pH, Am. J.
    Physiol., 205:337-340.
O'Riordan, J.F., Goldstick, T.K., Ditzel, J., and Ernest, J.T.,
    1982, Characterization of oxygen-hemoglobin equilibrium curves
    using nonlinear regression of the Hill equation: Parameter
    values for normal adults, in press.
Perutz, M.F., 1970, Stereochemistry of cooperative effects on
    hemoglobin, Nature, 228:726-739.
Robinson, B., 1981, NLREG-nonlinear Regression Subroutine Package,
    (Evanston IL: Vogelback Computer Center Document No. 328 (Rev.
    B), Northwestern University).
Rossi-Bernardi, L., Roughton, F.J.W., Pace, M., and Coven, E., 1972,
    The effects of organic phosphates on the binding of CO_2 to
    human hemoglobin and on CO_2 transport in the circulating blood,
    in: "Oxygen Affinity of Hemoglobin and Red Cell Acid Base
    Status", M. Rorth, P. Astrup, eds., Academic Press, New York,
    pp. 224-235.
Roughton, F.J.W., 1964, Transport of oxygen and carbon dioxide, in:
    "Handbook of Physiology", Sec. 3 "Respiration", 3 vols., W.
    Fenn, H. Rahn, eds., American Physiological Society,
    Washington.
Roughton, F.J.W., DeLand, E.C., Kernohan, J.C., and Severinghaus,
    J.W., 1972, Some recent studies of the oxyhemoglobin dissocia-
    tion curve of human blood under physiological conditions and
    the fitting of the Adair equation to the standard curve, in:
    "Oxygen Affinity of Hemoglobin and Red Cell Acid Base Status",
    M. Rorth, P. Astrup, eds., Academic Press, New York, pp. 73-81.

Severinghaus, J.W., 1966, Blood gas calculator, J. Appl. Physiol.,
    21:1108–1116.
Winslow, R.M., Morrissey, J.M., Berger, R.L., Smith, P.D., and
    Gibson, C.C., 1978, Variability of oxygen affinity of normal
    blood: An automated method of measurement, J. Appl. Physiol.:
    Respir. Environ. Exercise Physiol., 45:289–297.
Woodson, R.D., 1979, Physiological significance of oxygen dissocia-
    tion curve shifts, Crit. Care Med., 7:368–373.

IMPROVED O_2 TRANSFER TO TISSUES DURING DEEP HYPOXIA IN RATS WITH

A LEFT-SHIFTED BLOOD O_2 DISSOCIATION CURVE.

Z. Turek[1], F. Kreuzer[1], B.E.M. Ringnalda[2], and
P. Scotto[2]

[1]Dept. of Physiology, Univ. of Nijmegen, Nijmegen
The Netherlands
[2]Dept. of Human Physiology, Univ. of Naples, Naples
Italy

INTRODUCTION

Oxygen is transported in blood while bound to a highly specialized carrier, the hemoglobin in the red cells. The loading of hemoglobin with O_2 in the lung depends on the partial pressure of O_2 in
the alveolar air. During hypoxic hypoxia this pressure remains lowered in spite of hyperventilation, so that if the total transfer of O_2
from the lung to tissues is to be guaranteed, either the loading of
blood with O_2 in the lung must be enhanced, or blood flow through
the lung or O_2 extraction in tissues must increase. It is well known
that the extraction of O_2 from blood has its limits which are different in different organs. During hypoxic hypoxia, an increase of
the loading ability of blood in the lung or of blood flow through
the lung is required to compensate for the decrease of Po_2 in the
alveolar air. The loading ability of blood can be enhanced by an
increase of blood O_2 carrying capacity accompanied by polycythemia,
or by an increase of blood O_2 affinity, i.e., a shift of the blood
O_2 dissociation curve (ODC) to the left. The former is the usual
reaction of many mammalian species to chronic hypoxia, with the
potential danger that excessive polycythemia may increase blood
viscosity to such a degree that flow may become limited. An increase
of blood O_2 affinity improves the loading in the lung but impairs
the unloading in tissues. Depending on the degree of hypoxia, an
increase of blood O_2 affinity can improve or impair the net O_2 transfer to tissues. A shift of the ODC to the left occurs physiologically during exposure to acute hypoxic hypoxia, due to respiratory
alkalosis. Later this is counteracted by an increased concentration

of 2,3-diphosphoglycerate in the red cells. The reports about the position of the ODC in human natives of high altitude are contradictory (see e.g. Samaja et al., 1979). But many animals living for generations at high altitude have an ODC located to the left when compared with animals of similar body size living at sea level (Monge and Whittembury, 1976). An organism may succeed in maintaining the same transfer of O_2 from the lung to the tissues, in spite of a lower O_2 gradient between blood leaving and re-entering the lung, by decreasing the resistance to O_2 transfer within this section or, in other words, increasing the conductance.

Administration of sodium cyanate (NaOCN) to rats in drinking water induces a dramatic shift of the ODC to the left (Eaton et al., 1974). In our previous work (Turek et al., 1978a) we have shown that in NaOCN rats breathing room air or 14.9% O_2 in N_2 the mixed-venous Po_2 was lower than in control animals, whereas during breathing 8.0 or 5.6% O_2 in N_2 the mixed-venous Po_2 was higher in the NaOCN rats. The calculated (a-v) O_2 difference followed the same trend. Furthermore the conductance of O_2 from lung to tissue was lower in the NaOCN rats at normoxia and moderate hypoxia but larger during severe hypoxia when compared with control rats breathing the same inspiratory mixture. This was mainly due to an increase of the apparent (physical and chemical) solubility coefficient of O_2 in blood, the capacitance coefficient ß (Piiper et al., 1971). This defines the volume of O_2 transferred from the lung to the tissues per unit of blood volume and is, at least during hypoxic hypoxia, practically identical to the slope of the line connecting the arterial and venous points on the ODC. In the work of Turek et al. (1978a) no individual values of the capacitance coefficient were available and calculations were based on various assumptions. Besides, the lowest inspiratory mixture where the most significant differences occurred, was so low that resuscitation of many, mainly control, rats was necessary. Therefore new experiments were performed in order to estimate the capacitance coefficient and the conductance in individual animals at normoxia and several degrees of hypoxia as quantitative indexes of the efficiency of the O_2 transfer from the lung to the tissues. The lowest inspiratory mixture was chosen so that the majority of rats survived without apparent difficulties. In case of difficulties no resuscitation was performed. Rats with a normal and a left-shifted ODC were compared.

METHODS

Male rats of Wistar strain (initial weight about 200 g) were divided into two groups. One group received a 0.5% solution of NaOCN in drinking water for 14 days; control animals were given NaCl in water so that the concentration of Na^+ was identical. During the last 24 h before the measurements all animals were given tap water. This scheme follows that of Eaton et al. (1974). Five inspiratory mixtures were used during the measurements: room air, 14.8, 8.4, 7.0, and 5.9% O_2 in N_2. The rats breathed each mixture for 30 min. The animals were anesthetized with pentobarbital and heparinized.

The technique of the measurements was desribed in detail before (Turek et al., 1978a). The only difference was that not only the arterial and mixed-venous blood gases and pH were measured this time but also samples for blood O_2 content were taken to be analyzed according to Roughton and Scholander (1943). Measured data allowed us to calculate O_2 consumption, CO_2 production, (a-v) O_2, cardiac output, stroke volume, alveolar ventilation and the ideal alveolar Po_2 calculated from the alveolar gas equation using the arterial instead of alveolar Pco_2.

CALCULATION OF CONDUCTANCE AND CAPACITANCE COEFFICIENTS

Oxygen is transported by a pressure difference from the ambient air to the oxidizing enzymes in the tissues. This driving pressure is consumed along a series of transfer resistances, the O_2 flux being equal in all total cross-sections at steady state (Kreuzer, 1967). The transfer of O_2 in the blood from the lungs to the tissues involves only the section of the "cascade" between the pulmonary capillaries and the mixed-venous blood. The resistance of this part of the O_2 pathway is $(P_c'o_2-P_{\bar{v}}o_2)/\dot{V}o_2$, where $P_c'o_2$ is the pulmonary end-capillary Po_2, $P_{\bar{v}}o_2$ the mixed-venous Po_2 and $\dot{V}o_2$ the total body O_2 consumption. The conductance is the reciprocal of resistance and is equal to $\dot{V}o_2/(P_c'o_2-P_{\bar{v}}o_2)$. In the ideal lung $P_c'o_2$ equals P_ao_2, the arterial Po_2. In the nonideal lung during hypoxia these two values will be practically identical too. According to the Fick principle, $\dot{V}o_2 = \dot{Q}$ (a-v) O_2, where $\dot{Q}$ is cardiac output and (a-v) O_2 arterio-venous O_2 concentration difference $(C_ao_2-C_{\bar{v}}o_2)$. Thus, during hypoxia the conductance equals $\dot{Q}$ $(C_ao_2-C_{\bar{v}}o_2)/(P_ao_2-P_{\bar{v}}o_2)$. The second term of the product is the mean slope of the ODC and is equal to the apparent solubility coefficient of O_2 in blood, the capacitance coefficient.

At normoxia, however, $P_c'o_2$ is not close to P_ao_2 mainly due to the effect of the physiological shunt, but rather equals the ideal alveolar Po_2, $P_{Ai}o_2$. Therefore conductance equals $\dot{Q}$ $(C_ao_2-C_{\bar{v}}o_2)/(P_{Ai}o_2-P_{\bar{v}}o_2)$, the formula used in this paper. However, now the ratio on the right side does not equal the mean slope of the ODC, since $P_{Ai}o_2$ and C_ao_2 do not define any point on the ODC. On the other hand, $(C_ao_2-C_{\bar{v}}o_2)/(P_{Ai}o_2-P_{\bar{v}}o_2) = (1-Q_s/Q)(C_c'o_2-C_{\bar{v}}o_2)/(P_c'o_2-P_{\bar{v}}o_2)$, as derived from the shunt equation, where Q_s is the physiological shunt, $C_c'o_2$ is the pulmonary end-capillary O_2 content and $P_c'o_2$ equals $P_{Ai}o_2$. The ratio $(C_c'o_2-C_{\bar{v}}o_2)/(P_c'o_2-P_{\bar{v}}o_2)$ is the mean slope of the ODC between the "pulmonary-end capillary" and the mixed-venous point. The conceptual question or what actually should be called a capacitance coefficient cannot be solved here, not knowing the nature (due to $\dot{V}_A/\dot{Q}$ inequalities or anatomical), nor the location (intra-vs. extrapulmonary) of the shunt. In this paper we used the formula $(C_aO_2-C_{\bar{v}}O_2)/(P_{Ai}O_2-P_{\bar{v}}O_2)$, mainly for practical reasons. Therefore the capacitance coefficient at normoxia is not identical to the mean slope of the physiological ODC and also includes the physiological shunt. The mean physiological shunt was almost identical in NaOCN and control rats (2.9 and 3.0% of cardiac output, respectively) and thus could not be responsible for any difference between groups.

RESULTS

In order to check whether the administration of NaOCN indeed increased blood O_2 affinity, arterial and mixed-venous O_2 saturations were calculated. Blood O_2 capacity was estimated from hematocrit, assuming the same ratio between hematocrit and blood O_2 capacity as observed in previous measurements. From arterial and venous P_{O_2}, pH and saturation, P_{50} was calculated using the Hill plot (for details see Turek et al., 1978a). Only rats in which the arterial saturation was between 50 and 80% and the venous saturation was between 20 and 50% were used in the calculation of P_{50}. The mean P_{50} in NaOCN treated rats was (mm Hg, mean $\pm$ SD, 20 rats) 24.4 $\pm$ 2.8, the Hill number was 2.94 $\pm$ 0.51. In 9 control animals, the mean P_{50} and Hill number were 37.9 $\pm$ 2.9 and 2.86 $\pm$ 0.44, respectively. The difference in P_{50} is highly significant (P < 0.001), that in the Hill number is not.

The most important results of measurements at several levels of oxygenation (means $\pm$ SD) are shown in Tables 1 and 2. Because of the difference in body weight, $\dot{V}_{O_2}$, $\dot{Q}$ and conductance were calculated per kg of body weight, in order to make results in NaOCN and control rats comparable. From results not shown in the tables it appeared that the ideal alveolar-arterial O_2 pressure difference was significantly larger (at P < 0.01) in NaOCN rats during breathing room air and 14.8% O_2 in N_2. In the NaOCN and control rats it equalled 27.2 $\pm$ 4.6 (9) and 12.5 $\pm$ 5.1 (9), and at 14.8% O_2 the corresponding values were 13.0 $\pm$ 4.2 (12) and 6.7 $\pm$ 5.0 (11) mm Hg, respectively. Alveolar ventilation (not listed in the tables) and arterial hematocrit were not significantly different between groups breathing the same inspiratory mixture.

DISCUSSION

The results shown in Table 1 in general show the same trend as described before (Turek et al., 1978a). However, even though $P_{\bar{v}}O_2$ during breathing the two most hypoxic inspriratory mixtures tended to be higher in the NaOCN than in the control rats, no significant difference was found here. Also the results of O_2 content were similar to those in our previous paper (Turek et al., 1978b) with

Table 1. Body weight, O$_2$ consumption and blood gas values of NaOCN and control rats breathing various O$_2$ concentrations in N$_2$.

Inspir. O$_2$ conc.(%)	Group	Weight g	$\dot{V}_{O_2}$ (ml/(min.kg))	$P_{A_{i_{O_2}}}$ (mmHg)	Pa_{O_2} (mmHg)	$P_{\bar{v}_{O_2}}$ (mmHg)	$P_{a_{CO_2}}$ (mmHg)	$[H^+]_a$ (mmol/l)	$[H^+]\bar{v}$ (mmol/l)
20.9	NaOCN	219	16.51	103.1	75.8	34.9	42.2	3.86	4.31
		±11	±1.55	±3.7	±4.1	±2.8	±2.8	±0.16	±0.15
		(10)	(10)	(9)	(9)	(10)	(9)	(10)	(10)
		***)			***)	***)			
	CONTR.	275	16.75	103.9	91.4	43.2	41.3	3.81	4.25
		±13	±1.83	±4.1	±3.2	±2.9	±2.1	±0.29	±0.25
		(9)	(9)	(9)	(9)	(9)	(9)	(9)	(9)
14.8	NaOCN	210	18.21	65.8	52.8	25.7	36.5	3.24	3.86
		±17	±2.01	±4.4	±3.5	±3.9	±2.2	±0.25	±0.23
		(12)	(12)	(12)	(12)	(12)	(12)	(12)	(12)
		***)				**)			
	CONTR.	268	18.47	65.8	59.4	32.0	36.1	3.28	3.73
		±11	±2.94	±3.1	±5.1	±4.9	±1.5	±0.22	±0.25
		(12)	(12)	(11)	(12)	(12)	(11)	(12)	(12)
8.4	NaOCN	195	21.40	33.0	32.5	20.7	25.1	2.71	3.33
		±12	±2.10	±4.8	±4.3	±3.0	±2.6	±0.17	±0.25
		(10)	(10)	(10)	(10)	(10)	(10)	(10)	(10)
		***)	***)				**)		
	CONTR.	262	16.07	33.7	35.6	22.9	25.9	3.09	3.42
		±12	±3.40	±4.8	±2.2	±1.8	±2.3	±0.23	±0.27
		(8)	(8)	(8)	(8)	(8)	(8)	(8)	(6)
7.0	NaOCN	192	19.0	30.9	27.1	15.7	20.6	2.45	3.03
		±12	±4.7	±4.0	±3.5	±2.2	±2.3	±0.18	±0.33
		(8)	(8)	(7)	(8)	(8)	(7)	(8)	(8)
		***)	**)				**)	*)	
	CONTR.	247	11.7	34.3	27.7	13.5	20.7	2.96	3.50
		±24	±3.9	±3.1	±2.4	±4.3	±2.0	±0.34	±0.39
		(11)	(11)	(11)	(11)	(11)	(11)	(10)	(10)
5.9	NaOCN	183	18.1	26.8	22.3	13.7	17.6	2.56	3.27
		±18	±2.7	±3.0	±3.6	±2.3	±2.3	±0.27	0.40
		(9)	(9)	(9)	(9)	(9)	(9)	(9)	(9)
		***)	***)		**)		**)	*)	
	CONTR.	252	11.0	29.4	29.3	12.0	16.9	4.73	5.29
		±16	±3.4	±2.5	±2.5	±5.8	±1.3	±2.17	±2.17
		(6)	(6)	(6)	(6)	(6)	(6)	(6)	(6)

*) P < 0.05   **) P < 0.01   ***) P < 0.001

Table 2. Blood O_2 content, cardiac output, capacitance coefficient
and conductance of O_2 transfer in NaOCN and control rats.

Inspir. O_2 conc.(%)	Group	Ca_{O_2} (vol%)	$C\bar{v}_{O_2}$ (vol%)	$(a-v)_{O_2}$ (vol %)	$\dot{Q}$ (ml/min.kg)	Heart rate	Capac. coeff. (ml/(mmHg.l))	Conductance (ml/(mmHg. min.kg))
20.9	NaOCN							
		21.1	12.8	8.3	202	381	0.121	24
		±0.6	±1.3	±1.3	±32	±12	±0.016	±2
		(10)	(10)	(10)	(10)	(10)	(9)	(9)
	CONTR.	*)	*)			**)	**)	*)
		20.3	11.3	9.1	187	408	0.163	28
		±0.7	±1.3	±1.1	+32	±25	±0.031	±4
		(9)	(9)	(9)	(9)	(9)	(9)	(9)
14.8	NaOCN	20.1	12.1	8.0	235	400	0.302	70
		±1.2	±1.0	±1.3	±4.9	±22	±0.070	±18
		(12)	(12)	(12)	(12)	(12)	(12)	(12)
		***)	***)					
	CONTR.	16.8	7.8	9.0	211	404	0.341	78
		±185	±1.6	±1.62	±49	±27	±0.080	±33
		(12)	(12)	(12)	(12)	(12)	(12)	(12)
8.4	NaOCN	16.5	8.8	7.8	306	364	0.689	215
		±2.9	±3.2	±2.1	±136	±17	±0.203	±149
		(10)	(10)	(10)	(10)	(10)	(10)	(10)
		***)	***)	**)				
	CONTR.	7.7	2.7	5.1	322	372	0.437	139
		±1.7	±0.8	±1.2	±55	±38	±0.073	±34
		(8)	(8)	(9)	(8)	(8)	(8)	(8)
7.0	NaOCN	14.8	6.6	8.2	239	404	0.762	188
		±1.4	±1.7	±1.2	±72	±30	±0.193	±94
		(8)	(8)	(8)	(8)	(8)	(8)	(8)
		***)	***)	***)			***)	*)
	CONTR.	7.6	2.1	5.5	217	400	0.437	100
		±1.6	±1.3	±1.2	±70	±26	+0.142	±62
		(11)	(11)	(11)	(11)	(11)	(11)	(11)
5.9	NaOCN	12.0	4.4	7.7	250	409	0.922	232
		±2.2	±1.1	±2.2	±63	±42	±0.120	±70
		(9)	(9)	(9)	(9)	(9)	(9)	(9)
		***)	***)	**)			***)	***)
	CONTR.	5.8	1.6	4.3	268	368	0.293	81
		±1.3	±1.0	±0.9	±93	±53	±0.135	±58
		(6)	(6)	(6)	(6)	(6)	(6)	(6)

*) $P < 0.05$ **) $P < 0.01$ ***) $P < 0.001$

the exception that the differences in (a-v) O_2 were significant
here only during severe hypoxia. Again the administration of NaOCN
decreased heart rate at normoxia, the cause and significance of this
finding being unknown. For the lowest O_2 concentration, no difference in cardiac output between groups was found. It was confirmed
that in rats with a left-shifted ODC the alveolar-arterial O_2 pressure gradient is larger than in control animals but unlike our previous paper (Turek et al., 1978a), this was significant also during
breathing 14.8% O_2 in N_2. As the physiological shunt during normoxia
was almost identical in both groups, the most probable explanation
is that this change of alveolar-arterial O_2 gradients is due solely
to the effect of the shift of the ODC. A theoretical explanation has
been offered elsewhere (Frans et al., 1979; Turek and Kreuzer, 1981).

Capacitance coefficient and also conductance were larger in
control than in NaOCN rats at normoxia; the opposite occurred during
severe hypoxia. Therefore a shift of the ODC to the left seems to
impair O_2 transfer at normoxia but to improve it at severe hypoxia
(Turek et al., 1973; Turek and Kreuzer, 1976; Bakker et al., 1976).
In order to show the individual values of capacitance coefficient
and conductance, they were plotted as a function of the arterial P_{O_2}
(Figs. 1 and 2). In both groups capacitance coefficient and conductance increase during transition from normoxia to hypoxia substantially but this increase is much more marked in the NaOCN than in the

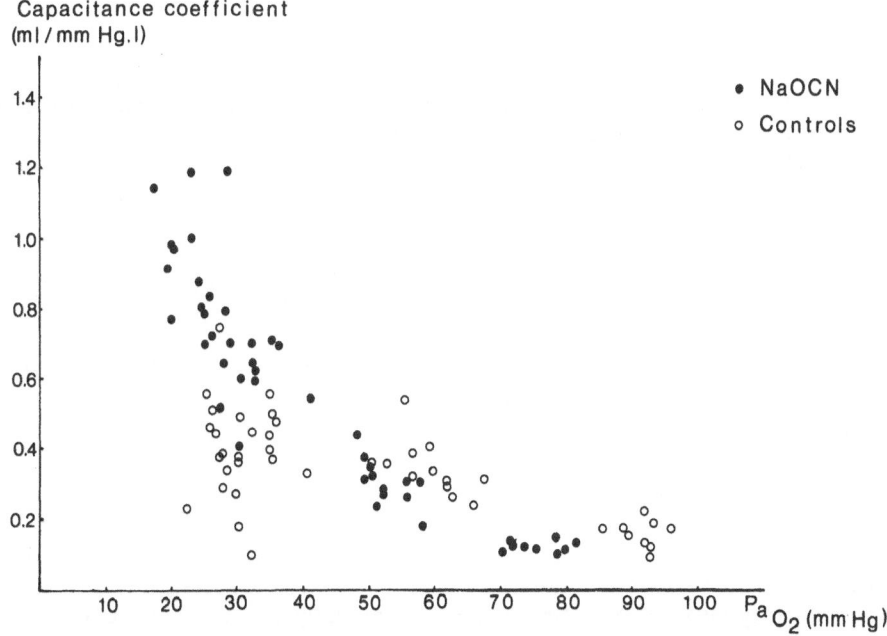

Fig. 1.   Capacitance coefficient (ml/(mm Hg·l) of rats with normal
          (controls) and left-shifted (NaOCN) blood O_2 dissociation
          curves plotted as a function of arterial P_{O_2}.

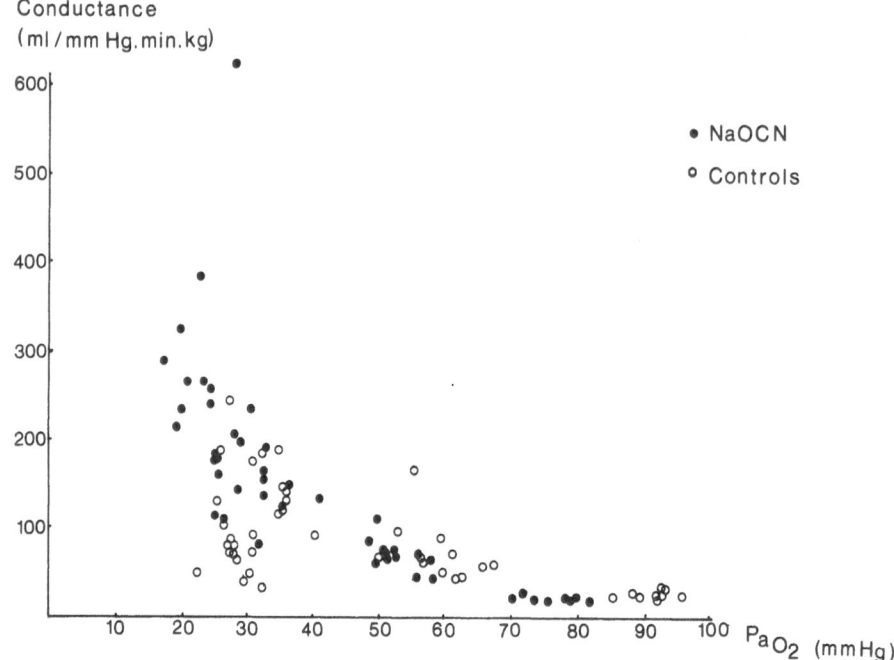

Fig. 2.   Conductance of O_2 transfer (ml/(mm Hg·min·kg)) from the
          lung to tissues in rats with normal (controls) and left-
          shifted (NaOCN) blood O_2 dissociation curves, plotted as a
          function of arterial Po_2.

control rats. The change of capacitance coefficient and conductance
at normoxia is very small when compared with that during hypoxia,
but nevertheless can have consequences for work performance, which
was found to be about 10% lower in rats with a left-shifted ODC
(Woodson et al., 1973). Indeed, at normoxia the NaOCN rats have not
only a lower $P_{\bar{v}}O_2$ but also a lower P_aO_2 due to an enlarged alveolar-
arterial gradient, so that their tissue Po_2 is expected to be lower
than in control animals.

     Both capacitance coefficient and conductance are substantially
larger in the NaOCN than in control animals during deep hypoxia, the
differences being significant at $P < 0.001$ at arterial Po_2 lower than
40 mm Hg. Similar results were obtained previously when the isolated
turtle heart was perfused with blood with low (left shift of the
ODC) or high (right shift of the ODC) concentration of 2,3-diphos-
phoglycerate in the red cells (Scotto et al., 1981). This probably
explains why control rats cannot maintain their O_2 consumption
during deep hypoxia while NaOCN rats manage quite well (Table 1).
In the absence of significant differences in alveolar ventilation
and cardiac output the most important difference between NaOCN and
control rats concerning their transfer of O_2 is the difference in

the capacitance coefficient. Its increase may explain the superiority
of NaOCN rats against controls in surviving at very severe hypoxia
(Eaton et al., 1974; Penney and Thomas, 1975; Turek et al., 1978b).
It may be concluded that the present experiments confirm that a
shift of the ODC to the left impairs O$_2$ transfer at normoxia but
improves it during deep hypoxia. However, when assessed quantitatively, this detrimental effect at normoxia is minor when compared
with the large beneficial effect during severe hypoxic hypoxia.
There is every reason to expect that a shift of the ODC to the
right would have an opposite effect. The conclusions derived from
the present results apply only to acute hypoxic hypoxia. The usefulness of a left shift of the ODC during chronic hypoxia is supported
by the observation that the blood of many species native to high
altitude has high O$_2$ affinity.

REFERENCES

Bakker, J.D., Gortmaker, G.C., Vrolijk, A.D.M., and Offerijns,
     F.J.G., 1976, The influence of the position of the oxygen dis-
     sociation curve on oxygen-dependent functions of the isolated
     perfused rat liver. I. Studies at different levels of hypoxic
     hypoxia, Pflügers Arch., 362:21-31.
Eaton, J.W., Skelton, T.D., and Berger, E., 1974, Survival at
     extreme altitude: protective effect of increased hemoglobin-
     oxygen affinity, Science, 183:743-744.
Frans, A., Turek, Z., Yokota, H., and Kreuzer, F., 1979, Effect of
     variations in blood hydrogen ion concentration on pulmonary
     gas exchange of artificially ventilated dogs, Pflügers Arch.,
     380:35-39.
Kreuzer, F., 1967, Transport of O$_2$ and CO$_2$ at altitude, in: "Exer-
     cise at Altitude", R. Margaria, ed., Excerpta Medica Foundation,
     Amsterdam.
Monge, C., and Whittembury, J., 1976, High altitude adaptations in
     the whole animal, in: "Environmental Physiology of Animals",
     J. Bligh, J.L. Cloudsley-Thompson, A.G. Macdonald, eds., Black-
     well Scientific Publications, Oxford.
Penney, D., and Thomas, M., 1975, Hematological alterations and
     response to acute hypobaric stress, J. Appl. Physiol., 39:
     1034-1037.
Piiper, J., Dejour, P., Haab, P., and Rahn, H., 1971, Concepts and
     basic quantities in gas exchange physiology, Respir. Physiol.,
     13:292-304.
Roughton, F.J.W., and Scholander, P.F., 1943, Micro gasometric esti-
     mation of the blood gases. I. Oxygen, J. Biol. Chem., 148:
     541-550.
Samaja, M., Veicsteinas, A., and Cerretelli, P., 1979, Oxygen affi-
     nity of blood in altitude Sherpas, J. Appl. Physiol.: Respir.
     Environ. Exercise Physiol., 47:337-341.

Scotto, P., Turek, Z., Licheri, D., and Ringnalda, B.E.M., 1981,
    Blood O_2 dissociation curve and O_2 transport to the isolated
    and perfused turtle heart, in: "Oxygen Transport to Tissue",
    Adv. Physiol. Sci. Vol. 25, A.G.B. Kovách, E. Dóra, M. Kessler,
    I.A. Silver, eds., Pergamon Press, Akadémiai Kiadó, Budapest.
Turek, Z., and Kreuzer, F., 1976, Effect of a shift of the oxygen
    dissociation curve on myocardial oxygenation at hypoxia, in:
    "Oxygen Transport to Tissue II", Adv. Exp. Med. Biol., Vol. 75,
    J. Grote, D. Reneau, G. Thews, eds., Plenum Press, New York-
    London.
Turek, Z., and Kreuzer, F., 1981, Effect of shifts of the O_2 disso-
    ciation curve upon alveolar-arterial O_2 gradients in computer
    models of the lung with ventilation-perfusion mismatching,
    Respir. Physiol., 45:133-139.
Turek, Z., Kreuzer, F., and Hoofd, L.J.C., 1973, Advantage or dis-
    advantage of a decrease of blood oxygen affinity for tissue
    oxygen supply at hypoxia, Pflügers Arch., 342:185-197.
Turek, Z., Kreuzer, F., and Ringnalda, B.E.M., 1978a, Blood gases
    at several levels of oxygenation in rats with a left-shifted
    blood oxygen dissociation curve, Pflügers Arch., 376:7-13.
Turek, Z., Kreuzer, F., Turek-Maischeider, M., and Ringnalda, B.E.M.,
    1978b, Blood O_2 content, cardiac output, and flow to organs at
    several levels of oxygenation in rats with a left-shifted
    blood oxygen dissociation curve, Pflügers Arch., 376:201-207.
Woodson, R.D., Wranne, B., and Detter, J.C., 1973, Effect of in-
    creased blood oxygen affinity on work performance of rats,
    J. Clin. Invest., 52:2717-2714.

BRAIN

CHANGES IN CEREBRAL OXYGEN TENSION AND RED CELL CONTENT ON SENSORY STIMULATION

I.S. Longmuir, J.A. Knopp, and J.L. Pittman

Department of Biochemistry, North Carolina State
University, Raleigh, North Carolina 27650-5050, USA

According to current studies, the cerebral cortical circulation shows well developed "oxygen autoregulation" to the extent that local increases in oxygen consumption in the sensory cortex produced by peripheral stimulation are accompanied by increases in blood flow so finely adjusted that no local changes in Po_2 can be detected (Leniger-Follert and Hossmann, 1979). If indeed there is no local fall in Po_2, then hypoxia cannot be the signal for increased blood flow, and it is necessary to postulate some other mechanism similar to the old "axon reflex" to account for this type of autoregulation. However, if the signal involves a change in Po_2 in a very small volume, perhaps a single neurone and occurring for a period of less than one second, then polarographic electrodes might not have sufficient temporal and spatial resolution to measure this. With our T.V. system we can measure changes in oxygen tension and microregional red cell mass in cylinders of tissue of 10 um diameter and approximately 150 um deep with a time resolution of 33 msecs.

We have applied this method to a study of the changes in oxygen tension and red cell mass in the cat's sensory cortex following cold paw stimulation.

MATERIALS AND METHODS

Windows are inserted in anesthetized cats' skulls by the method of Eke et al. (1979). One hundred mg/kg of pyrenebutyric acid is injected intraperitoneally and the cats are paralyzed with an adequate dose of Flaxedil. The cats are then ventilated with a minute volume sufficient to maintain normocapnia. The window is then placed under a 2.3 x objective of our television system

211

(Benson et al., 1980). An additional visible light source is brought in at the side to measure red cell mass changes by reflected light.

The images obtained from illumination by U.V. light used to measure local Po_2 and by visible light to measure red cells are recorded on tape for later data reduction. After a period of stabilization, data are recorded for ten seconds before immersing the contralateral paw in ice water for ten seconds and until fifteen seconds after drying.

The data are later digitized and the differences between control frames before stimulation and those after are calculated. The data can then be expressed numerically or by color contrast.

RESULTS

The data from light scattering give a measure of the changes in light absorption by changes in red cell absorption without further correction. The data from fluorescence measurement, however, must first be corrected for the change in red cell absorption before they can be expressed as a change in local Po_2.

The first step in the examination of the difference displays is to exclude changes due to movement of vessels. These are characterized by areas of increased density closely associated with similar areas of decreased density. After deleting these artifacts, which are not present in all difference plots, a number of areas of varying sizes remain showing decreases in light scattering and increases in fluorescence greater than 2%.

Thus, cold paw stimulation does result in transient, higly localized changes in blood flow and a fall in local Po_2 in tissue volumes which are not congruent.

ACKNOWLEDGEMENT

This work was supported in part by NIH Grant HL16828.

REFERENCES

Benson, D.W., Knopp, J.A., and Longmuir, I.S., 1980, Intracellular oxygen measurements of mouse liver cells using quantitative fluorescence video microscopy, Biochim. Biophys. Acta, 591:187.

Eke, A., Hutiray, G., and Kovách, A.G.B., 1979, Induced hemodilution detected by reflectometry for measuring microregional blood flow and blood volume in cat brain cortex, Am. J. Physiol., 236H:759.

Leniger-Follert, E., and Hossmann, K.-A., 1979, Simultaneous measurements of microflow and evoked potentials in the somatomotor cortex of the cat brain during specific sensory activation, Pflügers Arch., 380:85.

RELATIONS BETWEEN Po_2 AND NEURONAL ACTIVITY IN HIPPOCAMPAL SLICES[*]

D. Bingmann[1], G. Kolde[2], and H.G. Lipinski[1]

[1]Institute of Physiology and
[2]Institute of Cytobiology, University of Münster, FRG

It is well known that the oxygen supply of the brain is characterized by typical profiles of Po_2 (P_tO_2) in the tissue (Lübbers, 1969) and that the function of the CNS widely depends on an adequate distribution of P_tO_2. Hyperoxic P_tO_2 values, on the one hand, can induce seizures (cf. Bean, 1945; Lehmenkühler et al., 1978). During hypoxia, on the other hand, Speckmann and Caspers (1974) observed a depolarization of neurons in the spinal cord and in the brain cortex which was accompanied by a transient rise of neuronal activity until spike generation was blocked. Different mechanisms may contribute to this dependency of neuronal functions on Po_2. (i) Yamamoto and Kurokawa (1970) described that hypoxia modified the energy state of neurons in vitro. (ii) Vyskocil et al. (1972), Morris (1974), Silver (1977) and Lehmenkühler et al. (1981) have shown that the oxygen deficiency alters the ionic milieu of the extracellular space which may be due to neuronal and/or glial dysfunctions as well as to changes of transport rates of ions via microcirculation. (iii) Finally, synaptic inputs e.g. from chemoreceptors can alter the neuronal activity. As a whole, during changes of Po_2 neurons in vivo are affected by a great number of factors which can hardly be controlled simultaneously.

For a detailed analysis of the dependency of neurons on Po_2, better controlled experimental conditions appeared to be a prerequisite. Clearer conditions can be partly realized in hippocampal slices because (1) the excised tissue is superfused in vitro by a defined salt solution with constant ion concentrations. (2) Problems arising from changes of microcirculation are eliminated and (3) Po_2

[*]Supported by DFG (Bi 278/1-3)

dependent fluctuations of the synaptic input in slice preparations
probably are of minor importance. However, complex changes of P_tO_2
and of the ionic milieu may still exist in the slices when Po_2 in
the bath (P_BO_2) is altered. Therefore, in the present experiments
responses of hippocampal neurons to changes of Po_2 in vitro were
studied and the measurements were completed by investigations on
profiles of P_tO_2 and of the free potassium concentration in the
extracellular space $[K^+]_e$ of the excised tissue at low and high
P_BO_2 values.

METHODS

     The experiments were carried out on adult guinea pigs (300 –
500 g). In ether anaesthesia the brain was removed and the hippocampus was exposed. Slices (300 – 800 um thick) were cut almost perpendicular to the longitudinal axis of the hippocampus by means of a
handheld razor blade or with the aid of an Oxford Vibratome. The
slices were preincubated at a temperature of 28°C in a balanced salt
solution (BSS) and equilibrated with 5% CO_2 in O_2 (concentrations
in mmol·l^{-1}: NaCl 124, KCl 5, KH_2PO_4 1.24, $MgSO_4$ 1.3, $NaHCO_3$ 26,
Glucose 10, $CaCl_2$ 0.75 (cf. Yamamoto, 1972). After preincubation the
slices were transferred to a constant flow chamber mounted on an
inverted microscope. Within this chamber the excised tissue was
placed on the glass bottom and superfused by thermostabilized BSS
(30 – 36°C) at a rate of ca. 10 ml/min. The recording chamber contained 5 – 6 ml. In the superfusate the Ca^{++} concentration was elevated to 2.6 mmol·l^{-1} and bicarbonate was replaced by HEPES when
the BSS was equilibrated with air, N_2 or O_2. In an analyzer Po_2, pH
and Pco_2 of that BSS which had superfused the slices were measured
continuously. Neurons were impaled under visual control by glass
microelectrodes with tip diameter of less than 0.5 um. The resistances of these microelectrodes filled with 3 mmol·l^{-1} KCl amounted
to 60 – 100 megohms With the aid of a bridge circuit, current was
injected through the electrode which allowed either to stimulate the
neurons "electrically", to vary the membrane potential or to detect
changes of the membrane resistance.

     Profiles of Po_2 and of $[K^+]_e$ in hippocampal slices were
measured in a cylindrical incubation chamber (volume 6 ml), where
the excised tissue was placed on a nylon mesh. In this chamber the
lower surface was in direct contact with the BSS which perfused the
reservoir at a rate of 4 ml/min. The upper surface was exposed to
humidified gas which was also used for the equilibration of the incubation medium. For measuements of Po_2 in the tissue double barrelled platinum microelectrodes with tip diameters of 1 – 5 um were
used (cf. Lehmenkühler et al., 1976). At a polarizing voltage of
600 mV these electrodes passed currents of less than 1 nA when the
calibration solution was equilibrated with room air.

Potassium activity was measured by means of double barrelled microelectrodes (cf. Lux and Neher, 1973). The tip of one channel of the (theta-glass) microelectrode contained a K^+-selective ligand (Corning Nr. 477317). The other channel was filled with BSS and served as the reference electrode. Both channels were connected via Ag/AgCl wire to the inputs of a high impedance amplifier. The measured values were transformed into the concentration range. Thus, the free extracellular K^+ concentration $[K^+]_e$ was obtained.

RESULTS AND DISCUSSION

As outlined above, the results presented here were obtained in two experimental series. In a first series reactions of CA3 neurons to changes of Po_2 in the bath solution were recorded. The results of these experiments will be described in a first chapter. For a further clarification of the elementary mechanisms underlying the neuronal reactions, profiles of P_to_2 and of $[K^+]_e$ were measured at different P_BO_2 values. The results of this series are presented in a second chapter.

1a. Reactions of CA3 Neurons in vitro to an Increase of P_BO_2 from Normoxic to Hyperoxic Values

During the superfusion of hippocampal slices with air equilibrated balanced salt solution (BSS; P_BO_2 150 mm Hg, Pco_2 3 mm Hg, pH 7.4, 32°C) intracellular recordings from CA3 neurons rarely showed fluctuations of the resting membrane potential (RMP) resembling postsynaptic potentials (PSP) or spontaneous action potentials (AP). When the RMP amounted to at least −50 mV, P_BO_2 was elevated by perfusing the recording chamber with O_2-equilibrated BSS. This rise of P_BO_2 caused two different phenomena (Bingmann et al., 1982).

(i) With the rise of P_BO_2 the majority of neurons depolarized and the membrane resistance (RM) determined by current injections through the recording microelectrode increased as long as P_BO_2 was elevated. Typical examples of such long lasting depolarizations are given in Fig. 1A and 1B. The amplitudes of the RMP and RM shifts were found to be widely constant in single cells when the effects of further P_BO_2 elevations were tested. In different cells, however, the fluctuations of RMP and RM varied considerably. Thereby, the amplitudes of the RMP shifts were clearly related to changes of RM as shown in Fig. 1C. In this graph the maximum amplitudes of the long lasting depolarizations were plotted against the proper changes of RM. The RM of the neurons found at a P_BO_2 of 150 mm Hg was set 100%. The evaluation demonstrates a relation between both parameters which suggests that the permeability of K^+ and/or Cl^- decreased with the elevation of P_BO_2.

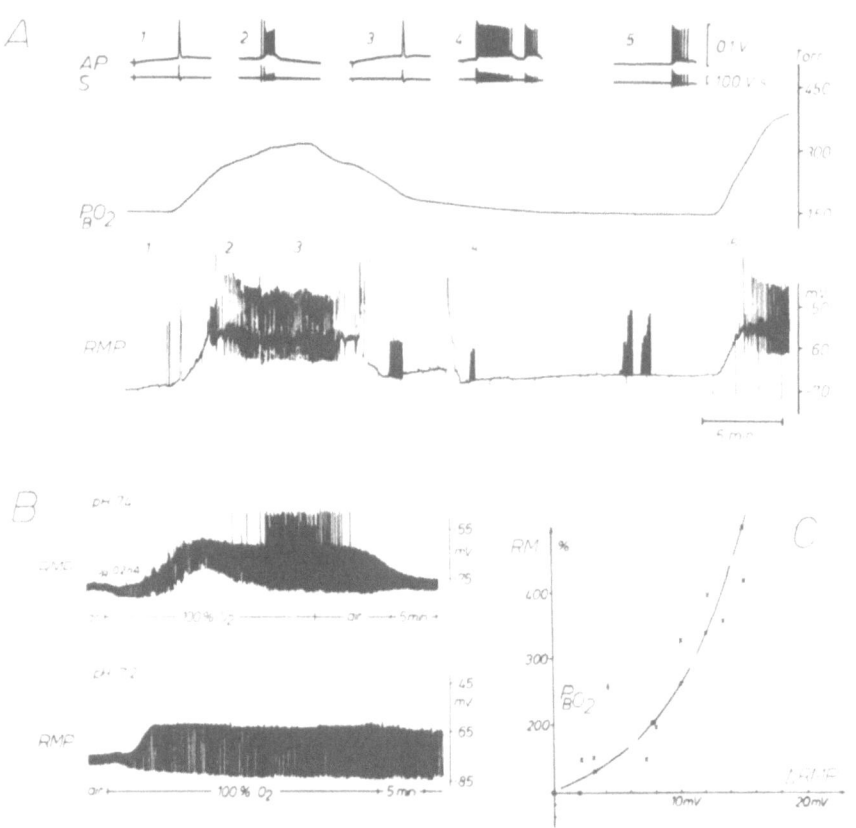

Fig. 1. Changes of the membrane properties of CA3 neurons after the
rise of Po_2 in the bath (P_BO_2). Fig. 1A, B: With the rise
of P_BO_2 the resting membrane potential (RMP) declined.
Simultaneously, the membrane resistance (RM) increased as
shown by the amplitudes of the hyperpolarizing pulses caused
by current injections (intensity: 0.2 nA. J_R duration:
0.6 s). Furthermore, burst activity occurred (Fig. 1A) which
was suppressed when the pH value of the bath was lowered
from 7.4 to 7.2 (Fig. 1B). AP = action potentials, S = time
derivation of AP. AP and S are related in time to the RMP
curve by numbers. Fig. 1C: Relation between maximum changes
of the RMP and of the concomitant changes of RM (in %) in
13 CA3 neurons observed after the rise of P_BO_2.

(ii) A few minutes after the rise of P_BO_2 a second phenomenon
appeared which did not depend on the amplitudes of the long lasting
depolarizations: PSP-like fluctuations of RMP increased in amplitude
and frequency. Suddenly, the CA3 neurons started to burst as demonstrated in Fig. 1A. The first AP during such bursts had full ampli-

tudes. They were followed by smaller ones riding on depolarization
waves. These bursts which were suppressed with a rise of Pco_2 or by
acidosis (cf. Fig. 1B) widely resembled penicillin-induced bursts
of CA3 neurons. When burst activity was fully developed at high P_BO_2
values, electrical stimulation through the recording microelectrode
evoked only single spikes which hardly differed from those elicited
at normoxic P_BO_2 values. Furthermore, hyperpolarization of the soma
membrane did not affect significantly the burst frequency. Therefore,
one may assume that the burst generator is located in dendrites or
axons remote from the soma. However, the elementary mechanisms underlying these bursts, are not clarified.

1b. Reactions of CA3 Neurons to Low P_BO_2 Values

        (i) Effects on RMP and RM: After lowering P_BO_2 from hyperoxic
to hypoxic values, burst activity stopped and the RMP initially returned to the level recorded at normoxic P_BO_2 (cf. Fig. 2). After
this transient repolarization which has previously been described
by Hansen et al. (1979) as well as by Misgeld and Frotscher (1982)

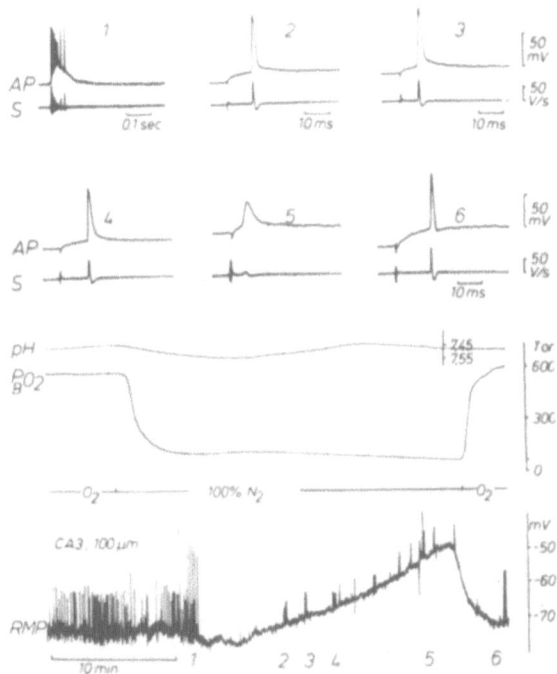

Fig. 2.   Alterations of the resting membrane potential (RMP) and of
          the action potentials (AP) as well as of their time deriva-
          tions (S) (related in time to the RMP curve by figures)
          after lowering the Po₂ in the bath (P_BO_2) from 600 to
          30 mm Hg. The tested CA3 neuron was found in a depth of
          100 um within a 300 um thick slice.

as a hyperpolarization shift, the neurons depolarized. Onset and
time course of these depolarization shifts varied considerably in
different neurons. Some neurons showed a steep depolarization which
started abruptly after a ten minutes exposition to hypoxia. Other
neurons depolarized slowly after latencies of 1 - 2 minutes. When
P_BO_2 was lowered from normoxic to hypoxic values, a transient hyperpolarization of CA3 neurons was missing.

Fig. 3 demonstrates that the pyramidal cell promptly depolarized
after the superfusion with N_2 - instead of air equilibrated BSS.
Simultaneously, RM decreased which indicates that this depolarization
is due to an increased leakage of the membrane. The RMP shift ceased
and turned over to a transient hyperpolarization when the recording
chamber was perfused with an O_2-equilibrated BSS. In parallel the RM
reincreased. 10 min later it exceeded the value found at a P_BO_2 of
150 mm Hg and the neuron slightly depolarized. A further exposition
to air - and then to N_2-equilibrated BSS again caused a decline of

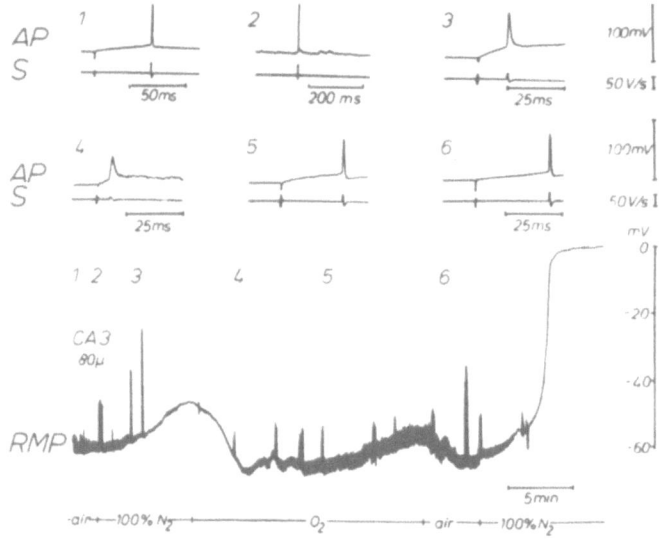

Fig. 3.   Changes of the resting membrane potential (RMP), of the
          action potentials (AP) and their time derivations (S) as
          well as of the membrane resistance of a CA3 neuron (found
          80 um below the surface of the slice) during superfusion of
          the tissue with air-, N_2-, O_2-, air - and N_2-equilibrated
          BSS. Changes of the membrane resistance are reflected in
          the RMP curve by the amplitudes of the hyperpolarizing
          pulses which resulted from repetitive intracellular current
          injections (intensity: 0.2 nA, duration: 600 ms, frequency:
          0.2 Hz). AP and S are related in time to the RMP curve by
          figures.

RM and a transient repolarization followed by a depolarization.
Finally, the intracellular recording was lost.

(ii) Effects of hypoxia on AP: With the lowering of P_BO_2 to
hypoxic values besides changes of RMP and of RM alterations of the
spike generator occurred. In Figs. 2 and 3 AP and their time derivations (S) elicited at different Po₂ values by intracellular current
injections are displayed. By means of numbers AP and S are related
in time to the RMP curves. Both figures demonstrate that the amplitude and the steepness of the AP declined while their duration and
the threshold increased until spike generation was blocked (Fig. 2:
AP 2-5, Fig. 3: AP 1-3). These alterations of the spike generator
were not clearly related to RMP shifts. Thus, in some experiments
AP could hardly be elicited by electrical stimulation although the
RMP still ranged between -55 mV and -65 mV. When the slices were re-
exposed to hyperoxic BSS, neurons regained their excitability within
some minutes (Fig. 2: AP 6, Fig. 3: AP 5, 6) and the amplitude,
duration and steepness of electrically evoked AP returned to control
values. Finally, spontaneous PSP-like RMP fluctuations and spontaneous AP reappeared which were the first to be suppressed after lowering P_BO_2 (Fig. 3: AP 2). This recovery was observed even when the
slices were exposed to the N_2-equilibrated BSS for 30 min. Misgeld
and Frotscher (1982) observed the same sequences of recovery phases
after superfusing hippocampal slices up to 20 min under hypoxic
conditions. Only when this hypoxic period was prolonged up to 1 h,
even a partial restoration of neuronal functions was missing.

The presented findings have shown that the neuronal behaviour
of CA3 neurons is widely influenced by changes of Po₂ as summarized
schematically in Fig. 4B. The long lasting depolarization and the
burst generation observed at high Po₂ values as well as the depolarization and the impairments of the AP under hypoxia may be attributed to changes of conductances in the membrane not only of the
soma but also of dendrites. This sensitivity of CA3 neurons in
slices to changes of Po₂ contrasted strikingly the poor dependency
of sensory spinal ganglion cells (Bingmann and Kolde, 1981) and of
petrosal ganglion neurons (Gallego and Belmonte, 1981) on Po₂
changes in vitro. Fig. 4A demonstrates schematically that shifts of
P_BO_2 which clearly modified the activity of CA3 neurons hardly affected either the RMP of sensory neurons or their threshold for AP.
Furthermore, significant alterations of the RM or of the amplitude,
duration, steepness and other parameters of AP were missing. Therefore, the question arises as to the reasons responsible for the
different sensitivities of CA3 neurons in slice preparations and
sensory spinal ganglion cells in vitro. If one assumes that changes
of Po₂ affect neurons by manipulating the distribution of ions in
the tissue, the different shapes of both neurons and their distinct
cellular organizations should contribute to these differences: On
the one hand, bipolar sensory neurons have spherical somata with
small surface/volume ratios. In vivo these nerve cells are surrounded

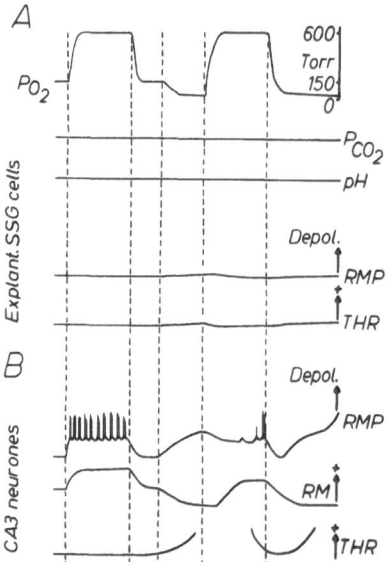

Fig. 4.   Schematic presentation of the effects with isolated changes
          of Po_2 in the bath exert on sensory spinal ganglion cells
          (Fig. 4A) and on CA3 neurons of hippocampal slices (Fig.
          4B). Pco_2 and pH were kept constant in the superfusate.
          Resting membrane potential: RMP. Threshold for action
          potentials: THR. Membrane resistance: RM.

by satellite cells which may stabilize the ionic milieu in the
tissue. In in vitro experiments the soma membrane of these neurons
is often directly exposed to bath solutions with constant ion concentrations. On the other hand, pyramidal cells in hippocampal slices
are characterized by extended dendrites which are embedded in a
densely packed neuronal tissue. These dendrites enlarge the surface/
volume ratio markedly. Due to this, high ratio changes of conductances will rapidly alter transmembraneous gradients of ions and thus
affect the RMP of dendrites and the soma. Unfortunately, little information is available about alterations of the ionic environment
in slices. Therefore, in a second series of experiments profiles
of tissue Po_2 (P_tO_2) and of $[K^+]_e$ in the extracellular space in
hippocampal slices were measured at different P_BO_2 values.

2. Profiles of P_tO_2 and of $[K^+]_e$ in Hippocampal Slices

        (a) Measurements of P_tO_2: Polarographic recordings with bipolar
microelectrodes showed that Po_2 declined already in the bath when
the tip of the electrode approached the surface of the slice (cf.
Ganfield et al., 1970). Inside vital slices P_tO_2 declined steeply
within the first 25 um below surface whereas the Po_2 gradients were
flat in those slices which proved to be poor in electrophysiological

experiments. Fig. 5A demonstrates that the Po_2 measured in an intact
slice at a depth of 25 um amounted to less than 50% of the P_BO_2 when
the excised tissue was superfused with air-equilibrated BSS. Minimum
P_tO_2 values were measured at a depth of 150 - 200 um when 300 - 400
um thick slices were tested. The minimum P_tO_2 was diminished and the
P_tO_2 gradients were steepened when temperature was elevated. The
oxygen supply of the innermost layers declined with the rise of tem-
P_tO_2 values were measured at a depth of 150 - 200 um when 300 - 400
um thick slices were tested. The minimum P_tO_2 was diminished and the
P_tO_2 gradients were steepened when temperature was elevated. The
oxygen supply of the innermost layers declined with the rise of temperature hardly improved, when the Po_2 in the bath increased from
150 to 600 mm Hg. With the rise of P_BO_2 tissue Po_2 initially increased proportionally. After forming a peak after 1 - 2 min, P_tO_2
in vital slices then declined within 5 - 10 min to a plateau although

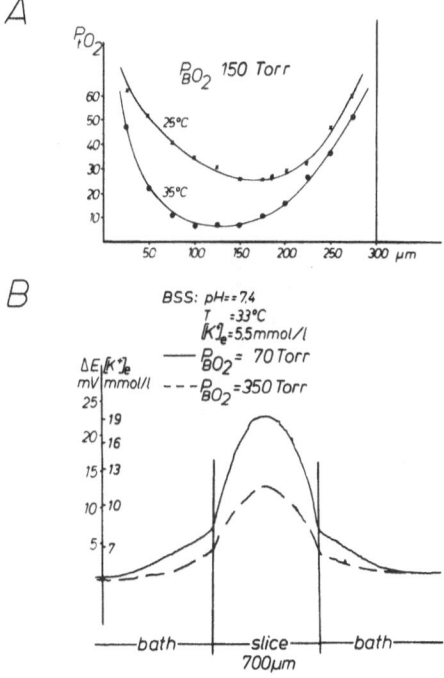

Fig. 5.   Depth-profiles of tissue Po_2 (P_tO_2) and of the concentra-
          tions of $[K^+]_e$ in slices. A: P_tO_2 gradients measured in
          300 um thick slices (at a Po_2 in the bath (P_BO_2) of
          150 mm Hg) at 25°C and 35°C. B: $[K^+]_e$ profiles (changes of
          the relative electrode potential) in a 700 um thick slice
          at P_BO_2 values of 350 Torr (--) and of 70 Torr (ΔE).

P_BO_2 remained constant at 600 mm Hg. This finding indicates that the O_2 consumption of the hippocampal slice increased with rising P_BO_2 (cf. Bingmann and Kolde, 1982). The steep P_tO_2 gradients in the tissue demonstrate that the CA3 neurons impaled in the present experiments at different depths were exposed to different degrees of hyper- or hypoxia when Po_2 in the bath was elevated or lowered. This may partly explain the great variability in the neuronal responses to changes of Po_2 in the superfusate. From the presented Po_2 profiles the question arose whether similar gradients exist concerning ion concentrations in hippocampal slices.

(b) Measurements of $[K^+]_e$: Recordings with liquid ion exchanger microelectrodes showed that $[K^+]_e$ profiles partly mirrored P_tO_2 distribution curves. However, at hyperoxic P_BO_2 values the gradients of $[K^+]_e$ were flat in 300 um thick slices. In 700 um thick slices, as sometimes used in neurophysiology, even at an elevated P_BO_2 of 350 mm Hg $[K^+]_e$ clearly increased from the surface to the innermost layers where the potassium concentration amounted to 11 mmol·1^{-1} (Fig. 5B). After lowering the P_BO_2 to 70 mm Hg, the gradients of the potassium concentrations were steepened and maximum concentrations of 20 mmol·1^{-1} were reached.

In the intact tissue of the CNS during hypoxia stable $[K^+]_e$ values between 12 and 50 mmol·1^{-1} usually are not maintained even for a few minutes (Staschen et al., 1981). In the central zones of slices, however, the potassium concentration ranged at about 20 mmol·1^{-1} throughout the hypoxic periods of 10 – 20 min. When P_BO_2 returned to hyperoxic values the potassium concentration in the slices promptly declined. It is likely that the Po_2-dependent shifts of $[K^+]_e$ were partly due to changes of leak currents across neuronal membranes because the RM of the pyramidal cells generally decreased when the neurons depolarized during hypoxia. Glial elements may also contribute to these changes of the potassium activity. However, the role played by these idle cells in hippocampal slices at low Po_2 values has to be studied in further experiments.

In summary, the observed changes of $[K^+]_e$ especially in the central zones of slices can alter not only the RMP of central neurons but also that of dendrites from remote cells and of afferent fibres. Thus, the presented findings suggest that Po_2 dependent shifts at $[K^+]_e$ markedly contribute to reactions of neurons in slice preparations to changes of Po_2.

REFERENCES

Bean, J.W., 1945, Effects of oxygen at increased pressure, Physiol. Rev., 25:1–147.
Bingmann, D., and Kolde, G., 1981, Reactions of neurons in vitro to changes of Po_2 in the bath solution, Pflügers Arch., 391:R32.

Bingmann, D., and Kolde, G., 1982, Po_2-profiles in hippocampal slices of the guinea pig, Exp. Brain Res., 48:89-96.

Bingmann, D., Kolde, G., and Speckmann, E.-J., 1982, Effects of elevated Po_2 values in the superfusate on the neural activity in hippocampal slices, in: "Physiology and Pharmacology of Epileptogenic Phenomena", M.R. Klee, H.B. Lux, E.-J. Speckmann, eds., Raven Press, New York, pp. 97-104.

Gallego, R., and Belmonte, C., 1981, Electrical properties of petrosal ganglion chemoreceptor cells, in: "Arterial Chemoreceptors", C. Belmonte, D.J. Pallot, H. Acker, S. Fidone, eds., Leicester University Press, Leicester, pp. 392-399.

Ganfield, R.A., Nair, P., and Whalen, W.J., 1970, Mass transfer, storage and utilization of O_2 in cat cerebral cortex, Am. J. Physiol., 219:814-821.

Hansen, A.J., Jahnsen, H., and Hounsgaard, J.B., 1979, Influence of hypoxia on hippocampal nerve cells in vitro, Acta Physiol. Scand., 473:55.

Lehmenkühler, A., Bingmann, D., Lange-Asschenfeldt, H., and Berges, D., 1978, Oxygen pressure and ictal activity in the cerebral cortex of artificially ventilated rats during exposure to oxygen high pressure, in: "Oxygen Transport to Tissue III", I.A. Silver, M. Erecinska and H.I. Bicher, eds., Plenum Press, New York, pp. 679-685.

Lehmenkühler, A., Caspers, H., and Speckamnn, E.-J., 1976, A method for simultaneous measurement of bioelectrical activity and local Po_2 in the CNS, in: "Oxygen Transport to Tissue II", J. Grote, D. Reneau and G. Thews, eds., Plenum Press, New York, pp. 3-7.

Lehmenkühler, A., Zidek, W., Staschen, M., and Caspers, H., 1981, Cortical pH and pCa in relation to DC potential shifts during spreading depression and asphyxiation, in: "Ion-Selective Microelectrodes and Their Use in Excitable Tissues", L. Vyklicky, E.S. Ykova, P. Huik, eds., Plenum Press, New York, pp. 225-230.

Lübbers, D.W., 1969, The meaning of the tissue oxygen distribution curve and its measurements by means of Pt-electrodes, in: "Oxygen Pressure Recording in Gases, Fluids and Tissues", F. Kreuzer and H. Herzog, eds., Karger, Basel, pp. 112-123.

Lux, H.D., and Neher, E., 1973, The equilibration time course of $[K^+]_o$ in cat cortex, Exp. Brain Res., 17:190-205.

Misgeld, U., and Frotscher, M., 1982, Dependency of the viability of neurons in hippocampal slices on oxygen supply, Brain Res. Bull., 8:95-100.

Morris, M.E., 1974, Hypoxia and extracellular potassium activity in the guinea pig cortex, Can. J. Physiol. Pharmacol., 52:872-882.

Silver, I.A., 1977, Ion fluxes in hypoxic tissues, Microvasc. Res., 13:409-420.

Speckmann, E.-J., and Caspers, H., 1974, The effects of O_2 and CO_2 tensions in the nervous tissue on neuronal activity and DC-potentials, in: "Handbook of Electroencephalography and Clinical Neurophysiology II", A. Rémond, ed., Elsevier, Amsterdam-London-New York, pp. 71-89.

Staschen, M., Zidek, W., Lehmenkühler, A., and Caspers, H., 1981,
    Changes of extracellular ion activities (K^+, Na^+, Ca^{++}, H^+, Cl^-)
    in relation to cortical DC-potential shifts during reversible
    asphyxia, Pflügers Arch., 389:R33.
Vyskocil, F., Kriz, N., and Bures, J., 1972, Potassium selective
    microelectrodes used for measuring the extracellular brain
    potassium during spreading depression and anoxic depolarization
    in rats, Brain Res., 39:255-259.
Yamamoto, C., 1972, Intracellular study of seizure-like afterdis-
    charges elicited in thin hippocampal sections in vitro. Exp.
    Neurol., 35:154-164.
Yamamoto, C., and Kurokawa, M., 1970, Synaptic potentials recorded
    in brain slices and their modification by changes in the level
    of tissue ATP, Brain Res., 10:159-170.

RESPONSE OF GERBIL CEREBRAL UNIT ACTIVITY TO SLOWLY DECLINING

TISSUE PO_2

R.M. Martin and J.H. Halsey

Department of Neurology, University of Alabama in
Birmingham, Birmingham, AL, USA

Within the central nervous system, regional differences may
exist in oxygen tension due to the variation in such factors as
blood flow, capillary density, and rates of tissue respiration.
Also, cells located near the venous portion of a capillary should
be exposed to a lower oxygen tension than cells at the arterial end
as suggested by modeling based on the Krogh cylinder (Krogh, 1919).
Such PO_2 differences, in fact, have been demonstrated experimentally (Silver, 1965; Metzger and Heuber, 1977). During an ischemic
or hypoxic insult, certain cells may be the first casualties in
that the PO_2 may reach zero more quickly for those cells normally
existing in low PO_2 areas.

The term "critical PO_2" has been used for that PO_2 below which
cellular dysfunction occurs. Is this level nearly the same for all
neurons, or can cells within regions of low PO_2 better tolerate
hypoxia because they may have undergone some adaptive process permitting them to function at lower oxygen levels?

We performed a series of experiments in awake, mechanically
ventilated gerbils, simultaneously recording PO_2 and action potentials from the oxygen microelectrode to observe the unit response
to slowly declining PO_2 to determine if the critical PO_2 is clearly
a function of resting PO_2 levels.

METHODS

Adult male gerbils were anesthetized with halothane for the
surgical procedures. After tracheostomy, the animal was placed into
a headholder. The scalp was removed, and a 3 mm x 3 mm craniectomy

227

was made in the parietal region. The dura was carefully removed, and a small Ag–AgCl reference wire was inserted between the skull and brain tissue and secured on top of the skull with dental acrylic. Bilateral EEG electrodes were similarly secured and placed into the skull frontally. Lidocaine was injected into pressure points and incision sites for local anesthesia.

The animal was moved to a Faraday cage where curare was administered, and the animal was placed onto a ventilator supplying a mixture of 25% O_2, 75% N_2. The temperature was maintained at 37°C $\pm$ 0.5°C by a heating blanket.

The oxygen microelectrode was of the "needle" design with a tip diameter of 1 um. Po_2 and unit activity were amplified through a modification of the circuit described by Kunke et al. (1972). These data were recorded on tape for subsequent computer analysis of Po_2 and spike frequencies.

Utilizing a motorized microdrive, the oxygen microelectrode was moved into the parietal cortex at a rate of 1 um/sec until an isolated unit was acquired. In four experiments, the electrode was driven into hippocampal structures. Once a satisfactory unit was found, a baseline period of 5 min was allowed. Hypoxia was then induced by systematically decreasing the oxygen concentration from the respirator so that the inspired Po_2 would reach zero in

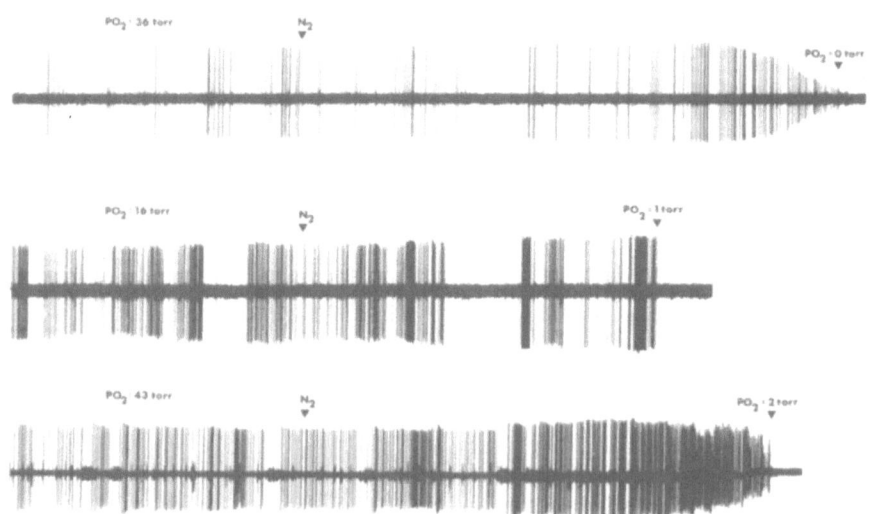

Fig. 1. Responses of three units at different resting Po_2 values
         to a hypoxic onset (arrow). All fail at very low Po_2 values
         that are not proportional to resting levels.

approximately 90 sec. As soon as a tissue Po_2 of zero was reached, the oxygen was restored. A 15-min monitoring period ensued to note unit recovery, if any. The electrode was re-calibrated as preexperimentally in saline equilibrated with oxygen concentrations of 0%, 5%, and 100%. In all, 13 units from 11 animals were recorded.

RESULTS AND DISCUSSION

     Fig. 1 displays the response of 3 units at various resting Po_2 values to hypoxia. Although the response in spike frequency varies somewhat, each failed at very low, though not identical, Po_2 values. The most typical response seen was a large enhancement in spike frequency with an abrupt cessation of activity. Interestingly, only 2 cells of the 13 recorded showed any sign of recovery within the post-hypoxia monitoring period. We have seen different responses in units exposed to step-changes in oxygen concentration, i.e. a gradual fall in activity with recovery. It appears, thus, that this type of slow hypoxic onset is most damaging to the cell.

     Fig. 2 is the plot of resting Po_2 versus the Po_2 at which spike failure occurred. No relationship is evident between the two. As mentioned previously, the Po_2 of failure was not the same for all cells, in that some cells maintained some activity until a Po_2 of zero where others which had a lower resting Po_2 failed sooner.

     The lack of correlation between resting Po_2 and the critical Po_2 is in contrast to data reported by Erdmann et al. (1973).

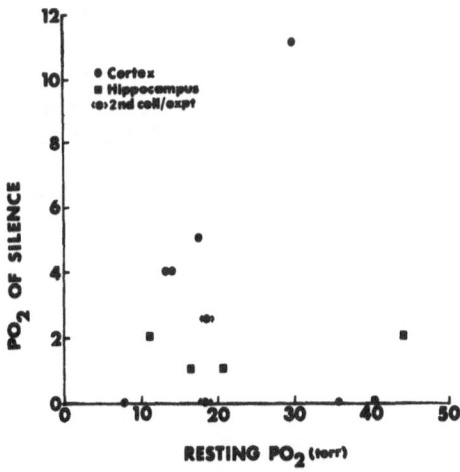

Fig. 2.   Plot of resting Po_2 versus Po_2 of silence (critical Po_2).
          No relationship can be seen between these parameters.

This difference is unexplained, although in this earlier work the authors used N_2O for analgesia. Since the oxygen electrode is sensitive to this compound, the Po_2 values they measured are somewhat suspect.

Our data tend to indicate that a cell's resistance to hypoxia may depend on many factors other than the proximity to the capillary. Such factors may include cell function, size, mitochondrial density, and amount of endoplasmatic reticulum. Longmuir and Pashko (1977) have suggested that the endoplasmatic reticulum may serve as an intracellular oxygen carrier, so if neurons possess more of this organelle they may have a higher survival rate. All such possibilities will have to be examined before a clear picture of neuronal susceptibility becomes available.

REFERENCES

Erdmann, W., Kunke, St., and Krell, W., 1973, Tissue Po_2 and cell function - An experimental study with multimicroelectrodes in the rat brain, in: "Oxygen Supply", M. Kessler, D.F. Bruley, L.C. Clark, D.W. Lübbers, I.A. Silver, J. Strauss, eds., University Park Press, Baltimore.

Krogh, A., 1919, The number and distribution of capillaries in muscles with the calculation of the oxygen pressure head necessary for supplying the tissue, J. Physiol. (London), 52: 409-415.

Kunke, St., Erdmann, W., and Metzger, H., 1972, A new method for simultaneous Po_2 and action potential measurements in microareas of tissue, J. Appl. Physiol., 32:436-438.

Longmuir, I., and Pashko, L., 1977, The role of facilitated diffusion of oxygen in tissue hypoxia, Int. J. Biometeor., 21: 179-187.

Metzger, H., and Heuber, S., 1977, Local oxygen tension and spike activity of the cerebral grey matter of the rat and its response to short intervals of O_2 deficiency or CO_2 excess, Pflügers Arch., 370:201-209.

Silver, I., 1965, Some observations in the cerebral cortex with an ultramicro, membrane-covered, oxygen electrode, Med. Electron. Biol. Engn., 3:377-387.

PROPERTIES OF THE SPONTANEOUS FLUCTUATIONS IN CORTICAL OXYGEN

PRESSURE

J. Manil[1], R.H. Bourgain[1], M. Van Waeyenberge[1],
F. Colin[2], E. Blockeel[3], B. De Mey[1], J. Coremans[1],
and R. Paternoster[4]

[1]Laboratory of Physiology (VUB)
[2] Faculté de Medicine (ULB)
[3]Medical Informatics Dept. (VUB)
[4]Electronics Laboratory (VUB)
Free Universities of Brussels, Laarbeeklaan 103
1090 Brussels, Belgium

INTRODUCTION

In previous publications (Manil et al., 1978; 1981; Bourgain et al., 1980) the spontaneous fluctuations of the cortical tissue pressure in oxygen (Po_2) in the awake rabbit were described in detail. The mean amplitude of these irregular oscillations was 5% of the normal local Po_2 where 100% is defined as the value recorded in normal control conditions and 0% as the value recorded when death occurred after sustained pure nitrogen breathing. Analysis of the frequency domain demonstrates that in control conditions the power of the spectrum decreased from 0.06 to 0.5 Hz.

Following the administration of α-blocking agents the oscillations increased in amplitude and became pseudosinusoidal with the power concentrated in two peaks at 0.1 Hz and 0.2 Hz, respectively; this phenomenon is highly reproducible and quite typical for α-blocking.

Oscillations affecting the oxygen tissue pressure as well as local microflow have already been described by Lübbers and Leniger-Follert (1978).

Two hypotheses can be advanced to explain this pattern: firstly the fluctuations result from variations in the tissue consumption

231

of oxygen, secondly the local Po$_2$ is controlled by a local feed-back
mechanism of vascular origin as advocated by Silver (1978). In other
words the fluctuations in oxygen pressure in the cortical tissue are
due to changes either in supply or in consumption of oxygen, the
interaction of both phenomena is also possible.

It had already been observed that the oscillations recorded in
the parietal area of the cortex disappeared during short anoxic anoxia episodes of 1 to 2 min as well as during the overshoot following
reoxygenation; the oscillations reappear 3 min later in these experimental conditions (Manil et al., 1981).

In order to investigate these phenomena further it was thought
essential to simultaneously record the oxygen tissue pressure in
different cortical areas (frontal, parietal and occipital) together
with the arterial blood pressure, local microflow and the electrocortical actvity.

The effect of α- and ß-blocking agents, as well as drugs affecting cortical function such as barbiturates has also been further
investigated.

METHODS AND MATERIALS

Dutch rabbits were anesthetized with Hypnorm (Duphar) (0.5 ml/
kg I.M.) and electrodes for the derivation of electrical potentials
were chronically implanted. They consisted of small silver discs
(500 um diameter) terminated by a wire (50 um diameter). These electrodes were introduced onto the dura mater through small holes
drilled into the skull. The electrodes were situated rostrocaudally
at a distance of 5 mm from the midline with a distance of 3 mm between each electrode; the frontal, central, parietal, occipital and
temporal regions of the cortex were thus explored over one hemisphere. These electrodes were implanted in order to record the evoked
somesthetic potentials and the electrocorticogram. Potentials were
measured with the ear used as a reference.

The oxygen electrodes consisted of a 100 um diameter platinum
wire insulated except at the tip (1 mm). The tip was covered with a
layer of cellulose acetate in order to allow free diffusion of oxygen molecules while avoiding poisoning from the platinum surface.
These electrodes were implanted in the frontal, parietal and occipital regions through a large opening drilled in the skull. The dura
was incised and the tip of the electrode implanted in the superficial layer of the cortex. On the day of the experiment the electrodes were negatively polarized with a 0.65 volt D.C. current. The
variations of the current due to local Po$_2$ variations were measured
through a 100 kΩ resistance with a 6.8 uF capacitor in parallel with
the input of a D.C. Tektronix amplifier.

All electrodes were imbedded into a layer of dental resin with free ends protruding and were thus protected against mechanical wear.

Antibiotic therapy was performed for eight successive days with chloramphenicol (R.I.T.) (100 mg I.M. daily).

Suitable rotameters were used for the administration of the respiratory gas mixtures when anoxia experiments were performed; the animal was cannulated with a tracheal tube under light and short halothane anesthesia, later curarized and artificially ventilated.

The evoked potentials were obtained by electrical stimulation of the contralateral forepaw at a value of twice the motoric threshold with a duration of 0.1 msec and a frequency of 0.5 Hz. The evoked potentials were obtained by an averaging procedure and represent the mean of 10 to 30 individual responses.

Spectral analysis was performed with a suitable spectrum analyser (Hewlett-Packard) allowing visual recording of the spectrum and determination of the value of the coherence function between simultaneously recorded sets of parameters.

The arterial blood pressure was recorded from the femoral artery through a polyethylene catheter and with a pressure transducer.

Local microflow was determined according to the method described by McCaffrey and McCook (1975).

As an α-blocking agent phenoxybenzamine was used at a dosage of 5 mg/kg body weight, and propranolol as a ß-blocking agent was used at a dosage of 0.4 mg/kg body weight; both drugs were administered intravenously. Nembutal was administered in shots of 10 mg/kg up to a total of 60 mg/kg body weight.

RESULTS

A group of eight animals were investigated in this study. In Fig. 1 the blood pressure and the tissue Po_2 values in the frontal, parietal and occipital zones are demonstrated from one animal. Spectral analysis of the Po_2 fluctuations demonstrates small peaks at 0.06 Hz, 0.17 Hz and at 0.36 Hz. The values of the coherence function, however, are very poor between these regions: they do not exceed 0.6 (for a maximal value of 1).

When the animal was in stress conditions (Fig. 2) the large fluctuations in blood pressure were accompanied by similar fluctuations in the local Po_2 curves. Peak values were then recorded at 0.06 Hz, 0.13 Hz, 0.21 Hz and 0.27 Hz, respectively. In these conditions the value of the coherence function becomes significant

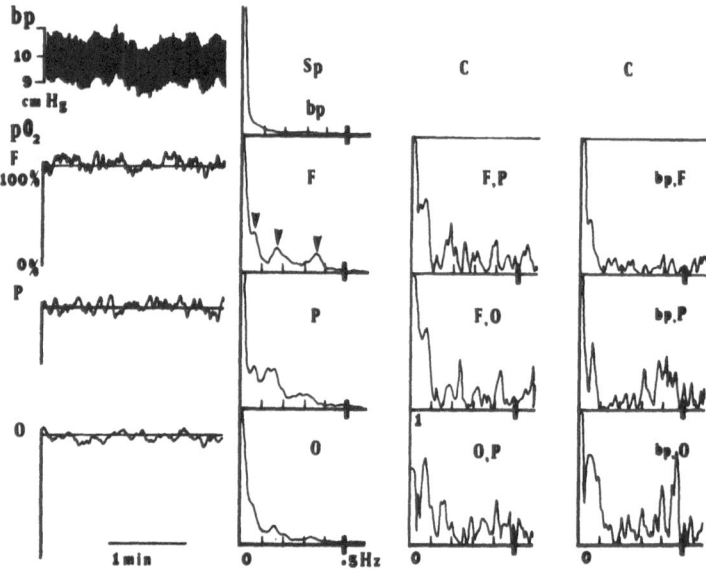

Fig. 1.   The blood pressure (bp) and the oxygen tissue pressure (Po₂)
in frontal (F), parietal (P) and occipital (O) areas of the
cortex from one animal. The normal oscillation pattern of
the Po₂ in the different derivations is clearly visualized.
The spectra of the frontal, parietal and occipital zones
demonstrate the different peak values (spectral analysis
is performed on 16 samples of 125 seconds each). A poor
value for the coherence function is observed when the frontal, parietal and occipital recordings are compared.

particularly for the recordings in the frontal and the occipital
areas. These values are then at 0.06 Hz, 0.98; at 0.13 Hz, 0.93 and
at 0.21 Hz, 0.91, respectively. The coherence function between the
arterial blood pressure and local Po₂ reaches the highest value in
the occipital cortical area, at 0.06 Hz, 0.91; and at 0.13 Hz,
0.21 Hz and 0.27 Hz a value of 0.98.

   Following the administration of phenoxybenzamine the blood pressure stabilized at a lower mean level (70 to 80 mm Hg) while the Po₂
oscillations became pseudosinusoidal presenting an increase in amplitude (30%) as demonstrated in Fig. 3. Correspondingly the spectra
also changed markedly while the value of the coherence function between the different cortical areas remained low. In one animal four
Po₂ electrodes were implanted rostrocaudally in the parietal cortex
at distances of respectively 2.5 mm, 2.5 mm and 1.5 mm. The value
of the coherence function between the Po₂ curves at oscillation
frequencies obtained from electrodes 2.5 mm apart is very low but
increases markedly (>0.8) when the coherence function is calculated for electrodes located at a distance of 1.5 mm from each other.

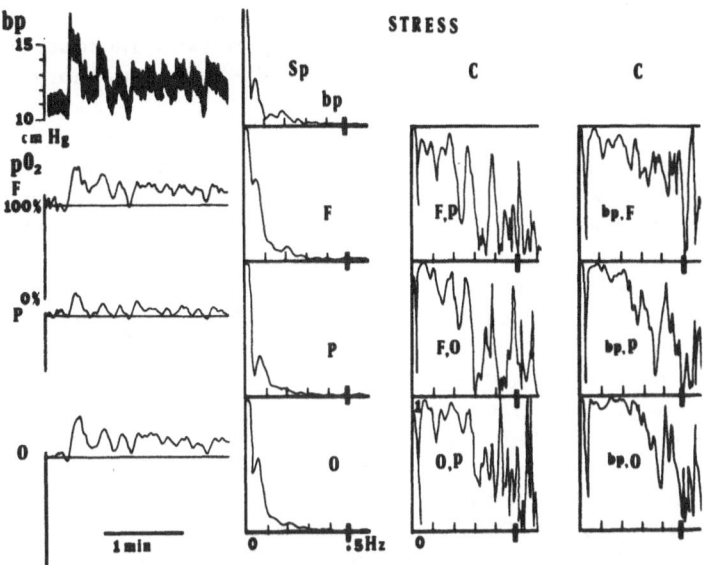

Fig. 2. In stress-like conditions variations in the arterial blood
pressure occur accompanied by similar variations in the Po_2
curves. The value of the coherence function increases markedly particularly for the frequencies between 0.1 and 0.3 Hz.

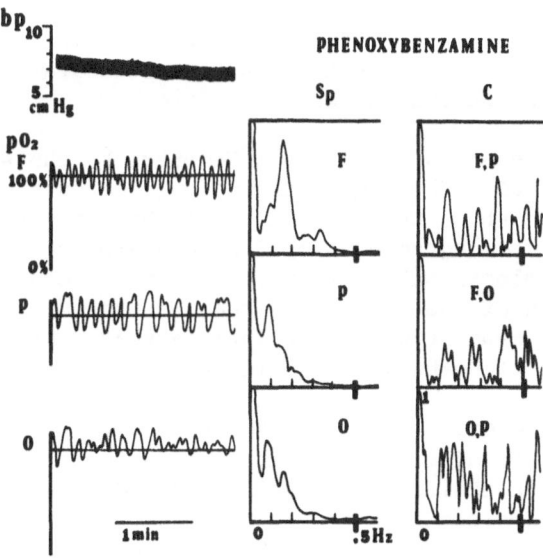

Fig. 3. Following the administration of α-blocking agents (phenoxybenzamine 5 mg/kg body weight) the arterial blood pressure
stabilizes and the local tissue Po_2 curves become pseudosinusoidal. As compared to Fig. 1 marked changes occur in
the power spectrum. The values of the coherence functions,
however, remain low.

The somesthetic evoked potential changes markedly after administration of α-blocking agents; there is a marked increase both in the positive P_2 wave and the late negative N wave, moreover the peak values are slightly delayed.

The electrocorticograms demonstrate definite changes after α- and ß-receptor blocking. In normal experimental control conditions the electrocorticogram is either of the awake type or of the slow wave type I and II, characterized respectively for the awake or active state in a low voltage recording (300 uV) at a wave frequency of 6 - 7/sec, for type I 5/sec; in type II bursts of spindles appear (10 - 12/sec) with slow waves 1.5 to 3/sec. Following the administration of phenoxybenzamine the electrocorticogram demonstrates a deeper sleep with more slow waves (1.5 - 3/sec) and some spindles (stage III) and ultimately very slow waves (< 1.5/sec) are observed (stage IV). The slow waves always present a high voltage of approximately 500 to 600 uV. A constant finding is the appearance of prolonged episodes (15 to 30 min) of rapid eye movement sleep (REMS) when ß-blocking agents are administered after α-blocking agents were given. During the REM sleep episodes a continuous dyssynchronized type of electrocortical recording of low voltage and a frequency of 6 - 7/sec is observed in all explored cortical areas. The value of the coherence function (Fig. 4) of the electrocorticogram recordings in the parietal and occipital area is high (0.9). However, the value of the coherence function between the Po_2 recordings in these areas is very low. Furthermore, when during the REM episode the animal demonstrates marked twitches and contractions of the facial muscles, the Po_2 recordings in the different areas show important slow oscillations in amplitude while the electrocorticogram remains stable and typical of the active state.

Following the administration of Nembutal by shots of 10 mg/kg up to a total of 60 mg/kg a consistent lowering of the blood pressure was observed while the mean value of the local Po_2 decreased and the electrocorticogram developed a very slow wave pattern, eventually becoming electrically silent. Under these conditions the oscillation pattern of the Po_2 curves changed markedly. Oscillations increased as the mean arterial pressure decreased being maximal with the largest dose of Nembutal when electrical activity disappeared.

The local microflow recorded simultaneously with the Po_2 recorded in the parietal and occipital area demonstrated a marked increase when CO_2 breathing (6%) was induced and as such emphasizes the role of local blood flow in establishing the tissue pressure in oxygen. In this situation, vasodilation seems to be maximal and oscillations of the local Po_2 are no longer observed.

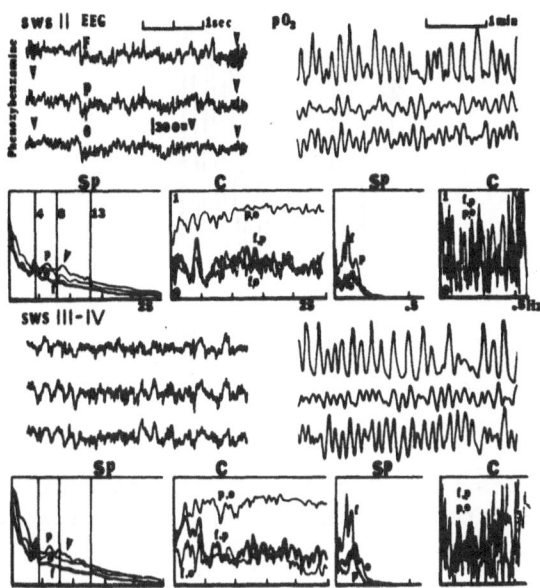

Fig. 4.   Following the administration of α-blocking agents the elec-
          troencephalogram (EEG) of the frontal (F), parietal (P) and
          occipital (O) regions demonstrates patterns of slow wave
          sleep (SWS II and SWS III-IV). The spectral analysis (Sp)
          up to 25 Hz and the coherence function between these EEGs
          are performed on an average value of 64 samples of 5 sec
          each. The simultaneously recorded Po_2 of these regions are
          also represented with their spectral analysis up to .5 Hz
          and also the coherence function between these Po_2 on an
          average value of 8 samples of 125 sec each.

DISCUSSION AND CONCLUSIONS

     These experimental findings clearly demonstrate that the re-
corded tissue Po_2 of the cortex remains quite constant in control
conditions and that the small amplitude oscillations vary from one
cortical area to another. When these oscillations recorded simultaneously in the frontal, parietal and occipital region are compared,
the value of the coherence function is very low suggesting a local
control mechanism. An increase in value of the coherence function
occurs only when two adjoining areas less than 1.5 mm distance apart
from each other are explored, giving an indication of the approximate dimension of a microcirculating unit.

     Marked increases in the blood pressure however, such as those
induced by stress-like conditions are accompanied by similar changes
in the local Po_2 recordings; in these conditions the value of the
coherence function between the blood pressure recording and the local
Po_2 becomes very significant. This phenomenon very probably results

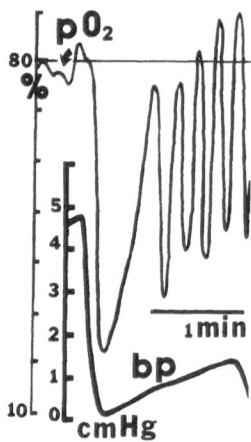

Fig. 5. Blood pressure (bp) in cm Hg and intracortical parietal
         Po$_2$ in % after I.V. injection at the arrow of 20 mg/kg
         Nembutal.

from a breakdown of the autoregulation when passive cortical vasodilation due to the hypertensive phenomenon occurs.

     Following the blocking of the α-receptors the blood pressure
stabilizes at a lower mean level while the Po$_2$ recordings in all
explored cortical areas display large pseudosinusoidal oscillations;
the value of the coherence function between different records, however, remains low demonstrating a local control.

     The pattern of oscillation following α-blocking can be observed
in the Po$_2$ recordings as well as in local flow. Admittedly this observation pleads for a vascular origin to the oscillations of the
tissue Po$_2$.

     Although it is quite clear that α-blocking not only affects the
local tissue Po$_2$ but also the electrical activity of the cortex
(spontaneous as well as evoked), it is not readily apparent how these
modifications of the functional state of the cortex could explain
quasi sinusoidal fluctuations of the O$_2$ consumption.

     Even more interesting is the following of the observations (Fig.
5) after the administration of Nembutal: oscillations are maximal
when the electrical activity has disappeared but when the arterial
pressure is minimal. So oscillations of Po$_2$ cannot be related to
periodic fluctuations of the cortical activity nor to fluctuations
of the systemic blood pressure.

     Extreme vasodilation due to CO$_2$ breathing is followed by an
increase in local flow and local tissue Po$_2$; this demonstrates the

importance of the vascular component in establishing the local oxygen pressure. Under these conditions oscillations of the Po_2 are no longer observed as if the local control was saturated.

ACKNOWLEDGEMENT

We wish to thank Fernand Vereecke and Hendrik De Backer for their skilled technical assistance.

This work was supported in part by grant No. 3.0036.80 and grant No. 3.0053.81 from the FGWO (National Foundation for Medical Scientific Research), Belgium.

REFERENCES

Bourgain, R.H., Colin, F., Vermarien, H., Maes, L., and Manil, J., 1981, Control mechanisms involved in the regulation of cerebral tissue pressure in oxygen, Adv. Physiol. Sci., 25:207-213.
Lübbers, D.W., and Leniger-Follert, E., 1978, Capillary flow in the brain cortex during changes in oxygen supply and state of activation, in: "Cerebral Vascular Smooth Muscle and its Control", Ciba Foundation Symposium 56, Elsevier/Excerpta Medica/North Holland, Amsterdam, pp. 21-47.
Manil, J., Bourgain, R., Colin, F., Maes, L., and Vermarien, H., 1981, The effect of α-blocking on the normal cortical cerebral tissue oxygen pressure and on the recovery from anoxic anoxia, Arch. Int. Physiol. Biochim., 89:P8.
Manil, J., Colin, F., and Bourgain, R.H., 1978, Modifications of somatosensory evoked cortical potentials during hypoxia in the awake rabbit, Adv. Exp. Med. Biol., 94:509-516.
McCaffrey, T.V., and McCook, R.D., 1975, A thermal method for the determination of tissue blood flow, J. Appl. Physiol., 39:170-173.
Silver, I.A., 1978, Cellular microenvironment in relation to local blood flow, in: "Cerebral Vascular Smooth Muscle and its Control", Ciba Foundation Symposium 56, Elsevier/Excerpta Medica/North Holland, Amsterdam, pp. 49-61.

INTRAOPERATIVE MONITORING OF CORTICAL SURFACE OXYGEN IN SUB-

ARACHNOID HAEMORRHAGE

M.P. Powell[1], I.A. Silver[3], H.B. Coakham[1] and
F.J.M. Walters[2]

[1]Department of Neurosurgery, Frenchay Hospital
[2]Department of Anaesthesia, Frenchay Hospital
[3]Department of Pathology
University of Bristol, Bristol, Great Britain

INTRODUCTION

We report the initial result of a series of clinical measurements of cerebral cortical oxygen tension in patients undergoing
surgery for intracranial arterial aneurysms. This group of patients, the majority of whom present with sub-arachnoid haemorrhage
(s.a.h.), frequently suffer from delayed cerebral artery vasospasm
which may not manifest itself clinically, yet may seriously decrease blood flow and thus cortical oxygenation may become compro-
mised. Once established, vasospasm is at present refractory to all
forms of treatment, and unfortunately anaesthesia and surgery
themselves may lead to further serious and irreversible vasospasm,
and hence to neurological deficits, despite the advances in both
operating and anaesthetic equipment and techniques of the last
decade.

The measurement of cortical oxygen has allowed us to observe
the effects of various established anaesthetic and surgical manipulations during aneurysm surgery, in particular sodium nitroprusside
(s.n.p.) induced hypotension (which is a standard manouevre during
surgery allowing safer manipulation of the aneurysm and the blood
vessels that surround it) and brain retraction; it has also allowed
us to measure the effect of new experimental drugs, in particular,
Nimodipine (Bayer BAY e 9736), a Ca^{++} antagonist whose therapeutic
use in man as an antispasm agent is of exciting great interest.

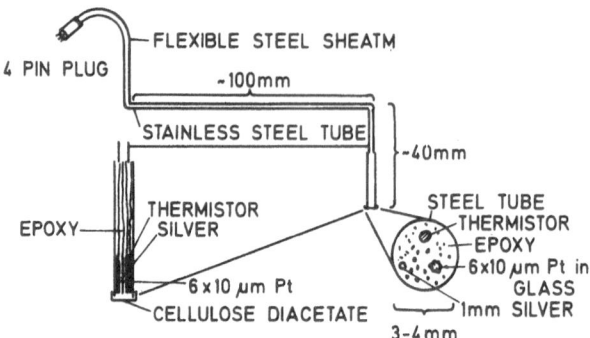

Fig. 1. Oxygen electrode.

In conjunction with this clinical work, a series of experiments are under way in the sheep to validate the measurement of cortical surface oxygen measurement by relating it to cerebral blood flow estimation using hydrogen clearance in closely adjacent areas of brain. Although the early results of these experiments are encouraging, they are not yet sufficiently advanced to allow us to make any conclusions.

METHOD

The measurement of oxygen tension at the cortical surface is made with a polarographic platinum/silver "weightless" electrode that was developed by Silver and Austin in 1981. The electrode used at present is shown in Fig. 1. It is constructed in our laboratory and contains six or seven 10 um platinum wires fused into glass and set in epoxy alongside a 1 mm silver reference electrode. The electrodes are shielded in a stainless steel tube that also serves as a convenient mount, and can be seen in Fig. 2. It has been modified recently to include a thermistor at the tip, so that cortical surface temperature may be measured, thus allowing for correction during calibration (q.v.) as well as observations of temperature changes during surgery. The tip is coated with a thin membrane of cellulose acetate.

The electrode is held in a clamp at one end of a 30 cm balance arm which pivots at the other end on a frictionless axle held in a gantry which allows movement in three dimensions (Fig. 2). The gantry is mounted firmly alongside the patient's head. A counter-balance is adjusted to allow the electrode tip to make contact with the cortex with minimal pressure (approximately 0.5 g cm^{-2}). The gantry and pivot system allow great choice in electrode position and also permit the electrode to rise and fall with the brain movements which occur with respiration and arterial pulse. The "S" shape of the electrode reduces intrusion into the operating field,

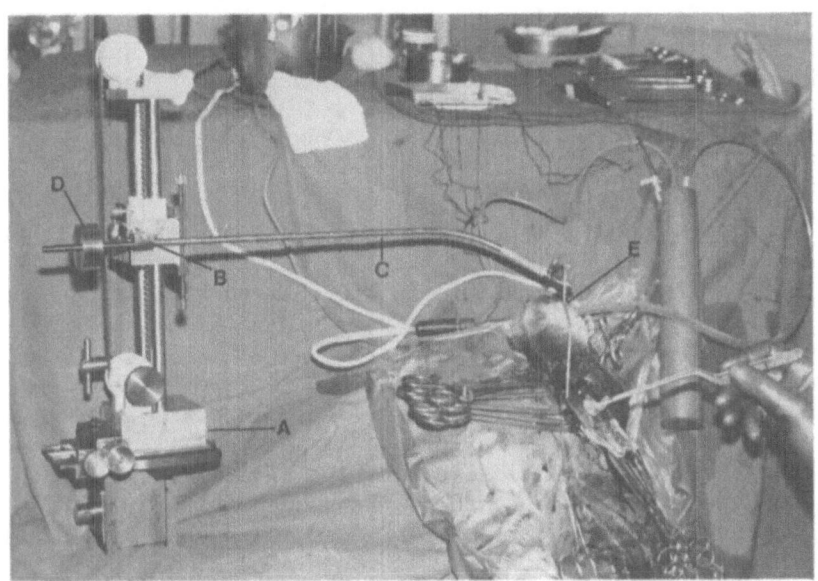

Fig. 2.   Gantry and balance arm details: A) Gantry. B) Pivot. C)
          Balance arm. D) Counter balance. E) Electrode (old design).

minimises interference with the operator, and technically allows
continuous monitoring. In practice, however, surgical manouevres
such as brain retraction interrupt the recording. Great care is
taken to ensure that the face of the electrode is parallel to the
surface of a gyrus. A position is chosen away from feeding arteries
and draining veins as these tend to give anomalously high readings.

The electrode is connected to an isolated high impedance
preamplifier via a sterile screened cable. This equipment protects
the patient from electrical inputs up to 1,500 V. The signal is fed
into a driver amplifier. A –0.75 V polarising voltage is applied to
the platinum electrode with reference to the silver anode. The
output of the amplifier is recorded on one channel of two-channel
recorder, the other being used for the continuous recording of
arterial blood pressure taken from the output of a Rigel Research
Ltd. 702 Blood Pressure monitor.

The electrode and electrical cables are sterilized in ethylene
dioxide and the gantry and balance arm in high pressure steam
(100 kPa).

PATIENTS

Measurements are at present made from those patients undergoing craniotomy to clip saccular intracranial artery aneurysms.

They have usually suffered s.a.h. at some time prior to surgery (usually 9-14 days). Patients who have suffered s.a.h. are in grades I and II (Hunt and Hess, 1968). A variety of anaesthetic agents are used with intermittent positive pressure ventilation with an endotracheal tube. The agents and dosages used are noted on each patient record (some anaesthetic agents such as halothane being profoundly vasoactive).

The equipment is set up prior to surgery and starts with the polarization of the electrode in sterile normal saline contained in a glass tonometer. The amplifier is switched on at least 1 hour before measurements start in order to minimize baseline drift. Once the electrode has been polarized, nitrogen gas (N_2) is bubbled through the tonometer so as to obtain a zero reading which can be checked against electrical zero. Once a stable nitrogen zero is obtained, the electrode is calibrated in 10% oxygen in nitrogen which covers the physiological range. The electrodes have been shown to have a linear response in the range of 0-120 mm Hg (0-15.8 kPa) oxygen. Any electrode that does not have a predictable response is rejected.

Once the brain is exposed the electrode can be placed on a suitable gyrus. An arterial blood sample is taken for blood gas analysis (for P_aO_2, P_aCO_2, and pH) and repeated at any stage if the anaesthetic agents or physiological parameters are altered. Blood pressure, E.C.G., C.V.P. and tidal carbon dioxide are continuously monitored.

Measurements are usually made before hypotension, during induction of hypotension with s.n.p. and during restoration of blood pressure following s.n.p. withdrawal. We have also observed the effect of the known vasodilator papaverine and the experimental vasodilator Nimodipine, applied topically and to the basal cisterns, both before and after s.n.p.

At the conclusion of the operation the electrode is once again checked for drift, nitrogen zero and 10% oxygen deflection in the tonometer.

A certain amount of difficulty was anticipated from artifacts that frequently occur in the course of physiological monitoring during operations. Apart from large artifacts which occur with both uni- and bi-polar diathermy and movement, we have been surprised at their absence; in particular there have not been problems with microphonics from long leads. Following diathermy the electrode settles down to its original level within a few seconds.

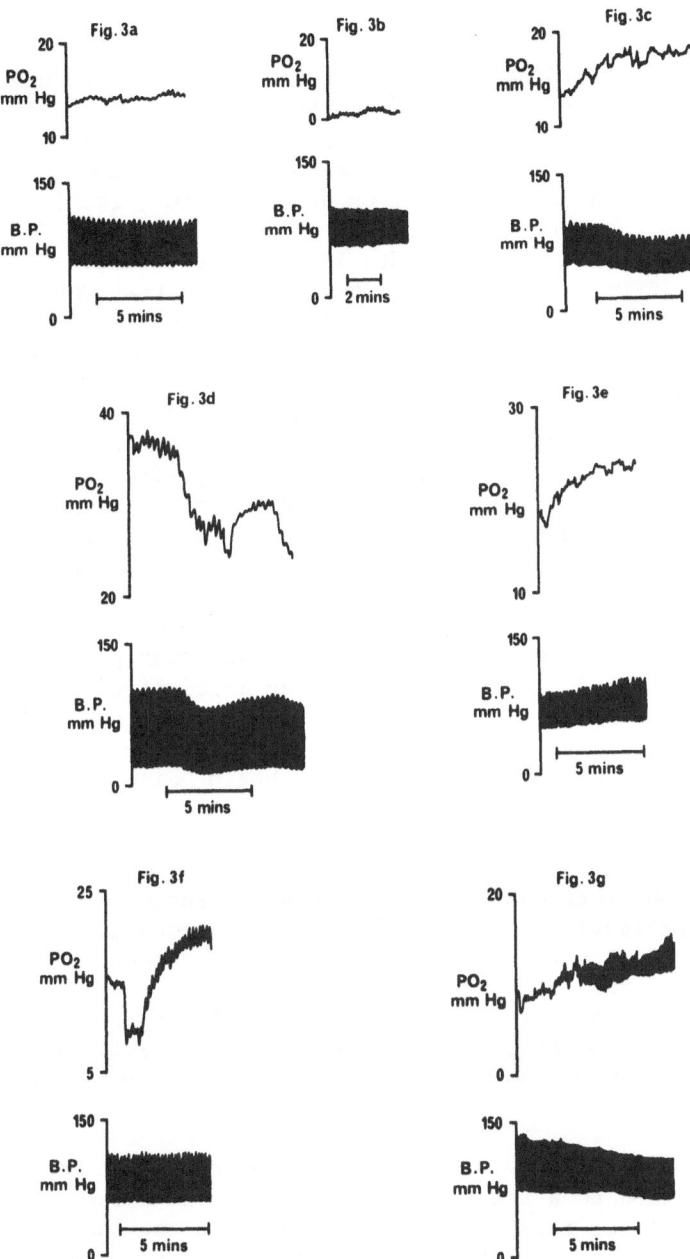

Fig. 3.   a) Normal oxygen level without vasoactive drugs. b) Zero
          oxygenation when P_aO_2 = 60 mm Hg. c) Increase in oxygena-
          tion to s.n.p. hypotension. d) Fall in oxygenation to
          s.n.p. hypotension. e) Increase in oxygenation to cister-
          nal Nimodipine following s.n.p. f) Increase in oxygenation
          to topical Nimodipine, note fluctuations with respiration
          and small temperature artifact. g) Increase in oxygenation
          to cisternal Nimodipine before s.n.p.

RESULTS AND DISCUSSION

To date, recordings have been made in 16 patients undergoing craniotomy for aneurysm surgery. We have found an initial cortical Po_2 range of 12-15 mm Hg (Fig. 3a) although the level has occasionally been lower (7-12 mm Hg). On one notable occasion when arterial P_aO_2 was 60 mm Hg, the cortical oxygen was zero (Fig. 3b). Various authors have reported the normal brain oxygen as 25 mm Hg (Silver, 1965; Whalen et al., 1970; Sundt, 1979). Observations of cerebral blood flow (CBF) in s.a.h. patients have shown that CBF is markedly reduced even in patients in good clinical grades (Hunt and Hess grades I & II). Our figures support these observations.

Three types of response have been observed during s.n.p. hypotension. Most commonly there is a sharp increase in cortical oxygen tension, usually rising to 25-35 mm Hg (Fig. 3c), despite a drop in mean arterial blood pressure (MABP) of 30-40 percent. With this rise in oxygen tension there are often small periodic fluctuations which correspond with respiration and which we believe show that the oxygen supply has exceeded tissue demands and the local tension now reflects the minor fluctuations in arterial Po_2 that occur in the blood leaving the lungs at different phases of respiration. The second group shows no change in cortical Po_2 despite the drop in MABP, which can be interpreted as successful autoregulation. In the final and so far small group, the cortical oxygen tension falls with blood pressure, the brain being unable to maintain oxygenation at prehypotensive levels (Fig. 3d). We believe that this represents failure of autoregulation.

These observations are in accord with studies of regional CBF in s.a.h. patients using xenon[133] clearance techniques. It has been observed that the majority of patients increase their CBF in response to vasodilator hypotensing agents such as s.n.p., trimetaphan and halothane but a small group fail to autoregulate and CBF falls (Heilbrun et al., 1972; Pickard et al., 1979). In one report a small group showed no change in CBF to hypotension (Griffiths et al., 1974).

Although we have yet to show experimentally the reduction of CBF in relation to a fall in oxygen tension, we believe that our tissue recordings indicate for the first time the local changes that occur in blood supply to human cerebral cortex following sub-arachnoid haemorrhage. Furthermore, our data supports the contention that the changes that occur in those patients that fail to autoregulate are not due to a resetting of autoregulation as suggested by Nornes et al. (1977) but rather to a "global failure of autoregulation"; a term used by Heilbrun et al. (1972).

Our results with vasodilator agents have also proved of interest. We have used Nimodopine, a calcium antagonist designed to

decrease cerebral vasospasm, in doses of 100-400 ug cisternally.
Following s.n.p. hypotension the drug may cause an increase in
local oxygen tension of 5-10 mm Hg sustained over 10 minutes (Fig.
3e). This has been observed in patients who increase their cerebral
oxygenation during s.n.p. hypotension but not in those patients who
fail to "autoregulate" or those who just maintain a steady Po_2
during hypotension. Topically in the vicinity of the electrode a
dose of 50-100 ug causes a temporary but repeatable increase in
local oxygen tension, which also shows periodic fluctuations with
respiration (Fig. 3f). Interpretation is more difficult with
topical application because of the artifact that tends to occur
from the temperature difference between the drug and the brain. The
only cases in which the increase fails to occur are those who are
not autoregulating. In general there appears to be some correlation
between those patients who have a substantial increase in oxygen
levels to vasodilators, and good clinical condition. However, in
our most recent measurements, we discovered an increase in oxygen
to Nimodipine in a patient in whom our present experience would
suggest little or no response would occur (Fig. 3g). The increase
was to a cisternal dose of 100 ug and was prior to s.n.p. hypotension (to which there was no response in cerebral Po_2).

Our measurements have shown for the first time a direct
beneficial effect from Nimodipine on the human brain following
s.a.h. It is planned to extend our experience of cisternal and
topical Nimodipine in different doses.

Our experience with papaverine is limited in as much as the
doses used are several orders of magnitude greater than Nimodipine.
Athough increases in oxygen tension have frequently been observed
with papaverine, it is interesting that Nimodipine can block this
effect.

Routine brain retraction has no effect on the cortical oxygen
measured either in the brain directly adjacent to the retractor
blade or in the territory of the feeding artery on which the
retractor may be exerting pressure. We have not measured oxygen
levels directly under the retractor in man in this series of
experiments, but our initial observations in sheep suggest that
oxygenation is reduced in proportion to the pressure directly under
the retractor blade and that when the remainder of the brain
autoregulates oxygen against s.n.p. hypotension, the brain under
the retractor fails to do so.

CONCLUSIONS

Microsurgical techniques in aneurysm surgery have reduced
operative mortality from 30 percent (McKissock et al., 1960) to
well below 10 percent in many centres (Krayenbuhl et al., 1972;

Yashimoto et al., 1979) but despite the technical advances of the last two decades, the complications of late ischaemia caused by cerebral vasospasm remain formidable. This risk can be reduced by delaying surgery until after the high risk period for spasm (one month) but the price is paid in a greater number of patients dying of aneurysmal rebleeds. Vasospastic ischaemia may cause progressive neurological deficits and even death in patients awaiting surgery, and although the operative results are excellent in those surviving the delay, the overall mortality may be in the region of 25 percent. Because of this dilemma, the neurosurgical trend is towards early surgery in s.a.h., despite the risk of exacerbating the dangers of vasospastic ischaemia with surgery.

A large amount of work has been performed to improve our understanding of vasospasm in s.a.h. but measurement and monitoring have presented great difficulties in the human. Much has been done to try to correlate angiographic spasm with the clinical condition, but as vasospasm is a dynamic process, this is frequently impossible or of little relevance. Others have measured in vitro preparations or constructed animal models of sub-arachnoid haemorrhage in order to assess physiological and pharmacological effects and then extrapolate to the human disease. Unfortunately, the exact conditions in s.a.h. are virtually impossible to mimic experimentally and any conclusion must be carefully qualified and its validity questioned, and it is for these reasons that a cure of vasospasm remains elusive (see Wilkins et al., 1980).

It would seem that local intraoperative monitoring of patients will give us the best insight into cerebrovascular dynamics in s.a.h. We hope that this study will continue to give useful information about autoregulation as affected by different pharmacological agents, and applying this to specific patients in relation to their progress, we will gain important knowledge of the complex and multifactorial features that govern the outcome in this ill understood and dangerous disease.

ACKNOWLEDGEMENTS

This work was supported by funds from the South West Regional Health Authority, U.K., Program Project N.S. 10939 of the National Institute of Health U.S.A., and Bayer U.K. Ltd. Pharmaceutical Division.

REFERENCES

Griffiths, D.P.G., Cummins, B.H., Greenbaum, R., Griffith, H.G., Staddon, G.E., Wilkins, D.G., and Zorab, J.M., 1974, Cerebral blood flow and metabolism during hypotension induced with sodium nitroprusside, Br. J. Anaesth., 46:194-198.

Heilbrun, M.P., Olesen, J., and Lassen, N.A., 1972, Regional cerebral blood flow in sub-arachnoid haemorrhage, J. Neurosurg., 37:36-44.

Hunt, W.E., and Hess, R.M., 1968, Surgical risk as related to time of intervention on the repair of intracranial aneurysms, J. Neurosurg., 28:14-20.

Krayenbuhl, H.A., Yasargil, M.G., Flamm, E.S., and Tew, J.M., 1972, Microsurgical treatment of intracranial saccular aneurysms, J. Neurosurg., 37:678-686.

McKissock, W., Paine, K.W.E., and Walsh, L.S., 1960, An analysis of the results of treatment of ruptured intracranial aneurysms, J. Neurosurg., 17:762-766.

Nornes, H., Knutzen, H.B., and Wikeby, P., 1977, Cerebral arterial blood flow at aneurysm surgery, Part 2: Induced hypotension and autoregulatory capacity, J. Neurosurg., 47:819-827.

Pickard, J.D., Matheson, M., Patterson, J., and Wiper, D., 1980, The prediction of late ischaemic complications after cerebral aneurysm surgery by the intraoperative measurement of cerebral blood flow, J. Neurosurg., 37:36-44.

Silver, I.A., 1965, Some observations on the cerebral cortex with an ultramicro membrane covered oxygen electrode, Med. Electron. Biol. Eng., 3:377.

Silver, I.A., and Austin, G., 1981, Oxygen measurements in human brain, Adv. Exp. Med. Biol., (in press).

Sundt, T.M., 1979, "Blood Flow Regulation in Normal and Ischaemic Brain", Current Concepts, Upjohn, London.

Whalen, W.J., Ganfield, R., and Nair, P., 1970, Effects of breathing O_2 or O_2 and CO_2 and of the injection of neurohumours on the P_{O_2} of the cat cortex, Stroke, 1:194-197.

Wilkins, R.H., 1980, Cerebral artery vasospasm, Proc. Second Int. Workshop, Amsterdam, 1979, Williams and Wilkins, Baltimore and London.

Yashimoto, T., Uchida, K., Kaneyko, U., Kayama, T., and Suzuki, J., 1979, An analysis of follow up results of 1.000 intracranial saccular aneurysms with definitive surgical treatment, J. Neurosurg., 50:152-157.

FOCAL EPICEREBRAL ISCHEMIA: POST-ISCHEMIC TISSUE OXYGENATION WITH AND WITHOUT RECIRCULATION

N. Wiernsperger

Preclinical Research, Sandoz Ltd., Basle, Switzerland

According to statistical analyses of stroke cases in intensive care units, about 70% of cerebrovascular accidents are the results of vascular occlusive diseases of various origins (mainly embolic or thrombotic) (Hachinski and Norris, 1981). The majority of brain infarcts occur in the territory of the middle cerebral artery (MCA), the pathological consequences being found in the insular cortex and basal ganglia. Whereas it is commonly accepted that the primary ischemic focus is irreversibly damaged, increasing interest is being devoted to the surrounding tissue, which is still viable. In fact, maintaining the collateral circulation should be of benefit to the infarct area and restrict the extension of ischemia into adjoining tissue.

The only experimental model presently available which provides the possibility of inducing focal brain infarcts is the occlusion of the MCA. However, this model requires neurosurgical procedures which are painstaking and time-consuming. In addition, ethical considerations prevent the use of this technique in many countries since it involves ocular enucleation. We therefore attempted to overcome these problems by using neurosurgical microclips in order to occlude small branches of the MCA which run over the cerebral cortex. The choice of the vessels to be occluded was based on macroscopic estimation of the tissue area supplied by this vessel. Tissue Po_2 measurements were performed to provide an indication of microcirculatory disturbances following the arterial occlusion.

MATERIAL AND METHODS

All experiments were performed using adult cats weighing 2 - 2.5 kg. They were anesthetized with pentothal for surgical

251

procedures, i.e., tracheostomy, catheterization of femoral arteries and veins, and trepanation of the skull in the parietal region. The animals were then artificially respirated with N_2O/O_2 (70%:30%, v/v) and immobilized with gallamine triiodoethylate. Blood pressure, heart rate, end-expiratory Pco_2 and body temperature were continuously monitored. After careful removal of the dura, a suitable area for clamping and measuring tissue Po_2 was selected. Small branching arteries from the MCA were clamped by a specially sharpened neurosurgical microclip. This procedure was used on vessels as small as 200 um diameter under microscopic control, extreme care being taken not to damage the surrounding tissue. Tissue Po_2 was measured with surface platinum micro-electrodes having a tip diameter of 2 - 5 um. The oxygen pressures were recorded in areas of approximately 10 mm^2.

Two different approaches were used: In one group of animals, the microclip was removed after 60 min of ischemia, while tissue Po_2 was continuously recorded throughout the experiment. In another group of cats, ischemia was maintained for up to 180 min. The Po_2 distribution histogram was determined by picking up local oxygen pressures at different sites over the infarcted area. In the latter case, statistical analysis was based on the determination of the 95%-confidence interval around the median (Wilcoxon-Mann-Whitney test). Post-treatment values were compared with ischemic ones.

Drugs were administered intravenously over a 10 min period, beginning after 50 min of ischemia as follows:

- aspirin: 10 mg/kg
- PY 108-068: 20 ug/kg/10 min (acute or continuous infusion)

PY 108-068 is a dihydropyridine calcium antagonist. In control experiments, Evans-blue was infused for testing possible disturbances of the blood-brain barrier. Before sacrifice, indian ink was given i.v. for testing the speed of blood-filling into the infarcted area.

RESULTS

Control Experiments

Immediately following clamping of the artery, observation through the operating microscope revealed an inversion of the direction of blood flow distal to the occlusion. After a few minutes, blood flow clearly becomes slower and the blood appears to be more viscous. Accompanying these macroscopic observations were Po_2 measurements which progressively decreased (Fig. 1). Ischemia in this model was not accompanied by any change in arterial blood pressure or heart rate.

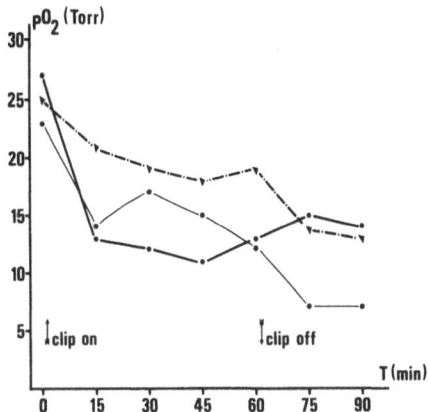

Fig. 1.   Three representative, typical Po_2 tracings recorded in the
          cerebral cortex after inducing focal epicerebral ischemia.
          Occlusion of the artery leads to variable decreases in
          tissue Po_2. By removing the microclip, most sites exhibit
          a further decrease in tissue oxygenation.

      a) Focal ischemia with recirculation: Fig. 1 shows character-
istic continuous Po_2 tracings obtained for the duration of the experiment. After an initial first stage decrease, most sites exhibit
a further, slight decrease in oxygen partial pressure. Removal of
the clip leads to a secondary decrease in tissue oxygenation. The
Po_2 measurements performed around the estimated area of infarction
reveal no, or only moderate changes (Fig. 2). From these measurements, we estimate the size of the focal ischemic area to be on the
order of 10 mm^2.

      b) Focal ischemia without recirculation: Fig. 3a shows the Po_2
distribution in cortical tissue before and after 2 hours of focal

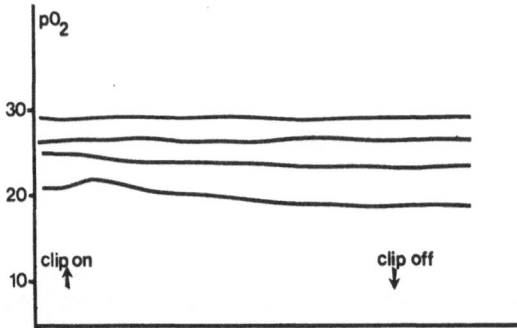

Fig. 2.   Representative Po_2 tracings recorded in the intact surround-
          ing tissue in the immediate vicinity of the infarcted area.
          Here, tissue oxygenation is not, or only slightly influenced
          by placing or removing the microclip.

ischemia in 33 untreated cats. The Po$_2$ histogram is shifted to the
left, the center of gravity being decreased from 27 to 17.5 Torr.
There is no accumulation of low, i.e., nearly anoxic values. Fig. 3b
represents the modification to the 95% confidence interval around the

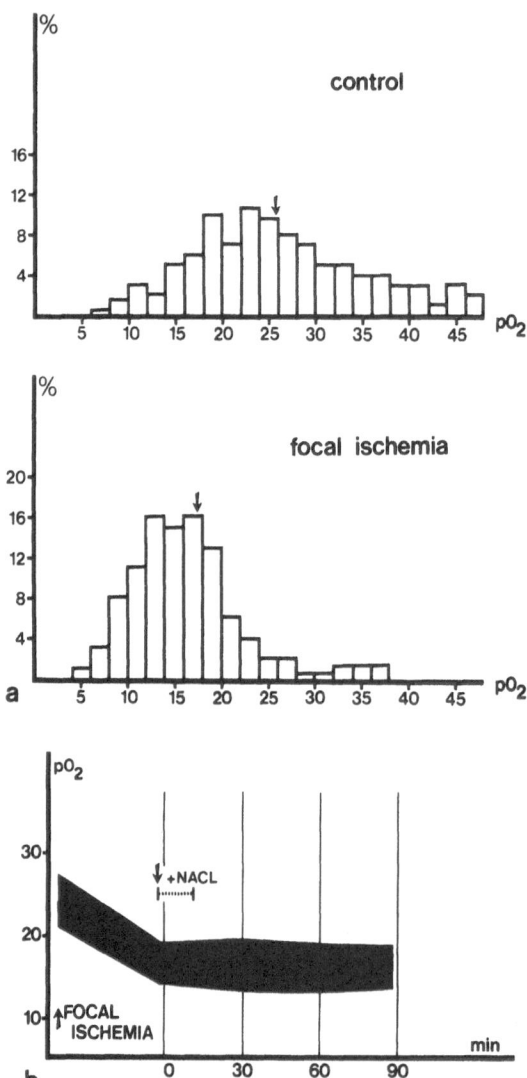

Fig. 3. a) Po$_2$ histograms of the cortical oxygenation in 33 un-
            treated cats before and at the end of 90' focal epi-
            cerebral ischemia. 424 Po$_2$ values were recorded before
            and 579 values after ischemia. Arrows indicate the center
            of gravity of the histogram.
         b) Evolution of the 95% confidence interval during the
            whole duration of the focal ischemia in the control
            group.

median over the whole duration of the experiment in the same animal group.

Indian ink administration before sacrifice revealed an 8 - 10 sec delay in the dye filling of the ischemic area as compared with the intact surrounding tissue. These observations confirmed the previous estimate obtained with Po_2 measurements that the size of the infarct was about 10 mm^2. At no time could any Evans-blue extravasation be seen in this group.

Pharmacological Experiments

a) Aspirin: Fig. 4 shows the effect of a large dose of aspirin in experiments with a permanent occlusion. As compared with ischemic values, the tissue Po_2 is significantly increased after aspirin administration. This effect is of variable duration in individual experiments, which explains the enlargement of the 95% confidence band at the end of the 90' period of ischemia.

b) PY 108-068: The effect of a single, short-lasting infusion (10') of the calcium antagonist PY 108-068 (total dosage: 20 ug/kg) on tissue oxygenation is shown in Fig. 5a. Following drug administration, there is a highly significant increase in cortical tissue oxygenation, which tends to diminish as a function of time. In Fig. 5b, experiments using a continuous infusion of PY 108-068 (20 ug/ kg/10 min for 3 hours) show that this procedure overcomes the problem of short-lasting drug efficacy.

DISCUSSION

The results obtained here compare favorably with those reported where the classical MCA occlusion (MCAO) was utilized. Bremer et al.

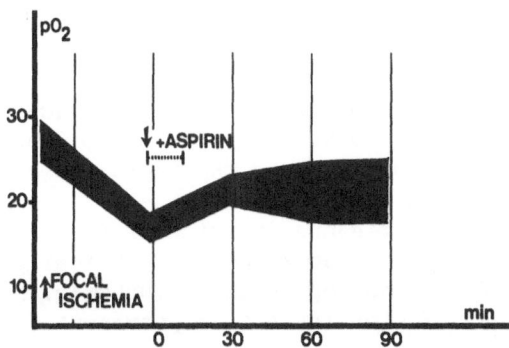

Fig. 4.   Effect of aspirin (10 mg/kg), administered during ischemia, on the 95% confidence interval of cortical tissue Po_2.

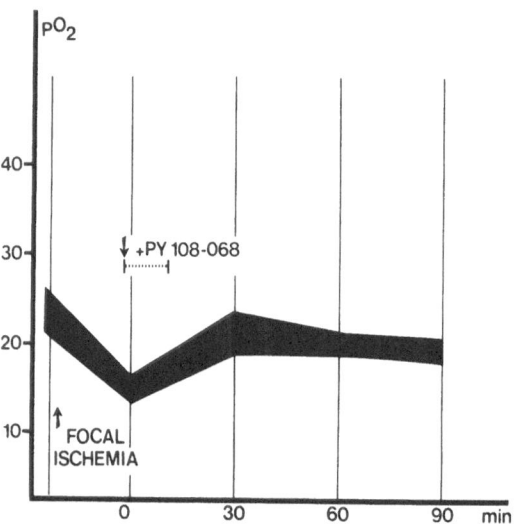

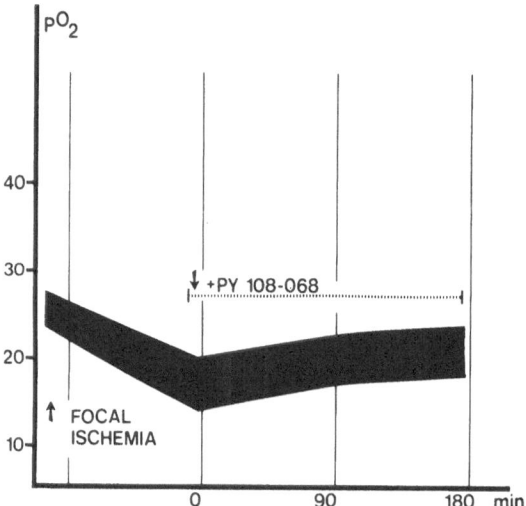

Fig. 5.   a) Effect of the calcium antagonist PY 108-068, administer-
             ed as a single dosage, on tissue Po_2 during focal epi-
             cerebral ischemia. Note the transient efficacy of a
             single infusion.
          b) Effect of continuous infusion of PY 108-068 (20 ug/kg/
             10 min) during 3 hours on the cortical tissue oxygena-
             tion after focal epicerebral ischemia. Note that con-
             tinuous infusion leads to a permanent efficacy of the
             drug.

(1979) found a mean control tissue Po_2 of 26 Torr. A decrease of approximately 30% was observed after MCAO. Ingvar et al. (1979) reported similar results. The data presented here using the epicerebral technique fall into the same range (27 and 17.5 Torr before and during ischemia respectively). In a similar model in the dog, Ravvin et al. (1977) reported a 20% decrease in local CBF and a perfusion delay of 8 seconds, which correlates well with the delay in the dye filling observed in this study.

MCAO induces ischemia over a large area of cortical tissue, which explains the usually higher values described in the literature for decreases in CBF and tissue Po_2 (Bremer et al., 1979; Cahn et al., 1980). In the model used here, cortical infarction is of a much more local type. The capacity of small vessels lying in the immediate vicinity of the ischemic core can be recruited, which probably explains why the microcirculatory changes are quantitatively less pronounced than in classical MCAO. Thus, only very few decidedly low Po_2 values appear on the histogram.

The detailed analysis of cortical oxygenation as a function of time reveals a progressive decrease with duration of ischemia. We did not find a biphasic behaviour of tissue oxygenation as described by Cahn et al. (1980). During the same period, macroscopic observation of the brain surface shows that blood flowing through the vessels within the ischemic area becomes progressively more viscous in appearance. The phenomenon, beginning in veins, is illustrated by the slowing of, and/or aggregate-like blood flow. Such observations have been made by Little et al. (1981), who measured a definite increase in erythrocyte and plasma transit time through the ischemic tissue. Morphological studies by the same author confirmed the obstruction of capillaries by erythrocytes (Little and O'Shaughnessy, 1979). In a sequential study of the influence of MCAO on microcirculation in primates, Theodore et al. (1980) found a progressive decrease in the number of perfused capillaries with time after the induction of ischemia. All these data support the concept of a low-flow state accompanied by, or due to, hemorheological changes caused by aggregation of platelets and/or erythrocytes. Pharmacological studies have shown the transient efficacy of therapeutic measures such as hemodilution with low molecular weight dextran (Crowell and Olsson, 1972), hemodilution combined either with hypothermia (Blöink et al., 1980) or with anticoagulant therapy (Sundt and Waltz, 1967), or oxygen carriers like Fluosol-DA (Peerless et al., 1981).

Our results show that, at least in most cases, large doses of aspirin are of benefit to the microflow in the ischemic tissue and confirm many studies showing the importance of platelets in cerebral infarction. In this model of epicerebral occlusion, we did not observe a Po_2 overshoot after clip removal, as has been described by others (Crockard et al., 1976). In contrast, we usually observed a secondary fall in tissue Po_2. We suggest that blood cell aggregates are released by reopening the vessel, which then clog the distal capillary bed. An alternative explanation could be steal effect by post-ischemic hyperperfusion in the border zone.

Similarly, the absence of Evans-blue extravasation in our experiments fits in with the observations made by others during the first 2 hours of MCAO. Neither Kamijyo et al. (1977) nor Schuier and Hossmann (1980) found the blood brain barrier to be permeable following the early phase of MCAO, the latter authors finding a most cytotoxic edema.

The beneficial effect observed by use of the calcium antagonist PY 108-068 confirms that calcium entry into vessel walls is another important factor in the pathophysiology of brain infarction. The disappearance of extracellular Ca^{++}, following an increase in extracellular K^{+}, has been described by Harris et al. (1981). Tamura and Teasdale (1980), have also reported relaxation of arterioles after perivascular microapplication of Nifedipine in a model of MCAO. However, they found the action of Nifedipine to be short-lasting, an observation we also made with PY 108-068 (a drug of the same class, namely dihydropyridine). We could overcome this problem by continuous infusion of the drug without potentiating negative cardiovascular effects.

In summary, this investigation has shown that for acute studies, epicerebral focal ischemia is a useful modification of the classical MCAO technique. It is less time-consuming, allows the study of small infarcted areas and has the same pathophysiological characteristics observed after MCAO. In addition, the model has ethical advantages. In contrast to many other CNS ischemia models, tissue oxygenation levels are not dramatically decreased because of a developing collateral circulation (Wiernsperger and Gygax, 1981). The latter avoids the occurence of strongly hypoxic or even anoxic sites in the cortical tissue, which possibly guarantees the reversibility of the ischemic damage over time. Accordingly, we have shown that during the early phase after the insult, oxygen supply in the ischemic area can be ameliorated by improving hemorheological complications involving platelets and erythrocytes, or by counteracting the increase in concentration of Ca^{++} in the vessel wall. Whether the improvement in tissue Po_2 is due to a better collateral circulation or due to additional factors cannot be answered by our measurements. This would require further studies on circulatory and metabolic events using iodoantipyrine or deoxyglucose techniques.

REFERENCES

Blöink, M., Hossmann, V., and Hossmann, K.A., 1980, Treatment of experimental infarcts following middle cerebral artery occlusion in cats, in: "Circulation Cérébrale", Proc. Congrès International de Circulation Cérébrale, 1979, A. Bès, G. Géraud, eds., Toulouse.

Bremer, A., West, C.R., and Yamada, K., 1979, Alterations in corti-
    cal oxygen tension during the development of ischemic cerebral
    edema in primates, Neurosurgery, 4:233.
Cahn, R., Foncin, J.F., and Pawelec, C., 1980, Brain cortical Po$_2$,
    EEG activity, cerebral (A-V) lactate and ultrastructural
    changes in temporary middle cerebral artery occlusion in
    chloralose anaesthetized mongrel dogs, in: "Pathophysiology
    and Pharmacotherapy of Cerebrovascular Disorders", A. Betz,
    J. Grote, D. Heuser, R. Wüllenweber, eds., Witzstrock, Baden-
    Baden, p. 187.
Crockard, H.A., Symon, L., Branston, N.M., and Juhasz, J., 1976,
    Changes in regional cortical tissue oxygen tension and cerebral
    blood flow during temporary middle cerebral artery occlusion
    in baboons, J. Neurol. Sci., 27:29.
Crowell, R.M., and Olsson, Y., 1972, Impaired microvascular filling
    after focal cerebral ischemia in the monkey. Modification by
    treatment, Neurology, 22:500.
Hachinski, V., and Norris, J.W., 1981, Intensive care of stroke, in:
    "Drugs and Methods in CVD", Pergamon Press, Paris, p. 375.
Harris, R.J., Symon, L., Branston, N.M., and Bayhan, M., 1981,
    Changes in extracellular calcium activity in cerebral ischemia,
    J. Cereb. Blood Flow Metab., 1:203.
Ingvar, M.C., Feustel, P.J., Brenneman, L., and Severinghaus, J.W.,
    1980, Local cerebral blood flow and oxygen availability measured
    with the same 25 micron electrodes in cat cortex before and
    during middle cerebral artery ligation at altered Pco$_2$ levels,
    in: "Cerebral Blood Flow and Metabolism", F. Gotoh, H. Nagai,
    Y. Tazaki, eds., Munksgaard, Copenhagen, pp. V-7.
Kamijyo, Y., Garcia, J.H., and Cooper, J., 1977, Temporary regional
    cerebral ischemia in the cat, J. Neuropathol. Exp. Neurol.,
    36:337.
Little, J.R., Cook, A., Cook, S.A., and McIntyre, W.J., 1981, Micro-
    circulatory obstruction in focal cerebral ischemia: albumin and
    erythrocyte transit, Stroke, 12:218.
Little, J.R., and O'Shaughnessy, D., 1979, Treatment of acute focal
    ischemia with continuous CSF drainage and mannitol, Stroke,
    10:446.
Peerless, S.J., Ishikawa, R., Hunter, I.G., and Peerless, M.J., 1981,
    Protective effect of Fluosol-DA in acute cerebral ischemia,
    Stroke, 12:558.
Ravvin, L.J., Feindel, W., Yamamoto, Y.L., and Hodge, C.P., 1977,
    Epicerebral arterial occlusion in the dog, in: "Cerebral Func-
    tion, Metabolism and Circulation", D.H. Ingvar, N.A. Lassen,
    eds., Munksgaard, Copenhagen, p. 363.
Schuier, F.J., and Hossmann, K.A., 1980, Experimental brain infarcts
    in cats. II. Ischemic brain edema, Stroke, 11:593.
Sundt, T.M., and Waltz, A.G., 1967, Hemodilution and anticoagulation.
    Effects on the microvasculature and microcirculation of the
    cerebral cortex after arterial occlusion, Neurology, 17:230.

Tamura, A., and Teasdale, G., 1980, Effects of perivascular micro-
     application of Nifedipine on cat pial arteriolar and venular
     caliber, in: "Pathophysiology and Pharmacotherapy of Cerebro-
     vascular Disorders", E. Betz, J. Grote, D. Heuser, R. Wüllen-
     weber, eds., Witzstrock, Baden-Baden, p. 19.
Theodore, D., and Abraham, J., 1980, A sequentious study of capil-
     laries in the infarcted area of primate brain, Ind. J. Med.
     Res., 71: 821-828.
Wiernsperger, N., and Gygax, P., 1981, Determination of tissular Po_2
     in brain microvascular research, Eur. Neurol, 20:200.

THE EFFECT OF GLUCOSE ON THE OXYGEN SUPPLY OF THE BLOOD-FREE PERFUSED GUINEA PIG BRAIN AS MEASURED BY REFLECTION SPECTRA AND Po_2 HISTOGRAMS

U. Heinrich, B. Yu, J. Hoffmann, and D.W. Lübbers

Max-Planck-Institut für Systemphysiologie, Rheinlanddamm
201, 4600 Dortmund 1, F.R.G.

INTRODUCTION

Earlier experiments with the guinea pig brain had shown that at a temperature of 37° with a substrate-free macrodex solution, equilibrated with 95% O_2 and 5% CO_2 and with a large flow cytochrome aa_3 (cyt aa_3) was only about 60% oxidized (Schwickardi, 1968; Heinrich et al., 1981). Under similar conditions the heart muscle behaves differently: In the blood-free perfused Langendorff heart the cyt aa_3 is about 95% oxidized (Figulla et al., 1979). Since in this case the perfusion medium was different, the question arises, whether the difference in the composition of the perfusion medium (for example the addition of substrate) influences the oxygen supply and can - at least in part - be responsible for the observed difference in the redox state of the respiratory chain. To have a better insight in the oxygen supply, in the following experiments, we combined the measurements of the redox states of the cytochromes with the measurements of Po_2 histograms on the brain surface. The redox state of cytochrome aa_3 serves as an indicator of the intracellular mitochondrial O_2 supply, whereas the Po_2 histogram indicates the Po_2 distribution in the extracellular space. The measurements were performed with two different perfusion media and at two different temperatures.

METHODS

Guinea pigs weighing 200-400 g were pretreated with 0.5 ml Liquemin 30 min before starting the experiments. Nembutal (60 mg/ml) in a concentration of 0.1 ml/100 g b.w. with an addition of 0.1 ml atropine (1 mg/ml) was used for anaesthesia.

The animals were tracheotomized at the first step. Then axillary vessels were ligated and the afferent catheter was fixed in the abdominal aorta as well as the efferent catheter in the inferior vena cava. At last the skull cap was trephinated by drilling a hole of 5 mm in diameter into the skull (Knaust, 1967).

The composition of the perfusion media was the following:

Perfusion medium A: Macrodex 6%, $NaHCO_3$ (20 nmol/l), papaverine (4.0 mg/100 ml perfusion solution).

Perfusion medium B: Macrodex 6%, glucose (5.5 $mmol \cdot l^{-1}$) + KCl (4.7 $mmol \cdot l^{-1}$) + CaCl (1.25 $mmol \cdot l^{-1}$) + KH_2PO_4 (1.2 $mmol \cdot l^{-1}$) + $NaHCO_3$ (24.9 $mmol \cdot l^{-1}$) + $MgSO_4$ (0.3 $mmol \cdot l^{-1}$) + pyruvate (2.0 $mmol \cdot l^{-1}$) + papaverine (4.0 mg/100 ml perfusion solution).

All perfusates were heated up to temperatures of 18°C and 37°C and aerated with carbogen (95% O_2 + 5% CO_2) by a disk oxygenator. pH value of perfusates amounted to 7.3. The perfusion rate was 30 ml/min.

Before and during the perfusion local cortical tissue Po_2 was measured polarographically with multiwire Pt surface electrodes according to Kessler and Lübbers (1966) covered by membranes of 12 um cellophane and 12 um Teflon. In these experiments we used an electrode with a silver oxide reference electrode and a special electrolyte (Yu et al., this volume). The advantage of this electrode is its relatively small drift, particularly at high Po_2 values. Calibration of the electrodes was performed with gas mixtures before and after each measuring series. The drift amounted almost to zero.

All recordings of reflection spectra of the perfused brain were performed by the rapid spectrometer T13/3 (Kieler Howaldswerke, Kiel, FRG) according to Lübbers and Niesel (1957). It is a double beam photometer with a measuring frequency of 25 kHz. In our case the desired wavelength range was 505-640 nm in 10 ms, so that movements of the organs could be easily cancelled. The light beam of the photometer was coupled to the brain by a solid light guide. The size of the light on the brain surface was about $4mm^2$, the depth of penetration about 1.0-1.5 mm.

The signals of the rapid spectrometer (wavelength and extinction) were digitized and stored in a computer (Honeywell, Interdata 7132). The reflection spectra were evaluated by a nonlinear multicomponent analysis according to Hoffmann et al. (this volume).

RESULTS

Figs. 1 and 2 show, as examples, the reflection spectra of the brain surface during a brain perfusion with the perfusion medium B

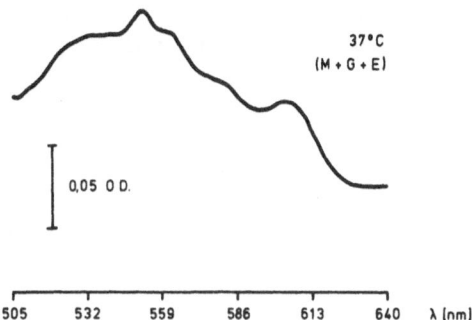

Fig. 1.   Reflection spectra of the brain surface measured during
          brain perfusion with medium B (temp. = 37°C). OD = optical
          density, λ = wavelength.

at two different temperatures (18°C and 37°C), in the wavelength
range of 505–640 nm. In Fig. 1 the spectra shows the characteristic
α-bands of reduced cytochromes: the peak at λ = 550 nm corresponds
to cyt c, the shoulder at λ = 562 nm to cyt b and the peak at λ =
605 to cyt aa_3. The spectra of the oxidized cytochromes (Fig. 2)
are rather uncharacteristic. From such spectra the absolute value
of the redox state of cytochrome aa_3, c and b were calculated.
Table 1 shows the calculated redox states at two different temperatures (18°C and 37°C) and with two different perfusion media A and
B. The error in the calculation of the percentage of oxidation
amounts to 3%. At 18°C cyt aa_3 and cyt c are almost oxidized. With
medium B a 100% oxidation of cyt aa_3 was measured. By increasing
the temperature to 37°C all cytochromes are becoming more reduced:
cyt aa_3 with medium A by 35%, with medium B only by 16%. Cyt c is
reduced with both media by about the same amount (A 30%, B 27%). Cyt
b is always more reduced than cyt aa_3 and c (Heinrich et al., 1981).
It decreases by 50% with medium A and only by 10% with medium B. As

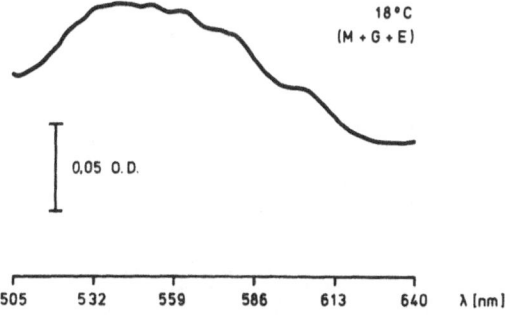

Fig. 2.   Reflection spectra of the brain surface measured during
          brain perfusion with perfusion medium B (temp. 18°C). OD =
          optical density, λ = wavelength.

Table 1. Degree of oxidation of cytochromes (cyt) in % in the
cortex of blood-free perfused guinea pig brain under
different conditions.

Temperature	cyt aa₃		cyt c		cyt b	
	med A	med B	med A	med B	med A	med B
18°C	95%	100%	95%	91%	70%	60%
37°C	60%	84%	65%	64%	20%	50%

med A,B: perfusion media; composition see methods

compared to medium A, medium B increases the degree of oxidation of
cyt aa₃ and b. At 37° the increase amounts to 24% for cyt aa₃, to
30% for cyt b. The redox state of cyt c does not change.

Fig. 3 shows the P_{O_2} histogram of the brain surface of a guinea
pig with normal blood flow, spontaneous respiration and Nembutal
anaesthesia at body temperature (37°C) as it is measured before the
blood-free perfusion starts. The P_{O_2} classes (abscissa) have a size
of 5 mm Hg. About 4-5% of the P_{O_2} values are below 5 mm Hg, but very
few values close to zero. The maximum lies between 15 and 20 mm Hg
P_{O_2}. In the blood-free perfused brain the surface P_{O_2} increases
up to 630 mm Hg. This corresponds to the high P_{O_2} of the inflowing
medium (P_aO_2 = 630 mm Hg (83.9 kPa)). The class size in the P_{O_2}
histogram of Figs. 4 and 5 is seven times larger, i.e. 35 mm Hg. In

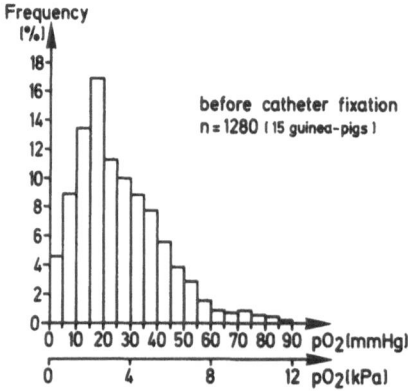

Fig. 3. P_{O_2} histogram of the blood perfused guinea pig brain
surface (temp = 37°C, Nembutal anaesthesia, spontaneous
respiration).

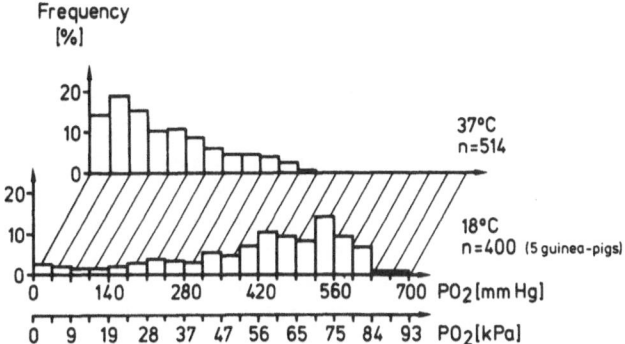

Fig. 4. Mean Po_2 histograms of the blood-free perfused brain
         cortex at two different temperatures (perfusion medium A).

Fig. 4 at 18°C corresponding to the degree of oxidation of cyt aa_3
there are only a few Po_2 values in the lowest class. The maximum
lies between 525-560 mm Hg Po_2. By increasing the temperature to
37°C the Po_2 histogram moves to the left, i.e. the number of low Po_2
values increases. The maximum is in the class between 35-70 mm Hg
Po_2.

       The perfusion medium B distinctly improves the oxygen supply.
At 18°C there are no Po_2 values smaller than 70 mm Hg any more.
The maximum lies in the class of 315-350 mm Hg Po_2, but the number
of high Po_2 values is increased. Even at 37°C the lowest Po_2 class
is empty. The Po_2 histogram is more uniform than all the others in
Fig. 4 and 5.

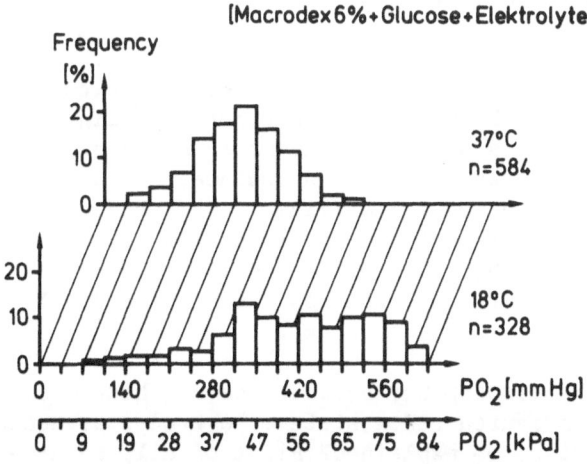

Fig. 5. Mean Po_2 histograms of the brain cortex determined from
         the blood-free perfused brain at two different temperatures
         (perfusion medium B).

DISCUSSION

The nonlinear multicomponent analysis according to Hoffmann et al. (this volume) allows a quantitative evaluation of the reflection spectra of the blood-free perfused brain. With 16 averaged spectra (i.e. 3 or 6 spectra/second) the absolute measuring error lies in the range of 0.0015 O.D.. From a single spectra the redox state of cyt aa_3, c and b is obtained. The degree of oxidation can be determined with an error of 3%. The Po_2 measurements could be improved by a Po_2 electrode with a special silver oxide reference electrode and an adapted electrolyte (borax-biphosphate buffer). This electrode was remarkably stable, even at high Po_2 values (Yu et al., this volume).

By combination of these methods it was possible to measure the redox state of the cytochromes and the Po_2 histogram simultaneously. Both data clearly showed that as compared to medium A, medium B considerably improves the oxygen supply to the brain (Table 1, Figs. 4 and 5). The effect at a temperature of 18°C is small, but visible. In the Po_2 histograms (Figs. 4 and 5) the number of low Po_2 values decreases and cyt aa_3 - as the reaction partner of molecular oxygen - becomes more oxidized. From the Po_2 histogram the conclusion is allowed that with medium B the O_2 supply to the brain, even at 37°C, is sufficient. After the general experience the uniform appearance of the Po_2 histogram at 37°C could be taken as a sign of intact microcirculation and perfect oxygen offer. Whether the addition of substrate or of other components is the main factor for this improvement and whether further improvements by better adaptation of the perfusion medium are possible has to be tested in additional experiments.

The degree of oxidation of cyt aa_3 in the brain cortex perfused at 37°C with medium B amounts to 80%. This was somewhat in contrast to our expectations since in the Langendorff heart cyt aa_3 is at least 95% oxidized. In respect to the reaction with oxygen there is no difference known between the behaviour of brain and heart mitochondria. Experiments with isolated brain mitochondria respiring in state 3 (oxygen and substrate in abundance; ADP and inorganic phosphate sufficient; Chance and Williams, 1956) had shown that at a temperature of 23-25°C cyt aa_3 is almost 100% oxidized, when the Po_2 in the medium is higher than 0.1 mm Hg (Starlinger and Lübbers, 1973). In these experiments the Po_2 was simultaneously measured with the redox state of cytochromes.

How can the discrepancy between the Po_2 histogram and the simultaneously measured redox state of cyt aa_3, as measured in our experiment in vivo, be explained? First of all, we have to consider that reflection photometry and surface Po_2 measure different parameter of the oxygen supply: the redox state of cyt aa_3 mirrors the mitochondrial oxygen supply whereas by the surface Po_2 the Po_2 in

the extracellular space is measured. Thus, it may be that there is
a Po_2 gradient across the cell membrane and/or within the cell. To
test this possiblity we will try to estimate for experiments with
the perfusion medium B at 37°C of which order of magnitude the Po_2
gradient between the extracellular space and the mitochondria would
be.

Assuming that at state 3 respiration cyt aa_3 is oxidized 95%
and that the relationship between the degree of oxidation and the
Po_2 in the surrounding medium measured in isolated brain mitochondria is also valid in vivo, we find for a 80% oxidation that a
tissue volume of about 15% has an oxygen pressure less than 0.1 mm
Hg. For such a percentage we find in the Po_2 histogram Po_2 values
up to about 140 mm Hg, so that the 3 lowest classes of the Po_2
histogram would roughly depict the existing Po_2 gradients. With
intact microcirculation such large Po_2 gradients have never been
observed. Thus, Po_2 gradients may contribute to the observed difference, but can hardly be the single cause.

Therefore, another difference between the measurements may be
of importance: whereas the Po_2 is measured directly on the brain
surface, the redox state is measured in the volume below the surface.
The penetration depth of the light beam is about 1-1.5 mm. The contribution of different depth follows about an exponential function,
so that the upper layers contribute most of the signal. Therefore,
it has to be discussed, whether the Po_2 histogram of the brain
surface is valid for deeper layers. The Po_2 histogram which was
measured before the perfusion started (Fig. 3) is similar to the
histograms published in the literature (see for example Baumgärtl
and Lübbers, 1983). It was experimentally proved that for these
conditions the Po_2 histogram measured by Po_2 needle electrodes
in deeper layers of the cortex showed no systematic difference as
compared to the simultaneously measured surface Po_2 histogram.
However, with the blood-free perfused brain much larger Po_2 decreases along the capillary occur. Using the extreme Po_2 values of
the Po_2 histogram we obtain a Po_2 difference of about 600 mm Hg
whereas in blood-perfused brain this difference amounts only to
90 mm Hg Po_2, i.e. it is about 7 times smaller.

In Hb-free perfused capillary of 600 um length the Po_2 would
approximately linearly decrease along a distance of 10 um by about
10 mm Hg. Within a blood perfused capillary the O_2 dissociation
curve produces a nonlinear decrease of Po_2, so that at the venous
end the Po_2 decrease along the same distance is less than 1 mm Hg.
Taking into account that the vasculature penetrates from the pial
surface into the brain, i.e. that the brain cortex is supplied with
oxygen from the pial surface and that large Po_2 gradients along the
capillary exist, it is possible that the Po_2 histograms of deeper
layers are more left shifted than those of the surface. We like to
follow that in our experimental conditions the state of oxygen

supply of the brain cortex is better measured by the reflection spectra than by the surface Po$_2$ histogram.

Since Jöbsis and Rosenthal (1978) have found a rather low degree of oxidation of cyt aa$_3$ in the normal blood perfused brain, it was discussed whether the kinetic of the isolated mitochondria is different to that of the mitochondria in vivo, but at the moment there is no strong evidence for such an assumption (Chance, 1981). However, one has to consider that from the methodological point of view the measurements of Jöbsis et al. are difficult to evaluate quantitatively. They used reflected light at two wavelengths according to Kramer (1934) and Millikan (1942) - an isosbestic and a measuring wavelength - and developed a correcting method for hemoglobin. The method assumes 1. that well-defined isosbestic wavelengths exist and 2. that the overall system is linear. In general, both presuppositions cannot be taken for guaranteed, since only the simultaneous changes of the redox state (and state of oxygenation) isosbestic points exist and since Hoffman et al. (this volume) showed that in brain a linear analysis of reflection spectra may produce erroneous results. Further experiments will be necessary to settle this question.

REFERENCES

Baumgärtl, H., and Lübbers, D.W., 1983, Microcoaxial needle sensor for polarographic measurement of local O$_2$ pressure in the cellular range of living tissue. Its construction and properties, in: "Polarographic Oxygen Sensors", E. Gnaigner, H. Forstner, eds., Springer, Berlin, Heidelberg, pp. 37-65.

Chance, B., Haselgrove, J., and Barlow, C., 1981, Redox gradients in oxygen delivery to tissue, in: "Oxygen Transport to Tissue", A.G.B. Kovách, E. Dorá, M. Kessler, eds., Adv. Physiol. Sci., Vol. 25, Pergamon Press, Akadémiai Kiadó, Budapest, pp. 13-17.

Chance, B., and Willians, G.R., 1956, The respiratory chain and oxidative phosphorylation, Adv. Enzym., 17:65.

Figulla, H.R., Wodick, R., and Hoffmann, J., 1979, The O$_2$-saturation of myoglobin (MYO) and cytochrome aa$_3$ (CYT) during high-flow hypoxia (HFH) and low-flow hypoxia (LFH) in the beating hemoglobin-free perfused Langendorff guinea-pig heart, Pflügers Arch., 379, R3

Heinrich, U., Hoffmann, J., and Lübbers, D.W., 1981, Quantitative analysis of reflection spectra on the perfused brain in different states of oxygen supply, in: "Oxygen Transport to Tissue IV", D.F. Bruley, H.I. Bicher, eds., Plenum Press, New York, in print.

Hoffmann, J., Wodick, R., Hannebauer, F., and Lübbers, D.W., 1982, Quantitative analysis of reflection spectra of the surface of the guinea pig brain, in: "Oxygen Transport to Tissue V", D.W. Lübbers, H. Acker, T.K. Goldstick, E. Leniger-Follert, eds., Plenum Press, New York, (this volume).

Jöbsis, F.F., and Rosenthal, M., 1978, Behaviour of the mitochondria
    respiratory chain in vivo, Ciba Foundation Symposium 56:149.
Kessler, M., and Lübbers, D.W., 1966, Aufbau und Anwendungsmöglich-
    keiten verschiedener Po_2-Elektroden, Pflügers Arch. ges.
    Physiol., 291: R82.
Knaust, K., 1967, Sauerstoffversorgung des hämoglobinfrei perfun-
    dierten Meerschweinchengehirns bei Normo- und Hypothermie,
    Dissertation, Marburg.
Kramer, K., 1934, Fortlaufende Registrierung der Sauerstoffsättigung
    im Blut an uneröffneten Blutgefäßen, Klin Wschr., 13:379.
Lübbers, D.W., and Niesel, W., 1957, Ein Kurzzeit-Spektralanalysator
    zur Registrierung rasch verlaufender Änderungen der Absorption,
    Naturwissenschaften, 4: 59-60.
Millikan, G.A., 1942, The oximeter, an instrument for measuring
    continuously the oxygen saturation of arterial blood in man,
    Rev. Sci. Instrum., 13:434.
Schwickardi, D., 1968, Konzentration und Kinetik der Atmungsfermente
    am isoliert perfundierten Meerschweinchengehirn in vivo und in
    Hypothermie von 18°C, Dissertation, Marburg.
Starlinger, H., and Lübbers, D.W., 1973, Polarographic measurements
    of the oxygen pressure performed simultaneously with optical
    measurements of the redox state of the respiratory chain in
    suspensions of mitochondria under steady state conditions at
    low oxygen tensions, Pflügers Arch., 341:15-22.
Yu, Bi., Baumgärtl, H., and Lübbers, D.W., 1982, An improved
    polarographic multiwire surface Po_2 electrode, particularly
    for measurement of high Po_2 values, in: "Oxygen Tansport to
    Tissue V", D.W. Lübbers, H. Acker, T.K. Goldstick, E. Leniger-
    Follert, eds., Plenum Press, New York (this volume).

OXYGEN SUPPLY TO THE BRAIN CORTEX IN SHR AND NORMOTENSIVE RATS

K. Skolasinska[1], H. Günther[1], J. Höper[1], and R. Funk[2]

[1]Institut für Physiologie und Kardiologie der Universität
Erlangen-Nürnberg, Waldstr. 6, 8520 Erlangen, FRG
[2]Institut für Anatomie der Universität Erlangen-
Nürnberg, Krankenhausstr. 9, 8520 Erlangen, FRG

INTRODUCTION

It is postulated that there is a structural adaptation of the
brain blood vessels to high blood pressure and that therefore
spontaneously hypertensive stroke-resistant rats (SHR-SR) are less
prone to develop blood-brain barrier (BBB) disturbances than stroke-
prone rats (SHR-SP). As previously reported there is evidence that
also in SHR-SR the early stage of hypertension is accompanied by
very early signs of BBB disorder such as dislocation of ATPase and
CTPase inside the endothelial cytoplasm of brain capillaries. In
the well-established phase of hypertension further changes such as
an increased number of empty pinocytic vesicles or floculent dense
material in the astrocyte processes were reported (Skolasinska et
al., 1981).

We studied whether these very basic changes in the endothelial
wall and/or other pathomorphological changes have an influence on
the oxygen supply to the brain and the behaviour of cortical Po_2 in
the well-established phase of hypertension.

METHODS

Eight to nine month old SHR-SP and normotensive (NR) rats were
anesthetized with Trapanal[R], tracheotomized, paralyzed and ventilated
with N_2O (30% O_2). Femoral artery (for blood pressure recording and
blood gas analysis) and femoral vein were exposed. Local tissue Po_2

was measured with a multiwire surface electrode (Kessler and Lübbers, 1966) according to the technique described by Leniger-Follert et al. (1975). The electrode, consisting of 8 single platinum wires, was positioned within the area supplied by the medial cerebral artery (MCA). To obtain the Po_2 histograms the electrode was gently moved 13 times over the organ surface (Kessler et al., 1982). The data were analyzed using a DEC LSJ 11/2 computer with a Plessey double-density floppy disc drive. After the data for Po_2 histograms had been obtained the animals were ventilated with 100% O_2 for 60 s, 10% O_2 for 40 s and 0% O_2 for 30 s with suitable recovery periods between each step.

Scanning electron microscopic (SEM) studies were carried out on vascular resin casts (modified technique of Castenholz et al., 1982). Technovit[R], prepared according to the manufacturer's instructions, was injected at about arterial pressure into the common carotid artery. The injection of 2 ml Technovit[R] took about 2 min.

Immediately after the injection the animal died. The heads were removed after 24 h and the tissue was macerated with 25% KOH. Bone structure remained principally unchanged so that a correct localization of the examined area was possible. The cast was cleaned under running and desalinated water and air dried. The vascular region from the medial cerebral artery was dissected out of the cast, fixed on the specimen holder with conductive silver coated with gold, and examined in a Cambridge scanning electron microscope at 10 - 20 kV.

RESULTS AND CONCLUSIONS

The two Po_2 histograms (Figs. 1 and 2) were obtained from NR (mean arterial blood pressure (MABP) = 135 $\pm$ 12 mm Hg) and SHR-SP (MABP = 190 $\pm$ 18 mm Hg) animals. The respiratory conditions were

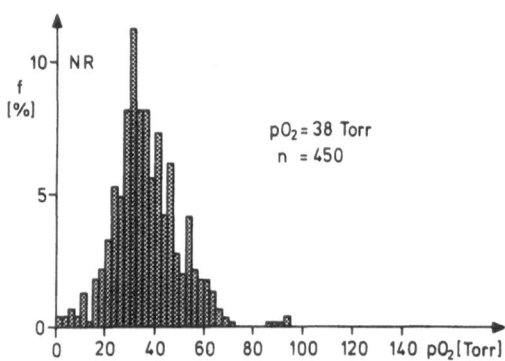

Fig. 1. Surface Po_2 histogram obtained from the brain cortex of normotensive rats under ventilation with 30% O_2. (mean Po_2 = 38 Torr)

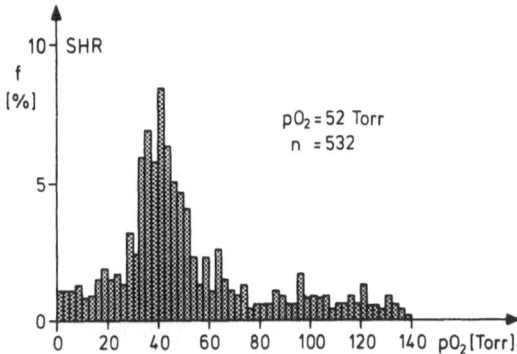

Fig. 2.   Surface Po_2 histogram obtained from the brain cortex of
          spontaneously hypertensive rats under ventilation with 30%
          O_2. (mean Po_2 = 52 Torr)

comparable: P_aco_2 = 32.5 $\pm$ 2.2 Torr for both groups. The two histograms differ in their mean Po_2 values (38 Torr for the NR, 52 Torr
for the SHR) as well as in the distribution of the Po_2 values.

     The cortical Po_2 behaviour during short-lasting drastic changes
in arterial oxygen is illustrated by the following examples.

     Fig. 3 shows a typical response of NR to an increase of the
inspiratory O_2 concentration. Tissue Po_2 shows a delayed increase
reaching normally maximum levels of around 100 Torr (higher values
were exceptions) and decreases only slowly after inspiratory O_2
concentration was changed back to the control level. In the SHR-SP
the response is much faster (Fig. 4), tissue Po_2 increasing immediately after inspiratory O_2 concentration of 100% was reached and

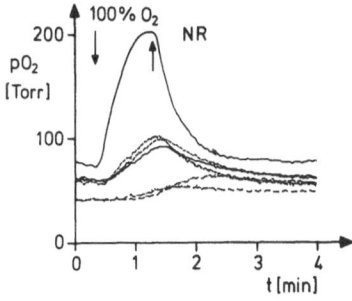

Fig. 3.   Tissue surface Po_2 changes on the brain cortex during 60 s
          ventilation with 100% O_2 in a normotensive rat. The gas
          mixture was changed at t = 0 min and the desired value was
          reached as indicated by the first arrow. The same procedure
          was used for the reactions shown in Figs. 4, 5, 6, 7 and 8.

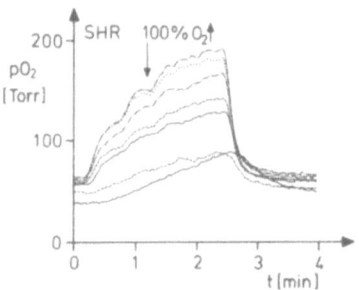

Fig. 4.   Tissue surface Po$_2$ changes on the brain cortex during 60 s
         ventilation with 100% O$_2$ in a spontaneously hypertensive
         rat.

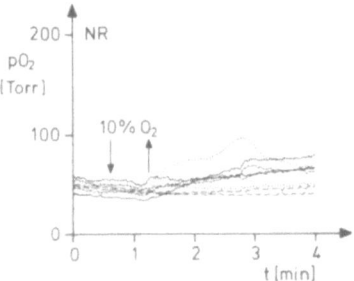

Fig. 5.   Tissue surface Po$_2$ changes on the brain cortex during 40 s
         ventilation with 10% O$_2$ in a normotensive rat.

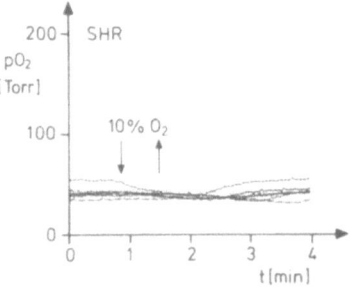

Fig. 6.   Tissue surface Po$_2$ changes on the brain cortex during 40 s
         ventilation with 10% O$_2$ in a spontaneously hypertensive
         rat.

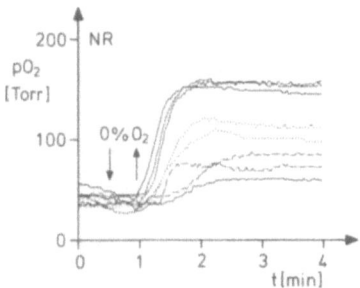

Fig. 7.   Tissue surface Po$_2$ changes on the brain cortex during 30 s
          ventilation with 0% O$_2$ in a normotensive rat.

decreasing abruptly with a return to control level. Maximum cortical
Po$_2$ levels reached around 200 Torr.

       Reduction of the inspiratory O$_2$ concentration from 30% to 10%
resulted in a decrease in MABP by about 20% in both the NR and
SHR-SP. The response of cortical Po$_2$ is shown in Figs. 5 and 6.
While the NR showed an overshoot of tissue Po$_2$ after hypoxia
(Fig. 5), this behaviour was rarely observed in the SHR-SP (Fig. 6).
The same was found for severe hypoxia with 0% O$_2$. NR showed a sharp
rise in tissue Po$_2$ above control levels immediately after the period
of hypoxia (Fig. 7). No such behaviour (Fig. 8) was observed in the
SHR-SP. Both groups of animals showed a decrease in MABP of 30%
below their control levels, and bradycardia, during severe hypoxia.

       The SEM examinations of the innermost layer of the arteries,
arterioles and capillaries belonging to the MCA are shown in Figs.

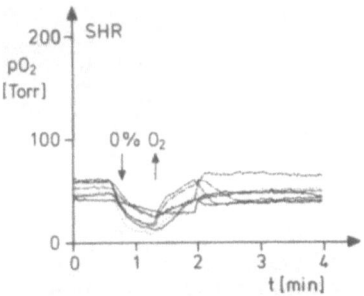

Fig. 8.   Tissue surface Po$_2$ changes on the brain cortex during 30 s
          ventilation with 0% O$_2$ in a spontaneously hypertensive
          rat.

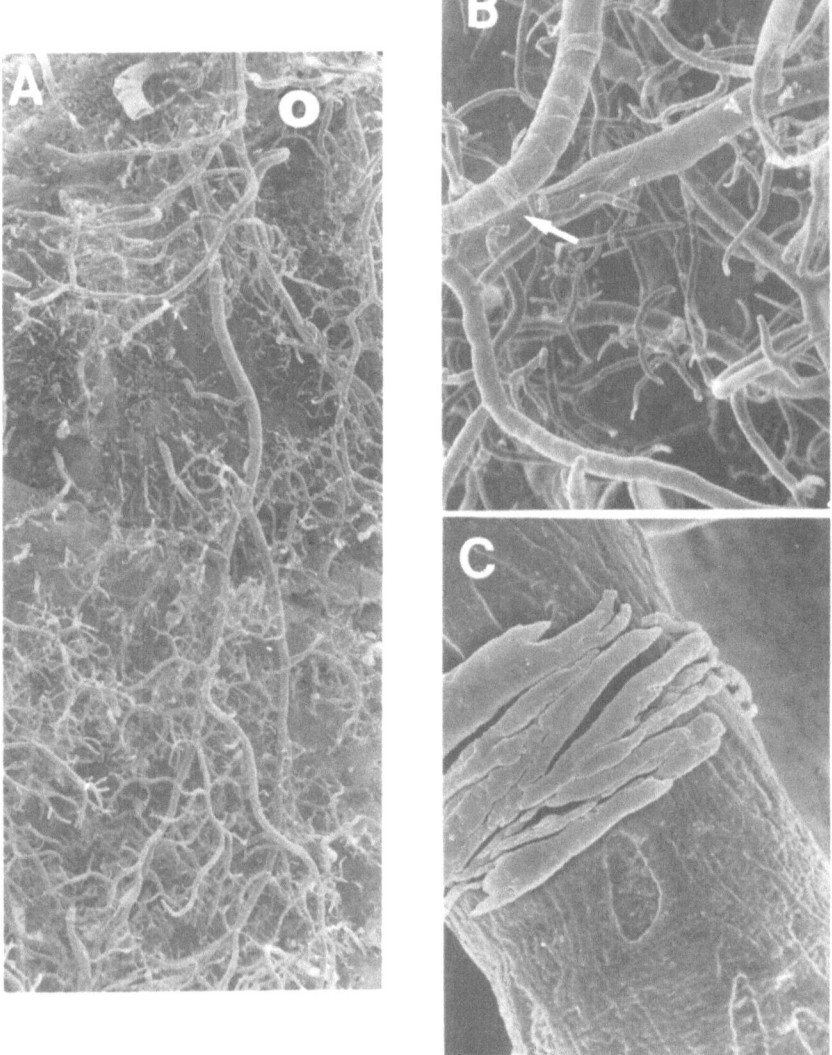

Fig. 9.   SEM examination (NR). A - view of the area supplied by MCA,
          mag x 47; B - homogeneous distribution of arterioles and
          capillaries, scantily located "myocytes" (arrow), mag x
          250; C - smooth surface of endothelial wall, impressions
          of nuclei, mag x 2460 (see text).

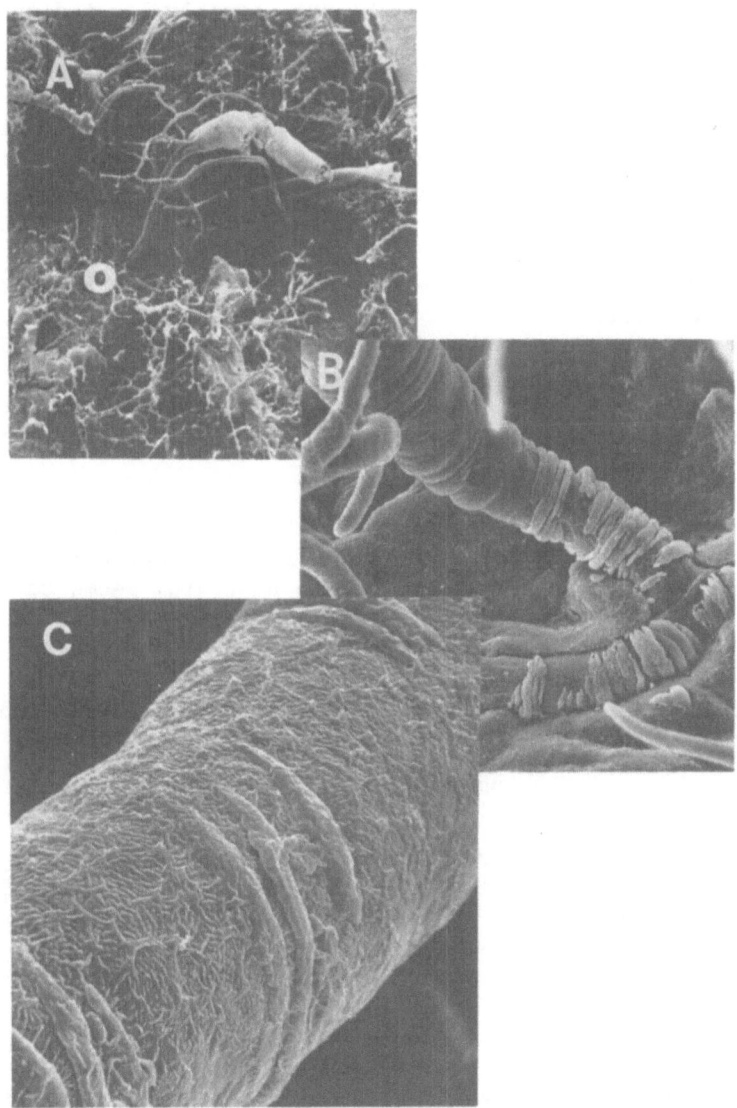

Fig. 10.   SEM examination (SHR-SP). A – view of the area supplied
            by MCA, mag x 39; B – "myocytes" are located on the
            arterioles, mag x 1200; C – very coarse surface of the
            endothelial wall, mag x 4480 (see text).

9 and 10. The ring (0) indicates the convergence of the MCA branches.
When compared to the NR, the SHR-SP shows a very inhomogeneous distribution of arterioles, extravasation of the plastic mass, ruptured
arteries, and "voids" covered mainly by arterioles of $\emptyset$ 15 - 30 um
diameter but only a few capillaries (4 - 6 um). Structures imitating
myocytes (Castenholz et al., 1982) are scantily located on the 20 um
arterioles in NR (Fig. 9B, C). On the other hand, in the SHR-SP most
of the arterioles below 14 um are covered with these structures
(Fig. 10B). The frequency of these structures increases in the SHR
proportionally to the narrowing of the lumen, but they were not seen
on capillaries. The endothelial walls differ significantly between
the two groups, in the NR they are smooth with impressions of nuclei
(Fig. 9C), but they are very coarse in the SHR-SP (Fig. 10C).

Surface electrodes for measuring Po_2 on the brain cortex were
used for the first time by Leniger-Follert et al. (1975). Under our
experimental conditions the shape of the histogram obtained in NR
is comparable to that obtained by Grote et al. (1981) for the brain
cortex of the cat. This suggests that there are no substantial differences in the technique used in our experiments. The conclusions
are, therefore:

1. The mean and distribution of the brain cortex surface Po_2
histogram obtained in SHR-SP differs from NR despite identical P_aO_2
values. The shape and range of the Po_2 histogram seems to be determined by the arrangement of the arterioles and capillaries supplying
the area examined.

2. The behaviour of the brain cortex Po_2 in SHR-SP indicates
that some of the regulatory mechanisms are absent. Changes in cortical
Po_2 occur passively in response to increases or decreases in arterial
Po_2.

3. Basic morphological changes in the surfaces of the endothelial
cells were recognized when the MCA vascular system of the SHR-SP was
compared with NR.

REFERENCES

Castenholz, A., Zöltzer, H., and Erhardt, H., 1982, Structures
     imitating myocytes and pericytes in corrosion casts of terminal
     blood vessels. A methodical approach to the phenomenon of
     "plastic strips" in SEM, Mikroskopie, 39:95.
Grote, J., Zimmer, K., and Schubert, R., 1981, Effects of severe
     arterial hypocapnia on regional blood flow regulation, tissue
     Po_2 and metabolism in the brain cortex of cats, Pflügers
     Arch., 391:195.

Kessler, M., Höper, J., and Pohl, U., 1982, Monitoring of local Po_2
    in skeletal muscle of critically ill patients, in: "Handbook
    of Critical Care", J.L. Berk, J.E. Fampliner, eds., Little,
    Brown and Comp., Boston, pp. 600-609.
Kessler, M., and Lübbers, D.W., 1966, Aufbau und Anwendungsmöglich-
    keiten verschiedener Po_2-Elektroden, Pflügers Arch. Ges.
    Physiol., 291:R32.
Leniger-Follert, E., Lübbers, D.W., and Wrabetz, W., 1975, Regula-
    tion of local tissue Po_2 of brain cortex at different arterial
    O_2 pressures, Pflügers Arch., 359:81.
Skolasinska, K., Kostrzewska, M., Ostenda, M., and Rudczynski, M.,
    1981, Cerebral blood flow and ultrastructural studies of the
    brain capillaries in the early state of hypertension in SHR,
    Clin. Exp. Hypertension, 3:319.

DISCUSSION

Question:
Could you comment on your findings that there appears to be very
little change with hypoxic gas mixtures in Figs. 5, 6 and 7 (even
with 0% O_2) compared with Fig. 8?

Answer of the authors:
At the time no explanation can be given because neither total nor
local blood flow were examined in these experiments. Also the values
for arterial Po_2 are not measured in this first series of experiments. In contrast to the brain, the same procedure causes very
pronounced changes in Po_2 measured at the liver surface.

LOCAL CYTOCHROME OXIDASE ACTIVITY IN THE CEREBRAL CORTEX OF THE

RAT, HISTOCHEMICALLY DETECTED WITH THE DAB-METHOD:

A MICRO-DENSITOMETRIC AND ELECTRON MICROSCOPE STUDY

A.W. Budi Santoso, and Th. Bär

Max-Planck-Institut für Systemphysiologie
Rheinlanddamm 201, 4600 Dortmund 1, FRG

SUMMARY

For demonstration of the local cytochrome oxidase activity, coronal sections through the frontal cortex of four Sprague-Dawley rats were incubated with DAB medium after Seligman et al. (1968).

The pattern of distribution of the DAB reaction product (DAB-RP) was measured at a wavelength of 460 nm. Cryostat sections were scanned with a microdensitometer. The cytochrome reaction product was accumulated in the laminae I and IV.

A comparative morphometric study of mitochondrial profiles in ultra-thin sections through equivalent cortical regions showed an approximately similar distribution in percentage of areal density of the mitochondria labeled by DAB-RP. On average 30% of the mitochondrial profiles were labeled with DAB-RP which was localized in the intracristate spaces and the outer mitochondrial compartment. The marked mitochondria mainly occurred in dendrites. Furthermore, the mean and median value of the size distribution of marked mitochondrial profiles was shifted to greater values when compared with non-marked ones.

INTRODUCTION

A lot of diamine reagents have been prepared to demonstrate cytochrome oxidase activity both in the light and electron micro-scope (Burstone, 1959; Plapinger et al., 1968) since Seligman et al. (1968) discovered that the best results could be reached on the

281

ultrastructural level by application of 3,3'-diaminobenzidine
(DAB), which had been previously introduced by Graham and Kasnovsky
(1966). The method was based upon the oxidative polymerization and
oxidative cyclization of DAB to an osmiophilic polymer which occurs
in a non-droplet form between the inner and outer mitochondrial
membranes. A similar localization also could be dectected for the
demonstration of succinodehydrogenase by using several osmiophilic
tetrazolium salts as electron acceptors (Seligman et al., 1967;
Seligman et al., 1971). Wong-Riley et al. (1979, 1980a, 1980b) considered a strict relationship existing between the neuronal activity
and intensity of staining by using the DAB-method after several experimental treatments.

The purpose of this study is to check how the light microscopic detectable density of the reaction product correlates with
the ratio of reactive and non-reactive mitochondria in order to
apply the DAB-method for a semi-quantitative approach by microdensitometry.

MATERIALS AND METHODS

Four female Sprague-Dawley rats were fixed by a transcardial
perfusion with 2% depolymerized paraformaldehyde and 0.5% glutar-
aldehyd in 0.1 mol·l^{-1} phosphate buffer, pH = 7.4. 4% sucrose was
added to the buffer solution. Two animals were prepared for the
morphometry and two animals for microdensitometry. They were perfused under i.p. Pentobarbital anaesthesia (35 mg/kg body weight).
The perfusion was stopped after 10 minutes. 1 hour after perfusion
the brains were removed. Coronal sections were cut with a vibratome
(thickness varied from 100 - 200 um). Frontal sections were cut at
the level of the anterior commissure. They were washed over night
at 4°C in phosphate buffer (pH = 7.4) with increasing concentrations
of sucrose added (10%, 20% and 30%). Sections were then incubated
for 1 hour at 37°C in DAB incubation medium containing 50 mg DAB,
25 mg cytochrome c, type IV (Sigma Chemical Co.), 4% sucrose and
200 ug/ml catalase in 90 ml 0.1 M phosphate buffer, pH = 7.4 (modified after Seligman et al., 1968). For control incubation 0.01
mol·l^{-1} KCN was added.

For the morphometric study, sections were dehydrated in ethanol,
postfixed in 1% OsO$_4$ and embedded in araldit. Systematic random
samples of TEM pictures running equidistantly from pial surface to
the white matter were taken from the edge zone (not more than 20 um
away from the surface) of ultrathin sections. the area of "marked"
(i.e., with osmiophilic DAB-RP within the mitochondria) and "non-
marked" mitochondrial profiles were measured by planimetry using a
computer assisted digitizer (Videoplan, Kontron Ltd., Munich).

For the microdensitometry, the optical density (ΔOD) on cryosections (thickness 30 um; Mod-2700-Frigo-cut, Jung-Reichert) of the
frontal section after incubation was measured using a microscope
densitometer M85A (Vickers Instruments, London). The sections were
mounted in glycerin gelatine (Sigma Chemical Co.). ΔOD was determined at 460 nm against reference measured by setting the mask in a
clear area of background adjacent to the object. The clear background area close to the measured specimen area was taken to have
zero density. The results obtained by measurements were in arbitrary units, expressed as ΔOD, presented on a digital display. The
following microdensitometer settings were used: objective 10x, spot
size 3 (d = 6 um), scanning size 1 (d = 100 um), bandwidth 50,
gating mask A3 (d = 60 um). For illustration see Fig. 1.

RESULTS

Microdensitometry

The absorption spectrum of the DAB-RP in comparison to a
tissue blanck measured on cryosection showed a spreading absorption
band at 450-480 nm (Fig. 2) by setting the gating mask A over the
laminae I, IV and V, respectively.

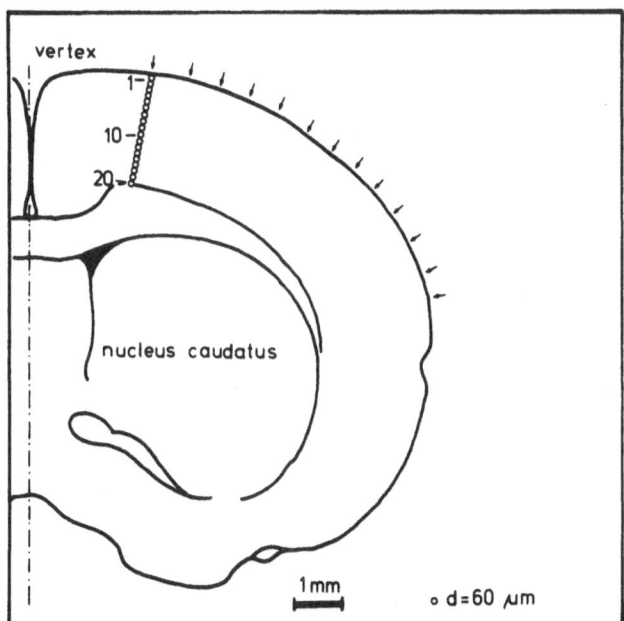

Fig. 1.   Schematic drawing shows the sequence of microdensitometric
          measurement steps in the frontal section plane. Thirty um
          thick cryosections were used.
          Arrows: measuring paths; Circles: gating masks of the
          microdensitometer.

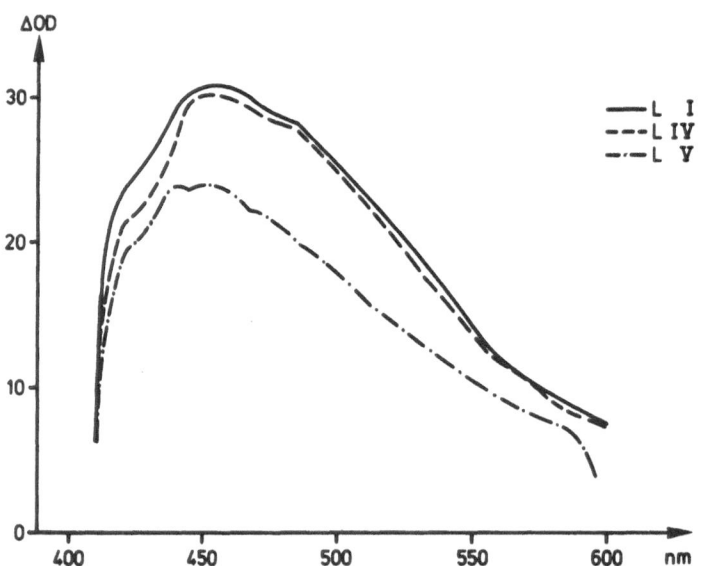

Fig. 2. Absorbance spectrum of oxidized DAB in different layers as
reaction product after incubation of 30 um thick cryosection measurements were carried out at wavelength between
420 and 700 nm.
(Roman numerals refer to cytoarchitectonic lamina = L).

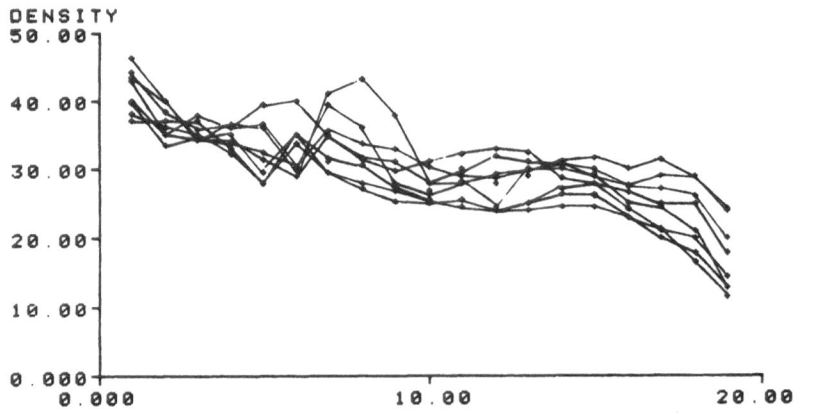

Fig. 3. Optical density. (ΔOD_{460} nm) is taken in 19 steps (sequences) as shown in Fig. 1.
Diagramms from eight measuring paths are plotted.

The measurements show that there is a high level of $OD_{460 \text{ nm}}$ in lam. I. In lam. II and III, the $\Delta OD_{460 \text{ nm}}$ is about 20-35% lower than in lam. I. The white matter showed the lowest $\Delta OD_{460 \text{ nm}}$. The value varied between 15% and 30% of the $\Delta OD_{460 \text{ nm}}$ in lam. I (Fig. 3).

Electron microscopy

The endoplasmatic reticulum did not show DAB-RP. Two populations of mitochondria could be detected, namely mitochondria with DAB-RP ("marked") and mitochondria without DAB-RP ("non-marked"), i.e. with histochemically non-detectable cytochrome oxidase activity (Fig. 4). The marked mitochondria occurred mainly within the dendrites. In neuronal pericarya, about 20-30% of mitochondrial profiles were provided with DAB-RP.

The size distribution of the area of marked mitochondria measured through the whole cortical depth is shifted to greater values as compared to the distribution of non-marked mitochondria. The median value of the area of marked mitochondria is 43-75% higher than the median value of the non-marked ones (Fig. 5).

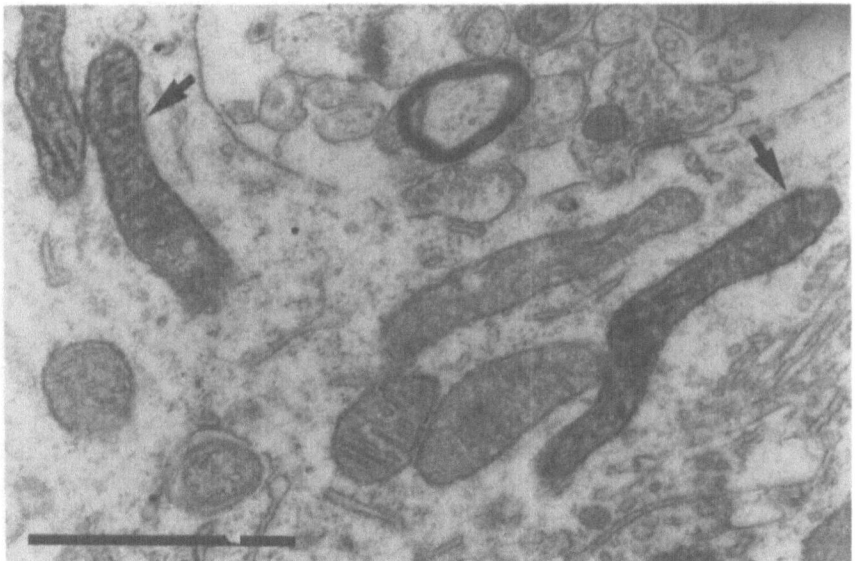

Fig. 4. Electron micrograph demonstrates the localization of the DAB-reaction product in the intracristate spaces of mitochondria (arrows).
a = myelinated axon; b = presynaptic element containing round clear vesicles. Bar represents 1 micron.

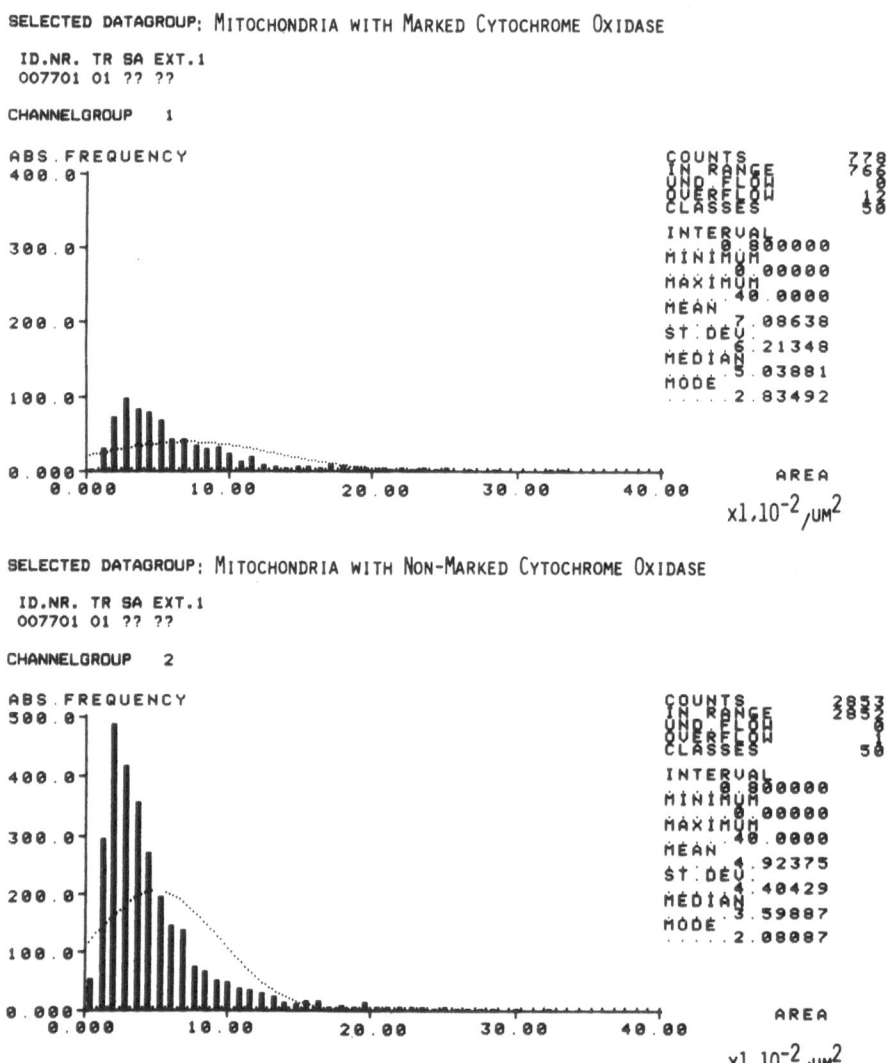

Fig. 5.   Frequency distribution of the profile area of mitochondria.
          Channel group 1 represents all mitochondria with marked
          cytochrome oxidase. Note the shift of the size of the
          profile area to greater mean and median values, compared
          to the histogramm of non-marked mitochondria (channelgroup
          2). Note the deviation of the histogramm from the Gaussion
          distribution (dotted curve).

     High numbers of mitochondrial profiles with positive cyto-
chrome oxidase were found in lamina I, upper part of lamina IV
and in lamina V (Fig. 6). Small mitochondria with DAB-RP could be
seen in lamina V. Large mitochondrial profiles were seen in
lamina IV. Furthermore, the mean areal density of all mitochondrial profiles has been estimated at about 5.42 ± 1.33%. The mean

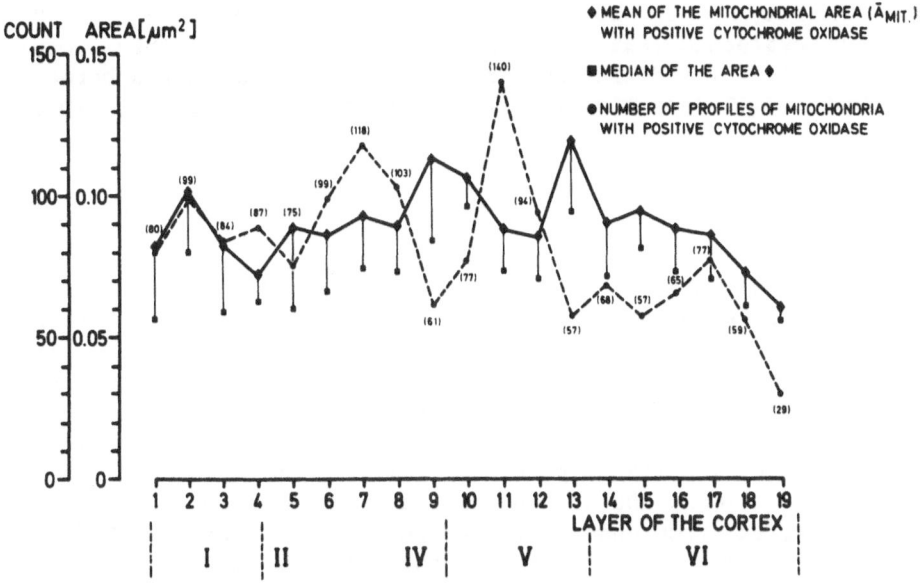

Fig. 6. Distribution of the area and number (count) of mitochondrial profiles with positive cytochrome oxidase activity
in different layer of the cortex.
Abscissa: The measurements were done in 19 equidistant
steps through the whole cortical depth (compare to Fig.
1). Roman numerals indicate the cytoarchitectonic laminae.

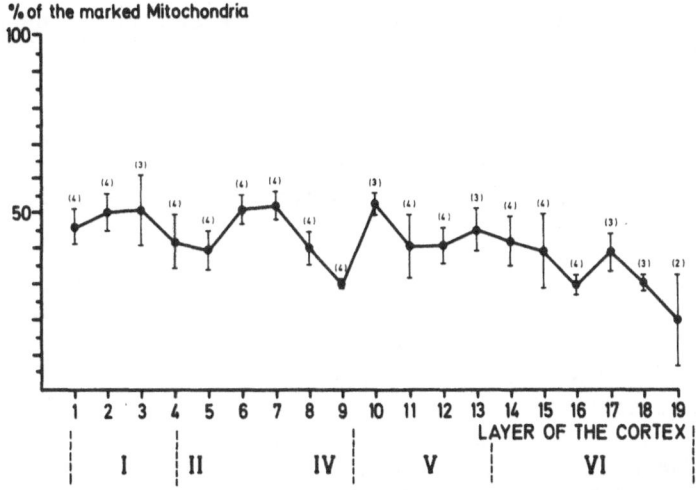

Fig. 7. Percentage of the marked mitochondria (Areal density of
the mitochondria with positive cytochrome oxidase/areal
density of total mitochondria) x 100%. Compare the diagram
with Fig. 3. Number of the abscissa see Fig. 6.

areal density of mitochondrial profiles with DAB-RP (1.75 $\pm$ 0.69%) has been evaluated (68% lower than the areal density of the total mitochondria). The differences between layers 7 and 8 and between layers 9 and 10 (see Fig. 7) were statistically significant (p < 0.05 with U-test).

DISCUSSION

Two populations of mitochondria seem to exist in the CNS: mitochondria with reactive or marked cytochrome oxidase and mitochondria with non-reactive or non-marked cytochrome oxidase. Several authors have also reported, that the dendritic processes contain very reactive mitochondria, whereas the mitochondria within axonal terminals are much less reactive (Wong-Riley et al., 1979; Ribak, 1981). Hajos and Kerpel-Fronius (1969) provided similar results for succinodehydrogenase in CNS. The reason for these differences in the mitochondria is far from understood at the moment. We assumed that it may be influenced by pentobarbital anaesthesia. As shown by several authors barbiturates can cause, for example, about 30% reduction of the $CMRo_2$ and CMR_{gluc} measured in the frontal and parietal region compared to the corresponding metabolic rates in animals ventilated with 40% N_2O (Nilsson and Siesjö, 1975; Berntman et al., 1979; Ingvar et al., 1980).

Similar as Hirai (1968) who had measured the absorption maximum of the oxidized DAB in solution, we found a spreading absorption band at 450-480 nm on cryosections. In fact, microscope densitometric examination revealed a relatively good coincidence between the distribution of the $\Delta OD_{460\ nm}$ and the distribution of the percentage of the marked mitochondrial areal density (see Figs. 3 and 7).

The shift in the size distribution of the marked mitochondria to greater mean and median values (Fig. 5) may be related to different states in the cycle of mitochondrial biogenesis. The maturity of mitochondria may be connected to its metabolic activity.

REFERENCES

Berntman, L., Carlsson, C., and, Siesjö, B.K., 1979, Cerebral oxygen consumption and blood flow in hypoxia: Influence of sympatho-adrenal activation, Stroke, 10:20-25.
Burstone, M.S., 1959, Histochemical demonstration of cytochrome oxidase with new amine reagents, J. Histochem., 7:112-122.
Graham, R., and Karnovsky, M.J., 1966, The early stages of absorption of injected horseradish peroxidase in the proximal tubules of mouse kidney: ultrastructural cytochemistry by a new technique, J. Histochem. Cytochem., 14:291-302.

Hajòs, F., and Kerpel-Fronius, S., 1969, Electron histochemical observations of succinic dehydrogenase activity in various parts of neurons, Exp. Brain Res., 8:66-78.

Hirai, K., 1968, Specific affinity of oxidized amine dye (radical intermediator) for heme enzymes: Study in microscopy and spectrophotometry, Acta Histochem. Cytochem., 1(N1):43-55.

Ingvar, M., Abdul-Rahman, A., and Siesjö, B.K., 1980, Local cerebral glucose consumption in the artificially ventilated rat: Influence of nitrous oxide analgesia and of phenobarbital anaesthesia, Acta Physiol. Scand., 109:177-185.

Nilsson, L., and Siesjö, B.K., 1975, The effect of phenobarbitone anaesthesia on blood flow and oxygen consumption in the rat brain, Acta Anaesth. Scand., 19, Suppl. 57:18-24.

Plapinger, R.E., Linus, S.L., Kawashima, T., Deb, C., and Seligman, A.M., 1968, Preparation and structure-activity relationship of reagents for cytochrome oxidase activity: Potential for light and electron microscopy, Histochemie, 14:1-16.

Ribak, C.E., 1981, The histochemical localization of cytochrome oxidase in the dentate gyrus of the rat hippocampus, Brain Res., 212:169-174.

Seligman, A.M., Karnovsky, M.J., Wasserkrug, H.L., and Hanker, J.S., 1968, Nondroplet ultrastructural demonstration of cytochrome oxidase activity with a polymerizing osmiophilic reagent, diaminobenzidine (DAB), J. Cell Biol., 38:1-14.

Seligman, A.M., Nir, I., and Plapinger, R.E., 1971, An osmiophilic distryl ditetrazolium salt (DS-NBT) for ultrastructural dehydrogenase activity, J. Histochem. Cytochem., 19(N5):273-285.

Seligman, A.M., Ueno, H., Morizono, Y., Wasserkrug, H.L., Katzoff, L., and Hanker, J.S., 1967, Electron microscopic demonstration of dehydrogenase activity with a new osmiophilic ditetrazolium salt (TC-NBT), J. Histochem. Cytochem., 15:1-13.

Wong-Riley, M., 1979, Changes in the visual system of monocularly sutured or enucleated cats demonstrable with cytochrome oxidase histochemistry, Brain Res., 171:11-28.

Wong-Riley, M.T.T., Walsh, S.M., Leake-Jones, P.A., and Merzenich, M.M., 1980, Maintenance of neuronal activity by electrical stimulation of unilaterally deafened cats demonstrable with cytochrome oxidase technique, Ann. Otol. Rhinol. Laryngol., Suppl. 90(N2):30-32.

Wong-Riley, M.T.T., and Welt, C., 1980, Histochemical changes in cytochrome oxidase of cortical barrels after removal in neonatal and adult mice, Proc. Natl. Acad. Sci., 77(N4):1333-1337.

RELATIONSHIP BETWEEN MICROFLOW, LOCAL TISSUE Po_2 AND EXTRACELLULAR

ACTIVITIES OF POTASSIUM AND HYDROGEN IONS IN THE CAT BRAIN DURING

INTRAARTERIAL INFUSION OF AMMONIUM ACETATE

J. Gronczewski, and E. Leniger-Follert

Max-Planck-Institut für Systemphysiologie
Rheinlanddamm 201, 4600 Dortmund 1, FRG

INTRODUCTION

In various pathophysiological states, i.e. hyperglycemia, hypoglycemia, hepatic coma, seizures and ischemia, the concentration of ammonia (NH_3/NH_4^+) is increased in both blood and tissue. One might suspect that ammonia could participate in the regulation of flow under these conditions. However, the reports in the literature about the effects of ammonia on flow are controversial. During infusion of ammonium salts in liver, a fall in regional blood flow was recorded by Barey et al. (1980), in brain, Schieve and Wilson (1953), and Gjedde et al. (1978) reported a decrease in cerebral blood flow, whereas Altenau and Kindt (1977) found an increase in cerebral blood flow.

The mechanism, for the effect of ammonia on cerebral blood flow is unknown. In rat brain slices, Benjamin et al. (1978) have shown that ammonium chloride causes an outward flow of K^+ into the extracellular space. The aim of our present study was to investigate the changes in cerebral microflow and local Po_2 during the infusion of ammonium acetate. In order to examine a possible K^+ and/or H^+ related ionic mechanism for the action of ammonia on cerebral blood flow, the extracellular H^+ and K^+ activities were also recorded.

METHODS

Our experiments were performed on 27 adult cats, anesthetized with Evipan$^{(R)}$ during the preparation. The animals were paralyzed

with Imbretil[R] and artificially ventilated with a mixture of
N_2O (70%) and O_2 (30%) during the measurements. Arterial blood
pressure, endtidal CO_2 content and blood gases were controlled
and were within the normal range. Rectal temperature was stabilized
at 37,5 $\pm$ 0.5°C by a heating pad. The right lingual artery was
cannulated for ammonium acetate infusion. After trepanation of the
calvarium on the right side, the dura mater was removed. From the
parietal cortex, microflow was continuously and qualitatively
measured in four sites of the gyrus suprasylvius by the local
hydrogen clearance method according to Lübbers and Stosseck (1970).
For these measurements we used the multiwire surface element
described by Leniger-Follert and Lübbers (1976). Local tissue Po_2
was recorded polarographically by a multiwire surface electrode
according to Kessler and Lübbers (1966). Extracellular pH was
monitored by using a commercial Ingold surface glass element with a
diameter of 2 mm.

Extracellular K^+ activity was recorded by a surface solid
contacted Valinomycin electrode as described by Kessler et al.
(1974). The electrocorticogram (ECoG) was recorded by a unipolar
silver ball electrode located immediately beside the pH or K^+
sensitive electrode. All measuring elements were either elastically
suspended or mounted on a counter balanced holder and thereby
followed the movements of the cortex. Ammonium acetate at pH 7.4 in
a dose of 35 uM/(kg·min) was infused for 10 minutes. In 20 cats the
change in microflow and local tissue Po_2 were simultaneously
recorded. In 7 cats, flow and extracellular K^+ and H^+ activities
were recorded.

RESULTS

Ammonium acetate infusion caused an increase in cerebral
microflow, local Po_2 and extracellular K^+ activity in all
experiments. The increase began about 30 secs after the start of
infusion. In some experiments, an initial transient decrease in
microflow was observed.

The extracellular H^+ activity was constant for the first
2 minutes of the infusion and then increased significantly. A fall
in ECoG activity during the ammonium infusion was noted.

In Fig. 1 a series of records taken simultaneously are shown,
demonstrating the effects of ammonium acetate infusion on: ECoG,
local tissue Po_2 and microflow.

In Fig. 2 the distribution of local tissue Po_2 values under
control conditions and at the end of infusion of ammonium acetate
is presented (20 cats = 120 Po_2 measuring sites).

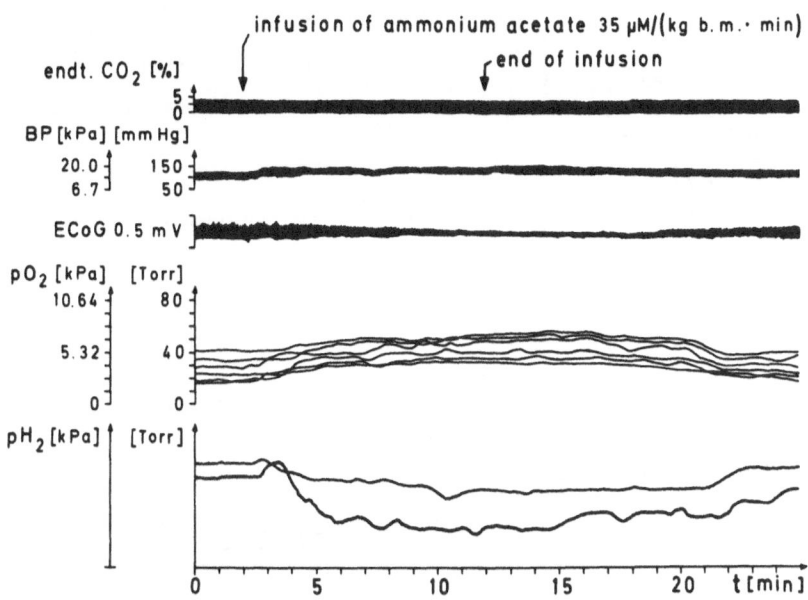

Fig. 1.   Original recordings of the effects of ammonium acetate
          infusion on cerebral local tissue Po_2, ECoG and microflow
          (pH_2 clearance) measured simultaneously.

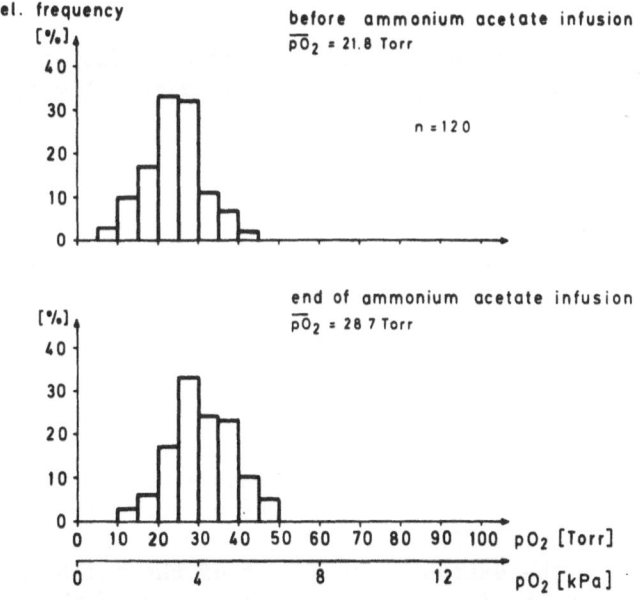

Fig. 2. Po_2 histograms in the brain cortex before and at the end
          of a 10 min infusion of ammonium acetate (35 umol·1^{-1}/
          (kg bm·min)). (20 cats, n = 120 Po_2 measuring sites).

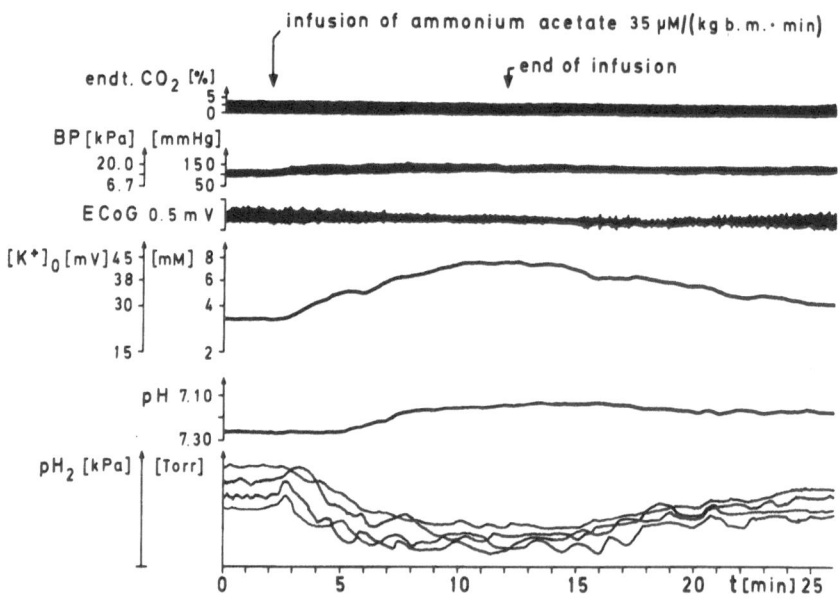

Fig. 3.   The effect of ammonium acetate infusion on ECoG, extracel-
         lular K^+ and H^+ activities (pH) and microflow (pH$_2$
         clearance).

Fig. 3 shows the effect of ammonium acetate infusion on ECoG,
microflow and K^+ and H^+ activities (pH).

Within thirty minutes after the end of infusion, extracellular
K^+ and H^+ activities (pH), local Po$_2$ and cerebral microflow returned
to control values.

DISCUSSION

As our results have shown, ammonium acetate infusion causes an
increase in microflow, local Po$_2$ and extracellular K^+ activity all
of which occur at about the same time. It seems probable that the
elevated extracellular K^+ activity is partially responsible for
the rise in local cerebral blood flow. The increase in extracellular
K^+ activity that we observed was in the range which is known to
cause relaxation of smooth muscle. Previously it has been shown
that in some tissues hyperammonemia causes changes in intracellular
pH, energy store and permeability to ions (Bessman and Pal, 1976;
Boron and De Weer, 1976; Aickin et al., 1977). Under conditions of
hyperammonemia the extrusion of K^+ out of the cells in studies in
vitro (Benjamin et al., 1978) and an increase in K^+ activity in
blood have also been reported (Benzi et al., 1977). The role of
H^+ activity in the regulation of cerebral blood flow seems to be
more limited under our experimental conditions. We were not able to
see any significant changes in extracellular H^+ activity during the

first 2 minutes of infusion. However, it is possible that the subsequently observed decrease in extracellular pH could participate
in a further increase in microflow.

From our results we suggest that the observed increase in
cerebral microflow, extracellular K^+ and H^+ activities during late
severe hypoglycemia as reported elsewhere in this volume may be
due partly to the effect of increased ammonium ion activity in
the brain.

The depressive action of ammonia on ECoG activity observed in
our study is in accordance with the reports of others (Benjamin et
al., 1978). The increase in local Po_2 can be explained by an increase in flow and possibly by a decrease in oxygen consumption.
The reports concerning oxygen uptake under hyperammonemic conditions are however controversial. A decrease in oxygen uptake has
been reported by McKhan and Tower (1961), no change by Schieve and
Wilson (1953), and an increase in oxygen consumption was reported
by both, Hawkins et al. (1973) and Benzi et al. (1977). Further
investigations on this subject and on the measurement of local
oxygen consumption should be performed.

REFERENCES

Aickin, C., and Thomas, R.C., 1977, Microelectrode measurement of
    the intracellular pH and buffering power in mouse soleus
    muscle fibers, J. Physiol., 267:791.
Altenau, L.L., and Kindt, G.W., 1977, Cerebral vasomotor paralysis
    produced by ammonia intoxication, Acta Neurol. Scand., 56:346.
Barey, W., Siwecka, B., Gronczewski, J., and Skolasinska, K., 1980,
    The depressive action of ammonium chloride on the hepatic
    blood flow in sheep, Q. J. Exp. Physiol., 65:99.
Benjamin, A.M., Okamoto, K., and Quastel, J.H., 1978, Effects of
    ammonium ions on spontaneous action potentials and on contents
    of sodium potassium, ammonium and chloride ions in brain in
    vivo, J. Neurochem., 30:131.
Benzi, G., Arrigoni, E., Strada, P., Pastoris, O., Villa, R.F.,
    and Agnoli, A., 1977, Metabolism and cerebral energy state:
    Effect of acute hyperammonemia in beagle dog, Biochem.
    Pharmacol., 26:2397.
Bessmann, S.P., and Pal, N., 1976, The Krebs cycle depletion
    theory of hepatic coma, in: "The Urea Cycle", S. Grisolia, R.
    Baguena, F. Mayor, eds., John Wiley and Sons, New York,
    pp. 83-89.
Boron, W.F., and De Weer, P., 1976, Intracellular pH transients in
    squid giant axions caused by CO_2, NH_3 and metabolic inhibi-
    tors, J. Gen. Physiol., 67:91.
Gjedde, A., Lockwood, A.H., Duffy, T.E., and Plum, F., 1978, Cere-
    bral blood flow and metabolism in chronically hyperammonemic
    rats: Effect of an acute ammonia challenge, Ann. Neurol.,
    3:325.

Hawkins, R.A., Miller, A.L., Nielson, R.C., and Veech, R.L., 1973,
    The acute action of ammonia in rat brain metabolism in vitro,
    Biochem. J., 34:1001.
Kessler, M., Höper, J., and Simon, W., 1974, Methodology and
    application of a multiple ions selective surface electrode
    (pH, pK, pNa, pCa, pCl) for tissue measurements, Fed. Proc.,
    33:279.
Kessler, M., and Lübbers, D.W., 1966, Aufbau und Anwendungsmöglich-
    keiten verschiedener Po_2-Elektroden, Pflügers Arch. Ges.
    Physiol., 291:R32.
Leniger-Follert, E., and Lübbers, D.W., 1976, Behavior of microflow
    and local Po_2 of the brain cortex during and after direct
    electrical stimulation. A contribution to the problem of
    metabolic regulation of microcirculation in the brain, Pflügers
    Arch., 366:39.
Lübbers, D.W., and Stosseck, K., 1970, Quantitative Bestimmung der
    lokalen Durchblutung durch elektrochemisch im Gewebe erzeugten
    Wasserstoff, Naturwissenschaften, 57:311.
McKhan, G.M., and Tower, G., 1961, Ammonia toxicity and cerebral
    oxidative metabolism, Am. J. Physiol., 200(3):420.
Schieve, J.F., and Wilson, W.P., 1953, The changes in cerebral
    vascular resistance of man in experimental alkalosis and
    acidosis, J. Clin. Invest., 32:33.

REGULATION OF MICROFLOW IN THE CAT BRAIN DURING INSULIN INDUCED

HYPOGLYCEMIA

E. Leniger-Follert, J. Gronczewski, and C. Danz

Max-Planck-Institut für Systemphysiologie
Rheinlanddamm 201, 4600 Dortmund 1, FRG

INTRODUCTION

It is well established that severe hypoglycemia is accompanied by changes in cerebral functional activity and by a decrease in the cerebral metabolism of glucose. The results published on cerebral blood flow (CBF) during hypoglycemia are, however, controversial. Whereas most authors reported constant CBF, Norberg and Siesjö (1976) demonstrated in rats that CBF increased significantly both when the EEG showed a pattern of slow waves and polyspikes and when electrical activity ceased.

In our investigations in rats (Krolicki and Leniger-Follert, 1980) we were able to show that local cerebral Po_2 increased in parallel with an increase in cerebral microflow during severe hypoglycemia at a mean arterial glucose concentration of 1.4 $mmol \cdot l^{-1}$. Whereas, in hypoglycemic cats Cilluffo et al. (1982) reported no change in CBF, we have also recently found in cats increases in cerebral tissue Po_2 during late severe hypoglycemia and during isoelectricity. After glucose injection Po_2 again decreased (unpublished results).

From these measurements, we assume that CBF also increases in cats during hypoglycemia, although electrical activity decreases. Up to now the mechanism for this increase has remained unknown. Astrup et al. (1978) reported that extracellular K^+ and H^+ activity, which we think to be important for flow regulation during enhanced activation, did not change in the rat brain. However, those authors only recorded the behaviour of H^+ and K^+ activities during a short period of hypoglycemia. The aim of our present studies was to investigate the behaviour of extracellular H^+ and K^+ activities in the

297

brain cortex of cats continuously during the entire course of the
hypoglycemia, that is, during normoglycemia, early hypoglycemia,
severe hypoglycemia, isoelectricity, and after glucose injection.

Furthermore, we simultaneously and continuously measured the
changes in microflow together with extracellular H^+ or K^+ activities
to see whether a temporal relationship exists between those factors
and the changes in microflow.

METHODS

Experiments were performed in 25 cats, weighing 2 to 3 kg. The
animals were anaesthetized initially with Evipan[R] and ventilated
with 70% nitrogen-suboxide and 30% oxygen after relaxation. Arterial
blood pressure, body temperature, endtidal CO_2 content and blood
gases were controlled. Extracellular pH was measured on the surface
of the gyrus suprasylvius in 8 cats with an Ingold pH surface electrode that had a flat glass membrane of 2 mm in diameter. In 3 cats
local extracellular H^+ activity was recorded in the deeper layers
of the gyrus suprasylvius by means of H^+ sensitive glass microelectrodes according to Saito et al. (1976).

In six cats extracellular H^+ activity and the dynamic changes
in microflow were simultaneously measured. Microflow was continuously and qualitatively recorded with the local hydrogen cleareance
method by Lübbers and Stosseck (1970) with a surface element. As the

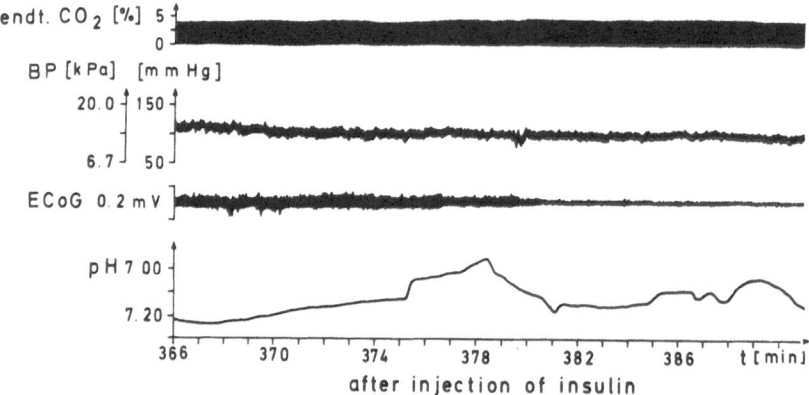

Fig. 1. Original registration of the changes in cerebral extracel-
         lular pH by means of a pH needle electrode. Note the acidic
         shift with irregular oscillations which occurs spontaneous-
         ly in severe late hypoglycemia with reduced ECoG activity.
         (endt. CO_2 = endtidal CO_2 content; BP = arterial blood
         pressure; ECoG = electrocorticogram)

experiments lasted for at least 6 to maximally 12 hours it was
necessary to avoid drifting of the measuring electrodes. Extracellular K^+ activity and microflow were recorded in 8 cats. For the
registration of K^+ activity we used a solid contacted Valinomycin
surface electrode according to Kessler et al. (1974).

In all experiments ECoG was monitored with an unipolar silver
ball electrode. The measuring electrodes were mounted on a counter
balanced holder so that the electrode could follow the movements of
the brain throughout the whole measuring time.

After a control period of 20 to 30 min with normal arterial
control parameters insulin (Novo Actrapid) was injected intravenously with a dose of 60 I.U. per kg body mass. After the ECoG had been
isoelectric for about 15 to 20 minutes, 5 ml of a 25% glucose solution was injected intravenously. Then the recordings were continued
for another hour. Arterial glucose concentration was determined by
using the hexokinase test every 30 min.

We obtained the following results:
1. Microflow in the brain remained constant during the control phase
of normoglycemia with a mean arterial glucose concentration of 5.8
$mmol \cdot l^{-1}$. During early hypoglycemia arterial pH and arterial blood
gases were kept in the normal range by correcting the metabolic acidosis which occurred in most cases after insulin application. At an
arterial glucose concentration between 1.5 and 0.5 $mmol \cdot l^{-1}$ micro-

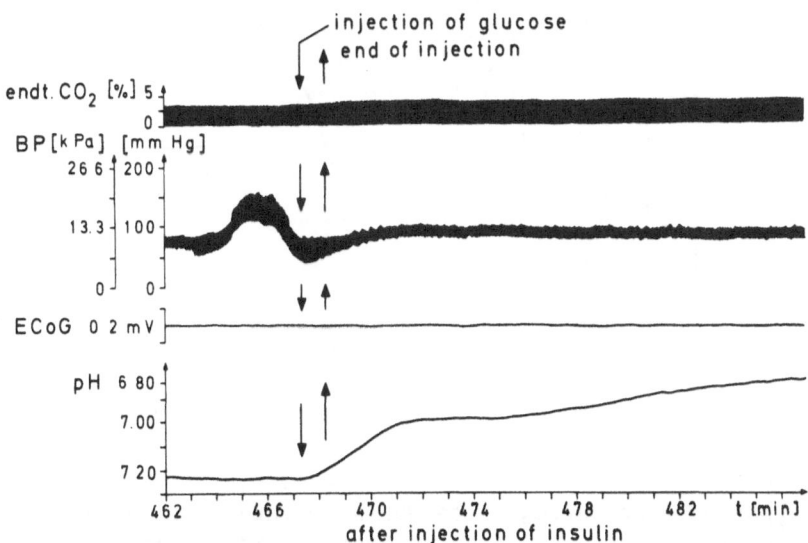

Fig. 2.   Original registration of the changes in extracellular pH.
          After a short period of isoelectricity glucose is adminis-
          tered. Note the strong acidosis which develops.

flow began to increase and showed irregular oscillations although
arterial pH showed a spontaneous tendency to increase and it remained elevated during isoelectricity as far as arterial blood pressure did not decrease below a critical level of about 40 - 50 mm Hg.
Immediately after glucose injection microflow decreased rapidly
within a few minutes.

2. The extracellular pH of the brain cortex in the control phase
was about 7.25 to 7.3 on the surface and in the deeper layers of
the brain cortex. It also remained constant after insulin application with a correction being applied to the pH. During severe hypoglycemia at a mean glucose level of about 1 $mmol \cdot l^{-1}$, a slight
acidic shift of (maximally) 0.2 to 0.3 pH units developed. During
that period irregular oscillations in pH appeared (Fig. 1). With
the onset of an isoelectric ECoG, pH returned to normal or even to
alkalotic values. After glucose injection (Fig. 2) a strong acidosis,
with a pH between 6.6 to 7.0 developed after which pH continued to
gradually decrease over the rest of the experiment.

3. Extracellular K^+ activity remained constant during normoglycemia
and early hypoglycemia. During severe hypoglycemia at a glucose

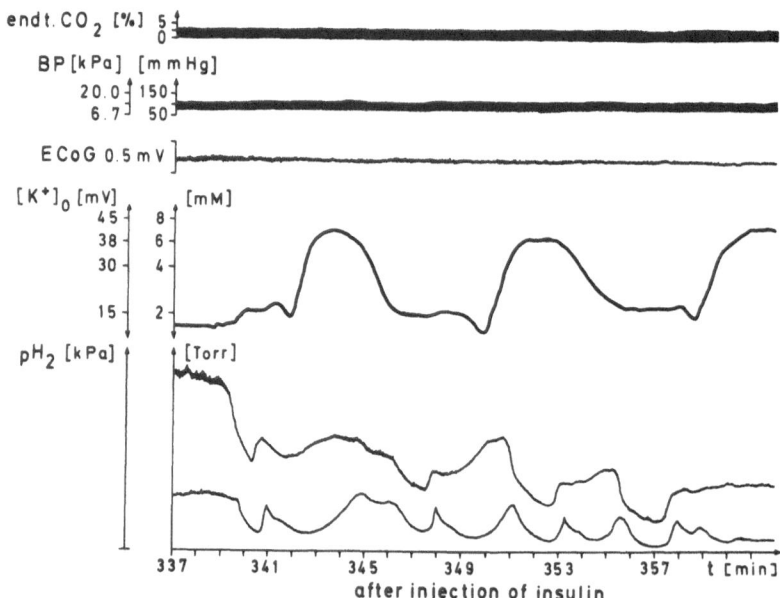

Fig. 3. Simultaneous recordings of extracellular K^+ activity and
         pH_2 at 3 measuring sites during severe late hypoglycemia.
         A decrease in pH_2 means an increase in microflow. Note the
         spontaneous changes in K^+ activity and in microflow with
         constant arterial blood pressure and constant endtidal CO_2
         contents. ECoG activity is diminished in that phase.

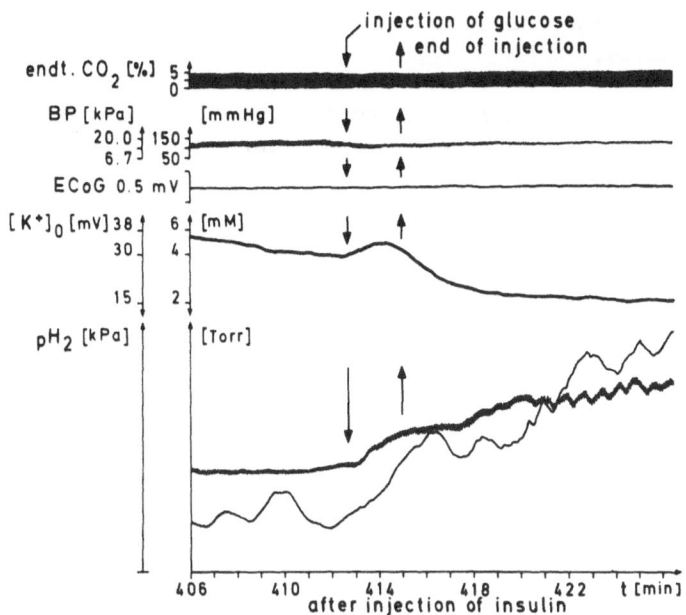

Fig. 4.   Original recordings of extracellular K^+ and pH_2 during and
after glucose administration. Note the decrease in K^+
activity and the concomitant decrease in microflow.

concentration less than 1.5 mmol·l^{-1}, K^+ increased rapidly to values
between 6 to 10 mmol·l^{-1}.

Additionally, irregular waves in K^+ activity were observed as
shown in Fig. 3. During isoelectricity K^+ activity remained elevated
and in most cases it remained constant. After glucose injection a
rapid decrease was recorded, which occurred in parallel with the
decrease in flow as shown in Fig. 4.

DISCUSSION

Our results agree with the findings of Norberg and Siesjö (1976)
in rats in that during a severe hypoglycemia of about 1 mmol·l^{-1}
arterial glucose concentration cerebral flow increases. We think
that the discrepancy with the results of other authors could be due
to methodological problems. As the extracellular K^+ and H^+ activities
change at about the same arterial glucose concentration and at the
same time when microflow increases we conclude that the increase in
flow during late severe hypoglycemia could be caused at least partly
by the increase in K^+ and H^+ activities.

During isoelectricity, H^+ activity obviously cannot be involved
in maintaining an increase in flow whereas K^+ activity remains

elevated in a range where it has a clear dilatory effect on the cerebral vessels.

After glucose injection microflow shows the same dynamic behaviour as K^+ activity. It decreases together with the decrease in K^+ in spite of the rapid and strong acidosis in extracellular pH. In this period the cerebral vessels are obviously not sensitive to the increased extracellular H^+ activity. The reason for this dissociation between decreased flow and increased H^+ activity is unknown.

We assume that the increase in extracellular K^+ and H^+ activities during severe hypoglycemia could be partly due to changes in the permeability of the cell membrane caused by increased production of ammonium ion from amino acids. Under the influence of the increased ammonium ion concentration, K^+ and H^+ ions could leave the cell partly in exchange with Na^+ (Gronczewski and Leniger-Follert, this volume).

REFERENCES

Astrup, J., Heuser, D., Lassen, N.A., Nilsson, B., Norberg, K., and Siesjö, B.K., 1978, Evidence against H^+ and K^+ as main factors for the control of cerebral blood flow: a microelectrode study, in: "Cerebral Vascular Smooth Muscle and its Control", Ciba Foundation Symposium 56 (new series), Elsevier/Excerpta Medica/North Holland, Amsterdam, Oxford, New York.

Cilluffo, J.M., Anderson, R.E., Michenfelder, J.D., and Sundt, Th.M. Jr., 1982, Cerebral blood flow, brain pH, and oxidative metabolism in the cat during severe insulin-induced hypoglycemia, J. Cereb. Blood Flow Metab., 2:337-346.

Gronczewski, J., and Leniger-Follert, E., 1982, Relationship between microflow, local tissue Po_2 and extracellular activities of potassium and hydrogen ions in the cat during intra-arterial infusion of ammonium acetate, this volume.

Kessler, M., Höper, J., and Simon, W., 1974, Methodology and application of multiple ion-selective surface electrode (pH, pK, pNa, pCa, pCl) for tissue measurements, Fed. Proc., 33:279.

Krolicki, L., and Leniger-Follert, E., 1980, Oxygen supply of the brain cortex (rat) during severe hypoglycemia, Pflügers Arch., 387:121-126.

Lübbers, D.W., and Stosseck, K., 1970, Quantitative Bestimmung der lokalen Durchblutung durch elektrochemisch im Gewebe erzeugten Wasserstoff, Naturwissenschaften, 57:311-312.

Norberg, K., and Siesjö, B.K., 1976, Oxidative metabolism of the cerebral cortex of the rat in severe insulin-induced hypoglycemia, J. Neurochem., 26:345-352.

Saito, Y., Baumgärtl, H., and Lübbers, D.W., 1976, The RF-sputtering
    technique as a method for manufacturing needle-shaped pH-micro-
    electrodes, in: "Ion and Enzyme Electrodes in Biology and
    Medicine", M. Kessler, L.C. Clark Jr., D.W. Lübbers, I.A.
    Silver, eds., Urban & Schwarzenberg, München-Berlin-Wien, pp.
    103-109.

GLYCOLYSIS AND REGULATION OF CEREBRAL BLOOD FLOW AND METABOLISM

E. Dóra, and A.G.B. Kovách

Exp. Res. Dept. and 2nd Institute of Physiology
Semmelweis University Medical School, Budapest, Hungary

INTRODUCTION AND METHODS

In past decades several humoral agents and neural mechanisms were suggested for the coupling between cerebral blood flow (CBF), metabolism and functional activity in the brain (Siesjö, 1978). To explore the role of glycolysis in the regulation of CBF, and also to assess the importance of glucose as a substrate in the energy homeostasis of the brain, glycolysis was inhibited topically in the brain cortex by iodoacetate (IAA).

The experiments were performed on 35 cats anaesthetized by chloralose, immobilized by flaxedil and respired artificially. The respiration was adjusted so as to have the arterial Po_2 and Pco_2 around 100 and 32 mm Hg, respectively. Various concentrations of IAA were dissolved in artificial CSF prepared according to Wahl and Kuschinsky (1976). Each dose of IAA (0.01, 0.1, 1 mg/kg) was administered for 30 min at a rate of 1 ml/min onto the surface of the brain cortex using a Harvard infusion pump and our cranial window technique. To exclude nonspecific effects, the brain cortex was also superfused with mock CSF and 1 mg/ml Na-acetate in the same way as with IAA. The cerebral vascular volume (CVV) and NADH fluorescence of the exposed brain cortex were measured by fluororeflectometry. In the majority of the experiments the brain cortex was also photographed and its ECoG was routinely recorded. In addition, in a couple of experiments the 1 mg/ml IAA-induced changes in rCBF and rCMR glucose were determined with the autoradiographic techniques of Sakurada et al. (1978) and Sokoloff et al. (1977). To see how IAA altered the vascular and metabolic reactivity of the brain cortex N_2-anoxia and metrazol-evoked epileptic seizures as tests were applied before and after IAA treatment.

305

RESULTS

    30 min superfusion of the brain cortex with mock CSF, or
1 mg/ml Na-acetate, or 0.0.1 mg/ml IAA did not alter CVV and NAD/
NADH redox state. 0.1 and 1 mg/ml IAA led to a dose-dependent increase in CVV and shifted the redox state toward a more oxidized
state (Figs. 1 and 2). Though 1 mg/ml IAA caused a very marked
NADH oxidation it did not affect the local electrical activity of
the exposed brain cortex discernibly (Fig. 3). In this experiment
IAA increased CVV by approximately 40% and shifted the NAD/NADH

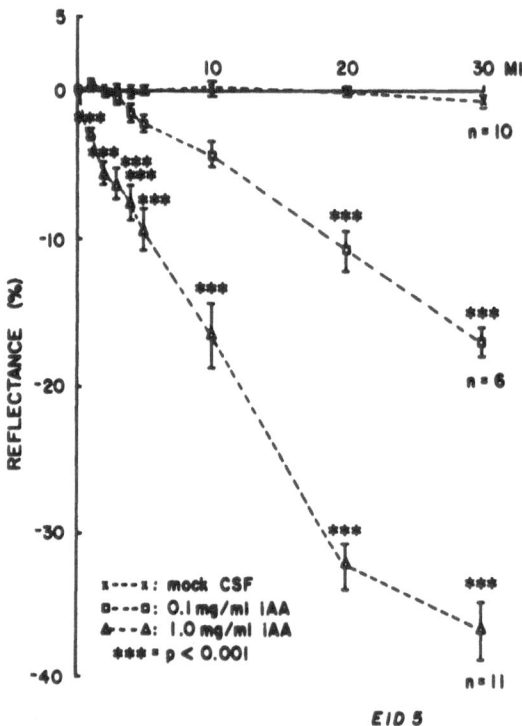

Fig. 1. Effect of the superfusion of the brain cortex with mock
         CSF and 0.1 and 1 mg/ml IAA on cerebrocortical vascular
         volume. The standard error of the mean values is repre-
         sented by vertical lines closed at both ends, n shows the
         number of experiments evaluated. The degree of significant
         changes as compared to the mock CSF superfused group is
         marked by asterisks. Note that the reflectance is invers-
         ly related to the changes in cerebrocortical vascular
         volume. As the figure shows 0.1 mg/ml iodoacetate increas-
         ed CVV only after 20 min in superfusion. 1 mg/ml iodoace-
         tate increased CVV already after 1 min superfusion and the
         vessels became nearly maximally dilatated after 20 min su-
         perfusion.

redox state by about 20% toward oxidation. During IAA superfusion
the correction factor (Harbig et al., 1976), that is used to eliminate the "hemodynamic artifact" in NADH fluorescence measurements,
was determined frequently. As the corrected NADH (CF) fluorescence
trace shows, the correction was made adequately. During IAA superfusion the arterial blood pressure and the ECoG of the non-superfused
brain cortex remained unaltered indicating the topicality of the
IAA treatment.

Corresponding to the reflectometrically revealed increase in
CVV 1 mg/ml IAA dilatated the pial arteries overwhelmingly (Fig. 4)
and increased rCBF 7 - 8-fold (Dora et al., unpublished).

A bolus injection of 8-10 mg/kg metrazól into the lingual artery elicited epileptiform activity in the ECoG which lasted for

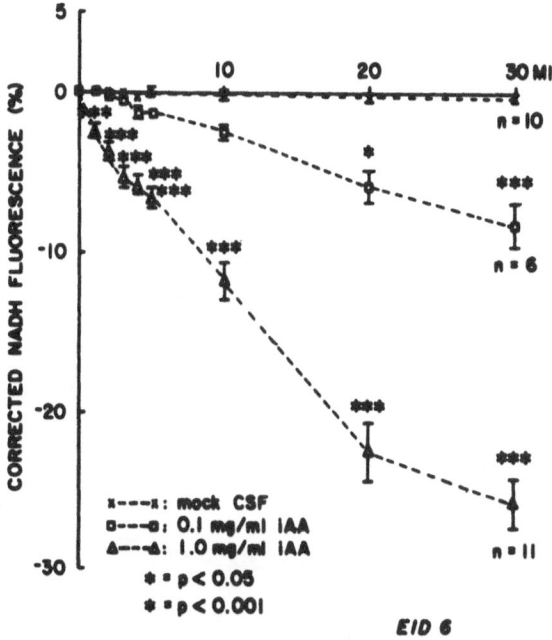

Fig. 2. Effect of the superfusion of the brain cortex with mock
         CSF and 0.1 and 1 mg/ml IAA on cerebrocortical NAD/NADH
         redox state. The standard error of the mean values is re-
         presented by vertical lines closed at both ends, n shows
         the number of experiments averaged. The degree of signifi-
         cant changes as compared to the mock CSF superfused group
         is marked by asterisks. Note that a decrease in corrected
         NADH fluorescence means NADH oxidation. As the figure shows
         0.1 mg/ml iodoacetate induced slight NADH oxidation, how-
         ever, the NADH oxidation at 1 mg/ml iodoacetate became very
         pronounced.

1-2 min. In the untreated brain cortex CVV and NAD reduction increased by approximately 30% and 8-10% during seizures, respectively
(Table 1). After pretreating the brain cortex with 0.1 and 1 mg/ml
IAA, instead of NAD reduction, NADH oxidation was obtained during
seizures. Because the cerebrocortical vessels have been already
dilatated with IAA, the seizures-induced CVV response was markedly
diminished at 0.1 mg/ml IAA, and vanished at 1 mg/ml IAA.

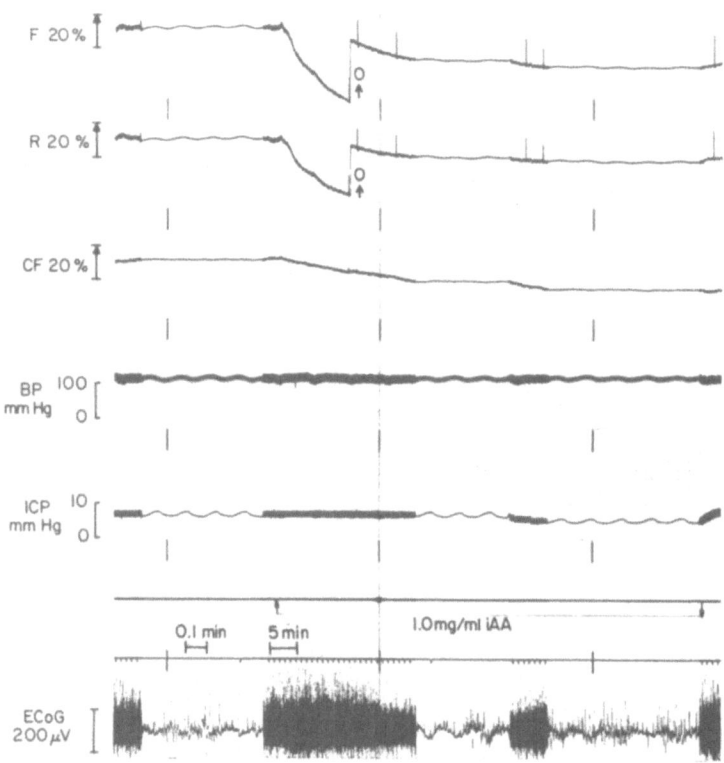

Fig. 3.   Effect of 1 mg/ml IAA on the routinely recorded parameters
          in a typical experiment. The duration of IAA superfusion
          and zero shifts are marked by arrows. The calibrations and
          direction of increases are marked before each trace. Time
          scale is shown above the local ECoG of the exposed brain
          cortex. Spikes on F and R traces were evoked by injecting
          0.1 ml saline into the lingual artery. Abbreviations: F and
          CF = uncorrected and corrected NADH fluorescence, R = re-
          flectance, BP = arterial blood pressure, ICP = intracranial
          pressure. Note that IAA did not discernibly alter the ECoG
          of the superfused brain cortex.

Fig. 5 shows the effect of 0.1 mg/ml IAA pretreatment of the brain cortex on the metrazol-induced CVV and redox state responses in a typical experiment. In the case of untreated brain cortex epileptic seizures resulted in an approximately 40% increase in CVV (R) and 10% NAD reduction (CF). When the epileptiform electrical activity spontaneously ceased the redox state transiently overshot toward a more reduced state. The recovery of the CVV was much faster. Though metrazol evoked similar epileptiform electrical response in the IAA treated brain cortex than in the control condition, CVV increased much less, and instead of NAD reduction NADH oxidation occurred during seizures.

Nitrogen anoxia applied during the control period increased CVV and NAD reduction by about 30% and 40%, respectively (Table 2). When the brain cortex was pretreated with 0.1 and 1 mg/ml IAA these reactions became markedly depressed. At 1 mg/ml IAA N$_2$-anoxia did not bring about a decrease in the intensity of the reflected light which clearly shows that variations in hemoglobin oxygen saturation do not result in discernible changes in the intensity of reflected light measured at 366 nm.

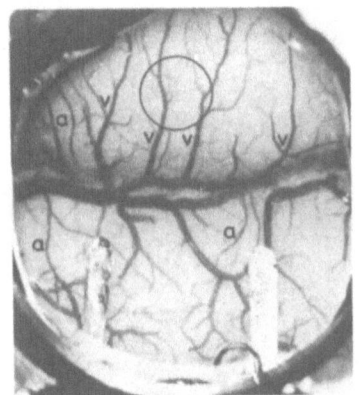
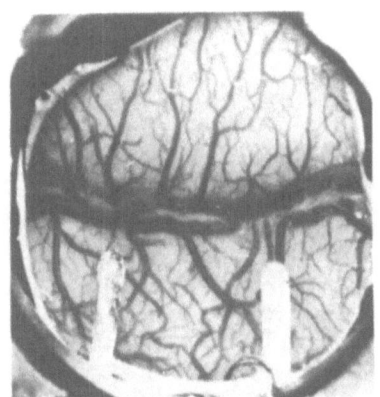

mock CSF   30min monoiodoacetate super-

0 500 1000µm   fusion (1mg/ml)

Fig. 4.   Effect of 30 min superfusion of the brain cortex with 1 mg/ml IAA on the diameter of the pial vessels in a typical experiment. The pictures on the left and right sides of the figure were taken when the brain cortex was superfused with mock CSF and IAA containing CSF, respectively. The arterial and venous pial vessels are marked by a and v letters. The circle drawn in the control picture shows the size and location of the cortical area monitored by fluororeflectometry. The silver plates that were used to record local ECoG can be seen in the bottom of the pictures. Note the very marked dilatation of the pial arterial vessels following IAA superfusion.

DISCUSSION

        According to in vitro experiments performed on various iso-
lated tissues, tissue extracts and purified enzymes, iodoacetate
inhibits glyceral-aldehyde-phosphate dehydrogenase (3-PGDH) quite
selectively (Webb, 1966), and consequently reduces tissue glucose
consumption (Heald, 1953; Webb, 1966). At higher concentrations
than 0.1 mg/ml it might react with other enzymes also containing
SH groups (creatine kinase), but even in this case it has negligible effect on the enzymes of Krebs-cycle, mitochondrial electron
transport and oxidative phosphorylation (Webb, 1966). Because of
this the topical inhibition of cerebrocortical glycolysis by IAA

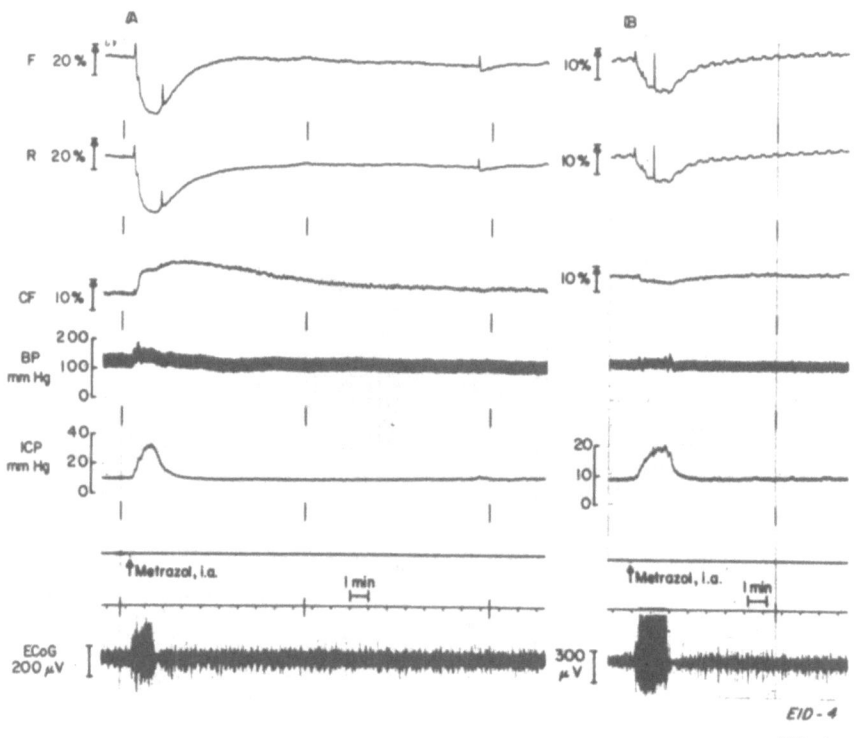

Fig. 5.   Effect of 0.1 mg/ml IAA pretreatment of the brain cortex
           on the metrazol-induced cortical vascular and redox state
           responses in a typical experiment. The reactions before
           and after IAA administration are marked by A and B let-
           ters. Spikes on F and R traces were evoked by the injec-
           tion of 0.1 ml saline into the lingual artery. Other marks
           and abbreviations are the same as in Fig. 3. Note that the
           vasodilatatory effect of epileptic seizures was consider-
           ably diminished after IAA treatment, and instead of the
           control NAD reduction (Fig. 5A) NADH oxidation (Fig. 5B)
           was obtained.

seemed to be a good experimental model for a better understanding of the relationships between glucose metabolism, blood flow and electrical activity of the brain. Compared to other experimental models (starvation, insulin-induced arterial hypoglycemia) it has the great advantage that systemic effects do not occur, therefore these relationships can be studied more clearly. In addition to this some of the possible coupling factors (H^+, lactate) are excluded since IAA inhibits lactate production (Webb, 1966).

Table 1. Effect of 0.1 and 1 mg/ml IAA on the metrazol-induced cerebrocortical vascular and redox state reactions.

	Control R %	Control CF %	0.1 mg/ml IAA R %	0.1 mg/ml IAA CF %	Control R %	Control CF %	1 mg/ml IAA R %	1 mg/ml IAA CF %
$\bar{x}$	-28.3	8.5	-8.3	-4.1	-28.3	9.0	1.3	-3.0
+SE	2.2	1.3	1.8	1.2	1.7	1.3	0.9	1.4
n	6	6	6	6	6	6	6	6
p	-	-	***	**	-	-	***	***

Abbreviations: R = reflectance, CF = corrected NADH fluorescence, IAA = iodoacetate, $\bar{x}$ +SE = mean value ± standard error, n = number of experiments, p = degree of significant changes (** = p < 0.01, *** = p < 0.001). The effects of 0.1 and 1 mg/ml IAA were tested in separate series of experiments. Note that reflectance is inversely related to changes of cerebrocortical vascular volume. Increase and decrease in CF means NAD reduction and NADH oxidation, respectively.

Table 2. Effect of 0.1 and 1 mg/ml IAA on the nitrogen anoxia-induced cerebrocortical vascular and redox state reactions.

	Control R %	Control CF %	0.1 mg/ml IAA R %	0.1 mg/ml IAA CF %	Control R %	Control CF %	1 mg/ml IAA R %	1 mg/ml IAA CF %
x	-28.0	40.7	-9.3	31.4	-29.9	42.4	0	20.5
+SE	1.1	0.8	1.2	2.1	1.1	2.1	1.3	1.2
n	7	7	7	7	8	8	8	8
p	-	-	***	***	-	-	***	***

Abbreviations and other marks are the same as in Table 1. The respiration of nitrogen gas was maintained until the ECoG became isoelectric (90-120 sec). The effects of 0.1 and 1 mg/ml IAA were tested in separate series of experiments.

Our results show that iodoacetate (0.1 and 1 mg/ml) dilates
cerebrocortical vessels and since cerebral perfusion pressure is
not altered, increases rCBF considerably. 1 mg/ml IAA led to maximal
dilatation of the vessels because neither epileptic seizures nor
nitrogen anoxia could induce further increase in CVV. At the same
time this concentration of IAA shifted the NAD/NADH redox state by
approximately 30% toward a more oxidized state and depressed the
anoxia and epileptic seizures-induced NAD reduction. These metabolic
effects of IAA indicate that the application of IAA should have reduced cerebrocortical glucose consumption considerably. Supporting
this indirect conclusion our pilot studies with ^{14}C labelled 2-
deoxyglucose indeed showed that 30 min pretreatment of the brain
cortex with 1 mg/ml IAA dminished CMR glucose to 1/3-1/4 of its
normal value. Since, in spite of the great reduction of glucose metabolism the ECoG and excitability of the IAA exposed brain cortex
was maintained at normal, it is supposed that some other substrates
besides glucose were consumed by the cortex. This finding agrees
with the corresponding data from literature obtained on isolated
brain slices (Heald, 1953; McIlwain, 1953), on isolated perfused
brain (Dirks et al., 1980), and on in vivo intact brain during
starvation and severe insulin-induced arterial hypoglycemia (Agardh
et al., 1981; Christensen et al., 1981).

Furthermore because recent studies of Chace and Odessey (1981)
and Odessey and Chace (1982) unequivocally proved that isolated
aorta readily uses other substrates than glucose, and the cerebrocortical small and microvessels contain all the necessary enzymes
to metabolize lipids, ketone bodies and amino acids (Cook et al.,
1978), it is very unlikely that IAA dilatated cerebrocortical
vessels via the reduction of energy-rich phosphates.

Winn et al. (1981; 1982) demonstrated that cerebral adenosine
concentration can be increased by as much as 10-15 times during
arterial hypoxia and epileptic seizures, and 60 times during the
slow wave activity phase of insulin-induced arterial hypoglycemia.
Since adenosine is a potent dilatator of pial vessels (Wahl and
Kuschinsky, 1976) - and if one accepts Winn's et al. (1981) presumption, namely adenosine stays only extracellularly -, adenosine
could be a very likely candidate of IAA-elicited vasodilatation.
However, the following results speak against this possibility: IAA
did not depress the ECoG of the exposed brain cortex in our experiments; a great part of cerebral adenosine might stay intracellularly and not extracellularly: adenosine deaminase treatment of
the brain cortex did not constrict the cerebrocortical vessels,
failed to influence the arterial hypoxia-evoked increase in CVV,
and only slightly diminished the augmentation of CVV associated to
epileptic seizures (Dora et al., submitted); superfusion of the
brain cortex with 10^{-3} M adenosine (Kovach and Dora, in this volume) led to half of the CVV response obtained with 1 mg/ml IAA.

Considering the above arguments it is unlikely that the generally accepted coupling factors (H^+, lactate, adenosine) or an energetic failure of the vascular smooth muscle would be responsible for the marked vasodilatatory effect of IAA. It should be due to some other mechanism.

Since IAA, arterial hypoxia and epileptic seizures stimulate phosphofructokinase and consequently increase the concentration of 3 glycolytic intermediates (fructose-1-6-di-P, glyceral-aldehyde-P, dihydroxyacetone-P) (Siesjö, 1978; Webb, 1966), it could be speculated that these intermediates or some associated changes adjust CBF to the actual functional activity of the brain. This would be the basic mechanism and the various other coupling factors would be secondary to this in the rapid adjustment of CBF. This presumption would also reasonably explain the reduction of CBF during deep barbiturate anaesthesia and hypothermia. However, to substantiate this hypothesis further experimental work is needed.

CONCLUSIONS

1. In the cerebral vasodilatatory mechanism of iodoacetate the generally accepted coupling factors (H^+, lactate, adenosine) do not take part.
2. Some substrates unidentified in the present study, can substitute for glucose as energy donors to maintain cortical electrical activity at a normal level.
3. A considerable part of the nitrogen anoxia-evoked cerebrocortical NAD reduction is not related to oxygen lack per se, but to increased glucose mobilization.
4. Epileptic seizure-induced cortical NAD reduction is not due to a relative hypoxia but to an elevated substrate influx from glucose mobilization.

REFERENCES

Agardh, C.D., Chapman, A.G., Nilsson, B., and Siesjö, B.K., 1981, Endogenous substrates utilized by rat brain in severe insulin-induced hypoglycemia, Acta Physiol. Scand., 36:490.
Chace, K.V., and Odessey, R., 1981, The utilization by rabbit aorta of carbohydrates, fatty acids, ketone bodies, and amino acids as substrates for energy production, Circ. Res., 48:850.
Christensen, T.G., Diemer, N.H., Laursen, H., and Gjedde, A., 1981, Starvation accelerates blood-brain glucose transfer, Acta Physiol. Scand., 112:223.
Cook, B.H., Granger, H.J., Granger, D.N., Taylor, A.E., and Smith, E.E., 1978, Metabolic profiles of canine cerebrovascular tree: A histochemical study, Stroke, 9:165.

Dirks, B., Hanke, J., Kriegelstein, J., Stock, R., and Wickop, G., 1980, Studies on the linkage of energy metabolism and neuronal activity in the isolated perfused rat brain, J. Neurochem., 35:311.

Dora, E., Koller, A., and Kovach, A.G.B., 1982, Possible role of extracellularly released adenosine in the regulation of cerebral blood flow, submitted to Circ. Res.

Harbig, K., Chance, B., Kovach, A.G.B., and Reivich, M., 1976, In vivo measurement of pyridine nucleotide fluorescence from cat brain cortex, J. Appl. Physiol., 41:480.

Heald, P.J., 1953, The effect of metabolic inhibitors on respiration and glycolysis in electrically stimulated cerebral-cortex slices, Biochem. J., 55:625.

Kovach, A.G.B., and Dora, E., 1982, Contribution of adenosine to the regulation of cerebral blood flow. The role of calcium ions in the adenosine-induced cerebrocortical vasodilatation, this volume.

McIlwain, H., 1953, Glucose level, metabolism, and response to electrical impulses in cerebral tissues from man and laboratory animals, Biochem. J., 55:618.

Odessey, R., and Chace, K.V., 1982, Utilization of endogenous lipid, glucogen, and protein by rabbit aorta, Am. J. Physiol., 243: H128.

Sakurada, O., Kennedy, C., Jehle, J., Brown, J.D., Carbin, G.L., and Sokoloff, L., 1978, Measurement of local cerebral blood flow with (^{14}C) iodoantipyrine, Am. J. Physiol., 234:H59.

Siesjö, B.K., 1978, "Brain Energy Metabolism", John Wiley & Sons, Chichester-New York-Brisbrane-Toronto.

Sokoloff, L., Reivich, M., Kennedy, C., Des Rosiers, M.H., Patlak, C.S., Pettigrew, K.D., Sakurada, O., and Shonohara, M., 1977, The (^{14}C)deoxyglucose method for the measurement of local cerebral glucose utilization: theory, procedure, and normal values in the conscious and anaesthetized albino rat, J. Neurochem., 28:897.

Wahl, M., and Kuschinsky, W., 1976, The dilatatory action of adenosine on pial arteries of cats and its inhibition by theophylline, Pflügers Arch., 362:55.

Webb, J.L., 1966, Iodoacetate, in: "Enzyme and Metabolic Inhibitors", Vol. III, J.L. Webb, ed., Academic Press, New York-London, pp.:1-283.

Winn, H.R., Rubio, G.R., and Berne, R.M., 1981, The role of adenosine in the regulation of cerebral blood flow, J. Cereb. Blood Flow Metab., 3:239-244.

Winn, H.R., Weaver, D.D., Ngai, A.C., and Berne, R.M. (Abstract), 1982, Changes in brain adenosine concentrations during hypoglycemia in the rat, Stroke, 13:123.

CONTRIBUTION OF ADENOSINE TO THE REGULATION OF CEREBRAL BLOOD FLOW:

THE ROLE OF CALCIUM IONS IN THE ADENOSINE-INDUCED CEREBROCORTICAL

VASODILATATION

A.G.B. Kovách, and E. Dóra

Experimental Research Department and 2nd Institute of
Physiology, Semmelweis University Medical School
Budapest, Hungary

INTRODUCTION AND METHODS

It is well established that cerebral blood flow (CBF) and
adenosine concentration in the brain can be elevated severalfold
during arterial hypoxia and epileptic seizures (Winn et al., 1981).
Since perivascularly administered adenosine dilatates pial arteries
(Wahl and Kuschinsky, 1976) and the stable analog of adenosine,
chloroadenosine, is an even more efficient dilatator of cerebrocortical vessels (Winn et al., 1981), it was suggested that adenosine
plays a central role in the regulation of CBF (Winn et al., 1981).
However, since the adenosine concentration of the normal brain is
very low (Winn et al., 1981: around 10^{-7} $mol\cdot l^{-1}$), this assumption
is valid only if adenosine is present exclusively in the extracellular fluid. Besides this, it is also not yet clear whether the vasodilatatory action of adenosine involves changes in calcium availability of the vascular smooth muscle (Dutta et al., 1980; Fenton et
al., 1982), or adenosine dilates the vessels by some other mechanism
(Kukovetz et al., 1978). In the present study the following questions were addressed: a) How efficient is topically administered
adenosine when compared with arterial hypoxia and epileptic seizures
in altering cerebrocortical vascular volume (CVV)? b) Is there a
possible role of adenosine-induced cortical metabolic changes in
the vasodilatatory mechanism? c) What is the importance of calcium
availability in the vascular action of adenosine? To answer these
questions the brain cortex was superfused with various concentrations of adenosine. The adenosine-, arterial hypoxia- and epileptic
seizure-induced cerebrocortical vascular and NAD/NADH redox state

315

responses were compared. Finally, the effect of organic calcium antagonists on these responses was studied.

    The experiments were performed on 40 cats anaesthetized by chloralose, immobilized by flaxedil, and artificially respired. The respiration was adjusted to keep the arterial Po_2 and Pco_2 at

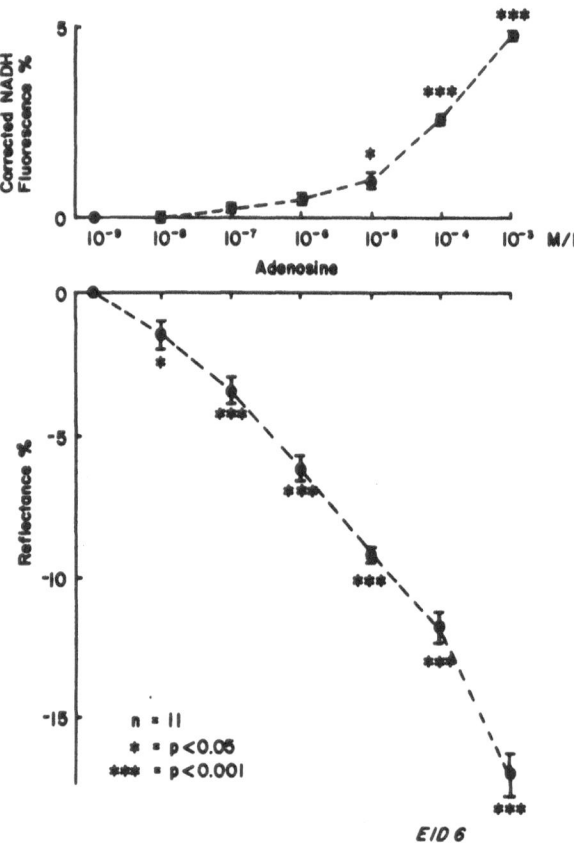

Fig. 1.   Effect of the superfusion of the brain cortex with various concentrations of adenosine (10^{-9}–10^{-3} M) on cerebrocortical vascular volume (CVV) and NAD/NADH redox state. The standard error of the mean values is represented by vertical lines closed at both ends, n is the number of experiments evaluated. The degree of significant changes is marked by the number of asterisks. Note that reflectance is inversely related to CVV, decrease in reflectance means increase in CVV. Increase in corrected NADH fluorescence represents NAD reduction. In the concentration range of 10^{-8}–10^{-3} M two to three min topical administration of adenosine resulted in a dose-dependent increase in CVV. NAD reduction occurred at concentrations of adenosine higher than 10^{-6} mol·l^{-1}.

approximately 100 and 32 mm Hg, respectively. The rectal temperature
of the animals was maintained at 37°C. Cerebrocortical vascular
volume and NADH fluorescence were measured through a cranial window
by a micro fluororeflectometer. Adenosine and organic calcium antagonists (verapamil, D-600) were dissolved in artificial CSF, prepared
according to Wahl and Kuschinsky (1976) and applied topically to the
surface of the brain cortex using a Harvard infusion pump and a
cranial window-superfusion technique. The artificial CSF was thermo-
stated at 38°C and equilibrated with 5% CO_2 in N_2. The exposed brain
cortex was photographed in some of the experiments but its local
electrocorticogram (ECoG) was recorded routinely. Arterial hypoxia
was induced by ventilating the animals with N_2 gas until the ECoG
ceased. Epileptic seizures, lasting for 1-2 min, were evoked by
bolus injection of 8-10 mg/kg metrazol into the lingual artery.

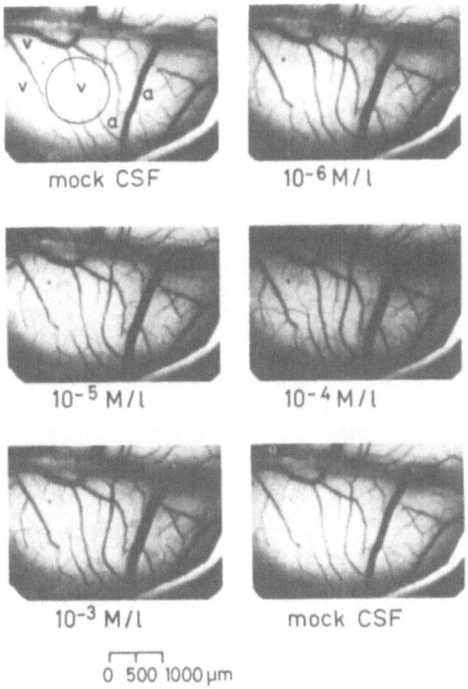

Fig. 2. Effect of topically administered adenosine on the diameter
         of pial arterious (a) and venous (v) vessels in a single
         experiment. The circle drawn in the control picture (mock
         CSF) shows the size and location of the cortical area mon-
         itored by the fluororeflectometer.

RESULTS

In agreement with previous data from the literature (Wahl and Kuschinsky, 1976), topically administered adenosine resulted in a dose-dependent increase in CVV (Fig. 1) and markedly dilatated the pial arteries (Fig. 2). The NAD/NADH redox state of the brain cortex was not altered significantly by low concentrations (10^{-8}–10^{-6} M) of adenosine but at higher adenosine concentrations (10^{-5} and 10^{-3} M)

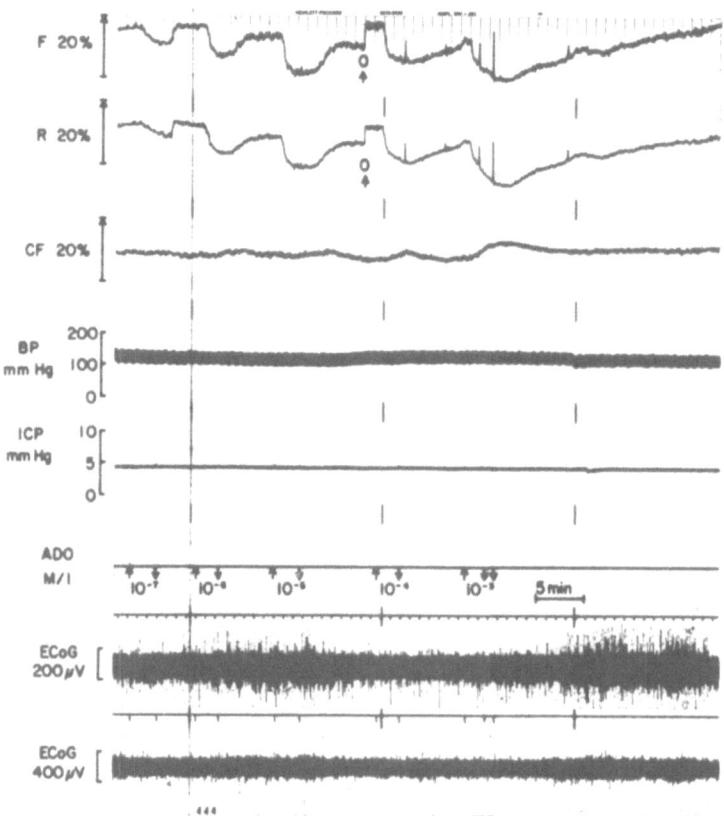

Fig. 3.   The effects of adenosine superfusion of the brain cortex in a single experiment. The duration of adenosine administration is marked by arrows. Before and between adenosine applications the brain cortex was superfused with mock CSF. The calibrations and direction of increase are shown before each parameter. Abbreviations: F and CF = uncorrected and corrected NADH fluorescence, R = reflectance, BP = arterial blood pressure, ICP = intracranial pressure. Time scale is shown above the local ECoG of the superfused brain cortex. Decrease in R and increase in CF mean increase in CVV and NAD reduction, respectively.

significant NAD reduction was obtained. The NAD reduction occurred in spite of the considerable increase of rCBF (adenosine increased CVV but did not alter cerebral perfusion pressure).

In Fig. 3 the effects of topically administered adenosine on various routinely recorded parameters are demonstrated in an original recording. In this experiment the dilatatory action (R decreased) of adenosine started at 10^{-7} M concentration. Discernible NAD reduction occurred at 10^{-5} M adenosine concentration. The arterial blood pressure (BP), intracranial pressure (ICP), and the ECoG of

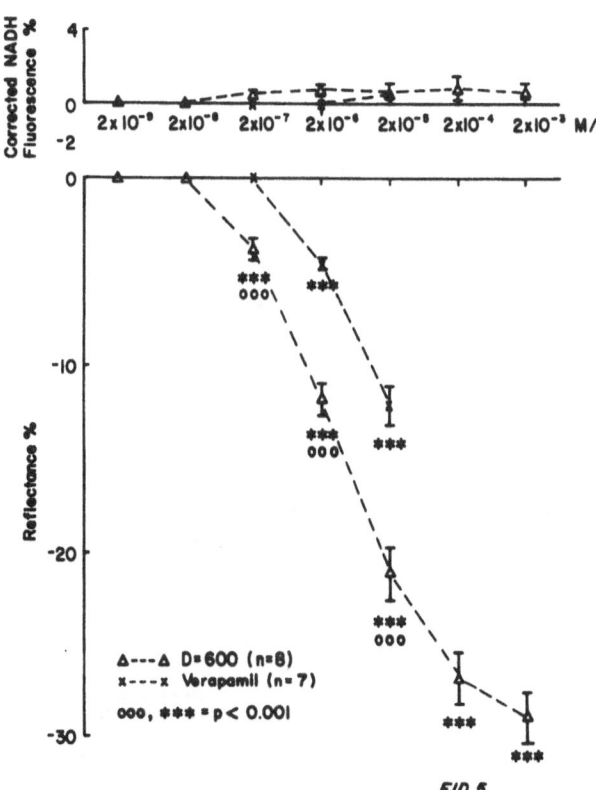

Fig. 4. Effect of various concentrations of D-600 and verapamil on cerebrocortical vascular volume and NAD/NADH redox state. The standard error of the mean values is represented by vertical lines closed at both ends, n is the number of experiments averaged. The degree of significant changes is marked by the number of asterisks. The number of small "o" letters show the significant changes between verapamil and D-600 treated groups. Note that D-600 was approximately 10 times more efficient than verapamil in altering CVV but neither of these calcium antagonists had significant effect on NAD/NADH redox state.

the adenosine superfused and non-superfused brain cortex were not altered at any of adenosine doses tested.

Superfusion of the brain cortex with organic calcium antagonists (verapamil, D-600) increased CVV dose-dependently and markedly but none of their doses altered NAD/NADH redox state significantly (Fig. 4). A concentration of 2×10^{-6} M D-600 increased CVV approximately as much as 10^{-4} M adenosine (Fig. 1), and 2×10^{-5} M D-600 increased CVV twice as much as 10^{-4} M adenosine. In spite of the considerable increase of CVV obtained with D-600, the lack of NADH oxidation indicates that the brain of normoxic animals is well oxygenated.

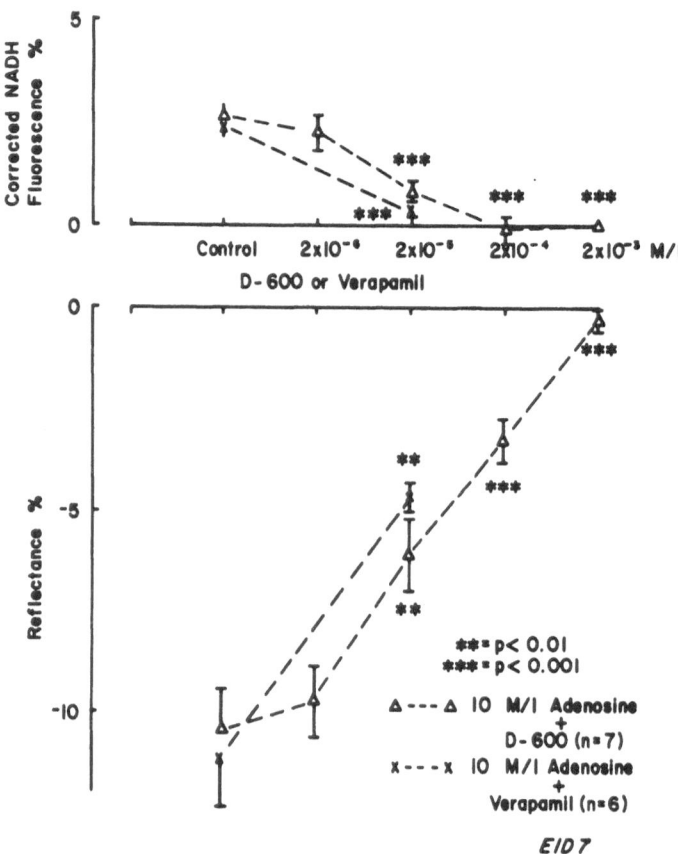

Fig. 5. Individual effects of D-600 and verapamil on the cortical vascular and redox state responses evoked by 10^{-4} M adenosine. The standard error of the mean values is marked by vertical lines closed at both ends, n is the number of experiments evaluated. The degree of significant alterations in adenosine-induced cortical vascular and redox state responses is shown by the number of asterisks.

When 10^{-4} M adenosine was superimposed on various concentrations of D-600 the vascular and metabolic effects of adenosine were not altered by $2x10^{-7}$ and $2x10^{-6}$ M concentrations of this organic calcium antagonist (Fig. 5). At $2x10^{-5}$ M concentrations of D-600 and verapamil the vascular and metabolic effects of adenosine were significantly depressed, however it should be remembered that the vessels were already considerably dilatated by these drugs. At $2x10^{-3}$ M D-600 concentration, when the cerebrocortical vessels were almost maximally dilatated, adenosine had no effect on CVV and NAD/NADH redox state.

In Fig. 6, besides the effects of D-600 on adenosine-induced vascular and NAD/NADH redox state responses, it can also be seen that D-600 did not influence the local electrical activity of the exposed brain cortex. This shows that calcium channels of the cerebral vascular smooth muscle are much more sensitive to this organic calcium antagonist than those calcium channels of the brain paren-

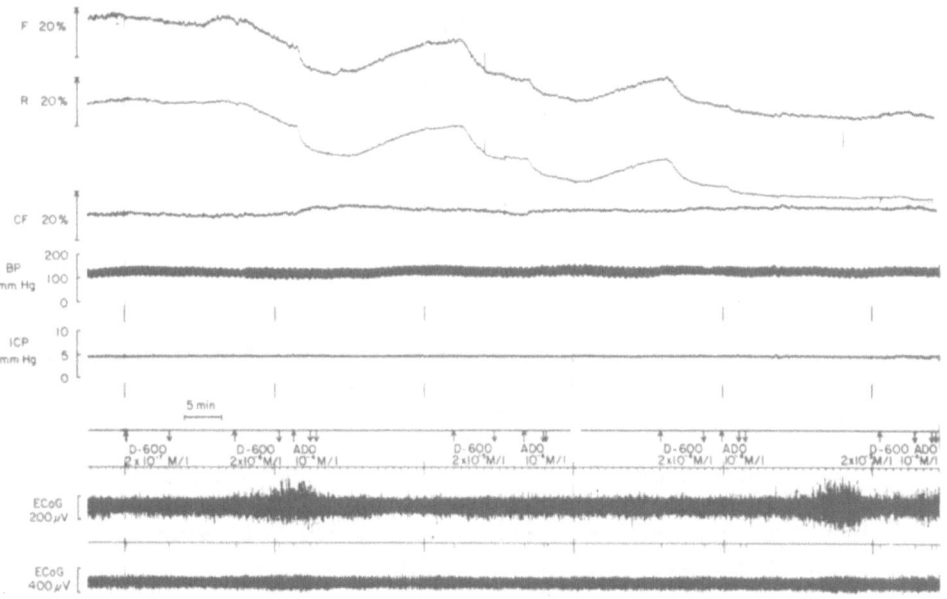

Fig. 6.   Effect of various concentrations of D-600 on 10^{-4} M adenosine-evoked cortical vascular and NAD/NADH redox state responses in the same experiments as in Fig. 3. Drug administrations are marked by arrows. Other marks and abbreviations are the same as in Fig. 3. Note that none of the doses of D-600 affected the local electrical activity of the superfused brain cortex (see ECoG below the time scale). Note also that though D-600 increased CVV - and because of the constancy of cerebral perfusion pressure - rCBF very markedly, the NAD/NADH redox state was not shifted toward a more oxidized state.

Table 1.   Effect of 2×10^{-3} M D-600 pretreatment of the brain cortex
on the N_2-anoxia and epileptic seizures-evoked cortical
vascular volume and NAD/NADH redox state responses.

	Nitrogen anoxia				Epilepsy			
	Mock CSF		D-600		Mock CSF		D-600	
	R %	CF %	R %	CF %	R %	CF %	R %	CF %
$\bar{x}$	-28.0	40.7	-4.5	22.8	-28.3	8.5	-6.1	1.3
+SE	1.1	0.8	0.7	2.2	2.2	1.3	0.8	0.5
n	7	7	6	6	6	6	6	6
p	-	-	***	***	-	-	***	**

Abbreviations: R = reflectance, CF = corrected NADH fluorescence,
$\bar{x} \pm$ SE = mean value $\pm$ standard error, n = number of experiments
averaged, p = degree of significant changes between "mock CSF" and
"D-600" treated groups (** = $p < 0.01$, *** = $p < 0.001$). Note: since
during anoxia the autoregulation of CBF is lost and the arterial
blood pressure initially increases and later decreases, only those
anoxic reactions were evaluated where the arterial blood pressure
at the end of N_2 respiration did not differ markedly ($\pm$ 15 mm Hg)
from its normoxic value.

chyma cells which are connected to the synaptic transmission. A
finding similar to this was reported previously on isolated brain
slices (Ververken et al., 1982).

In normal brain cortex, N_2-anoxia and epileptic seizures resulted in approximately 28% increase in CVV, and 40% and 8% increase
in NAD reduction, respectively (Table 1). Following 2×10^{-3} M D-600
pretreatment of the brain cortex the same experimental interventions
led only to slight increase in CVV and markedly depressed NAD reduction.

DISCUSSION

It is well recognized that cerebral glycogenolysis and glycolysis are considerably accelerated by epileptic seizures, arterial
hypoxia and various depolarizing agents (Ibrahim, 1975; Siesjö, 1978;
Nimit et al., 1981; Ververken et al., 1982). It is also known from
in vitro studies performed on isolated brain slices and cerebral
small vessels that externally added adenosine stimulates the enzyme
of adenylate cyclase and cAMP production, and consecutively glyco-

genolysis and glycolysis (Ibrahim, 1975; Huang and Drummond, 1979; Nimit et al., 1981). Increased glucose metabolism on other hand will lead to accumulation of lactate and acidosis, and hence dilatation of the vessels (Siesjö, 1978). Though these metabolic effects of adenosine could dilatate cerebral vessels indirectly, the possible importance of adenosine in the regulation of CBF was solely attributed to extracellularly released adenosine (Winn et al., 1981).

Our present data clearly showed that adenosine stimulates glucose metabolism also in the in vivo brain cortex since it shifted cerebrocortical NAD/NADH redox state toward a more reduced state, though rCBF was increased. Because of this finding it can be presumed that those experimental conditions (hypoxia, activations of the brain) which result in an increase in cerebral adenosine concentration, elevate CBF partially via glucose mobilization due to accumulation of adenosine and, possibly calcium in the intracellular space.

Since the extracellular concentration of adenosine in the brain is not known and the adenosine concentration of the whole brain is quite low compared to the effective doses of adenosine determined by perivascular micro-injection (Wahl and Kuschinsky, 1976) and superfusion (present data), the importance of extracellularly released adenosine in the regulation of CBF is difficult to estimate. However, several findings speak against the presumption of Winn et al. (1981), namely extracellularly released adenosine is a major factor in the adjustment of CBF to the actual needs of the brain. The main objections against this presumption are as follows: 1) In our experiments 10^{-3} M adenosine increased CVV only by about 17% while epileptic seizures and anoxia augmented CVV approximately by 28%. 2) The reuptake mechanisms and adenosine deaminase inactivate free adenosine very rapidly (Nimit et al., 1981). 3) Adenosine deaminase superfusion of the brain cortex did not constrict the pial arteries (Wei and Kontos, 1981), and did not depress the CVV responses associated to various kinds of brain hypoxia and activations (Dora et al., submitted).

How adenosine elicits relaxation of vascular smooth muscle is obscure (Kukovetz et al., 1978; Huang and Drummond, 1979; Dutta et al., 1980; Fenton et al., 1982). To clear the importance of calcium availability in the mechanism of adenosine-induced vasodilatation the usage of organic calcium antagonists seemed to be reasonable. According to Fleckenstein (1977) these drugs, at a certain concentration range, selectively inhibit the inward movement of calcium into the vascular smooth muscle cells. However, our results showed that the organic calcium antagonist, D-600, inhibited adenosine-induced cerebrocortical vasodilatation only at those concentrations which themself dilatated the vessels near to maximally. Since according to recent data from literature (Church and Zsoter, 1980) organic calcium antagonists at high concentrations inhibit not only the inward movement of calcium but also its release from cytoplasmic binding sites, and they might effect directly the contraction-relaxation process itself, for the vasodilatatory mechanism of adenosine, the following might be suggested: inhibition of transmembrane calcium influx plays no or minor role in the vascular action of adenosine; adenosine probably dilatates cerebrocortical vessels partly by stimulation of intracellular binding of calcium, partly by some other mechanism, like increased cAMP production as proposed by Kukovetz et al. (1978), partly via glucose mobilization (see discussion above).

CONCLUSIONS

1) Extracellularly located adenosine plays a minor role in the regulation of CBF. Intracellularly accumulated adenosine might dilatate the cerebrocortical vessels indirectly during arterial hypoxia and epileptic seizures via substrate (glucose) mobilization.

2) In the cerebral vasodilatatory mechanism of adenosine the inhibition of the transmembrane calcium influx plays negligible role.

3) A considerable part of the cerebrocortical NAD reduction obtained in vivo by fluororeflectometry during epileptic seizures and nitrogen anoxia is not released to the relative or absolute lack of oxygen but to increased substrate influx due to glucose mobilization.

REFERENCES

Church, J., and Zsoter, T.T., 1980, Calcium antagonistic drugs. Mechanism of action, Can. J. Physiol. Pharmacol., 58:254.

Dora, E., Koller, A., and Kovach, A.G.B., 1982, Possible role of extracellularly released adenosine in the regulation of cerebral blood flow, submitted to Circ. Res.

Dutta, P., Mustafa, S.J., and Jones, A.W., 1980, Effect of adenosine on the uptake and efflux of calcium by coronary arteries of dog, Fed. Proc., 39:530.

Fenton, R.A., Bruttig, S.P., Rubio, R., and Berne, R.M., 1982, Effect of adenosine on calcium uptake by intact and cultured vascular smooth muscle, Am. J. Physiol., 242:H797.

Fleckenstein, A., 1977, Specific pharmacology of calcium in myocardium, cardiac pacemaker, and vascular smooth muscle, Ann. Rev. Pharmacol. Toxicol., 17:149.

Huang, M., and Drummond, G.I., 1979, Adenylate cyclase in cerebral microvessels: action of guanine nucleotides, adenosine, and other agonists, Mol. Pharmacol., 16: 462.

Ibrahim, M.Z.M., 1975, Glycogen and its related enzymes of metabolism in the central nervous system, Adv. Anat. Embryol. Cell. Biol., 52:1.

Kukovetz, W.R., Posch, G., Holzmann, S., Wurm, A., and Rinner, I., 1978, Role of cyclic nucleotides in adenosine-mediated regulation of coronary flow, Adv. Cyclic Nucleotide Res., 9:397.

Nimit, Y., Skolnick, P., and Daly, J.W., 1981, Adenosine and cyclic
    AMP in rat cerebral cortical slices: effects of adenosine up-
    take inhibitors and adenosine deaminase inhibitors, J. Neuro-
    chem., 36:908.
Siesjö, B.K., 1978, "Brain Energy Metabolism", John Wiley & Sons,
    Chichester-New York-Brisbane-Toronto.
Ververken, D., van Veldenhofen, P., Proost, C., Carton, H., and
    de Wulf, H., 1982, On the role of calcium ions in the regula-
    tion of glycogenolysis in mouse brain cortical slices,
    J. Neurochem., 38:1286.
Wahl, M., and Kuschinsky, W., 1976, The dilatatory action of adeno-
    sine on pial arteries of cats and its inhibition by theophyl-
    line, Pflügers Arch., 362:55.
Wei, E.P., and Kontos, H.A., 1981, Role of adenosine in cerebral
    arteriolar dilation from arterial hypoxia, J. Cereb. Blood
    Flow Metab., 1 (Suppl. 1):S395.
Winn, R.H., Rubio, R.G., and Berne, R.M., 1981, The role of adeno-
    sine in the regulation of cerebral blood flow, J. Cereb. Blood
    Flow Metab., 1:239.

EFFECT OF DMSO AND BARBITURATES ON BRAIN OXYGEN DISTRIBUTION

H.I. Bicher[1], H. Dujovny[2], C. Codas[2], and S. Honig[2]

[1]Valley Cancer Institute, 5222 Sepulveda Blvd.
Van Nuys, CA 91411, USA
[2]Henry Ford Hospital, Detroit, Michigan, USA

ABSTRACT

Brain oxygen distribution in the anesthetized cat's brain was determined using oxygen ultramicroelectrodes. The animals were anesthetized using Ketalar. Vein, artery and tracheal cannulation were performed. Oxygen distribution was expressed in histogram fashion for brain cortex, white matter and nucleus caudatus. The average P_tO_2 distribution was higher in cortex and nucleus caudatus than in the white matter. There was no difference between right or left side of the brain.

Three different groups of 12 cats each were experimented upon. Group I was control, anesthesia only. Group II received Thiopental, 30 mg/kg bolus followed by 30 mg/kg the second and third hours to a total of 90 mg/kg. Group III received DMSO 2 mg/kg, in addition to the procedures of Group II.

There was no difference in the histograms for cortex, white matter or nucleus caudatus between groups I and II. There was, however, a significant increase in P_tO_2 in Group III, especially at the level of the nucleus caudatus.

327

HEART

LOCAL OXYGEN SUPPLY AND REGIONAL WALL MOTION OF THE DOG'S HEART

DURING CRITICAL STENOSIS OF THE LAD[*]

M. Kessler[1], W.P. Klövekorn[2], J. Höper[1], F. Sebening[2],
M. Brunner[1], K.H. Frank[1], D.K. Harrison[1], C. Kernbach[1],
W. Anderer[1], H. Richter[1], and R. Ellermann[1]
[1]Institut für Physiologie und Kardiologie der
Universität Erlangen-Nürnberg, 8520 Erlangen, FRG
[2]Deutsches Herzzentrum, 8000 München, FRG

INTRODUCTION

When open heart surgery is performed in order to bypass ste-
nosed coronary arteries by a vein graft, the heart surgeon often
discovers hypokinetic zones in the poorly perfused area of the myocardium.

So far, the biological factors which induce these hypokinetic
zones in the beating heart are unknown. For quite a long time it was
supposed that a local depletion of energy-rich phosphates might be
the cause of disturbances of wall motion in the critically perfused
myocardium. However, systematic measurements of the local content of
creatine phosphate and ATP performed by several investigators (Opie,
1976; Kannengießer et al., 1979) did not give any evidence to prove
this working hypothesis.

Recently, we initiated experiments in the beating heart of the
anesthetized dog with the aim of investigating the mechanism that
may induce hypokinetic zones. In order to do this, direct and continuous measurements of several local tissue parameters were performed.

---

[*] With support of Thyssen-foundation

METHODS

Stepwise stenosis of the left anterior coronary artery (LAD) was induced by a micrometer occluding device consisting of a non expandable Teflon cable fixed to a micrometer screw.

The different tissue sensors used for local or regional measurements in the beating heart of 25 dogs were kept in position by a thin and highly flexible silicone disc attached to the tissue by atraumatic sutures.

The following local, regional and global parameters were measured continuously.

1. Intracapillary hemoglobin oxygenation with a micro light-guide spectrophotometer (Frank et al., this volume).
2. Local Po_2 with multiwire platinum electrodes according to Kessler and Lübbers (1966).
3. NAD(P)H-fluorescence with a micro light-guide spectrofluorometer (Ji et al., 1979; Frank et al., this volume).
4. Local K^+ activity with valinomycin surface electrodes (Kessler et al., 1974).
5. Regional wall motion by measurement of an ultrasonic transient time with piezo crystals (Heimisch et al., 1975).
6. Aortic and coronary blood flow with electromagnetic flow-meters.
7. Ventricular and aortic blood pressure.
8. Blood gases, acid-base status, hematocrit and hemoglobin concentration.

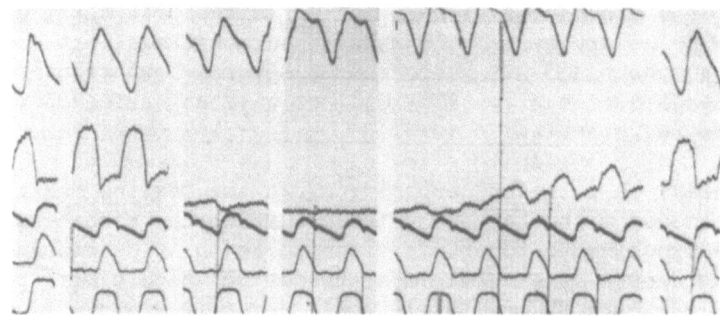

Fig. 1.   Regional wall motion and hemodynamic parameters of the myocardium during coronary occlusion (top to bottom): 1. wall motion of the myocardium, 2. coronary blood flow (LAD), 3. aortic blood pressure, 4. aortic blood flow, 5. ventricular pressure.
LAD (trace 2) was occluded between the 2. and 3. part of the record.

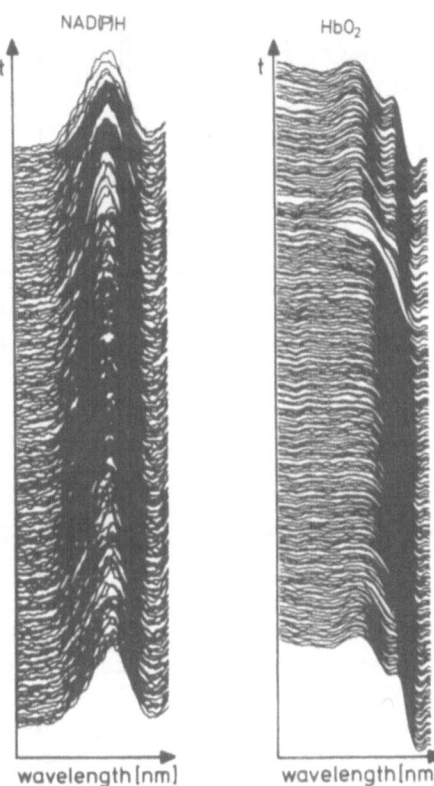

Fig. 2.   Right: Kinetic changes of intracapillary hemoglobin oxygen-
          ation during a period of occlusion of the LAD.
          Left: Reduction of pyridine nucleotides of the myocardium
          during a period of coronary occlusion.
          Time: from bottom to top.

     The anesthetized mongrel dogs were ventilated mechanically with
a Starling pump using a 1:1 mixture of oxygen and nitrous oxide.
Relaxation was induced by pancuroniumbromide and for analgesia piri-
tramid was used.

     A typical example of a simultaneous recording of regional myo-
cardial wall motion, blood flow in the LAD, aortic blood pressure,
aortic blood flow and ventricular pressure is shown in Fig. 1.

     The local change in myocardial length of the myocardium is
characterized by the slope and the amplitude of the curve of wall
motion found during the time interval between the opening and
closing of the aortic valve. As shown in Fig. 1 regional dL/dt
decreases rapidly when coronary occlusion is induced (see the flow
trace). After reopening of the clamp regional wall motion recovers
after a few heart beats.

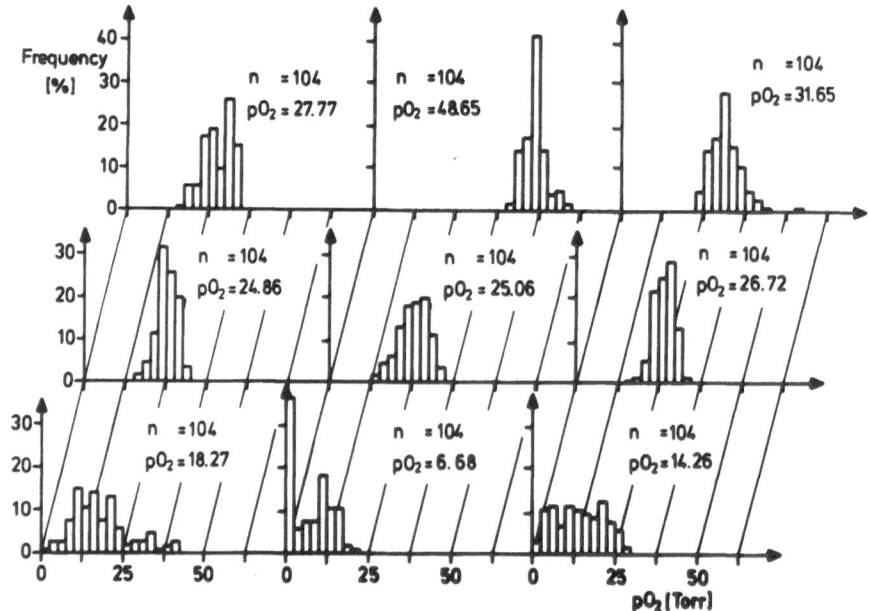

Fig. 3.   Po₂ histograms of the beating heart in dogs: 1. normal
          tissue oxygenation (top), 2. local oxygenation after iso-
          volemic hemodilution (middle), 3. when critical stenosis
          causes a decrease in contractility the lowest Po₂ values
          approach the region of 0-2.5 Torr (bottom).
          The results of the Po₂ measurements demonstrate that pro-
          nounced heterogeneity exists in the beating heart muscle.
          Aside of the Po₂ histogram mean tissue Po₂ values are given.

     Typical kinetic changes of intracapillary hemoglobin oxygena-
tion and of the redox state of intramyocardial pyridine nucleotides
(NAD(P)H) induced by coronary occlusion can be seen in Fig. 2. The
spectra of both pigments, recorded at a rate of one spectrum per
second reveal the fast kinetics of the on- and off-reactions.

RESULTS

     The local oxygen supply in the myocardium of dogs was investi-
gated before and during critical stenosis. In order to be able to
adjust the critical stenosis more precisely with our mechanical
occluding device we decreased the oxygen transport capacity by isovolemic hemodilution.

     The Po₂ histograms measured in three different areas of the
myocardium supplied by the LAD demonstrate a rather pronounced
heterogeneity of local oxygen supply (Fig. 3).

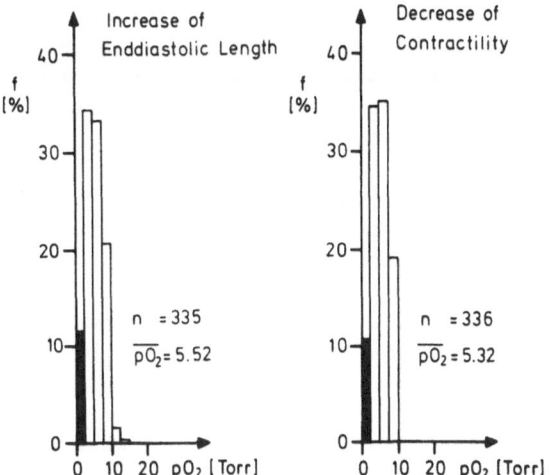

Fig. 4. Left: Po₂ histogram when the enddiastolic length of the
myocardium starts to increase.
Right: local oxygenation at the point when critical stenosis causes a decrease of contractility (dL/dt).

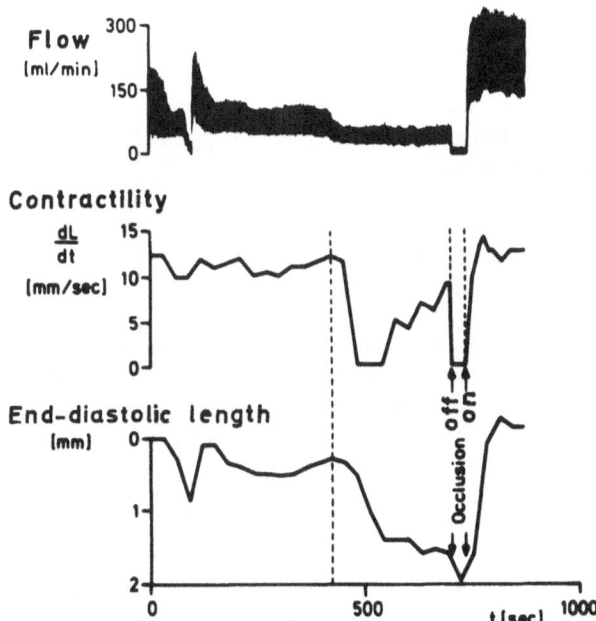

Fig. 5. Correlation between coronary blood flow, enddiastolic
length and dL/dt ("contractility") during critical stenosis
and coronary occlusion (dashed line). Enddiastolic length
reacts first when local anoxia starts to develop at the
"lethal corner".
For further explanation see Figs. 6 and 7.

Hemodilution down to a hematocrit of 10% causes a moderate shift of the distributions towards lower Po_2 values. When critical stenosis is induced it becomes evident that the deterioration of local oxygen supply does not show uniform changes and cannot be predicted from the initial control histograms.

The concomitant measurements of myocardial wall motion show that regional changes in myocardial length starts to decrease when the first Po_2 values approach zero Torr.

Typical examples of local oxygen supply under the condition of critical stenosis of the LAD are characterized by the histograms in Fig. 4.

The Po_2 histogram on the left was found when a pronounced increase in enddiastolic length was observed whereas the distribution curve on the right corresponds to a situation when dL/dt distinctly decreases. The two histograms do not show any significant differences.

Therefore, simultaneous measurements of the redox state of pyridine nucleotides and of intracapillary hemoglobin oxygenation were performed in order to assess the threshold for a decrease in dL/dt more precisely.

Fig. 5 shows the behaviour of coronary blood flow, contractility and enddiastolic length during critical stenosis and occlusion of the LAD. The corresponding relation between the dynamic parameters of the myocardium and the intracapillary hemoglobin oxygenation as well as the intracellular redox state of pyridine nucleotides can be seen in Figs. 6 and 7.

When a critical stenosis and a subsequent short occlusion are induced a transient increase in enddiastolic length is observed whereas only minor variations in regional wall motion are found.

The intracapillary hemoglobin oxygenation and the redox state of NAD(P)H undergo characteristic changes under these conditions. When the oxygen saturation of hemoglobin falls below 10% and the levels of NAD(P)H increase by 5 to 20%. Mechanical performance then starts to deteriorate (Figs. 5 and 6).

A second stenosis, induced after a short period of recovery causes similar changes in enddiastolic length, HbO_2 and NAD(P)H. A more pronounced stenosis leads to a sudden fall of active wall motion to zero and regional dilatation of the myocardium. When the enddiastolic length has increased by 1.5 mm, active wall motion starts to recover (Figs. 5 and 7). A close correlation seems to exist between the mechanical parameters of the myocardium and the local oxygen supply as measured by spectrophotometry.

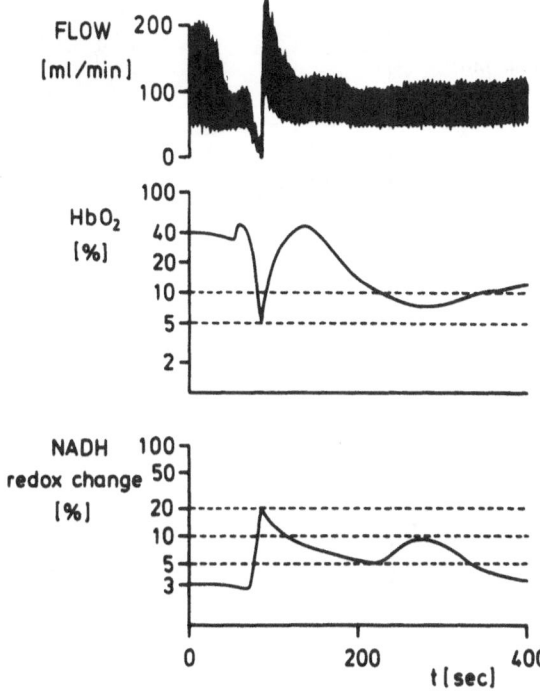

Fig. 6. Correlation between coronary blood flow, intracapillary
HbO$_2$ and redox state of NADH(P)H. When HbO$_2$ falls below
10% and NAD(P)H reaches a reduction between 5 and 20%
mechanical performance starts to deteriorate.

A very interesting result may be that the levels of NAD(P)H
decrease when the enddiastolic length increases and that, subsequently, the wall motion starts to recover, even though the coronary
blood flow and the oxygenation of hemoglobin do not show any significant change.

The relation between extracellular potassium activity, and the
wall motion during stenosis of the LAD shows a biphasic behaviour
when plotted semi-logarithmically (Fig. 8).

Minor changes in extracellular K$^+$ (less than 1 mmol·l^{-1}) are
accompanied by changes in dL/dt of up to 20%. This means that this
change in contractility is not caused by depolarization of the sarcolemma.

When extracellular potassium activity rises by more than
1 mmol·l^{-1}, inducing a depolarization of the myocardial cell
membrane, a pronounced decrease in wall motion is observed.

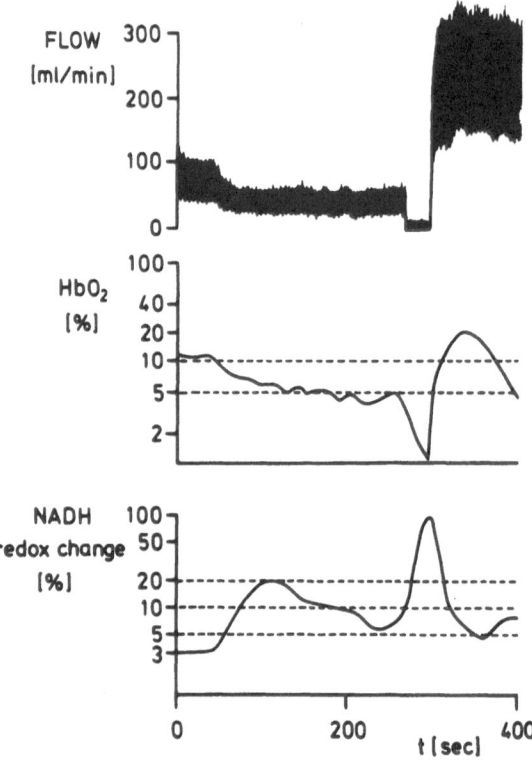

Fig. 7. Correlation between coronary blood flow, intracapillary
HbO$_2$ and redox state of NADH(P)H. When HbO$_2$ falls below 10%
and reaches a reduction between 5 and 20% mechanical performance starts to deteriorate.

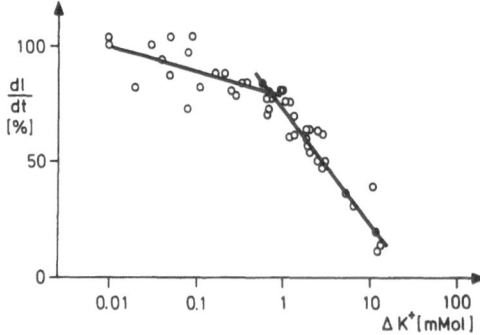

Fig. 8. Correlation between extracellular potassium activity and
myocardial wall motion (dL/dt). The very minor changes of
K$^+$ activity down to a 20% decrease of contractility do not
indicate a significant depolarization of the muscle cells.

DISCUSSION

Koyama (1978) investigated the relation between O_2 availability and pulsatile intramyocardial pressure. He assumed that during coronary occlusion an unknown factor causes a rapid decrease of local myocardial contractile force and that the local O_2 consumption of myocardial tissue is reduced in accordance with the rapid change in contractile force.

The Po_2 measurements indicated that a great heterogeneity of local oxygen supply exists in the myocardium under control conditions. Under pathological conditions this heterogeneity increases unpredictably.

In our experiments we found that regional wall motion starts to decrease when some of the measured local Po_2 values approach zero Torr.

However, the precise threshold of local oxygen supply to tissue could not be determined by means of the Po_2 electrode.

Only the simultaneous measurements of intracapillary hemoglobin oxygenation and of intracellular redox state of pyridine nucleotides allowed a precise determination of this threshold. The results show that under the given experimental conditions regional hypokinetic zones start to appear when intracapillary hemoglobin oxygenation falls below 10% and when the redox state of NAD(P)H increases locally by 5-10%. In this context it should be emphasized that, due to the great heterogeneity of tissue oxygen supply, regional decrease in mechanical performance is already induced when the tissue oxygenation reaches critical thresholds only very localized.

Furthermore, it seems to be of great interest that a 20% range of regulation of contractility exists which may have a protective function for the beating heart. As followed from the measurements of extracellular K^+ activity within the "regulatory range" of contractility no significant change in the resting potential of the myocardium may occur.

Our experimental observations raise the question as to whether or not "signal oxidases" may influence the local myocardial contractile force during critical stenosis.

Experiments performed in the liver (Kessler et al., 1981; Kessler et al., this volume) revealed evidence that a regulatory system affecting oxygen uptake rate and energy consumption may exist.

ACKNOWLEDGEMENT

We thank A. Brehm, G. Kerl and G. Schuster for skillful technical assistance.

REFERENCES

Frank, K.H., Schabert, A., Friedl, A., Brunner, M., Höper, J., Kerl, G., and Kessler, M., 1982, Correlation between tissue Po_2 and intracapillary Hb-spectra, this volume.

Heimisch, W., Hagl, S., Meisner, H., Franklin, D., and Kemper, W.S., 1975, Aufzeichnung lokaler Myokardfunktion nach dem Ultraschall-Laufzeit-Prinzip, Biomed. Tech. (Berlin), 20 (Suppl.):135.

Ji, S., Chance, B., Nishiki, K., Smith, T., and Rich, T., 1979, Micro-light guides: a new method for measuring tissue fluorescence and reflectance, Am. J. Physiol., 52:C144–C156.

Kannengießer, G.J., Opie, L.H., and van der Werft, T.J., 1979, Impaired cardiac work and oxygen uptake after reperfusion of regionally ischaemic myocardium, J. Mol. Cell Cardiol., 11:197.

Kessler, M., Höper, J., Lübbers, D.W., and Ji, S., 1981, Local factors affecting regulation of microflow, O_2 uptake and energy metabolism, in: "Oxygen Transport to Tissue", Adv. Physiol. Sci., Vol. 25, A.G.B. Kovách, E. Dóra, M. Kessler, I.A. Silver, eds., Akadémiai Kiadó, Budapest, pp. 155-162.

Kessler, M., Höper, J., and Simon, W., 1974, Methodology and application of a multiple ion selective surface electrode for tissue measurements, Fed. Proc., 33:279.

Kessler, M., Klövekorn, W.P., Höper, J., Sebening, F., Brunner, M., Frank, K.H., Harrison, D.K., Kernbach, C., Anderer, W., Richter, H., and Ellermann, R., 1982, Local oxygen supply and regional wall motion of the dog's heart stenosis of the LAD, this volume.

Kessler, M., and Lübbers, D.W., 1966, Aufbau und Anwendungsmöglichkeiten verschiedener Po_2-Elektroden, Pflügers Arch. Ges. Physiol., 291:82.

Koyama, T., Sasajima, T., Yagi, T., Miki, N., and Kikuchi, Y., 1978, Effects of coronary occlusion on myocardial oxygen and pulsatile intramyocardial pressure, in: "Oxygen Transport to Tissue", I.A. Silver, M. Erecińska, H.I. Bicher, eds., Plenum Press, New York-London, pp. 429-432.

Opie, L.H., 1976, Effects of regional ischemia on metabolism of glucose and fatty acids, Circ. Res., 38 (Suppl.):52-68.

INTRAMYOCARDIAL OXYGEN PRESSURE AND CORONARY BLOOD FLOW DURING

EXPERIMENTAL CORONARY STENOSIS

W. Menke, S. Schuchhardt, and H. Fritz

Physiologisches Institut der Freien Universität Berlin
1000 Berlin, FRG

INTRODUCTION

The present study was carried out in order to examine the effect of experimental coronary artery stenoses of various degrees on intramyocardial oxygen pressure and coronary blood flow.

Especially under periods of increased oxygen demand coronary stenoses can be the reason for the alteration of perfusion pressure or changes in myocardial oxygen delivery and may cause ischemia. As resting coronary blood flow is an insensitive index for the assessment of coronary stenosis, we quantified stenosis severity by determining hyperemic flow responses to temporary total occlusions as described by Gould et al. (1975). While the hemodynamics of coronary stenoses are well understood, there exists no exact knowledge of the influence of intramyocardial oxygen pressure on compensatory mechanisms such as blood flow autoregulation and reactive hyperemia. Therefore, we determined in open-chest dogs intramyocardial oxygen pressure ($imPo_2$) and coronary blood flow (CBF) during partial occlusions on the left descending coronary artery.

METHODS

Experiments were performed on 11 adult mongrel dogs of either sex with a body weight of 24-30 kg. Anesthesia was induced and maintained with sodium pentobarbital (30 mg/kg Nembutal i.v. at the beginning and about 10 mg/kg Nembutal i.p. on the average every two hours, or as needed). Two 15-cm plastic catheters were introduced via the femoral arteries into the arterial system to measure blood pressure (BP) (Statham P-23 pressure transducer) and to sample blood

341

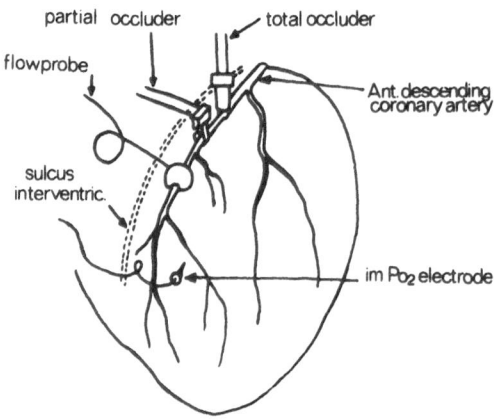

Fig. 1.   Measurement of intramyocardial oxygen pressure ($imPo_2$) and
          coronary blood flow in the anterior wall of the canine left
          ventricle.

for arterial Po_2 (Radiometer D616, Clark-type electrode). Loss of
fluid was compensated for by infusion of Macrodex.

        Artificial respiration was maintained with a Starling ventila-
tory pump through a tracheotomy tube. Supplemental oxygen (21% - 50%
O_2 in N_2) was adjusted during each experiment to stabilize Po_2 at
the initial value. These initial values were in the range 75 - 100
mm Hg. Minute volume was adjusted to keep end-expiratory CO_2
(Hartmann & Braun Uras 4 infrared gas-analyzer) constant in a range
from 5.0% to 5.6%. Left thoracotomy was performed in the 4th intercostal space, pericardium was opened and proximal segments of the
left descending coronary artery were dissected free. Fig. 1 shows
the instrumentation of the beating dog heart with the total occluder,
partial occluder, electromagnetic flowmeter probe and $imPo_2$ electrode (from proximal to distal).

        The total occluder was of the snare type and was used to produce
10 sec complete occlusions and consisted of an approximately 2 mm
wide stretched teflon band, which was passed around the dissected
vessel segment and through the two slits of a plastic tube.

        The partial occluder allowed the precise production of graded
stenoses by means of either a special screw or a digital micrometer.
The partial occluder permitted transient narrowing by compressing
the vessel between a plunger and a hook-shaped plate both made
either out of stainless steel or plastic. Before the partial occluder
was placed, we determined the zero setting by transitory total closure and measured external vessel diameter with sliding callipers
(values ranged from 2.5 to 3.5 mm; mean value was 3.3 mm). Percent
stenosis of the external vessel diameter was then calculated for

each intervening micrometer or screw setting by the following equation:

$$\% \text{ stenosis} = \frac{\text{baseline setting} - \text{actual setting}}{\text{baseline setting}} \cdot 100$$

where "baseline setting" was the point at which the plunger of the partial occluder came into contact with the surface of coronary vessel wall. We defined "zero setting" by complete closure of the partial occluder before it was put around the artery.

The flowmeter probe (Statham SP-type electromagnetic blood flow transducer) with a lumen slightly smaller than the vessel was snuggly fitted around the coronary artery distal to both occluders.

The imPo$_2$ electrodes of the riding type (for further details see Schuchhardt et al., 1978) with tip diameters between 15 um and 25 um were inserted into the beating heart by hand and placed about 3 mm deep (range 1 - 5 mm) in the ramification area of the occluded left descending coronary artery. Microelectrodes of the size generally used for Po$_2$ measurements in other organs proved to be too fragile for insertion into the beating heart (Lösse et al., 1973). Though puncture was made at a specially prepared small region of

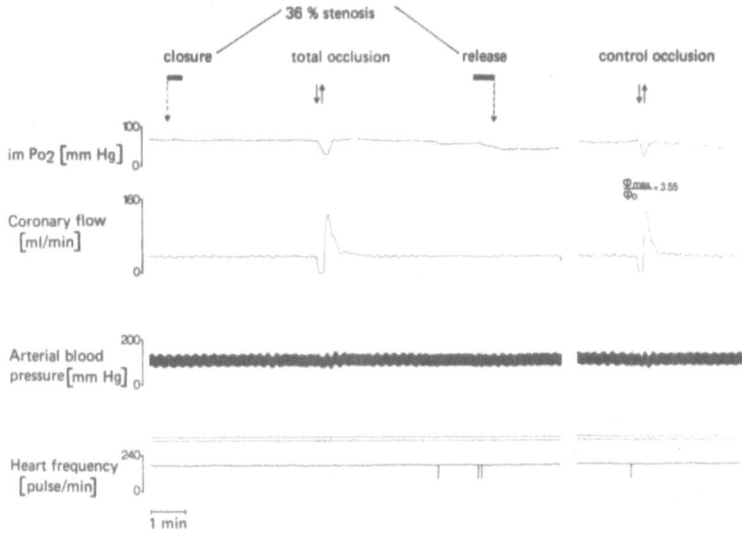

Fig. 2. Recording of arterial coronary flow and intramyocardial oxygen pressure (imPo$_2$) during intermittant 36% stenosis of the coronary artery. Total occlusion during partial stenosis shows no significant reduction compared to control occlusion.

less than 20 mm^2 where the epicardium was scraped off, more than 50% of our electrodes broke when they were inserted and had to be discarded.

ImPo$_2$, mean CBF, BP and PF were recorded continuously; all values are expressed as the mean $\pm$ standard deviation of the mean.

RESULTS

Mean imPo$_2$ in the wall of the left ventricle (40 $\pm$ 20 mm Hg) and mean CBF (33 $\pm$ 11 ml/min) were measured simultaneously during 64 experimental stenoses in 11 dogs. At the beginning of each experiment we performed 10 sec total occlusions which were repeated from time to time, because peak flow response tended to decrease slightly

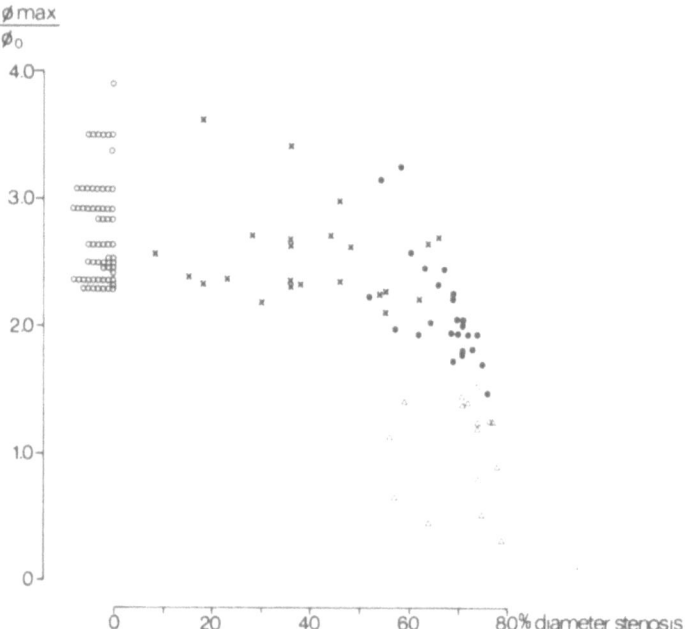

Fig. 3.   Relationship of reactive hyperemia following 10 sec occlusions to the degree of experimental coronary stenosis ($\emptyset_{max}/\emptyset_0$ = ratio of hyperemic peak flow $\emptyset_{max}$ to resting flow $\emptyset_0$; % stenosis = degree of narrowing of the external coronary artery diameter).
o = control values; x = stenoses group 1: no reduction of resting flow and reactive hyperemia; ● = stenoses, group 2: reduction of reactive hyperemia, but no reduction of the resting flow; Δ = stenoses, group 3: reduction of resting flow and reactive hyperemia.

from the beginning to the end of the experiment. Reactive hyperemia
following control occlusion was characterized by ratio of hyperemic
peak flow $\emptyset_{max}$ to resting flow $\emptyset_O$ ($\emptyset_{max}/\emptyset_O$ = 2.82 $\pm$ 0.46, n = 27).

Fig. 2 shows a typical record of an occlusion cycle during 36%
stenosis. Four minutes after the start of a partial occlusion
reactive hyperemia was tested by producing a total occlusion for
10 sec. About 3 min later the partial occluder was released. There
were no significant systemic changes during the 4-min period of
partial occlusion. Initial values of BP (145 $\pm$ 10/95 $\pm$ 10 mm Hg) and
PF (173 $\pm$ 15 min^{-1}) remained constant.

The ratio between hyperemic peak flow $\emptyset_{max}$ and resting flow $\emptyset_O$
is a sensitive index for the extent of partial coronary occlusion.
Results obtained from 64 occlusion cycles are summarized in Fig. 3
which shows the relationship of the index $\emptyset_{max}/\emptyset_O$ to % stenosis. As
% stenosis is only a rough parameter calculated without consideration
of vessel wall thickness we used the more sensitive index $\emptyset_{max}/\emptyset_O$ to
quantify stenoses. Three groups of stenoses were distinguished by
the index $\emptyset_{max}/\emptyset_O$ which are labelled by crosses (group 1), closed
circles (group 2) and triangles (group 3). During stenoses of group
1 and group 2 resting flow was not affected while stenoses of group
3 restricted coronary blood flow.

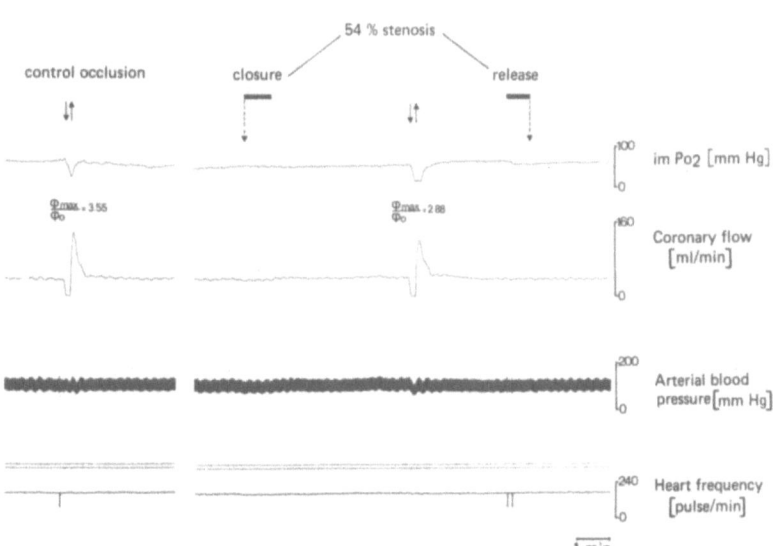

Fig. 4.   Recording of arterial coronary flow and intramyocardial
          oxygen pressure (imPo$_2$) during intermittant 54% stenosis
          of the coronary artery. Compared to control occlusion hyper-
          emia during stenosis was reduced. Corresponding values are
          written on top of the hyperemic peak. ($\emptyset_O$ = resting flow,
          $\emptyset_{max}$ = peak flow).

Stenoses, Group 1: No Reduction of Resting Flow and Reactive
Hyperemia

Stenoses up to about 50 - 60% did not significantly influence
CBF, $imPo_2$ and $\emptyset_{max}/\emptyset_O$ ($\emptyset_{max}/\emptyset_O$ = 2.58 $\pm$ 0.41, n = 23; control:
$\emptyset_{max}/\emptyset_O$ = 2.67 $\pm$ 0.48, n = 11). Stenoses of this group had no
significantly hemodynamic effect nor did they elicit compensatory
vasodilatory mechanisms of the distal vascular bed ($\emptyset_{max}/\emptyset_O$ not
reduced). Stenoses of group 1 allowed us to verify the range of
scattering of our experiments.

Stenoses, Group 2: No Reduction of Resting Flow, but Reduction of
Reactive Hyperemia

Stenoses from about 50 - 60% to about 70 - 80% diminished reac-
tive hyperemia ($\emptyset_{max}/\emptyset_O$ = 2.13 $\pm$ 0.41, n = 25; control: $\emptyset_{max}/\emptyset_O$ =
2.85 $\pm$ 0.43, n = 21). CBF was kept at its control resting flow by
vasodilatation of the distal vascular bed as indicated by reduced
reactive hyperemia. Adjustment of the peripheral resistance was
very quick and exact; no reaction of $imPo_2$ or CBF could be observed
either during low (54% stenosis, Fig. 4) or during high (73%
stenosis, Fig. 5) degree of stenosis.

During 24 of the 25 stenoses in group 2 compensatory mechanisms
of distal vascular bed maintained $imPo_2$ at or near normal control
values. Histograms of Fig. 6 show changes of intramyocardial oxygen

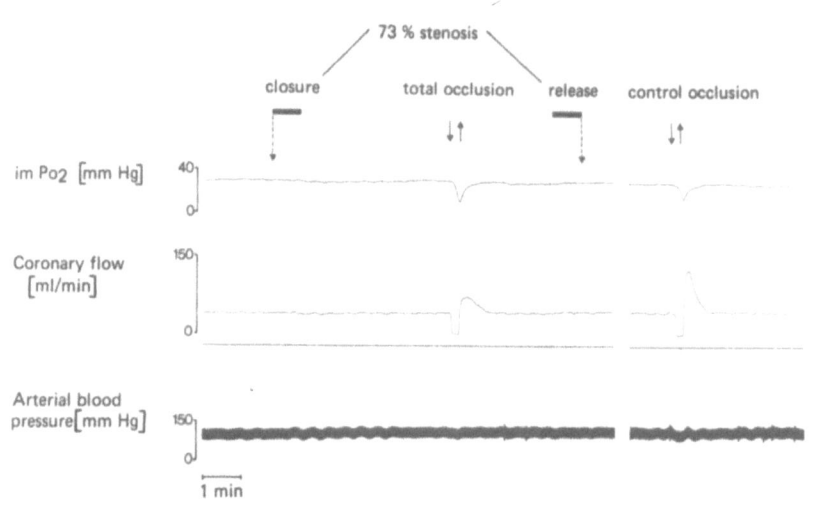

Fig. 5.   Recording of arterial coronary flow and intramyocardial
          oxygen pressure ($imPo_2$) during intermittant 73% stenosis
          of the coronary artery. Reactive hyperemia was determined
          during and after release of stenosis.

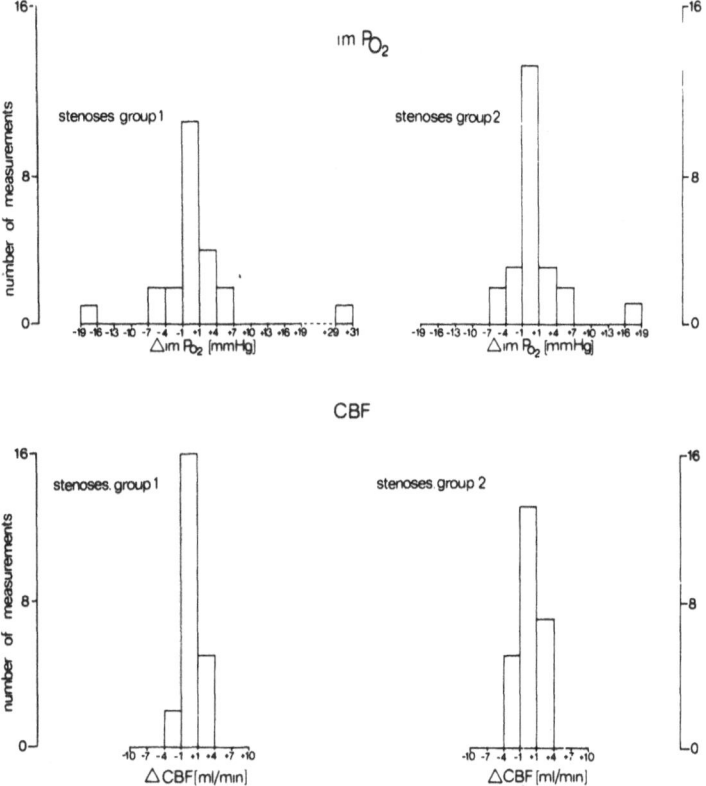

Fig. 6.   Changes of intramyocardial oxygen pressure ($imPo_2$) and
          coronary blood flow (ΔCBF) during 4 min experimental
          coronary stenoses of various degree.
          Stenoses, group 1: no reduction of reactive hyperemia,
                             8% - 66% stenosis of coronary artery.
          Stenoses, group 2: reduction of reactive hyperemia,
                             52% - 76% experimental stenosis.
          In all experiments partial stenoses did not affect resting
          flow.

pressure ($\Delta imPo_2$) and of coronary blood flow (ΔCBF) during 4 min
experimental stenoses. Histograms demonstrate uniform ranges of
scattering in both groups.

Stenoses, Group 3: Reduction of Resting Flow and Reactive Hyperemia

        In the range between 70 - 80% stenosis, even very small inten-
sifications of the stenosis caused considerable decrease of CBF and
$\emptyset_{max}/\emptyset_0$ ($\emptyset_{max}/\emptyset_0$ = 1.35 $\pm$ 0.36, n = 16; control: $\emptyset_{max}/\emptyset_0$ = 2.78 $\pm$
0.31, n = 17); first resting flow dropped and then reactive hyper-
emia was abolished. Fig. 5 (73% stenosis, group 2) and Fig. 7 (74%
stenosis, group 3) show the change-over from group 2 to group 3 of

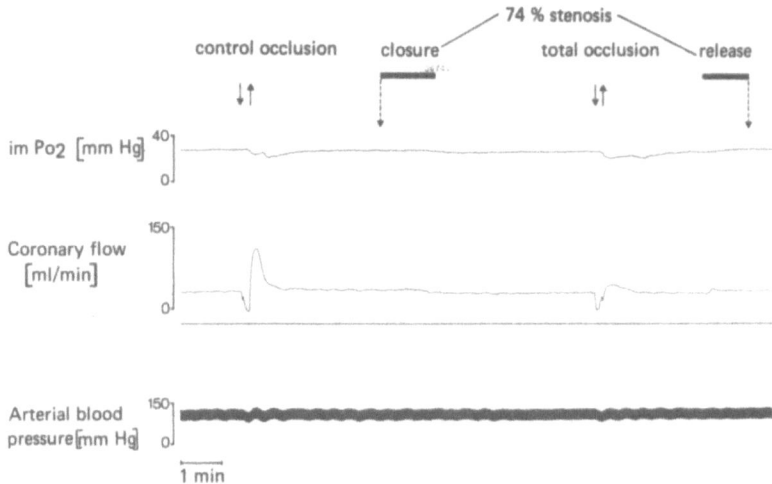

Fig. 7.   Recording of arterial coronary flow and intramyocardial
          oxygen pressure (imPo$_2$) during intermittant 74%% stenosis
          of the coronary artery. Reactive hyperemia was determined
          before and during stenosis.

experiments in the same dog. During 74% stenosis (Fig. 7) resting
flow and imPo$_2$ decreased slightly. The peripheral vasodilatation
was not maximal as detectable by hyperemia test.

In 6 of these 16 experiments the drop in CBF was maintained at
the new level. In the other 10 experiments flow returned toward but
not to the previous control level. During 12 experiments imPo$_2$ fell
significantly and then increased while in 4 cases imPo$_2$ did not
change (these 4 experiments represented stenoses of group 3 with
relatively high values of $\emptyset_{max}/\emptyset_0$ and very slight drop of resting
flow).

The release of partial occluder after occlusion cycles with
$\emptyset_{max}/\emptyset_0 < 1$ (n = 6) caused significantly reactive hyperemia. After
occlusion cycles with $\emptyset_{max}/\emptyset_0 > 1$ (n = 58) hyperemia occurred in
only 2 cases, although 10 stenoses (of group 3, $\emptyset_{max}/\emptyset_0 > 1$) produced flow deficit.

DISCUSSION

Over a wide range of coronary stenoses blood flow is closely
matched to myocardial oxygen requirements. Our experiments performed
to define the role of oxygen in regulating coronary blood flow
provide data about the relationship between imPo$_2$ and mean CBF under
conditions of experimental coronary stenosis.

Tissue Po_2 is assumed to be the central parameter in local control of blood flow. Our results support this hypothesis only for very severe stenoses with significant drop of CBF (group 3), but our observations are in conflict with this hypothesis when control steady-state flow was perfectly autoregulated: $imPo_2$ remained constant in spite of decreased perfusion pressure (group 2).

Therefore, we have to conclude that reduction of perfusion pressure by coronary artery stenosis activates two different compensatory vasodilatory mechanisms of the microvascular bed:

The first (probably myogenic) mechanism is independent of tissue Po_2, very fast and very sensitive to changes of perfusion pressure, while O_2 has probably a very important effect on the second mechanism.

It is useful to keep in mind some methodical aspects. Temporal and spatial resolution of Po_2 measurement is limited. Needle electrodes with relatively large diameters needed for the beating heart preparation may compress microvessels reducing the measured Po_2 values (Lösse et al., 1973). In the present experiments with thinner electrodes (tip diameter 15 - 25 um) we obtained relatively high steady-state values (mean value 40 mm Hg, range from 5 mm Hg to 88 mm Hg). Although the number of $imPo_2$ values is not large enough for histograms, we can say that disturbances of microcirculation seem to be minimized using thin electrodes with tip diameters of about 20 um. The magnitude of our $imPo_2$ values compared to former investigations (Lösse et al., 1973) are in part also caused by higher arterial Po_2 in the present study, but this fact does not entirely explain the probably too high mean value of $imPo_2$ of 40 mm Hg.

Our $imPo_2$ measurements were performed in the subepicardium. This is another point we must discuss as many authors suppose that stenoses may cause a redistribution of blood flow from endocardial to epicardial layers (for details see Bache and Schwartz, 1982). Therefore, we have to expect a reduction of flow reserve and $imPo_2$ first endocardially and then epicardially.

Evidence for collateral O_2 supply after acute experimental stenoses was found in only a few of our experiments: e.g., when $imPo_2$ increased while the reduced CBF remained constant, or when $imPo_2$ remained stable at a steady-state level while CBF dropped. However, even in these few cases the effect seemed to be very small.

Our 10-sec total control occlusions resulted in an increase in flow of about 280%. Lipscomb and Gould (1975) obtained an increase of about 390%. This higher value is partially explainable by very high blood Po_2 values not kept in the physiological range as in our experiments. Arterial blood Po_2 values up to about 150 mm Hg probably reduced resting flow and thus allowed a higher flow reserve.

Our calculated % stenosis values are in good agreement up to
about 60 - 70% stenosis with those of others (Gould et al., 1975;
Buss and Wüsten, 1978). For severe degrees of stenosis our values
are about 10 - 20% lower. It should be pointed out that in our experiments narrowing of the coronary artery was not concentric, that
vessels often assumed an oval shape in the partial occluders and
that % stenosis was calculated as the degree of narrowing of external vessel diameter not considering the thickness of the vessel
wall. We gave, moreover, preference to the functional index $\emptyset_{max}/\emptyset_0$
over % stenosis, which is only a rough parameter for classification
of stenoses.

ACKNOWLEDGEMENT

We thank Mr. Horst Ewald, chief of the precision mechanics
workshop of the institute, for the construction of the occluders.

REFERENCES

Bache, R.J., and Schwartz, J.S., 1982, Effect of perfusion pressure
     distal to a coronary stenosis on transmural myocardial blood
     flow, Circulation, 65:928-935.
Buss, D., and Wüsten, B., 1978, Koronardurchblutung und Dilatations-
     reserve bei Koronarkonstriktion, Verh. Dtsch. Ges. Kreislauf-
     forsch., 44:187.
Gould, K.L., Lipscomb, K., and Calvert, C., 1975, Compensatory
     changes of the distal coronary vascular bed during progressive
     coronary constriction, Circulation, 51:1085-1094.
Lipscomb, K., and Gould, K.L., 1975, Mechanism of the effect of
     coronary artery stenosis on coronary flow in the dog, Am.
     Heart J., 89:60-67.
Lösse, B., Schuchhardt, S., Niederle, N., and Benzing, H., 1973,
     The histogram of local oxygen pressure (Po_2) in the dog myo-
     cardium and the Po_2 behaviour during transitory changes of
     oxygen administration, Adv. Exp. Med. Biol., 37A:535-540.
Schuchhardt, S., Schuster, J., and Ryzlewicz, Th., 1978, "Untersu-
     chungen zur Sauerstoffversorgung des Warmblütermyocards mit
     Platin-Nadel-Elektroden", Westdeutscher Verlag, Opladen.

MITOCHONDRIAL OXIDATIVE PHOSPHORYLATION: TISSUE OXYGEN SENSOR FOR REGULATION OF CORONARY FLOW

E.M. Nuutinen, D.F. Wilson, and M. Erecińska

Department of Biochemistry and Biophysics, University of Pennsylvania, Philadelphia, PA 19104, USA

INTRODUCTION

The heart is able to match its metabolic need for oxygen with alterations in oxygen delivery via the coronary blood flow (see for example Alella et al., 1955; Ball et al., 1975; Nuutinen et al., 1982). The mechanisms which elicit this autoregulatory response in coronary flow have been the subject of intensive investigation for many years but they remain incompletely understood. In earlier studies we observed that in suspensions of intact cells the cellular energy state (expressed as $[ATP]/[ADP][Pi]$ is dependent on oxygen tension throughout the physiological range (Wilson et al., 1977; 1979a; 1979b). This dependence was manifested by a progressive increase in the reduction of cytochrome $\underline{c}$ and lowering of the cytosolic $[ATP]/[ADP][Pi]$ as oxygen tension was lowered. The changes in these two parameters allowed the rate of oxygen consumption, and thus the rate of ATP synthesis, to remain almost undiminished to oxygen tensions below 5 Torr. This oxygen dependence was an expression of the reaction of oxygen with cytochrome $\underline{c}$ oxidase of the mitochondrial respiratory chain and suggested that oxidative phosphorylation may be the primary tissue "oxygen sensor". In cardiac tissue the mitochondria sustain continually high rates of ATP production and consume most (>95%) of the oxygen utilized by this tissue. Since ATP is required for muscular work and it and its metabolic products (ADP, Pi, AMP, adenosine) are involved in essentially every cellular metabolic pathway as reactants and regulators, changes in the cellular energy level arising from alterations in tissue oxygen tension can readily be transmitted to the rest of cellular metabolism. In this paper we will summarize some of the data which support the concept that mitochondrial oxidative phosphorylation is the major oxygen sensor for regulation of coronary blood flow.

351

MATERIALS AND METHODS

Male Sprague Dawley rats (250-300 g) were used in the study. The animals were given a lab chow diet and tap water ad libitum. They were anesthetized with sodium pentobarbital (Veterinary Laboratories, Lenexa, KA) administered intraperitoneally at a concentration of 10 mg/100 g body weight. Immediately before the excision of the heart 500 IU of heparin were injected into the caudal vena cava.

Isolated hearts were perfused at 37°C with Krebs-Henseleit buffer (Krebs and Henseleit, 1932), equilibrated with O_2/CO_2 (19:1) and containing 10 $mmol \cdot l^{-1}$ glucose and 12 IU of insulin per liter, at an aortic pressure of 70 cm H_2O (6.87 kPa) essentially by the method of Langendorff (1895), without recirculation. The hearts were paced electrically at a frequency of 5 Hz to maintain a constant work load. Coronary flow and oxygen concentration in the venous effluent were monitored continuously. Oxygen consumption was calculated from the arterio-venous difference in O_2 concentration multiplied by the coronary flow.

Experimental manipulations were commenced after an equilibrating perfusion period of 20 min at an aortic pressure of 6.87 kPa. Perfusions with adenosine, Amytal and 2,4-dinitrophenol (DNP) were performed by either using media containing the desired concentration of each substance or by using an accessory pump to infuse a more concentrated solution at a rate less than 0.4% of the main flow. Work load and oxygen consumption of the heart were altered either by elevating the perfusion pressure (to 9.81 kPa or 12.8 kPa) or by lowering the free Ca^{2+} concentration in the medium from 2.5 $mmol \cdot l^{-1}$ to 0.5 $mmol \cdot l^{-1}$. A Ca^{2+} of 0.5 $mmol \cdot l^{-1}$ was chosen because the changes in coronary flow and oxygen consumption were observed to be fully reversible.

Metabolite assays were carried out on hearts which were freeze-clamped at the indicated times with aluminum clamps precooled in liquid N_2. The assays were carried out as previously described (Nuutinen et al., 1982).

The cytoplasmic phosphorylation potential $[ATP]/[ADP][Pi]$ was calculated by assuming that the creatine phosphokinase reaction was near equilibrium. The equilibrium constant for the reaction:

Creatine + ATP = Creatine-Phosphate + ADP

was taken to be 6.3×10^{-3} for an intracellular free Mg^{2+} concentration of 1 $mmol \cdot l^{-1}$ (Lawson and Veech, 1979) and the intracellular pH was assumed to be 7.0. Since neither the intracellular Mg^{2+} nor pH are accurately known at this time, it is essential to be sure that the same values for these and the equilibrium constant were used when comparing data from different laboratories.

RESULTS

Several observations made recently have supported the idea that mitochondrial oxidative phosphorylation plays a central role in sensing tissue oxygen tension and thus is important in the regulation of coronary flow.

1. Coronary flow increases parallel to an increase in oxygen consumption as the work load is increased (constant A-V oxygen difference).
2. Coronary flow increases with decreasing oxygen tension in the influent medium (hypoxia).
3. Coronary flow increases when isolated perfused rat hearts are perfused with Amytal, an inhibitor of mitochondrial electron transport, although oxygen consumption decreases and the effluent oxygen tension increases (Nuutinen et al., 1982).
4. Coronary flow increases when isolated rat hearts are perfused with 2,4-dinitrophenol (DNP), an uncoupler of mitochondrial oxidative phosphorylation where oxygen consumption increases (Nuutinen et al., 1983).

The results using Amytal show that coronary flow is not related to the tissue oxygen tension per se because the flow increased with either increased (Amytal), unchanged (increased work load) or decreased (hypoxia) oxygen tensions.

Tissue energy metabolism has been evaluated by measuring metabolite levels in hearts which were freeze-clamped at systematically varied work loads, levels of Amytal infusion and levels of DNP infusion. In all three methods of vasodilation quantitatively the same correlation was observed between the tissue energy level (expressed as $[Cr-P]/[Cr][Pi]$ or as $[ATP]_f/[ADP]_f[Pi]$ assuming an equilibrium of creatine phosphokinase) and coronary flow (Table 1 and Nuutinen et al., 1982). Since oxidative phosphorylation is responsible for maintaining the metabolic energy supply of cardiac tissue this correlation implies that oxidative phosphorylation is involved in vasodilation and specifically points out that the tissue $[ATP]_f/[ADP]_f$ $[Pi]$ has an important function in regulating flow through the coronary blood vessels.

Evidence that the cardiac muscle $[ATP]_f/[ADP]_f[Pi]$ is a major determinant of coronary flow establishes it as a link between tissue oxygen metabolism on one hand and the response of the vascular smooth muscle on the other. It does not necessarily mean that the tissue $[ATP]_f/[ADP]_f[Pi]$ is the vasodilatory effector. It is possible that for some conditions the $[ATP]_f/[ADP]_f[Pi]$ of the smooth muscle cells declines in parallel with that of the cardiac muscle cells and this causes vasodilation but this does not appear likely, although such a mechanism could explain the effects of Amytal.

Table 1. Relationship between tissue energy metabolism and coronary flow in isolated perfused rat hearts.

Perfusion conditions	$\dfrac{[Cr-P]}{[Cr]}$	$[Pi]$ (mM)	$\dfrac{[ATP]}{[ADP][Pi]}$ $(\times 10^4 \ M^{-1})$	Coronary flow	O_2 consumption
A. Work					
0.5 mM Ca^{2+}					
6.87 kPa	1.93	2.38+.11	7.47+.48	4.12+.65	2.10+.42
6.87 kPa	1.63	2.57+.48	5.41+1.03	5.45+.69	3.64+.44
9.81 kPa	1.10	3.81+.23	2.69+.46	6.93+.73	4.79+.23
12.8 kPa	0.94	4.06+.20	2.13+.16	8.38+.64	5.75+.49
B. Amytal infusion					
0.5 mM Ca^{2+}					
0.22 mM Amytal	1.45	3.00+.52	4.64+1.41	5.70+.48	2.5 +.08
0.22 mM Amytal	1.10	3.62+1.13	2.92+.61	7.23+1.06	3.31+.50
0.44 mM Amytal	0.77	4.35+.84	1.68+.37	8.40+.55	2.82+.35
0.88 mM Amytal	0.50	4.64+.16	0.99+.13	9.40+.66	1.76+.33
C. DNP infusion					
0.5×10^{-5} M	1.17	3.66+.20	3.20+.30	7.02+.32	5.28+.24
1.0×10^{-5} M	1.03	4.20+.41	2.44+.23	7.41+.16	5.59+.13

Isolated rat hearts were perfused by the Langendorff (1895) procedure for 20 minutes and then perfused for 10 minutes (at an aortic
pressure of 6.87 kPa and a Ca^{2+} concentration of 2.5 mmol·1^{-1} unless
otherwise noted) and the indicated concentration of Amytal or DNP.
Data are taken from Nuutinen et al. (1982; 1983). Coronary flow is
given as ml/(min·g ww) and oxygen consumption as umole/(min·g ww).

There is no evidence that increasing the cardiac work rate
causes a parallel increase in work by the vascular smooth muscle.
This makes it more reasonable to propose that changes in the energy
state of the cardiac muscle cells give rise to "second messengers"
which are the affectors of the vascular smooth muscle.

The possible identity of the "second messenger(s)" remains a
matter of considerable speculation. It has been proposed that adenosine is an important regulator of local flow in cardiac tissue (see
for example Berne, 1963; 1974; Rubio and Berne, 1975; Olsson and
Patterson, 1976). We have examined this possibility by two methods:
A. Precise measurement of the effluent concentrations of adenosine
and its metabolites at different levels of vasodilation induced by
work changes, Amytal infusion and DNP infusion (Table 2). B. Comparison of the ability of theophylline, a methyl xanthine which com-
petes with adenosine for the adenosine receptor, to block vasodilation by increased work, Amytal infusion and DNP infusion with its
ability to block vasodilation by adnosine infusion.

Table 2.   Effluent concentrations of adenosine, inosine and hypoxanthine from rat hearts perfused under conditions of
various coronary flow values.

Perfusion pressure (kPa)	Hypoxanthine nM	Inosine nM	Adenosine nM	Total nM
6,87	37 $\pm$ 5	23 $\pm$ 6	36 $\pm$ 7	99 $\pm$ 9
9.81	33 $\pm$ 12	<20	38 $\pm$ 2	83 $\pm$ 13
12.8	38 $\pm$ 3	21 $\pm$ 13	40 $\pm$ 5	100 $\pm$ 10
0.7 x 10^{-6} M Adenosine	120 $\pm$ 11	123 $\pm$ 23	88 $\pm$ 14	331 $\pm$ 12
control	63 $\pm$ 9	31 $\pm$ 4	25 $\pm$ 2	119 $\pm$ 8
control	67 $\pm$ 6	30 $\pm$ 2	21 $\pm$ 1	117 $\pm$ 4
5 x 10^{-6} M DNP	67 $\pm$ 17	39 $\pm$ 17	26 $\pm$ 4	130 $\pm$ 17
1 x 10^{-5} M DNP	48 $\pm$ 2	23 $\pm$ 3	17 $\pm$ 2	88 $\pm$ 3

Values are means $\pm$ SEM for from 2 to 4 experiments (see Nuutinen et
al., 1982; 1983). Separate controls are given for the adenosine and
DNP experiments. Controls are the values for effluent in the last 1
minute of the 20 minute preperfusion and the experimental values are
for the effluent 4 minutes after beginning perfusion with adenosine
or DNP when coronary flow had been stable for at least 2 minutes.

The level of adenosine and its metabolites (inosine and hypoxanthine) in the effluent of perfused rat heart is sufficiently low
(20-40 nM) that it was necessary to develop a new assay for these
compounds. A modified high pressure liquid chromatography (HPLC)
capable of measuring adenosine, inosine and hypoxanthine at levels
down to 15 nM (Nuutinen et al., 1983) was applied to the various
conditions shown in Table 1 (see Table 2). For comparative purposes
hearts were also perfused with a medium containing adenosine at the
concentration needed to increase coronary flow by 37%, one-half the
maximum obtainable by adenosine under these conditions. The infusion of 0.7 x 10^{-6} M adenosine caused a 3 fold increase in the
effluent concentration of adenosine and a 2-4 fold increase in the
effluent concentrations of inosine and hypoxanthine. No significant
increase in the effluent concentrations of these compounds occurred
for increased work load, Amytal infusion of DNP infusion even when
these conditions induced larger increases in coronary flow than did
the adenosine.

Theophylline is known to compete with adenosine for binding to
the adenosine receptor and to block adenosine receptor mediated
responses. In the present work 1.4 x 10^{-5} M theophylline was found
to completely block vasodilation induced by 0.7 x 10^{-6} M adenosine
but it had not effect on that induced by increasing work load,

Amytal infusion or DNP infusion. Moreover it had no effect on the
uptake, release or metabolism of adenosine by the heart tissue as
evidenced by the absence of changes in the amount of adenosine and
adenosine metabolites in the perfusion effluent.

SUMMARY

The observation that mitochondrial oxidative phosphorylation
in vivo is dependent on oxygen tension throughout the physiological
range (Wilson et al., 1979a, 1979b) has made this metabolic pathway
the most probable candidate for the tissue oxygen sensor in the regulation of local blood flow. We have utilized the oxygen dependent
regulatory system for coronary blood flow to examine this possiblity.
Alterations in coronary flow were induced by: 1. Varied work load;
2. Infusion of Amytal (an inhibitor of mitochondrial respiration);
3. Infusion of DNP; 4. Hypoxia. Increased work load caused increased
coronary flow with no decrease in effluent oxygen tension while
Amytal infusion and hypoxia caused vasodilation with increased and
decreased O_2 tension respectively. This indicates that oxygen tension per se cannot be responsible for the observed vasodilation.
Tissue energy metabolism was evaluated by measuring metabolite
levels in hearts which were freeze-clamped in each state of perfusion. In all four methods of vasodilation, a decrease in cellular
energy state ratio ($[ATP]_f/[ADP]_f[Pi]$) expressed as the calculated
ratio of free adenine nucleotides, was observed for conditions which
increased flow. Systematic variation of work load, Amytal or DNP
concentration resulted in quantitatively the same correlation between tissue $[ATP]_f/[ADP]_f[Pi]$ and coronary flow. It is concluded
that mitochondrial oxidative phosphorylation is the oxygen sensor
for the regulation of coronary blood flow by tissue oxygen tension.

Infusion of adenosine, a known coronary vasodilator, induced
vasodilation which was completely blocked by theophylline. Adenosine
is not the active compound for increased flow with work load, Amytal
infusion or DNP infusion, however, as: 1. In none of these cases
did the effluent level of adenosine or its metabolites inosine and
hypoxanthine increase with increasing flow. 2. In none of these
cases was the vasodilation sensitive to theophylline. It is concluded that adenosine is unlikely to be the "second messenger" in
regulation of coronary blood flow. (Supported by GM-21524).

REFERENCES

Alella, A., Williams, F.L., Bolene-Williams, C., and Katz, L.N.,
    1955, Interrelation between cardiac oxygen consumption and
    coronary blood flow., Am. J. Physiol., 183:570-582.
Ball, R.M., Bache, R.J., Cobb, F.R., and Greenfield, J.C. Jr., 1975,
    Regional myocardial blood flow during graded treadmill exercise
    in the dog, J. Clin. Invest., 55:43-49.

Berne, R.M., 1963, Cardiac nucleotides in hypoxia: possible role in regulation of coronary blood flow, Am. J. Physiol., 204:317-322.

Berne, R.M., 1974, The coronary circulation, in: "The Mammalian Myocardium", G.A. Langer, A. Brady, eds., Wiley, New York, pp. 251-281.

Krebs, H.A., and Henseleit, K., 1932, Untersuchungen über die Harnstoffbildung im Tierkörper, Hoppe-Seyler's Z. Physiol. Chem., 210:33-36.

Langendorff, O., 1895, Untersuchungen am überlebenden Säugetierherzen, Pflügers Arch. Ges. Physiol., 61:291-322.

Lawson, J.W., and Veech, R.L., 1975, Effects of pH and free Mg^{2+} on the K_{eq} of creatine phosphokinase and other phosphate hydrolyses and phosphate transfer reactions, J. Biol. Chem., 254:6528-6537.

Nuutinen, E.M., Nelson, D., Wilson, D.F., and Erecińska, M., 1983, Regulation of coronary blood flow: effects of 2,4-dinitrophenol and theophylline, Am. J. Physiol., in press.

Nuutinen, E.M., Nishiki, K., Erecińska, M., and Wilson, D.F., 1982, Role of mitochondrial oxidative phosphorylation in regulation of coronary blood flow, Am. J. Physiol., 243:H159-H169.

Olsson, R.A., and Patterson, R.E., 1976, Adenosine as a physiological regulator of coronary blood flow, Prog. Mol. Subcell. Biol., 4:227-248.

Rubio, R., and Berne, R.M., 1975, Regulation of coronary blood flow, Prog. Cardiovasc. Dis., 18:105-122.

Wilson, D.F., Erecińska, M., Drown, C., and Silver, I.A., 1977, Effect of oxygen tension on cellular energetics, Am. J. Physiol., 233(5):C135-C140.

Wilson, D.F., Erecińska, M., Drown, C., and Silver, I.A., 1979a, The oxygen dependence of cellular energy metabolism, Biochem. Biophys., 195:485-493.

Wilson, D.F., Owen, C.S., and Erecińska, M., 1979b, Quantitative dependence of mitochondrial oxidative phosphorylation on oxygen concentration: a mathematical model, Arch. Biochem. Biophys., 195:494-504.

RESPIRATORY CHAIN O_2 REQUIREMENTS AND THE METABOLIC ANSWER TO

DIFFUSE ISCHEMIA OF MECHANICALLY OVERLOADED LEFT VENTRICULAR

MYOCARDIUM[*]

J. Moravec, J. Nzonzi, C. Bowe[1], and D. Feuvray[1]

I.N.S.E.R.M. U2, Hopital Léon Bernard
94450 Limeil-Brévannes, France
[1]Department of Physiology, Faculté des Sciences
 91405 Orsay, France

INTRODUCTION

Cardiac adaptation to mechanical overload proceeds in a three step manner (Meerson, 1969). After a short transitional period, a new steady state is usually attained before the heart fails. This period of enhanced metabolic activity (compensatory cardiac hypertrophy) can sometimes last over several weeks. In rats with a surgically induced aorto-caval communication (Hatt et al., 1980a), the compensated cardiac hypertrophy can persist over several months. Morphologically, the hearts from rats with a prolongated volume overload exhibit a decreased vascularization of the left ventricle (Rakusan et al., 1980). At the cellular level, the persistence of an activation of protein synthesis was suggested (Hatt et al., 1980a), the size of the left ventricular myocytes are increasing (Hatt et al., 1980b) and quantitative changes in intracellular organization appear (Anversa et al., 1971). The most striking modification is the increase in numerical density of the mitochondria resulting in an improved surface/volume ratio of mitochondria and decreased oxygen requirements for mitochondrial function (decreased cytochrome oxidase apparent K_M $[O_2]$; Moravec et al., 1981). In this work we tried to quantify the range of intracellular Po_2's compatible with the unimpaired mitochondrial function (full oxidation of the cytochrome oxidase). The hemodynamic and metabolic effects of an acute ischemia (one-way valve, working heart preparation) were

[*]This work was supported by INSERM (CRL no. 81 50 44).

also tested in order to evaluate the resistance of the left ventricular myocardium to a decreased oxygen supply.

METHODS

     An aorto-caval communication was surgically created in young
female Wistar rats. Sham-operated animals were used as controls.
Three experimental and two control rats were caged together. The
animals were sacrificed 3 months after the surgery. The majority of
these animals exhibited a compensatory heart hypertrophy. The
animals having peripheral signs of congestive heart failure were
rejected from further study. One group of hearts, half with a
fistula and half without (total number 40) was atrially perfused
with bicarbonate buffer containing 10 $mmol \cdot l^{-1}$ glucose. In some
cases 1.5 $mmol \cdot l^{-1}$ palmitate was added to the glucose-containing solution (Feuvray et al., 1981). The perfusate (200 ml) was recircu-
lated. All hearts were preperfused for 15 min. Then, in some of
them, a 50% reduction of coronary flow was induced by the insertion
of a one-way valve into the aortic outflow tract (Neely et al.,
1973). Left ventricular function was assessed by monitoring the
aortic pressure and the heart rate. The aortic flow was measured by
means of a rotameter. At the end of the perfusion, the hearts were
deeply frozen in liquid nitrogen and the ventricular fragments processed for ATP, CP, ADP, Pi determinations. An aliquot of frozen
tissue was used for CoA and long chain acyl-CoA (Garland et al.,
1965) determinations. Tissue carnitine and long chain carnitine were
determined by a radio isotope procedure (Mc Garry and Foster, 1976).
A separate group of hearts was used to assess the oxygen requirements
of the mitochondrial function (Moravec et al., 1981). These hearts
were perfused at 37°C and 70 Torr according to Langendorff (non-
working heart preparation). Two parallel columns of the perfusion
fluid containing 10 $mmol \cdot l^{-1}$ glucose were equilibrated with either
95% O_2 and 5% CO_2 (Po_2 620 Torr) or 95% N_2 and 5% CO_2 (40 Torr)
and connected to the aortic cannula by means of a miniature solenoid
valve (General Valve Corporation) with negligible dead space. This
allowed abrupt changes in the rate of oxygen delivery to the tissue.
All hearts were pre-perfused for 10 min with the oxygenated solution,
then they were subjected to several 2 min anoxic periods followed
by reoxygenation (state 5-to-state 3 transition). During the reoxygenation phase, the cytochrome oxidase reduction and the myoglobin
O_2 saturation were monitored by means of a rapid scanning spectroscope according to Lübbers and Niesel (1959). Care was taken to
alternate the cytochrome aa_3 and myoglobin recordings and only those
hearts whose individual responses were reproducible were considered.
The superposition of successive cytochrome aa_3 and myoglobin double
wavelengths recordings allowed an evaluation of the range of intracellular Po_2's necessary to maintain the cytochrome oxidase full
oxidation using the Theorell's equation for the myoglobin dissociation curve read at 37°C (Leniger-Follert and Lübbers, 1973). In some

hearts, the fluorescence emission from the reduced pyridine nucleotide was also recorded (Moravec et al., 1974) in order to assess the
O_2 requirements of reduced pyridine nucleotide reoxidation (Chance,
1976).

RESULTS

The concentrations of intracellular oxygen necessary to maintain cytochrome c oxidase and pyridine nucleotide fully oxidized

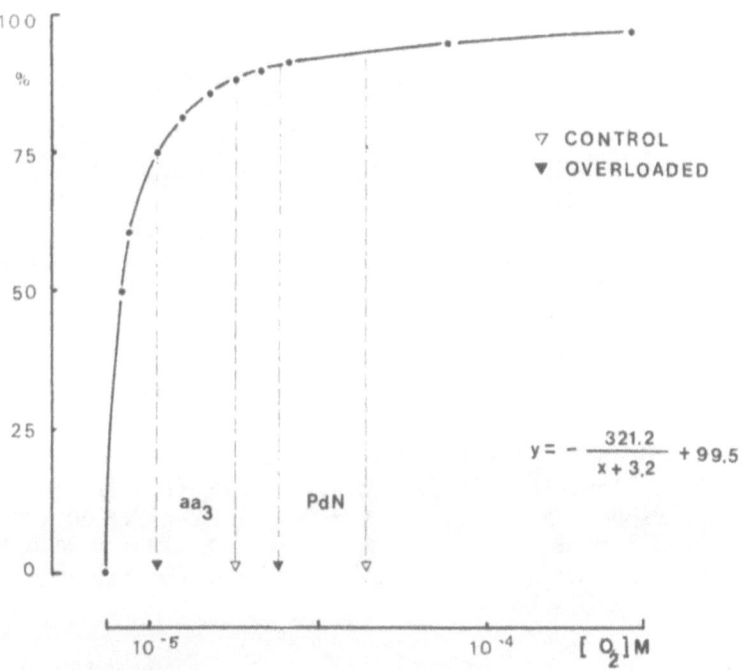

Fig. 1. The evaluation of intracellular O_2 concentrations necessary
to maintain the cytochrome c oxidase (aa_3) and total tissue
pyridine nucleotide (PdN) fully oxidized in control and
overloaded hearts. The hearts were perfused at 37°C according to Langendorff with a bicarbonate buffer containing
2.5 mmol·l⁻¹ pyruvate and exposed to repetitive anoxic
periods. Myoglobin and cytochrome oxidase absorption
changes were obtained from double wavelength runs (rapid
scanning spectroscope T 13/3); the reduced pyridine nucleotide fluorescence emission was monitored by a home made
fluorimeter. Note the apparent decrease of O_2 concentrations at which the respiratory chain components are still
fully oxidized. The equation for the myoglobin dissociation
curve observed at 37°C is that used by Lübbers (Leniger-
Follert and Lübbers, 1973).

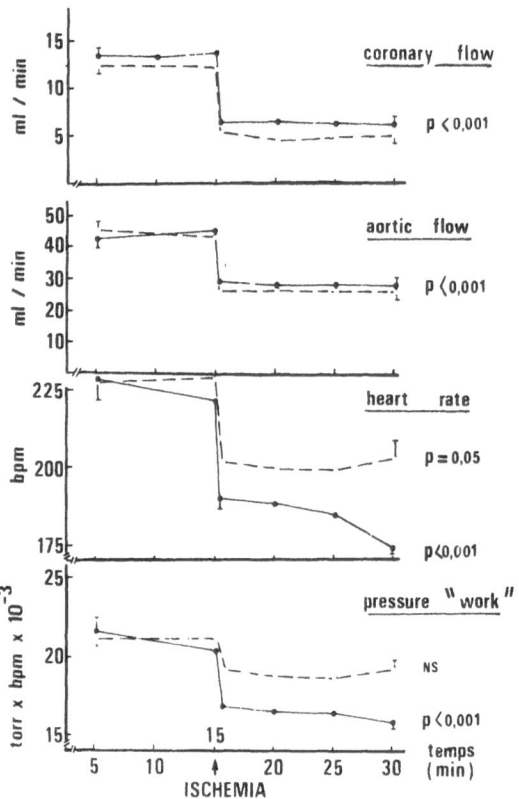

Fig. 2.   Hemodynamic effects of reduced coronary flow rate (50%
          reduction, 15 min) as observed in control and mechanically
          overloaded hearts. Note better preservation of the heart
          rate and of resulting isometric work' (HR x peak AoP) in
          hearts from animals with the aorto-caval fistula (o----o)
          as compared to controls (●———●).

are indicated in Fig. 1. According to the equation of myoglobin
dissociation curve read at 37°C (Leniger-Follert and Lübbers, 1973),
the critical intracellular $[O_2]$'s for cytochrome aa_3 and pyridine
nucleotide oxidation are decreased by about 1/3 of order. In the
control hearts, these O_2 concentrations can be estimated as 4 x
10^{-5} M and 7 x 10^{-5} M respectively. In volume overloaded hearts they
equal to 1 x 10^{-5} M and 3 x 10^{-5} M (Fig. 1).

     The effects of the ischemia, as induced by the insertion of a
one-way valve into the aortic outflow tract of the atrially perfused hearts, are shown in Fig. 2. The 50% reduction of the coronary
flow rate is followed by similar changes of left ventricular ejec-
tion in both groups of hearts. However, the pressure work (peak
aortic pressure (AoP) x heart rate) is significantly more depressed
in hearts from control animals; the main reason being the decrease

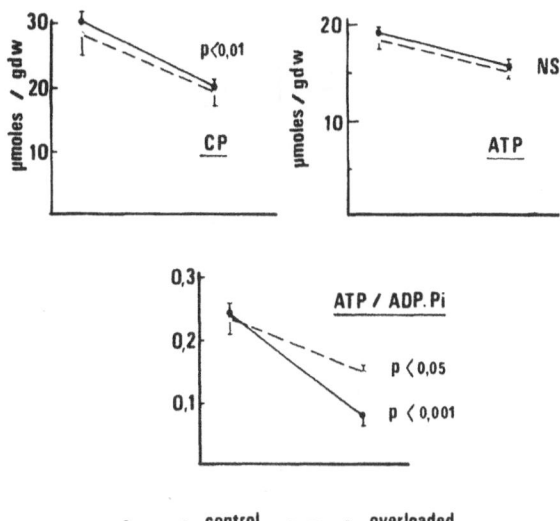

Fig. 3. The effects of a 15 min cardiac ischemia on tissue levels
of creatine phosphate and of ATP. While there is no difference between the ischemic levels of macroergic compounds as
found in control and overloaded hearts, the ATP/[ADP]x[Pi]
ratio is significantly more depressed in the former: the
ischemic level of CP and that of ATP in the overloaded
hearts seems to be maintained only at the expense of higher
tissue levels of their hydrolysis products (NS = nonsigni-
ficant).

of the heart rate (Fig. 2). In hearts perfused with glucose and
palmitate this deceleration is even more significant. Some of them
became contractured before the end of the experiment.

The above relative protection of left ventricular activity, as
observed in mechanically overloaded hearts submitted to acute
ischemia, does not seem simply related to a better preservation of
intracellular ATP or creatine phosphate (CP). These two compounds
are affected to the same extent in both ischemic groups (Fig. 3).
However, in the control hearts, the ischemic levels of ATP and CP
are reached only at the expense of higher concentrations of ADP and
Pi. The resulting total tissue ATP/[ADP.Pi] ratio is significantly
lower in the control hearts (Fig. 3). This ADP and Pi accumulation
might reflect a disturbed adenine nucleotide cycling between the
cytosol and the mitochondrial matrix (Moravec, 1980).

Another difference between the reaction of control and overloaded hearts to a decreased coronary perfusion concerns the lipid
intermediates. While in control hearts a large accumulation of both
long chain acyl CoA and long chain acyl carnitine regularly occurred,
no such change could be detected in mechanically overloaded hearts
perfused with 10 mmol·l⁻¹ glucose. In the presence of glucose and

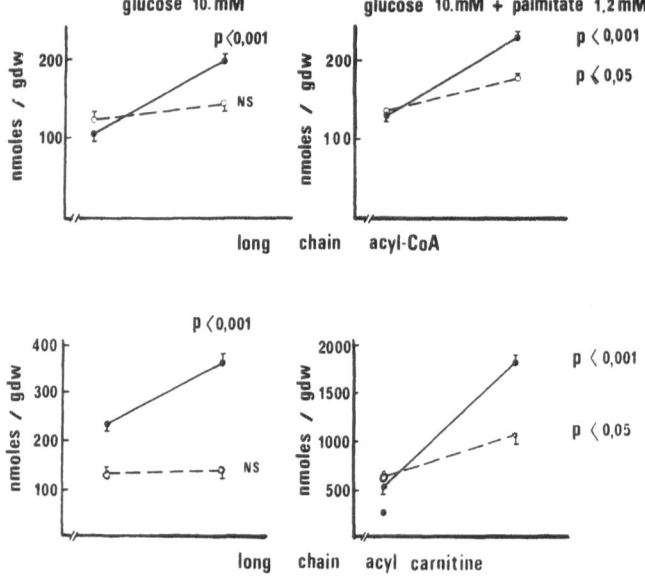

Fig. 4. The effects of a 15 min cardiac ischemia on tissue levels
         of long chain acyl CoA and long chain acyl carnitine. Note
         the absence of any accumulation of lipid intermediates in
         hearts from rats with the aorto-caval fistula, when glucose
         10 mmol·l^{-1} is used. In those perfused with glucose 10
         mmol·l^{-1} + palmitate 1.2 mmol·l^{-1} an attenuation of long
         chain acyl ester accumulation can be noted. (•——• control;
         o----o overloaded; NS = nonsignificant).

1.2 mmol·l^{-1} palmitate, the ischemia-induced accumulation of long
chain acyl esters was considerably attenuated in the overloaded
hearts (Fig. 4).

DISCUSSION

        The acceleration of reduced pyridine nucleotide and cytochrome
aa$_3$ reoxidation which appears in volume overloaded hearts exposed
to a short period of anoxia was already described in our recent
papers (Moravec, 1980; Moravec et al., 1981). This acceleration was
found unrelated to improved oxygen supply to the myocardium, since
neither coronary flow rate nor the speed of myoglobin reoxygenation
in the epicardial layer of the overloaded hearts were changed. The
reoxidation of pyridine nucleotide and that of cytochrome oxidase
seemed to be completed at lower intracellular Po$_2$'s, indicating a
decrease in O$_2$ requirements for respiratory chain function (decreased apparent K_M [O$_2$]; Mela et al., 1976; Jöbsis, 1977). Such an
adaptation may result from a disinhibition of the cytochrome oxidase
which, in intact tissues, was found to be partially inhibited
(Jöbsis, 1977; Wilson et al., 1979). The decrease in O$_2$ requirements

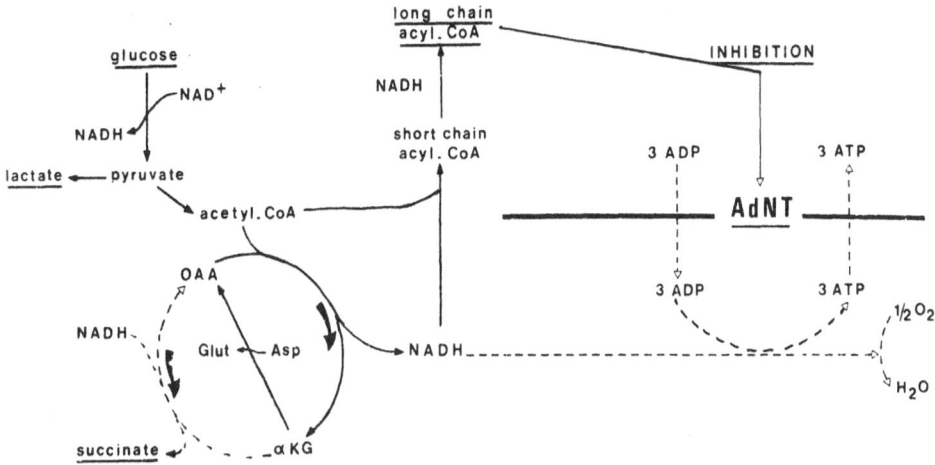

Fig. 5.   Schematic representation of the interference between the
respiratory chain and lipid metabolism at the level of reduced pyridine nucleotide reoxidation. Unter normal conditions, most of the NAD⁺ necessary for the maintenance of
metabolic activity is regenerated by the respiratory chain.
Under conditions of decreased oxygenation, the respiratory
chain function becomes O₂ limited. Small amounts of NAD⁺
can then be regenerated by enhanced transaminations and by
the intramitochondrial synthesis of free fatty acids
(Whereat, 1967; Hochachka, 1980). If, during the development of cardiac hypertrophy, the oxygen requirements of
the respiratory chain function decreased, the accumulation
of long chain acyl intermediates of lipid metabolism could
be attenuated. Glut: glutamate; Asp: arspartate; αKG: keto-
gentarate; DAA: oxaloacetate.

of the mitochondria from mechanically overloaded hearts is further
suggested by the absence of any long chain acyl ester accumulation,
which regularly occurs in control hearts under conditions of decreased oxygenation (Neely, 1979; Hochachka, 1980). In fact, it has
been suggested that under conditions of impaired respiratory chain
function, the regeneration of NAD⁺, necessary for the maintenance
of metabolic fluxes, proceeds via the incorporation of acetyl CoA
into the mitochondrial fatty acids (Whereat et al., 1967). Succinate,
as well as citrate promote this process. If, in the overloaded
hearts exposed to ischemia, most of the NADH reoxidation proceeded
effectively via the respiratory chain, the neoformation of long
chain acyl CoA esters, normally occurring during hypoxia, could be
avoided (Fig. 5). This, in turn, would have a beneficial effect on
mitochondrial energy production, since long chain acyl CoA were
shown to specifically inhibit the adenine nucleotide translocase
system (AdNT) (Chua and Shrago, 1977). In the absence of any long
chain acyl ester accumulation, the resistance of the mechanically
overloaded myocardium to a transient hypoxia could therefore be

improved. The data presented in this paper are consistent with such a possibility.

ACKNOWLEDGEMENT

The technical assistance of Mr. A. Corsin, P. Pocholle and of Mrs. M.-T. Dronne is deeply appreciated.

REFERENCES

Anversa, P., Vitali-Mazza, L., Visioli, O., and Marchetti, G., 1971, Experimental cardiac hypertrophy: a quantitative ultrastructure study, J. Mol. Cell. Cardiol., 3:213.

Chance, B., 1976, Pyridine nucleotide as an indicator of the oxygen requirements for energy-linked functions of the mitochondria, Circ. Res., 38(Suppl. 1):31.

Chua, B., and Shrago, E., 1977, Reversible inhibition of adenine nucleotide translocase in bovine heart mitochondria by long chain acyl CoA esters. Comparison with atractyloside and bonkrekik acid, J. Biol. Chem., 252:6711.

Feuvray, D., 1981, Structural, functional and metabolic correlates in ischemic hearts: effect of substrates, Am. J. Physiol., 240:391.

Garland, P.E., Shepherd, D., and Yates, D.W., 1965, Steady state concentrations of coenzyme A, acetyl coenzyme A and long chain fatty acyl CoA in rat liver mitochondria oxidizing palmitate, Biochem. J., 97:587.

Hatt, P.Y., Rakusan, K., Gastineau, P., Laplace, M., and Cluzeaud, F., 1980a, Aorto-caval fistula in rat: an experimental model of heart overloading, Bas. Res. Cardiol., 75:105.

Hatt, P.Y., Rakusan, K., Gastineau, P., and Laplace, M., 1980b, Morphometry and ultrastructure of heart hypertrophy induced by chronic volume overload, J. Mol. Cell. Cardiol., 11:989.

Hochachka, P.W., 1980, "Living Without Oxygen", Harvard University Press, Cambridge, London.

Jöbsis, F.F., 1977, What is molecular oxygen sensor: What is a transduction process, in: "Tissue Hypoxia and Ischemia", M. Reivich, R. Coburn, S. Lahiri, B. Chance, eds., Plenum Press, New York.

Leniger-Follert, E., and Lübbers, D.W., 1973, Determination of local myoglobin concentration in the guinea pig heart, Pflügers Arch., 341:271.

Lübbers, D.W., and Niesel, W., 1959, Der Kurzzeit-Spektralanalysator. Ein schnellarbeitendes Spektralphotometer zur laufenden Messung von Absorptions- bzw. Extinktionsspektren, Pflügers Arch. Ges. Physiol., 268:286.

Mc Garry, J.D., and Foster, D.W., 1976, An improved and simplified radio isotope assay for the determination of free and esterified carnitine, J. Lipid Res., 17:277.

Meerson, F.Z., 1969, The myocardium in hyperfunction hypertrophy and failure, Circ. Res., 25(Suppl. 2):1.

Mela, L., Goodwin, C.W., and Miller, L.D., 1976, In vivo control of mitochondrial enzyme concentrations and activity by oxygen, Am. J. Physiol., 231:1811.

Moravec, J., 1980, Possible relationship between tissue levels of long chain acyl CoA and the ability of the overloaded myocardium to oxidize an excess of reduced pyridine nucleotide, FEBS Lett., 113:134.

Moravec, J., Corsin, A., Owen, P., and Opie, L.H., 1974, Effect of increased aortic pressure on fluorescence emission of isolated rat heart, J. Mol. Cell. Cardiol., 6:187.

Moravec, J., Moravec, M., and Hatt P.Y., 1981, Rate of pyridine nucleotide oxidation and cytochrome oxidase interaction with intracellular oxygen in hearts from rats with compensated volume overload, Pflügers Arch., 392:106.

Neely, J.R., Garber, D., Mc Donough, K., and Idell-Wenger, J., 1979, Relationships between ventricular function and intermediates of fatty acid metabolic during myocardial ischemia, in: "Ischemic Myocardium and Antianginal Drugs", M.M. Winsburg, ed., Plenum Press, New York.

Neely, J.R., Rovetto, M.J., Whitmar, J.T., and Morgan, H.E., 1973, Effect of ischemia on ventricular function and metabolism in the isolated working rat hearts, Am. J. Physiol., 225:651.

Rakusan, K., Moravec, J., and Hatt, P.Y., 1980, Regional capillary supply to the normal and hypertrophied rat heart, Microvasc. Res., 20:319.

Whereat, A.F., Mull, F.E., and Orishimo, M.W., 1967, The role of succinate in the regulation of fatty acid synthesis of heart mitochondria, J. Biol. Chem., 242:4013.

Wilson, D.F., Owen, C.S., and Erecinska, M., 1979, Quantitative dependence of mitochondrial oxidative phosphorylation on oxygen concentrations, Arch. Biochem. Biophys., 195:495.

DISTRIBUTION OF MYOCARDIAL GLUCOSE CONSUMPTION UNDER NORMAL CONDITIONS AND DURING ISOPRENALINE AND DOBUTAMINE INFUSION

W. Breull, and M. Rubart

Department of Physiology, University of Bonn
Nussallee 11, 5300 Bonn 1, FRG

INTRODUCTION

Tracer microsphere studies on regional myocardial blood flow revealed an inhomogeneous distribution within the left ventricular wall under various experimental conditions. In the normal and the anesthetized closed chest dog as well as in isolated heart preparations blood flow to the subendocardial layers exceeds that to the subepicardial layers; ratios, calculated from subendocardial and subepicardial flow ranged from 1.1 to 1.8 (Flohr et al., 1973; Ypintsoi et al., 1973; Domenech et al., 1980) depending largely on the size of the particles applied. While it is a matter of discussion, which particle size can delineate nutritional blood flow in the myocardium, it is commonly accepted that under normal conditions a blood flow gradient between the subepi- and subendocardial layer exists. This blood flow pattern may be induced by a comparable pattern of myocardial metabolic rate. Measurements of sarcomere length in diastole and in systole (Sonnenblick et al., 1967), calculations of tension developed (Streeter et al., 1970) and the analysis of tissue oxygen clearance in acute ischemia (Winbury et al., 1981) in different layers of the left ventricular wall seem to support this hypothesis. By measuring local capillary oxygen saturation and local blood flow, Holtz et al. (1977) calculated peak values of oxygen consumption in the subendocardial region.

Since methodological difficulties occur, when estimating local myocardial oxygen consumption, we determined local myocardial substrate utilization. 2-Deoxyglucose (2 DG) was used to study regional myocardial metabolic rate of glucose (MMR Gl). Applying this tracer, regional glucose consumption of brain tissue has been measured in rats (Sokoloff et al., 1977) and in man (Reivich et al., 1979).

Even though free fatty acids (FFA) and lactic acid (LA) are the
major fuel of myocardial energy metabolism, and even though these
substrates can substitute glucose (Neely et al., 1974), local
MMR Gl values reflect the distribution pattern of myocardial energy
metabolism.

Applying the 2-DG technique we investigated the transmural
distribution of MMR Gl in the left ventricular free wall of the
anesthetized closed chest dog. As an increase in cardiac performance and in cardiac energy metabolism can be obtained by myocardial
β_1-adrenoceptor stimulation, additional investigations of regional
MMR Gl were performed under conditions of an isoprenaline infusion.
Since with isoprenaline the increase in cardiac energetics was
accompanied by a decrease in cardiac afterload, dobutamine infusion
was chosen in a third experimental series.

METHODS

a) Determination of MMR Gl: Metabolic rate of glucose was
determined according to Sokoloff et al. (1977). According to several
investigators the assumption on which the Sokoloff model is based,
seems to be valid for myocardial tissue (Opie et al., 1967; Takala
et al., 1981; Schelbert et al., 1982). Previous studies on isolated
perfused non-working hearts demonstrated a good correlation between
values determined by direct measurement of glucose extraction and
values determined by means of the 2-DG method (Breull et al., 1981).
We therefore used the operational equation developed by Sokoloff and
co-workers.

$$R_i = \frac{C_i^*(T) - k_1^* e^{-(k_2^*+k_3^*)T} \int_0^T C_p^* e^{(k_2^*+k_3^*)t} \, dt}{\left(\frac{\lambda \cdot V_{max}^* \cdot K_m}{\phi \cdot V_{max} \cdot K_m^*}\right)\left[\int_0^T (C_p^*/C_p)\,dt - e^{-(k_2^*+k_3^*)T} \int_0^T (C_p^*/C_p)\,e^{(k_2^*+k_3^*)t}\,dt\right]}$$

R_i = MMR Gl, $C_i^*(T)$ = tissue content of 2-DG and 2-DG-6-P at the
end of the experiment, k_1^*, k_2^*, k_3^* = rate constants,

$$\left(\frac{\lambda \cdot V_{max}^* \cdot K_m}{\phi \cdot V_{max} \cdot K_m^*}\right) = \text{"lumped constant"}$$

C_p^* = 2-DG concentration in the arterial plasma,
C_p = glucose concentration in the arterial plasma
(for details see Sokoloff et al., 1977).

Since k_1^*, k_2^*, and k_3^* have not yet been determined for myocardial tissue, values determined for the gray matter of brain were
used (k_1^* = 0.189, k_2^* = 0.245, k_3^* = 0.052). For the "lumped constant"
in the denominator a value (0.39) determined by Bünger and Kuschinsky
(pers. com.) was used.

b) Experimental protocol: 27 mongrel dogs with a body weight
ranging from 8 to 12 kg were investigated. The animals, which were
fasted for about 16 hours prior to the experiment, were anesthetized with pentobarbital (20-30 mg·kg^{-1}), relaxed by succinylcholine
and artificially ventilated with room air. Arterial Po_2, Pco_2 and
pH were monitored by means of a blood gas and blood pH analyzing
system (Eschweiler). Polythene catheters were placed in the right
atrium, the inferior caval vein, a femoral vein and in a femoral
artery. A thermistor probe was placed in the aortic arc and a Millar
catheter tip pressure transducer with a separate injection channel
in the left ventricle. The ECG was recorded throughout the experiment. Body temperature could be adjusted by means of a water heating
device.

The following parameters were measured: left ventricular
systolic (LVSP) and enddiastolic (LVeDP) pressures as well as
maximum rate of left ventricular pressure rise (dp/dt$_{max}$) by means
of the Millar tip transducer; mean arterial blood pressure (MABP)
and right atrial pressure by means of Statham P23 DB transducer;
cardiac output (CO) and enddiastolic volume (EDV) by means of a
thermodilution method and heart rate (HR) by electrocardiography.
The arterial plasma concentration of glucose was determined in all
experiments (Schmidt, 1961), while the concentrations of FFA
(Howorth et al., 1966) and LA were measured in 11 experiments.

In the control group (n = 12) MMR Gl was measured as soon as
steady state conditions were reached. In the isoprenaline (n = 8)
and in the dobutamine (n = 7) group, isoprenaline (mean dosage:
1.3 ug·kg^{-1}·min^{-1}) or dobutamine (mean dosage: 32 ug·kg^{-1}·min^{-1})
was infused in ascending doses until constant hemodynamic conditions
were attained. 0.1-0.3 mCi 1-14-C-2-deoxyglucose, spec. activity
50-60 mCi/mmol, diluted in 2 ml isotonic saline was injected i.v.
within 15 s. Timed arterial blood samples of 0.8 ml each were drawn
through the femoral artery catheter at 30 s and 1, 2, 3, 5, 7, 10,
15, and at each consecutive fifth minute. 60 minutes after injection
the animal was sacrificed without excitation by means of T 61^R,
Hoechst. The heart was quickly removed, cleaned, weighed and cut
into 6 blocks (basal, middle and apical), which were frozen in
isopentane chilled to -50°C. Tissue specimen from the subepicardial,
middle and subendocardial layers of the anterior, lateral and
posterior part of the left ventricular wall were dissected, weighed,
dissolved, and counted for 14-C activity in a Packard liquid
scintillation counter. Plasma from the blood samples was counted in
a similar way. Glucose concentration was determined in the same

samples. From the tissue 2-DG and 2-DG-6-P content at the end of the experiment $(C_i^*(T))$ as well as from the time course of plasma 2-DG (C_p^*) and glucose (C_p) concentration throughout the experiment, myocardial glucose consumption was calculated.

Autoradiographic studies: From the different tissue blocks 200 um sections were cut. After freeze-drying the sections were exposed

Table 1. Values for cardiac output, stroke volume and enddiastolic volume are relative to the corresponding body weights (CO/BW, SV/BV, EDV/BW).
Experimental data (mean, $\pm$ SD)

Table 1.	Experimental data (mean, $\pm$ SD)		
	Control	Isoprenaline $1.3 \, \mu g \cdot kg^{-1} \cdot min^{-1}$	Dobutamine $32 \, \mu g \cdot kg^{-1} \cdot min^{-1}$
n	12	8	7
Body weight (kg)	11.0 ± 1.4	9.8 ± 1.4	9.4 ± 1.0
Heart weight (g)	89.7 ± 25.8	82.3 ± 16.4	78.8 ± 24.8
PaO_2 (mm Hg)	95.3 ± 7.8	87.3 ± 7.7	77.7 • ± 13.1
LVSP (mm Hg)	123.0 ± 15.3	114.8 ± 23.7	117.1 ± 16.1
LVeDP (mm Hg)	4.64 ± 1.94	4.25 ± 2.11	3.01 ± 0.95
dp/dt_{max} (mm Hg·s^{-1})	3012 ± 960	5188 •• ± 1123	6785 •• ± 1627
HR ($1 \cdot min^{-1}$)	165.9 ± 25.3	209.4 • ± 23.8	207.3 • ± 32.5
MABP (mm Hg)	106.3 ± 12.1	78.2 • ± 19.2	98.6 ± 19.2
CO/BW ($ml \cdot kg^{-1} \cdot min^{-1}$)	174 ± 60	209 ± 57	220 ± 29
SV/BW ($ml \cdot kg^{-1}$)	1.054 ± 0.342	1.025 ± 0.355	1.050 ± 0.229
EDV/BW ($ml \cdot kg^{-1}$)	3.561 ± 1.765	2.657 ± 0.560	4.400 ± 2.011

• $p \leqslant 0.05$; •• $p \leqslant 0.01$

with a Kodak Definix X-ray film in a dry chamber at room temperature. Exposition times varied from 7 to 40 days.

RESULTS

Table 1 summarizes mean values of parameters influencing
cardiac performance in the three experimental groups. Isoprenaline
and dobutamine increased heart rate, dp/dt_{max} and cardiac output,
demonstrating an increase in cardiac work. On the other hand isoprenaline induced a decrease in aortic blood pressure and therefore
in cardiac afterload. Stroke volume remained constant. The interpretation of the observed enddiastolic volumes is restricted due to
the limitations of the thermodilution method applied (Holt, 1956).
With dobutamine infusion arterial plasma concentrations of LA increased from 1.62 to 2.4 $mmol \cdot l^{-1}$ and from 0.6 to 1.2 $mmol \cdot l^{-1}$ for
FFA.

Regional MMR Gl: In the control group MMR Gl of the subendocardial region - except for the apical parts of the ventricle -
exceeded that of the subepicardial by about 10-60% (Table 2). With
isoprenaline only slight differences in local MMR Gl were found,
demonstrating a more or less homogeneous distribution pattern
(Table 3). With dobutamine in 7 out of 9 regions MMR Gl decreased
from the subepi- to the subendocardial region (Table 4). However,
these differences could not be statistically verified.

Table 2. MMR glucose l.v. free wall ($umol \cdot g^{-1} \cdot min^{-1}$, mean $\pm$ SEM)
          control, n = 12.

Region	anterior			lateral			posterior		
	Epi	m	Endo	Epi	m	Endo	Epi	m	Endo
basal	0.208 ±0.052	0.210 ±0.063	0.312 ±0.060	0.185 ±0.041	0.178 ±0.043	0.249* ±0.066	0.220 ±0.060	0.234 ±0.054	0.234 ±0.050
middle	0.161 ±0.042	0.145 ±0.037	0.231* ±0.052	0.144 ±0.033	0.117 ±0.025	0.239* ±0.058	0.199 ±0.046	0.177 ±0.041	0.291** ±0.053
apical	0.157 ±0.036	0.231 ±0.075	0.140 ±0.033	0.146 ±0.034	0.156 ±0.034	0.191** ±0.047	0.186 ±0.038	0.165 ±0.036	0.212 ±0.064

* $p < 0.05$

** $p < 0.01$

Table 3. MMR glucose l.v. free wall (umol$\cdot$g$^{-1}\cdot$min^{-1}, mean $\pm$ SEM) isoprenaline 1.27 ug$\cdot$kg$^{-1}\cdot$min^{-1}, n = 8.

Region	anterior			lateral			posterior		
	Epi	■	Endo	Epi	■	Endo	Epi	■	Endo
basal	0.265 ±0.050	0.237 ±0.047	0.245 ±0.051	0.261 ±0.061	0.207 ±0.036	0.238 ±0.049	0.244 ±0.046	0.247 ±0.052	0.226 ±0.036
middle	0.240 ±0.051	0.226 ±0.049	0.240 ±0.046	0.220 ±0.048	0.232 ±0.048	0.251 ±0.053	0.218 ±0.039	0.223 ±0.044	0.252 ±0.046
apical	0.219 ±0.045	0.206 ±0.041	0.194 ±0.034	0.189 ±0.033	0.210 ±0.038	0.208 ±0.040	0.215 ±0.041	0.204 ±0.039	0.210 ±0.040

All results are summarized in Fig. 1. MMR Gl significantly (p < 0.01) increased from the subepicardial (0.178 umol$\cdot$g$^{-1}\cdot$min^{-1}) and middle (0.174 umol$\cdot$g$^{-1}\cdot$min^{-1}) to the subendocardial layer (0.238 umol$\cdot$g$^{-1}\cdot$min^{-1}) under control conditions. A mean value of 0.197 umol$\cdot$g$^{-1}\cdot$min^{-1} was calculated. Isoprenaline caused a rise in mean MMR Gl to 0.227 umol$\cdot$g$^{-1}\cdot$min^{-1}. MMR Gl of the subepicardial region amounted to 0.238 and to 0.229 umol$\cdot$g$^{-1}\cdot$min^{-1} for the subendocardial region. Under dobutamine infusion a mean value of 0.21 umol$\cdot$g$^{-1}\cdot$min^{-1} was calculated. MMR Gl was 0.23 umol$\cdot$g$^{-1}\cdot$min^{-1} in the subepicardial and 0.207 umol$\cdot$g$^{-1}\cdot$min^{-1} in the subendocardial region. With isoprenaline and dobutamine no significant differences in

Table 4. MMR glucose l.v. free wall (umol$\cdot$g$^{-1}\cdot$min^{-1}, mean $\pm$ SEM) dobutamine 32.5 ug$\cdot$kg$^{-1}\cdot$min^{-1}, n = 7.

Region	anterior			lateral			posterior		
	Epi	■	Endo	Epi	■	Endo	Epi	■	Endo
basal	0.242 ±0.054	0.187 ±0.036	0.227 ±0.041	0.192 ±0.037	0.202 ±0.042	0.206 ±0.037	0.287 ±0.089	0.219 ±0.05?	0.200 ±0.042
middle	0.224 ±0.027	0.193 ±0.034	0.195 ±0.043	0.193 ±0.037	0.180 ±0.034	0.183 ±0.037	0.312 ±0.066	0.195[*] ±0.039	0.345 ±0.063
apical	0.204 ±0.026	0.178 ±0.019	0.155 ±0.028	0.191 ±0.039	0.171 ±0.031	0.175 ±0.035	0.232 ±0.035	0.171 ±0.031	0.174 ±0.032

● p ≤ 0.05

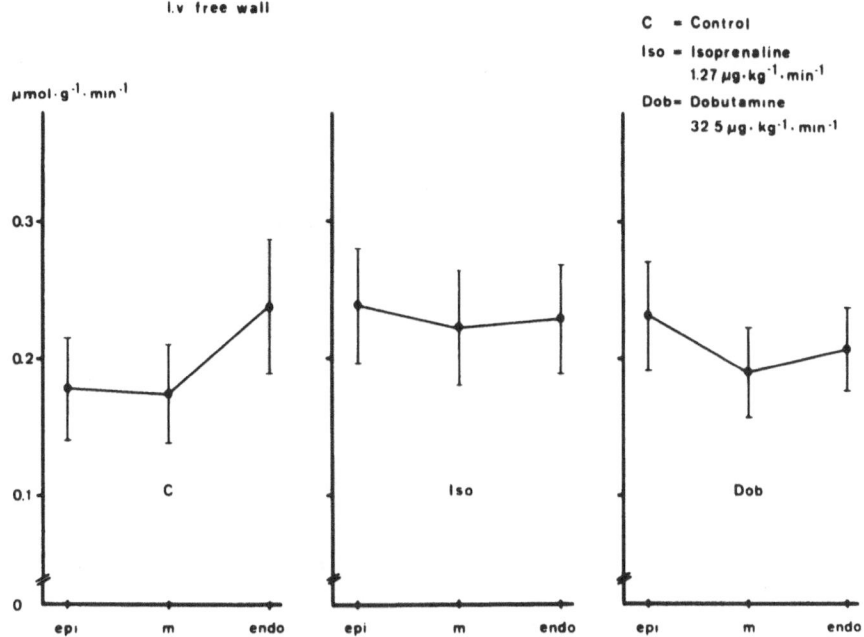

Fig. 1.   Regional myocardial metabolic rate (glucose) (MMR Gl) of
the subepicardial (epi), middle (m) and subendocardial
(endo) layer of the left ventricular (l.v.) free wall.
Control group (C), and two groups with isoprenaline (Iso)
or dobutamine (Dob) infusion, respectively. Mean values ±
SEM.

MMR Gl between the different layers of the ventricular wall could
be found.

        These results are supported by autoradiographic studies on
myocardial uptake of 2-DG (Fig. 2). Under control conditions peak
values of 2-DG metabolism were found in the subendocardial region
with special reference to the papillary muscles. A homogeneous
distribution can be observed in the isoprenaline as well as in the
dobutamine group. With dobutamine small regions of higher activity
were found in different layers of the ventricular wall.

DISCUSSION

        The 2-DG technique, for measuring regional glucose consumption
of brain tissue, was established by Sokoloff et al. (1977). Since
the glycolytic enzyme pattern of the heart does not substantially
differ from that of the brain, glucose-6-phosphatase activity is low
(Opie et al., 1967; Takala et al., 1981) and biological half life of
2-DG-6-P is in the order of hours, the Sokoloff model can be applied

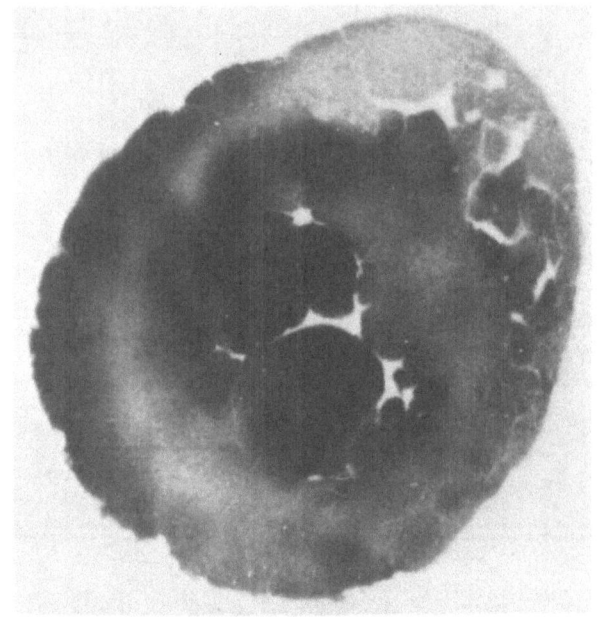

(a)

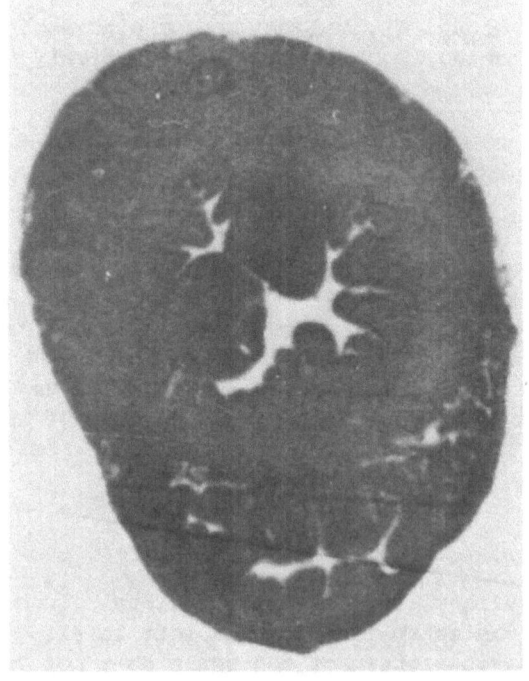

(b)

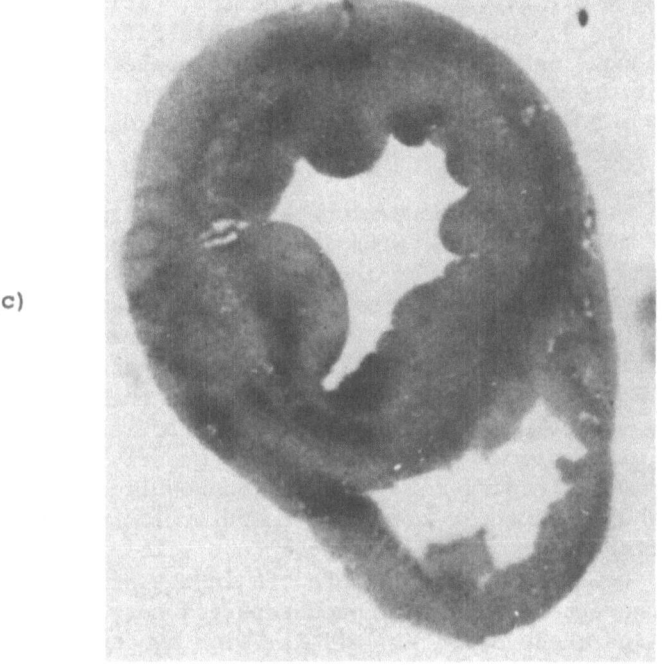

(c)

Fig. 2.   3 typical autoradiographs of 14-C-2-DG and -2-DG-6-P
          distribution in 200 um sections of the canine heart under
          control conditions (a) and under isoprenaline (b) and
          dobutamine (c) infusion.

to determine regional myocardial glucose consumption. Since in
myocardial tissue glucose can be substituted by FFA and LA, a
pattern of MMR Gl can only reflect a pattern of total myocardial
metabolism, if steady state conditions with regard to oxygen, FFA
and glucose supply are achieved.

     Mean values of MMR Gl, determined in the present study, are in
agreement with values measured in the human heart by means of the
Fick principle (Bernsmeier et al., 1962). Regional MMR Gl demonstrated an inhomogeneous distribution under control conditions; MMR
Gl of the subendocardial region exceeded that of the subepicardial
by 30%. These inhomogeneities may result from those functional inhomogeneities described by Sonnenblick et al. (1967). Similar inhomogeneities have also been observed in myocardial blood flow
(Flohr et al., 1973; Domenech et al., 1980). Since a close connection between myocardial blood flow and metabolism is commonly
accepted, subendocardial flow should exceed subepicardial by a
comparable proportion (30%). We therefore believe, that 12-15 um
microspheres will most accurately measure blood flow distribution
within the myocardium.

Isoprenaline and dobutamine caused only a slight (7-15%) increase in MMR Gl while myocardial energy demand, estimated from
data shown in Table 1, increased by 30-50%. This difference can be
explained by changes in myocardial metabolism, preferring LA and
FFA under conditions of an adrenoceptor stimulation. This hypothesis
is supported by a 50-100% increase in arterial plasma concentration
of FFA and LA, observed under dobutamine infusion.

Under both experimental conditions MMR Gl revealed a significant pattern of redistribution. With isoprenaline, the homogeneous
pattern observed, results from an increase in MMR Gl of the subepicardial and middle layer, whereas with dobutamine a net decrease in
MMR Gl of the subendocardial layer was measured. A redistribution
of local metabolic rate can result from altered ventricular working
conditions due to altered ventricular volumes. In the different
layers of the ventricular wall, changes in ventricular volumes will
influence diastolic sarcomere length and consequently systolic
tension developed, to a different degreee (Sonnenblick et al., 1967)
On the other hand a different amount of muscular activation can
result from different amounts of adrenoceptors within the layer of
the ventricular wall; until now differences in the number of myocardial β_1-adrenoceptors have only been reported between atrial and
ventricular tissue (Hedberg et al., 1979). With myocardial blood
flow, adrenoceptor stimulation induced a pattern of redistribution,
comparable to that of MMR Gl (Buckberg et al., 1973). Blood flow
redistribution under these conditions, therefore, originates from a
redistribution of local metabolic rate, but not from restrictions
in local blood flow.

SUMMARY

The regional myocardial metabolic rate of glucose (reg. MMR Gl)
of the left ventricular free wall was determined applying the
2-deoxyglucose method (Sokoloff, 1977) in the anesthetized closed
chest dog. Under control conditions an inhomogeneous distribution
of MMR Gl was observed. Isoprenaline or dobutamine infusions resulted
in a redistribution of MMR Gl. With isoprenaline a homogeneous
distribution pattern was observed, while with dobutamine a slight
decrease in MMR Gl of the subendocardial region was measured. This
redistribution may result from altered ventricular working conditions, due to changes in ventricular volumes or from differences in
the frequency of myocardial adrenoceptors within the ventricular
wall.

ACKNOWLEDGEMENT

We thank Mrs. Ch. Pusch and Mrs. B. Schreiber for their skillful technical assistance.

REFERENCES

Bergmeyer, H.U., 1974, "Methoden der enzymatischen Analyse", Verlag
    Chemie, Weinheim.
Bernsmeier, A., and Rudolph, W., 1962, Neue Ergebnisse über Durch-
    blutung und Substratversorgung des menschlichen Herzens,
    Münch. Med. Wochenschr., 104:46-50.
Breull, W., Flohr, H., Schuchhardt, S., and Dohm, H., 1981,
    Transmural gradients in myocardial metabolic rate, Basic Res.
    Cardiol., 76:399-403.
Buckberg, G.D., and Ross, G., 1973, Effects of isoprenaline on
    coronary blood flow: its distribution and myocardial perform-
    ance, Cardiovasc. Res., 7:429-437.
Domenech, R.J., and Maclellan, P.R., 1980, Transmural distribution
    of coronary blood flow during coronary β_2-adrenergic receptor
    activation in dogs, Circ. Res., 46:29-36.
Flohr, H., Breull, W., Redel, D., and Dahners, H., 1973, Regional
    myocardial blood flow, Bibl. Anat., 11:158-163.
Hedberg, A., Minneman, K.P., and Molinoff, P.B., 1979, Regional
    distribution of β_1- and β_2-adrenoceptors in the right atrium
    and left ventricle of the cat and guinea pig heart, Brit. J.
    Pharmacol., 66:505P.
Holt, J., 1956, Estimation of the residual volume of the ventricle
    of the dog's heart by two indicator-dilution techniques, Circ.
    Res., 4:187.
Holtz, J., Grunewald, W.A., Manz, R., Restorff, W.V., and Bassenge,
    E., 1977, Intracapillary hemoglobin oxygen saturation and
    oxygen consumption in different layers of the left ventricular
    myocardium, Pflügers Arch., 370:253-258.
Howorth, P.J.N., Gibbard, S., and Marks, V., 1966, Evaluation of a
    colorimetric method (Duncombe) of determination of plasma none-
    esterified fatty acids, Clin. Chim. Acta, 14:69-73.
Neely, J.R., and Morgan, H.E., 1974, Relationship between carbo-
    hydrate and lipid metabolism and the energy balance of heart
    muscle, Ann. Rev. Physiol., 36:413-459.
Opie, L.H., and Newsholme, E.A., 1967, The activities of fructose
    1,6-diphosphatase, phosphofructokinase and phosphoenolpyruvate
    carbokinase in white muscle and red muscle, Biochem. J., 103:
    391-399.
Reivich, M., Kuhl, D., Wolf, A., Greenberg, J., Phelps, M., Ido, T.,
    Casella, V., Fowler, J., Hoffmann, E., Alavi, A., Som, P., and
    Sokoloff, L., 1979, The (18 F) fluorodeoxyglucose method for
    the measurement of local cerebral glucose utilization in man,
    Circ. Res., 44:127-137.
Schelbert, H.R., Henze, E., Phelps, M.E., and Kuhl, D.E., 1982,
    Assessment of regional myocardial ischemia by positron-emission
    computed tomography, Am. Heart J., 103:588-597.
Schmidt, F.H., 1961, Die enzymatische Bestimmung von Glukose und
    Fructose nebeneinander, Klin. Wochenschr., 39:1244-1247.

Sokoloff, L., Reivich, M., Kennedy, C., des Rosiers, M.H., Patlak,
    C.S., Pettigrew, K.D., Sakurada, O., and Shinohara, M., 1977,
    The (14-C)-deoxyglucose method for the measurement of local
    cerebral glucose utilization: Theory, procedure and normal
    values in the conscious and anesthetized albino rat, J.
    Neurochem., 28:897.
Sonnenblick, E.H., Ross, J., Covell, J.W., Spotnitz, H.M., and
    Spiro, D., 1967, Ultrastructure of the heart in systole and
    diastole: changes in sarcomere length, Circ. Res., 21:423-431.
Streeter, D.D., Vaishnav, R.N., Patel, D.J., Spotnitz, H.M., Ross,
    J., and Sonnenblick, E.H., 1970, Stress distribution in the
    canine left ventricle during diastole and systole, Biophys.
    J., 10:345-363.
Takala, T.E.S., and Hassinen, I.E., 1981, Effect of mechanical work
    load on transmural distribution of glucose uptake in the
    isolated perfused rat heart, studied by regional deoxyglucose
    trapping, Circ. Res., 49:62-69.
Winbury, M., Howe, B.B., and Weiss, H.R., 1971, Effect of nitro-
    glycerine and dipyridamole on epicardial and endocardial oxygen
    tension - further evidence for redistribution of myocardial
    blood flow, J. Pharmacol. Exp. Therap., 176:184-199.
Ypintsoi, T., Dobbs, W.A., Scanlon, P.D., Knopp, T.J., and
    Bassingthwaighte, J.B., 1973, Regional distribution of diffus-
    ible tracers and carbonized microspheres in the left ventricle
    of isolated dog hearts, Circ. Res., 33:573-587.

CARDIOVASCULAR AND METABOLIC RESPONSES TO CAROTID CLAMPING IN

ANEMIC DOGS

S.M. Cain[1], and C.K. Chapler[2]

[1]University of Alabama in Birmingham, USA
[2]Queen's University, Kingston, Ontario, Canada

INTRODUCTION

It has been observed that blood flow to the hindlimb was maintained or even increased in proportion to cardiac output during
isovolemic hemodilution of anesthetized dogs with dextran (Grupp et
al., 1972; Cain and Chapler, 1978; 1981). This was true even when
the decrease in total O_2 transport limited whole body O_2 uptake. In
one study, we infused norepinephrine to increase vasoconstrictor
tone. The results suggested that the reduction in viscosity prevented any measurable increase in hindlimb resistance during anemia
so that blood flow could not be redistributed to areas more essential than resting skeletal muscle. In these new experiments we have
used a more physiologic cardiovascular challenge, clamping of the
carotid arteries, to see if lowered blood viscosity alters the response to baroreceptor stimulation. As will be shown, the buffering
effect of aortic baroreceptors prevented much change in the peripheral resistance but carotid clamping did alter O_2 uptake.

METHODS

Eight dogs were anesthetized with sodium pentobarbital (30 mg/
kg). The venous circulation of the left hindlimb was redirected
through the femoral vein by tourniquets placed at the thigh. The
paw circulation was excluded by another tourniquet. At the end of
each experiment, the isolation of hindlimb blood flow was verified
by blue dye collection and the mass of perfused muscle was stained,
dissected, and weighed. All values for blood flow and O_2 uptake for
the hindlimb are given per kg of muscle. The animals were paralyzed
with succinylcholine chloride and ventilated at a constant rate to

keep arterial Pco$_2$ near 40 mm Hg. Blood samples were obtained simultaneously from the abdominal aorta, left femoral vein, and pulmonary
artery. Expired gas was collected and analyzed every 10 min.

The experimental protocol is explained in Fig. 1. In the top
panel, the mean systemic arterial blood pressure is shown for each
10 min period. The mean heart rate is shown in the middle panel and
stroke volume of the heart in the bottom panel. Following a 20-min
control period, both carotid arteries were occluded in the midcer-
vical region of the neck by means of previously placed ties. The
occlusion was maintained for 20 min and then released. This was
followed by a 20-min recovery period. From 60 to 70 min in the experiment, blood was isovolemically exchanged with warm 6% dextran
in Tyrode solution until an average hematocrit of 15% was reached.
The sequence of 20-min preclamp, 20-min clamp, and 20-min postclamp
observation periods was repeated while the dogs were anemic. At 130
min of the experiment, recovered red blood cells were returned to
the animal in exchange for anemic blood. Blood samples were taken

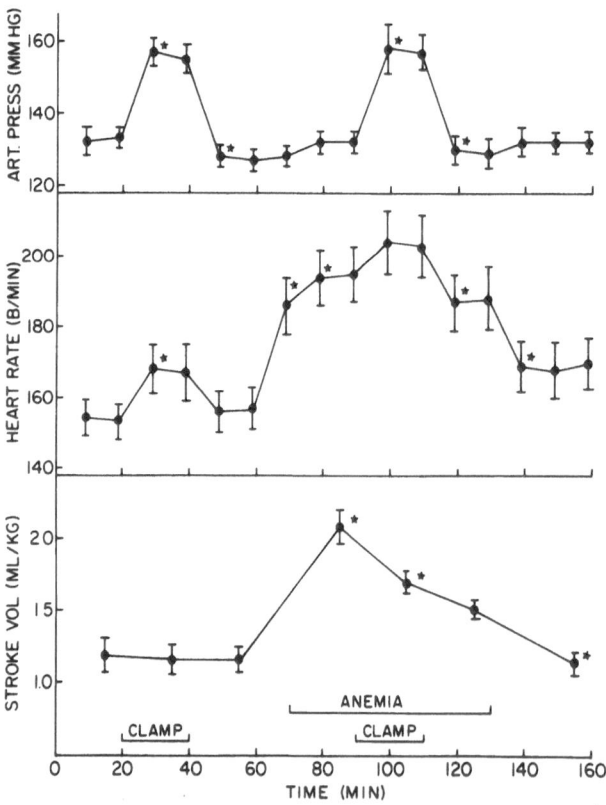

Fig. 1. Mean (+SE) arterial pressure, heart rate, and stroke volume.

midway through the last 10-min period of each step in the experimental sequence.

RESULTS

The asterisks in this and the following figures denote a significant (p < 0.05) change from the preceding measurement. Carotid clamping caused a blood pressure increase of approximately 25 mm Hg which was well maintained throughout the 20-min period of clamping. The isovolemic exchange with dextran did not alter pressure and the response to clamping was the same. Heart rate tended to be elevated by clamping but a more significant change occurred with anemia when heart rate increased about 25%. Stroke volume which was unchanged by clamping, nearly doubled with anemia at first, and then declined steadily until returning back to the control level with the reinfusion of red cells.

Cardiac output and limb blood flow are shown in Fig. 2. Carotid clamping did not significantly affect either one. With anemia, both cardiac output and limb blood flow approximately doubled. In contrast to the significant decrease in cardiac output during carotid clamping in anemia, limb blood flow did not change but then decreased with release of clamping while cardiac output declined even further.

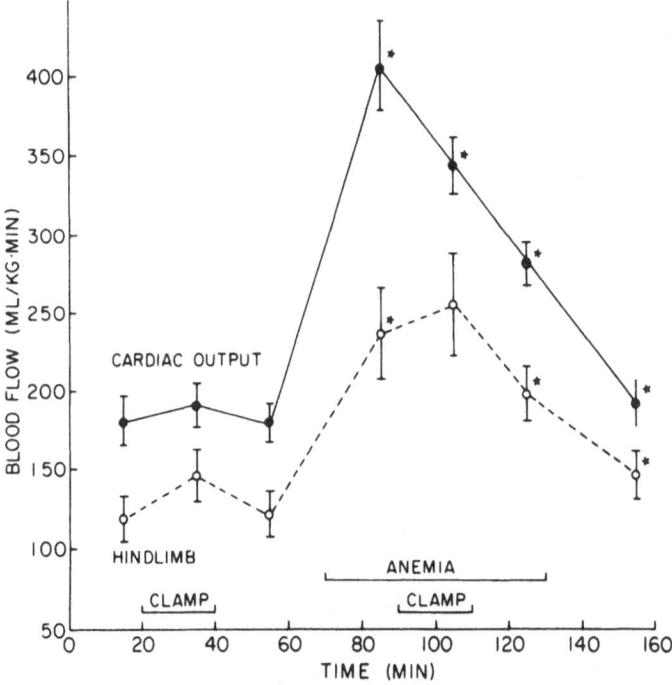

Fig. 2.   Cardiac output and limb blood flow (mean ± SE).

The ratio of blood pressure to flow is shown in the next figure
(Fig. 3) as peripheral resistance. In spite of a marked and sustained increase in blood pressure, the changes in cardiac output and
limb blood flow were sufficient during carotid clamping to cancel
out any significant effect on resistance while the animals were
normocythemic. There was an expected and significant decrease in
both whole body and limb resistance during anemia. To see if any of
that decrease was attributable to resistance vessel diameter changes,
the viscosity was estimated from the hematocrit using the data of
Fan et al. (1980) and vascular hindrance was calculated. There was
no significant change in hindrance with anemia so the entire effect
on resistance can be assigned to the viscosity change. During anemia,
there was a slight but significant increase in whole body peripheral
resistance with carotid clamping that was not seen in the hindlimb.
Although it may have not been meaningful physiologically, aortic
body chemoreceptor stimulation by anemia may have added to vasoconstrictor tone in non-muscle organ systems to cause a change in
total resistance. Most of the effects of carotid clamping on resistance were buffered out by the intact aortic baroreceptors, however.
Although it is not shown here, the initial response of the limb
circulation to carotid clamping was the expected one of no change
or a decrease in flow while pressure rose sharply during the first
minute or two. A reflex relaxation of resistance vessels was then
seen as blood flow began to parallel the increase in pressure. With
pressures approaching 200 mm Hg in some cases, aortic baroreceptor
stimulation would have been nearly maximal. The ability of aortic
baroreceptors to decrease limb vascular resistance is even more
marked than their effect on heart rate, for example, because they

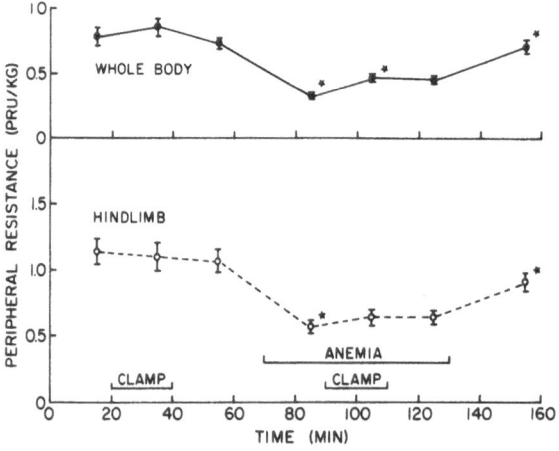

Fig. 3. Mean values (± SE) of systemic vascular resistance for
         whole body and hindlimb.

inhibit medullary vasoconstrictor tone (Carswell et al., 1968;
Donald and Edis, 1971). This may explain the maintenance of limb
blood flow while cardiac output declined and the lack of significant
resistance change with carotid clamping during anemia.

The metabolic consequences of carotid clamping and anemia are
shown in Fig. 4. Whole body O_2 uptake increased about 6%, a slight
but significant effect ($p < 0.05$) when the carotids were clamped
while the animals were normocythemic. Limb O_2 uptake was not changed.
With anemia, there was a 10% increase in whole body O_2 uptake which
again was not matched by the hindlimb musculature. The effect of
carotid clamping was additive to that of anemia and whole body O_2
uptake went up another 7%. With release of carotid occlusion, O_2 uptake returned to the preclamp level, whether normocythemic or anemic.
With reinfusion of red cells, the O_2 uptake decreased to the initial
control level. Limb O_2 uptake never changed.

One factor that might have increased the measured whole body O_2
uptake was increased work of the heart. That would have been consistent with the absence of any effect on skeletal muscle. Stroke work,
the product of stroke volume and mean arterial pressure, was calculated as an index to work of the heart. The results are shown in
Fig. 5. Carotid clamping did significantly increase stroke work
during normocythemia. Was this the reason that whole body O_2 uptake
increased? Although increased work of the heart may have contributed,
it could not have accounted entirely for the increased O_2 uptake.
This becomes clear when the results obtained during anemia are examined.

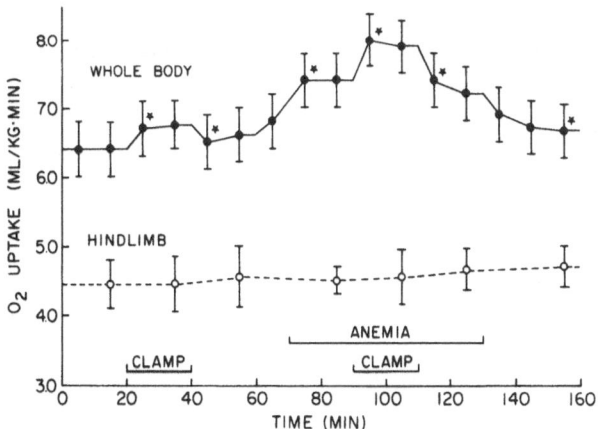

Fig. 4.   Mean values ($\pm$ SE) of whole body and hindlimb O_2 (in ml O_2/
          (kg·min)) uptake.

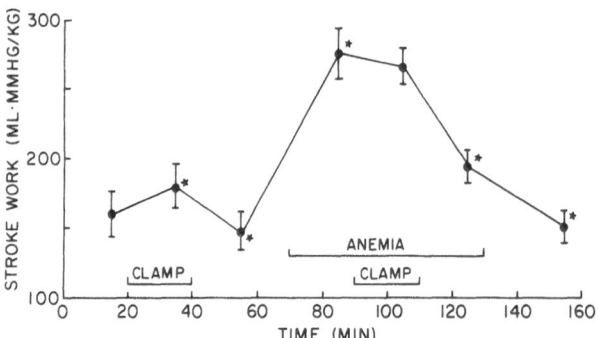

Fig. 5. Average (± SE) stroke work of the heart.

There was nearly a doubling of stroke work with hemodilution, primarily as a result of the increase in stroke volume. At the same time, O_2 uptake increased about 0.7 ml/(kg·min). Using data from Von Restorff et al. (1975) who measured the heart's O_2 uptake at a similar level of dilutional anemia, the heart could have accounted for 75% of the total increase in whole body O_2 uptake. The additional increase of 7% seen with carotid clamping in anemia, however, occurred with no change in stroke work of the heart so some other factor must have been responsible. We believe that this was probably the calorigenic action of catecholamines whose levels were raised by both baroreceptor and chemoreceptor stimulation. We have shown previously that norepinephrine infusions do not increase hindlimb O_2 uptake but do increase whole body O_2 uptake (Cain and Chapler, 1981; Chapler and Cain, 1981). The effect on whole body O_2 uptake, therefore, must have its origin in organ systems other than skeletal muscle. That would fit the results presented here. If the metabolic action of catecholamines is responsible for the increased O_2 uptake seen with carotid clamping, then it should be eliminated by ß-adrenergic receptor blockade. We are now looking to see if that was indeed the case and the first results indicate that our explanation will prove to be correct.

ACKNOWLEDGEMENT

    This work was supported by NIH Grant HL 14693 to S.M. Cain and by a grant from MRC of Canada to C.K. Chapler.

REFERENCES

Cain, S.M., and Chapler, C.K., 1978, O_2 extraction by hindlimb
    versus whole dog during anemic hypoxia, J. Appl. Physiol.:
    Respir. Environ. Exercise Physiol., 45:966.
Cain, S.M., and Chapler, C.K., 1981, Effects of norepinephrine and
    α-block on O_2 uptake and blood flow in dog hindlimb, J. Appl.
    Physiol.: Respir. Environ. Exercise Physiol., 51:1245.
Carswell, F., Hainsworth, R., and Ledsome, J.R., 1968, The effects
    of varying pressures in the aortic arch on limb resistance and
    heart rate, J. Physiol. (London), 196:38P.
Chapler, C.K., and Cain, S.M., 1981, Blood flow and O_2 uptake in
    dog hindlimb with anemia, norepinephrine, and propranolol, J.
    Appl. Physiol.: Respir. Environ. Exercise Physiol., 51:565.
Donald, D.E., and Edis, A.J., 1971, Comparison of aortic and carotid
    baroreflexes in the dog, J. Physiol. (London), 215:521.
Fan, F.-C., Chen, R.Y.Z., Schuessler, G.B., and Chien, S., 1980,
    Effects of hematocrit variations on regional hemodynamics and
    oxygen transport in the dog, Am. J. Physiol., 238 (Heart Circ.
    Physiol., 7):H545.
Grupp, I., Grupp, G., Holmes, J.C., and Fowler, N.O., 1972, Regional
    blood flow in anemia, J. Appl. Physiol., 33:15.
Restorff, W. v., Hofling, B., Holtz, J., and Bassenge, E., 1975,
    Effect of increased blood fluidity through hemodilution on
    coronary circulation at rest and during exercise in dogs,
    Pflügers Arch., 357:15.

FUNCTIONAL ALTERATION OF MEMBRANE INTEGRITY DURING GLOBAL ISCHEMIA IN PERFUSED WORKING RABBIT HEARTS

H. Rhee, and J. Cooper

Department of Pharmacology, Oral Roberts University
School of Medicine, Tulsa, Oklahoma 74171, USA

INTRODUCTION

Ischemic myocardial injury is certainly reversible following
an early reperfusion of the ischemic region. A prolonged ischemic
insult however, produces irreversible myocardial injury leading to
permanent cell death and myocardial infarction. We now have an im-
pressive body of knowledge about the characteristics of myocardial
ischemic injury and cell death in terms of cardiac biochemistry
(Allison et al., 1977; Idell-Wenger and Neely, 1977; Jennings et
al., 1978; Chua et al., 1979), pathophysiology (Jennings et al.,
1969; Neely et al., 1973; Lucchesi et al., 1976), and histology
(Jennings et al., 1975; Schaper et al., 1979).

Biochemical investigation indicates that a decrease in the
cellular energy level due to an experimental ischemia may be a key
event leading to irreversible cardiac injury and ultimate myocardial infarction. However, in many animal models cardiac contractility
decreases well before (within seconds after ischemia) any detectable changes in high-energy compounds after the experimental ischemia. Histological and pathophysiological demonstrations of structural damage to cardiac tissue after an experimental ischemia requires a rather long period of ischemia (Neely et al., 1973; Jennings
et al., 1974; Lucchesi et al., 1976). Thus it is important to determine the biochemical nature of irreversible cardiac injury and
to identify the dominant factor determining irreversibility. We know
little about the molecular mechanism of cardiac ischemic damage.
Especially, the role of the cardiac membrane in early irreversible
cardiac injury is unknown. Therefore, the primary objective of this
work was to test the hypothesis that the irreversible ischemic injury
is a secondary effect of myocardial membrane defects induced by an
experimental ischemia in the perfused, working rabbit heart.

METHODS

Rabbit Heart Preparation

White New Zealand rabbits of either sex, weighing 2 to 4 kg, were sacrificed by cervical dislocation after an injection of 400 u/kg of heparin I.V. The isolated hearts were perfused by a Krebs-Henseleit (K-H) solution (Winegrad and Shanes, 1962). For the working heart preparation the hearts were initially perfused by a modified Langendorff technique (Huang et al., 1979) until the left atrium was cannulated in order ot operate the heart in a working mode according to Neely et al. (1973). The temperature of the perfusion medium was maintained at $37^\circ C$ and oxygenated with a mixture of 95% O_2 - CO_2, pH 7.4.

The hearts were paced at 20% above their intrinsic rate by a stimulator (Grass Instruments Co., Quincy, MA, Model SS88) which delivered square wave pulses of 5 msec duration at twice the threshold voltage. The left intraventricular pressure was monitored by a teflon catheter inserted in the chamber. The first derivative of the intraventricular pressure (dp/dt) and other hemodynamic parameters were recorded continuously (Gould Inc., Cleveland, OH, Model ES1000). After a 10-min equilibration period, the heart was switched to work against a pressure of 80 cm H_2O by turning the stop-cock of the left atrium (Fig. 1). Cardiac output was estimated by measuring the aortic flow to a reservoir and coronary flow was measured by a flow probe (I.D. 1.5 mm, Gould, Inc.) which was applied to the pulmonary artery. The initial perfusion pressure was maintained at 12 cm H_2O by a variable-speed pump. In some experiments an accurate control of coronary flow was achieved by blocking the cardiac output so that the left atrial perfusion rate was equal to the coronary flow. By this procedure, the coronary flow can be reduced exactly to the level that may produce the critical reversible or irreversible ischemic injury.

Ischemic Procedure

After obtaining the control record, an ischemia was produced by reducing the perfusion rate from 7 ml/min to 0.7 ml/min. A 15-min ischemic perfusion definitely produced reversible hemodynamic changes (Fig. 2, top three panels) upon reperfusion. However, when the ischemic period continued for another 15 min, for a total of 30 min, reperfusion did not return cardiac function (dp/dt) to even 50% of the control value. This clearly indicates that there must be permanent damage in the ischemic heart (Fig. 2, bottom three panels) because the cardiac parameters did not revocer even after a reperfusion for several hours with fully oxygenated K-H solution (data not shown).

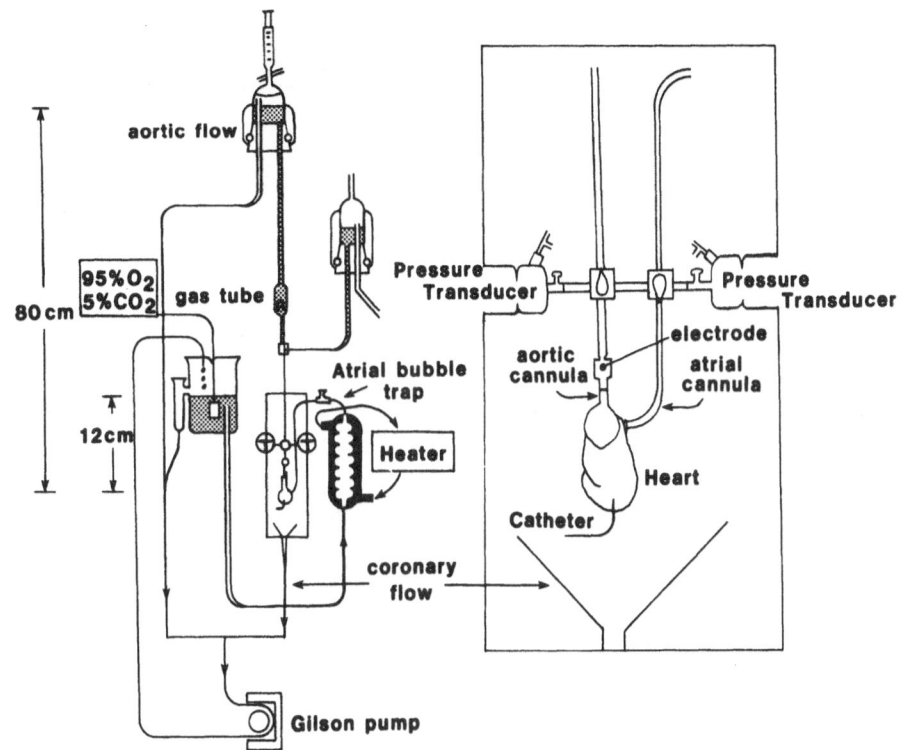

Working Heart Apparatus

Fig. 1. Perfusion apparatus for the working rabbit heart. The rabbit heart was perfused via a catheter inserted into the left atrium using a constant-speed pump at 7 ml/min. The heart had to work against an 80-cm H_2O pressure gradient and the onset of ischemia was produced by a reduction of the speed of the pump to 0.7 ml/min.

Determination of $^{86}Rb^+$ uptake and binding of ^3H-ouabain to cardiac tissue. A Rb^+ uptake experiment was carried out as described by Rhee (1981). One modification to this method was that the left ventricular tissue blocks were incubated in a final 10 ml of K-H solution instead of K^+-free K-H solution (Rhee, 1981). ^3H-ouabain binding was carried out as described by Rhee (1982) after the completion of normal and ischemic perfusion.

Preparation of cardiac sarcoplasmic reticulum (CSR): Assay of Ca^{++}- ATPase activity and $^{45}Ca^{++}$ uptake. The modified method of Harigaya and Schwarz (1969) was used to prepare CSR. Samples of the left ventricle were homogenized in 9 volumes of cold 250 $mmol \cdot l^{-1}$ sucrose by a Brinkmann polytron (PT 20 generator) for 15 seconds at half maximal speed. This procedure was repeated once more and the homogenate was centrifuged in a refrigerated centrifuge for 10 min

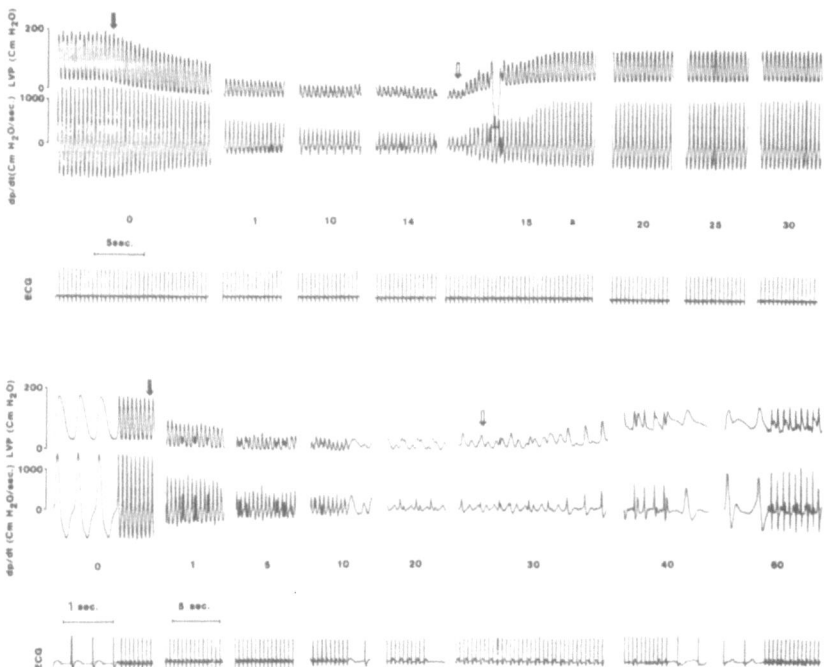

Fig. 2. Typical experiment of reversible and irreversible ischemic
         protocols in the paced, working rabbit heart. At time 0
         (closed arrow), the speed of the pump was reduced from 7
         ml/min to 0.7 ml/min. Left ventricular pressure (LVP) and
         its dp/dt were recorded during the ischemia after restora-
         tion of the pump speed to 7 ml/min at 15 min (open arrow).
         Numbers indicate time after the onset of ischemic procedure.
         The irreversible ischemic protocols (lower 3 panels) were
         identical to the above except the ischemic period was for
         30 min instead of 15 min, when a reperfusion was begun at
         open arrow.

at 14,000 x g to remove cell debris and mitochondria. The supernatant was taken up by a 0.6 $mmol \cdot l^{-1}$ KCL solution to dissolve the
contractile protein and the mixture was subsequently centrifuged at
30,000 x g for 60 min. The final pellet was homogenized in a teflon
glass homogenizer with 10 ml of cold 250 $mmol \cdot l^{-1}$ sucrose. The entire
procedure was conducted at $4^{\circ}C$.

    Ca^{++}-ATPase activity was calculated from the difference between
ATPase activity assayed in the presence of 50 $umol \cdot l^{-1}$ $CaCl_2$ in a
standard assay medium and the ATPase activity assayed in the presence

of 1 mmol·l^{-1} EDTA in the standard medium at 37°C. The standard assay medium contains 50 mmol·l^{-1} Tris (pH 7.4), 5 mmol·l^{-1} ATP and MgCl$_2$, and 100 mmol·l^{-1} KCL in a final volume of 1 ml. The reaction was initiated by the addition of 5 mmol·l^{-1} ATP with 15 to 25 ug of the CSR preparation in the presence or in the absence of 50 umol·l^{-1} CaCl$_2$. The reaction was terminated by the addition of 0.5 ml of 10% trichloroacetic acid in a mixture of CHCl$_3$ and CH$_3$OH (1:1) after a 10 min incubation period. The incubation mixture was centrifuged at 3000 x g for 10 min after the vigorous shaking on the vertex. An aliquot of the supernate was assayed for its Pi content according to the method of Fiske and SubbaRow (1925). In each assay, tubes for blank and standard inorganic phosphate (1 um) were included. Based on a protein assayed by Lowry et al. (1951), the Ca^{++}-ATPase activity was expressed in um Pi/mg CSR/hr.

For the Ca^{++}-uptake experiment, 10 to 15 ug of the CSR were incubated in a 50 mmol·l^{-1} Tris buffer solution (pH 7.4) containing 0.1 mmol·l^{-1} CaCl$_2$ and ATP, 100 mmol·l^{-1} KCl, and a trace of ^{45}Ca^{++} (S.A. 0.8 Ci/mmol) in a final volume of 2 ml at 37°C. After 2 to 5 minutes incubation, an aliquot of the incubation mixture was filtered through a pre-wet Millipore filter (HAWP, 0.45 um) under a light vacuum. After washing the filter paper with 4 ml of 250 mmol·l^{-1} cold sucrose, the CSR on the filter paper was dissolved and counted for ^{45}Ca^{++}. Based on the specific activity of ^{45}Ca^{++} and the protein content of CSR, Ca^{++} uptake was expressed as umol·l^{-1} Ca^{++} bound per mg protein for 2 min in the presence of 3 mmol·l^{-1} Tris-oxalate.

RESULTS

Effect of Ischemia and Reperfusion on Cardiac Contractility

A reduction of coronary flow from the control 7 ml/min to 0.7 ml/min (closed arrow in Fig. 2) immediately reduced the left ventricular pressure (LVP). Within a minute after the ischemic procedure, the LVP was reduced from 180 cm H$_2$O to 70 cm H$_2$O. The first derivative of the LVP (dp/dt) was also reduced after one minute from 1300 cm H$_2$O/sec to 600 cm H$_2$O/sec. Cardiac contractility under this condition was continuously reduced without any gross abnormality of the electrocardiogram for up to 15 min when reperfusion was begun again at 7 ml/min. The reperfusion partially restored the LVP and its dp/dt. Within 5 min after the reperfusion, the parameters returned to 60% and 70% of the control values, respectively, which indicates that the 15-min ischemic procedure did not leave a permanent impairment of the contractile machinery.

When the ischemic procedure was continued for 30 min, however, there were marked cardiac arrhythmias, which persisted even after the reperfusion for 30 min. Contractility (LVP and dp/dt) was usually below 50% of the respective control levels (Fig. 2),

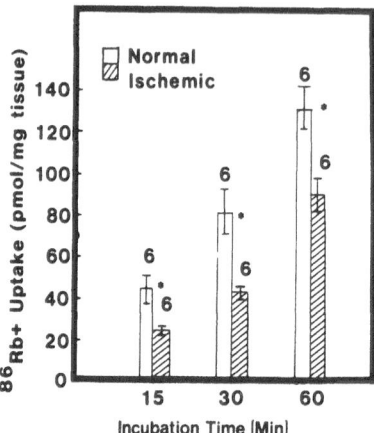

Fig. 3. Effects of an irreversible ischemia on the uptake of $^{86}Rb^+$
         in working rabbit hearts. The ischemic procedure was
         applied to the heart for 30 min when the ventricular
         slices were incubated for 15, 30, and 60 min with 0.1
         $mmol \cdot 1^{-1}$ Rb^+ and 10 uCi of $^{86}Rb^+$ at 30°C as described
         under "Methods". Vertical bars indicate standard errors
         and numbers on the bars indicate number of animals used.
         * indicates p is less than 0.05.

suggesting that the ischemic protocol for 30 min was sufficient to
cause a permanent change in the contractile process.

$^{86}Rb^+$ Uptake Experiments

      Cardiac sarcolemmal membrane activity during the early irre-
versible ischemic injury was assessed by the measurement of $^{86}Rb^+$
transport. In both control and ischemic experiments Rb^+ uptake was
time dependently increased <u>in vitro</u> (Fig. 3). Rb^+ uptake was significantly reduced after the experimental ischemic procedure in this
specific working rabbit heart.

^3H-Ouabain Binding Study and Scatchard Analysis

      Uptake of ^3H-ouabain to the heart after the control and irre-
versible ischemic procedure is summarized in Table 1. In intact
tissues uptake of ouabain was a slow process because a 120-min
incubation provided a further increase in the uptake of ouabain
compared to the value of 60 min. As the concentration of ouabain
was increased from 5.0 x 10^{-8} $mol \cdot 1^{-1}$ to 2.5 x 10^{-7} $mol \cdot 1^{-1}$, ^3H-
ouabain binding was increased in both the control and irreversibly
ischemic hearts. Specific binding of ouabain was significantly
greater (p 0.001) in the control heart than the binding of ouabain
in the irreversibly ischemic heart preparation. The nonspecific
binding of ouabain to tissues prepared from the control heart was

Table 1. Specific binding of ^{3}H-ouabain to cardiac tissue prepared from the control and irreversibly ischemic rabbit hearts[a].

Concentration of Ouabain	N[c]	Specific Binding of Ouabain (fm/mg tissue)[b]	
		Control	Irreversibly Ischemic
5.0×10^{-8} mol·l^{-1}	21	3.66 $\pm$ 0.14	3.38 $\pm$ 0.08[d]
7.5×10^{-8} mol·l^{-1}	21	5.79 $\pm$ 0.26	3.24 $\pm$ 0.12[d]
1.0×10^{-7} mol·l^{-1}	21	6.36 $\pm$ 0.21	4.06 $\pm$ 0.13[d]
2.5×10^{-7} mol·l^{-1}	21	8.21 $\pm$ 0.52	5.76 $\pm$ 0.18[d]

[a]All values represent mean and SEM after 120 min incubation.
[b]Specific binding was calculated from the difference between the total binding and nonspecific binding assayed in the presence of 5 mmol·l^{-1} nonlabeled ouabain as described in Methods.
[c]N indicates number of determination.
[d]Indicates probability is less than 0.05.

not significantly different from the nonspecific binding of ouabain in the irreversibly ischemic heart preparation (data not shown).

In order to characterize the reduced ouabain binding by the ischemia, a Scatchard (1949) analysis was performed, using the data in Table 1. The ratio of ouabain bound to free ouabain was plotted with respect to bound ouabain (Fig. 4). Both tissue blocks prepared from the control (o) and irreversibly ischemic (x) hearts showed straight monophasic Scatchard plots, which indicates there is a single homogeneous binding site in the two hearts. The number of binding sites in the control heart was 14.2 $\pm$ 1.9 fm of ouabain per mg tissue-wet weight. The dissociation constant (Kd) was 1.03 $\pm$ 0.03 nmol·l^{-1} ouabain. However, in case of the irreversibly ischemic preparation, the number of binding sites was 10.4 $\pm$ 1.0 fm/mg tissue, which is not significantly different from the number of binding sites in the control heart. The dissociation constant (Kd) in this heart was 1.39 $\pm$ 0.16 nmol·l^{-1}, which is significantly greater than the Kd of the control-heart preparation.

Effects of Ischemia on Ca^{++}-ATPase Activity and Ca^{++} Uptake in Cardiac Sarcoplasmic Reticulum (CSR)

After the completion of the specific perfusion schedule the control and ischemic hearts were subjected to prepared CSR. In the control nonischemic hearts, the Ca^{++}-ATPase activity was as high as

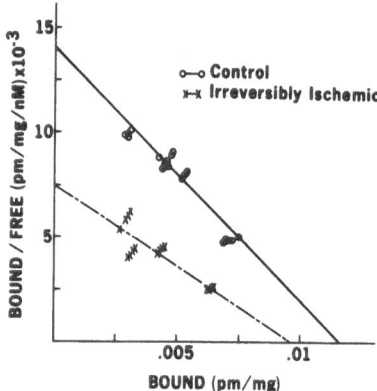

Fig. 4.   Scatchard plots of ouabain binding to tissue slices pre-
          pared from the control and irreversibly ischemic hearts.
          Using data in Table 1, Scatchard (1949) plots were con-
          structed. The control heart (solid line, with open circles)
          and irreversibly ischemic heart (broken line) have regres-
          sion lines Y = 0.014 - 1.12X and Y = 0.008 - 0.78X, respec-
          tively. nm = nmol·l^{-1}.

37 uM Pi/(mg protein/hr). The enzymatic activity was reduced significantly by the specific ischemic procedure (Fig. 5). Likewise, the
ischemia significantly reduced the Ca^{++} uptake from the control
700 nm per mg in 2 min to 320 nm/mg.

     Reduced Ca^{++} pumping activity induced by ischemia may result
in an increased cytoplasmic Ca^{++} concentration with the possible
increased Ca^{++} influx as suggested by Fleckenstein (1970). Thus, we
also examined the possible detrimental effect of isoproterenol by
promoting the influx of Ca^{++}, which may cause damage to the myo-

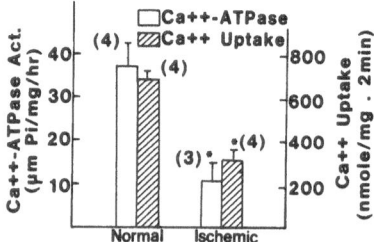

Fig. 5.   Effect of irreversible ischemia on Ca^{++}-ATPase activity and
          Ca^{++} uptake in cardiac sarcoplasmic reticulum (CSR). CSR
          was prepared as in Methods and assayed for Ca^{++}-ATPase and
          for Ca^{++}-uptake ability in the control and ischemic hearts.
          Numbers in parentheses indicate number of animals used and
          vertical bars indicate standard errors.

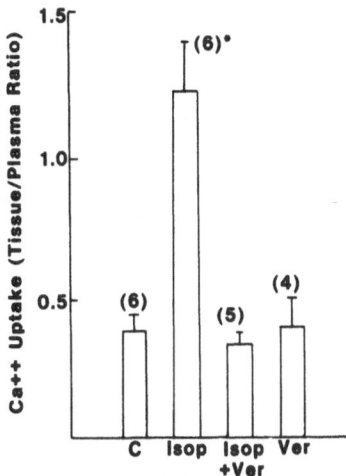

Fig. 6.   Effects of isoproterenol (ISOP) and verapamil (Ver) on
$^{45}Ca^{++}$ tissue to plasma ratio. $^{45}Ca^{++}$ was injected into
rats with or without ISOP (3 mg/kg, body weight), 6 hours
before sacrifice. Other rats were received ISOP plus Ver
(5 mg/100 g) or Ver alone. The ratio of $^{45}Ca^{++}$ in heart to
$^{45}Ca^{++}$ in the plasma was calculated. * indicates p is less
than 0.05. Other keys were as used in Fig. 5.

cardium. The ratio of $^{45}Ca^{++}$ in cardiac tissue to $^{45}Ca^{++}$ in the
plasma was under 0.5 after a 6-hour injection of $^{45}Ca^{++}$ in rats
(Fig. 6). The ratio was increased by more than 300% after an injection of isoproterenol, 3 mg/100 g, body weight. Verapamil, a potent
Ca^{++} channel blocker, inhibited the effect of isoproterenol at the
dose of 5 mg/100 kg, although the drug alone had no effect on the
$^{45}Ca^{++}$ tissue to plasma ratio.

DISCUSSION

     In these experiments, the 15 min ischemia which was followed
by a reperfusion (see Methods and Fig. 2) initially produced a
time-dependent cardiac hypofunction. A subsequent reperfusion
usually allowed a speedy recovery without severe disturbance in
cardiac rhythmicity. Had there been any irreversible damage at any
level of cardiac cellular activity in the heart, restoration of
cardiac contractility would not be possible upon reperfusion of the
globally ischemic heart. This indicates that this experimental
model and protocol are suitable for studying the cellular and biochemical events associated with the reversible myocardial ischemic
injury. However, the ischemic protocol for 30 min produced a definite
irreversible myocardial ischemic injury since reperfusion with the
fully oxygenated K-H solution for 30 min (or even for an hour, data
hour, data not shown) did not restore the heart to even 50% of the

preischemic contractile force. Therefore, this protocol also serves
as an acceptable model for the study of irreversible ischemia.

In the process of excitation and contraction coupling, the
cardiac membrane plays an important role in the regulation of Ca^{++},
Na^+, and K^+ fluxes. The effect of a specific degree of ischemia on
the integrity of the myocardial membrane is poorly understood. Sodium
and potassium activated adenosine triphosphate phosphohydrolase
(Na^+, K^+-ATPase, E.C.3.6.1.3.) is known to be a membrane-bound
enzyme (Lee and Klaus, 1971; Hokin et al., 1973; Schwartz et al.,
1975) and it is responsible for Na^+ and K^+ counter transport across
a variety of animal membranes (Skou, 1957; Hilden et al., 1974).
Therefore, it is not unreasonable for many investigators to examine
the binding of ouabain to the enzyme during the ischemia as an index
of myocardial membrane function. Bellar et al. (1976) observed that
an in vitro binding of ouabain in the infarcted canine myocardium
was considerably lower than the ouabain binding in the normal myocardium, suggesting that the characteristics of the enzyme have been
altered. The assay of Na^+, K^+-ATPase activity in vitro from the affected myocardium was well correlated with the reduced ouabain
binding, from which they suggested the possible alteration of the
plasma membrane structure. Although other studies in the same species
under comparable experimental conditions did not confirm this finding
(Kohama et al., 1971; Ku and Lucchesi, 1979), a decrease of ouabain
accumulation in ischemic tissue in comparison to the normal healthy
tissue is well documented (Beller et al., 1972, 1975; Hopkins and
Taylor, 1973).

Unfortunately, in most of the published studies, the changes
in Na^+, K^+-ATPase activity or ^3H-ouabain binding to the enzyme in
vitro were not detectable until at least a few hours after the
ischemic insult. According to our rabbit experimental protocol
(Fig. 2), the myocardium already entered permanent injury by this
time. Furthermore, in dogs undergoing circumflex coronary ligation,
Schwartz et al. (1973) did not observe a decrease in enzyme activity in the first 24 hours of ischemia. This indicates that either
there is a change in this enzyme only after severe irreversible
damage or that the sensitivity of the enzyme assay technique is not
able to detect any small changes in the enzyme activity even though
there may be such a change.

In rabbit hearts, the specific ouabain binding to the control
hearts was greater than ouabain binding to irreversibly ischemic
hearts (Table 1). According to an analysis by the Scatchard method,
the nature and characteristics of the specific binding of ouabain
in the two hearts are qualitatively different (Fig. 4). That is,
the affinity of ouabain to the irreversibly ischemic hearts was
lower (i.e., a high Kd) than the affinity of ouabain to the control
hearts. This suggests that the property of the former membrane was
altered by the irreversible ischemic procedure. Since the number of

binding sites for ouabain in the irreversibly ischemic heart was not significantly different from the control preparation, the fact that the reduction in the affinity of ouabain binding and in cation pump activity (Fig. 3) is solely responsible for the initiation of the irreversible myocardial injury remains to be tested. However, an impairment of ion pumping activity induced by ischemia suggests that there must be an irreversible physical alteration in the structure of the sarcolemmal membrane.

This interpretation may be in agreement with Jennings et al. (1975) who considered that a possible sarcolemmal defect is a primary event leading to the irreversible myocardial injury. They predicted that the sarcolemmal defect might allow the massive influx of Ca^{++} to the injured cells. They further consolidate their speculation of the involvement of $Ca^{++}{}_i$ in the process of myocardial cell injury utilizing verapamil, a calcium antagonist as shown by numbers of investigators (Fleckenstein, 1970; Caroni and Carafoli, 1981). Recently, Trump et al. (1974) have conducted an investigation on the reversibility of myocardial ischemic injury after reperfusion of the myocardium following an occlusion of the coronary artery. They observed a good correlation between an irreversible injury and the inability of mitochondrial function. They concluded that the mitochondrial damage could result from the possible activation of mitochondrial phospholipase by an influx of calcium. Of course, in order to substantiate this hypothesis, it is a prerequisite to demonstrate the early high influx of Ca^{++}, which will activate not only mitochondrial phospholipase but also many Ca^{++}-dependent proteolytic enzymes (Raab, 1953; Reimer et al., 1977). In addition to the activation of the proteases, considerable evidence suggests the direct detrimental effect of Ca^{++} in the evolution of myocardial injury leading to myocardial infarctions (Gazes et al., 1959; Fleckenstein, 1970). Our study with Ca^{++}-ATPase and Ca^{++} uptake experiments supports the above contention because reduced Ca^{++} uptake by CSR will result in an increase in the cytosolic Ca^{++} level (Fig. 6). The detrimental effect of adrenergic drugs by promoting the massive influx of Ca^{++} (Raab, 1953) is also confirmed in the present study (Fig. 6) using verapamil. From this study it suggests that ischemic cardiac injury might be due to a potential defect in both the sarcolemmal and sarcoplasmic reticular membranes, although a further molecular characterization of the ischemic injury remains to be established.

ACKNOWLEDGEMENT

This research was supported by an intramural grant from Oral Roberts University School of Medicine.

REFERENCES

Allison, T.B., Ramey, C.A., and Holsinger, J.W., Jr., 1977, Trans-
    mural gradients of left ventricular metabolites after circum-
    flex artery ligation in dogs, J. Mol. Cell. Cardiol., 9:837-852.
Beller, G.A., Conroy, J., and Smith, T.W., 1976, Ischemia-induced
    alterations in myocardial $(Na^+ + K^+)$-ATPase and cardiac glyco-
    side binding, J. Clin. Invest., 57:341-350.
Beller, G.A., Hood, W.B., Jr., and Smith, T.W., 1975, Effects of
    ischemia and coronary reperfusion on myocardial digoxin uptake,
    Am. J. Cardiol., 36:902-907.
Beller, G.A, Smith, T.W., and Hood, W.B., Jr., 1972, Altered distri-
    bution of titrated digoxin in the infarcted canine left ven-
    tricle, Circulation, 46:572-579.
Caroni, P., and Carafoli, E., 1981, Regulation of Ca^{++}-pumping ATPase
    of heart sarcolemma by a phosphorylation-dephosphorylation
    process, J. Biol. Chem., 256:9370-9373.
Chua, B., Kao, R.L., Rannels, D.E., and Morgan, H.E., 1979, Inhibi-
    tion of protein degradation by anoxia and ischemia in the per-
    fused rat heart, J. Biol. Chem., 254:6617-6623.
Fiske, C.H., and SubbaRow, Y., 1925, The colorimetric determination
    of phosphorus, J. Biol. Chem., 66:375-400.
Fleckenstein, A., 1970, in: "Calcium and the Heart", Proc. of the
    Meeting of the European Section of the International Study Group
    for Research in Cardiac Metabolism, Academic Press, London,
    pp. 135-189.
Gazes, P.C., Richardson, J.A., and Woods, E.F., 1959, Plasma catecho-
    lamine concentrations in myocardial infarction and angina pec-
    toris, Circulation, 19:657-661.
Harigaya, S., and Schwartz, A., 1969, Rate of calcium binding and
    uptake in normal animal and failing human cardiac muscle, Circ.
    Res., 25:781-794.
Hilden, S., Rhee, H.M., and Hokin, L.E., 1974, Sodium transport by
    phospholipid vesicles containing purified sodium and potassium
    ion-activated adenosine triphosphatase, J. Biol. Chem., 249:
    7432-7440.
Hokin, L.E., Dahl, J.L., Deupree, J.D., Dixon, J.F., Hackney, J.F.,
    and Perdue, J.F., 1973, Studies on the characterization of the
    sodium-potassium transport adenosine triphosphatase, J. Biol.
    Chem., 248:2593-2605.
Hopkins, B.E., and Taylor, R.R., 1973, Digoxin distribution in the
    dog's left ventricle in the presence of coronary artery liga-
    tion, J. Mol. Cell. Cardiol., 5:197-203.
Huang, W., Rhee, H.M., Chiu, R.H., and Askari, A., 1979, Re-evalua-
    tion of the relationship between the positive inotropic effect
    of ouabain and its inhibitory effect on $(Na^+ + K^+)$-dependent
    adenosine triphosphatase in rabbit and dog hearts, J. Pharmacol.
    Exp. Ther., 211:571-582.

Idell-Wenger, J.A., and Neely, J.R., 1977, Effects of ischemia on myocardial fatty acid oxidation, in: "Pathophysiology and Therapeutics of Myocardial Ischemia", Spectrum Publication, New York, pp. 227-238.

Jennings, R.B., and Ganote, C.E., 1974, Structural changes in myocardium during acute ischemia, Circ. Res., 34, 35 (Suppl. III): 156-172.

Jennings, R.B., Ganote, C.E., and Reimer, K.A., 1975, Ischemic tissue injury, Am. J. Pathol., 81:179-198.

Jennings, R.B., Hawkins, H.K., Lowe, J.E., Hill, M.L., Klotman, M.S., and Reimer, K.A., 1978, Relation between high energy phosphate and lethal injury in myocardial ischemia in the dog, Am. J. Pathol., 92:187-214.

Jennings, R.B., Sommers, H.M., Herdson, P.B., and Kaltenback, J.P., 1969, Ischemic injury of myocardium, Annal. N. Y. Acad. Sci., 156:61-78.

Kohama, A., Boyd, W.A., Ballinger, C.M., and Veda, I., 1971, Adenosine triphosphatase activities of subcellular fractions of normal and ischemic muscles, J. Surg. Res., 11:297-302.

Ku, D.D., and Lucchesi, B.R., 1979, Ischemic-induced alterations in cardiac sensitivity to digitalis, Eur. J. Pharmacol., 57:135-147.

Lee, K.S., and Klaus, W., 1971, The subcellular basis for the mechanism of inotropic action of cardiac glycosides, Pharmacol. Rev., 23:193-261.

Lowry, O.H., Rosebrough, N.J., Farr, A.L., and Randall, R.J., 1951, Protein measurement with the folin phenol reagent, J. Biol. Chem., 193:265-275.

Lucchesi, B.R., Burmeister, K.E., Lomas, T.E., and Abrams, G.D., 1976, Ischemic changes in the canine heart as affected by the dimethyl quaternary analog of propranolol UM-272 (SC-27761), J. Pharmacol. Exp. Ther., 199:310-328.

Neely, J.R., Rovetto, M.J., Whitmer, J.T., and Morgan, H.E., 1973, Effect of ischemia on ventricular function and metabolism, Am. J. Physiol., 225:651-658.

Raab, W., 1953, "Hormonal and Neurogenic Cardiovascular Disorders", Williams and Wilkins Co., Baltimore, Maryland.

Reimer, K.A., Lowe, J.E., and Jennings, R.B., 1977, Effect of the calcium antagonist verapamil on necrosis following temporary coronary artery occlusion in dogs, Circulation, 55:581-587.

Rhee, H., 1981, Accumulation of ^{3}H-ouabain in functionally different canine cardiac tissues: differential Rb^+ uptake, Br. J. Pharmacol., 73:81-86.

Rhee, H., 1982, Ouabain sensitivity of Rb^+ uptake in canine Purkinje fiber and ventricular muscle, Naunyn-Schmiedeberg's Arch. Pharmacol., 318:344-348.

Scatchard, G., 1949, The attractions of proteins for small molecules and ions, Ann. N. Y. Acad. Sci., 51:660-672.

Schaper, J., Mulch, J., Winkler, B., and Schaper, W., 1979, Ultrastructural, functional, and biochemical criteria for estimation
of reversibility of ischemic injury: A study on the effects of
global ischemia on the isolated dog heart, J. Mol. Cell.
Cardiol., 11:521-541.

Schwartz, A., Lindenmayer, G.E., and Allen, J.C., 1975, The sodium-
potassium adenosine triphosphatase: pharmacological, physiological, and biochemical aspects, Pharmacol. Rev., 27:3-134.

Schwartz, A., Wood, J.M., Allen, J.C., Bornet, E.P., Entman, M.L.,
Goldstein, M.A., Sordahl, L.A., and Suzuki, M., 1973, Bio-
chemical and morphologic correlates of cardiac ischemia. 1.
Membrane systems, Am. J. Cardiol., 32:46-61.

Skou, J.C., 1957, The influence of some cations on an adenosine
triphosphatase from peripheral nerves, Biochim. Biophys. Acta,
23:394-401.

Trump, B.F., and Mergner, W.J., 1974, Cell injury, in: "Inflammatory
Process", Vol. 1, Academic Press, New York, pp. 115-257.

Winegrad, S., and Shanes, A.M., 1962, Calcium flux and contractility
of guinea pig atria, J. Gen. Physiol., 45:371-394.

EFFECTS OF PROPRANOLOL AND EPINEPHRINE ON CAPILLARY PLASMA FILLING

AND MINIMAL INTERCAPILLARY DISTANCES IN THE RAT HEART

P. Vetterlein, and G. Schmidt

Institut für Pharmakologie und Toxikologie
Robert-Koch-Str. 40, 3400 Göttingen, FRG

Many factors may influence substrate exchange between blood
and tissue in the heart. One of these is variations in the density
of the perfused capillaries (Honig and Gayeski, in press). In this
study we examined the density of capillaries which are open to
plasma flow at rest compared to decreased or increased cardiac work.

The method is based on the infusion of fluorescent, protein-
conjugated dyes (RB 200, FITC) into the vascular system of anesthetized rats (Vetterlein et al., 1982). By rapidly freezing the heart
after defined periods of time those capillaries which had been perfused with the plasma marker during the respective period could be
identified histologically. In this way dye containing capillaries
became clearly discernible due to their intense red fluorescence in
the plasma.

The durations of the arterial dye infusion were increased
stepwise in the different experimental groups. This procedure
allowed us to gain a dynamic description of the plasma filling of
the myocardial microcirculation.

In each experiment the maximal attainable density of capillaries was also determined. Ten minutes before the fixation of the
heart FITC globulin was applied as a fluorochrome, which could be
separated from RB 200 due to its different fluorescence spectrum.

The influence of a decrease or increase in cardiac work on the
pattern of capillary plasma filling was studied on three different
groups: controls, propranolol and epinephrine treated rats, respectively.

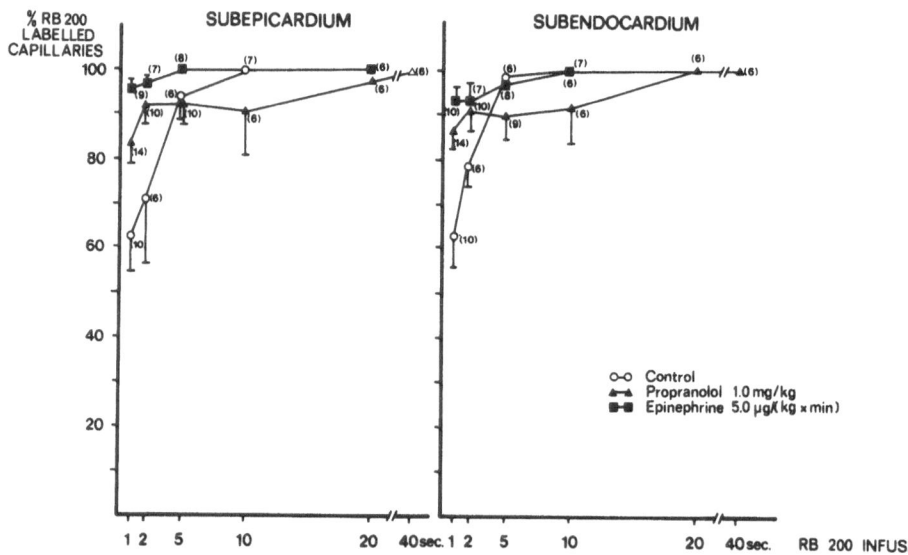

Fig. 1. Filling of the myocardial microcirculation in the rat left
ventricle with a plasma label during application of propranolol and epinephrine. Densities of RB 200-labelled
capillaries are expressed in percent of capillaries marked
with FITC for 10 min. x̄ ± SEM, number of experiments
indicated in brackets.

The effects of the drugs on coronary blood flow were also
measured in separate experiments using the microsphere technique.
In the controls a mean flow rate of 5.2 ml·min^{-1}·g^{-1} was measured,
the corresponding value of propranolol treated rats was 2.3 ml·
min^{-1}·g^{-1} and of epinephrine-treated ones, 12.6 ml·min^{-1}·g^{-1}.

Fig. 1 shows the results of the plasma labelling experiments.
When RB 200 had been infused in the control group for 1 sec, about
60% of the maximal attainable density was found labelled with this
dye. When the time of staining was increased this fraction rose in
parallel. With a marking period of 10 sec all capillaries stained
with FITC were found to contain RB 200 also. The density obtained
with plasma labelling for 10 sec had become identical with that
found with a labelling period of 10 min.

When propranolol was applied the filling was observed to be
delayed. However, when the time of labelling exceeded 20 and 40 sec,
respectively, complete filling could be observed, too. Epinephrine
on the other hand led to an acceleration of capillary plasma filling.
With both dyes identical values were found when the hearts had been
exposed to RB 200 for only 5 sec.

The density of short time labelled capillaries was expressed
as a percent of the maximal attainable one. The following evalua-

Table 1. Maximal capillary density (capillaries labelled with
          FITC-globulin for 10 min) and mean intercapillary distan-
          ces (ICD) in the myocardium of the rat treated with pro-
          pranolol or epinephrine. $\bar{x} \pm$ SEM; * significant difference
          from the control group, $p < 0.05$ (U-test).

	subepicardium		subendocardium		
	cap./mm^2	mean ICD (um)	cap./mm^2	mean ICD (um)	(n)
controls	3.530 $\pm$ 90	18.1 $\pm$ 0.2	3.260 $\pm$ 90	18.8 $\pm$ 0.2	(40)
propranolol 1.0 mg/kg	* 3.890 $\pm$ 90	* 17.2 $\pm$ 0.2	3.410 $\pm$ 100	18.4 $\pm$ 0.3	(42)
epinephrine 5.0 ug/(kg x min)	* 3.930 $\pm$ 110	* 17.1 $\pm$ 0.2	* 3.510 $\pm$ 70	* 18.1 $\pm$ 0.1	(32)

tion concerns these maximal values of capillary density, which had
been taken as 100%. Mean intercapillary (center to center) distances
(ICD) were derived from these values, based on a hexagonal model
(e.g. Bassingthwaighte et al., 1974).

     Differences were observed between the absolute, maximal den-
sities of the controls and the drug treated rats (Table 1).

     Significantly higher capillary densities were obtained in the
propranolol as well as the epinephrine treated rat subepicardium.
A mean ICD of 18.1 and 18.8 um was found in the controls, the corresponding values amounted to 17.2 and 18.4 in the propranolol and
to 17.1 and 18-1 in the epinephrine treated rats.

     These results made us wonder whether additional capillaries
had become available to the plasma marker or whether some other
effects had been responsible as a result of drug application. When
considering alternative explanations the following mechanism proved
to be of great value. In the propranolol as well as in the epinephrine treated animals the mean degree of myocardial dilation was
found to be somewhat higher than that of the controls.

     In the first attempt to study the influence of differences in
muscular contraction the cases were grouped according to their
degree of ventricular dilation. Densities were calculated for those
hearts where the luminal area of the left ventricle was less than
15%, 15% to 30% and more than 30% of total area. A presentation of
these data is given in Fig. 2. The upper part of the graph shows
the subepicardial values, the lower part the subendocardial ones.
The arrangement of the data according to the area of the left ven-

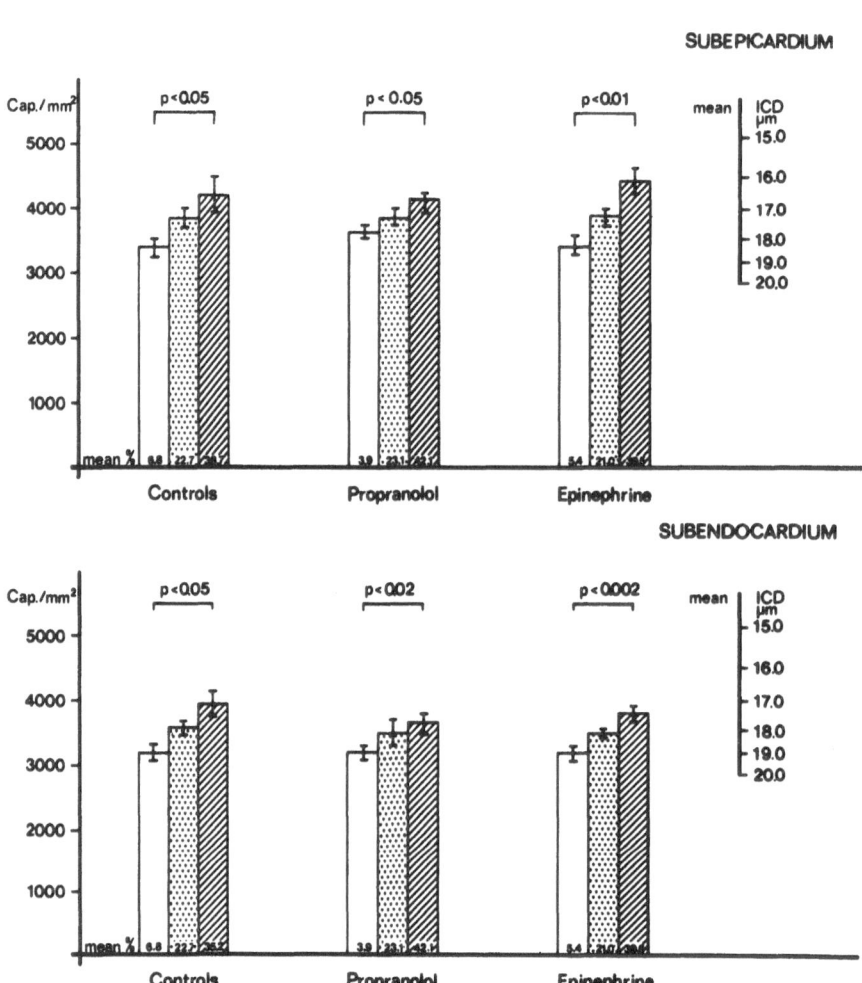

Fig. 2.   Maximal, absolute capillary densities in the subepicardium
          and in the subendocardium of the rat left ventricle in
          control experiments and during application of propranolol
          and epinephrine. The data are arranged according to three
          classes of degrees in left ventricular dilation. The error
          bar represents ± SEM.

tricular lumen was performed separately for the controls, the propranolol and the epinephrine treated rats. It is evident from this
graph that in each group the contracted hearts showed the lowest
capillary densities whereas the dilated ones showed significantly
higher densities. This trend could be observed in all groups.

The most likely explanation of this effect involves changes in
fiber dimensions due to distension of the ventricular wall. Wright
and Hudlicka (1981) and Rakusan (1971) pointed out that differences
in the degree of contraction may influence the capillary density
and may be responsible for discrepancies between reported values of
density in muscular tissue. Ventricular fibers are expected to
lengthen and thin with increasing wall dilation (Hort, 1960;
Sonnenblick and Skelton, 1973). Thinning of fibers on the other hand
would cause the surrounding capillaries to move closer to each other,
thereby decreasing their ICD. If this conception is correct fiber
densities would have increased in parallel, and capillary/fiber
relations would have remained constant.

Further experiments were performed in which the intravasal
space was again labelled with FITC-globulin. In addition the intravital application of RB 200, conjugated with myoglobin permitted us
to demonstrate the volume of extracellular space. Due to its low
molecular weight this protein rapidly leaves the capillaries and
diffuses into the extracellular, but not into the intracellular
space. In this way the borderlines of the muscular fibers become
clearly visible and capillaries as well as fibers can be counted in
the same area.

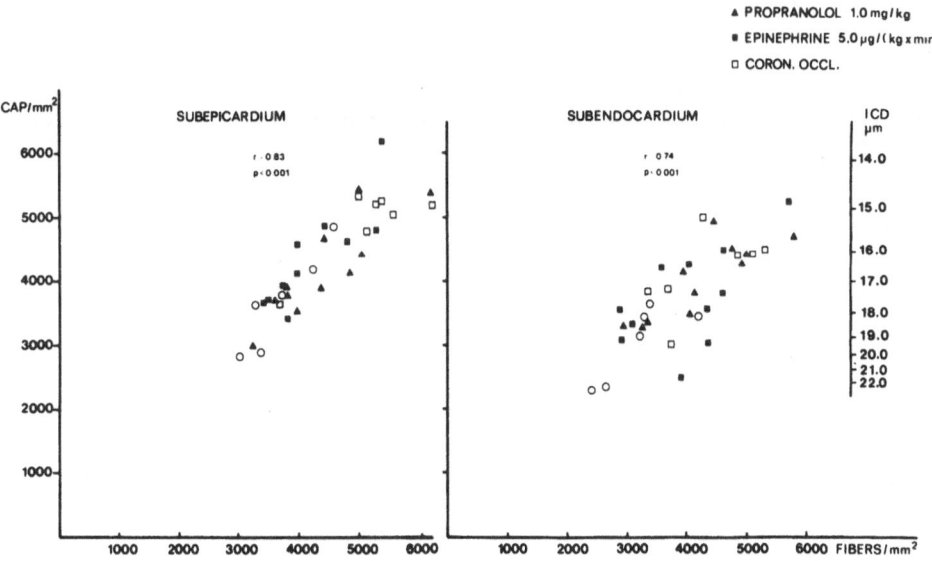

Fig. 3.   Correlation of capillary vs. fiber density in the rat left
          ventricle during control conditions, application of pro-
          pranolol and epinephrine and coronary occlusion, as indi-
          cated for each experiment.

The density of these structures was determined additionally in hearts which showed a rather extreme dilation of the left ventricle due to the occlusion of a coronary artery.

The results of these experiments are summarized in Fig. 3. This graph shows the correlation of capillary versus fiber density, indicated separately for each experiment.

Capillary densities covered the wide range of 3.000 to 5.000 cap./mm^2 in the subepicardium and of 2.500 to 5.000 cap./mm^2 in the subendocardium. Controls were at the lower end of the range and the drug-treated hearts at the higher end. The highest values were obtained in the coronary occluded organs.

Fiber densities which were concomitantly counted showed corresponding changes. The hearts with low capillary counts revealed low fiber densities and those organs with high capillary counts showed high densities of muscular fibers. The correlation coefficient for the pooled data was significantly different from zero (see Fig. 3). This result strengthens the view that the differences in maximal capillary densities observed in this study were due to changes in the degree of contraction of the muscular fibers.

In conclusion we have found that in the rat heart at rest, capillary plasma filling is completed within a few sec and so filling was observed to be delayed in propranolol and accelerated in epinephrine treated rats. The absolute, maximal attainable density was found to be higher in the drug treated rats than in the controls. This difference could be attributed to changes in the degree of fiber lengthening, which influence the distances between the surrounding capillaries.

ACKNOWLEDGEMENT

This work was supported by the "Deutsche Forschungsgemeinschaft, SFB 89, Kardiologie Göttingen".

REFERENCES

Bassingthwaighte, J.B., Yipintsoi, T., and Harvey, R.B., 1974, Microvasculature of the dog ventricular myocardium, Microvasc. Res., 7:229.

Honig, C.R., and Gayeski, T.E.J., Capillary reserve and tissue O_2 transport in normal and hypertrophied hearts, Basic Res. Cardiol., in press.

Honig, C.R., and Gayeski, T.E.J., in press, Capillary reserve and tissue O_2 transport in normal and hyertrophied hearts, NIH Symposium, Workshop on "Left Ventricular Hypertrophy", Raven Press, New York.

Hort, W., 1960, Makroskopische und micrometrische Untersuchungen am
    Myocard verschieden stark gefüllter linker Kammern, Virchows
    Arch. path. Anat., 333:523.
Rakusan, K., 1971, Quantitative morphology of capillaries of the
    heart. Number of capillaries in animals and human hearts under
    normal and pathological conditions, in: "Methodological
    Achievements of Experimental Pathology", E. Bajusz and G.
    Jasmin, eds., Karger, Basel.
Sonnenblick, E.H., and Skelton, C.L., 1974, Reconsideration of the
    ultrastructural basis of cardiac length-tension relations,
    Brief review, Circ. Res., 35:517.
Vetterlein, F., dal Ri, H., and Schmidt, G., 1982, Capillary
    density in rat myocardium during time plasma staining, Am. J.
    Physiol., 242:H133.
Wright, A.J.A., and Hudlicka, O., 1981, Capillary growth and
    changes in heart performance induced by chronic bradycardial
    pacing in the rabbit, Circ. Res., 49:469.

A COMPARISON OF THE METHODS FOR ASSESSMENT OF MYOCARDIAL CAPIL-

LARITY

K. Rakusan[1] and Z. Turek[2]

[1]Department of Physiology, School of Medicine, University
of Ottawa, Ottawa, Ontario, Canada
[2]Department of Physiology, Faculty of Medicine
University of Nijmegen, Nijmegen, Netherlands

INTRODUCTION

In all models of the O_2 supply to the tissue, starting from
the classical model of Krogh (1919), one of the principal input
variables is the maximal diffusion distance, i.e. the longest
distance from the capillary. This is equal in the Krogh model to
the radius of the tissue cylinder. In most models, it is assumed
that the maximal diffusion distance is equal to half of the intercapillary distance.

Intercapillary distance has been estimated on the surface of
the beating heart by cinematographic methods. These measurements,
however, are limited to the most superficial layer of the heart
which is hardly representative of the cardiac tissue. An alternative method involves the evaluation of capillary supply by histometric methods (Rakusan, 1971).

The purpose of this presentation is to critically evaluate the
available histometric methods for the estimation of the capillary
supply to the heart.

METHODS

The study is based on morphometric analysis of 18 rat hearts
with a wide range of cardiac weight obtained from 10 normal rats
and 8 rats with cardiac hypertrophy (SHR, rats with spontaneous
hypertension, see Fig. 1). Standard histological techniques were
used for preparation of the tissue samples: the heart was fixed by

411

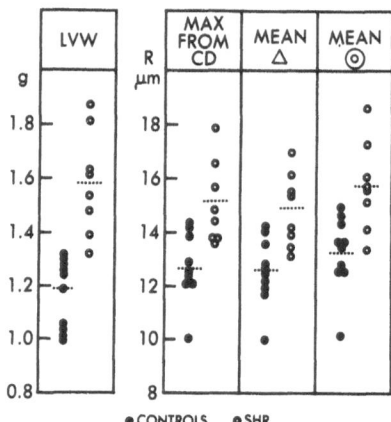

Fig. 1.   Left ventricular weight and radius of the Krogh cylinder
          (R). R was calculated from the capillary density (CD) and
          from measurements based on the triangular net method ($\triangle$)
          or method of concentric circles ($\odot$) (basic data). SHR =
          rats with spontaneous hypertension.

retrograde perfusion with 1.5% glutardehyde in cacodylate buffer.
It was removed after 5 min of perfusion, its ventricles were separated from atria and weighed. The right ventricles were discarded and
from each left ventricular wall a few cubical samples were taken in
its midportion. They were oriented in order to obtain the sections
perpendicular to the long axis of the muscle fibers. Subsequently,
the tissue samples were postfixed for 1 hour in 2% O_sO_4 and
150 mmol·l^{-1} sodium cacodylate, washed overnight, dehydrated and
finally embedded in Epon.

     Photomicrographs were taken of 2 um toluidine blue-stained
sections originating from the subendocardial region of the left
ventricular free wall. From each heart, 5-7 photomicrographs were
taken each corresponding to 28500 um^2. Fields used for morphometric
analysis contained cross-sectioned muscle cells and capillaries and
they were free of larger vessels. All the distances were measured
using a MOP 3 (Zeiss) image analyzer, coupled with an Apple II
computer for subsequent numerical analysis.

     First, traditional indicators of the capillary supply were
estimated, namely the capillary density (CD), based on direct
counting of the capillaries in tissue cross-sections and the
percentage of the tissue area occupied by capillaries as derived by
the point counting method. The average capillary radius (r) was
then calculated from the total area of tissue occupied by capillaries and the number of capillaries. The mean radius of the tissue
Krogh's cylinder (R) was estimated from the capillary density. A
minimal value of R is obtained if it is calculated as half of the

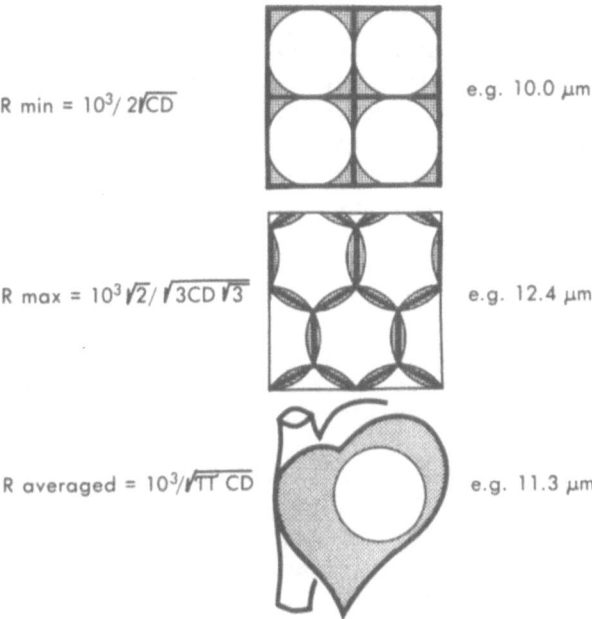

$R \min = 10^3/ 2\sqrt{CD}$ e.g. 10.0 μm

$R \max = 10^3 \sqrt{2}/ \sqrt{3CD \sqrt{3}}$ e.g. 12.4 μm

$R \text{ averaged} = 10^3/\sqrt{\pi \ CD}$ e.g. 11.3 μm

Fig. 2.   Three concepts of calculation of the radius of the Krogh
          cylinder (R) as calculated from the capillary density (CD).

distance between the capillaries arranged in a regular square array
(R min = $10^3/2 \sqrt{CD}$). In this case, however, a sizeable tissue area
is neglected (see top portion of Fig. 2). An alternative method is
to calculate the maximal values of the radius of the Krogh cylinder
when each capillary is considered to be a center of regular hexagon
(R max = $10^3\sqrt{2/} \sqrt{3 \ CD\sqrt{3}}$, Fig. 2 middle portion). In this case, an
overlapping occurs. Finally, it is possible to assume that the total
area of the tissue cross-section consists of circles with radii
equal to those of the Krogh cylinder. Then, after dividing the total
surface R area by the number of capillaries, the average circle
surface is obtained, from which the average radius can be calculated.
(R ave = $10^3/\sqrt{\pi CD}$, bottom part of Fig. 2).

     Tissue oxygenation, however, is not influenced only by the
mean values of the radius of the Krogh's cylinder but also by its
variability, i.e. by the inhomogeneity of the capillary spacing
(Turek and Rakusan, 1981). Two methods for the estimation of the
variability of the capillary spacing are currently available. In
the first approach, the distribution of the tissue in different
distances from the nearest capillary is first determined by the
method of concentric circles (Loats et al., 1978) from which the
degree of inhomogeneity of the capillary net is subsequently calculated (Turek and Rakusan, 1981). A second approach is a direct
measurement of the distances between the capillaries connected in a

triangular net as proposed by Renkin and coworkers (1981). Both
methods (method of concentric circles and triangular net method)
were applied to the same micrographs on which the capillary density
and the total area of capillaries were previously determined. In
contrast to our previous computations (Turek and Rakusan, 1981),
the mean radius of the Krogh cylinder was not estimated from the
capillary density but calculated from the data obtained by the
method of concentric circles and from the mean capillary radius.

To summarize: in each micrograph from hearts of various weights,
the following indices of the tissue capillarization were determined:

a) Capillary density (CD), i.e. the number of capillaries per
mm^2 from which the radius of the Krogh cylinder was subsequently
calculated in its three possible variants (minimal, averaged and
maximal, see Fig. 2),

b) the relative area of the tissue occupied by capillaries and
which knowing the capillary density, allows the calculation of the
mean capillary cross-section and mean capillary radius.

c) The distribution of intercapillary distances as derived from
the results obtained by the method of concentric circles and the
mean capillary radius.

d) The distribution of intercapillary distances as by the
method of triangular nets.

RESULTS

The mean values of the radius of the Krogh cylinder calculated
as R max from the capillary density are compared with similar values
obtained by the triangular net method and by the method of concentric circles in Fig. 1. The distribution of intercapillary distances, however, may be approximated by the log normal distribution
(Turek and Rakusan, 1981; Renkin et al., 1981). This distribution
is better described by its median values and by the log standard
deviation. Fig. 3 summarizes both parameters on the same micrographs obtained by the triangular net method and by the method of
concentric circles. The triangular net yields lower median and mean
values but a higher degree of inhomogeneity (log standard deviation) than the results obtained from the same hearts by the method
of concentric circles. This difference may be explained by the geometrical considerations inherent in these methods. Estimates based
on the method of concentric circles are programmed as though the
capillaries were distributed in a net of equilateral triangles. The
most distant point from the nearest capillary would then lie in the
center of this triangle. This distance is therefore longer than
half of the site of the triangle, i.e., half of the distance between
capillaries connected into a triangle net. Consequently, it should
equal half of the site of triangle multiplied by $2/3\sqrt{3}$.

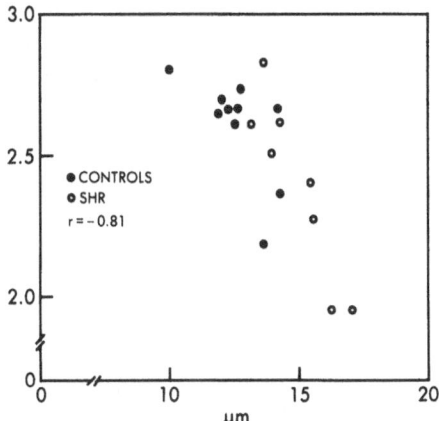

Fig. 3. Characteristics of log-normal distributions. Median values
of R and σ log from the measurement based on the triangular net method and method of concentric circles.
For explanation of the symbols see Fig. 1.

The difference in the values of inhomogeneity may be explained
by the deviations from the equilaterality of triangles and also a
certain degree of subjectivity in a construction of triangular net.

The following three graphs summarize the relationship among
the various indicators of tissue capillarity. Fig. 4 depicts the
correlation between half of the intercapillary distance obtained by
the method of triangular nets and the linear index derived from the
measurements of capillary area by the point counting; Fig. 5
compares the results obtained by the method of triangular net and
the results based on the mean capillary density and, finally Fig. 6

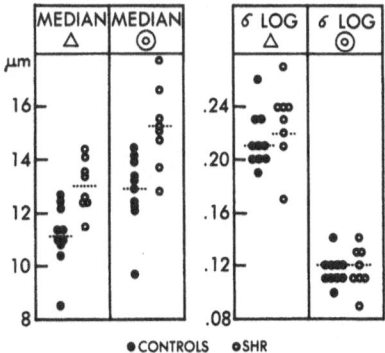

Fig. 4. Correlation between the linear factor derived from the %
of tissue occupied by capillaries and half of the intercapillary distance measured by the triangular net method
($\sqrt{A\ Cap\ (\%)}$ versus 1/2 ICD).

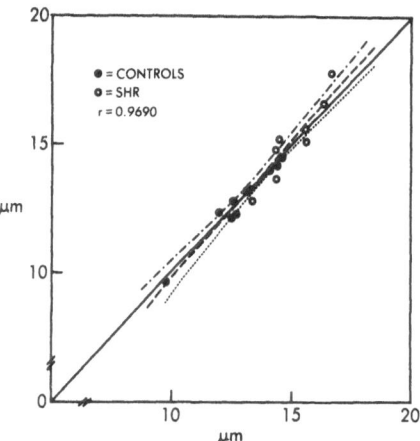

Fig. 5.   Correlation between the R's calculated from the capillary
          density and half of the intercapillary distance measured
          by the triangular net method. (R as calculated from CD
          versus 1/2 ICD (▲).

depicts the comparison between the medians obtained by the method
of triangular net after the correction for a geometrical factor and
the medians calculated from the data obtained by the method of
concentric circles. It thus follows that the degree of correlation
between the percentage of tissue occupied by capillaries and directly
measured intercapillary distances is much lower than the remaining
correlations. This probably reflects variation in the size of capillary lumen as well as artefacts due to histological tissue processing. On the other hand, the remaining methods yield values which are
closely correlated.

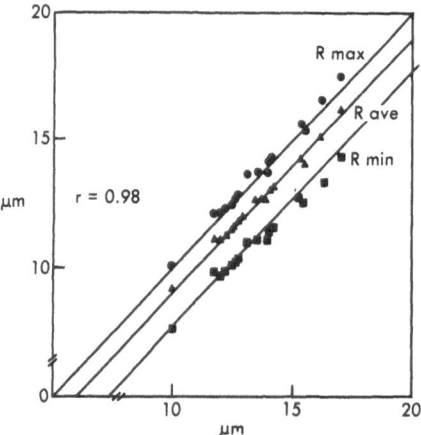

Fig. 6.   Correlation between R derived from the method of concen-
          tric circles and half of the intercapillary distance
          measured by the triangular net method and corrected by the
          geometrical factor. (R_θ versus 1/2 ICD x 2/3 $\sqrt{3}$ (medians)).

DISCUSSION AND CONCLUSION

The results seem to indicate that measuring the tissue area occupied by capillaries yields more variable results than the remaining methods, which provide comparable values. However, measuring the capillary density enables us to calculate the mean radius of the Krogh's cylinder only and provides no estimate of the inhomogeneity of the capillary spacing which is an additional oxygen determinant. This may be calculated from the remaining two methods, i.e. methods based on triangular net and concentric circles. The former method yields higher values of variability (σ log) for the reasons discussed above. In addition, the method of triangular net requires the direct measurement of the intercapillary distances. This is more tedious and time-consuming than the method of concentric circles, which is a special case of the point counting method.

It is our opinion that measuring only the capillary density does not provide sufficient data to study the effect of capillarization on the oxygen supply to the tissue. Some index of the inhomogeneity of capillary spacing, i.e. variability of the radii of tissue cylinders, provides important additional information. Another important feature, up to now neglected, is the character of the diffusion distance itself. So far, we did not distinguish between the portion of the diffusion distance occupied by the myocytes and the portion lying in the extracellular space. Furthermore, at the same time, it is well known that the oxygen consumption of the two tissue components differs considerably. Finally, even within the portion of the diffusion distance occupied by myocytes, special attention should be paid to the spatial distribution of the oxygen sinks, i.e. mitochondria, which may play a role in the terminal energy supply (Mainwood and Rakusan, 1982).

In conclusion, we compared several means of evaluation of the capillary supply to the heart. The following methods, arranged in order of preference, were assessed: method of concentric circles, method of triangular net, counting the capillary density and estimating the relative surface area occupied by capillaries.

REFERENCES

Krogh, A., 1919, The number and distribution of capillaries in muscles with calculations of the oxygen pressure head necessary for supplying the tissue, J. Physiol. (Lond.), 52:409.

Loats, J.T., Sillau, A.H., and Banchero, N., 1978, How to quantify skeletal muscle capillarity, in: "Oxygen Transport to Tissue III", I.A. Silver, M. Erecinska, H.I. Bicher, eds., Plenum Press, New York and London, p. 41.

Mainwood, G., and Rakusan, K., 1982, A model for the intracellular energy transport, Can. J. Physiol. Pharmacol., 60:98.

Rakusan, K., 1971, Quantitative morphology of capillaries of the
    heart, in: "Functional Morphology of the Heart", E. Bajusz, G.
    Jasmin, eds., Karger, Basel, p. 272.
Renkin, E.M., Gray, S.D., Dodd, L.R., and Lia, B.D., 1981, Hetero-
    geneity of capillary distribution and capillary circulation in
    mammalian skeletal muscles, in: "Underwater Physiology VII",
    A.J. Bachrach, M.M. Matzen, eds., Undersea Medical Soc.,
    Bethesda.
Turek, Z., and Rakusan, K., 1981, Log-normal distribution of inter-
    capillary distance in normal and hypertrophic rat heart as
    estimated by the method of concentric circles: its effect on
    tissue oxygenation, Pflügers Arch., 391:17.

Supported by the Medical Research Council of Canada.

OXYGEN RADICALS

DELETERIOUS EFFECTS OF OXYGEN RADICALS ON REOXYGENATED MYOCARDIAL

CELLS

Y. Gauduel, and M.A. Duvelleroy

Laboratoire de Biophysique, Fernand Widal Hospital, 200
rue du Faubourg, St. Denis, 75010 Paris, France

INTRODUCTION

Oxygen deprivation induces severe damage in cardiac muscle.
The ultrastructural lesions, which include mitochondrial and cellular swellings and structural defects in cell plasma membranes, are
accentuated by the duration of the hypoxic period (Hearse et al.,
1973).

An oxygen paradox phenomenon is observed when the hypoxic myocardium is suddenly submitted to reoxygenation (Hearse et al., 1973;
1978; Gauduel et al., 1979; Ganote et al., 1979). Irreversible
ultrastructural lesions are the consequence of the toxic effects of
molecular oxygen (Feuvray and de Leiris, 1975; Ganote et al., 1979).
The mechanisms of these cytotoxic effects are still imperfectly
understood but are of fundamental importance in cardiology.

The purpose of the present study is to examine a hypothesis
which states that the survival of the hypoxic heart submitted to
reoxygenation involves a complicated interplay between the biological generation of very reactive oxygen free radicals and the ability,
by the cells, to control the production of these chemical species.
Selective extrinsic scavengers of free oxygen radicals were used:
glutathione reduced form (GSH), catalase, superoxide dismutase (SOD)
and α-tocopherol. These compounds are powerful tools for use in
dissecting some of the complexities which characterize the relations
between free oxygen radicals and cellular components.

METHODS

Male rats of the Whistar strain, weighing 300 to 350 g and fed ad libitum, are used. Hearts, rapidly excised from ether anesthetized animals, are washed and arrested with cold isotonic saline solution. Retrograde perfusion through the aorta at 90 cm H_2O of after load is performed according to the technique of Langendorff (1895). The pulmonary artery is incised to allow complete coronary drainage. The standard perfusion fluid is a modified Krebs-Henseleit buffer. During the hypoxic perfusion, the hearts are K^+ arrested by increasing the concentration of potassium from 5.5 to 16 $mmol \cdot l^{-1}$ with a corresponding decrease in Na^+ concentration. During all experimental conditions, the high potassium concentration prevented the occurrence of cardiac arrhythmias. The aerobic perfusate is equilibrated with 95% O_2 and 5% CO_2. Arterial Po_2 and pH are respectively 670 mm Hg and 7.4. The anaerobic phase is produced with an electrolyte solution with N_2/CO_2 (95%/5%). Aortic oxygen pressure is less than 5 mm Hg. The whole apparatus is included in a thermostatic chamber at 37°C.

The hearts are perfused in normoxic conditions for a 15 minutes equilibration period with Krebs-Henseleit buffer containing glucose (11 $mmol \cdot l^{-1}$). The hearts are then submitted to hypoxia for 60 minutes. After this period, the cardiac muscle is reoxygenated for 10 or 20 minutes.

Coronary flow is monitored and samples of the coronary effluent are collected at specified intervals for analysis. At the end of experiments, the hearts are blotted and weighed.

Drug treatment is given during hypoxia or reoxygenation. Glutathione, reduced form (Sigma G 4251) is administered in the perfusion medium at 0.5 $mmol \cdot l^{-1}$. Catalase (EC 1,11,1,6 Boehringer) from beef liver is pulse-administered during hypoxia (20 10^3 Iu every 6 minutes). Superoxide dismutase (EC 1,11,1,1 Sigma) is pulse-administered through the aortic cannula at 280 Iu every 6 minutes during the entire hypoxic period (total injection 1 mg). α-tocopherol (Sigma) is administered in the perfusion medium during the reoxygenation (total perfusion 70 mg).

Anaerobic energy production (ATP) in the absence of exogenous substrate is estimated from the lactate production, assuming that all lactate is derived from endogenous glycogen and that one glucose equivalent will yield 3 ATP molecules.

Lactate production, expressed as $nmol \cdot min^{-1} \cdot g^{-1}$ wet weight, is calculated from coronary flow and lactate concentration in the venous effluent. Lactate is measured according to the method of Gutmann and Wahlefeldt (1974).

The release of macromolecules in the venous effluent is also
studied to define the integrity of the cardiac cells. Creatine
kinase activity (CK) is determined with the optimized method of
Siegel and Cohen (1974), using a Perkin Elmer dual beam spectrophotometer equipped with thermostable cuvettes and an enzymatic calculator (Perkin-Elmer 5-100). The catalase activity is assayed according to the method of Aebi (1974) and expressed as Bergmeyer units.

The enzymes and chemical reagents used are of the first grade
and obtained from Boehringer Mannheim and Sigma.

All the results are expressed as mean $\pm$ standard error of the
mean (SEM). P values are calculated by an unpaired Student's t-test
and $P < 0.05$ is taken as the level of significance.

RESULTS

The effect of oxygen deprivation on lactate production is shown
in Fig. 1. This production is quickened in the first 30 minutes of
hypoxia and secondarily depressed between 30 and 60 minutes of perfusion.

In comparison with untreated hearts, glutathione reduced form
and superoxide dismutase limit lactate production during all the
continuance of the oxygen deprivation.

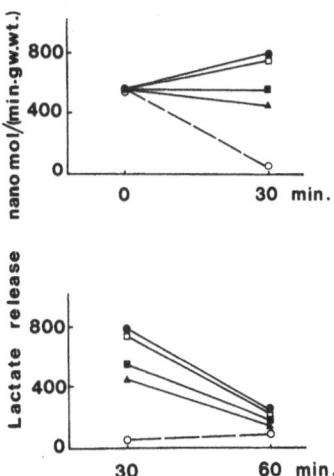

Fig. 1. Lactate production under a variety of hypoxic conditions.
Each point represents the mean of five or six experiments.
O Normoxic hearts (6). Hypoxic hearts: ● untreated (6),
▲ + GSH (5), □ + catalase (5), ■ + SOD (5) (in nmol/
(min·g w.wt.)).

Catalase does not influence the glycolytic activity of hypoxic hearts.

The results of anaerobic ATP production are summarized in Table 1. In different groups of hearts, glycolytic ATP production is increased under hypoxic conditions. But the energetic production is severely depressed between 30 and 60 minutes of hypoxia. In consequence, net rate of ATP utilization is depressed if hypoxia is longer than 30 minutes. Superoxide dismutase and GSH limit the glycolytic ATP production and ATP utilization during hypoxic period. Exogenous catalase does not influence the energy production of oxygen deprivated myocardium.

Table 1. Total glycolytic ATP production and net rate of ATP utilization under a variety of hypoxic conditions.

Conditions	Total glycolytic ATP production (umol ATP/g· wet weight)	Net rate ATP utilization (umol/min/g· wet weight)
Hypoxia (30 min)		
1 Untreated n = 6	31 + 3	1 + 0.09
2 GSH n = 5	22 + 2 *	0.75 + 0.07 *
3 SOD n = 5	25 + 2 *	0.83 + 0.05 *
4 Catalase n = 5	29 + 4 ns	0.98 + 0.08 ns
Hypoxia (60 min)		
1 Untreated n = 6	56 + 6	0.57 + 0.06
2 GSH n = 5	36 + 4 *	0.30 + 0.03 *
3 SOD n = 5	41 + 3 *	0.36 + 0.02 *
4 Catalase n = 5	51 + 5 ns	0.47 + 0.05 ns

(2), (3), (4) vs (1); * $p < 0.05$; ns: not significant

The integrity of cardiac cells can be estimated by the level of enzyme released in the venous effluent. Table 2 shows the patterns of creatine kinase and catalase released during 60 minutes of norm-oxic and hypoxic perfusions.

Untreated hypoxic hearts exhibit a biphasic CK release. An initial peak appears after 30 minutes and a second phase after 60 minutes of oxygen deprivation (Fig. 2).

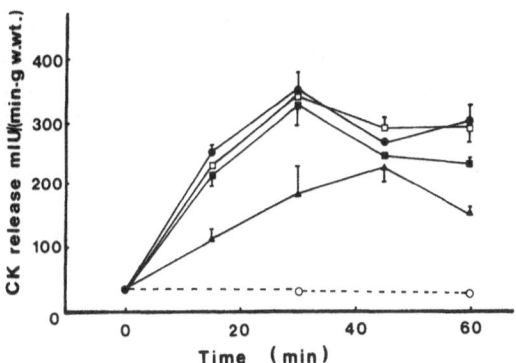

Fig. 2. Creatine kinase release kinetics from isolated perfused rat
       hearts following the onset of hypoxia. Each point represents
       the mean ± SEM of five or six experiments.
       ○ Normoxic hearts (6). Hypoxic hearts: ● untreated (6),
       ▲ + GSH (5), □ + catalase (5), ■ + SOD (5) (in m IU/
       (min·g w.wt.)).

Table 2. Enzymes release in hearts perfused under normoxic and
         hypoxic conditions.
         (Each value represents the mean ± SEM of six hearts.)

Conditions	Enzymes $(mIU \cdot min^{-1} \cdot g^{-1})$	Time (minutes)				
		0	15	30	45	60
Normoxia	1 CK	35 ±2	–	38 ±4	–	30 ±2
	2 catalase	15 ±3	13 ±2	17 ±3	14 ±1	15 ±3
		ns		*		*
Hypoxia	3 CK	35 ±2	253 ±6	354 ±27	270 ±14	304 ±34
	4 catalase	25 ±3	113 ±7	146 ±24	176 ±32	187 ±37
		ns	*	*	*	*

(3) vs (1) and (4) vs (2); *: p < 0.05; ns: not significant

     Hypoxia also increases the pattern of catalase released in the
coronary sinus but this kinetic is monophasic (Table 2).

During hypoxia, pulsed administration of catalase and superoxide dismutase does not significantly modify the kinetc of CK
release (Fig. 2). In contrast, glutathione reduced form delays the
first phase of CK release up to 45 minutes of oxygen deprivation.

In untreated hearts submitted to reoxygenation after 60 minutes
of hypoxic perfusion, CK release is greatly increased (Fig. 3), an
effect previously reported by several authors (Hearse et al., 1973;
Ganote et al., 1975; Gauduel et al., 1979). This paradoxal effect of
molecular oxygen is Po_2 dependent and is completely prevented when
arterial Po_2 is lower than 150 mm Hg (Fig. 4).

Glutathione reduced form, catalase and superoxide dismutase are
effective in delaying or decreasing the CK release induced by high
arterial Po_2 (670 mm Hg).

These scavengers of oxygen free radicals limit significantly
the total CK released during the 10 minutes of reoxygenation (Fig. 4).

The oxygen paradox phenomenon can be largely prevented if
α-tocopherol is present during the reoxygenation (Fig. 5). When the
reoxygenation is continued for 20 minutes, the creatine kinase
release increases progressively, but α-tocopherol inhibits the
development of abrupt alteration of myocardial cells during the post
anoxic period.

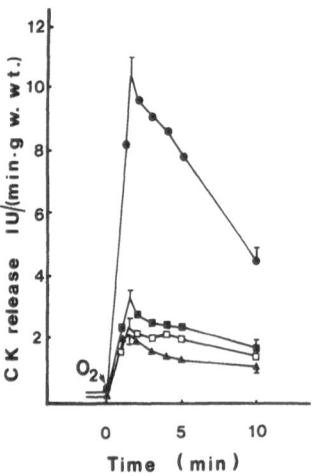

Fig. 3. Effect of reoxygenation on creatine kinase release from iso-
        lated perfused hearts previously submitted to hypoxia (60
        minutes). Each point represents the mean of five or six
        experiments (in IU/(min·g w.wt.).
        ● untreated (6), ▲ GSH (5), □ catalase (5), ■ SOD (5).

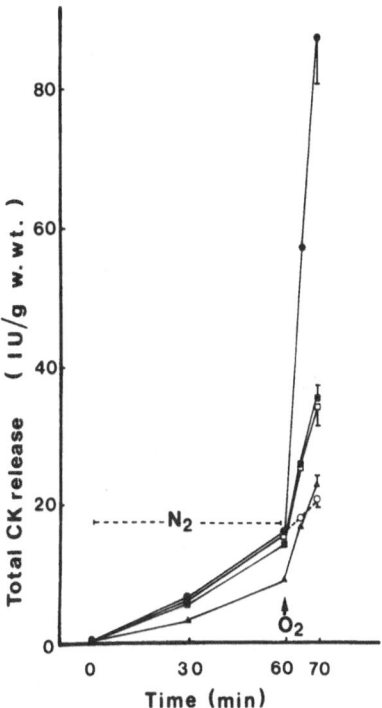

Fig. 4. Cumulated levels of creatine kinase released during hypoxia
(60 minutes) and reoxygenation (10 minutes) at high Po_2
(670 mm Hg) or low Po_2 (150 mm Hg).
● untreated hearts, Po_2: 670 (6) - ○ untreated hearts,
Po_2: 150 (6) - ▲ GSH, Po_2: 670 (5) - □ catalase, Po_2: 670 (5),
■ SOD, Po_2: 670 (5).

DISCUSSION

This study, in accordance with previous works (Hearse et al.,
1978; Gauduel et al., 1979; Ganote et al., 1979) provides evidence
that a first phase of CK release appears after 30 minutes of oxygen
deprivation and that a broad plateau occurs in the first 60 minutes
of hypoxic perfusion. The release of catalase, which is an enzyme
located in the matrix of the mitochondria (Nohl and Hegner, 1978a),
indicates that the cellular alterations produced by hypoxia are not
only localized in cytosol but also in mitochondrial matrix and
membranes.

The glutathione reduced form is used in this study with a concentration which cannot induce swelling or lysis of mitochondria
(Hunter et al., 1964). This thiol compound which may have a scavenging potential in ionic environment of cells (Pryor, 1976; Kosower
and Kosower, 1976), presents a hydrophylic protective mechanism
during the oxygen deprivation. The GSH exerts a direct effect on

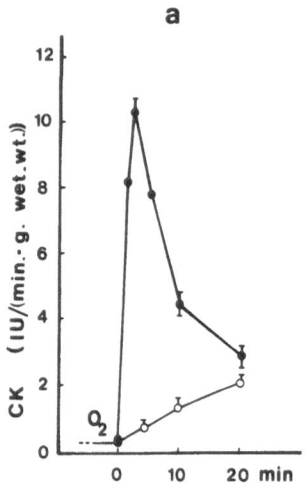 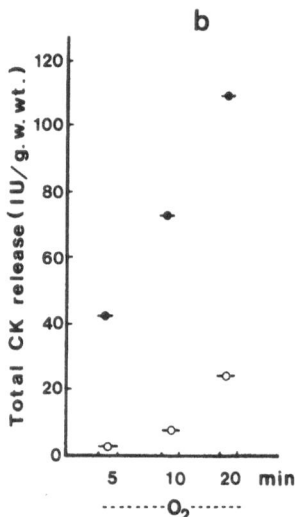

Fig. 5: Effect of α-tocopherol on CK release induced by the reoxygenation (Po_2 = 670 mm Hg) of hypoxic hearts. Each point
represents the mean $\pm$ SEM of five experiments.
a: kinetics of creatine kinase release; b: cumulated levels;
$\bullet$ untreated hearts, $\ominus$ hearts treated with α-tocopherol.

membrane permeability of hypoxic cells. This protective activity is
probably enhanced by the limitation of the glycolytic activity and
glycogen depletion in myocardium.

The exogenous catalase and superoxide dismutase do not protect
the cardiac cells against the hypoxic injury. These enzymes have no
effect of glycolytic activity of oxygen deprived myocardium. In
accordance with study of Pryor (1966), these results show that free
oxygen radicals are unlikely to play prominent role during hypoxic
period since oxygen free radicals generation is limited in the absence of molecular oxygen.

One of the first manifestations of hypoxic cardiac muscle injury
is the sensitization of cells to molecular oxygen. The contribution
of the univalent pathway to oxygen reduction has been investigated
during the reoxygenation period. The elucidation of the relationship
between the scavengers of oxygen free radicals and the oxygen induced
enzyme release may aid in the understanding of the mechanisms which
occur during the development of the oxygen paradox phenomenon. This
work shows that the toxicity of oxygen can be significantly decreased during the reoxygenation if the oxygen free radicals scavenging
mechanisms are preserved during hypoxia.

In unpathological conditions, the majority of the oxygen reduced
by the normoxic cells is carried out tetravalently by cytochrome

oxidase, without the release of intermediates (Antonini et al., 1970). But the contribution of the univalent pathway to oxygen reduction is practicable in pathological conditions of normoxia. In these conditions, the reduction of molecular oxygen stimulates the formation of powerful free oxygen radicals: hydrogen peroxide (H_2O_2), superoxide anion radical (O_2^-), hydroxyl radical ($OH^\bullet$).

Different protective and controlling mechanisms can be identified to prevent the release of H_2O_2, O_2^- and $OH^\bullet$ in the cells.

The oxidation-reduction state of the reduced glutathione-oxidized glutathione couple is of major importance in the defense of cells against oxygen toxicity. When the intracellular level of reduced glutathione (GSH) is maintained by the perfusion of exogenous GSH, the oxygen paradox phenomenon is forbided. This protective effect of GSH is probably dependent on several mechanisms. During the reoxygenation, GSH which is more easily available to oxygen than are thiol groups of other enzymes, can inhibit the irreversible inactivation of enzymes such as glyceraldehyde 3 phosphate dehydrogenase (Haugaard, 1968). The GSH level which determines the activity of GSH peroxidase in the rat heart mitochondria (Nohl and Jordan, 1980), can also increase the damage of hydrogen peroxide (H_2O_2) and so reduce the toxic effect of this radical in the reoxygenated cells.

The ability of heart mitochondria to produce hydrogen peroxide is evident with the studies of Loschen et al., (1973), Nohl and Hegner (1978b), Nohl and Jordan (1980). The catalase plays a fundamental role in the cellular metabolism of hydrogen peroxide (Chance et al., 1979). The matrix of mitochondria is altered during hypoxia and when the cardiac cells are reoxygenized the integrity of catalase system is inadequate to control the production of hydrogen peroxide in the mitochondria. This hypothesis is confirmed by the protective action of exogenous catalase obtained during the reoxygenation. The hydrogen peroxide is also produced during the spontaneous dismutation of superoxide anion (Boveris, 1977). Actually, the punctual origin of O_2^- is unknown. But the alteration of mitochondrial function during hypoxia can reduce the detoxification of O_2^- by the cytochrome oxidase. In these conditions, the formation of superoxide anion by the univalent pathway will be likely (Del Maestro, 1980). In our study, the role of superoxide anion radical in the development of oxygen paradox phenomenon can be reserved. Indeed, the superoxide dismutase (SOD), which accelerates the dismutation of O_2^- (Fridovich, 1975, 1978), prevents the appearance of severe oxygen toxicity during the reoxygenation period.

The hydrogen peroxide and the superoxide anion are necessary intermediates in the initiation of lipid peroxidation (Fong et al., 1973; Halliwell, 1978). Although the role of Haber-Weiss reaction is debated in the biological systems (Czapski and Ilan, 1978), lipid peroxidation results in the interaction of unsaturated fatty acids

with superoxide anion, hydroxyl radical and singlet oxygen ($O_2{}^1\Delta g$),
(Halliwell, 1978). The α-tocopherol (vitamine E), a powerful scaven-
ger of singlet oxygen, prevents chain propagating reactions in lipid
membranes (Pryor, 1976; Halliwell, 1978). In our study, α-tocopherol
impedes the toxic effect of molecular oxygen during reoxygenation.
This result suggests that the oxygen paradox includes the alteration
of membrane components and the activation of lipid hydroperoxides
formation (Narabayashi et al., 1982).

So, the development of paradoxal effect of molecular oxygen on
the hypoxic cardiac musle is dependent on cumulative processes
including the alteration of mitochondrial function, the oxygen
derived free radicals production, the chain propagating reactions
in lipid membranes and the disruption of cell integrity.

REFERENCES

Aebi, H., 1974, in: "Methods of Enzymatic Analysis", H.U. Bergmeyer,
     ed., Academic Press, New York.
Antonini, E., Brunori, M., Greenwood, C., and Malmström, B.G., 1970,
     Catalytic mechanism of cytochrome oxidase, Nature, 228:936.
Boveris, A., 1977, in: "Oxygen and Physiological Funtion", F.F.
     Jöbsis, ed., Professional Information Library, Washington.
Chance, B., Sies, H., and Boveris, A., 1979, Hydroperoxide metabo-
     lism in mammalian organs, Physiol. Rev., 59:527.
Czapski, G., and Ilan, Y.A., 1978, On the generation of the hydroxy-
     lation agent from superoxide radical. Can the Haber-Weiss
     reaction be the source of OH· radicals?, Photochem. Photobiol.,
     28:651.
Del Maestro, R.F., 1980, An approach to free radicals in medicine
     and biology, Acta Physiol. Scand., Suppl. 492:153.
Feuvray, D., and de Leiris, J., 1975, Ultrastructural modifications
     induced by reoxygenation in the anoxic isolated rat heart
     perfused without exogenous substrate, J. M. C. C., 7:307.
Fong, K.L., Mc Cay, P.B., Poyer, J.L., Leele, B.B., and Misra, H.,
     1973, Evidence that peroxidation of lysosomal membranes is
     initiated by hydroxyl free radicals produced during flavin
     enzyme activity, J. Biol. Chem., 248:7792.
Fridovich, I., 1975, Superoxide dismutases, Ann. Rev. Biochem.,
     44:147.
Fridovich, I., 1978, Superoxide radicals, superoxide dismutases and
     the aerobic life style, Photochem. Photobiol., 28:733.
Ganote, C.E., and Kaltenbach, J.P., 1979, Oxygen-induced enzyme
     release: early events and a proposed mechanism, J. Mol. Cell.
     Cardiol., 11:389.
Ganote, C.E., Seabra Gomes, R., Nayler, W.G., and Jennings, R.B.,
     1975, Irreversible myocardial injury in anoxic perfused rat
     hearts, Am. J. Pathol., 80:419.

Gauduel, Y., Karagueuzian, H.S., and de Leiris, J., 1979, Deleterious
    effects of endogenous catecholamines on hypoxic myocardial
    cells following reoxygenation, J. Mol. Cell. Cardiol., 11:717.
Gutmann, I., and Wahlefeld, A.W., 1974, in: "Methods of Enzymatic
    Analysis", H.U. Bergmeyer, ed., Academic Press, New York.
Halliwell, B., 1978, Biochemical mechanisms accounting for the toxic
    action of oxygen on living organisms: the key role of super-
    oxide dismutase, Cell Biol. Int. Rep., 2:113.
Haugaard, N., 1968, Cellular mechanisms of oxygen toxicity, Physiol.
    Rev., 48:312.
Hearse, D.J., 1978, The oxygen paradox and the calcium paradox: two
    facets of the same problem?, J. Mol. Cell. Cardiol., 10:641.
Hearse, D.J., Humphrey, S.M., and Chain, E.B., 1973, Abrupt reoxy-
    genation of the anoxic potassium arrested perfused rat heart:
    a study of myocardial enzyme release, J. Mol. Cell. Cardiol.,
    5:395.
Hunter, F.E., Scott, J.A., Hoffsten, P.E., Gebicki, J.M., Weinstein,
    J., and Schneider, A., 1964, Studies on the mechanism of swel-
    ling lysis and disintegration of isolated liver mitochondria
    exposed to mixtures of oxidized and reduced glutathione, J.
    Biol. Chem., 239:614.
Kosower, N.S., and Kosower, E.M., 1976, The glutathione-glutathione
    disulfide system, in: "Free Radicals in Biology", W.A. Pryor,
    ed., Academic Press, London.
Langendorff, O., 1895, Untersuchungen am überlebenden Säugetierher-
    zen, Pflügers Arch. Ges. Physiol., 61:251.
Loschen, G., Azzi, A., and Flohe, L., 1973, Mitochondrial H_2O_2 for-
    mation: relationship with energy conservation, FEBS Lett., 33:89.
Narabayashi, H., Takeshige, K., and Minakami, S., 1982, Alteration
    of inner membrane components and damage to electron transfer
    activities of bovine heart submitochondrial particles induced
    by NADPH dependent lipid peroxidation, Biochem. J., 202:97.
Nohl, H., and Hegner, D., 1978a, Evidence for the existence of
    catalase in the matrix space of rat heart mitochondria, FEBS
    Lett., 89:126.
Nohl, H., and Hegner, D., 1978b, Do mitochondria produce oxygen
    radicals "in vivo"?, Eur. J. Biochem., 82:563.
Nohl, H., and Jordan, W., 1980, The metabolic fate of mitochondrial
    hydrogen peroxide, Eur. J. Biochem., 111:203.
Pryor, W.A., 1966, "Free Radicals", Mc Graw-Hill, New York.
Pryor, W.A., 1976, in: "Free Radicals in Biology", W.A. Pryor, ed.,
    Academic Press, New York.
Siegel, A.L., and Cohen, P.S., 1974, in: "Methods of Enzymatic
    Analysis", H.U. Bergmeyer, ed., Academic Press, New York.

SPONTANEOUS LIPID PEROXIDATION IN RABBIT SPERMATOZOA: A USEFUL

MODEL FOR THE REACTION OF O_2 METABOLITES WITH SINGLE CELLS

J.G. Alvarez, M.K. Holland, and B.T. Storey

Departments of Physiology and Obstetrics and Gynecology
University of Pennsylvania, School of Medicine
Philadelphia, PA 19104, USA

INTRODUCTION

Mature mammalian spermatozoa comprise a relatively homogeneous population of cells. In suspension, their motility provides a readily visualized index of cell viability. Suspensions of spermatozoa are, therefore, convenient and useful systems for studying the reactions of O_2 and its partially reduced metabolites which cause cell damage. It has been known for nearly 40 years that high O_2 tensions are deleterious to the motility of human sperm (McLeod, 1943). Bull and ram spermatozoa were shown to lose motility on storage at $4^{\circ}C$ in concert with loss of phospholipid, particularly plasmalogen, suggesting that this loss was due to lipid peroxidation leading to plasma membrane damage (Jones and Mann, 1973; Mann and Lutwak-Mann, 1981). Since most mammalian sperm contain little or no catalase (Mann, 1964), O_2-induced damage to spermatozoa has been attributed to H_2O_2 (Wales et al., 1959). Another source of O_2-induced damage could be the superoxide anion O_2^{-}: Menella and Jones (1980) demonstrated superoxide dismutase (SOD) activity in spermatozoa from a variety of mammalian species, suggesting that O_2^{-} might be produced in these cells. But no documentation of O_2^{-} production was provided.

We have been studying the production by and reaction with mature rabbit epididymal spermatozoa of O_2 and its partially reduced metabolites, H_2O_2 and O_2^{-}. We have shown that rabbit spermatozoa produce H_2O_2 and lack detectable catalase activity; they contain detectable glutathione peroxidase and oxidized glutathione reductase activity but no detectable glutathione (Holland and Storey, 1981). Rabbit sperm are deficient in two of the enzymatic defence systems utilized by other cells against partially reduced metabolites

433

of O_2 (Chance et al., 1979). Production of O_2^- as well as SOD activity in these cells was also documented (Holland et al., 1983). Spontaneous lipid peroxidation, as measured by malondialdehyde (MDA) production, was also demonstrated to take place in rabbit epididymal spermatozoa (Alvarez and Storey, 1983). The production of MDA by a sperm suspension correlated with the increasing number of inert, non-motile sperm in the suspension. The rate of MDA production was 8-fold greater in a suspending medium of high K^+ content than in a medium of high Na^+ content. We have further observed that, concomitant with MDA production and loss of sperm motility, there is a loss of CN^--sensitive SOD activity and an increase in the apparent reactivity of the cells with O_2^- (Alvarez and Storey, ms. in preparation).

Taken together, these observations implicated superoxide as the principal dioxygen species leading to spontaneous lipid peroxidation and so to loss of motility in mature rabbit epididymal spermatozoa. Two questions remained unanswered, however. The first was: what are the sources of superoxide production in the sperm cells? The second was: which superoxide species, O_2^- or $HO_2\cdot$, is the more potent in inducing lipid peroxidation? In this paper we report evidence showing both mitochondria and cytosol as sources of superoxide and $HO_2\cdot$ as a very potent inducer of peroxidation.

THEORETICAL ANALYSIS

The assay for the rate of production of O_2^- from both intact sperm and hypotonically treated rabbit epididymal spermatozoa (HTRES) (Holland et al., 1983) involves reduction of exogenous acetylated ferricytochrome c (AC_{ox}) by the O_2^- released from the cells (Azzi et al., 1975); this assay is a modification of the ferricytochrome c assay of McCord and Fridovich (1969). In the absence of a reaction of O_2^- with the sperm cells, the rate of reduction of AC_{ox} under a given set of experimental conditions should be linear with sperm cell concentration, expressed as cells/ml. Contrary to this expectation, the reduction rate of AC_{ox} by O_2^- produced from both intact rabbit epididymal spermatozoa and HTRES was shown (Holland et al., 1983) to have the following double reciprocal dependence on sperm cell concentration at a given concentration of AC_{ox}:

$$(V_{ac})^{-1} = v_{int}^{-1} (SP)^{-1} + k_s \left\{ v_{int}\, k_c\, (AC_{ox}) \right\}^{-1} \qquad (1)$$

where V_{ac} is the rate of reduction of AC_{ox}, (SP) is the concentration of sperm cells, v_{int} is the intrinsic net rate of O_2^- production from the sperm cells, k_s is the second order rate constant for reaction of sperm cells and O_2^- and k_c is the second order rate constant for reaction of AC_{ox} and O_2^-. The form of the equation is a consequence of competition for the steady state concentration of

($O_2^{\overline{\cdot}}$) by both AC_{ox} and the cells; the derivation of the equation is detailed in Holland et al. (1983). The double reciprocal plot of $(V_{ac})^{-1}$ versus $(SP)^{-1}$ yields v_{int}^{-1} as the slope. The value of k_c is determined in the absence of sperm (Holland et al., 1983) from plots of V_{ac} versus (AC_{ox}) at different rates of $O_2^{\overline{\cdot}}$ generation from xanthine and O_2 catalyzed by xanthine oxidase (Azzi et al., 1975); k_c in the high K^+ medium KTP (see Experimental Procedures) at 24°C and pH 7.4 was found to be 7.8 x 10^4 M^{-1}min^{-1}. Since $(AC)_{ox}$ is set by the assay conditions and v_{int} is calculated from the slope of the double reciprocal plot, k_s is determined by the intercept of the plot; it is given in the empirical units (cells/ml)$^{-1}$min^{-1}. The values for v_{int} found from the double reciprocal plots for intact sperm in medium KTP at 24° and pH 7.4 (Holland et al., 1983) were 0.20 nmol/min-10^8 cells in the absence of CN$^-$ and 1.80 nmol/min-10^8 cells in the presence of 10 mmol·l^{-1} CN$^-$ to inhibit sperm SOD activity. The values of v_{int} for HTRES were 0.24 nmol/min-10^8 cells in the absence and 0.58 nmol/min-10^8 cells in the presence of 10 mmol·l^{-1} CN$^-$. In medium KTP at 24° and pH 7.4, the values of k_s in the absence and presence of 10 mmol·l^{-1} CN$^-$ were determined from the double reciprocal plots to be 12.5 x 10^{-8} (cells/ml)$^{-1}$min^{-1} and 22.9 x 10^{-8} (cells/ml)$^{-1}$min^{-1} respectively. For HTRES, the corresponding values of k_s were 8.2 x 10^{-8} (cells/ml)$^{-1}$min^{-1} and 10.8 x 10^{-8} (cells/ml)$^{-1}$min^{-1} (Holland et al., 1983).

EXPERIMENTAL PROCEDURES

Two different media were used for sperm suspensions in this study. One was a modification of the high potassium medium of Keyhani and Storey (1973), designated KTP, with the composition: 113 mmol·l^{-1} KCl, 12.5 mmol·l^{-1} KH$_2$PO$_4$, 2.5 mmol·l^{-1} K$_2$HPO$_4$, 3 mmol·l^{-1} MgCl$_2$, 20 mmol·l^{-1} Tris, 0.4 mmol·l^{-1} EDTA, 1.5 mmol·l^{-1} D-glucose, 0.6% penicillin/streptomycin, adjusted with HCl to pH 7.4. The second was a high sodium medium, designated NTP, which contained 10 mmol·l^{-1} KCl, 103 mmol·l^{-1} NaCl, and 15 mmol·l^{-1} NaH$_2$PO$_4$, but was otherwise identical in composition to KTP. In experiments not involving motility determination or long-term incubations: e.g. in determining rates of $O_2^{\overline{\cdot}}$ production, the glucose and penicillin/ streptomycin were omitted.

Spermatozoa were obtained from the caudae of excised epididymides of mature male New Zealand White rabbits by retrograde flushing through the vas deferens with the appropriate medium, followed by washing (Holland and Storey, 1981). The final stock suspensions ranged from 1-5 x 10^8 cells/ml. Care was taken to prevent contamination of the sperm sample with hemoglobin from epididymal blood vessels, which interferes with the spectrophotometric procedure used to determine lipid peroxidation. HTRES, which lack much of the plasma membrane, were prepared as described by Keyhani and Storey (1973).

The rate of O_2^- production was measured by the method of Azzi et al. (1975) using AC_{ox} at room temperature (24 $\pm$ 1^{O}C with the DW-2A dual wavelength spectrophotometer (American Instrument Co.) using the wavelength pair 550-540 nm and the difference extinction coefficient $\Delta\varepsilon$ = 19.1 mM^{-1}cm^{-1} (Margoliash and Frohwirt, 1959).

Spontaneous lipid peroxidation was induced by exposure of the spermatozoa to O_2 during aerobic incubation of sperm supsensions containing 1-5 x 10^7 cells/ml. Production of MDA was determined by a modification of the thiobarbituric acid (TBA) assay (Barber and Bernheim, 1967), in which the TBA concentration was increased to 28 uM and the absorbance of the TBA/MDA chromogen was determined with the DW-2A dual wavelength spectrophotometer using the wavelength pair 534-570 nm. Full details of the procedure have been described by Alvarez and Storey (1983).

Sperm motility was estimated by the modification of the method of Heffner and Storey (1982) as described by Alvarez and Storey (1983), in which duplicate aliquots of the sperm suspension were taken and the average of the percentage motile in both aliquots estimated. Variation between duplicate aliquots was within $\pm$ 5%. The assay was extended to include a second mode of flagellar motion in rabbit sperm, that of flagellar beating. In this mode, there is active flagellar motion and the sperm move, but forward progress is close to nil in this group of sperm.

RESULTS

Boveris (1977) had pointed out that one important subcellular source of O_2^- production is the mitochondrial compartment. The contribution of sperm mitochondria to spermatozoal O_2^- production was assessed by comparing production rates in intact sperm compared to HTRES in the absence and presence of the mitochondrial respiratory chain inhibitor, antimycin A, and the respiratory chain plus SOD inhibitor, cyanide. The HTRES preparation has lost much of the plasma membrane but retains full mitochondrial integrity and function (Keyhani and Storey, 1973). The production rate of O_2^-, as assayed by the rate of reduction of exogenous AC_{ox}, V_{ac}, for these conditions is given in Table 1. From the measured rate V_{ac} one calculates a value for the intrinsic O_2^- production rate, v_{int}, using the appropriate k_s value (Holland et al., 1983). The values of v_{int} so calculated in the absence of inhibitor and presence of CN$^-$ agree well with the values of v_{int} obtained from analysis of the double reciprocal plots representing equation (1) as described in Theoretical Analysis. This indicates that the other values of v_{int} calculated from V_{ac} may be accepted with confidence.

In the presence of CN$^-$, v_{int} from HTRES was 0.6 nmol/min-10^8 cells, while in the presence of antimycin A it was 0.5 nmol/min-10^8

cells. With the latter inhibitor, production of H_2O_2 from HTRES was shown to be maximal (Holland and Storey, 1981) and all the H_2O_2 so produced could be accounted for by dismutation of O_2^- (Holland et al., 1983). Antimycin A has been shown to elicit maximal O_2^- production from mitochondria isolated from other mammalian tissues (Boveris and Chance, 1973). We therefore assign the value of 0.6 nmol/min-10^8 cells to the intrinsic mitochondrial O_2^- production, which can be realized when the cytosolic SOD activity is inhibited by CN^-. Comparison of the mitochondrial v_{int} with the value of v_{int} = 1.6 nmol/min-10^8 cells for intact cells in the presence of 10 mmol·1^{-1} CN^-, indicates that the cytosolic v_{int} is 1.0 nmol/min-10^8 cells.

The effect of mitochondrial substrates on O_2^- production was also tested with HTRES, yielding the results shown in Table 1. Lactate plus malate are preferred mitochondrial oxidative substrates in rabbit spermatozoa (Storey and Kayne, 1978); these increase v_{int} to nearly double that observed with endogenous substrate.

The rate of lipid peroxidation of intact rabbit sperm was examined in medium KTP as a function of pH, with the results shown in Fig. 1. At pH 7.0 and above in this medium, the rate of peroxidation, as measured by MDA formation, is very low, amounting to 0.093 x 10^{-3} nmol/min-10^8 cells. This is 8-fold slower than in medium KTP (Alvarez and Storey, 1983), and so NTP is the medium in which to look for increase of lipid peroxidation rate with decreasing pH, the result expected if $HO_2\cdot$, with pK_a = 4.7 (Bielski, 1978), were the active species in this reaction. This is what was observed: as the pH decreased, the rate increased markedly, as shown in Fig. 1. At pH 4.7, the accumulation of MDA became linear with time after a short lag; at pH 5.0 and above, the accumulation was linear with time (Fig. 1). The rates of lipid peroxidation determined at pH values between 7.4 and 5.0 are collected in Table 2.

Above pH 7.0, there was little further decrease in lipid peroxidation rate with increasing pH. This rate we attribute to radical chain oxidation induced by O_2^-; between 7.4 and 7.0, the H^+ and $HO_2\cdot$ concentrations would rise 2.5-fold, but the rate of lipid peroxidation increased less than 10% (Table 2). We take the rate of lipid peroxidation to be the sum of two reactions, one with O_2^- and one with $HO_2\cdot$. The simplest expression for the rate equation is:

$$v_{ma} = k_A(O_2^-) + k_H(HO_2\cdot) \tag{2}.$$

In equation (2), v_{ma} is the specific rate of MDA production, k_A and k_H are composite second order rate constants for the reaction of sperm cell lipids with O_2^- and $HO_2\cdot$, respectively, to produce MDA. Values of k_A = 0.26 x 10^{-11} $(cells/ml)^{-1}min^{-1}$ and k_H = 81 x 10^{-11} $(cells/ml)^{-1}min^{-1}$ were estimated from the measured values in Table 2

Table 1. Intrinsic rates of O_2^- production from intact and hypotonically treated rabbit epididymal sperm calculated from
rates of reduction of acetylated ferricytochrome c.

Expt.	Sperm Preparation	Additions[c]	V_{ac}[a]	v_{int}(calc)[b]
			nmol/min - 10^8 cells	
I	Intact	None	0.07 ± 0.02	0.20
		Anti. A (4 ug)	0.14 ± 0.02	0.39
		CN- (10 mmol·l^{-1})	0.36 ± 0.02	1.60
II	HTRES	None	0.11 ± 0.01	0.24
		Anti. A (4 ug)	0.23 ± 0.02	0.50
		CN- (10 mmol·l^{-1})	0.24 ± 0.02	0.61
III	HTRES	None	0.09 ± 0.02	0.20
		+Lac (5 mmol·l^{-1})	0.14 ± 0.02	0.30
		+Mal (5 mmol·l^{-1})	0.17 ± 0.01	0.37
		+ADP (0.5 mmol·l^{-1})	0.18 ± 0.02	0.39

a. Measured rate of reduction in medium KTP of acetylated ferricytochrome c (AC_{ox}) by O_2^- produced from sperm cells (see Theoretical Analysis). Sperm cell concentration was 1 x 10^8 cells/ml.
AC_{ox} was 90 uM. Rates are means of three experiments ± S.D.

b. Intrinsic net rate of O_2^- production by sperm cells calculated
from equation (1), using measured value of V_{ac} and k_s = 12.5 x 10^{-8}
$(cells/ml)^{-1}min^{-1}$ in the absence of CN^- and 22.9 x 10^{-8} (cells/
$ml)^{-1}min^{-1}$ in the presence of 10 mmol·l^{-1} CN^- for intact cells. Corresponding values of k_s for HTRES are 8.2 x 10^{-8} $(cells/ml)^{-1}min^{-1}$
and 10.8 x 10^{-8} $(cells/ml)^{-1}$. The values of v_{int} obtained by analysis of double reciprocal plots of equation (1) are (see Theoretical
Analysis): for intact cells, 0.20 nmol/min-10^8 cells in the absence
and 1.80 nmol/min-10^8 cells in the presence of CN^-; corresponding
values for HTRES are 0.24 nmol/min-10^8 cells in the absence and
0.58 nmol/min-10^8 cells in the presence of CN^-.

c. In experiments I and II, antimycin A and CN^- were added in
different experiments. In experiment III, additions of lactate,
malate, and ADP were sequential.

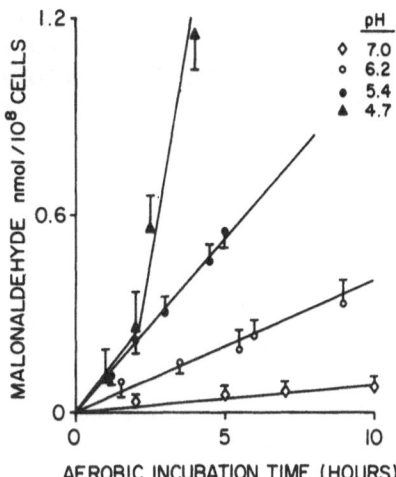

Fig. 1.   MDA production by intact rabbit epididymal spermatozoa as
          a function of aerobic incubation time in medium NTP at dif-
          ferent pH. Procedures for incubation and determination of
          MDA are given in Experimental Procedures. The sperm concen-
          tration ranged between 0.4 and 0.9 x 10^8 cells/ml. Each
          point is the mean of 3 determinations; error bars are
          standard deviations.

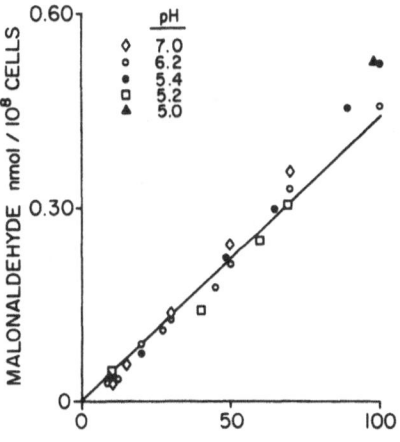

Fig. 2.   MDA accumulated in the medium during aerobic incubation of
          intact rabbit spermatozoa at different pH in NTP as a func-
          tion of percent inert spermatozoa. The determinations were
          made under the same conditions as those in Fig. 1. Each
          point is the mean of three determinations. The linear re-
          gression line calculated through the origin has the form
          y = 0.0045x (r = 0.927).

(see Appendix). The estimated values of v_{ma} calculated from equation
(2) using these values of k_A and k_H at each pH are also listed in
Table. 2. The agreement is fair but satisfactory, given the simple
mode of calculation. Comparison of the values of k_H and k_A show that
$HO_2\cdot$ is a far more potent inducer of lipid peroxidation in rabbit
spermatozoa than is $O_2^{\cdot-}$.

Table 2. Measured specific rates of malondialdehyde (MDA) production, v_{ma} (meas.)
during spontaneous lipid peroxidation
of intact rabbit sperm in medium NTP
at different pH; comparison with
values estimated from equation (2),
v_{ma} (est.).

pH	v_{ma}(meas.)	v_{ma}(est.)[a]
	10^3 x nmol/min-10^8 cells	
7.4	0.093	0.071
7.0	0.098	0.110
6.8	0.105	0.151
6.6	0.107	0.209
6.5	0.217	0.258
6.2	0.55	0.46
5.8	1.08	1.12
5.4	1.83	1.87
5.2	2.33	2.37
5.0	3.67	3.70

a. See Appendix for calculation of estimated
values of v_{ma}.

Inreased lipid peroxidation results in loss of motility. This
is shown in Fig. 2 for spontaneous lipid peroxidation carried out
over the pH range 5.0 to 7.4. The sperm population loses all motility at the point where 0.5 nmol MDA/10^8 cells has been produced,
regardless of the rate of peroxidation. This is the same value
obtained at pH 7.4 with media of different salt composition (Alvarez
and Storey, 1983). This indicates that the mechanism of motility
loss is the same in both sets of experimental conditions.

DISCUSSION

The first question posed in this study: what are the sources of superoxide production in rabbit sperm? yielded a surprising answer. Only about 35% of the O_2^- comes from the mitochondria, which are usually considered a major source of this species (Boveris, 1977). The rest comes from the cytosol. Mammalian sperm contain an unusually high content of plasmalogen in their phospholipid component (Mann, 1964), which have vinyl ether links prone to autoxidation by O_2 abstraction of H to yield $HO_2\cdot$ (Walling, 1957). We attribute the high cytosolic rate of O_2^- production to this source, in accord with the observation by Jones and Mann (1973) of plasmalogen loss during lipid peroxidation.

The second question: which of O_2^- or $HO_2\cdot$ is the most potent agent for induction of lipid peroxidation? was clearly answered. The perhydroxyl radical is some 2 orders of magnitude more potent, in accord with the results of Gebicki and Bielski (1981) on chain oxidation of linoleic acid. The values of k_H and k_A may also be compared with the value of k_S in NTP, which is 1.2×10^{-8} (cells/ml)$^{-1}$min^{-1}. The ratios: $k_A/k_S = 0.3 \times 10^{-3}$ and $k_H/k_S = 68 \times 10^{-3}$ reflect the composite nature of k_A and k_H and the low yield of MDA from the many reactions occurring during lipid peroxidation. The potency of $HO_2\cdot$ in these reactions suggest that the 8-fold difference observed in lipid peroxidation rate between the media NTP and KTP could be due to perturbation of the equilibrium $HO_2\cdot \leftrightarrows H^+ + O_2^-$ or that it might be due to formation of neutral ion pair species of the type $\{M^+O_2^-\}$. Formation of the latter might be expected to occur at the membrane surface where lipid peroxidation is presumed to take place.

APPENDIX

The specific rate of MDA production is given by the equation:

$$v_{ma} = k_A(O_2^-) + k_H(HO_2\cdot) \tag{2}$$

The following simplifying assumptions have been made:
a) At pH = 7.4, O_2^- production rate per 10^8 cells is given by v_{int}; its consumption rate is given by $k_S(O_2^-)$.
b) The steady state (O_2^-) plus $(HO_2\cdot)$ varies little with pH.
c) At pH values at which v_{ma} is 10-fold or greater than v_{ma} at pH 7.4, the $HO_2\cdot$ reaction dominates and the O_2^- reaction can be neglected. From assumption a, the steady state (O_2^-) is:

$$(O_2^-) = v_{int}(k_S)^{-1} \tag{3}.$$

In medium NTP for intact cells at pH 7.4, v_{int} is 0.17 nmol/min-10^8 cells, and k_S is 1.2×10^8 (cells/ml)$^{-1}$min^{-1}, yielding

0.14 uM for the steady state (O_2^-·) at pH 7.4. From the assumption b, we assume that the steady state (O_2·) + (HO_2·) = 0.14 uM over the pH range studied, and that (HO_2·) can be calculated simply from this value and its pK_a. From assumption c, we consider v_{ma} from pH 5.8 to pH 5.0 to be dominated by k_H(HO_2·) and neglect k_A(O_2·). the calculation yields the following values for k_H in units of 10^{11} (cells/ml)$^{-1}$min^{-1}:

pH	5.8	5.4	5.2	5.0	
k_H	88.2	73.7	76.2	86.4	Ave: 81

This average value of k_H is then used to calculate the contribution of k_H(HO_2·) at the higher pH values, from which k_A can then be estimated. This yields the following values of k_A in units of 10^{11} (cells/ml)$^{-1}$min^{-1}:

pH	7.4	7.0	6.8	6.6	6.5	6.2	
k_A	0.39	0.19	0.00	0.00	0.32	0.66	Ave: 0.26

Note that determination of k_A requires the subtraction of two numbers close in magnitude with resulting large error in the difference. For this reason, we have included in the average the values at pH 6.8 and 6.6, where k_H term appears to account for entire measured rate.

ACKNOWLEDGEMENT

    This work was supported by NIH grants HD-06274 and HL-19737. We thank Mrs. Dorothy Rivers and Ms. Dawn Cesarini for technical help and Ms. Neisha Son for secretarial assistance.

REFERENCES

Alvarez, J.G., and Storey, B.T., 1983, Spontaneous lipid peroxidation in rabbit epididymal spermatozoa: its effect on sperm motility, Biol. Reprod., 28: (in press).
Azzi, A., Montecucco, C., and Reichert, C., 1975, The use of acetylated ferricytochrome c for the detection of superoxide radicals produced in biological membranes, Biochem. Biophys. Res. Comm., 65:597.
Barber, A.A., and Bernheim, F., 1967, Lipid peroxidation: its measurement, occurrence, and significance in animal tissues, Adv. Gerontol. Res., 2:355.
Bielski, B.H.J., 1978, Re-evaluation of the spectral and kinetic properties of HO_2 and O_2-free radicals, Phytochem. Phytobiol, 28:645.

Boveris, A., 1977, Mitochondrial production of superoxide radical
    and hydrogen peroxide, in: "Tissue Hypoxia and Ischemia",
    M. Reivich, R. Coburn, S. Lahiri, and B. Chance, eds., Plenum
    Press, New York.
Boveris, A., and Chance, B., 1973, The mitochondrial generation of
    hydrogen peroxide. General properties and effect of hyperbaric
    oxygen, Biochem. J., 134:707.
Chance, B., Sies, H., and Boveries, A., 1979, Hydroperoxide metab-
    olism in mammalian organs, Physiol. Rev., 59:527.
Gebicki, J.M., and Bielski, B.H.J., 1981, Comparison of the capaci-
    ties of the perhydroxyl and superoxide radicals to initiate
    chain oxidation of linoleic acid, J. Am. Chem. Soc., 103: 7030.
Heffner, L.J., and Storey, B.T., 1982, Cold lability of mouse sperm
    binding to zona pellucida, J. Exp. Zool., 219:155.
Holland, M.K., Alvarez, J.G., and Storey, B.T., 1983, Production of
    superoxide and activity of superoxide dismutase, Biol. Reprod.
    28: (in press).
Holland, M.K., and Storey, B.T., 1981, Oxygen metabolism of mamma-
    lian spermatozoa. Generation of hydrogen peroxide by rabbit
    epididymal spermatozoa, Biochem. J., 198:273.
Jones, R., and Mann, T., 1973, Lipid peroxidation in spermatozoa,
    Proc. Roy. Soc. London B., 184:103.
Keyhani, E., and Storey, B.T., 1973, Energy conservation capacity
    and morphological integrity of mitochondria in hypotonically
    treated rabbit epididymal spermatozoa, Biochim. Biophys. Acta
    305:557.
Mann, T., 1964, "The Biochemistry of Semen and the Male Reproduc-
    tive Tract", pp. 265-292, Methuen and Co., London.
Mann, T., and Lutwak-Mann, C., 1981, "Male Reproductive Function
    and Semen", Springer, Berlin/Heidelberg/New York.
Margoliash, E., and Frohwirt, N., 1959, Spectrum of horse heart
    cytochrome c, Biochem. J., 71:570.
McCord, J.M., and Fridovich, I., 1969, Superoxide dismutase. An
    enzymic function for erythrocuprein (hemocuprein), J. Biol.
    Chem., 244:6049.
McLaed, J., 1943, The role of oxygen in the metabolism and motility
    of human spermatozoa, Am. J. Physiol., 138:512-518.
Mennella, M.R.F., and Jones, 1980, Properties of spermatozoal super-
    oxide dismutase and lack of involvement of superoxide in metal-
    ion-catalyzed lipid-peroxidation reactions in semen, Biochem.
    J., 191:289.
Storey, B.T., and Kayne, F.J., 1978, Energy metabolism of sperma-
    tozoa. VII. Interactions between lactate, pyruvate and malate
    as oxidative substrates for rabbit sperm mitochondria, Biol.
    Reprod., 18:527.
Walling, C.W., 1957, "Free Radicals in Solution", John Wiley and
    Sons, New York.

MUSCLE

A COMPARISON OF Po_2 HISTOGRAMS FROM RABBIT HIND-LIMB MUSCLES

OBTAINED BY SIMULTANEOUS MEASUREMENTS WITH HYPODERMIC NEEDLE

ELECTRODES AND WITH SURFACE ELECTRODES

W. Fleckenstein, and Ch. Weiss

Institut für Physiologie der Medizinischen Hochschule
Lübeck, Ratzeburger Allee 160, 2400 Lübeck, FRG

INTRODUCTION

Compression of the tissue in the vicinity of a Po_2 sensitive hypodermic needle electrode leads to low Po_2 values which do not represent the oxygenation of the tissue. Therefore, instead of hypodermic needle probes for measurements of tissue oxygenation, multiwire surface probes are widely used (Hauss et al., 1982). However, as will be shown, if the response time of needle probes is sufficiently short it is possible to record "true" tissue Po_2 values if the probe is moved in a stepwise fashion within the tissue, and if only the initial Po_2 value indicated shortly after the end of each step movement is accepted.

In the present study Po_2 histograms from resting skeletal muscle in the rabbit, obtained by surface probe measurements, were compared with histograms obtained by fast responding hypodermic needle probes which were inserted stepwise into the muscle.

The following three questions have been studied:

1. How fast does the Po_2 decrease at the tip of a Po_2 needle probe (shaft diameter of 400 um) in resting skeletal muscle?

2. Are Po_2 histograms measured with multiwire surface probes similar to simultaneously recorded Po_2 histograms obtained with needle probes?

3. Is it possible with Po$_2$ needle probe histograms to follow
   up transient changes in tissue Po$_2$ such as are elicited by
   sudden changes of respiratory gas composition or by drug
   administration?

METHODS

The hypodermic Po$_2$ needle probe used in this study (Fig. 1)
was of the fast responding polarographic type (Fleckenstein, 1982),
and had the following mechanical and electrical specifications: A
stainless steel tube containing a silica fiber (Ø 40 um), imbedded
in Al$_2$O$_3$ filled epoxy resin. The shaft of the silica fiber carries
at its elliptically ground end the polarographic thin film Au-cathode
of 2000 Å thickness. The cathode was recessed by an iodine etch and
covered by a phenolic resin (Formvar[R] F 15, Monsanto Inc.) membrane.
Mechanical data: 60 mm overall length; outer diameter of the stainless steel tubing: 400 um; tip lancet ground (25°/40°/40°); Electrical data: response time (T$_{90}$, 37°C): < 300 ms; current sensitivity:
10 - 15 pA/mm Hg Po$_2$; current in oxygen-free physiological NaCl
solution: < 10 pA; stirring effect: < 3%; drift per hour < 5%; N$_2$O
sensitivity (physiological NaCl solution 37°C, 730 mV polarisation):
< 0.1 pA/mm Hg pN$_2$O.

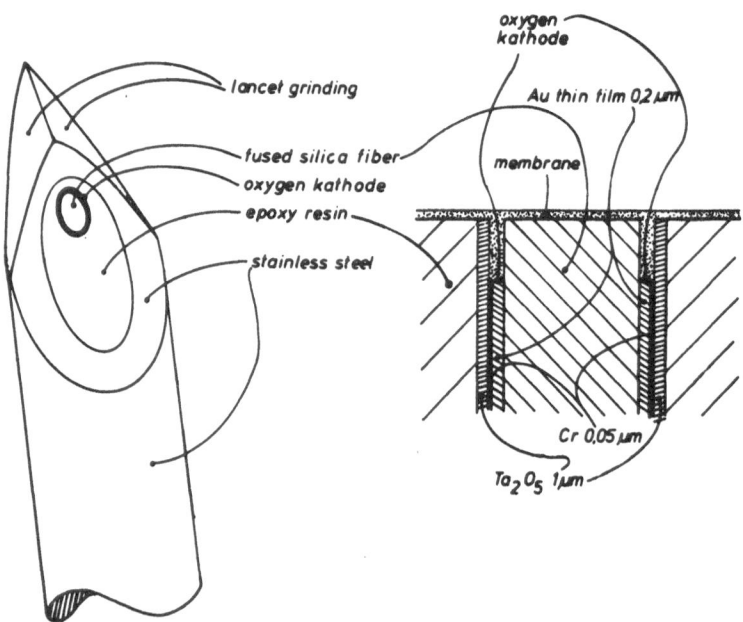

Fig. 1.   left : view of the tip of a hypodermic needle probe;
           right: cross-section of the polarographic cathode region
                  of the probe.

The anode potential was applied by an Ag-AgCl-KCl skin electrode of 1 cm^2 surface. In order to reduce friction within the tissue and in order to minimize electrical artifacts during movement the surface of the stainless steel tubing is covered by polytetrafluorethylen (PTFE).

The multiwire probe was of the Clark-type (similar to MDO probe) and contained 12 Po$_2$ gold microwire (∅ 10 um) cathodes isolated by fused silica and imbedded in epoxy resin. The wires were recessed and covered by a 10 um thick membrane of PTFE. Mechanical data: total outer diameter of the multiwire probe: 8 mm; diameter of the disk shaped cathode array: 4 mm; hight of the electrode housing: 6 mm. Electrical data: response time (37°C) < 3 sec; sensitivity: 20 pA/mm Hg Po$_2$; current in oxygen-free NaCl solution: < 20 pA; stirring artifact < 1.5%; drift < 2%/h.

As a measuring setup, a newly developed KIMOC^R 400 system (G.M.S.m.b.H. Kiel, FRG) was used to apply the hypodermic needle probes. The calibration of the probe, the probe movements and the on-line histogram display were controlled by a microcomputer. The probe was moved into the muscle while protected by a steel-covered needle. Once within the tissue the probe was moved stepwise out of its protective steel cover. The histogram setup time (100 Po$_2$ values) is about 2 minutes. White New Zealand rabbits of 1.2 - 4.3 kg body weight were used.

In a first group, the animals were sedated by the i.v. administration of 0.25 mg droperidol/kg b.w. and of 1 mg diazepam/kg b.w. In a second group, animals were narcotized by the administration of 0.5 mg droperidol/kg b.w. and of 2.5 mg diazepam/kg b.w.; pentobarbital was administered as required. The animals breathed spontaneously. Their blood pressure was measured in a branch of the a. brachialis. For the simultaneous measurement with surface and hypodermic needle probes the muscles of the thigh were dissected free and kept humid and at constant temperature (35°C).

RESULTS

Concerning the first question: In order to determine the slope of decrease in the tissue Po$_2$ in the vicinity of the probe, i.e., the compression zone, the probe signal during the halting of the probe following a rapid step movement of 2 mm was recorded (Fig. 2). The electrical movement artifact (* in Fig. 2) which occurs during the step movements (st→←in Fig. 2) is over 400 ms after the end of the movement. The probe current at this moment is defined as the "initial value" (see ∠ in Fig. 2). Subsequently the probe current decreases in a nonlinear fashion. Depending on the localization of the tip of the probe the slope of the decrease varies. In order to estimate the possible error due to probes with slow response times,

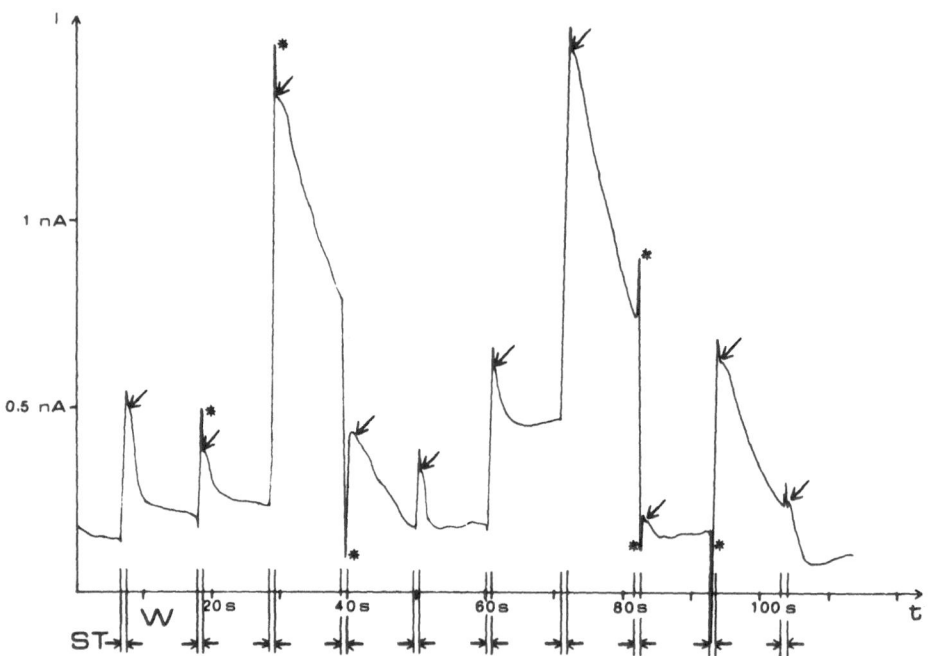

Fig. 2. Probe signal in relation to periods of rapid step movement
(st→ ←); * electrical "movement artifact"; signal decreases
(w) during standstill of the probe.

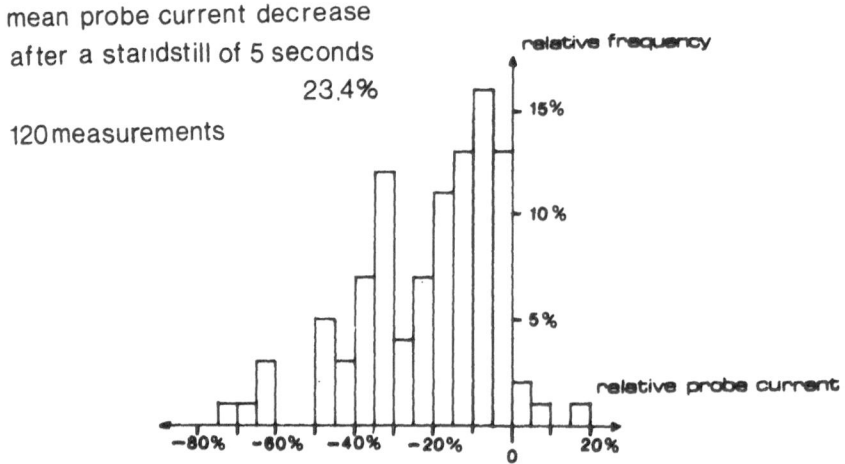

Fig. 3. Frequency distribution of the probe signal measured during
standstill of the probe 5 secs after the initial value
(see ↙ in Fig. 2) expressed as a percentage of the initial
value.

in Fig. 3 the percentage decreases from the initial values after
5 secs, measured at 120 tissue points has been plotted as a histogram. It can be seen that in a number of measurements very steep
decreases of local Po$_2$ occur.

Concerning the second question: In the second group of experiments data obtained simultaneously from hypodermic needle- and surface multiwire probes were compared. The narcotized animals breathed
air spontaneously. The data obtained from five animals were pooled
and plotted as histograms. On the left side of Fig. 4 there is a
histogram from surface probe measurements and the corresponding
histogram from hypodermic needle data is shown on the right. In
both cases all data obtained were pooled regardless of the animals
blood pressure. However, the mean blood pressure varied considerably
from animal to animal and also within one experiment, and exerted a
significant effect on the shape of the individual histograms. The
mean blood pressure of the rabbits of group 2 remained within
physiological ranges for only 15 - 30 minutes. It fell continuously
within the following 60 - 90 minutes to values of 30 mm Hg.

Since in this study we were interested primarily in a comparison of the Po$_2$ values indicated by the two different measuring
methods we continued our measuring at low blood pressures. This
explains the "pathological" form of the pooled histograms of Fig. 4.
However, histograms derived from data obtained from rabbits with
blood pressures above 65 mm Hg MAP have a physiological form as
shown in Fig. 5 (left: surface probe data; hypodermic needle probe
data on the right).

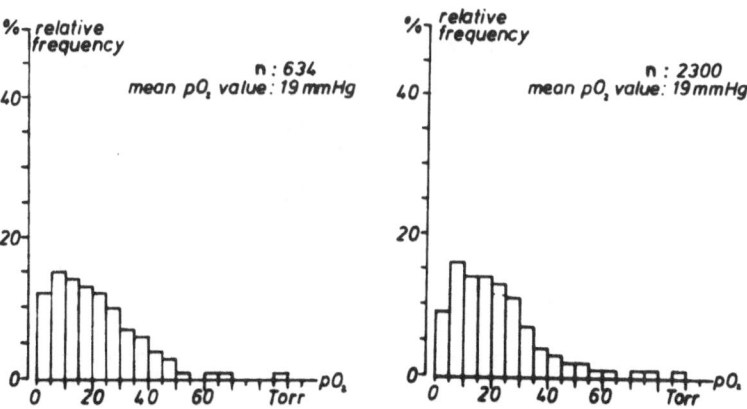

Fig. 4.   Frequency distribution of pooled Po$_2$ values measured
          simultaneously in skeletal muscle of 5 rabbits.
          left : surface probe data
          right: hypodermic needle probe data

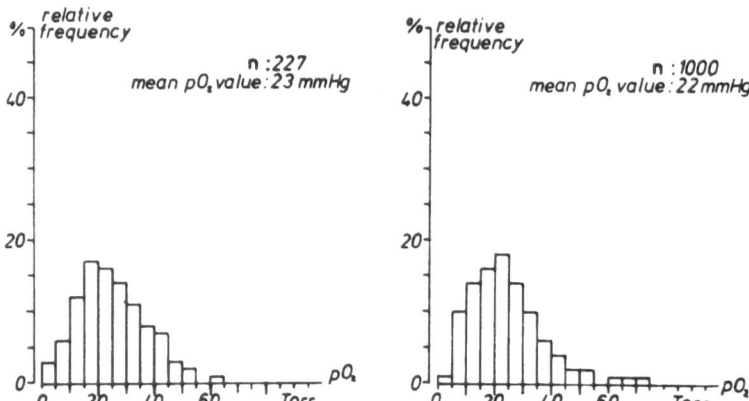

Fig. 5. Frequency distribution of Po_2 values measured simultaneously in skeletal muscle of rabbits with a MAP > 65 mm Hg.
left : surface probe measurements
right: hypodermic needle probe measurements

While the maximum difference of the mean Po_2 of the data
obtained with the two measuring methods lies at 2 mm Hg, even in
the range of very low mean blood pressure, i.e., < 50 mm Hg the
difference of the mean Po_2 values reaches only 4 mm Hg as shown in
Fig. 6 (left: surface probe data; right: hypodermic needle probe
data).

Concerning the third question: In 3 rabbits of the sedated
group the Po_2 in the respiratory gas has been suddenly raised

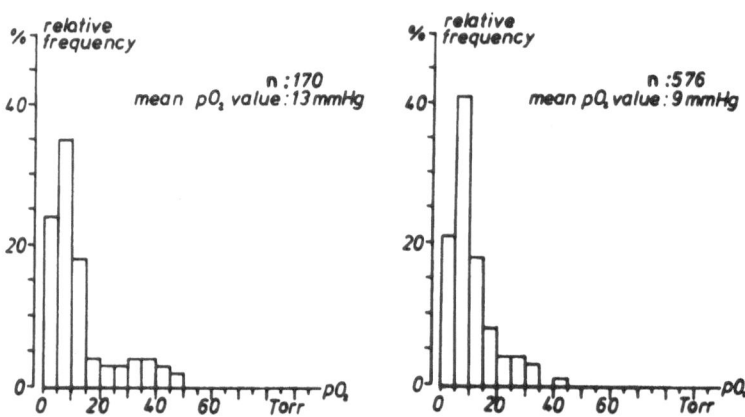

Fig. 6. Frequency distribution of Po_2 values measured simultaneously in skeletal muscle of rabbits with a MAP < 50 mm Hg.
left : surface probe measurements
right: hypodermic needle probe measurements

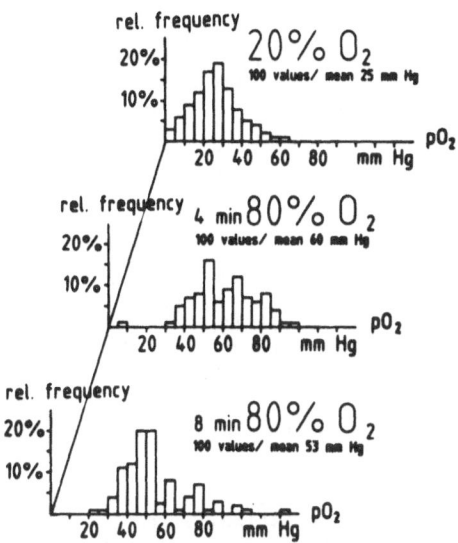

Fig. 7.   Frequency distribution of Po₂ values from rabbit skeletal
          muscle measured with hypodermic needle probes. The animal
          breathed spontaneously.
          top histogram   : air
          middle histogram: for 4 min 80% O₂; 20% N₂
          bottom histogram: for 8 min 80% O₂; 20% N₂

from 20% to 80%. In Fig. 7 the effect on the muscle Po₂ histogram
of an individual rabbit is shown after 4 and 8 min respiring the
altered gas mixture. Only hypodermic needle Po₂ probes were used.
A typical maldistribution of the type observed in the human under
analogous conditions could only be observed for the first 20 minutes,
subsequently the histograms slowly reattained a more physiological
form. The mean Po₂ after 30 min lay at 46 mm Hg.

     In another group of experiments hypodermic needle Po₂ measure-
ments were carried out to demonstrate the effects of the administration of phentolamine on tisue Po₂ histograms 2, 8 and 14 minutes after administration of 1 mg/kg b.w. phentolamine (Fig. 8). In
both groups of experiments, those with changes of the respired gas
(Fig. 7), and those with administration of phentolamine (Fig. 8)
individual histograms were plotted in shortest intervals of only
two minutes. Thus fairly rapid changes of tissue oxygenation could
be followed up.

CONCLUSIONS

     If a hypodermic needle Po₂ probe of a shaft diameter of 400 um
is inserted in resting skeletal muscle the local Po₂ decreases

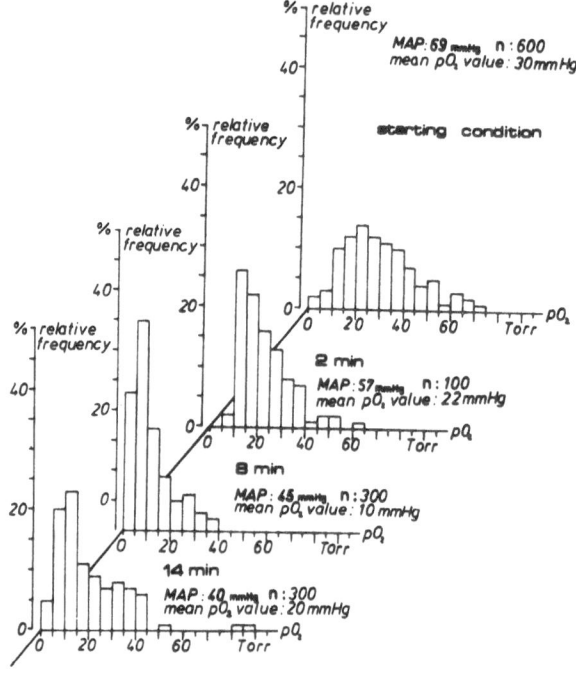

Fig. 8.   Frequency distribution of Po$_2$ values from rabbit skeletal
          muscle measured with hypodermic needle probes. The animal
          breathed air spontaneously. Po$_2$ histograms 2, 8, 14 minutes
          after i.v. administration of 1 mg phentolamine/kg b.w.

relatively fast in the vicinity of the probe, probably due to tissue
compression. Therefore, information about the oxygenation of such
tissue can only be obtained with probes with a response time ≪ 1 sec.
These probes must be moved in sufficiently large and quick steps and
only a Po$_2$ value indicated shortly after the standstill of the probe
can be evaluated. In this way Po$_2$ tissue histograms are obtained
which are similar in form, and mean, to histograms obtained simultaneously on the same tissue by multiwire surface Po$_2$ probes.

     The results of the experiments with sudden changes in the
respired gas and results obtained after the administration of phentolamine demonstrate that the hypodermic needle measurements are
fast enough to follow swift changes in tissue oxygenation. A special
advantage of the hypodermic needle method besides its swiftness
lies in the fact that the probes are easily sterilized and produce
a negligible wound.

REFERENCES

Fleckenstein, W., 1982, In vivo measurements of Po_2 histograms
      using a hypodermic needle electrode system, <u>Pflügers Arch.</u>,
      392, Suppl.:R209.
Hauss, J., Schönleben, S., and Spiegel, H.K., 1982, "Therapiekon-
      trolle durch Überwachung des Gewebe-Po_2", Hans Huber, Bern.

DIFFUSION-PERFUSION RELATIONSHIPS IN SKELETAL MUSCLE: MODELS AND

EXPERIMENTAL EVIDENCE FROM INERT GAS WASHOUT

J. Piiper, and M. Meyer

Abteilung Physiologie, Max-Planck-Institut für
experimentelle Medizin, Göttingen, FRG

INTRODUCTION

The aim of the present research was to study the basic gas
exchange mechanisms - diffusion and perfusion - and their interaction in order to better understand O_2 and CO_2 exchange in muscle
tissue. Instead of the respiratory gases O_2 and CO_2, inert gases
were used because their transfer behaviour is more easily analyzed
due to absence of chemical combination and of comsumption or production.

To investigate the diffusion aspect, two inert gases of differing diffusivity, CH_4 and SF_6, were analyzed simultaneously in
tissue washout experiments.

METHODS

The experiments were performed on 12 artificially ventilated
dogs, body weight 22-29 kg, anaesthetized with chloralose (80 mg/
kg) and urethane (250 mg/kg) intravenously. The isolated-perfused
gastrocnemius muscle preparation of the dog was used. The venous
outflow of the gastrocnemius muscle was obtained by cannulation of
the femoral vein after clamping of all branches from other areas.
The outflow was returned to the jugular vein by a roller-pump. The
muscle was wrapped in tissue soaked with Ringer solution and kept
at the animal's body temperature by a heating pad. Two different
inert gas washout procedures were used.

Series I

In this series two experimental animals, A and B, were used. The dog A (with isolated gastrocnemius muscle) was ventilated with room air. The muscle was perfused by arterial blood from dog B ventilated with a gas mixture containing 45% CH_4, 35% SF_6 and 20% O_2. After equilibration of at least 60 minutes, the perfusion of the isolated muscle was switched back to dog A, whereby test gas washout was started. During 60 minutes of washout at least 12 spot samples of venous outflow blood were withdrawn and analyzed for CH_4 and SF_6 content using gas chromatographic analysis. The wash-out rate constants were obtained from the time course of the venous CH_4 and SF_6 concentrations plotted on logarithmic scale, applying the 'curve peeling' technique.

Series II

In this series only one experimental animal was used. Similar-ly to series I, the dog was equilibrated by ventilation with the test gas mixture for at least 60 min. Washout was initiated by changeover to ventilation with room air and arterial and muscle venous samples were simultaneously withdrawn and analyzed for CH_4 and SF_6. From the time course of veno-arterial concentration dif-ferences the washout rate constants of CH_4 and SF_6 were obtained.

There was no significant difference between the results ob-tained by the two procedures.

Table 1. Ratios of component rate constants (k) and component amplitudes (A) of washout of CH_4 and SF_6. Mean values $\pm$ SD; n, number of measurements

	k_1/k_2	k_3/k_2	A_1	A_2	A_3
Rest (n = 10)					
CH_4	4.4 $\pm$2.0	0.16 $\pm$0.04	0.37 $\pm$0.14	0.32 $\pm$0.11	0.31 $\pm$0.15
SF_6	4.4 $\pm$1.9	0.13 $\pm$0.04	0.36 $\pm$0.18	0.27 $\pm$0.18	0.37 $\pm$0.20
Exercise (n = 10)					
CH_4	2.4 $\pm$0.5	0.13 $\pm$0.07	0.51 $\pm$0.17	0.29 $\pm$0.11	0.20 $\pm$0.16
SF_6	2.3 $\pm$0.8	0.16 $\pm$0.06	0.40 $\pm$0.21	0.29 $\pm$0.13	0.31 $\pm$0.21

RESULTS

The mean values of simultaneous washout measurements are shown in Tables 1 and 2. In spite of considerable scatter of the individual measurements the following features are apparent.

1. The time course of CH_4 and SF_6 washout was markedly non-monoexponential. Three components characterized by their respective rate constants could be identified when the washout curves were analyzed to 5% of the initial value.

2. The shape of the washout curves was remarkably similar for both CH_4 and SF_6 as documented by the similar values of relative amplitudes and rate constant ratios of the three exponential components.

3. The washout rate constants for CH_4 were on the average by a factor 1.10 to 1.25 higher than those for SF_6.

4. The results in resting and exercising muscles were similar except that the k_1/k_2 ratio for both CH_4 and SF_6 was reduced at exercise.

Table 2. Ratio of exponential washout rate constants (k) of CH_4 and SF_6 together with the respective ratios predicted for perfusion-limited transfer (λ^{-1}), diffusion-limited transfer (d) and transfer limited by veno-arterial back diffusion $[(\lambda^2 \cdot d)^{-1}]$. Mean values $\pm$ SD; $\bar{k} = (k_2+k_3)/2$; n, number of measurements.

	k_2	k_3	$\bar{k}$	λ^{-1}	d	$(\lambda^2 \cdot d)^{-1}$
			CH_4/SF_6			
Rest						
	1.18 ± 0.33	1.32 ± 0.38	1.25 ± 0.36			
(n = 10)						
				1.46	3.02	0.71
Exercise						
	1.09 ± 0.50	1.11 ± 0.17	1.10 ± 0.33			
(n = 10)						

THEORY

To analyze the kinetics of inert gas washout from tissue use will be made of extremely simplified analog models in an attempt to reduce the number of variables as far as possible.

The particular aspects considered are:
1. homogeneity vs. inhomogeneity
2. diffusion vs. perfusion limitation
3. diffusion shunt

Homogeneous Tissue Model

The washout of inert gas, measured as decrease of concentration in venous outflow, C_v, from its initial pre-washout value, $C_{v(0)}$ is monoexponential for a homogeneous (i.e. efficiently 'stirred') tissue:

$$C_v/C_{v(0)} = A \cdot exp \; (-k \cdot t) \tag{1}$$

If washout of a homogeneous tissue is perfusion-limited, i.e. an equilibrium between tissue and outflowing blood is always attained, the rate constant k is proportional to the specific blood flow $\dot{q}$ (= blood flow per tissue volume) and inversely proportional to the tissue/blood partition coefficient λ (Kety, 1951):

$$k = \dot{q}/\lambda \tag{2}$$

Thus for perfusion-limited washout of several inert gases the k values should be proportional to $1/\lambda$.

For evaluation of the role of diffusion limitation it is appropriate to use a model in which capillary blood is separated from a homogeneous tissue compartment by a thin diffusion barrier of uniform thickness and zero capacitance. The diffusion characteristics of the barrier are given by the surface/thickness ratio, γ, and the diffusion coefficient, d, of the gas in the barrier material. The washout is again monoexponential, but the washout rate constant k depends additionally on diffusion parameters:

$$k = \; 1 - exp(- \gamma \cdot d \cdot \lambda/\dot{q}) \; \cdot \dot{q}/\lambda \tag{3}$$

For $\gamma \cdot d \cdot \lambda/\dot{q} \to 0$, i.e. exclusive diffusion limitation, eq. (3) yields:

$$k = \gamma \cdot d \tag{4}$$

When test gases of differing diffusivity, d, are used in washout experiments, the presence and the extent of diffusion limitation may be derived from the relationship between k and d.

Inhomogeneous Tissue Model

It follows from the general structure of the vascular system that an organ or a tissue can be likened to a multitude of parallel capillary-tissue units. When the units are unequal with respect to gas exchange parameters, washout of the whole system will be multi-exponential:

$$C_v/C_{v(0)} = A_1 \cdot \exp(-k_1 \cdot t) + A_2 \cdot \exp(-k_2 \cdot t) + \ldots \qquad (5)$$

For the perfusion-limited case the rate constants, k_n, are proportional to the specific perfusion:

$$k_n = (\dot{q}/\lambda)_n \qquad (6)$$

In the case that there is additional diffusion limitation, one obtains (assuming constant d and λ):

$$k_n = \left[1 - \exp(-\gamma_n \cdot d \cdot \lambda/\dot{q}_n)\right] \cdot \dot{q}_n/\lambda \qquad (7)$$

It follows that for an inhomogeneity based on unequal distribution of blood flow to tissue volume in a perfusion-limited system, the k ratios of given compartments (e.g. k_1/k_2) should be equal for all test gases. But if the inhomogeneity is mainly due to differing diffusion conditions (γ), the k values of slow components are expected to be relatively smaller for less diffusible gases.

Diffusion Shunt

In densely vascularized organs like muscle the highly developed diffusion conditions may lead to diffusive shunting between arterial and venous vessels (or between arterial and venous ends of capillaries). The result for inert gas washout is 'trapping' of test gas by veno-arterial back diffusion which would preferentially delay the washout of gases of higher diffusivity.

For a simple model in which inert gas diffusion between arterial and venous vessels, arranged in counter-current manner, occurs through a barrier with area/thickness ratio per unit tissue volume γ', the washout rate constant is given by the relationship:

$$k = \frac{\dot{q}/\lambda}{\left[1 - \exp(\gamma \cdot d \cdot \lambda/\dot{q})\right]^{-1} + \gamma' \cdot d \cdot \lambda/\dot{q}} \qquad (8)$$

This relationship shows the following features:
1. With increasing $\gamma' \cdot d$, k decreases.
2. With increasing d, k first increases (due to enhanced blood-tissue equilibration) and thereafter decreases (due to enhanced veno-arterial back diffusion).

DISCUSSION

Inhomogeneity

The markedly non-monoexponential washout pattern shows that
the muscle is functionally inhomogeneous. In exercise the inhomogeneity is slightly reduced as indicated by the decrease of the
k_1/k_2 ratio from 4.4 to 2.4. The simplest and most likely explanation is unequal tissue perfusion.

If the inhomogeneity was mainly due to presence of varied diffusion conditions, the k_1/k_2 ratio should be higher and the k_3/k_2
ratio should be smaller for the less diffusible gas, SF_6. The experimental results reveal no such tendency.

The variance of blood flow distribution in muscle tissue has
been shown in several previous studies. Paradise et al. (1971)
found considerable regional variations of tritiated water in the
extensor digitorum longus muscle of the dog following 2-3 min perfusion with labeled blood. Sparks and Mohrman (1977) displayed the
heterogeneity of muscle blood flow by autoradiography after intra-
arterial bolus injection of ^{85}Kr in skinned calf muscle preparations
of the dog. Pendergast et al. (1982) showed by radioactive microsphere embolization and by local ^{133}Xe clearance that a large variation of $\dot{q}$ existed, both at rest and during exercise, in a dog gastrocnemius preparation similar to that used in this study. It is
remarkable, that the same degree of variance was found in intact
gastrocnemii of the dog.

Diffusion Limitation

If tissue/blood equilibration had been limiting the washout,
the CH_4/SF_6 rate constant ratio should have been in the range
between exclusive perfusion limitation (eq. 2) and exclusive diffusion limitation (eq. 4).

Ohta et al. (1978) found in whole dogs that simultaneous washout of N_2 and Ar was inversely proportional to their λ values, in
accordance with perfusion-limited gas transport. The same result
was obtained in the dog brain using CH_4 and Ar as test gases (Ohta
et al., 1979).

In our experiments washout of the less diffusible gas SF_6 was
faster than expected for a simple perfusion-limited model. Such
behaviour is unexplainable by the simple diffusion-perfusion models
considered above, but may be tentatively attributed to a 'diffusion
shunt' mechanism.

Veno-Arterial Back Diffusion (Diffusion Shunt)

Veno-arterial back diffusion provides an explanation for slower washout of a more diffusible gas. Of interest is the theoretically greatest effect of veno-arterial back diffusion. It follows from eq. (8) that with no tissue/blood diffusion limitation ($\gamma \cdot d = \infty$) and high γ' ($\gamma' \cdot d \cdot \lambda/\dot{q} \gg 1$), $k = (\dot{q}/\lambda)^2/(\gamma' \cdot d)$ and therefore

$$\frac{k_y}{k_x} = \frac{(\lambda_x)^2 \cdot d_x}{(\lambda_y)^2 \cdot d_y} \tag{9}$$

The predicted value for kCH_4/kSF_6 is 0.71 (Table 2). Thus the experimental kCH_4/kSF_6 value is in the explainable range.

The 'diffusion shunt', i.e. transfer between arterial and venous vessels by-passing capillaries, is well known from heat exchange in tissue (Bazett et al., 1948; Piiper, 1959; Aukland, 1967).

The non-monoexponential washout curves measured by Sejrsen and Tønnesen in the cat gastrocnemius after intra-arterial bolus injection, prolonged intra-arterial infusion (corresponding to our method) and after atraumatic local application (Tønnesen and Sejrsen, 1967; Sejrsen and Tønnesen, 1968; Sejrsen, 1970) have been interpreted to be due to diffusion shunting. However, diffusion shunting does not necessarily lead to non-monoexponential washout, as was clearly pointed out also by Kruhøffer (1970).

The effects of diffusion shunt on the behaviour of diffusible indicators in perfused organs have been studied by various groups (e.g. Bassingthwaighte et al., 1970) and its effects on O_2 supply to tissue have been investigated by Grunewald and Sowa (1977). We believe that veno-arterial back diffusion was effective in our experiments, producing the unexpectedly slow washout of CH_4 as compared to SF_6. Further support for this hypothesis could be provided by simultaneous, and more accurate, measurement of more gas species with widely varied diffusion properties and by experimental changes of blood flow over a wide range.

SUMMARY

In order to study the dependence of blood-tissue gas exchange upon diffusion, the simultaneous washout of two inert gases of differing diffusivity was investigated in isolated-perfused dog gastrocnemius preparations. The muscles were equilibrated with CH_4 and SF_6 via arterial blood. The washout kinetics were determined from venous blood samples analyzed by gas chromatography. The results revealed the following features:

1. The washout of the test gases was pronouncedly multi-exponential, and could be described by three exponential components when analyzed to 5% of the initial value. The non-exponential washout was attributed to unequal distribution of capillary blood flow to tissue volume.

2. The mean ratio of washout rate constants CH_4/SF_6 was within 1.10 - 1.25 and was even smaller than the ratio expected for pure perfusion limitation (1.46). Therefore, no evidence for effective tissue-blood diffusion limitation was obtained.

3. The observed washout rate constant ratio could be explained by a model with veno-arterial back diffusion which more strongly retards washout kinetics of the better diffusible gas (CH_4) as compared to the less diffusible gas (SF_6).

REFERENCES

Aukland, K., 1967, Renal medullary heat clearance in the dog, Circ. Res., 20:194-203.

Bassingthwaighte, J.B., Strandell, T., and Yipintsoi, T., 1970, Flow limited washout of diffusible solutes from the heart, in: "Benzon Symposium II: Capillary Permeability", C. Crone, N.A. Lassen, eds., Munksgaard, Copenhagen, pp.580-585.

Bazett, H.C., Love, L., Newton, M., Eisenburg, L., Day, R., and Forster, R., 1948, Temperature changes in blood flowing in arteries and veins in man, J. Appl. Physiol., 1:3-19.

Grunewald, W.A., and Sowa, W., 1977, Capillary structures and O_2 supply to tissue, Rev. Physiol. Biochem. Pharmacol., 77:149-209.

Kety, S.S., 1951, Theory and applications of the exchange of inert gas at the lungs and tissues, Pharmacol. Rev., 3:1-41.

Kruhøffer, P., 1970, Discussion remark, in: "Benzon Symposium II: Capillary Permeability", C. Crone, N.A. Lassen, eds., Munksgaard, Copenhagen, pp.597-598.

Ohta, Y., Song, S.H., Groom, A.C., and Farhi, L.E., 1978, Is inert gas washout from tissues limited by diffusion? J. Appl. Physiol., 45:903-9o7.

Paradise, N.F., Swayze, C.R., Shin, D.H., and Fox, I.J., 1971, Perfusion heterogeneity in skeletal muscle using tritiated water, Am. J. Physiol., 220:1107-1115.

Pendergast, D., Cerretelli, P., Heisler, N., Marconi, C., Meyer, M., and Piiper, J., 1982, Muscle blood flow distribution in resting and exercising dog gastrocnemius, Fed. Proc., 41:1680.

Piiper, J., 1959, Durchblutung der arterio-venösen Anastomosen und Wärmeaustausch an der Hundeextremität, Pflügers Arch. Ges. Physiol., 268:242-253.

Sejrsen, P., 1970, Convection and diffusion of inert gases in cutaneous, subcutaneous, and skeletal muscle tissue, in: "Benzon Symposium II: Capillary Permeability", C. Crone, N.A. Lassen, eds., Munksgaard, Copenhagen, pp. 586-596.

Sejrsen, P., and Tønnesen, K.H., 1968, Inert gas diffusion method
    for measurement of blood flow using saturation technique: com-
    parison with directly measured blood flow in isolated gastroc-
    nemius muscle of the cat, Circ. Res., 22:679-693.
Sparks, H.V., and Mohrman, D.E., 1977, Heterogeneity of flow as an
    explanation of the multiexponential washout of inert gas from
    skeletal muscle, Microvasc. Res., 13:181-184.
Tønnesen, K.H., and Sejrsen, P., 1967, Inert gas diffusion method
    for measurement of blood flow: comparison of bolus injection
    to directly measured blood flow in the isolated gastrocnemius
    muscle, Circ. Res., 20:552-564.

OXYGEN SUPPLY OF SKELETAL MUSCLE IN EXPERIMENTAL ENDOTOXIC SHOCK[*]

K.H. Kopp[1], E. Sinagowitz[2], and H. Müller[1]

[1]Dept. of Anesthesiology, Universitätsklinik Freiburg
7800 Freiburg, FRG
[2]Dept. of Urology, Städtisches Krankenhaus
7990 Friedrichshafen, FRG

The hypodynamic form of septic shock is a disease with a high grade of mortality. It is characterized by severe disturbances of hemodynamics, pulmonary gas exchange and metabolism. Additionally, severe disturbances of microcirculation and a distinct cellular hypoxia are assumed (Lasch, 1979).

As shown in experimental hemorrhagic shock (Sinagowitz et al., 1973) microcirculatory disorders can be detected by measuring tissue Po_2 in skeletal muscle, even before changes in systemic hemodynamics are found. Similar observations are made in patients (Schönleben et al., 1978). In human hyperdynamic septic state increased muscle Po_2 due to increased cardiac output, could be found (Kopp et al., 1982). However, in hypodynamic septic shock tissue Po_2 measurements are not practicable, since early treatment of those patients is imperative. Thus, hypodynamic septic shock was induced experimentally in pigs by infusion of endotoxin to study oxygen supply and microcirculation in skeletal muscle. The goal of these investigations was to answer the following two questions:

1. Are there early microcirculatory disorders in skeletal muscle during endotoxic shock, similar to those observed in hemorrhagic shock and

2. which are the relationships between hemodynamic pulmonary and metabolic changes and local tissue Po_2?

[*]Supported by grants of the Deutsche Forschungsgemeinschaft, 5300 Bonn-Bad Godesberg/FRG

METHOD

The experiments were done in pigs with a mean body weight of
17 kg. In 8 animals endotoxic shock was induced by continuous infusion of highly purified salmonella endotoxin over a period of
4 hours (dose 50 ug/kg body weight). Anesthesia was induced with
Nembutal and maintained with Methomidate and Fentanyl under controlled respiration, 7 pigs served as control group infused with
saline only.

After tracheostomy a Swan-Ganz catheter was placed in the
pulmonary artery for measurement of cardiac output (thermodilution),
pulmonary arterial pressure and pulmonary capillary wedge pressure.
Two catheters in the jugular vein provided for the infusion of endotoxin or saline and the measurement of central venous pressure.
Arterial blood pressure was monitored in the aorta and arterial P_{O_2}
was measured continuously with a catheter electrode in the same
vessel. Local tissue P_{O_2} in the gracilis muscle was measured by means
of the multiwire surface electrode according to Kessler and Lübbers
(1966). The experimental protocol was as follows:

After a stabilization period of 2 hours continuous infusion of
endotoxin (or saline) was started for a period of 4 hours. Measurements were done before infusion and then in 1 hour-intervals:

MAP   mean arterial blood pressure
CO   cardiac output
HR   heart rate
PAP   pulmonary arterial pressure (mean)
PCwP   pulmonary capillary wedge pressure (mean)
CVP   central venous pressure
P_aO_2 arterial P_{O_2}
P_aCO_2 arterial P_{CO_2}

acid-base state of arterial blood
lactate
catecholamines
P_mO_2 mean tissue P_{O_2} of skeletal muscle
tissue P_{O_2} histograms

RESULTS

1. Hemodynamic Parameters

Fig. 1 represents MAP, HR and CO during the course of the experiment. In the endotoxic group MAP decreased after 1 hr from 73
to 41 mm Hg (4. hr), HR increased continuously to 153 b/min and CO
dropped from 1.8 to 0.85 l/min, close to 50%.

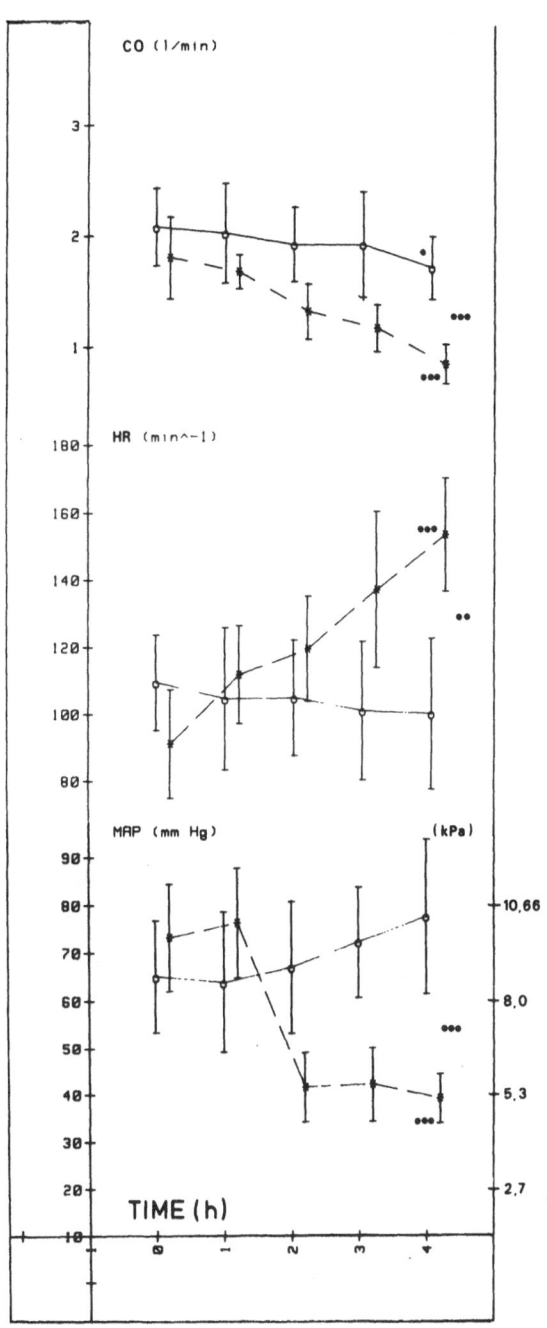

Fig. 1. MAP, HR and CO
        (full line: control group, dashed line: endotoxic group,
        abszissa: time in hours (h)); $\bar{x} \pm x_s$; the small stars
        on the left side of the 4. hr values indicate the statis-
        tical difference between the 0. hr and 4. hr within each
        group, those on the right side of the figure the difference
        between the two groups to 4. hr. (* p < 0.05, ** p < 0.01,
        *** p < 0.001).

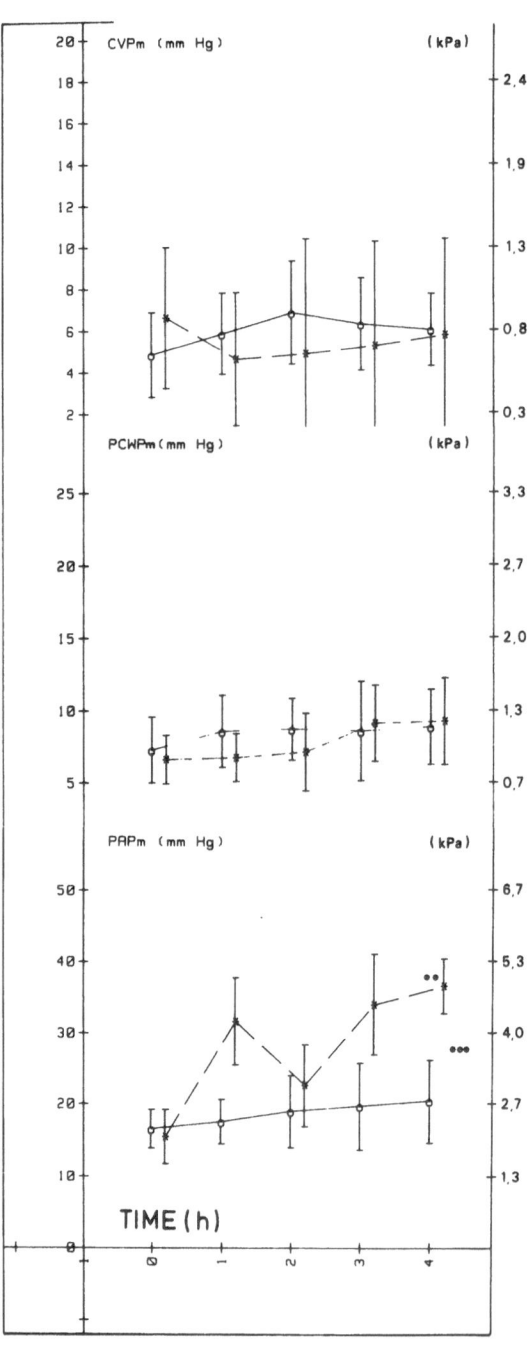

Fig. 2. PAP, PCwP and CVP
        (full line: control group, dashed line: endotoxic group).
        For further explanations see Fig. 1.

PAP (Fig. 2) increased biphasically in the endotoxic group from 15 to 36 mm Hg (4. hr) whereas PCwP and CVP remained nearly unchanged.

Fig. 3 demonstrates the very early excessive increase of PAP in one of the endotoxic pigs. It is shown, that this increase occurred about 11' after starting infusion of endotoxin. The pressure recorded increased to nearly systemic values (70/35 mm Hg). All other parameters measured simultaneously, except P_aO_2, remained stable. P_aO_2, however, showed a small decrease which paralleled the increase of PAP. This must be considered as an effect of the pronounced pulmonary vasoconstriction on the pulmonary ventilation/ perfusion. The increase of pulmonary resistance is presumably mediated by the endotoxic induced release of Thromboxan TxA_2 (Demling et al., 1981), the pronounced changes of the other hemodynamic parameters by additional release of substances (histamin, kinins, prostaglandin, complement fractions (C_{3A} and C_{5A}) which act by different ways: vasocontrictions, vasodilation or increasing permeability of capillaries.

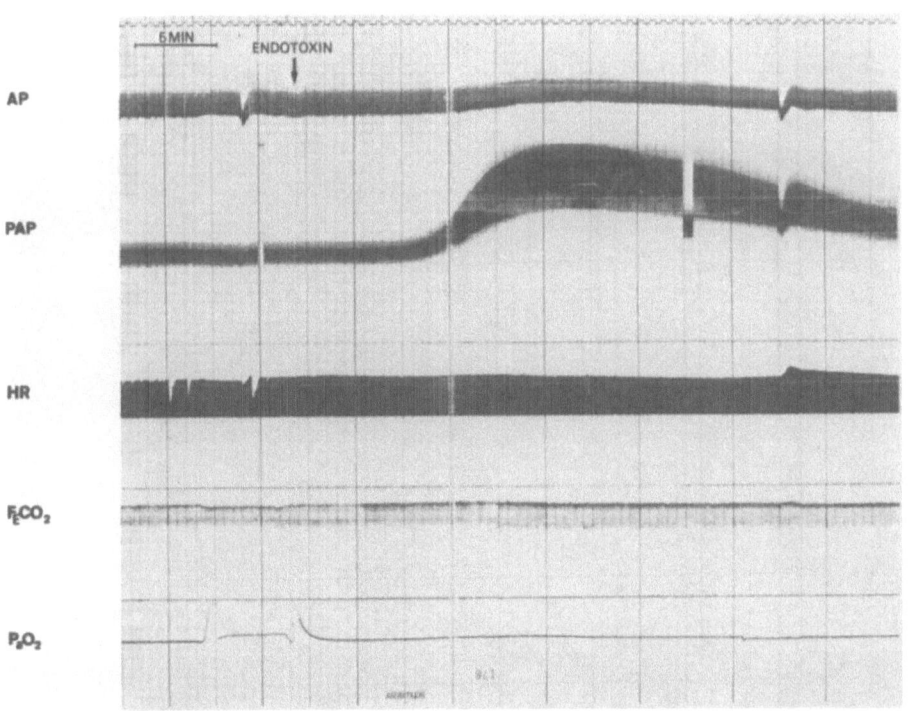

Fig. 3. Continuous recording of AP, PAP, HR, F_ECO_2 and P_aO_2 during infusion of endotoxin (started at ↓).

2. Pulmonary Gas Exchange

Whereas P_aCO_2 remained constant in both groups, arterial P_aO_2 continuously decreased from 77 to 61 mm Hg in the endotoxic group (Fig. 4), indicating a progressive development of a pulmonary failure with impaired gas exchange, which clinically appeared as a pulmonary edema.

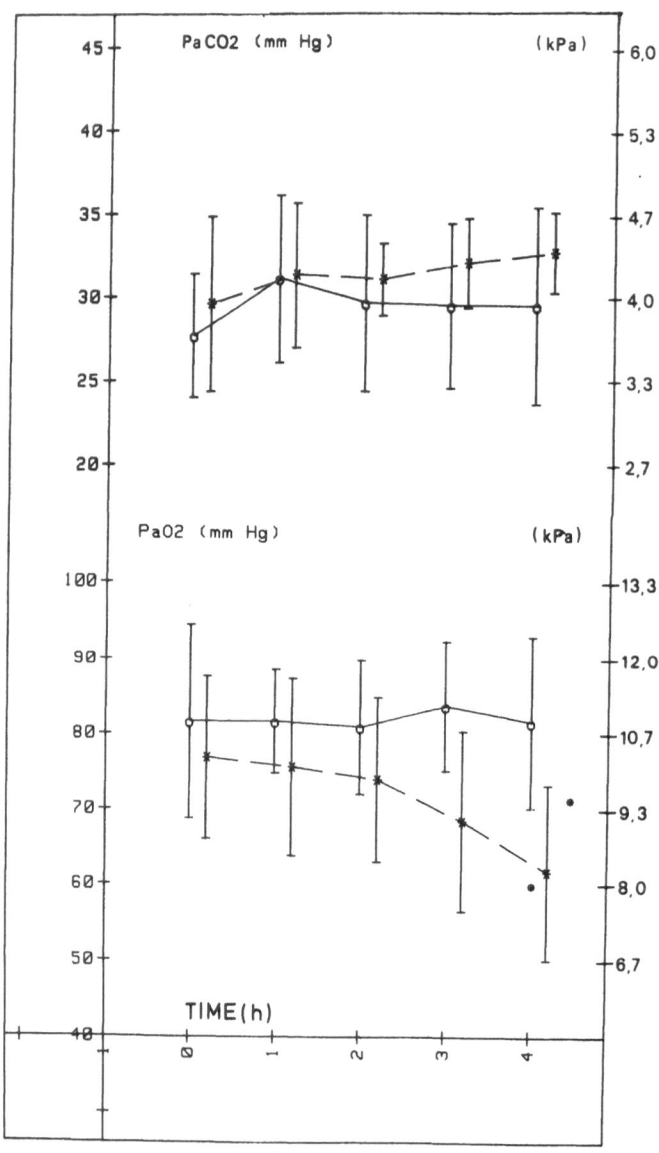

Fig. 4. P_aO_2
        (full line: control group, dashed line: endotoxic group).
        For further explanations see Fig. 1.

3. Metabolism

During the infusion of endotoxin a progressive metabolic acid-osis developed: pH decreased from 7.51 to 7.32 and standard bicar-bonate from 25 to 15 $mmol \cdot l^{-1}$. At the end of the experiment base excess was lowered from 1 to -7.7 $mmol \cdot l^{-1}$. This acidosis seems to be a lactatcidosis since lactate in the blood of the endotoxic pigs increased 3-fold from 2.2 to 6.9 $mmol \cdot l^{-1}$. A concomitant elevation of plasma epinephrine and n-ephinephrine levels also could be ob-served (epinephrine 0. hr: 50 pg/ml, 4. hr: 200 pg/ml; n-epinephrine 0. hr: 25 pg/ml, 4. hr: 250 pg/ml).

4. Muscle Po_2 Measurements

Mean tissue Po_2 of skeletal muscle (Fig. 5) increased in both groups during the first hour. The low initial values must be con-sidered as a long lasting effect of Nembutal with preferential per-fusion of non-nutritive capillaries, as shown by Franke et al. (1982).

While mean tissue Po_2 remained stable in the control group, tissue Po_2 in the endotoxic group decreased again during the fol-lowing hours to the initial value of 19 mm Hg. These changes, how-ever, were, at least in the early phase of the experiment, not as

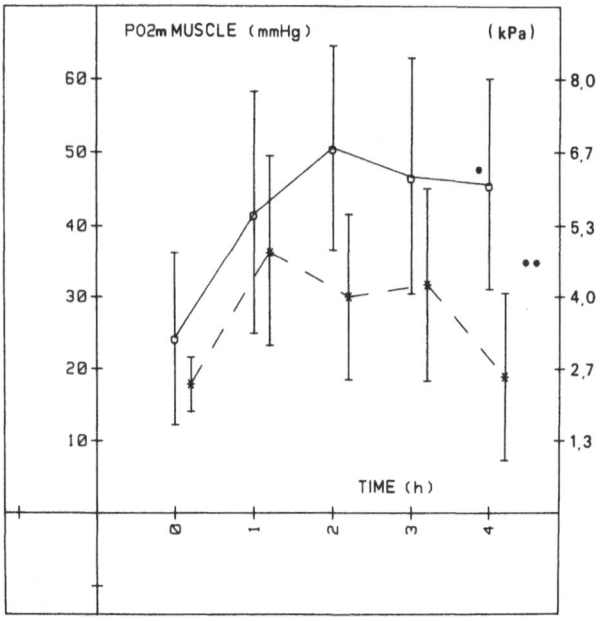

Fig. 5.   Mean tissue Po_2 of skeletal muscle
         (full line: control group, dashed line: endotoxic group).
         For further explanation see Fig. 1.

pronounced as the other mentioned hemodynamic, pulmonary and metabolic parameters; a deep decrease only seemed to occur between the
third and the fourth hour.

Fig. 6 demonstrates the summarized Po_2 histograms of the
control group (left) and the endotoxic group (right). In contrast
to the control group the Po_2 histograms during endotoxic shock show
a left shift after 1 hr with an increased amount of hypoxic values
up to 17% at the end of the experiment. However, these hypoxic
values did not reach the extent observed in hemorrhagic shock
(Sinagowitz et al., 1973).

Our findings demonstrate, that microcirculatory changes in the
muscle took place very slowly and in a very late phase during the
development of hypodynamic endotoxic shock. This may be due to individual but normally well-functioning compensating mechanism to
low flow - or hypoxic hypoxia as described by the model of local
redistribution of microflow according to Kessler et al. (1976). The

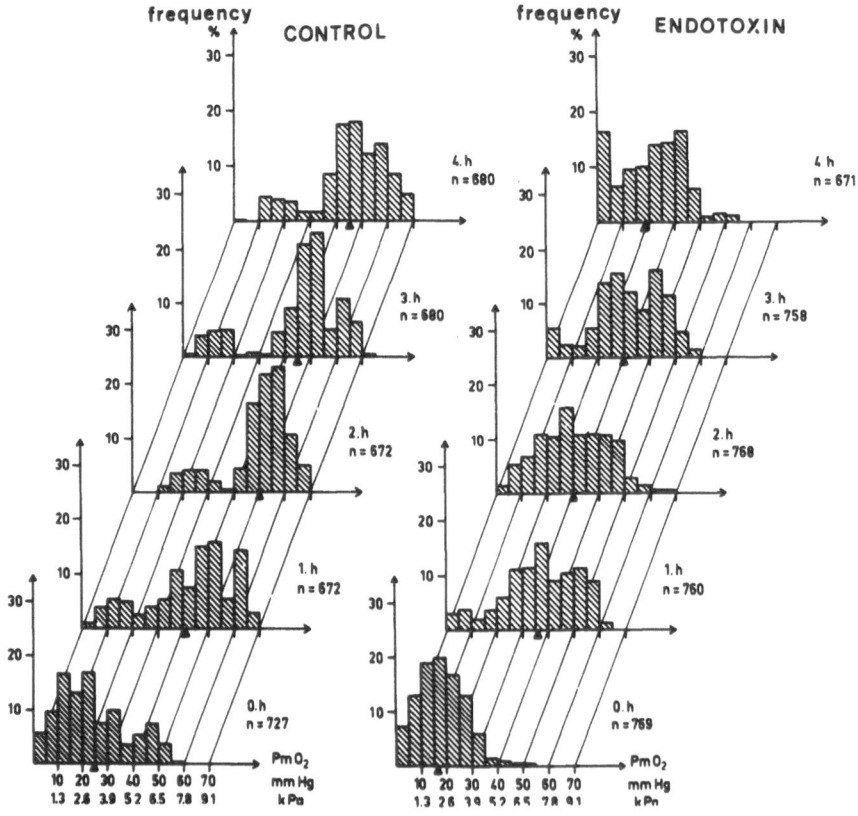

Fig. 6.   Summarized Po_2 histograms of the control group (left) and
the endotoxic group (right). P_mo_2 = muscle Po_2.

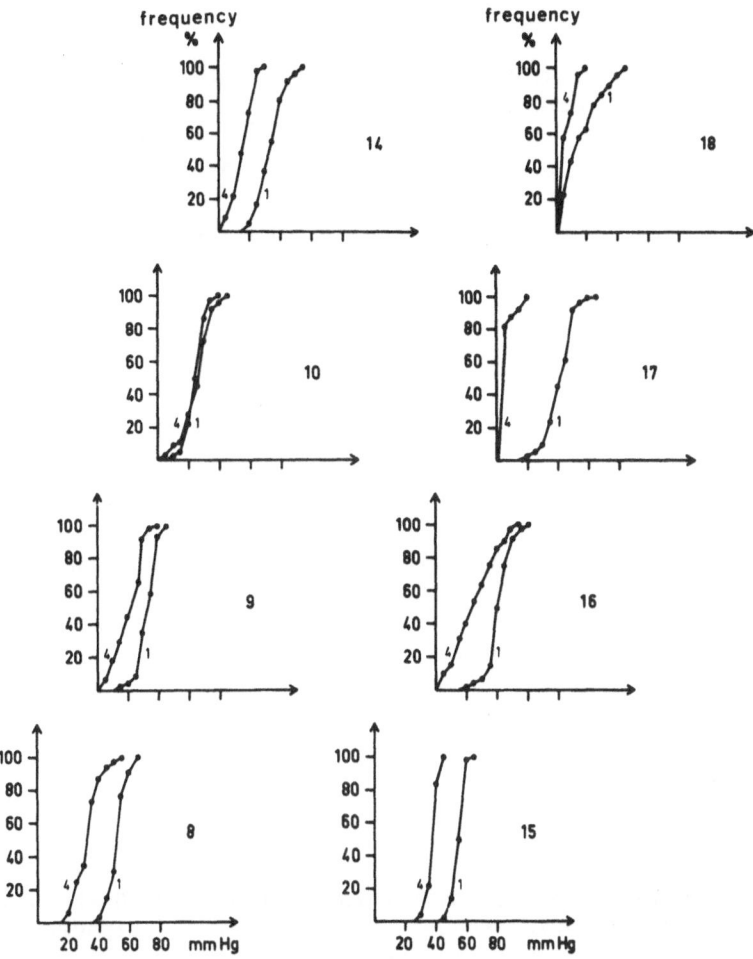

Fig. 7.   Cumulative Po_2 histograms of each endotoxic pig, compared
          at the 1. and 4. hour.

individual responses of skeletal muscle Po_2 to endotoxic shock
support this suggestion (Fig. 7).

     Distinct differences can be seen in the animals no. 10 and no.
17. Whereas muscle oxygen supply remained unchanged in no. 10, it
was disturbed dramatically in no. 17. But both animals show similar
changes in the general parameters.

     From the development of an early lactacidose it must be conclud-
ed that microcirculatory alterations during endotoxic shock may
occur earlier in other organs for example intraabdominal organs or
skin, which can also be suspected by the appearance of an early and
pronounced cyanosis.

SUMMARY

In our model of hypodynamic endotoxic shock systemic pulmonary and metabolic alterations such as

- hypotension, tachycardia, low output syndrom
- pulmonary hypertension
- pulmonary failure with arterial hypoxemia and
- lactacidosis

developed earlier than disturbances in oxygen supply of skeletal muscle. This points to the different responses of the organ systems in endotoxic shock.

REFERENCES

Demling, R.H., Smith, M., Gunther, R., Flynn, J.F., and Gee, M.H., 1981, Pulmonary injury and prostaglandin production during endotoxemia in conscious sheep, Am. J. Physiol., 240:349.

Franke, N., Endrich, B., Laubenthal, H., Peter, K., and Messmer, K., 1982, Einfluß von Pentobarbital auf die Mikrozirkulation von Skelettmuskulatur und Subcutis. Eine tierexperimentelle Studie, Anaesthesiol. Intensivther., Notfallmed., 17:11.

Kessler, M., Höper, J., and Krumme, A., 1976, Monitoring of tissue perfusion and cellular function, Anesthesiology, 45:184.

Kessler, M., and Lübbers, D.W., 1966, Aufbau und Anwendungsmöglich-keiten verschiedener Po_2-Elektroden, Pflügers Arch. Ges. Physiol., 291:82.

Kopp, K.H., Klieser, H.P., and Würdinger, B., 1982, Muskelsauer-stoff-Partialdruckmessungen zur Klärung der peripheren Sauer-stoffversorgung hyperdynam-septischer Intensivpatienten, Anaesthaesist, 31:500.

Lasch, H.G., 1979, Klinik und Pathophysiologie des Schocks, in: "Schock und Intensivmedizin", W. Sandritter, G. Dohm, eds., Fischer, Stuttgart-New York, p. 2.

Schönleben, K., Hauss, J.P., Spiegel, U., Bünte, H., and Kessler, M., 1978, Monitoring of tissue Po_2 in patients during intensive care, Adv. Exp. Med. Biol., 94:583.

Sinagowitz, E. Palmer, H., Rink, R., Görnandt, L., and Kessler, M., 1973, Local oxygen supply in intraabdominal organs and in skeletal muscle during hemorrhagic shock, Adv. Exp. Med. Biol., 37A:505.

MICROCIRCULATION AND Po$_2$ IN SKELETAL MUSCLE DURING RESPIRATORY HYPOXIA AND STIMULATION

D.K. Harrison, J. Höper, H. Günther, H. Vogel,
K.H. Frank, M. Brunner, R. Ellermann, and M. Kessler

Institut für Physiologie und Kardiologie der Universität
Erlangen-Nürnberg, Waldstr. 6, 8520 Erlangen, FRG

INTRODUCTION

The regulation of blood flow to skeletal muscle has, for many years, been the subject of much research (Hudlická, 1973) with the result that nowadays many phenomena can be explained by the presence of local oxygen sensors which serve to regulate the flow in order to maintain an adequate oxygen supply to the tissue (Granger and Shepherd, 1973). For example, the vasodilation which occurs during hypoxia almost certainly takes place as a result of signals from such sensors, and there is strong evidence to suggest that the majority of these sensors may be located in the tissue cells (Kessler et al., 1983).

In order to examine this regulatory mechanism further, we have investigated not only the level of Po$_2$ in the tissue, but also the capillary blood flow (referred to here as microcirculation) and intracapillary haemoglobin oxygenation under various conditions in the intact sartorius muscle in anaesthetized dogs.

Our first model was that of the reaction to systemic hypoxia of the blood flow in resting muscle. The aim was to reduce the tissue oxygen supply to such an extent that we could determine the levels of capillary haemoglobin oxygenation and tissue Po$_2$ at which changes in flow both to and within the organ take place.

The second method used to induce changes in flow and oxygen uptake in the tissue was stimulation of the non-loaded muscle via the femoral nerve in order to find out how the oxygen supply was regulated under "working conditions".

477

METHODS

Techniques

    Tissue Po_2 measurements were made with a multiwire surface
electrode (Kessler and Lübbers, 1966) and intracapillary HbO_2 was
measured using a micro light-guide spectrophotometer constructed in
our laboratory, as described by Brunner et al. (1981). In order to
investigate changes in microcirculation, we applied the technique
of washout of a freely diffusible indicator, hydrogen, measured by
means of a multiwire surface pH_2 electrode (Krumme et al., 1975).
The catchment area of each wire of the electrode is a hemisphere of
about 30 um diameter, which means that each individual curve (e.g.,
see Fig. 1b) represents the flow only within this region.

    An improvement to the method of Baumgärtl and Lübbers (1973)
for palladinizing the platinum wires has produced a sensitive electrode with a drift of less than 5% per hour, and a response time
comparable to that of the Po_2 electrode. For in vivo experiments,
hydrogen was administered to the inspired gas mixture at a controlled

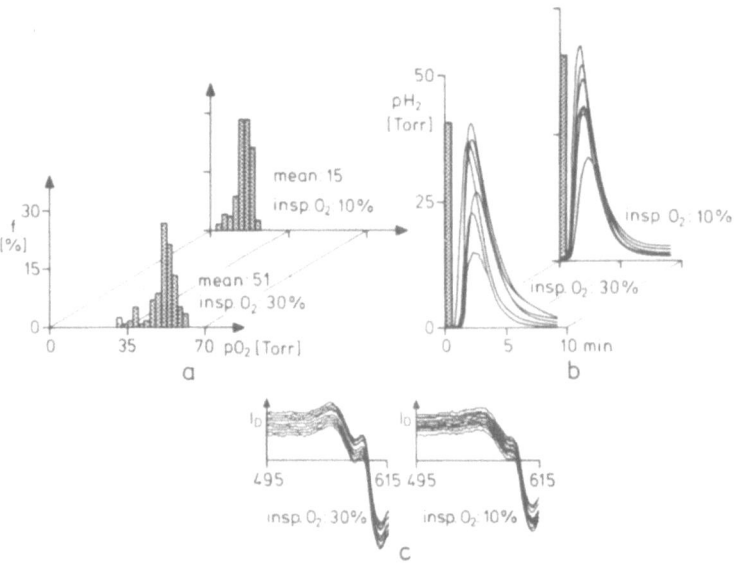

Fig. 1.   a) Tissue Po_2 histograms, b) hydrogen clearance curves and
          c) uncorrected HbO_2 spectra during control and hypoxic
          periods of ventilation. The shaded area in b) indicates
          the duration of administration of H_2 to the inspired gas
          mixture. In c) I_d is the recorded intensity plus a spacing
          factor. Each spectrum shown is the average from 100 single
          spectra recorded at a rate of 15 Hz.

rate for a period of 30 seconds (the partial pressure of oxygen being kept constant) and the washout curve (see Fig. 1) recorded in the muscle. Microflow values were compared using the reciprocal of the time taken for the pH_2 to fall from its maximum to the half-maximum value ($T_{1/2}$).

Procedure

Mongrel dogs (25 - 30 kg) were first anaesthetized with a short-acting barbiturate, and then maintained with piritrimide, fluonitrezapan and N_2O. Pancuronium was used as a relaxant (except when nerval stimulation was carried out) and the dogs were ventilated at an initial FiO_2 of 0.3 in order to obtain consistent control arterial Po_2 values (130 ± 12 mm Hg). ECG was monitored continuously, and arterial blood gases, Hb concentration and Hct were measured at regular intervals. Arterial pressure was measured via a catheter located in the iliac aorta and venous pressure via a catheter in the abdominal vena cava, both by means of a pressure transducer. A small area of the sartorius muscle was exposed in order that tissue measurements could be carried out before positioning an electromagnetic flow probe around the femoral artery.

In the hypoxia experiments (6 dogs) the following procedure was carried out. Control measurements were taken of all parameters before reducing the FiO_2 to 20%, 15%, 10% and then returning to 30% once again. Flow, arterial and venous pressure changes, together with tissue HbO_2 and Po_2 were monitored continuously after each change of inspired O_2 concentration, and steady-state values recorded concurrently with tissue Po_2 histograms and H_2 clearances at each stage of hypoxia.

Stimulation of the femoral nerve was applied with a pulse length of 0.5 ms at 1.3 volts in two further dogs. Stimulation frequency was increased stepwise (see Fig. 5) with no recovery period between, but at regular time intervals. At each stage, control measurements of all parameters were made.

RESULTS

Results from the two experimental treatments are presented. For comparison purposes, results for flow (F), pressure (P), resistance (P/F) and microcirculation are calculated in terms of percentage changes from the initial control value. Also for comparison purposes, HbO_2 spectra are presented in a form uncorrected for the absorption factor of tissue. The tissue Po_2 histograms presented here all contain at least 100 values and represent the frequency distribution of the actual values of oxygen tensions in the tissue.

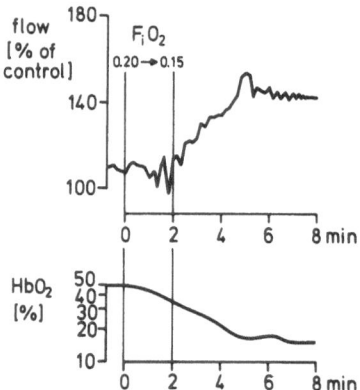

Fig. 2.   Response of femoral flow and intracapillary HbO_2 saturation
to a change in inspired O_2 content from 20% to 15%.

Fig. 1 shows tissue Po_2 values, hydrogen clearance and intracapillary HbO_2 during an inspired oxygen concentration of only 10%
(P_aO_2 33 mm Hg) as compared with control conditions. It can be
clearly seen that the hydrogen clearance is not only faster, but
more homogeneous in the presence of a relatively low capillary HbO_2,
despite the absence of anoxic tissue Po_2 values.

Fig. 2 illustrates how the femoral flow changes in response to
the decrease in FiO_2 from 0.2 to 0.15. Simultaneous intracapillary
HbO_2 values are also given, calculated from the change in intensity
of the 541.2 nm peak of the HbO_2 spectrum with relation to the
549.2 nm isosbestic point. The SO_2 values were then estimated from
the appropriate mean tissue Po_2 values.

The haemodynamic and microcirculatory changes during hypoxia
are summarized in Fig. 3. The error bars for iliac pressure, femoral
flow and resistance represent the standard deviation <u>between</u> experiments. The error bars for microflow represent the mean standard
deviation <u>within</u> experiments from the seven locations on the skeletal muscle. The results are presented in this way in order to
illustrate the increasing homogeneity of the capillary flow during
hypoxia - particularly at an FiO_2 of 0.1.

Results from our experiments with stimulation of the muscle via
the femoral nerve can be seen in Figs. 4 and 5. Examples of tissue
Po_2 histograms and pH_2 clearances are illustrated in Fig. 4 for
stimulation frequencies of 2, 4 and 20 Hz, together with the initial
control measurements. The haemodynamic changes at all five frequencies are summarized in Fig. 5 where, again, the error bars on the
microflow values represent the standard deviation of the flows at
the different wires of the electrode.

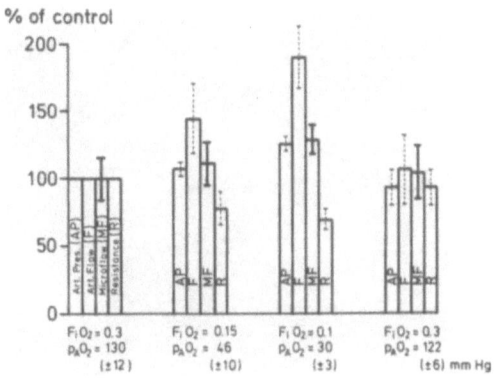

Fig. 3.   Summary of haemodynamic and microflow changes during
          hypoxia.

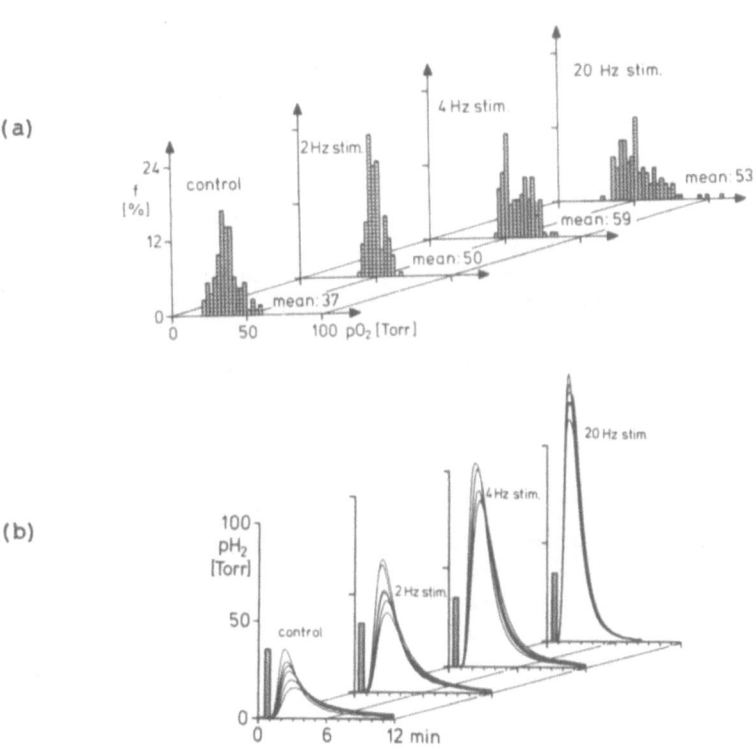

Fig. 4.   Po₂ histograms (a) and H₂ clearance curves (b) during
          stimulation at three different frequencies. As in Fig. 1,
          the shaded areas on the clearance curves indicate the
          period of inspiration of H₂.

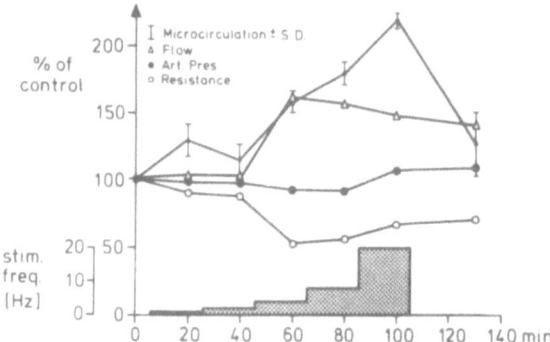

Fig. 5.   Haemodynamic and microflow changes during stimulation at
          different frequencies.

DISCUSSION

        It should, first of all, be noted that only changes in micro-
flow are quoted, since by taking the initial half-life value from
the clearance curve, a part of the inflow of hydrogen is also included in the calculation. Indeed, this is only one of the factors
of the multi-exponential function which comprised the clearance
curve (Hutton, 1970). However, experiments which we have carried
out using total equilibration of the dog with hydrogen yield exactly
the same percentage changes in flow as with the inspired bolus
method. We may assume, then, that the changes in flow which we have
measured are real, and that what we have observed are genuine re-
distributions of blood flow in the capillaries.

        Both the example of Fig. 1 and the summary of our results given
in Fig. 3 show that the capillary blood flow not only increased, but
that also the distribution of flow through the capillary network
became more homogeneous during acute hypoxia. Whilst it would seem
that a proportion of the increased capillary flow must be due to
the increase in systemic flow, it is interesting to note that the
proportional increase in microflow is considerably smaller than the
decrease in peripheral resistance at the FiO_2 = 0.15 level of
hypoxia. One explanation for this may be that the nutritional
demands of other muscles supplied by the femoral artery are higher
than those of sartorius muscle, and that the mean capillary blood
flow in these muscles has already increased in order to maintain a
sufficient tissue oxygen supply. The other explanation, assuming
that the sartorius muscle is typical of all muscles in the limb,
would be that the oxygen supply to the tissue is still sufficient
for some higher flow shunt capillaries to be operating.

        At the higher level of hypoxia (FiO_2 = 0.1) the relative
changes in peripheral resistance and microflow were the same, and

the distribution of flow values through the capillaries became much
narrower probably indicating that maximal oxygen transport to the
tissue may have been reached. However, even at such an acute level
of hypoxia there was no real evidence of tissue anoxia with only
4.5% of tissue Po_2 values lying within the potentially dangerous
0 to 5 mm Hg.

One answer to this apparent paradox may lie in the reaction
shown in Fig. 2 which shows quite clearly how fast intracapillary
HbO_2 falls after a decrease in inspired O_2, but it is not until it
reaches approximately 30% that the femoral flow starts to increase.
Thus given the absence of tissue anoxia, but the speed of the reaction of flow to changes in SO_2, it may be surmised that local muscle
blood flow may be regulated, at least in part, at this level of
oxygenation by means of intracellular oxygen sensors (Höper et al.,
1981).

The influence of stimulation on blood flow to and within the
muscle was dramatic as can be seen quite clearly from the hydrogen
clearance curves of Fig. 4. The results are, however, even more
interesting when the changes in microcirculation are compared with
those in femoral flow (Fig. 5). After an initial increase in mean
capillary flow at 1 Hz stimulation, a decrease is observed at 2 Hz
presumably because the tissue oxygen supply is sufficient to meet
energy requirements, as indicated by the normal, although slightly
narrower histogram. However, despite a very small increase in total
flow, a 30% increase in mean tissue Po_2 was achieved through what
was evidently an equivalent redistribution of microcirculation.

At 4 Hz the total flow to the organ increased by some 60% above
the control value, and the microflow increased by the same degree,
but further increase of the stimulation frequency actually produced
a decrease in femoral flow and an increase in peripheral resistance.
Meanwhile, the capillary flow continued to increase and became more
homogeneous. The tissue Po_2 histograms display an overall shift to
the right probably indicating an increase of tissue oxygen reserve.

These early results point to two interesting aspects of microflow regulation. The first is that the capillary network appears to
be capable of auto-redistribution, or redistribution as a result of
local stimuli, to the extent of increasing tissue blood flow by at
least 50%. The second feature is that local regulation of capillary
flow may be modulated not only by oxygen levels, but also by the
energy requirements of the tissue (Höper, 1983).

One of the key questions raised by the above results both from
the hypoxia and stimulation experiments is, how does the capillary
network reorganize itself? The results of our hydrogen clearance
studies indicate that under normal conditions a bimodal distribution
of flow is found in the type of muscle studied (Kessler et al.,

1983), comprising some 80% of "low flow" values and 20% "high flow" values. If these high flow channels (which could be termed shunt capillaries) have a surplus oxygen capacity, the question arises as to whether or not at times of increased oxygen requirement, the proportion of these channels actually decreases as the capillary flow becomes more homogeneous (Kessler et al., 1981). In the case of hypoxia our results are too few in number to be conclusive in this respect. However, if one considers the conditions existing in the tissue during stimulation up to 4 Hz as being within the normal range of capillary regulation, then it would seem from Fig. 4 that since faster control channels actually become proportionally slower at higher mean flows, the redistribution of microcirculation would uphold this hypothesis.

CONCLUSIONS

Whilst it is recognized that central regulation plays a significant role in determining blood flow to skeletal muscle, particularly during systemic hypoxia, it is clear from the above results that local regulation of the capillary blood flow is important in maintaining an adequate tissue oxygen supply. This local regulation can occur in the absence of tissue anoxia.

Our preliminary results have shown that large increases in capillary blood flow also occur in skeletal muscle during our conditions of stimulation, apparently due to modulation of the Po_2 regulatory mechanism by the redox state of the tissue cells.

ACKNOWLEDGEMENT

This work was supported by the Deutsche Forschungsgemeinschaft Ke 138/5. Dr. D.K. Harrison is an Alexander von Humboldt-Research Fellow.

REFERENCES

Baumgärtl, H., and Lübbers, D.W., 1973, Platinum needle electrode for polarographic measurement of oxygen and hydrogen, in: "Oxygen Supply. Theoretical and Practical Aspects of Oxygen Supply and Microcirculation of Tissue", M. Kessler, D.F. Bruley, L.C. Clark Jr., D.W. Lübbers, I.A. Silver, J. Strauss, eds., Urban & Schwarzenberg, München-Berlin-Wien, pp. 130-136.
Brunner, M., Kastner, N., Schabert, A., Höper, J., and Kessler, M., 1981, On-line Verarbeitung von Hämoglobin-Reflexionsspektren hoher Repetitionsraten, in: "Medizinische Informatik und Statistik 28", S. Koller, P.L. Reichertz, K. Überla, eds., Springer, Berlin-Heidelberg-New York.

Granger, H.J., and Shepherd, A.P., 1973, Intrinsic microvascular
    control of tissue oxygen delivery, Microvasc. Res., 5:49.
Höper, J., 1983, Correlation between redox-state of NADH(P)H and
    total flow in the perfused rat liver, this volume.
Höper, J., and Kessler, M., 1981, Po₂ and sodium dependent mechanism
    regulating liver blood flow, in: "Oxygen Transport to Tissue",
    Adv. Physiol. Sci., Vol. 25, A.G.B. Kovách, E. Dóra, M. Kessler,
    I.A. Silver, eds., Pergamon Press, Akadémiai Kiadó, Budapest.
Hudlická, O., 1973, "Muscle Blood Flow", Swets and Zeitlinger,
    Amsterdam.
Hutton, H., 1970, Untersuchung nichtstationärer Austauschvorgänge
    in gekoppelten Konvektions-Diffusions-Systemen, Abh. Akad.
    Wiss. Lit., Mainz.
Kessler, M., 1983, Tissue O₂ supply under normal and pathological
    conditions, this volume.
Kessler, M., Höper, J., Lübbers, D.W., and Ji, S., 1981, Local fac-
    tors affecting regulation of microflow, O₂ uptake and energy
    metabolism, in: "Oxygen Transport to Tissue", Adv. Physiol.
    Sci., Vol. 25, A. Kovách, E. Dóra, M. Kessler, I.A. Silver,
    eds., Pergamon Press, Akadémiai Kiadó, Budapest.
Kessler, M., and Lübbers, D.W., 1966, Aufbau und Anwendungsmöglich-
    keiten verschiedener Po₂-Elektroden, Pflügers Arch. Ges.
    Physiol., 291:R82.
Krumme, B.A., Strehlau, R., and Kessler, M., 1975, Hydrogen clear-
    ance measurements on the liver surface in situ with the
    multiwire electrode, Arzneim.-Forsch. (Drug Res.), 25:1666.

HETEROGENEITY OF CAPILLARY BLOOD FLOW IN SKELETAL MUSCLE: EVIDENCE

FROM LOCAL ^{133}Xe CLEARANCE AND MICROSPHERE TRAPPING

C. Marconi[1], P. Cerretelli[1], D. Pendergast[1], M. Meyer[2],
N. Heisler[2], and J. Piiper[2]

[1]Centro per lo studio del Lavoro Muscolare, C.N.R.
Milano, Italy
[2]Max-Planck-Institut für Experimentelle Medizin, Abtlg.
Physiologie, D-3400 Göttingen, FRG

ABSTRACT

Blood flow and its distribution was studied in isolated-perfused and intact dog gastrocnemius muscles at rest and at several
levels of rhythmic tetanic exercise. In each experiment, micro-
spheres 15 um in diameter labeled with ^{125}I, ^{141}Ce, ^{51}Cr, ^{85}Sr, ^{95}Nb
or ^{46}Sc, were injected into the arterial inflow while clearance of
^{133}Xe from several small intramuscular depots was monitored for
determination of local blood flow ($\dot{Q}_{Xe}$). For determination of blood
flow from microsphere trapping ($\dot{Q}_m$), the muscle was cut into 140 to
340 pieces, of about 700 mg each.

Both at rest and during exercise the $\dot{Q}_m$ values followed normal distribution around the mean value. Typical $\dot{Q}_m$ values were
0.14 $\pm$ 0.05 (mean $\pm$ SD) and 0.75 $\pm$ 0.30 ml g^{-1} min^{-1} for a resting
and maximally stimulated isolated-perfused muscle preparation. $\dot{Q}_m$
results for intact gastrocnemii were similar: 0.11 $\pm$ 0.08 at rest
and 0.75 $\pm$ 0.23 ml g^{-1} min^{-1} at maximal exercise. Thus, the coefficient of variation (SD/mean) of local blood flow averaged 45%% and
showed no clear correlation with rest vs. exercise or isolated-
perfused vs. intact state.

The total muscle blood flow was simultaneously determined by
three independent methods: direct collection of venous effluent
($\dot{Q}_v$), ^{133}Xe clearance ($\dot{Q}_{Xe}$) and microsphere trapping ($\dot{Q}_m$). The
average $\dot{Q}_m$ was not significantly different from $\dot{Q}_v$. Local washout
of ^{133}Xe was multiexponential, indicating perfusion or diffusion

inhomogeneity at the microlevel. Usually three components were found and the $\dot{Q}_{Xe}$ value therefore depended substantially on the component selected for analysis.

The $\dot{Q}_{Xe}$ values based on the middle component averaged 40% of $\dot{Q}_{V}$ at all blood flow levels. This finding, which is in agreement with previous measurements of muscle blood flow in running dogs, may be related to diffusion limitation of Xe washout, to veno-arterial diffusion shunt or to functional shunt perfusion.

BIOCHEMICAL AND ^{31}P-NMR STUDIES OF THE ENERGY METABOLISM IN

RELATION TO OXYGEN SUPPLY IN RAT SKELETAL MUSCLE DURING EXERCISE

J.-P. Idström[1], V. Harihara Subramanian[2],
B. Chance[2], T. Scherstén[1], and A.-C. Bylund-Fellenius[1]

[1]Surgical Metabolic Research Laboratory, Dept. of
Surgery I, Sahlgrenska Hospital, University of
Göteborg, S-413 45 Göteborg, Sweden
[2]Johnson Research Foundation, University of
Pennsylvania, Philadelphia, PA 19104, USA

INTRODUCTION

The energy metabolism of muscle tissue during exercise has
been extensively studied, yet there is no unambiguous candidate as
being the limiting factor for the exercise performance (c.f. di
Prampero, 1981). One possible factor of importance in this respect
is the oxygen supply to the muscle tissue.

We have used the perfused rat hind limb preparation to study
the influence of the oxygen delivery on the energy state in contracting muscle. Here data from two different studies are reported,
one in which we have used biochemical methods and the other where
we have used the ^{31}P-NMR technique, and the results are compared.

METHODS

Perfusion Technique

The rat hind limb perfusion technique as described in detail
elsewhere (Walker et al., 1982a) was used. The perfusate consisted
of Krebs-Henseleit buffer, pH 7.4, (Krebs and Henseleit, 1932) with
6% albumin, 5.5 $mmol \cdot l^{-1}$ glucose and 0.1 U/l insulin. Erythrocytes
were excluded and the oxygen content was varied by equilibrating the
perfusate with different gas mixtures (20, 50, 70 and 95% oxygen).
Isometric muscle contractions of maximal amplitude and a frequency

of 4 Hz were induced by electrical stimulation of the sciatic nerve.
The amplitude of the muscle contractions was recorded with a force
displacement transducer.

The ^{31}P-NMR Technique

A Johnson Foundation NMR spectrometer connected to an Oxford
Instruments 7 inch superconducting magnet was used for the ^{31}P-NMR
studies (Chance et al., 1980; 1981). The magnet was set at 1.5 T
and the spectrometer at 24.3 MHz for phosphorus and 60.1 MHz for
protons. The water-proton signal was used for shimming. A Helmholz
coil (∅ = 2.5 cm) was built to fit the calf muscle of a 250 g rat.
The perfusion apparatus was modified in order to maintain the temperature at 37° and to ensure good oxygenation of the perfusate.
The operative procedures were performed with the rat lying on a
specially designed probe, which was then moved into the magnet,
where the hind limb was connected to the perfusion system. The
magnetic field homogeneity was considered to be acceptable when the
half-width of the water peak was 0.5 - 1 ppm. The ^{31}P-NMR spectra
were obtained by accumulating 60 FIDs with a repetition rate of one
pulse per second. The accumulated FID was then Fourier transformed.
The inorganic phosphate (Pi), phosphocreatine (PCr), α-ATP, ß-ATP
and γ-ATP peaks were identified and the peak heights were measured.

The PCr/Pi ratio was calculated as a convenient expression for
the energy state of the muscle tissue since the NMR spectra just
give relative measurements of the concentrations.

Biochemical Methods

Muscle biopsies for biochemical analyses were rapidly excised
with a pair of scissors and immediately frozen in liquid nitrogen.
The biopsies were freeze dried, extracted and analyzed for PCr and
Pi concentrations according to previously described techniques
(Bylund-Fellenius et al., 1981). The PCr/Pi ratio was calculated to
make the biochemical data comparable to the NMR data.

Experimental Procedure

The experimental procedure was identical for the biochemical
studies and the NMR studies. The perfusions began with a 10 min
equilibration period during which the flow was gradually increased
to 0.85 ml·min^{-1}·g^{-1} muscle. In the NMR experiments the equilibration period was also used for proton shimming on the rat leg. A 10
min resting period proceeded hereafter during which three NMR
spectra were collected.

After the equilibration period and the additional resting
period, nerve stimulation was started. Simultaneously the perfusion
flow was increased to 1.7 ml·min^{-1}·g^{-1} muscle over a period of

1 min. NMR spectra were collected with 1.5 min intervals throughout
the 10 min exercise period and a 10 min recovery period. In the
experiments where metabolite levels were analyzed biochemically,
biopsies from the soleus and gastrocnemius muscles were taken at
the completion of the 10 min exercise period.

RESULTS

Representative ^{31}P-NMR spectra of a rat hind limb perfused at
an oxygen delivery of 30 umol·h^{-1}·g^{-1} at rest, exercise and 10 min
of recovery, are shown in Fig. 1. For comparison spectra obtained
at corresponding time points in a rat hind limb in vivo, stimulated
to contract by the same procedure are shown (Fig. 2). The signal-
to-noise level and the resolution was better in vivo compared to in
the perfused limb. The increase in Pi and decrease in PCr with exercise were more pronounced in the rat perfused at the lower oxygen
delivery compared to the in vivo preparation. A complete restoration
of the Pi and PCr levels was obtained in both conditions after cessation of exercise, but the recovery was more rapid in the in vivo
rats.

The PCr/Pi ratios derived with the NMR technique at the end of
exercise during perfusion with various oxygen deliveries, are shown
in Fig. 3. A significant relationship between the oxygen delivery,
expressed as the product of the arterial oxygen concentration and
the flow rate, and the PCr/pi ratio was obtained. For comparison

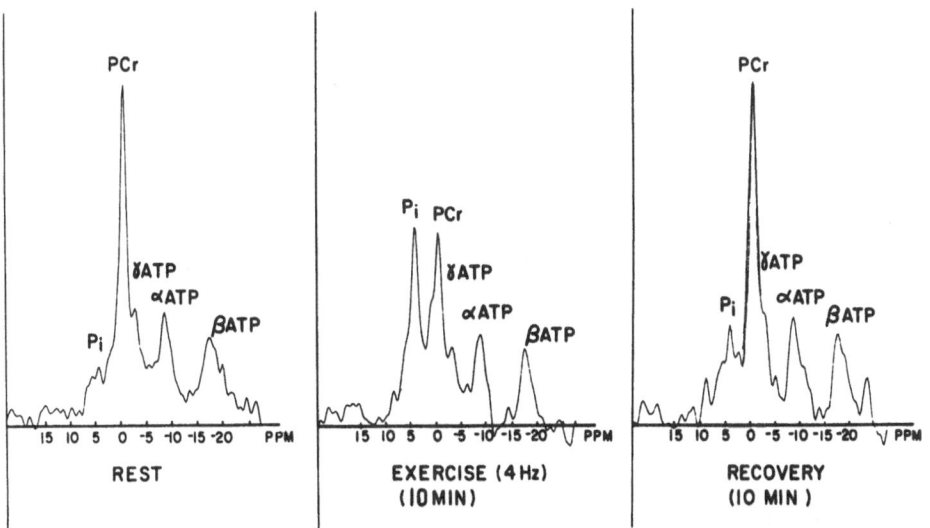

Fig. 1. Representative ^{31}P-NMR spectra from calf muscles of the perfused rat hind limb, at an oxygen delivery of 30 umol·h^{-1}·g^{-1}
muscle. Spectral accumulation time was 1 min, 60 scans.

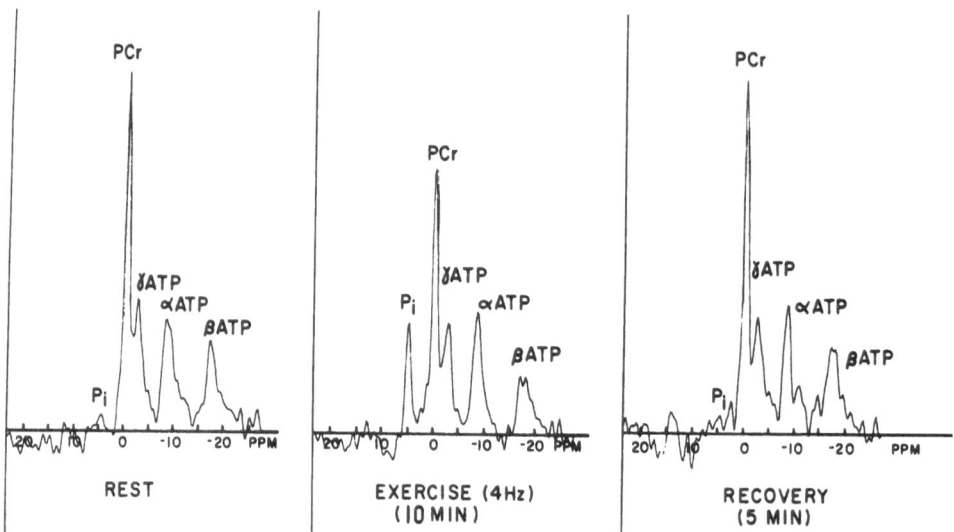

Fig. 2.   Representative ^{31}P-NMR spectra from rat calf muscles in
          vivo. Spectral accumulation time was 1 min, 60 scans.

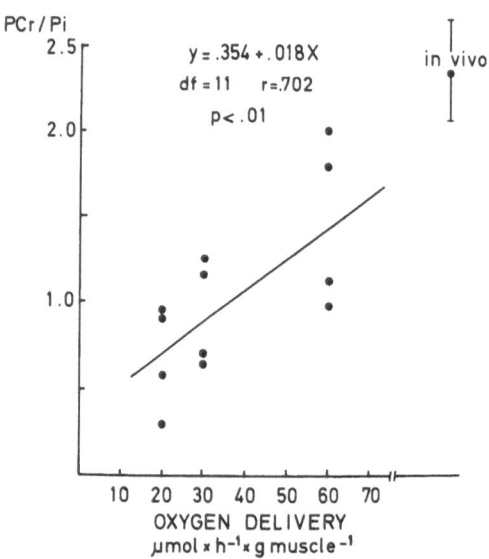

Fig. 3.   Relationship between the oxygen delivery and the phospho-
          creatine/inorganic phosphate ratio in the rat calf muscle
          after 10 min of contractions. PCr and Pi peak heights were
          obtained from NMR spectra.

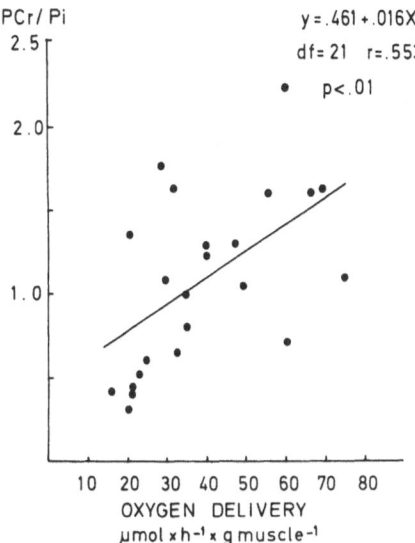

PCr/Pi

$y = .461 + .016X$

$df = 21$ $r = .553$

$p < .01$

OXYGEN DELIVERY

$\mu mol \times h^{-1} \times g\ muscle^{-1}$

Fig. 4. Relationship between the oxygen delivery and the phospho-
creatine/inorganic phosphate ratio in the soleus muscle
after 10 min of contractions. PCr and Pi concentrations
were obtained by biochemical analysis.

the 10 min exercise PCr/Pi ratio for the in vivo experiments (n = 7)
is also shown in this figure.

Biopsies from the soleus and the gastrocnemius muscles, taken
at the completion of the exercise period were analyzed for the
absolute concentrations of PCr and Pi. To make the biochemical data
comparable to the NMR data the PCr/Pi ratios were calculated for
the two muscles. The relationship between the oxygen delivery and
the PCr/Pi ratio in the soleus muscle and the gastrocnemius muscles
are shown in Figs. 4 and 5, respectively.

An increasing degree of fatigue, expressed as the percent
remaining contraction force at the end of the exercise period was
found with a decrease in the oxygen delivery. There was, however,
no significant difference in the total work performed between the
groups, as judged from integration of the work output recordings.

DISCUSSION

The present study was undertaken to gain information on the
metabolic response of contracting skeletal muscle to a reduced
oxygen supply. For this purpose the rat hind limb preparation was
used, a technique that offers the possibility to control the oxygen
delivery in a predictable way while all other variables can be kept
constant.

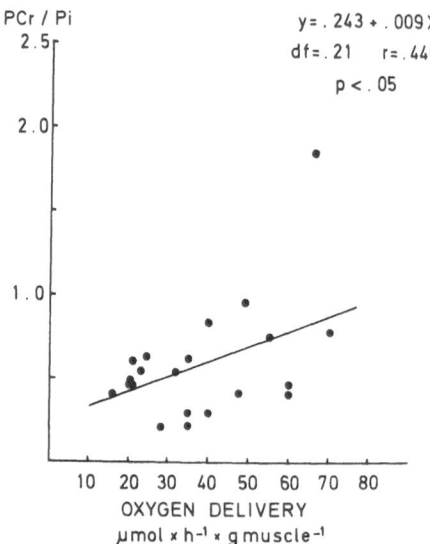

Fig. 5.   Relationship between the oxygen delivery and the phospho-
           creatine/inorganic phosphate ratio in the gastrocnemius
           muscle after 10 min of contractions. PCr and Pi concentra-
           tions were obtained by biochemical analysis.

       Two different approaches were used to study the energy metabo-
lism in skeletal muscle tissue - biochemical analysis of metabolite
concentrations in muscle biopsies and measurements of phosphate
compounds with the non-invasive ^{31}P-NMR technique.

       The NMR system developed for these experiments gave a signal-
to-noise ratio of more than 20 for the rat hind limb in vivo. The
resolution of the NMR spectra was good for a system run at 1.5
Tesla, 24.3 MHz, as judged from good separation of the PCr and
γ-ATP-peak (Fig. 2). Reliable NMR spectra could be obtained within
one minute of accumulation (60 scans) which shows the power of the
NMR technique for continuous measurements of the energy state under
various conditions.

       The signal-to-noise ratio and the resolution were not quite as
good for the perfused system as for the in vivo system due to perfusate leakage in the coil surroundings, but still reliable spectra
were obtained within one minute (Fig. 1).

       With both the biochemical and the NMR technique significant
relationships between the oxygen delivery to the tissue and the
energy state, expressed as the PCr/Pi ratio were obtained. This
demonstrates that the oxygen supply has a direct influence on the
energy metabolism in contracting muscle.

The slope of the relationship was different for the soleus and the gastrocnemius muscles. The fiber composition of these muscles differs and thereby also the biochemical characteristics. Thus, the soleus muscle is mainly composed of slow oxidative fibers, while the lateral portion of the gastrocnemius muscle is mainly composed of fast glycolytic fibers. The gastrocnemius muscle has a higher content of PCr than the soleus muscle, but responds with a more pronounced decrease in response to a certain workload (Walker et al., 1982b). The difference between the two muscles in the decrease of the PCr/Pi ratio with a reduction in the oxygen delivery can thus be explained by these differences. We have previously shown that the gastrocnemius muscle is more susceptible to a reduced blood flow during exercise than the soleus muscle, and the present data confirm that this is also true for a reduced oxygen supply.

The slope of the regression line obtained from the NMR data was in between the slopes obtained for the soleus and gastrocnemius muscle, respectively. This can be expected since the NMR technique gives an average value for all the muscles in the calf, where soleus and gastrocnemius represent two extremes with respect to the fiber composition.

In conclusion, this study demonstrates a relationship between the oxygen delivery to the tissue and the intramuscular energy state as reflected by the PCr/Pi ratio, measured with two independent techniques, in the perfused contracting rat hind limb. This suggests that the energy metabolism in a contracting muscle is directly dependent on the oxygen supply.

REFERENCES

Bylund-Fellenius, A.C., Walker, P.M., Elander, A., Holm, S., Holm, J., and Scherstén, T., 1981, Energy metabolism in relation to oxygen partial pressure in human skeletal muscle during exercise, Biochem. J., 200:247-255.
Chance, B., Eleff, S., and Leigh Jr., J.S., 1980, Noninvasive, nondestructive approaches to cell bioenergetics, Proc. Natl. Acad. Sci. USA, 77:12, 7430-7434.
Chance, B., Eleff, S., Leigh Jr., J.S., Sokolow, D., and Sapega, A., 1981, Mitochondrial regulation of phosphocreatine/inorganic phosphate ratios in exercising human muscle: A gated ^{31}P-NMR study, Proc. Natl. Acad. Sci. USA, 78:11, 6714-6718.
Krebs, H.A., and Henseleit, K., 1932, Untersuchungen über die Harn-stoffbildung im Tierkörper, Hoppe Seyler's Z. Physiol. Chem., 210:33-36.
Lowry, O.H., and Passonneau, J.V., 1978, "A Flexible System of Enzymatic Analysis", Academic Press, New York.
Prampero di, P.E., 1981, Energetics of muscular exercise, Rev. Physiol. Biochem. Pharmacol., 89:143-222.

Walker, P.M., Idström, J.-P., Scherstén, T., and Bylund-Fellenius,
    A.-C., 1982a, Glucose uptake in relation to metabolic state in
    perfused rat hind limb at rest and during exercise, Eur. J.
    Appl. Physiol., 48:163-176.
Walker, P.M., Idström, P.J., Scherstén, T., and Bylund-Fellenius,
    A.-C., 1982b, Metabolic response in different muscle types to
    reduced blood flow during exercise in perfused rat hind limb,
    Clin. Sci., 63:293-299.

DEPENDENCE OF O_2 UPTAKE ON TISSUE Po_2: EXPERIMENTS IN INTACT EXCISED RAT SKELETAL MUSCLE

T. Kawashiro*, and P. Scheid**

Max-Planck-Institut für experimentelle Medizin, Abtlg.
Physiologie, 3400 Göttingen, FRG

ABSTRACT

Oxygen uptake ($\dot{M}$) at different surface O_2 pressures (P_O) was measured in intact, excised rat skeletal muscle suspended in the gas phase. Above a critical value of P_O (P_C), $\dot{M}$ was apparently independent of P_O; below P_C, $\dot{M}$ declined in close proportion to the square root of P_O. This experimental relationship was compared with predictions based on the model of Warburg (1923) which assumes the tissue to be homogeneous in respect of its metabolic rate, diffusivity and geometry. The good agreement between the Warburg model and the experimental values suggests that the assumptions underlying this model are justified.

INTRODUCTION

In 1923, Warburg reported a model that allowed one to predict the critical boundary Po_2 (P_C) for a slice of respiring tissue of given thickness. This P_C constitutes the minimum Po_2 that must prevail at the tissue surface in order that the entire tissue is supplied with O_2 for aerobic metabolism. The Warburg model is based on a number of assumptions (see below). Measurements on liver, kidney

---

* Present address: Dept. of Medicine, School of Medicine, KEIO University, Shinano-machi 35, Tokyo, Japan
** Present address: Institut für Physiologie der Ruhr-Universität, Universitätsstr. 150, 4630 Bochum-Querenburg, FRG

and heart slices have yielded results that supported this model
(Longmuir and Bourke, 1960).

Recently, some authors have questioned the applicability of
Warburg's formula to respiring muscle tissue. In fact, Pichotka and
his colleagues (Claessen and Pichotka, 1975; Schmidt and Pichotka,
1977) have claimed that rat diaphragm at 37°C increases O_2 consumption with increasing Po_2 far beyond the value of P_C predicted from
Warburg's formula.

We have studied O_2 uptake in relation to Po_2 in intact rat
skeletal muscle in order to test the applicability of Warburg's
formula.

METHODS

The experiments were performed in intact muscle tissue from ten
rats (Sprague Dawley; average body weight 180 g). For measurements,
a sheet comprising the abdominal transverse and internal oblique
muscles was excised immediately after killing the animal with an
overdose of ethyl ether.

Principle

The intact muscle sheet was mounted in the diffusion chamber
of Kawashiro et al. (1975). In this apparatus, the muscle sheet
separated the chamber into two smaller chambers as would a membrane
(surface area, F = 2.01 cm^2). Both chambers were initially ventilated with humidified gas mixtures of identical composition (O_2
partial pressure, P_O). In one, the recording chamber (volume, V =
0.43 cm^3), ventilation could be discontinued and the rate of change
in O_2 partial pressure, dP_O/dt, could be recorded by an O_2 electrode. This recording period was kept small as to prevent significant changes in Po_2 within the recording chamber. Thus, the Po_2 at
both tissue surfaces (P_O) was nearly identical during the recording
interval and did not significantly vary with time of recording.

If the tissue is homogeneous in composition, one half of its
total O_2 consumption ($\dot{M}$) will be provided by each chamber and can
be calculated from the rate of change of Po_2 in the recording chamber
(dP_O/dt), its volume (V), and the capacitance coefficient in the gas
phase, β_g (= 0.0517 x 10^{-3} mmol·cm^{-3}·Torr^{-1} at 37°C; cf. Piiper et
al., 1971) as:

$$\dot{M} = -2\beta_g \cdot V \cdot dP_O/dt \qquad\qquad\qquad (1)$$

Measurements performed at different values of P_O yield $\dot{M}$ as a
function of surface Po_2.

Equipment

The gas mixtures (2 to 50% O_2, 7% CO_2 in N_2) were provided by precision gas mixing pumps (M 301 a-F, Wösthoff, Bochum / FRG) and were humidified at 37°C, the temperature at which all experiments were conducted. The Po_2 in the recording chamber was measured by an electrode (Type E 5046, Radiometer, Copenhagen / Denmark) covered with 6 um teflon and filled with an alkaline buffer as suggested by Hahn et al. (1975). Thus, 95% response time was less than 20 sec. The Po_2 was recorded at high sensitivity on a strip chart recorder. Electrode sensitivity was checked repeatedly but was found rather stable in the course of one experiment.

Procedure

After mounting the muscle sheet in the apparatus, a period of 30 min was allowed for the tissue and the apparatus to warm up to 37°C in the water bath in which it was kept during the entire experiment. In this warm-up period both chambers were flushed with a gas mixture containing 50% O_2, 7% CO_2 in N_2.

After warm-up, several measurements were conducted which differed only in the Po_2 of the ventilating gas mixture (P_0). In each measurement, about 15 min were allowed, after setting the Po_2 in the gas mixture to P_0, before the recording chamber was closed and dP_0/dt recorded for 3 min, while flow continued to the other chamber. Thereafter, flow to the recording chamber was resumed, and P_0 in the gas ventilating both chambers was changed for the next measurement.

Between 4 and 9 such measuring periods were performed with steps of decreasing Po_2. The last period was usually executed again at a high level of Po_2 in order to check if the tissue had suffered from exposure to sub-critical Po_2.

At the end of the experiment, the thickness of the muscle sheet was determined by weighing as described by Kawashiro et al. (1975). Wet-to-dry weight ratio of experimental and control muscles from the same animal did not differ significantly suggesting that shrinkage or swelling of the muscle membrane during the experiment was insignificant. The entire experiment, from anaesthetizing the animal until determination of tissue thickness was completed within 2 to 3 hours.

Model Analysis

In the theory underlying the model of Warburg (1923) a tissue slice is considered, the cells of which are supplied with O_2 by diffusion from the slice surface, Krogh's diffusion constant (K) being constant within the tissue. Above a critical cell Po_2 the cells are considered to consume O_2 at a maximum rate, independent of Po_2. This

O_2 consumption rate per unit tissue volume ($\dot{m}$) is considered constant throughout the slice.

This model predicts that the O_2 consumption of the muscle sheet (half-thickness, L; surface area, F) is constant provided the surface Po_2 remains above a critical value, P_C. The value for P_C may be obtained (cf. Kawashiro et al., 1975) from:

$$P_C = \frac{\dot{m} \cdot L^2}{2K} \tag{2}$$

For surface Po_2 (P_O) above P_C, O_2 uptake rate of the total tissue ($\dot{m}$) is constant at its maximum value, $\dot{M}_\infty$:

$$\dot{M}_\infty = \dot{m} \cdot F \cdot 2L \tag{3}$$

But when P_O is below P_C, the critical depth from the surface (L_C) that is supplied with O_2 is obtained from eq. (2) as

$$L_C = \sqrt{2K \cdot Po/\dot{m}} \tag{4}$$

Combining eqs. (2) and (3):

$$\dot{M}_\infty = 2F \cdot \sqrt{2K \cdot P_C \cdot \dot{m}} \tag{5}$$

Similarly, for $P_O < P_C$

$$\dot{M} = 2F \cdot \sqrt{2K \cdot P_O \cdot \dot{m}} \tag{6}$$

Combining eqs. (5) and (6) yields

$$\dot{M}/\dot{M}_\infty = \begin{cases} 1 & \text{for } P_O \geq P_C \\ \sqrt{P_O/P_C} & \text{for } P_O < P_C \end{cases} \tag{7}$$

If K is known as well as the tissue half-thickness, L and $\dot{M}_\infty$ (measured as O_2 uptake rate in the Po_2 range where there is no variation of $\dot{M}$ with Po_2), P_C may be obtained from eq. (2). Experimental data may then be plotted as $\dot{M}/\dot{M}_\infty$ against P_O/P_C according to eq. (7), which permits a comparison of results obtained in tissues of different geometry and different O_2 uptake rate.

RESULTS

Fig. 1 shows the O_2 uptake rate, $\dot{M}$, of one experimental tissue against the surface Po_2, P_O. Above P_O of about 100 Torr, $\dot{M}$ did not appear to depend on P_O. The average value of $\dot{M}$ in this range, $\dot{M}_\infty$, was $47 \cdot 10^{-6}$ mmol·min^{-1} with a standard deviation of about 8% of the

Table 1. Average values for experimental parameters and results. The
         symbol L is the half-thickness of tissue; $\dot{m}$, its metabolic
         rate per unit volume (when Po$_2$ is above critical); P_c, cri-
         tical surface Po$_2$ predicted from the Warburg model (eq. (2));
         n, number of measuring periods.

Rat	Body weight (g)	L (um)	$\dot{m}$ (umol·min^{-1}·cm^{-3})	P_c Torr	n
1	189	200	0.96	147	4
2	172	197	0.60	89	6
3	170	170	0.62	68	6
4	161	139	0.95	70	6
5	160	176	0.64	77	6
6	183	160	0.94	91	7
7	180	190	0.61	84	9
8	185	170	0.69	76	7
9	188	217	0.89	161	8
10	195	170	0.78	87	7
Mean	178 ± 12	179 ± 23	0.77 ± 15		

mean. With membrane half-thickness (L = 190 um in this example) and
the area of the tissue (F = 2.01 cm^2, see above), the O$_2$ uptake rate
per tissue volume, $\dot{m}$, can be calculated as 0.61 umol·min^{-1}·cm^{-3}.

        In all 10 experiments, $\dot{M}$ was found to attain a maximum value.
Table 1 is a summary of half-thickness (L) and specific O$_2$ uptake
rate ($\dot{m}$) found in all tissues. The mean value of $\dot{m}$, 0.77 umol·min^{-1}
·cm^{-3}, is similar to the value obtained by Kawashiro et al. (1975)
in the same preparation in the Warburg apparatus, when the tissue
slices were exposed to O$_2$ in a gaseous environment.

        The line in Fig. 1 corresponds to the Warburg model (eq. (7)).
For the calculation of P_c according to eq. (2), a value of 1.31 x
10^{-9} mmol·cm^{-1}·min^{-1}·Torr^{-1} was utilized for K as obtained by
Kawashiro et al. (1975) in the same preparation. It is apparent
that the experimental data fit the theoretical curve fairly well.

        Another comparison can be made by plotting $\dot{M}/\dot{M}_\infty$ against $\sqrt{P_0/P_c}$.
This is shown for all values from all 10 experiments in Fig. 2.
Despite a general agreement between experimental data and theory,
the data points below P_c tend to be below the theoretical line.
The regression line fitting the data for $P_0 < P_c$ deviates slightly
from the theoretical line and this difference is statistically
significant at the 1% level. The regression line intersects the
horizontal line of $\dot{M}/\dot{M}_\infty = 1$ at a somewhat higher value of $\sqrt{P_0/P_c}$,
corresponding to an apparent critical Po$_2$ of $(P_c)_{app} = 1.39\ P_c$.

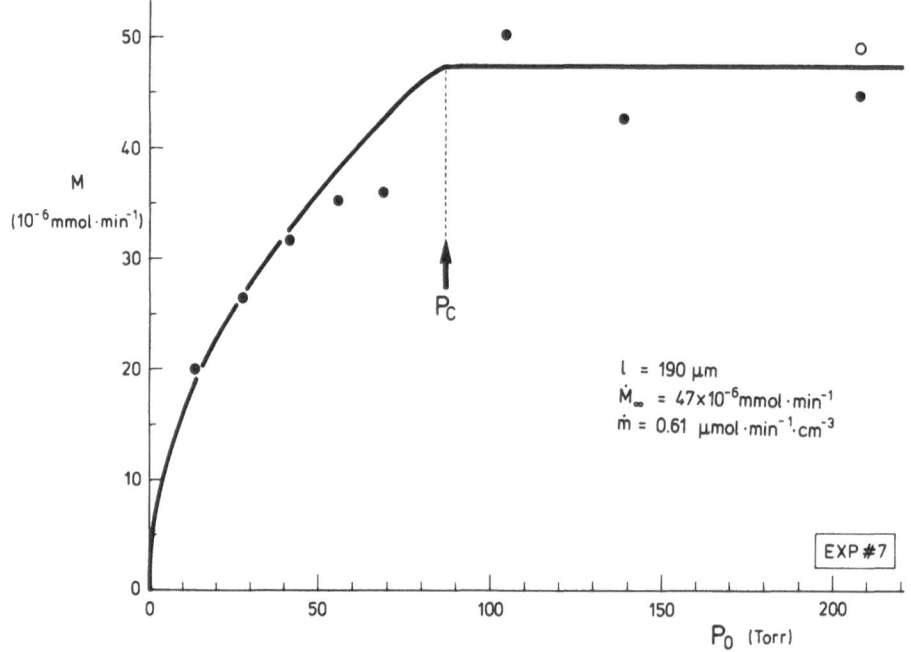

Fig. 1. Plot of O_2 uptake rate, $\dot{M}$, against surface Po_2, P_O. Nine
measuring periods in one experimental tissue of half-thick-
ness, L = 190 um. The open circle represents the last measurement, after the tissue had been exposed to Po_2 values at
which it did not respire at its maximum rate, $\dot{M}_\infty$. The symbol
$\dot{m}$ is the maximum O_2 uptake rate per tissue volume, P_C,
critical Po_2 according to the Warburg model (eq. (2)), to
which the lines correspond (eq. (7)).

DISCUSSION

        Our results suggest that above a critical Po_2 value, O_2 uptake
of the muscle sheet is independent of Po_2. This finding is in agreement with results obtained in mitochondrial preparations which show
that mitochondrial O_2 uptake is independent of Po_2 down to Po_2 levels
below 1 Torr (Jöbsis, 1964). This finding is in contradiction to
recent results of Pichotka and his co-workers (Claessen and Pichotka,
1975; Schmidt and Pichotka, 1977) who found O_2 uptake of excised
mouse diaphragm to increase with Po_2 well beyond the predicted critical Po_2, with no tendency to approach a final level. We have no
simple explanation to offer for this apparent discrepancy. It should,
however, be noted that these authors have used the classical Warburg
technique in which the muscle is suspended in liquid. The observation
that the O_2 uptake rate increased with increasing shaking rate of
the tissue in the apparatus (Pichotka et al., 1975) may be explained
by the existence of stagnant layers of medium adhering to the tissue

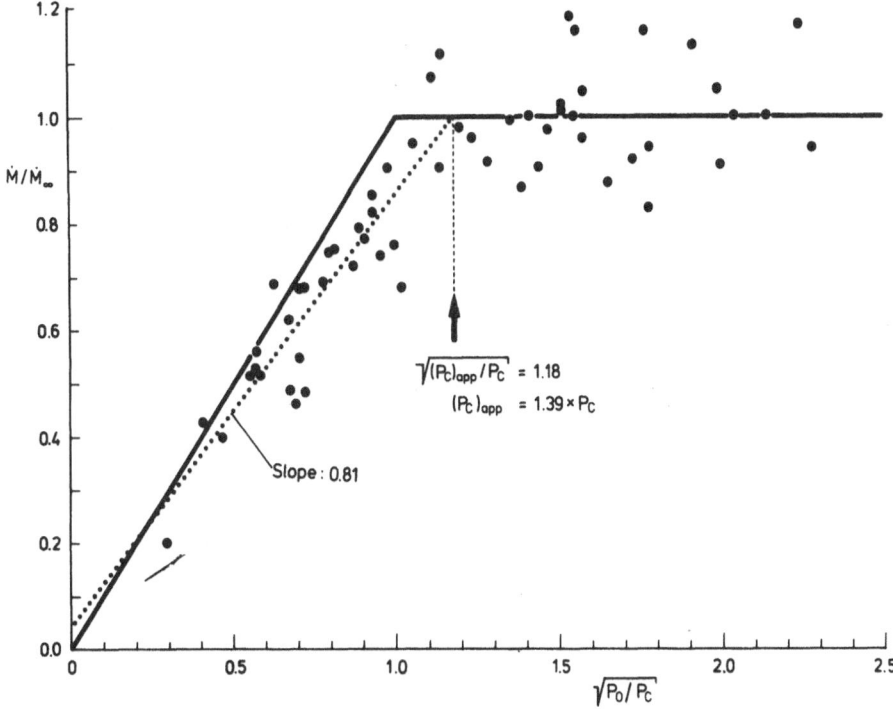

Fig. 2. Plot of all data obtained in 10 experimental tissues. The
linear regression line, calculated for values of $P_O < P_C$,
is drawn as a dotted line. The solid line is predicted by
the theory of eq. (7). For details see text.

and offering a diffusion barrier for O_2. Since in our preparation
the tissue was suspended in the gas phase, no mixing problems were
expected in our study.

Below a critical Po_2 level, O_2 uptake in our muscle preparation
dropped and this decline followed closely the square-root relationship predicted by the Warburg formula. The only deviation from the
theoretical prediction concerns the value for P_C which appears to be
about 40% larger when evaluated from the plot of Fig. 3 than when
calculated according to eq. (2).

A possible explanation for this deviation could be the biological variability which is apparent from the wide scatter in our
data points. It should, however, be noted that the majority of values
of $\dot{M}/\dot{M}_\infty$ in the range $P_O < P_C$ fall below the predicted line and that
thus the discrepancy in P_C between the experimental and the theoretical value may be real. To evaluate whether errors in determining the relationship of $\dot{M}$ with P_O might be responsible for the
apparent deviation, one can evaluate the relative importance of
each of the experimental parameters.

Using eqs. (1) to (3), the ratio of P_O/P_C can be re-written to contain only experimentally determined parameters:

$$P_O/P_C = \frac{2K \cdot F}{\beta_g \cdot V \cdot L} \cdot \frac{P_O}{(dP_O/dt)_\infty} \tag{8}$$

In this equation, $(dP_O/dt)_\infty$ is the value of dP_O/dt when the tissue is at its maximum metabolic activity $(\dot{M}_\infty)$. In addition to the constant β_g, eq. (8) contains parameters that are constant for all tissues (V and F) and those that are directly determined (L and $(dP_O/dt)_\infty$).

Krogh's diffusion constant, K, taken from Kawashiro et al. (1975), had been determined in the same muscle preparation using the same apparatus as in the present study. It is of interest to investigate whether a systematic error in the experiments of Kawashiro et al. (1975) could, via K, give rise to systematic errors in P_C.

In the experiments of Kawashiro et al. (1975), K was determined from steady state diffusion across the tissue membrane. At two levels of Po_2, P_O, the rate of change of Po_2, dP_O/dt, was observed in the transiently closed recording chamber. With suffixes 1 and 2 denoting the measurements at the two levels, the following relationship was used in their study:

$$\frac{K \cdot F}{2\beta_g \cdot V \cdot L'} = \frac{(dP_O/dt)_1 - (dP_O/dt)_2}{(P_O)_1 - (P_O)_2} \tag{9}$$

F and V were identical in their study and in the present investigation and the tissue half-thickness, L', which in general differs from tissue to tissue, was determined in the same way in both studies. Introducing eq. (9) into eq. (8) yields

$$P_O/P_C = - \frac{4L'}{L} \cdot \frac{(dP_O/dt)_1 - (dP_O/dt)_2}{(P_O)_1 - (P_O)_2} \cdot \frac{P_O}{(dP_O/d_t)_\infty} \tag{10}$$

If L and L' are assumed to be correct or to contain the same erroneous factor, then accuracy of P_O/P_C depends upon the relative accuracy of the Po_2 measurement within the recording chamber, which depends entirely on the accuracy of Po_2 in the gas mixture used for calibration and for measurement. Both were provided from gas mixing pumps which were checked to be accurate to about 0.2 Torr. The method used is thus likely to yield no systematic errors in the relationship of $\dot{M}/\dot{M}_\infty$ against P_O/P_C.

We conclude that the data are in basic agreement with the Warburg model. Further experiments are needed to confirm or reject the significance of deviations of experimental from predicted values which appear to suggest that the critical surface Po_2 of the muscle slice is larger than that predicted on the basis of the Warburg model.

REFERENCES

Claessen, J.U., and Pichotka, J., 1975, Oxygen uptake of whole diaphragms of mice in relation to oxygen pressure and temperature, Pflügers Arch., 355:R41.

Hahn, C.E.W., Davis, A.H., and Albery, W.J., 1975, Electrochemical improvement of the performance of Po_2 electrodes, Respir. Physiol., 25:109-133.

Jöbsis, F.F., 1964, Basic processes in cellular respiration, in: "Handbook of Physiology", Section 3, Respiration, Vol. I, W.O. Fenn, H. Rahn, eds., American Physiol. Soc., Washington, pp. 63-124.

Kawashiro, T., Nüsse, W., and Scheid, P., 1975, Determination of diffusivity of oxygen and carbon dioxide in respiring tissue: results in rat skeletal muscle, Pflügers Arch., 359:231-251.

Longmuir, I.S., and Bourke, A., 1960, The measurement of the diffusion of oxygen through respiring tissue, Biochem. J., 76:225-229.

Pichotka, J., Schmidt, H.J., Schmidt, I., and Sommer, H.M., 1975, The influence of the shaking rate on the oxygen uptake of isolated tissue, Pflügers Arch., 355:R41.

Piiper, J., Dejours, P., Haab, P., and Rahn, H., 1971, Concepts and basic quantities in gas exchange physiology, Respir. Physiol., 13:292-304.

Schmidt, H.J., and Pichotka, J., 1977, The relation of oxygen tension and oxygen uptake of isolated tissue under stable conditions, Pflügers Arch., 368:R18.

Warburg, O., 1923, Versuche an überlebendem Carcinomgewebe (Methoden), Biochem. Z., 142:317-333.

METABOLIC ADAPTATION IN RESPONSE TO INTERMITTENT HYPOXIA IN RAT SKELETAL MUSCLES

A. Elander, J.-P. Idström, S. Holm, T. Scherstén, and A.-C. Bylund-Fellenius

Surgical Metabolic Research Laboratory, Department of Surgery I, Sahlgrenska Hospital, University of Göteborg 413 45 Göteborg, Sweden

INTRODUCTION

Endurance training causes an increased activity of oxidative enzymes in skeletal muscle tissue (Holloszy and Booth, 1976; Bylund et al., 1977). A similar adaptation has also been described as a spontaneous phenomenon in patients with peripheral arterial occlusive disease (Holm et al., 1972; Bylund et al., 1976), who have an insufficient blood supply to the muscles even at mild exercise.

We have recently shown that the intramuscular oxygen tension (Po_2) decreases during exercise both in patients with arterial insufficiency and normal subjects. The decrease in the intramuscular Po_2 was associated with marked changes in the energy potential and the redox state in the muscle tissue (Bylund-Fellenius et al., 1981). It was suggested that these changes are an important link in the adaptation of the oxidative enzyme capacity.

The aim of this study was to examine acute as well as chronic effects of reduced blood flow on muscle metabolism during exercise.

MATERIAL AND METHODS

Chemicals

Substrates, enzymes and co-factors for metabolite and enzyme analyses were all purchased from Sigma Corporation (St. Louis, Missouri, USA).

507

Animals

Female Sprague-Dawley rats, weighing about 250 g (Anticimex, Stockholm, Sweden) were used. The rats were kept in separate cages with free access to food (Purina chow) and tap water during the experimental period.

Operative Procedure

The rats were anesthetized with Nembutal 30 mg x kg^{-1} (ACO, Solna, Sweden). Both common iliac arteries were exposed through a midline incision and the right artery was ligated. In the experiments involving nerve stimulation, an electrode was placed around the sciatic nerve to each hindlimb through a lateral incision in the thigh. The electrodes were made of multistrand stainless steal wire (Cooner Sales Co. Inc., Chatsworth, California, USA) and were tunnulated subcutaneously from the hindlimbs to the neck.

Nerve Stimulation

A Grass ST-5 stimulator connected to the electrodes by a flexible cable was used for the electric stimulation of the sciatic nerves. Square wave impulses, at a frequency of 4 Hz and a duration of 0.3 ms were used and the voltage was adjusted to give palpable contractions in the calf muscles.

Measurement of Intramuscular Po_2

The intramuscular Po_2 was monitored with the technique previously described for human skeletal muscle (Holm and Bylund-Fellenius, 1981). A flexible catheter (diam. 1.3 mm) with an oxygen transducer mounted in the tip (G.D. Searle and Co., High Wycombe, Bucks, UK) was inserted into the calf muscles of the rat hindlimb.

Enzyme Analysis

Phosphofructokinase (PFK), citrate synthase (CS) and cyto-chrome-c-oxidase (cyt-c-ox) activities in muscle tissue were determined with our previously described methods (Bylund et al., 1977).

The enzyme activities were expressed per g of protein in the fraction analyzed. The protein content was determined according to Lowry et al. (1951).

Metabolite Analyses

The muscle biopsies were immediately frozen in liquid nitrogen, freeze-dried at $-25^{o}C$, stored at $-80^{o}C$ and extracted as described previously (Walker et al., 1982) for determination of metabolite concentrations (ATP, ADP, phosphocreatine, glycogen, lactate and pyruvate).

Experimental Protocol

    Group I: In 10 rats the intramuscular Po_2 was recorded at rest, and during nerve stimulation. When the Po_2 had reached a constant level the common iliac artery was ligated and the Po_2 was monitored for 5 min, whereafter the ligature was removed.
    Group II: In 9 rats the right common iliac artery was ligated as described above. The day after, the rats were anesthetized and both hindlimbs were stimulated for 20 min. Biopsies were taken from the soleus and the extensor digitorum longus (EDL) muscles from both legs at the end of stimulation period and analyzed for metabolite concentrations.
    Group III: In 12 rats the right common iliac artery was ligated as described above. During the following 6 days both hindlimbs were stimulated for 20 min, 4 times a day. The rats were allowed to move freely in their cages during the stimulation periods. At the 7th day the rats were anesthetized and the soleus and the EDL muscles were excised and analyzed for the activities of PFK, CS and cyt-c-ox.

Statistics

    Standard statistical procedures were used to calculate the mean and SEM in the different groups. For comparison between the ligated and the control leg, Wilcoxons matched pair tests were used (Siegel, 1956).

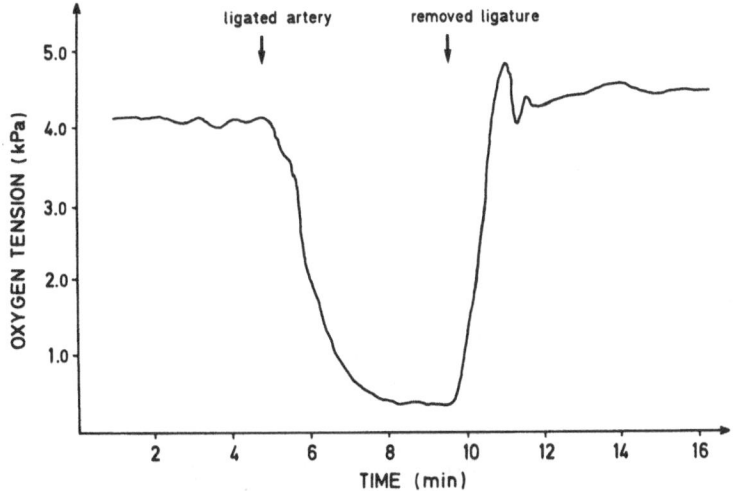

Fig. 1. A representative registration of the intramuscular oxygen tension in the calf muscles (rat), during muscle stimulation and common iliac artery ligation.

RESULTS

The intramuscular Po_2 was recorded in the calf muscles of 10 rats during muscle stimulation and ligation of the common iliac artery. A representative registration is shown in Fig. 1. Three min after the ligation the Po_2 had decreased from 4.6 ± 0.3 kPa to 0.5 ± 0.1 kPa (mean $\pm$ SEM). Removal of the ligature caused a rapid normalization of the Po_2.

The effect of the reduced blood flow on various metabolite levels in the soleus and the EDL muscles during muscle stimulation is shown in Fig. 2. Muscle biopsies were obtained at the end of a 20 min stimulation period 1 day after the ligation. Phosphocreatine, glycogen and the ATP/ADP ratio were lower and the lactate/pyruvate ratio was higher in both muscles in the ligated leg compared to the control leg.

Fig. 3 shows the results obtained after 6 days of intermittent stimulation. The activities of CS and cyt-c-ox had increased in both the soleus and the EDL muscles in the ligated leg compared to

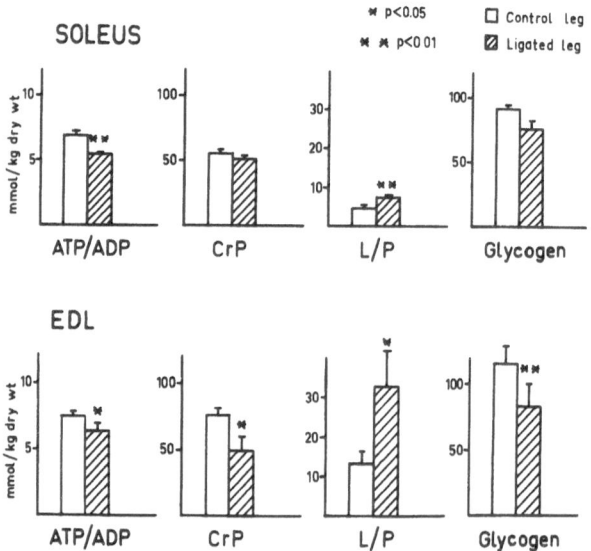

Fig. 2.   Metabolite levels during muscle stimulation in the ligated leg and the control leg, analyzed one day after ligation of the common iliac artery.

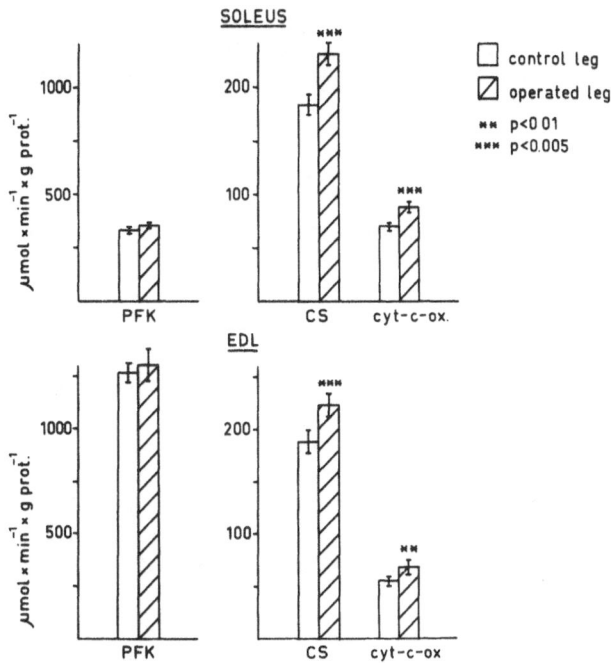

Fig. 3.   Enzyme activities in the ligated leg and the control leg
          analyzed after 6 days of intermittent sciatic nerve stimu-
          lation (rat).

the control leg. In the soleus muscle the increase amounted to an
increase in the mean values of 26 percent of control and in the EDL
to 20 percent for both enzymes. No change in PFK activity was
observed in either of the muscles.

DISCUSSION

        The aim of the present study was to investigate the acute and
chronic effects of reduced blood flow on skeletal muscle metabolism
during exercise. For this purpose an experimental rat model was
developed. The blood flow to the right hindlimb was restricted by a
ligature around the common iliac artery, leaving the left hindlimb
as a control. Measurements of muscle blood flow with the [133]Xenon
wash-out technique showed no difference in the resting flow between
the legs, but during muscle stimulation the flow was reduced by 50
percent in the ligated leg.

        The intramuscular Po_2 measurements demonstrated that this
reduction of the blood flow was associated with a decrease in the
oxygen tension during exercise.

To evaluate the acute effect of the reduced flow on the energy metabolism, muscle contractions were induced in both hindlimbs and muscle biopsies were taken during the stimulation and analyzed for various metabolite levels.

More pronounced alterations in the energy potential and the redox state were found in the muscles in the ligated leg compared to the control leg. This together with the higher glycogen ulitization demonstrates that the muscles exposed to the reduced blood flow have a more anaerobic metabolism - in line with a lower oxygen supply in this situation.

To evaluate the longer term effects of the reduced flow, muscle contractions were induced intermittenly for 6 days in both hindlimbs, and muscle biopsies were then taken for analysis of enzyme activities. Significantly increased activities of CS and cyt-c-ox were found in the muscles in the ligated leg as compared to the control leg, showing that repeated exposure to a reduced flow during exercise provokes an increase in the activities of oxidative enzymes.

It is concluded, that exercise at a reduced blood flow is associated with pronounced changes in the intramuscular Po_2 as well as the intracellular energy - and redox state in the working muscles. Furthermore, repeated exposure to a reduced flow during exercise is associated with chronic changes in the activities of oxidative enzymes in the muscle tissue. The results strengthen the hypothesis that changes in the energy - and redox state during periods of intermittent hypoxia trigger the adaptation of these enzymes.

REFERENCES

Bylund, A.-C., Bjurö, T., Cederblad, G., Holm, J., Sjöström, M., Ängquist, K.A., and Scherstén, T., 1977, Physical training in man. Skeletal muscle metabolism in relation to muscle morphology and running ability, Eur. J. Appl. Physiol., 36:151-169.

Bylund, A.-C., Hammarsten, J., Holm, J., and Scherstén, T., 1976, Enzyme activities in skeletal muscles from patients with peripheral arterial insufficiency, Eur. J. Clin. Invest., 6:425-429.

Bylund-Fellenius, A.-C., Walker, P.M., Elander, A., Holm, S., Holm, J., and Scherstén, T., 1981, Energy metabolism in relation to oxygen tension in human skeletal muscle during exercise, Biochem. J., 200:247-255.

Holloszy, J.-O., and Booth, F.W., 1976, Biochemical adaptation to endurance exercise in muscle, Ann. Rev. Physiol., 38:273-391.

Holm, J., Björntorp, P., and Scherstén, T., 1972, Metabolic activity in rat skeletal muscle. Effect of intermittent hypoxia, Eur. J. Clin. Invest., 2:279-283.

Holm, S., and Bylund-Fellenius, A.-C., 1981, Continuous monitoring
    of oxygen tension in human gastrocnemius muscle during exer-
    cise, Clin. Physiol., 1:541-552.
Lowry, O.H., Rosebrough, N.J., Farr, L.A., and Randall, R.J., 1951,
    Protein measurements with the Folin phenol reagent, J. Biol.
    Chem., 193:265-275.
Siegel, S., 1956, "Nonparametric statistics for behavioral sciences",
    McGraw-Hill, New York.
Walker, P.M., Idström, J.-P., Scherstén, T., and Bylund-Fellenius,
    A.-C., 1982, Glucose uptake in relation to metabolic state in
    perfused rat hindlimb at rest and during exercise, Eur. J.
    Appl. Physiol., 48:163-176.

LOCAL REGULATION OF BLOOD FLOW

G. Siegel, A. Walter, M. Thiel, and B.J. Ebeling

Institute of Physiology, Biophysical Research Group
The Free University of Berlin, 1000 Berlin 33, FRG

INTRODUCTION

In vascular smooth muscle as in other excitable structures the K^+ permeability is considerably higher than the Na^+ permeability ($P_K : P_{Na}$ = 1 : 0.024) (Siegel and Schneider, 1981). Therefore, the intra- and extracellular K^+ ion distribution play a decisive role in the passive potential genesis. Electromechanical coupling pro- vided, a change of internal and/or external K^+ concentration can thus influence vascular tone. With the aid of spectroscopic and mor- phometric data the K^+ fraction located in the intracellular space of smooth muscle cells is calculated to be 6/7 of the total potas- sium in a vessel wall, that located in the extracellular space to be 1/7. 43% of the extracellular potassium is distributed in the in- terstitial fluid space, while 57% is bound to connective tissue structures. The microdynamic K^+ binding properties of the latter fraction permit, under pH or concentration shifts of other cation species, a K^+ release from or adsorption to the polyanionic macro- molecules of vascular connective tissue, which can considerably alter the external K^+ concentration close to the cell membrane (Siegel et al., 1977a). Further, it is known from studies with nuclear magnetic resonance spectroscopy that a shift in proton con- centration not only changes the binding properties of cations to polyelectrolytes but also directly alters their conformation (Gustavsson et al., 1981). Thus, besides the effect of K^+ ions on vascular smooth muscle cells, we have also studied the influence of the external pH value on membrane permeability.

MATERIALS AND METHODS

Experimental Preparations and Solutions

     Common carotid arteries were surgically excised from anaesthe-
tized dogs within 5 min. Adventitial connective tissue was almost
completely removed from carotis segments (5 - 25 mg wet wt.) with
a microscopically controlled microdissection method. Both, vascular
connective tissue and the remaining media were equilibrated in a
modified Krebs solution having the following composition: Na^+ 151.16;
K^+ 4.69; Ca^{++} 2.52; Mg^{++} 0.11; Cl^- 143.42; HCO_3^- 16.31; $H_2PO_4^-$ 1.38;
glucose 7.77 $mmol \cdot l^{-1}$ (37^oC). For carotid vascular smooth muscle
different pH values were obtained by changing the O_2/CO_2 ratio of
the gas mixture with which the Krebs solutions were aerated. For adventitial connective tissue the pH value was varied between 4.0 and
9.0 whereby the pH steps 6, 7 and 8 were attained again by changing
the O_2/CO_2 ratio of the gas mixture. The pH steps 4 and 5 were ar-
rived at by titration of the Krebs solution with 65% HNO_3, the pH
step 9 by titration with 25% NH_4OH (Siegel et al., 1977a). The preparations were incubated for 3 hrs.

     Basilar arteries were surgically excised from anaesthetized
dogs after trepanation and equilibrated in modified Krebs solutions.
The pH steps 6.8, 7.3, and 7.8 were attained by changing the O_2/CO_2
ratio of the gas mixture with which the Krebs solutions were aerated.
The experiments were performed at the earliest after 3 hr incubation,
when the preparation was in equilibrium regarding ion concentrations
and fluxes (Siegel et al., 1976c, 1981a).

Electrical Recording

     Intracellular recordings of membrane potential were made with
glass microelectrodes filled with 3 M KCl. The resistances ranged
from 10 to 50 $M\Omega$ and the tip potentials from -5 to -20 mV. The electrodes were shielded to the tips; otherwise conventional recording
techniques were used. All intracellular recordings were corrected
for tip and junction potentials by a method described earlier (Haas
et al., 1966). Arteries with membrane potentials between -40 and
-70 mV were selected for the final averaging.

Mechanical Recording

     From the carotid arteries, 4 - 5 mm long cylinders were cut
lengthwise. The folded-out vessel ring, 10 - 15 mm long, was attached at the cut ends to an isometrically measuring tension device. The
specimens were equilibrated for 3 hrs in Krebs solution at an initial tension of 2 g. After this time the tension was 1.3 $\pm$ 0.1 g.
The extracellular H^+ or K^+ concentration was then varied. At each
concentration step the tension reached an equilibrium value after
1.5 - 2 hrs.

Determination of Ion Concentration

The total ion concentrations of the isolated tissues were determined using an atomic absorption spectrometer (Perkin-Elmer, model
403). After determination of wet and dry weight, the preparation was
placed in an ice-cold quartz Erlenmeyer flask containing pure oxygen
and Pt catalyst and burned ash-free. The gaseous products were dissolved in a defined volume of 1.3% HNO_3. Atomic absorption measurements were made directly from this fluid for Na, K and Mg; for Ca
after addition of a 2% lanthanum nitrate solution; and for Cl by
measurement of Ag after quantitative precipitation with $AgNO_3$. The
total method error is $< \pm 3\%$.

Isotope Flux Measurements

The preparations were loaded until saturation for at least 3 hrs
in a radioactive bath containing ^{24}Na, ^{28}Mg[1]) or ^{42}K (Siegel et al.,
1976b; 1976c). Washout was then performed in corresponding inactive
solutions. VIP series coaxial Ge(Li) detector systems with a ratio
of peak height to Compton plateau height of 30/1 were used (Ortec,
Oak Ridge, U.S.A.), which permitted adequate discrimination of the
γ-lines (system resolution 2.2 keV using 1.333 MeV photons) of ^{24}Na
at 1.37 MeV, ^{28}Mg at 1.342 MeV (Alburger and Harris, 1969) or of ^{42}K
at 1.52 MeV by means of stabilized single channel analyzers (SCA-N-3;
Elscint, Haifa, Israel). The γ-spectra of the preparation were measured continuously in a 2.3 ml chamber and printed out at intervals
of 10 s. The output was either on paper tape (Tally, Kent, U.S.A.),
magnetic tape (RG 23; Deutsche Intertechnik GmbH, Mainz, F.R.G.),
or as numbers from a Diehl Combitron S (Berlin, F.R.G.). During the
measurements, a strictly constant and laminar flow of the washout
medium was maintained. The flow rate was 2.7 ml/s, so that the fluid
in the measurement chamber was completely exchanged 1.17 times per
second to avoid an accumulation of radioactive ions in the preparation chamber, interstitium, or on the surface of the specimen.

For $^{24}Na^+$ and $^{42}K^+$ efflux measurements from the basilar artery,
Krebs solutions with the three pH steps, 6.8, 7.3, and 7.8, were
used as radioactive incubation media and as washout fluid. After
loading, in direct efflux experiments, the decay of radioactivity
in the preparation was recorded over 40 min at intervals of 10 s,
rinsing with inactive solutions of the same composition and pH. Solutions in 3 concentration steps of Mg^{++} or K^+ as the only counter
ion were taken for the radioactive incubation and the subsequent
washout of isolated adventitial connective tissue in the ^{28}Mg and
^{42}K flux measurements (Siegel et al., 1980b). The Mg^{++} solutions

[1] ^{28}Mg was produced via reaction $^{26}Mg(t,p)^{28}Mg$ using the cyclotron
of the Fachbereich Physik, Technical University of Munich (Prof.
Dr. H. Morinaga).

contained pure $MgCl_2$ in concentrations of 0.11, 1.1 and 11 $mmol \cdot l^{-1}$
Mg^{++}. A pH of 7.0 was obtained by addition of Trizma base (Sigma
Chemical Co., St. Louis, U.S.A.) as a buffer substance, where the
final Tris concentration within the various solutions was 2 $mmol \cdot l^{-1}$
at the most. The K^+ solutions contained K^+ in concentrations of
0.469 $mmol \cdot l^{-1}$ (KCl 0.269 $mmol \cdot l^{-1}$; $KHCO_3$ 0.1 $mmol \cdot l^{-1}$; $KH_2 PO_4$
0.1 $mmol \cdot l^{-1}$), 4.69 (KCl 2.69 $mmol \cdot l^{-1}$; $KHCO_3$ 1 $mmol \cdot l^{-1}$; KH_2PO_4
1 $mmol \cdot l^{-1}$) and 46.9 $mmol \cdot l^{-1}$ (KCl 29.21 $mmol \cdot l^{-1}$; $KHCO_3$ 16.31
$mmol \cdot l^{-1}$; KH_2PO_4 1.38 $mmol \cdot l^{-1}$). A pH of 7.0 was attained by changing
the O_2/CO_2 ratio of the gas mixture with which the K^+ solutions
were aerated. The temperature was adjusted to 37°C throughout the
experiments. Vascular connective tissue was pretreated before incubation in the radioactive bath for 1 hr with the corresponding Mg^{++}
or K^+ solutions which were exchanged several times to reduce the concentration of unwanted ion species to zero or a minimum. After loading, in direct efflux experiments the decay of radioactivity in the
preparation was registered over 40 min at intervals of 10 s, rinsing
with inactive solutions of the same composition. The analysis of
the efflux kinetics and the calculation of affinity constants using
a thermodynamic approach were described in an earlier publication
(Siegel et al., 1980b). After each experiment, the concentrations
of Na^+, K^+, Cl^-, Mg^{++} and Ca^{++} ions in basilar arteries or adventitial connective tissue were determined by atomic absorption spectroscopy.

[23]Na Nuclear Magnetic Resonance Measurements

A Bruker WH 90 Fourier transform spectrometer operating at a
resonance frequency of 23.8 MHz was used for the ^{23}Na NMR transverse relaxation rate, $R_2 = 1/T_2$, determinations. Each spectrum is
the result of at least 100 accumulations with an acquisition time
of 0.5 s, and each reported value is the average of at least 3 separate determinations. R_2 is calculated from $R_2 = 1/T_2 = \pi \cdot \Delta v_{1/2}$,
where $\Delta v_{1/2}$ is the line width at half height of the observed resonance signal. The longitudinal relaxation rate, $R_1 = 1/T_1$, measurements were performed on a Bruker BK 322s NMR spectrometer at 23.812
MHz, using the 180°-t-90° plot method. The corresponding T_2's were
obtained using the Meiboom-Gill-Carr-Purcell sequence with 8 ms
between the 180° pulses. In both cases, the signals were averaged
using a Varian CAT-1024 time average computer.

For the competition experiments the samples were prepared by dis-
solving an amount of chondroitin sulphate·Na salt (SERVA, Heidelberg,
F.R.G.), corresponding to 0.1 M concentration in both carboxylate
and sulphate groups in 2 g H_2O (pH = 5.6) or in 2 g Mg^{++}-free Krebs
solution (pH = 6.9). The chondroitin sulphate was found to be composed of 95% chondroitin-4-sulphate and 5% protein (Siegel et al.,
1977a). Therefore, the preparation was classified as a multichain
chondroitin sulphate-polypeptide complex (CS-P) (Gustavsson et al.,
1981). In the case of H_2O-dissolved samples, the pH was adjusted

to 3.6 (lower plateau of the pH titration curve, Siegel et al., 1977b) by the addition of a small amount of 1 M aqueous HCl. Then defined amounts of crystalline NaCl, KCl or $MgCl_2 \cdot 6H_2O$ were added to the samples. The pH value of the samples remained unchanged with these additions. For samples dissolved in Mg^{++}-free Krebs solution, a physiological pH value of 7.38 (plateau of the pH titration curve between pH 6.3 and 7.6 (Siegel et al., 1980b) was attained by changing the O_2/CO_2 ratio of the gas mixture with which the solutions were aerated before measurement. Then defined amounts of crystalline $MgCl_2 \cdot 6H_2O$ were added to the samples. pH determinations and readjustments to 7.38 were performed before and after running the NMR spectra.

RESULTS AND DISCUSSION

Membrane Properties of Vascular Smooth Muscle

Table 1 shows the known relation between membrane potential and tension development of vascular smooth muscle and the external K^+ concentration (Scott et al., 1961; Kuschinsky et al., 1972; Brace, et al., 1974; Siegel et al., 1974, 1976c, 1977b; Casteels et al., 1977; Kuriyama and Suzuki 1978; Bolton, 1979; Ito et al., 1979; Suzuki 1981). Proceeding from $[K^+]_o = 2.35$ mmol$\cdot l^{-1}$, a decrease and increase of the extracellular K^+ concentration led to depolarization and contraction. Comparing the changes in tension with these alterations in membrane potentials, it becomes obvious that the maximum relaxation nearly corresponds to the maximum hyperpolarization of

Table 1. Membrane potential (V) and tension development (T) in isolated carotid segments of the dog as a function of the external K^+ concentration $[K^+]_o$ (n = 20)

$[K^+]_o$ [mM]	V ± SEM [mV]	T ± SEM [g]
0.00	- 47.9 ± 2.4	2.63 ± 0.39
0.47	- 49.1 ± 1.6	2.69 ± 0.36
1.17	- 50.4 ± 1.5	2.35 ± 0.26
2.35	- 52.8 ± 1.3	1.61 ± 0.14
4.69	- 51.7 ± 1.4	1.29 ± 0.09
11.73	- 44.4 ± 1.4	1.44 ± 0.15
23.45	- 39.7 ± 0.7	1.95 ± 0.24
46.90	- 29.8 ± 1.5	4.43 ± 0.32
93.80	- 22.4 ± 2.2	4.84 ± 0.38

the membrane. The small difference between the extracellular K^+ concentrations, at which the most negative membrane potential and the most relaxed state were registered, may be due to the fact that membrane potentials were recorded during stable intracellular penetrations about 10 - 20 min after changing the K^+ concentration, while mechanical force was measured in the stationary equilibrium after 1.5 - 2 hrs. Further, we observed a hyperpolarization of 2.6 mV in 45% of the transitions from 4.7 to 11.7 $mmol \cdot l^{-1}$ $[K^+]_o$, and in a few cases a hyperpolarization of 2.3 mV during the transition to 23.5 $mmol \cdot l^{-1}$ $[K^+]_o$. In all cases relaxation corresponded with these hyperpolarizations. Besides a possible stimulation of an electrogenic Na^+ pump (Brace et al., 1974), our results indicate that these opposing effects depend on the instantaneous value of the membrane potential when the hyperkalemic solution is introduced during a slow wave (Siegel et al., 1976a).

The first two figures show the results of testing the effect of another physiological parameter, the exterior pH value of the Krebs solution, on the membrane potential and mechanical force development (Vanhoutte and Clement, 1968; Kuschinsky et al., 1972;

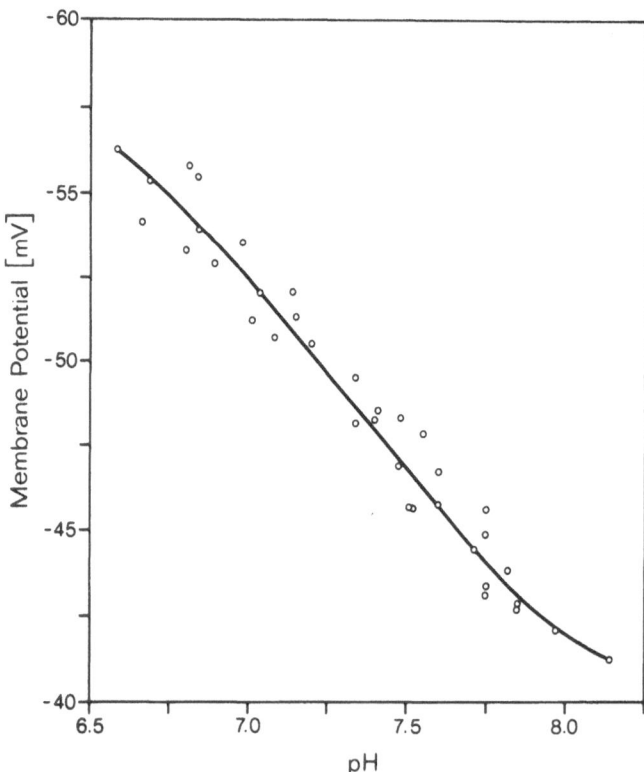

Fig. 1.   Membrane potential of vascular smooth muscle in relation to the external pH value of the Krebs solution.

Siegel and Schneider, 1981b). Fig. 1 represents the correlation between pH and membrane potential. The curve shows a sigmoid course in a wider pH range, between 6.8 and 7.8 it is linear. Acidification of the Krebs solution causes hyperpolarization of the vascular smooth muscle cells, alkalinization causes depolarization. A quantitative correlation between pH change and change in tension shows also a linear variation of the tension development in the pH range 6.8 to 7.8 (Fig. 2). This curve, too, has a sigmoid course in a wider pH range. A very similar relation between tension and respiratory changes in pH was found by Vanhoutte and Clement (1968) in saphenous vein strips.

Knowing the membrane potential and tension as a function of the extracellular H^+ and K^+ concentration allows an examination of electromechanical coupling under the given experimental conditions. The parameters $[K^+]_o$ and pH can be eliminated and the membrane potential plotted against the developed force. Fig. 3 illustrates the

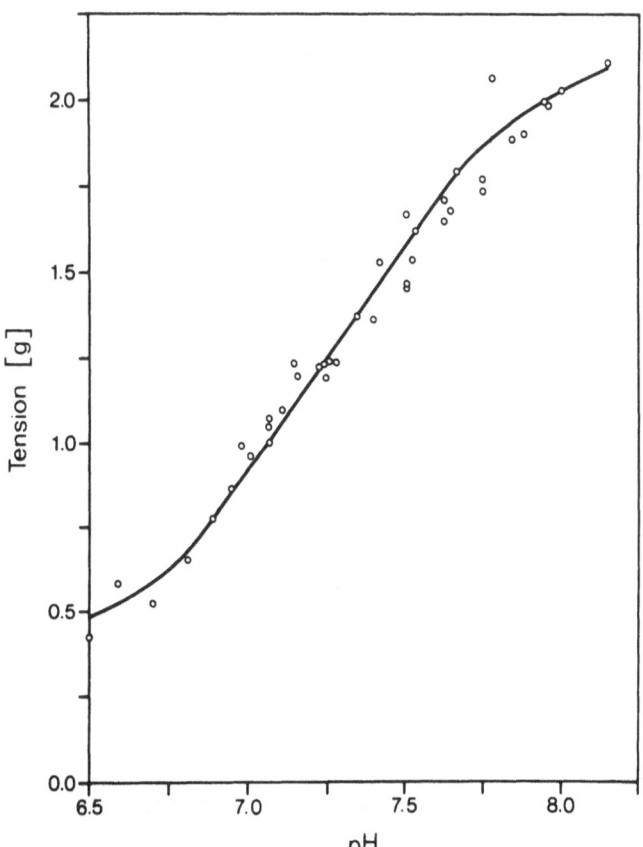

Fig. 2.   Tension developed in isolated carotid segments as a function of the extracellular H^+ concentration.

stationary activation curve of tension versus membrane potential in
isolated carotid segments. The membrane potential was changed by
variations in the external H^+ or K^+ concentration. It is immediately
obvious from the sigmoid activation curve and the measurement points
that the same change in tone occurs with a defined alteration of membrane potential, regardless whether the latter results from a variation of the extracellular K^+ concentration or the exterior pH (Siegel
and Schneider, 1981b). So, this curve gives strong evidence for the
existence of an electromechanical coupling in vascular smooth muscle,
at least under the physiological conditions described here. The point
marked on the curve indicates the membrane potential and the mechanical force under normal conditions. In the linear range of the characteristic curve, a potential change of 8.7 mV corresponds to an alteration of force of 1 g. A very similar characteristic curve, though
without pH-dependent potential changes, was observed by Ito et al.
(1979) in the coronary artery. The present curve demonstrates that
the effect of a change in H^+ or K^+ concentration is mediated via a

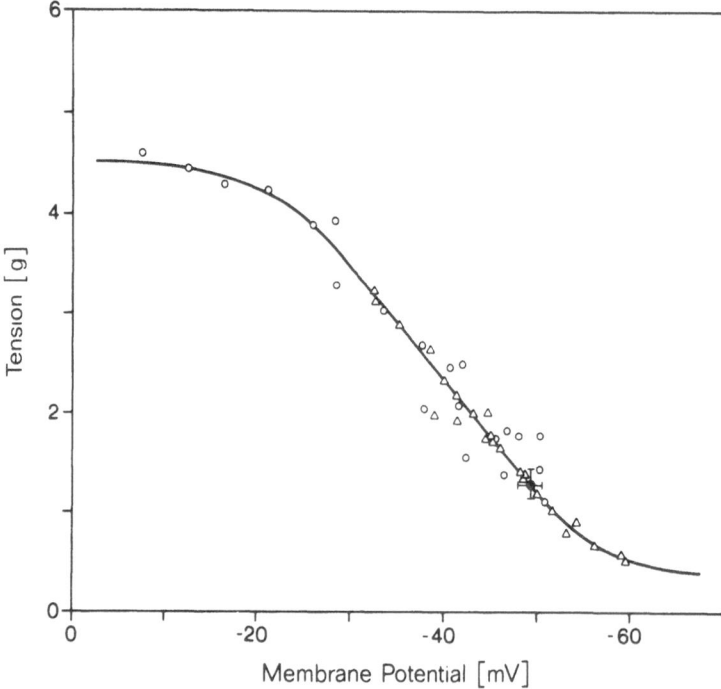

Fig. 3. Dependency of mechanical tension on the membrane potential
         in isolated carotid segments. The membrane potential was
         changed by variations of the extracellular H^+ (Δ) or K^+
         concentration (o). The point marked on the curve indicates
         the membrane potential and the mechanical force under
         normal conditions.

change in membrane potential. The force developed can be inferred
from the stationary activation curve.

The vasodilator hyperpolarization at acidic pH and the vasoconstrictor depolarization at basic pH demand an explanation of their
basis on the cell membrane level. Looking at the equilibrium potentials of the most important ion species in relation to the membrane
potential, a hyperpolarization with acidosis is possible by an increase in K^+ permeability of the cell membrane or by a decrease in
Na^+ permeability or by both. The opposite argumentation would be
valid for alkalosis. By means of $^{42}K^+$ and $^{24}Na^+$ flux experiments at
pH values 6.8, 7.3, and 7.8 under steady state conditions, the
question could be answered if, with a pH change, the origin of the
membrane potential shift and the related change of vascular tone
has to be attributed to a permeability change of the vascular
smooth muscle cell membrane for K^+ and/or Na^+ ions.

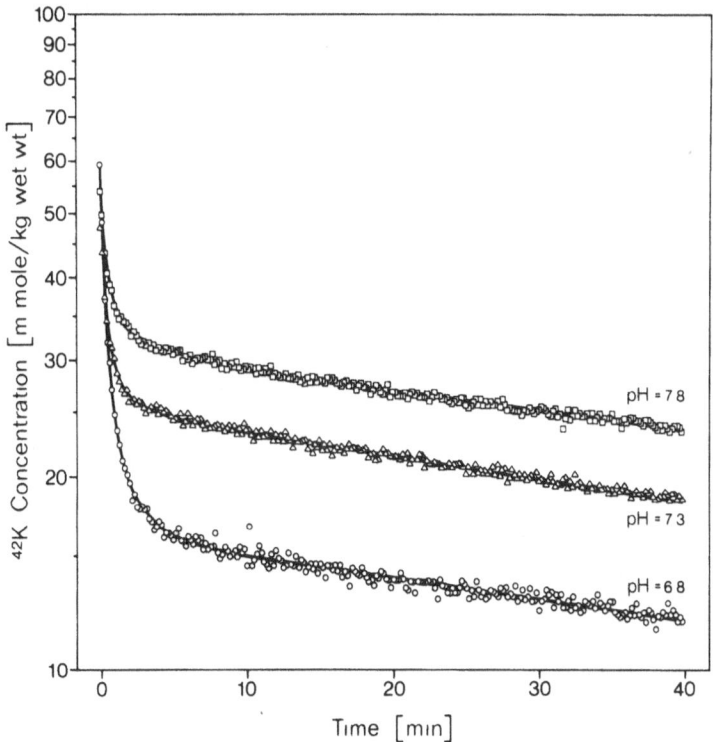

Fig. 4. $^{42}K^+$ efflux in the basilar artery of the dog according to
the direct method in Krebs solutions with pH 6.8, 7.3, and
7.8. The graph shows the time course of radioactive K^+ decrease within the preparations at intervals of 10 s in
single experiments. The superimposed lines represent the
optimal triple-exponential functions found by computer
fitting.

In Fig. 4 a semilogarithmic plot of $^{42}K^+$ efflux curves of the
basilar artery of the dog at pH steps 6.8, 7.3, and 7.8 with a time
resolution of 10 s is presented (Siegel et al., 1980a, 1981a). It
can be seen that the flux curves have a regularly graded, pH-dependent course. The fastest K^+ efflux is found at pH 6.8, the slowest
at pH 7.8. Triple-exponential decay curves could be fitted iteratively to the measured values with the aid of a nonlinear least square
fit (Marquardt, 1963). The three amplitudes and time constants of
the e-functions were calculated and a detailed compartment model was
applied (Siegel et al., 1980b, 1981a). Finally a statement on a pH-
dependent, passive K^+ net current has to be reserved for an exact
analysis of the effects of a pH change on the quantity of the various
compartments and the kinectic coefficients.

Fig. 5 shows ideal $^{24}Na^+$ efflux curves at pH steps 6.8, 7.3,
and 7.8. The amplitudes and time constants of five experiments each
were computed and averaged. The mean values served for calculating

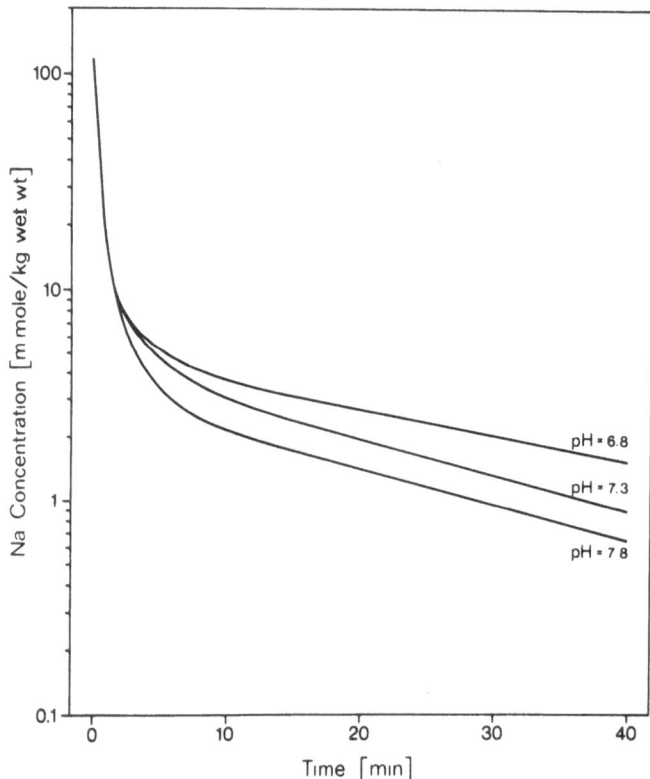

Fig. 5. Ideal $^{24}Na^+$ efflux curves at pH steps 6.8, 7.3, and 7.8 in
the Krebs solution. The amplitudes and rate constants of
five experiments each were computed and averaged. The mean
values served for calculating and drawing the ideal flux
curves.

the ideal flux curves. The Na$^+$ exchange is much faster than that of K$^+$ ions. Already in this figure, a rise in Na$^+$ flux is clearly seen at pH 7.8 versus pH 7.3 or 6.8, at least concerning the slowest exchange phase. Therefore, the relation of the Na$^+$ exchange to the external pH is inverse compared to K$^+$ ions.

The summarized results of a compartment analysis of the flux data are presented in Table 2, leading to the calculation of the ionic net currents I$_{Na}$ and I$_K$. Na$^+$ net current, but also Na$^+$ conductance and permeability, increase steadily from pH = 6.8 to 7.8. The corresponding K$^+$ values for I$_K$, g$_K$ and P$_K$ decrease continuously with increasing pH (Siegel et al., 1980 a; 1981a; Siegel, 1981; 1982). Therefore, the highest K$^+$ permeability and the lowest Na$^+$ permeability are found at pH = 6.8, which is in complete agreement with the hyperpolarization of the cell membrane at this pH value (cf. Harder, 1982). This shift to the K$^+$ side becomes particularly obvious in the quotients g$_K$: g$_{Na}$ and P$_K$: P$_{Na}$. These ratios decrease steadily with increasing pH.

These findings give a satisfactory explanation for the hyperpolarization and relaxation at acidic pH, and the depolarization and constriction at basic pH.

Ionic Interactions in Vascular Connective Tissue

So far we have dealt with the direct effect of ionic changes on the membrane properties of vascular smooth muscle cells. In the

Table 2. Effects of external pH changes on undirectional fluxes (ϕ), ionic currents (|), conductances (g), and permeabilities (P) in vascular smooth muscle of the basilar artery of the dog.

		Krebs solution		
		pH = 6.8	pH = 7.3	pH = 7.8
ϕ_K	[pmole/cm^2/s]	48.82	27.61	21.22
I$_K$	[μA/cm^2]	3.69	2.13	1.69
g$_K$	[Ω^{-1}cm^{-2}]	90.41 · 10^{-6}	49.51 · 10^{-6}	36.17 · 10^{-6}
P$_K$	[cm/s]	1.00 · 10^{-6}	0.56 · 10^{-6}	0.40 · 10^{-6}
ϕ_{Na}	[pmole/cm^2/s]	3.01	5.81	6.87
I$_{Na}$	[μA/cm^2]	· 0.28	· 0.52	· 0.62
g$_{Na}$	[Ω^{-1}cm^{-2}]	3.27 · 10^{-6}	7.17 · 10^{-6}	8.51 · 10^{-6}
P$_{Na}$	[cm/s]	8.80 · 10^{-9}	18.32 · 10^{-9}	23.41 · 10^{-9}
g$_K$: g$_{Na}$		27.6	6.9	4.3
P$_K$: P$_{Na}$		113.8	30.5	17.3

following we want to turn to the ionic interactions in vascular connective tissue and to their indirect influence on vascular tone. It
is well known from morphological and histochemical investigations
that vascular smooth muscle cells are covered by the basement membrane and a fine network of connective tissue fibres (Siegel et al.,
1980b; 1981a). Furthermore, we could demonstrate that these structures have a high accumulation rate of mono- and divalent cations
(Siegel et al., 1976c; 1977a; 1980b; 1981 c). Although quantitative
morphology assigns only 15% of the total volume of the vessel wall
to this compartment, about 65% of the cations in the vascular wall
are structurally bound to the proteoglycans which comprise only 1%
of the connective tissue (Siegel et al., 1981c). The basic significance of this large, extracellular cation pool became clear when acid
mucopolysaccharides, in the form of proteoglycans, were identified
as an essential component of basement membranes and connective tissue
fibres (Chamley-Campbell et al., 1979; Kennedy, 1979; Rodén, 1980).
The basal lamina builds a tight-fitting sheath around the plasma
membrane of the smooth muscle cell in a distance of 8 - 15 nm (Siegel
et al., 1980b; 1981c). Besides, vascular smooth muscle cells are also
fitted into a tight network of other connective tissue fibres. In
this way small pockets and narrow cleft spaces between cell membrane,
basal lamina and connective tissue fibres emerge (Siegel et al.,
1977a; 1977b; Bolton, 1979; Siegel et al.,1980b; 1981c). Under suitable, non-stationary conditions, even small amounts of released or
adsorbed ions can lead to measurable activity changes in these tiny
extracellular fluid spaces (Zidek and Lange-Asschenfeldt, 1980).
Well-known examples are the accumulation and depletion effects under
long-lasting voltage clamping in cardiac and smooth musculature (Cohen
et al., 1976; Baumgarten and Isenberg, 1977; Bolton et al., 1981;
Kleine and Kupersmith, 1982). They affect transmembrane ionic currents, membrane potential and mechanical tension development. As the
K^+ concentration in the immediate neighbourhood of the cell membrane
can be influenced by the microdynamic binding properties of the basement membrane and the other vascular connective tissue fibres, we
studied K^+ binding and exchange in dependence on proton and cation
concentration by tracer and NMR techniques. Isolated, adventitial
connective tissue of the canine carotid artery has been used as test
object. These investigations served as a model for the physico-chemical prerequisites of cation binding to medial connective tissue and
its susceptibility, especially by alterations of the extracellular
pH value and ionic environment. Hence, the question arises of specific
cation binding to discrete sites with ionic selectivity and possible
competitive and cooperative binding effects.

Fig. 6 shows K^+ binding curves for normal and Ca^{++}-free Krebs
solutions. In a normal Krebs solution, the K^+ binding increases
with increasing pH and attains a maximum at pH = 6.8. Surprisingly,
it decreases in the basic pH range, which is a most remarkable fact
for a pH-dependent cation binding curve (Siegel et al., 1977a; 1981c).
Before an explanation for this behavior is given, the significance

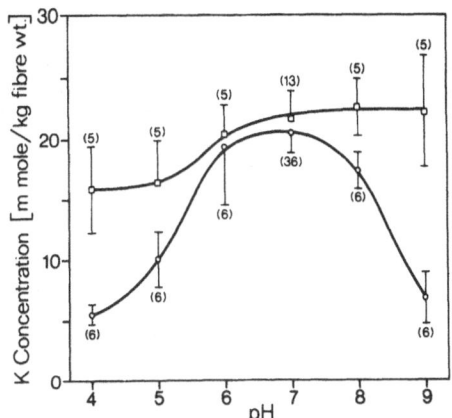

Fig. 6. K^+ binding of isolated adventitial connective tissue from
the canine carotid artery as a function of the pH of a
normal (o) and a Ca^{++}-free (□) Krebs solution.

of the K^+ binding curve shall be discussed. The actual K^+ binding
curve shows that with a displacement of the pH from 7.4 towards acidity the K^+ ions are bound to connective tissue, with a displacement
towards basicity they are released. The physiological meaning of this
mechanism could be that alkalosis increases the extracellular K^+ concentration in the immediate neighbourhood of the cell membrane, while
acidosis has the opposite effect. Extracellular accumulation of potassium would lead to membrane depolarization of smooth muscle cells
and sympathetic nerve fibres located at the adventitia-media boundary,
and thus to contraction, depletion of potassium to hyperpolarization
with vasodilatation. Bearing in mind the membrane hyperpolarization
of vascular smooth muscle cells with acidosis, also a decrease of
the external K^+ concentration close to the cell membrane can be
discussed as one of its causes besides an increase of K^+ permeability and a decrease of Na^+ permeability.

The K^+ binding characteristic in a Ca^{++}-free Krebs solution
explains the unusual course of the K^+ binding curve under normal
conditions (Siegel et al., 1977a; 1981c). Comparing both curves, one
can immediately conclude that the decrease of K^+ binding in the basic
pH range for normal Krebs solution must be attributed to Ca^{++} competition. The Ca^{++} competition is least pronounced between pH 6 and 7.
Therefore, the result of a Ca^{++}-free Krebs solution is a simple sigmoid K^+ titration curve. Furthermore, the regulating role of K^+ ions
on the membrane potential is abolished by flattening its binding characteristic, since the K^+ binding at pH = 7.4 has reached a steady
state value. In summary, K^+ and Ca^{++} ions compete for the same binding sites on connective tissue structures. Because of the higher
affinity of Ca^{++} ions, the K^+ ions are expelled from their binding
sites. A strong increase of Ca^{++} binding in the basic pH range can

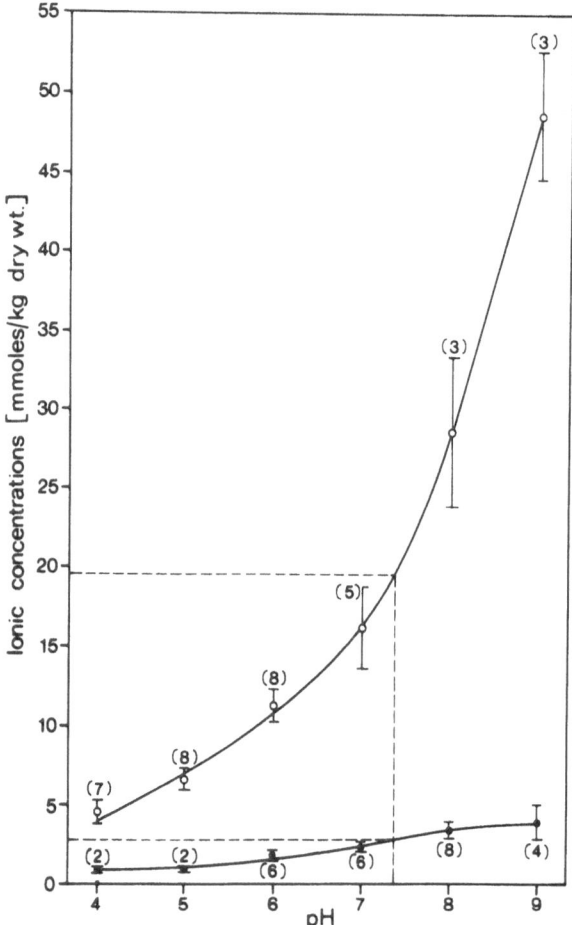

Fig. 7. Ca (o) and Mg binding (•) of isolated adventitial connec-
         tive tissue from the dog A. carotis as a function of the
         pH of the Krebs solution.

be supposed. This assumption is confirmed by Fig. 7. The very steep
Ca^{++} binding characteristic, especially in the basic pH range, confirms that K^+ and Ca^{++} ions compete for the same binding sites. Moreover, a very small pH change in the physiological range can lead to
a considerable release or binding of Ca^{++} ions (Siegel et al., 1977b).
In principle, the same argument holds for Mg^{++} ions, but their binding curve is flatter. However, it must be noted that the Mg^{++} concentration in our modified Krebs solution was only 1/10 that of a
normal blood plasma. Increasing the extracellular Mg^{++} concentration
to 1.1 $mmol \cdot l^{-1}$ does not change the binding characteristics for Na^+,
Mg^{++}, and Ca^{++} ions in principle, while the K^+ binding is considerably modified. This will be discussed later.

    The polyelectrolyte connective tissue that was used for the ex-
periments described before, consists of at least three very complicated macromolecules, collagen, elastin, and proteoglycan. Therefore,
we decided to study affinities and ionic interactions in substances
derived from connective tissue using ^{23}Na NMR relaxation rate measurements. To obtain a good basis for interpretation it was useful to
begin on relatively simple systems like chondroitin and dermatan sulphate and their proteoglycans (Siegel et al., 1977a; 1977b; 1980b;
1981c; Gustavsson et al., 1981).

    Fig. 8 demonstrates the result of a competition experiment with
a chondroitin sulphate polypeptide (CS-P). The ionic concentrations
are plotted on the abscissa, the transverse relaxation rate R_2 on
the ordinate. R_2 is a measure for the ratio bound/ free Na^+ ions. A
high value of R_2 results from a great bound Na fraction, while a decreasing R_2 means that Na^+ ions are more and more ionized. One recognizes that with the increasing concentration of Na^+, K^+, and Mg^{++}
the ^{23}Na relaxation rate decreases to differing but approximately
constant end values. If one looks at the initial parts of the curves
in the range 0 to 200 meq, which include also physiological ion concentrations, then one notices that the ^{23}Na relaxation rate decreases
fastest upon addition of K^+ or Mg^{++} ions, and distinctly slower
upon addition of Na^+ ions. This means that due to the high affinity
of K^+ or Mg^{++} ions for the binding sites of CS-P, these ion species
expel most of the Na^+ ions when all three ion species are at the
same normality (Siegel et al., 1980b; 1981). The nearly complete coincidence between the K^+ and Mg^{++} competition curves allows one to
expect the affinities of K^+ and Mg^{++} ions to CS-P to be almost identical.

    After these more qualitative conclusions we intended to measure
global affinity constants of the various ion species for in vivo

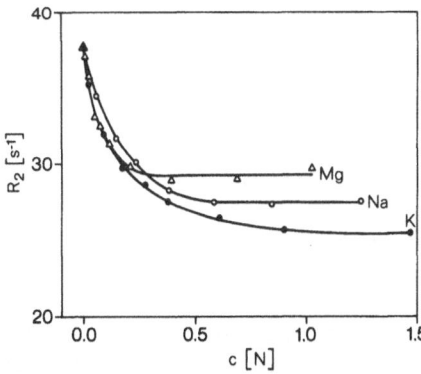

Fig. 8. $^{23}Na^+$ transverse relaxation rate as a function of Na^+, K^+
          and Mg^{++} concentrations in the presence of a 0.1 M CS-P
          solution at pH = 3.6.

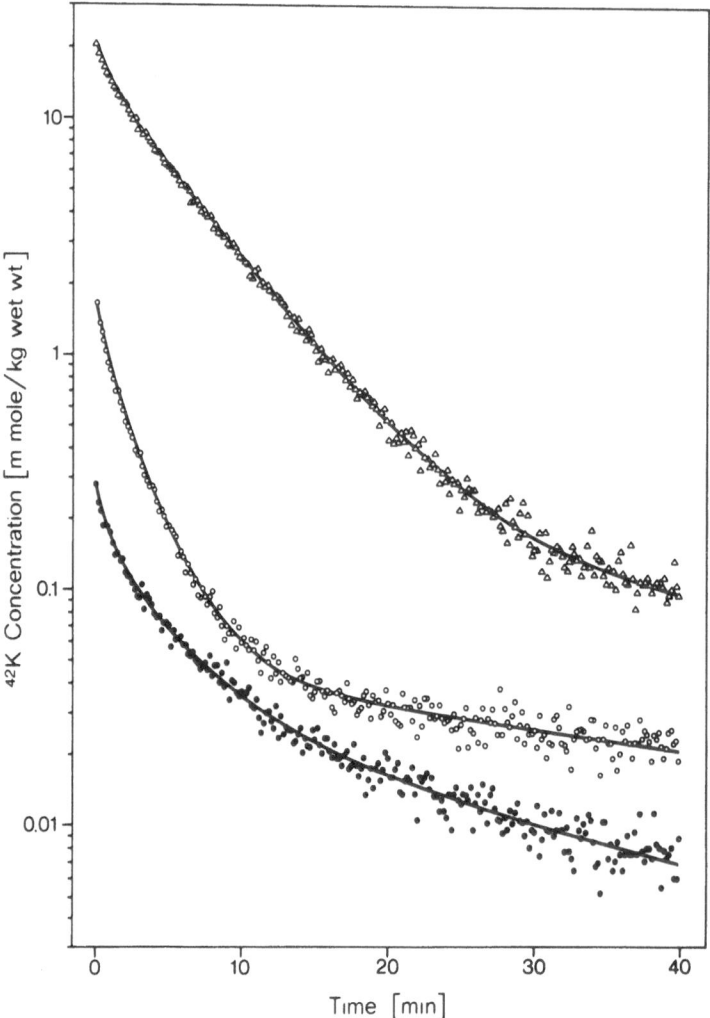

Fig. 9.   Steady state $^{42}K^+$ efflux from isolated adventitial connec-
          tive tissue of canine carotid artery according to the direct
          method. The graph shows the time course of radioactive K^+
          decrease within the preparation at intervals of 10 s in
          single experiments for 0.469 $mmol \cdot l^{-1}$ (●), 4.69 $mmol \cdot l^{-1}$ (o)
          and 46.9 $mmol \cdot l^{-1}$ (Δ) K^+ solutions at pH 7.0. The superim-
          posed lines represent the optimal triple-exponential func-
          tions found by computer fitting.

vascular connective tissue. Therefore, tracer efflux experiments were
designed, which would supply us with the necessary information (amplitudes, rate constants) for calculating affinity constants. Isolated adventitial connective tissue as a polyelectrolyte with at least
two distinct binding sites (carboxylate and sulphate groups) was used,

and $^{42}K^+$ or $^{28}Mg^{++}$ ions in three different concentrations each as ligands (cf. Siegel et al., 1981c). In order to exclude competition, pure KCl or $MgCl_2$ solutions without any other cation species were used as incubation media. Fig. 9 illustrates $^{42}K^+$ washout curves for 0.469, 4.69, and 46.9 $mmol \cdot l^{-1}$ potassium solutions, respectively. Under the assumption of three compartments (K^+ efflux from tissue water and from two binding sites) a triple-exponential decay curve has been fitted iteratively to the measured values. The three amplitudes and rate constants of the e-functions were calculated. From Table 3 it becomes evident that the rate constants (b_2, b_3) and the derived kinetic coefficients (k_{21}, k_{31}) for the slower exchange fractions are equal and mutually uniform in all series of experiments. The kinetic coefficients are an expression of the exchange deceleration by chemical binding and the affinity of binding sites for K^+ ions. The homogeneity of the kinetic coefficients under the various experimental conditions shows that the negatively charged groups in the preparations after K^+ accumulation can be subdivided into few classes of binding sites with uniform intrinsic affinity constants for potassium. The fraction of K^+ bound to the sites of one class corresponds to an experimentally defined compartment of the tracer exchange.

Table 3. Amplitudes, rate constants, compartments and kinetic coefficients of $^{42}K^+$ efflux kinetics in the isolated adventitial connective tissue of the carotid artery of the dog after equilibration in K^+ solutions at three concentration steps ($s_m \pm$ SEM of 5 experiments)

$\lvert K^+\rvert_0$ [mM]	Amplitude A_i [mmole/kg wet wt.]	Rate constant b_i [s^{-1}]	Compartment S_i [mmole/kg wet wt.]	Kinetic coefficient k_i [s^{-1}]
0.469	$A_1 = 0.138 \pm 0.028$	$b_1 = 2.619'\text{-}2 \pm 0.564'\text{-}2$	$S_1 = 0.202$	$k_{10} = 2.207'\text{-}2$
	$A_2 = 0.198 \pm 0.077$	$b_2 = 4.219'\text{-}3 \pm 0.767'\text{-}3$	$S_2 = 0.134$	$k_{21} = 5.008'\text{-}3$
	$A_3 = 0.054 \pm 0.027$	$b_3 = 4.619'\text{-}4 \pm 0.632'\text{-}4$	$S_3 = 0.054$	$k_{31} = 4.619'\text{-}4$
	$\Sigma A_i = 0.390 \pm 0.086$			
4.69	$A_1 = 0.765 \pm 0.268$	$b_1 = 2.292'\text{-}2 \pm 0.424'\text{-}2$	$S_1 = 1.206$	$k_{10} = 1.872'\text{-}2$
	$A_2 = 1.220 \pm 0.213$	$b_2 = 4.136'\text{-}3 \pm 0.707'\text{-}3$	$S_2 = 0.779$	$k_{21} = 5.062'\text{-}3$
	$A_3 = 0.167 \pm 0.057$	$b_3 = 4.620'\text{-}4 \pm 0.632'\text{-}4$	$S_3 = 0.167$	$k_{31} = 4.620'\text{-}4$
	$\Sigma A_i = 2.152 \pm 0.347$			
46.9	$A_1 = 8.982 \pm 0.981$	$b_1 = 2.000'\text{-}2 \pm 0.267'\text{-}2$	$S_1 = 14.834$	$k_{10} = 1.612'\text{-}2$
	$A_2 = 13.599 \pm 1.268$	$b_2 = 4.369'\text{-}3 \pm 0.650'\text{-}3$	$S_2 = 7.747$	$k_{21} = 5.423'\text{-}3$
	$A_3 = 0.265 \pm 0.038$	$b_3 = 3.660'\text{-}4 \pm 0.632'\text{-}4$	$S_3 = 0.265$	$k_{31} = 3.660'\text{-}4$
	$\Sigma A_i = 22.846 \pm 1.604$			

Table 4. Binding site $S_{i,j}$, affinity constant K_A and concentration
         of sites (S_i) of arterial vascular connective tissue of the
         dog for the cation species K^+ and Mg^{++}

$S_{i,j}$	K_A [l/mole]	$\log K_A$	$\Sigma[S_i]$ [meq/kg fibre wt.]
$S_{K,1}$	365	2.56	16.068
$S_{K,2}$	1,009	3.00	
$S_{Mg,1}$	559	2.75	15.672
$S_{Mg,2}$	3,177	3.50	

   After having computed a compartment model and solved a set of
reaction kinetic equations which describe the binding of the ligand
to compartments S_2 and S_3, the affinity constants could be calculated (Siegel et al., 1980b; 1981c) The result of these flux experiments
is summarized in Table 4. One notices that the log K_A values for $S_{i,1}$
and $S_{i,2}$, respectively, are nearly equal for both the ion species K^+
and Mg^{++}, where $S_{i,2}$ binds more strongly than $S_{i,1}$. Thus, these data
are in excellent agreement with the competition experiments from the
NMR investigations on CS-P. Also the total number of binding sites
for K^+ and Mg^{++} ions is nearly identical. This means that K^+ and Mg^{++}

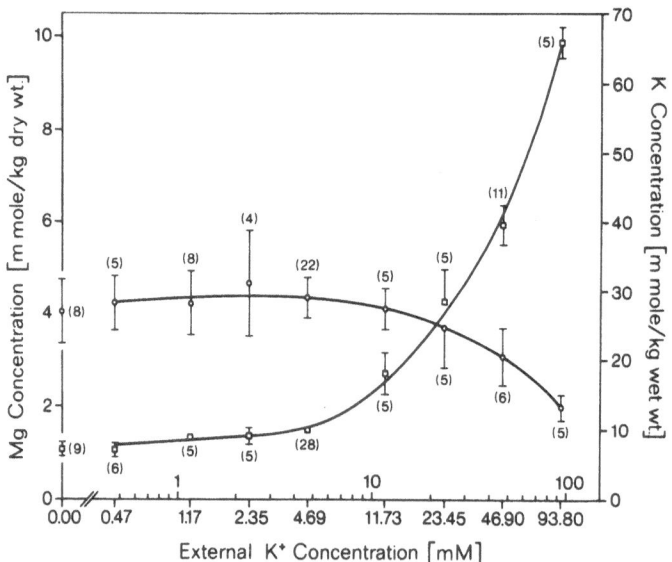

Fig. 10. K^+ (□) and Mg^{++} binding (o) in isolated adventitial con-
           nective tissue of the dog carotid artery as a function of
           the external K^+ concentration $[K^+]_o$ of a Krebs solution;
           semilogarithmic plot. $[K^+]_o$ and $[Na^+]_o$ vary in an opposite
           manner so that the sum is always 155.9 $mmol \cdot l^{-1}$.

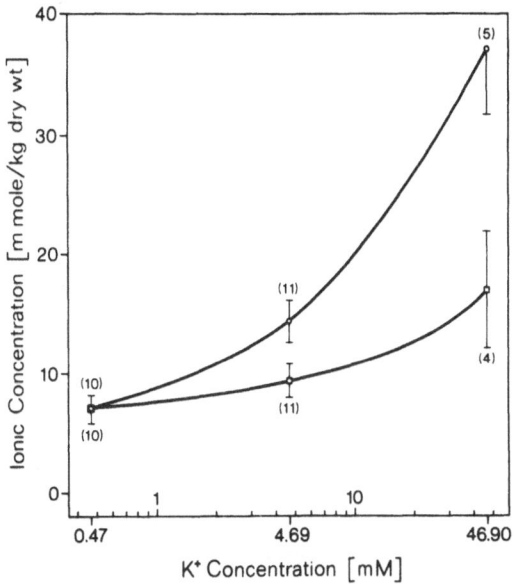

Fig. 11.   K$^+$ (o) and Cl$^-$ binding ($\square$) of isolated adventitial connec-
            tive tissue from the dog A. carotis as a function of the
            K$^+$ concentration in aqueous solutions.

ions with equal affinities strongly compete for the same molecular
sites.

     In order to prove this competition under in vivo conditions,
vascular connective tissue was equilibrated in physiological Krebs
solutions, which were varied in K$^+$ content in a wide range under opposite change in Na$^+$ content. All the other ion concentrations as
well as osmotic pressure and ionic strength were kept constant. Fig.
10 shows the K$^+$ and Mg^{++} binding in a semilogarithmic plot. While
the K$^+$ binding rises with increasing $[K^+]_o$ concentration, the Mg^{++}
binding distinctly decreases with constant Mg^{++} $_o$ concentration.
Since an ion analysis of the tissue was performed after the $^{42}K^+$
efflux experiments in pure KCl solutions, the competitive effect of
K$^+$ ions could easily be detected. From Fig. 11 one can see the expected increase of K$^+$ and Cl$^-$ binding with increasing KCl concentration.
And Fig. 12 illustrates the competitive effect of bound K$^+$ on the
remainder ions in the tissue. The Na$^+$ binding is reduced strongest
with increasing K$^+$ concentration. But Mg^{++} and Ca^{++} binding clearly decrease as well.

     After having proved the competitive effect of increasing K$^+$
concentrations on Na$^+$, Mg^{++}, and Ca^{++} binding, the cationic interaction with the anionic polyelectrolyte connective tissue was systematically studied. Fig. 13 represents the results of determinations
of ion concentrations from the $^{28}Mg^{++}$ flux experiments. Here, the

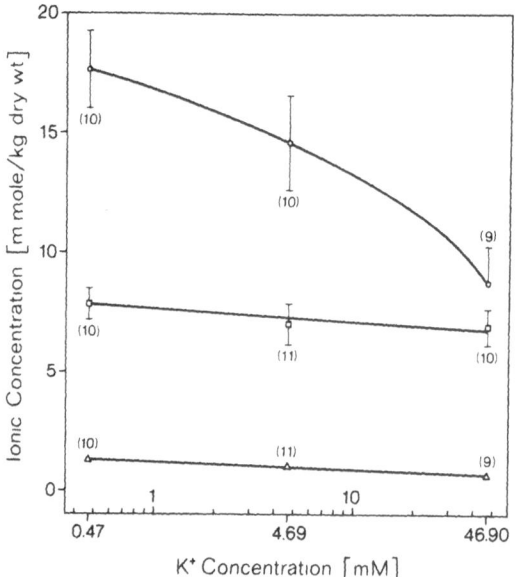

Fig. 12. Na$^+$ (o), Mg^{++} (Δ) and Ca^{++} binding ($\square$) of isolated adventitial connective tissue from the dog A. carotis as a function of the K$^+$ concentration in aqueous solutions.

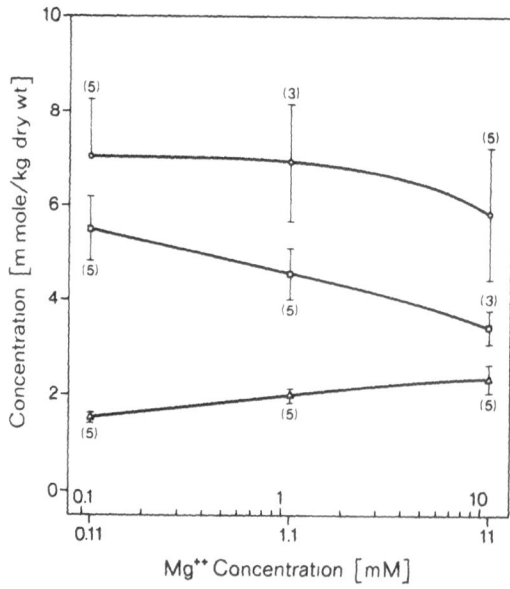

Fig. 13. Na$^+$ (o), K$^+$ (Δ), and Ca^{++} binding ($\square$) of isolated adventitial connective tissue from the dog A. carotis as a function of the MgCl$_2$ concentration in aqueous solutions.

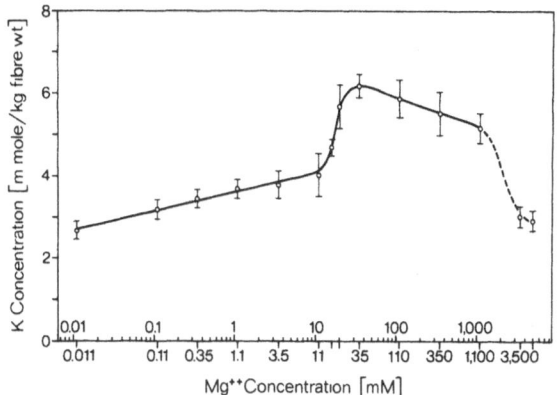

Fig. 14. K^+ binding of isolated adventitial connective tissue of the canine carotid artery after a 3 hr incubation in pure, aqueous $MgCl_2$ solutions at pH 7.0.

connective tissue was equilibrated in pure $MgCl_2$ solutions. The competitive effect of increasing the Mg^{++} concentration on Na^+ and Ca^{++} binding is easily recognizable. Na^+ and Ca^{++} binding curves decrease steadily. After having seen that K^+ and Mg^{++} ions compete for the same binding sites, the significant increase in K^+ binding in correspondence to the increase of Mg^{++} binding is at first a surprising finding (Siegel et al., 1981c). Consequently, the K^+ binding was inspected more closely for a wide range of Mg^{++} concentrations (Fig. 14). In pure $MgCl_2$ solutions the K^+ binding increases continuously in a low Mg^{++} concentration range (Siegel et al., 1981c), then increases steeply between 15 and 20 $mmol \cdot l^{-1}$, and finally decreases gradually after saturation. The Hill coefficient of cooperative K binding is 6.8, the $[Mg^{++}]_o$ concentration at half-maximal velocity 18 $mmol \cdot l^{-1}$. Yet it has to be considered that the ionic strength of these pure $MgCl_2$ solutions is by no means physiologic. Preliminary studies in Krebs solutions of increasing Mg^{++} concentration yielded also cooperative K^+ binding for an $[Mg^{++}]_o$ range between 0.3 and 0.8 $mmol \cdot l^{-1}$.

Cooperative binding is possible if a conformational change of the substrate takes place via the allosteric effect of a ligand (Siegel et al., 1977b). Monovalent cations are not suitable for such a change of proteoglycan structure, but rather cations like Mg^{++} and Ca^{++} with their two valences (Gustavsson et al., 1981). Stretched in between the side chains of these anionic polyelectrolytes, they can rearrange the whole macromolecular structure. With the help of NMR techniques, the ability of Mg^{++} ions in a low concentration range has been examined to cause such a change in configuration. In Fig. 15 the ^{23}Na excess transverse relaxation rate is plotted versus the Mg^{++} concentration of a physiological Krebs solution at pH = 7.38 in the presence of a physiological CS-P concentration. It is highly

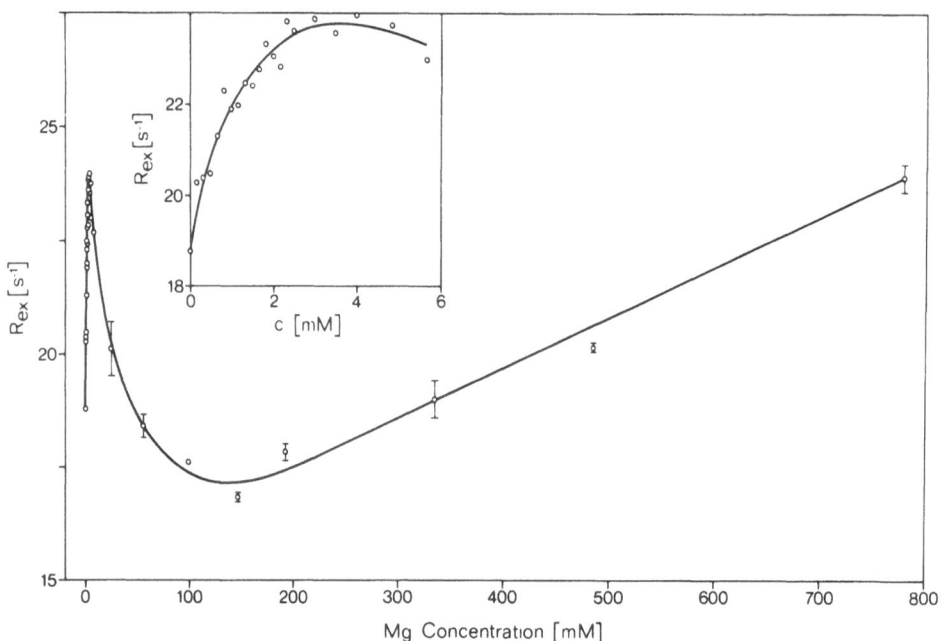

Fig. 15.   The ^{23}Na$^+$ excess transverse relaxation rate, R_{2ex}, in
relation to the Mg^{++} concentration of a Krebs solution
containing 0.1 M CS-P at pH = 7.38. The initial rise of
the relaxation rate in the range of physiological Mg^{++}
concentrations is supplementarily illustrated in a separate inset. This increase is indicative of a conformational change of the CS-P macromolecules.

surprising that R_{2ex} increases for small additions of MgCl$_2$ (Siegel
et al., 1981c), and decreases only with higher Mg^{++} concentrations,
that is Na$^+$ is expelled from its binding sites (normal competition).
With Mg^{++} concentrations >150 mmol·l^{-1} the ^{23}Na$^+$ excess transverse
relaxation rate rises steadily due to ion-ion and ion-solvent interactions (Lindman and Forsén, 1976). The remarkable initial increase
of R_{2ex} is drawn once again in a separate inset. The additional measurement of the longitudinal relaxation rate, R_{1ex}, in the range of
Mg^{++} concentrations between 0 and 3 mmol·l^{-1} gave further evidence
for a conformational change of the macromolecules, which is indicated by the R_{2ex} increase in the competition experiment. This means
that already physiological concentrations of Mg^{++} ions can induce a
specific change in configuration, which enables K$^+$ ions to bind co-
operatively to connective tissue structures (Siegel et al., 1981c).
Therefore, with extracellular Mg^{++} deficiency, not only less Mg^{++}
ions are bound to vascular connective tissue but also less K$^+$ ions,
and additionally a K$^+$ release is initiated. [K$^+$]$_o$ would increase near
the cell membrane of vascular smooth muscle cells and sympathetic

nerve terminals, depolarization and vasoconstriction would occur
(cf. Scott et al., 1961; Siegel et al., 1981c).

SUMMARY

1. H^+ and K^+ ions participate decisively in the local regulation of blood flow. Variation of their extracellular concentration
changes the membrane potential of vascular smooth muscle cells and
tension via electromechanical coupling. The effect of K^+ ions can
be primarily attributed to a change of K^+ equilibrium potential and
electrogenic pump rate, the effect of H^+ ions to a change of Na^+ and
K^+ permeability of the cell membrane.

2. Shifts of external proton and/or cation concentrations cause
changes of the binding properties of the polyanionic macromolecules
in vascular connective tissue. Thus, the extracellular concentration
of various cation species can vary fast and drastically in the tight
mesh-work of connective tissue fibres close to the membrane of vascular smooth muscle cells. Especially, the K^+ adsorption with extracellular acidification as well as the cooperative K^+ binding as consequence of a conformational change induced by Mg^{++} ions are of great
importance for membrane hyperpolarization and vasodilatation.

ACKNOWLEDGEMENTS

The authors thank Mrs. Ch. Fuhrmann for her skillful technical
assistance and Messrs. H. Dannenberg and J. Ofner from the precision
workshop for their help in the development of new analytical methods.
We are grateful to Mrs. M. Krawczynski for her outstanding work in
preparing the illustrations and to Mrs. E. Gaebel for the translation, editorial elaboration and typing of the manuscript. We would
also like to thank the staff of the "Großrechenzentrum für die Wis-
senschaft in Berlin" for their cooperation and for generously making
computer time available.

This work was supported by the Deutsche Forschungsgemeinschaft
(Si 182/1-4) within its overall program "Structure and Function of
Biological Membranes", the Stiftung Volkswagenwerk (Az. 111 327) and
the Erwin Riesch-Stiftung (ERS 7/8).

REFERENCES

Alburger, D.E., and Harris, W.R., 1969, Decay scheme of Mg^{28},
     Physiol. Rev., 185:1495.
Baumgarten, C.M., and Isenberg, G., 1977, Depletion and accumulation
     of potassium in the extracellular clefts of cardiac Purkinje
     fibers during voltage clamp hyperpolarization, Pflügers Arch.,
     368:19.

Bolton, T.B., 1979, Mechanisms of action of transmitters and other substances on smooth muscle, Physiol. Rev., 59:606.

Bolton, T.B., Tomita, T., and Vassort, G., 1981, Voltage clamp and the measurement of ionic conductances in smooth muscle, in: "Smooth Muscle: An Assessment of Current Knowledge", E. Bülbring, A.F. Brading, A.W. Jones, T. Tomita, eds., E. Arnold, London.

Brace, R.A., Anderson, D.K., Chen, W.-T., Scott, J.B., and Haddy, F.J., 1974, Local effects of hypokalemia on coronary resistance and myocardial contractile force, Am. J. Physiol., 227:590.

Casteels, R., Kitamura, K., Kuriyama, H., and Suzuki, H., 1977, Excitation-contraction coupling in the smooth muscle cells of the rabbit main pulmonary artery, J. Physiol. (Lond.), 271:63.

Chamley-Campbell, J., Campbell, G.R., and Ross, R., 1979, The smooth muscle cell in culture, Physiol. Rev., 59:1.

Cohen, I., Daut, J., and Noble, D., 1976, The effects of potassium and temperature on the pace-maker current, i_{K_2}, in Purkinje fibres, J. Physiol. (Lond.), 260:55.

Gustavsson, H., Siegel, G., Lindman, B., and Fransson, L.-Å., 1981, $^{23}Na^+$-NMR studies of cation binding to multi-chain and single-chain glycosaminoglycan peptides, Biochim. Biophys. Acta, 677:23.

Haas, H.G., Glitsch, H.G., Kern, R., Hantsch, F., and Siegel, G., 1966, Kalium-Fluxe und Membranpotential am Froschvorhof in Abhängigkeit von der Kalium-Außenkonzentration, Pflügers Arch. Ges. Physiol., 288:43.

Harder, D.R., 1982, Effect of H^+ and elevated Pco_2 on membrane electrical properties of rat cerebral arteries, Pflügers Arch., 394:182.

Ito, Y., Kitamura, K., and Kuriyama, H., 1979, Effects of acetylcholine and catecholamines on the smooth muscle cell of the porcine coronary artery, J. Physiol. (Lond.), 294:595.

Kennedy, J.F., 1979, "Proteoglycans - Biological and Chemical Aspects in Human Life", Elsevier Scientific Publ. Comp., Amsterdam-Oxford-New York.

Kline, R.P., and Kupersmith, J., 1982, Effects of extracellular potassium accumulation and sodium pump activation on automatic canine Purkinje fibres, J. Physiol. (Lond.), 324:507.

Kuriyama, H., and Suzuki, H., 1978, Electrical property and chemical sensitivity of vascular smooth muscles in normotensive and spontaneously hypersensitive rats, J. Physiol. (Lond.), 285:409.

Kuschinsky, W., Wahl, M., Bosse, O., and Thurau, K., 1972, Perivascular potassium and pH as determinants of local pial arterial diameter in cats, Cir. Res., 31:240.

Lindman, B., and Forsén, S., 1976, "Chlorine, Bromine and Iodine NMR: Physico-chemical and Biological Applications", Springer, Berlin-Heidelberg-New York.

Marquardt, D.L., 1963, An algorithm for least squares estimates of nonlinear parameters, J. Siam., 11:431.

Rodén, L., 1980, Structure and metabolism of connective tissue pro-
    teoglycans, in: "The Biochemistry of Glycoproteins and Proteo-
    glycans", W.J. Lennarz, ed., Plenum Press, New York-London.
Scott, J.B., Frohlich, E.D., Hardin, R.A., and Haddy, F.J., 1961,
    Na^+, K^+, Ca^{++}, and Mg^{++} action on coronary vascular resistance
    in the dog heart, Am. J. Physiol., 201:1095.
Siegel, G., 1981, The effect of membrane protonation on ionic per-
    meabilities in vascular smooth muscle, Pflügers Arch., 391:R36.
Siegel, G., 1982, The effect of external pH changes on Na^+ and K^+
    permeabilities in the smooth muscle fibres membrane of canine
    cerebral vessels, J. Physiol. (Lond.), 329:56P.
Siegel, G., Ehehalt, R., Gustavsson, H., and Fransson, L.-Å., 1977a,
    Ion binding properties of vascular connective tissue, in: "Ex-
    citation-contraction Coupling in Smooth Muscle", R. Casteels,
    T. Godfraind, and J.C. Rüegg, eds., Elsevier/North-Holland Bio-
    medical Press, Amsterdam-New York-Oxford.
Siegel, G., Gustavsson, H., Ehehalt, R., and Lindman, B., 1977b, The
    role of membrane potential in the regulation of vascular tone,
    in: Recent Advances in Basic Microcirculatory Research, Bibl.
    Anat., 15, D.W. Lewis, ed., Karger, Basel-München-Paris-London-
    New York-Sidney.
Siegel, G., Kämpe, Ch., and Ebeling, B.J., 1981a, pH-dependent myo-
    genic control in cerebral vascular smooth muscle, in: "Cerebral
    Microcirculation and Metabolism", J. Cervós-Navarro, E. Frit-
    schka, eds., Raven Press, New York.
Siegel, G., Niesert, G., Ehehalt, R., and Bertsche, O., 1976a, Mem-
    brane basis of vascular regulation, in: "Ionic Actions on Vas-
    cular Smooth Muscle", E. Betz, ed., Springer, Berlin-Heidelberg-
    New York.
Siegel, G., Rettig, W., Kämpe, Ch., Ebeling, B.J., and Walter, A.,
    1980a, Potassium fluxes in cerebral arteries, in: "Pathophysio-
    logy and Pharmacotherapy of Cerebrovascular Disorders", E. Betz,
    J. Grote, D. Heuser, R. Wüllenweber, eds., G. Witzstrock, Baden-
    Baden-Köln-New York.
Siegel, G., Roedel, H., and Hofer, H.W., 1976b, Basic rhythms in vas-
    cular smooth muscle, in: "Smooth Muscle Pharmacology and Physio-
    logy", INSERM Colloque, Vol. 50, M. Worcel, G. Vassort, eds.,
    Editions INSERM, Paris.
Siegel, G., Roedel, H., Jäger, R., and Bertsche, O., 1974, Relation-
    ship between membrane potential of vascular smooth muscle and
    external K^+ concentration, Pflügers Arch., 347:R14.
Siegel, G., Roedel, H., Nolte, J., Hofer, H.W., and Bertsche, O.,
    1976c, Ionic composition and ion exchange in vascular smooth
    muscle, in: "Physiology of Smooth Muscle", E. Bülbring, M.F.
    Shuba, eds., Raven Press, New York.
Siegel, G., and Schneider, W., 1981, Anions, cations, membrane po-
    tential, and relaxation, in: "Vasodilatation", P.M. Vanhoutte,
    I. Leusen, eds., Raven Press, New York.
Siegel, G., Walter, A., Gustavsson, H., and Lindman, B., 1981b, Mag-
    nesium and membrane function in vascular smooth muscle, Artery,
    9:232.

Siegel, G., Walter, A., Rettig, W., Kämpe, Ch., Ebeling, B.J., and
    Bertsche, O., 1980b, Sodium compartmens in the arterial wall,
    in: "Intracellular Electrolytes and Arterial Hypertension", H.
    Zumkley, H. Losse, eds., Thieme, Stuttgart-New York.
Suzuki, H., 1981, Effects of endogeneous and exogeneous noradrenaline
    on the smooth muscle of guinea-pig mesenteric vein, J. Physiol.
    (Lond.), 321:495.
Vanhoutte, P., and Clement, D., 1968, Effect of pH and Pco$_2$ changes
    on the reactivity of isolated venous smooth muscle, Arch. Int.
    Physiol. Biochim., 76:144.
Zidek, W., and Lange-Asschenfeldt, H., 1980, Continuous measurements
    of extracellular ion activities in rat carotid artery by liquid
    ion exchanger microelectrodes, in: "Intracellular Electrolytes
    and Arterial Hypertension", H. Zumkley, H. Losse, eds., Thieme,
    Stuttgart-New York.

DISCUSSION

Fig. 6 shows that in the physiological pH range the changes of
K^+ concentration with changing pH values are small and the error is
rather large. Is your statement that alkalosis increases while acidosis decreases the extracellular K^+ concentration justified?

When the K^+ binding of isolated adventitial connective tissue
is investigated for pH steps 4 to 9 in one and the same dog, the
course of the lower curve shown in Fig. 6 results with relatively
small deviations. The error in Fig. 6 is therefore mainly due to
variations in the dogs. The reason may be that dogs of different
breed and either sex, and of various nutrition, were used. Following
transition from pH 7.3 to 6.8, 0.64 mmole K^+/kg fibre wt. are bound
to connective tissue structures, while 1.52 mmole K^+/kg fibre wt.
are released upon transition from pH 7.3 to 7.8. Application of a
simple geometric model, which considers the narrow cleft spaces between cell membrane, basal lamina and connective tissue fibres, can
lead to a decrease of the K^+ concentration in the cleft spaces to
2/3 to 1/2 in the first case, in the latter case to an increase to
the double to triple of its normal value.

ABDOMINAL
ORGANS

FLOW DEPENDENCE AND INDEPENDENCE OF OXYGEN CONSUMPTION WITHIN THE INTESTINAL VASCULATURE

J. Lutz

Department of Physiology, University of Würzburg,
Röntgenring 9, 8700 Würzburg, FRG

INTRODUCTION

The curve presenting the relation between oxygen consumption ($\dot{V}o_2$) and blood flow (BF) can be divided into two parts (see Fig. 1). The first one shows a proportionality between the values, supporting the hypothesis of a metabolic control of blood flow. In the second portion a distinct saturation of oxygen consumption is reached. A variation in O_2 consumption is then possible independent of flow, thereby changing the amount of O_2 extraction. In this upper range a metabolic control of blood flow becomes insufficient to explain changes in the horizontal course of the pressure-flow curve, generally called "autoregulation of blood flow" (see Fig. 3).

Both parts of the $\dot{V}o_2$-BF curve will be analyzed using our own experimental results along with results reported in the literature with respect to the trajectory course, dispersion, and the calculated intersection of the two components. Additionally, the influence of hemodilution will be studied.

METHODS

The intestinal vascular bed supplied by the superior mesenteric artery of cats (n = 14) and Wistar rats (n = 38), intraperitoneally anesthetized with chloralose (60 mg/kg body weight) and urethane (260 mg/kg body weight), was perfused in situ after separation by ligation from adjoining tissue.

In cats perfusion was performed by means of an occlusive roller pump from the aorta as described earlier (Lutz et al., 1975).

Measurements of Po_2 and pH were continuously carried out in the
arterial and mesenteric venous blood and $\dot{V}o_2$ calculated from the
known O_2 dissociation curves for this species (Bartels and Harms,
1959) according to a computer subroutine.

In rats an oxygenator system was used filled with heparinized
bovine blood diluted 3-fold with Krebs-Henseleit solution and made
isooncotic by addition of hydroxyethyl starch (Fresenius KG, Bad
Homburg). The perfusing roller pump was fed from this source and
supported the intestine in a closed circuit. Small blood samples
were taken every 10 minutes from the arterial and venous side and
analyzed for pH by an external temperature regulated pH electrode
unit and for O_2 content by a Lex-O_2 Con Analyzer (Lexington Instr.,
Waltham, Ma.). Hemoglobin content was regularly determined from
microsamples from both cats and rats using a Hb-CN-method (Merck,
Darmstadt).

Changes in blood flow were induced solely mechanically by
variations of the speed of the roller pump and were recorded together
with the perfusion pressure (strain gauge P 23 Db, Statham, Puerto
Rico) on a multiple pen recorder (Linseis, Selb). $\dot{V}o_2$ versus blood
flow curves were drawn and regression lines for the ascending part
calculated according to the method of least squares with confidence
limits of 95% probability; the horizontal part of the curve was expressed by the mean maximal O_2 consumption (see results) $\pm$ 2 S.D..
Values given in the text are means $\pm$ 1 S.D..

RESULTS

For the cats the maximal O_2 consumption amounted to 1.86 ml
$O_2/(min \cdot 100\ g)$ under a blood flow of 52.5 ml/(min$\cdot$100 g). If all
values for $\dot{V}o_2$ in the upper range blood flow are taken together
(mean maximal $\dot{V}o_2$) a value of 1.80 $\pm$ 0.07 resulted. The intersection of the line through these values and the ascending regression
line occurred below a flow of 20.0 ml/(min$\cdot$100 g) (Fig. 1). The
importance of this value in regard to the interpretation of the
pressure-flow curve will be mentioned in the discussion.

A curve for the values in the rat is given in Fig. 2. Again the
two parts of the $\dot{V}o_2$-BF curve can be distinctly discriminated.
However, for the intersection the appreciably lower O_2 content of
the perfusing blood must be considered. Calculated for a normal
hemoglobin content of 14.7 g/dl the intersection lies below a flow
of 23.0 ml/(min$\cdot$100 g). This value is consistent with most of those
calculated from the literature and shows that even under a marked
hemodilution no qualitative change in the mutual interaction of $\dot{V}o_2$
and BF occurs, as can also be deduced from the similar relation of
$\dot{V}o_2$ and O_2 supply.

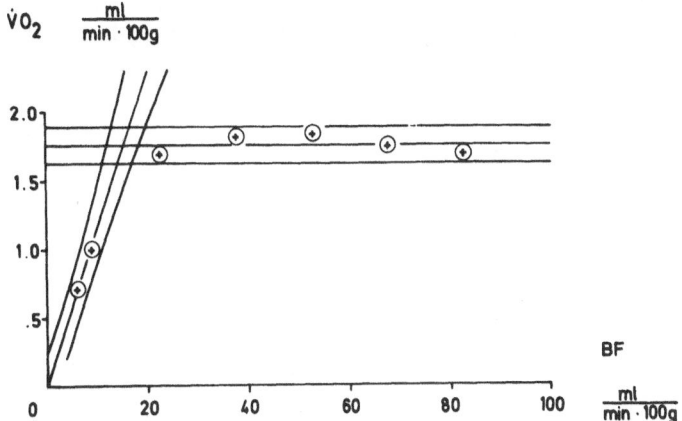

Fig. 1. Relation of oxygen consumption and blood flow in the intestine of the cat. Each cycle represents mean values of 10 cats. BF = blood flow, Vo_2 = O_2 consumption.

DISCUSSION

A separation of the Vo_2-BF curve into two linear sections appears at first view to be somewhat arbitrary. In the range of high blood flows, however, in most cases no distinct increase in Vo_2 can be found. What should be shown is the functional difference between both parts of the curve, which can also be demonstrated in examples taken from the literature (Table 1).

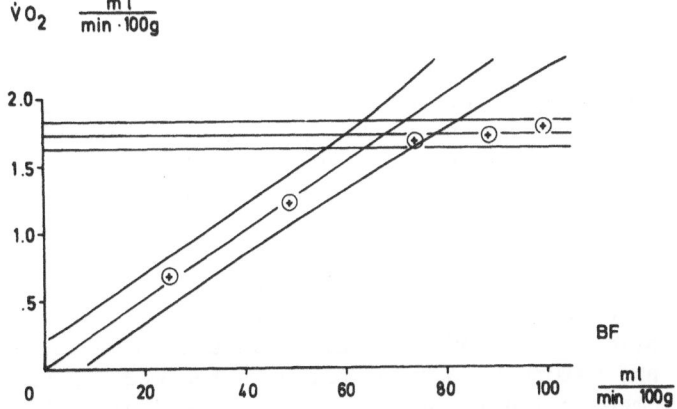

Fig. 2. Relation of oxygen consumption and blood flow in the intestine of the rat. Blood was diluted 3-fold (Hb = 4.9 $\pm$ 0.2 g/dl). Cycles represent mean values of 29 rats.

Table 1. Survey of $\dot{V}O_2$-BF relations given here and in the literature. Dispersions are calculated as $\pm$ 1 S.D..

Species	Intersection ml/(min·100 g)	Mean max. $\dot{V}O_2$ ml O_2/(min·100 g)	Author(s)
Dog	21.5 + 0.58	1.72 + 0.05	Kvietys & Granger 1982
Dog[1]	41.1 $\pm$ 3.60	2.40 $\pm$ 0.39	Shepherd 1982
Cat	15.2 $\pm$ 0.76	1.80 $\pm$ 0.07	Lutz
Cat	16.0 $\pm$ 0.96	1.12 $\pm$ 0.05	Hamar et al. 1975
Rat[2]	22.7 $\pm$ 0.62	1.73 $\pm$ 0.05	Lutz

[1] fasted in contrast to other values for fed animals
[2] calculated for Hb = 14.7 g/dl

Though many papers have been published dealing with the effects
of vasoactive agents on oxygen consumption and the attending flow
variations, only little data and few diagrams have appeared concerning the relation between oxygen consumption and blood flow under
constant - not increased - metabolic situations. The pressure-flow
curves with which they could be compared (Fig. 3) are obtained under
such stable conditions (Lutz, 1969).

The examples taken from the literature reveal different varian-
ces for the calculated intersections. These again depend on the
dispersion of the mean maximal $\dot{V}O_2$ and thus reflect the degree
to which the upper part of the curve approaches a horizontal course.
The values of Kvietys and Granger (1982) show a small dispersion and
indicate a curve with a course very similar to that for the cat
and rat as described here. That means that with a blood flow increase a reciprocal diminution of oxygen extraction occurs. On the
other hand, when a blood flow decrease threatens, oxygen extraction
in the intestine cannot be enhanced to values above 75% (Lutz et
al., 1975) and before this takes place, a decrease in resistance
(and thus a metabolically induced flow regulation) will appear.

The horizontal part of the $\dot{V}O_2$-BF curve led authors to
sometimes speak of an "autoregulation of O_2 uptake" (Öhman, 1976).

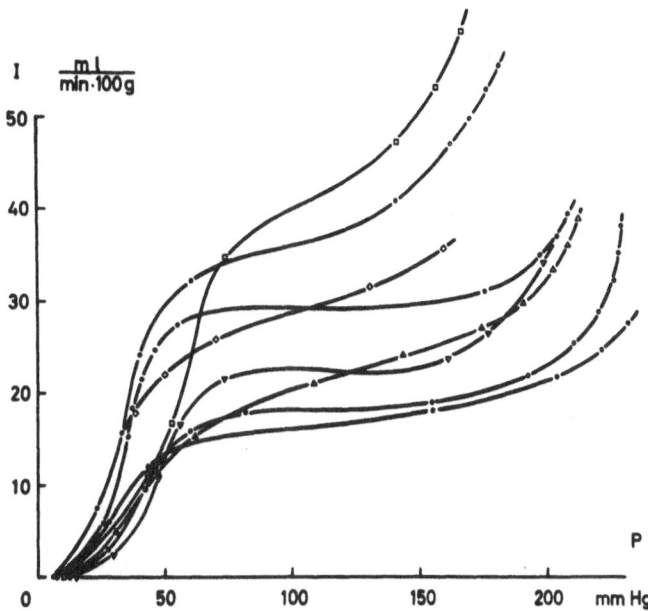

Fig. 3.   Pressure-flow autoregulation in the feline intestine. In
          most of the curves the regulated range lies above 20 ml/
          (min·100 g). I = blood flow.

However, it seems misleading to describe a value as being autoregulated in relation to flow when it is in fact constant due to other
reasons (e.g., a constant metabolic demand in the tissue).

The question raised here is, if metabolic processes solely
determine the whole course of the pressure-flow curve under normal
conditions (without a hypermetabolic state). The calculated intersections of the two lines in the described curves lie - varying with
the different O_2 capacity of the perfusing blood - in the range
of 20 ml/(min·100 g). Thus, only the lower range of blood flow
may be well-regulated by metabolites in unfed animals. For the rest
of the "autoregulation" in the pressure-flow curve other (e.g.,
myogenic) mechanisms seem to be responsible. This means there is a
major limitation in a purely metabolic model for predicting a
phenomenon like pressure-flow autoregulation (Granger and Shepherd,
1979; Shepherd, 1982).

The general expression "autoregulation of blood flow" should
be replaced by "autoregulation of vascular resistance" because it
is more the filtration pressure in the capillaries than the blood
flow which is regulated during arterial or venous pressure changes
(Johnson, 1964; Lutz, 1966; Mellander and Johansson, 1968).

Added during edition: After termination of this study an
article of Adams, Dieleman and Cain, J. Appl. Physiol., 53:660
(1982) appeared, revealing the independence of total Vo_2 of the rat

above a certain value of O_2 supply. The intersection between the linearly dependence of $\dot{V}o_2$ upon O_2 supply and the supply independent maximal $\dot{V}o_2$ became similarly distinct.

REFERENCES

Bartels, H., and Harms, H., 1959, Sauerstoffdissoziationskurven des Blutes von Säugetieren, Pflügers Arch. Ges. Physiol., 268: 334-365.

Granger, H.J., and Shepherd, A.P., 1979, Dynamics and control of the microcirculation, in: "Advances in Biomedical Engineering", Vol. 7, J.H.V. Brown, ed., Academic Press, New York, pp. 1-63.

Hamar, J., Ligeti, L., Kovach, A.G.B., Tkachenko, B.I., Ovsjannikov, V.I., and Tcherniavskaja, G.V., 1978, Blood supply and O_2 consumption of the small intestine in low flow, Acta Physiol. Acad. Sci. Hung., 52:351-390.

Johnson, P.C., 1964, Origin, localisation and homeostatic significance of autoregulation in the intestine, Circ. Res., 15, Suppl. 1:225-232.

Kvietys, P.R., and Granger, D.N., 1982, Relation between intestinal blood flow and oxygen uptake, Am. J. Physiol., 242:G202-208.

Lutz, J., 1966, Über veno-vasomotorische Gefäßreaktionen im Mesenterialkreislauf der Katze, Pflügers Arch. Ges. Physiol., 287:330-344.

Lutz, J., 1969, Hämodynamische Eigenschaften und Gefäßreaktionen der intestinalen Strombahn, Arch. Kreislaufforsch., 59:99-152.

Lutz, J., Henrich, H., and Bauereisen, E., 1975, Oxygen supply and uptake in the liver and the intestine, Pflügers Arch., 360:7-15.

Mellander, S., and Johansson, B., 1968, Control of resistance, exchange and capacitance functions in the peripheral circulation, Pharmacol. Rev., 20:117-196.

Öhman, V., 1976, Blood flow and oxygen consumption in the feline small intestine; responses to artificial distension and intestinal obstruction, Acta Chir. Scand., 142:329-333.

Shepherd, A.P., 1982, Metabolic control of intestinal oxygenation and blood flow, Fed. Proc., 41:2084-2089.

Question to the author:
What is the hemoglobin content?

Answer of the author:
The mean Hb content in the cat experiments amounted to 13.5+1.4 g/dl (n = 10) and was thus in the normal range like in other studies of Table 1, therefore not especially mentioned in the text.

Question to the author:
How many animals are used to obtain $\dot{V}o_2$ = 1.86?

Answer of the author:
V_{O_2} = 1.86±0.39 (S.D.) was received from 10 cats just as in the case
of all other circled values.

Question to the author:
Do you think that two measured points in the first stage of the
O_2 consumption – flow curve are sufficient to define the intersection point?

Answer of the author:
The flow dependent range of the $\dot{V}_{O_2}$/flow curve was not the chief
subject of these experiments, it should represent only the contrast
to the nearly horizontal course of the upper part of the curve. The
meaning of the intersection point of both parts of the curve might
be overestimated, the chief statement is that above a certain flow
value $\dot{V}_{O_2}$ becomes widely flow independent.

Question to the author:
You assume that only the arterial blood content determines the
slope of the first part and the intersection point. Is this experimentally proved?

Answer of the author:
The slope of the first part of the $\dot{V}_{O_2}$/flow curve might be influenced by some other factors, but chiefly it is determined by the
O_2 supply and thus the arterial blood (Hb) content. In spite of
the fact that thus the intersection of both parts of the curve
might differ somewhat, nothing is changed in the character of the
upper part of the curve.

EFFECT OF SOMATOSTATIN ON INTESTINAL MICROCIRCULATION AND METABOLISM

J. Hamar[1], S.S. Polenov[2], G.V. Tcherniavskaya[2],
L. Dézsi[1], and B.I. Tkachenko[2]
[1]Experimental Research Dept., Semmelweis University
School of Medicine, Budapest, Hungary
[2]Institut of Experimental Medicine, Academy of
Medical Sciences, Leningrad, USSR

INTRODUCTION

The gastrointestinal tract (GIT) besides its major activities
in digesting and absorbing food, has other important functions.
Fluid balance is delicately regulated across the intestinal mucosa.
The turnover rate of water in the GIT is about 8-10 l/day. The GI
mucosa also produces important proteins and lipids required by the
whole organism. These functions (fluid balance and metabolism) take
place within the intestinal mucosa. The GIT requires a significant
amount of O_2 and glucose for maintaining its integrity even during
fasting periods (Alteveer et al., 1973; Hamar et al., 1981). Several
neural and humoral factors take part in regulating GIT functions.
Much attention has been focused recently on the role of GI hormones.
They are produced by special endocrine like cells located within
the mucosa or close to the neural elements of the gut (Solcia et
al., 1980). Somatostatin produced by the D cells (Pollack et al.,
1975) is found all along the GIT (Meyers and Coy, 1980). It inhibits
the release of many GI hormones, like gastrin, VIP, GIP, glucagon,
etc. (Schally et al., 1978). It also inhibits other functions of
the GIT like electrolyte secretion or secretion of digestive enzymes
(Konturek, 1980). It has also been used in the treatment of patients
with bleeding ulcers (Mathes et al., 1975). On the basis of its
widespread presence and its widespread inhibitory effect on the
GIT, somatostatin is believed to have direct effects on elementary
physiological functions of the small intestine.

METHODS

22 cats of either sexes (2.5-3.5 kg body weight) were anesthetized by chloralose (40 mg/kg) and urethan (0.5 g/kg) after ether
induction. The small intestine (ileum and jejunum) were completely
isolated. The distal duodenum and the coecum were ligated, the colon
was removed, and the mesentery completely isolated by ligature. Only
one of the major lymphatic vessels running in the mesentery was left
intact in order to avoid edema formation. The superior mesenteric
artery (AMS) was autoperfused using one of the channels of a two-
channel perfusion pump. Arterial blood for perfusion was taken from
the femoral artery. The outflowing venous blood from the isolated
superior mesenteric vein was collected into a cylinder by a polythene cannula. The mobile end of the cannula was set at a hight
that established a venous pressure of 8-10 mm Hg (Fig. 1, manometer
I). Venous blood was returned to the femoral vein by the second
channel of the pump. The output of the second channel was identical
to the venous outflow. In this case a steady level of the blood
column was reached within the cylinder. Pressure in the column was
continuously measured by a manometer (manometer II). The isolated
small intestine was left within the abdominal cavity. The temperature of the animal and the arterial blood for perfusion were kept
at 37°C. No artificial ventilation was applied.

Systemic arterial blood pressure (manometer IV) and perfusion
pressure (manometer III) were also continuously recorded. The capillary filtration coefficient (CFC), venous capacitance (C) changes
to hydrostatic load (the mobile end of the catheter was lifted by 7
cm for 1.5-2 min) were determined by the method of Dvoretsky (1981).

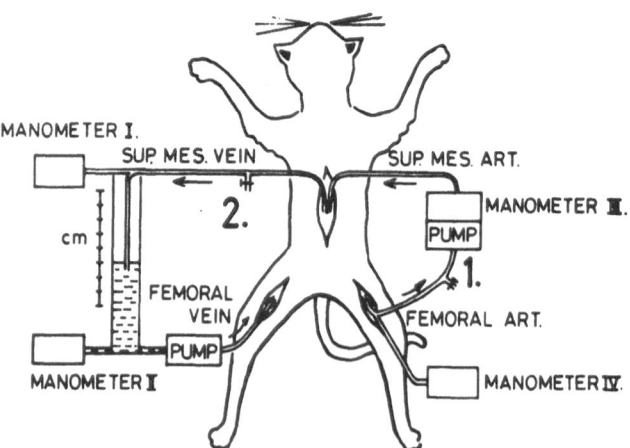

Fig. 1.   Schematic drawing of the experimental arrangement.
          Arrows indicate direction of flow. 1., 2. sites of blood
          sampling. For explanation see text.

Arterial and venous blood gases (Po_2, Pco_2) and pH were determined by a blood gas microanalyser (Radiometer). Hematocrit (Ht)
was measured. Bicarbonate solution (4.2%) was infused if necessary
to keep arterial pH between 7.35-7.40. Systemic arterial blood
pressure, perfusion and venous pressures and also changes of the
hydrostatic pressure of the blood column within the cylinder were
continuously recorded. Somatostatin (Sigma) 20, 50, and 100 ng/min,
dissolved in buffered saline (pH: 7.50) was infused at a rate of
0.2 ml/min into the AMS for 15 min. Atropine (0.1 mg/kg), propranolol (0.2 mg/kg), dihydroergotoxin (0.5 mg/kg), norepinephrine and
epinephrine as test drugs (1 ug each) were injected into the arterial side of the perfusion system in a single bolus. The extracor-
poreal system (cylinder and catheters) were filled by rheomacrodex
solution, 30 ml Heparin was given to avoid blood clotting.

EXPERIMENTAL PROTOCOL

Two series of studies were carried out. In the first, the dose
response effect of S and the effect of atropine, in the second, the
influence of adrenoreceptor blockers were analysed.

1. Dose Response Studies and the Effect of Atropine

The animals were given a 1 hour rest after surgical intervention and arterial pH was corrected. The perfusion pressure was set
equal to the systemic arterial blood pressure. A constant perfusion
volume was used afterwards. The perfusion pressure and venous outflow reached a steady level within 30 min. CFC, blood gases and Ht
were determined, then somatostatin (20 ng/min) was infused and the
above determinations repeated during the last 5 min of the infusion.
The whole procedure was repeated when 50 and 100 ng/min were given.
10-15 min elapsed between each procedure. Atropine was then given
and each step of the above procedure repeated.

2. Effect of Alpha and Beta Receptor Blocking Agents

Somatostatin (S) was infused at a dose of 100 ng/min. CFC,
blood gases and Ht were measured before and during somatostatin
infusion as described above. Epinephrine and norepinephrine were
injected in a bolus of 0.1 ml after each determination of CFC.
Propranolol or dihydroergotoxin (DHE) were injected. 30 min after
administration of the drug the same measurements were performed as
before blockade.

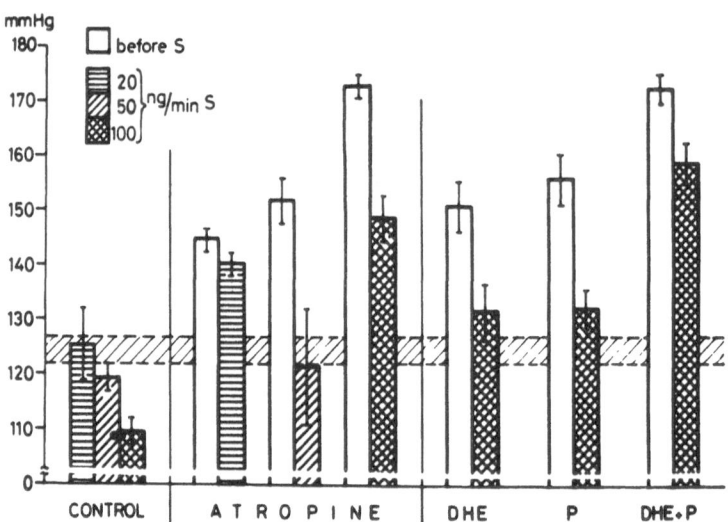

Fig. 2. Effect of the somatostatin (S) on the perfusion pressure.
DHE: dihydroergotoxin, P: propranolol. Horizontal shaded
area means perfusion pressure values (128 ± 5 mm Hg) before
any blocking agent (atropine, DHE, P) was applied.

RESULTS AND DISCUSSION

Effect of Somatostatin on Resistance Vessels

     The control perfusion pressure (PP) was 128 ± 5 mm Hg. Somato-
statin decreased it in a dose dependent manner: PP was 100, 94 and
85% of control when 20, 50, and 100 ng/min of S were infused,
respectively (Fig. 2).

     After administration of atropine PP increased to 149 ± 4 mm Hg.
It further increased in a stepwise manner after completing each
infusion of S. However, S itself significantly reduced PP to 94, 80
and 84% of each preinfusion value when 20, 50, and 100 ng/min were
infused, respectively. DHE increased PP to 151 ± 9 mm Hg. During
infusion of S (100 ng/min) it was reduced to 134 ± 7 mm Hg (88% of
preinfusion level). Propranolol also increased PP to 161 ± 9 mm Hg,
and the above dose of S decreased it to 83% of this preinfusion
value. A combined alpha and beta receptor blockade elevated PP to
174 ± 5 mm Hg and it was reduced to 90% of this value during infusion of S. Reactions of PP evoked by a bolus injection of epinephrine (E) and norepinephrine (NE) were not significantly influenced
by the infusion of S (100 ng/min) either before or after receptor
blockade (Table 1).

     Somatostatin exerts an immediate effect on the resistance
vessels by dilating them. This effect is reversible. As removal of

Table 1. Changes in PP (mm Hg), (both with and without the presence of S) and venous outflow (ml/100 g) following i.a. injections of NE and E.

Perfusion Pressure

	Control		S		DHE		DHE+S		Prop.		Prop.+S	
	NE	E	NE	E	NE	E	NE	E	NE	E	NE	E
x	52.5	37.9	60.8	35.4	-19.0	-64.0	-18.0	-68.0	53.8	71.2	56.2	72.5
SE	3.5	2.3	4.8	3.6	2.6	1.6	0.4	3.4	6.9	4.1	5.8	7.3
n	12	12	12	12	5	5	5	5	4	4	4	4

Venous outflow

	Control		S		DHE		DHE+S		Prop.		Prop.+S	
	NE	E	NE	E	NE	E	NE	E	NE	E	NE	E
x		–	-0.54	-0.39	0	+0.21	0	+0.22	-0.35	-0.45	-0.33	-0.47
SE			0.07	0.07	0	0.07	0	0.06	0.09	0.12	0.06	0.09
n			12	12	5	5	5	5	4	4	4	4

+ = increase   E = epinephrine   S = somatostatin

- = decrease   NE = norepinephrine

the polypeptide from the circulation is very quick, a dose dependent
level of relaxation of the vascular smooth muscle is reached within
2-3 min and it is kept constant whilst somatostatin is present
within the circulation. The effect of atropine by making S more
effective suggests that cholinergic mechanisms antagonise the vasodilatation elicited by S.

Alpha or beta adrenergic receptors do not seem to be involved
in the dilatatory mechanisms elicited by the polypeptide. Besides
the immediate effect of somatostatin, a longer term effect is also
present. In our earlier studies using the same intestinal preparation and perfusion technique, a decrease of perfusion pressure was
always found after administration of atropine or DHE. This was not
the case in the present series (see PP values before S, Fig. 2).

Effect of Somatostatin on Exchange Vessels

With the exception of the first determination following the
administration of atropine, CFC values before S were unchanged
throughout the experiments:

0.090-0.100 ml/(mm Hg·min·100 g) CFC increased significantly
during infusion of S at a dose of 100 ng/min, however, an increase
was already observed in 8 experiments out of 10 when 20 and 50 ng/min
were infused, respectively (Fig. 3). Atropine significantly increased the effect of S on CFC in each infusion given. A dose effect
response could not be observed.

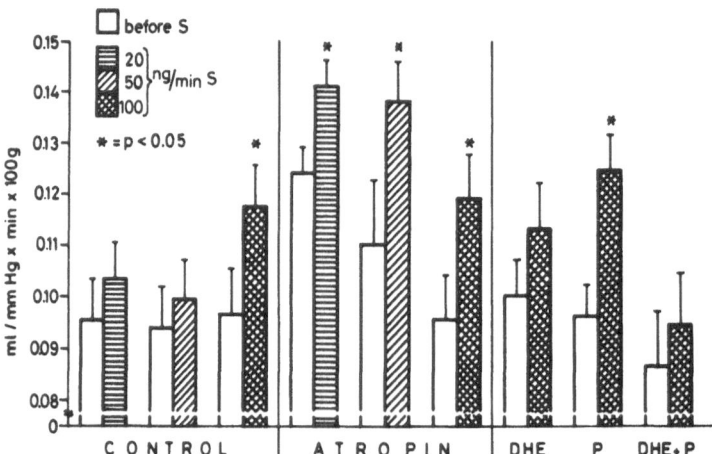

Fig. 3. Effect of somatostatin on the capillary filtration coeffi-
         cient. Significant differences between the values repre-
         sented by white and neighbouring shaded columns are indi-
         cated by *.

Propranolol also increased the effect of S and there was no change in CFC to the infusion of 100 ng/min following DHE and a combined alpha - beta receptor blockade.

Somatostatin increases the rate of capillary filtration. This effect is abolished by alpha receptor blockade. As is discussed below, venous filling pressure is increased by S. This results in an increased capillary filtration. However, other mechanisms, but only physical ones, may take part in the process that involve alpha receptors.

Effect of Somatostatin on Capacitance Vessels

Capacitance changes to hydrostatic load: Average C was 0.197 +0.006 ml/(mm Hg·100 g) before infusing somatostatin. It did not change throughout the experiment and showed no significant deviations compared to the above value after administration of any blocking agent. S significantly increased C. Dose effect relations to different concentrations of S infused could not be demonstrated. This increase of C was not influenced by either atropine or alpha .and beta blocking agents (Fig. 4).

Venous outflow: Venous outflow did not change when 20 ng/min of S was infused. It increased in 1, it did not change in 5 and it decreased in 4 experiments out of 10 when 50 ng/min was administered. The highest dose of S decreased venous outflow in 14, did not change in 3 and increased in 3 out of 20 cases. Following administration of atropine there was a decrease in venous outflow by 0.31,

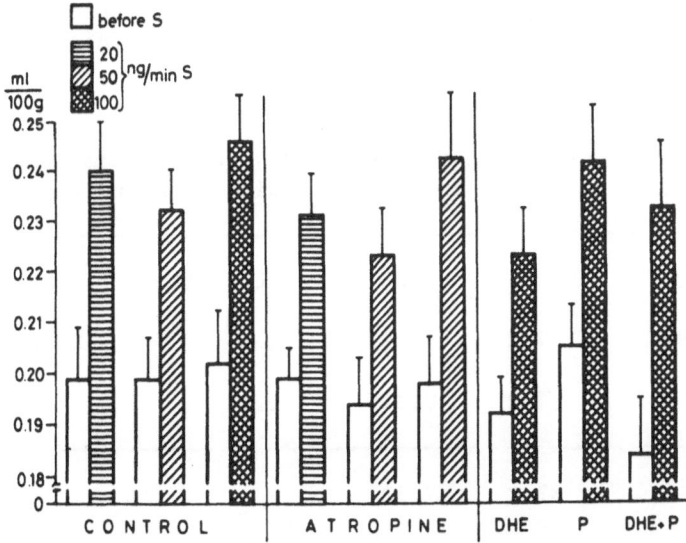

Fig. 4. Effect of somatostatin (s) on venous capacitance changes

0.52, and 0.86 ml/100 g when 20, 50, and 100 ng/min of S were infused, respectively. The same effect was observed after propranolol
(0.91) and after a combined alpha + beta receptor blockade (0.62
ml/100 g). Following DHE administration, venous outflow decreased
in 6 (0.38 ml/100 g), it did not change in 2 and it increased in 1
out of 9 cases when somatostatin (100 ng/min) was infused.

An i.a. bolus injection of NE increased venous outflow in 2
and decreased it in 8 experiments. A decrease (0.54 ± 0.07 ml/100 g)
was always observed during infusion of S. There was no change
following alpha receptor blockade and the outflow decreased by 0.35
$\pm$ 0.09 and 0.33 ± 0.06 ml/100 g before and during administration of
S respectively, after administration of proparanolol.

E increased outflow in 3 and decreased it in 6 experiments. In
3 cases there was a biphasic reaction, a decrease being always the
first. During infusion of S, venous outflow always decreased (0.39
$\pm$ 0.06 ml/100 g), however, in 3 experiments a moderate increase
followed this first reaction. Following DHE it increased ($0.21 \pm$
0.07 and 0.22 ± 0.09 ml/100 g) and after propranolol it decreased
(0.45 ± 0.09 and 0.47 ± 0.09 ml/100 g) both before and during S,
respectively (Table 1).

Somatostatin decreases venous outflow and increases venous
dilatation to a hydrostatic load. This means that more blood is
stored within the venous system. These effects are accounted for
the relaxation of the vascular smooth muscle. Blood content of the
small intestine is about 7 ml in 100 g tissue (Folkow et al., 1963;
Svanvick, 1973). 75-80% of blood volume is located in the intestinal veins in 100 g tissue. This volume is further increased by
15-20% when 100 ng/min of S is infused. Alpha receptor stimulation
increases the effect of S, while cholinergic mechanisms seem to
decrease it (see Table 1 and Fig. 4).

Table 2. GI arterial and venous blood, pre-drug-administration
control values for the following parameters.

	Po_2	Pco_2	pH	Ht	flow ml/(min·100 g)
Art.	92	27.2	7.362	32	32.6
$\pm$	2.3	1.2	0.015	2.1	1.4
Ven.	43	33.4	7.307	-	-
$\pm$	1.7	1.3	0.013		

Blood Gases

Control values (before administration of either S or any blocking agent) of blood gases, pH and Ht are shown in Table 2. Neither
arterial nor venous Po_2, Pco_2 and pH changed significantly when
somatostatin was infused in any dose. Atropine, DHE and propranolol
did not change blood gases either.

CONCLUSION

Somatostatin relaxes both arterial and venous vascular smooth
muscle. This relaxation may lead to an increase in the blood flow
of the small intestine. At the same time venous filling pressure is
also increased which favours filtration in the intestinal muscosa.

ACKNOWLEDGEMENT

This work has been supported by the Hungarian Department of
Health Grant No. TPB EüM 05.

REFERENCES

Alteveer, R.J., Goldfarb, R.D., Lau, J., Port, M., and Spitzer,
    J.J., 1973, Effect of acute severe hemorrhage on metabolism of
    the dog intestine, Am. J. Physiol., 224:197-201.
Dvoretsky, D.P., 1981, New method for measuring capillary filtra-
    tion coefficient and postcapillary vessel compliance in
    different organs and tissues under constant perfusion, Acta
    Physiol. Acad. Sci Hung., 57:395-397.
Folkow, B., Lundgren, O., and Wallentin, I., 1963, Studies on the
    relationship between flow resistance, capillary filtration
    coefficient and regional blood volume in the intestine of the
    cat, Acta Physiol. Scand., 57:270-283.
Hamar, J., Polenov, S.A., Tcherniavskaya, G.V., Berezina, T.P., and
    Dézsi, L., 1981, Effect of hypoxia on microcirculation and
    energy supply of the small intestine, Adv. Physiol. Sci.,
    25:173-174.
Konturek, S.J., 1980, Somatostatin and opiate peptides: Their action
    on gastrointestinal secretion, in: "Comprehensive Endocrinology,
    Gastrointestinal Hormones", G.B. Jerzy Glass, ed., Raven Press,
    New York, pp. 693-715.
Mathes, P., Heil, Th., Raptis, S., Rasche, H., and Scheck, R., 1975,
    Extended somatostatin treatment of a patient with bleeding
    ulcer, Horm. Metab. Res., 7:508-511.
Meyers, C.A., and Coy, D.H., 1980, Somatostatin, encephalins and
    endorphin, in: "Comprehensive Endocrinology, Gastrointestinal
    Hormones", G.B. Jerzy Glass, ed., Raven Press, New York, pp.
    363-385.

Pollack, J.M., Pearse, A.G.E., Grimelius, L., Bloon, S.R., and
    Arimura, A., 1975, Growth hormone release inhibiting hormone
    in gastrointestinal and pancreatic D cells, Lancet, I:1220-
    1222.
Schally, A.V., Coy, D.H., and Meyers, C.A., 1978, Hypothalamic
    regulatory hormones, Ann. Rev. Biochem., 47:89-128.
Solcia, E., Cappela, C., Ruffa, B., Frigerio, B., Usellini, L., and
    Fiocca, R., 1980, Morphological and functional classifications
    of endocrine cells and related growths in the gastrointestinal
    tract, in: "Comprehensive Endocrinology, Gastrointestinal,
    Hormones", G.B. Jerzy Glass, ed., Raven Press, New York, pp.
    1-17.
Svanvick, J., 1973, Mucosal blood circulation and its influence on
    passive absorption in the small intestine, Acta Physiol.
    Scand. Suppl., 385, 87:1-44.

PO_2 AND pH MEASUREMENTS WITHIN THE RABBIT OVIDUCT FOLLOWING TUBAL MICROSURGERY: REANASTOMOSIS OF PREVIOUSLY DISSECTED TUBES

D.H.A. Maas, B. Stein, and H. Metzger

Dept. of Gynaecology and Dept. of Physiology
Medical School Hannover, Karl-Wiechert-Allee 9
3000 Hannover 61, FRG

The oviduct as a transport system for the gametes and as "reactive tube" for the early development of the fertilized ovum needs an optimal oxygen and substrate supply. Functional and mechanical disturbances of the oviduct might cause generalized dysfunction. Microsurgical resection of such affected tubular segments and reanastomosis of the free ends have been applied in order to try to improve fertility, however, with contradictory results. In order to measure the physiological properties of the fallopian tube before and after microsurgery, catheter electrodes for PO_2 and pH measurements within the lumen of the oviduct have been developed. These measurements might be helpful in gaining further information concerning blood supply and oxygen transport to the oviduct before and after reanastomosis had been performed.

MATERIALS AND METHODS

Adult female rabbits (Dutch belted strain, n = 19, 3.5-4.5 kg b.w.) were investigated, all animals being in the oestrus phase. The animals were anesthetized by an i.m. application of pentobarbital (Nembutal[R], 20-30 mg/kg b.w.). The peritoneal cavity was opened and the oviduct on one side was dissected. An ampullary segment of 1 cm was resected and the ends reanastomosed according to the method described by Vammen et al. (1979). Microsurgical instruments designed for eye surgery were employed in order to reduce traumatic influences as much as possible. The free ends of the dissected oviduct were connected through the different layers by applying three ligatures (Ethilon 8-o, Ethicon). The defect within the mesosalpinx was closed by vicryl single button-connections (Ethicon).

561

After surgery the rabbits were kept under regular veterinary control
and treated by the use of antibiotics. Six to eight weeks later the
rabbits were relaparotomized and the pH and Po_2 values within the
oviducts were measured at both sides of the animals.

Four to five weeks later in 6 remaining rabbits, ovulation was
induced by 75 IE human choriogonadotropin and simultaneously artificial insemination was performed. Sixty-four hours later the animals
were sacrificed. The oviducts were isolated and flushed with Ringer
solution. The gametes were investigated by microscope and the corpora lutea of each ovary were counted. The oviducts were fixed in
Bouin's solution. The area of the reanastomosis was prepared and
analyzed histologically. Corresponding procedures were applied to
the contralateral undisturbed side, which served as a control.

Construction of the Catheter Electrodes

The pH measurements were performed by use of small catheter
electrodes with tip diameters of 2.2 mm (Ingold, Mainz, F.R.G., type
MS-142) inserted through the fimbrial end of the oviduct about 1.5-
2.0 cm into the lumen of the fallopian tube. The reference electrode
(Ingold, type NaCl 1 mol) was located within the abdominal cavity.
Measurements were performed using a pH meter (Knick, Berlin, F.R.G.),
calibrated with standardized buffer solutions (Merck, Darmstadt,
F.R.G.) in the range 4.0 to 10.0. The exact adjustment was done
immediately before and after the experiments by means of buffer
solutions of pH 6.88 and 7.43 (Radiometer, Copenhagen, Denmark). If
marked deviations in the calibration line were observed, the surface
of the electrode and the diaphragm were treated with cleaning solutions (Ingold).

The catheter Po_2 electrode was constructed from fine platinum
wires (15 micron in diameter) with an Ag/AgCl reference electrode
isolated by use of Hysol[R] (Resin R-8-2028, Dexter, N.Y.). The specific resistance of this material is comparable with that of glass
being in the range of $6 \cdot 10^{15}$ ohm/m. The polarization voltage was
adjusted according to the individual polarogram for each electrode,
being between 750 and 900 mV negative relative to the reference. In
order to obtain stable boundary layers at the liquid-metal interface
it was necessary to prepolarize the platinum cathode 6-8 hours prior
to the experiments. The tip of the catheter electrode was cleaned
and covered by a teflone membrane of 12 micron thickness. The fixation of the membrane was performed by use of a small piece of silicone tube (Silastic[R], Dow-Corning, Midland Michigan, USA). A buffer
solution (0.05 mol KCl, 0.05 mol H_2CO_3, 0.213 mol NaOH) was used
between the platinum surface and the teflone membrane. In addition,
Ficoll-400[R] (Pharmacie Fine Chemicals, Uppsala, Sweden) was added
for mechanical stabilization of the membrane. The electrode was
calibrated in 0.9% NaCl solutions maintained at 37°C by means of
precalibrated gas mixtures of known oxygen content.

RESULTS

Po$_2$ Measurements

From the pilot studies (n = 6) and the control as well as re-anastomosis series (n = 10) the Po$_2$ within the oviduct of the rabbit was determined. The mean Po$_2$ of the oviduct was found to be 53 ($\pm$ 16) mm Hg for the control series (mean $\pm$ SD) while for those fallopian tubes, which had undergone microsurgery the Po$_2$ was 56 ($\pm$ 13) mm Hg depending partially upon the degree of deformation of the reanastomosed tubular segments (mean $\pm$ SD) (see Table 1). The Po$_2$ records from the tubes showed typical Po$_2$ oscillations (Fig. 1) with either fast-small amplitudes of 1 mm Hg size or slow-large amplitudes of about 20 mm Hg size; combinations of both types were also observed.

A subdivision into four types of oscillations was possible.

Type 1. Superposition of large Po$_2$ oscillations of 20-40 mm Hg amplitude and additional small oscillations of 2-3 mm Hg in the frequency range of 2/sec.

Table 1. Results of the Po$_2$ and pH measurements within the oviduct of the rabbit. One side of the oviduct had undergone microsurgery with resection and reanastomosis of the fallopian tube while the other side of the oviduct served as control.
[+]) Two rabbits died due to circulatory collapse during laparatomy. Rabbit No. 330 and 349 have not been considered in the calculation of the mean values. $\bar{x} \pm$ SD = mean values and standard deviations.

rabbit No.	control side		reanastomosed side		
	Po$_2$ (mm Hg)	pH	Po$_2$ (mm Hg)	pH	adhesion
298	48	8.05	45	8.04	small
357	75	7.95	74	7.86	tender
342	65	7.93	52	7.80	no
299	34	7.89	35	7.78	small
363	49	8.01	62	7.75	no
324	72	8.04	70	7.68	pron.
292	41	7.92	49	7.60	pron.
348	40	7.73	63	7.51	pron.
330	26	7.21	33	6.80	– [+])
349	–	–	10	7.58	– [+])
$\bar{x} \pm$ SD	53 $\pm$ 16	7.94 $\pm$ 0.10	56 $\pm$ 13	7.75 $\pm$ 0.16	

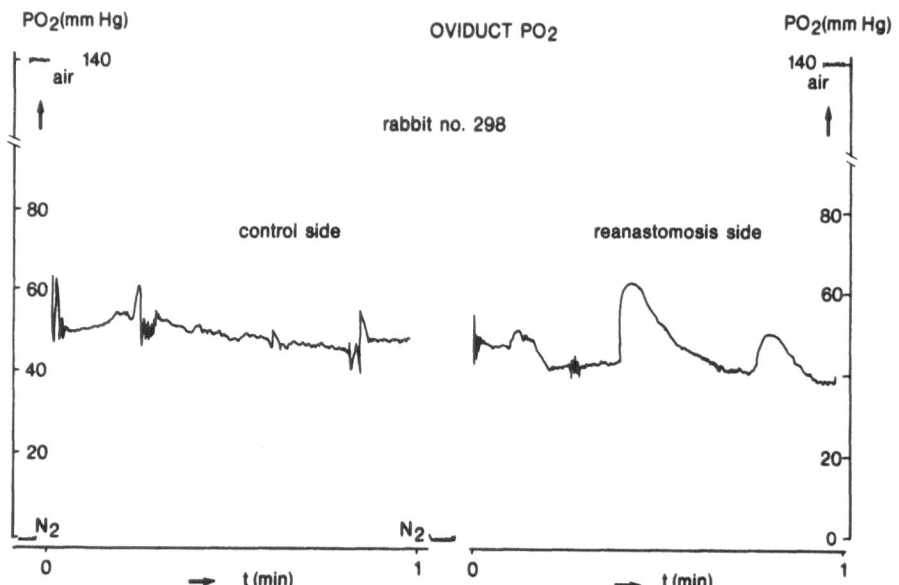

Fig. 1.   Po$_2$ registration obtained from the rabbits' oviduct:
          control and reanastomosed side. Note the irregular fluctua-
          tions of the Po$_2$.

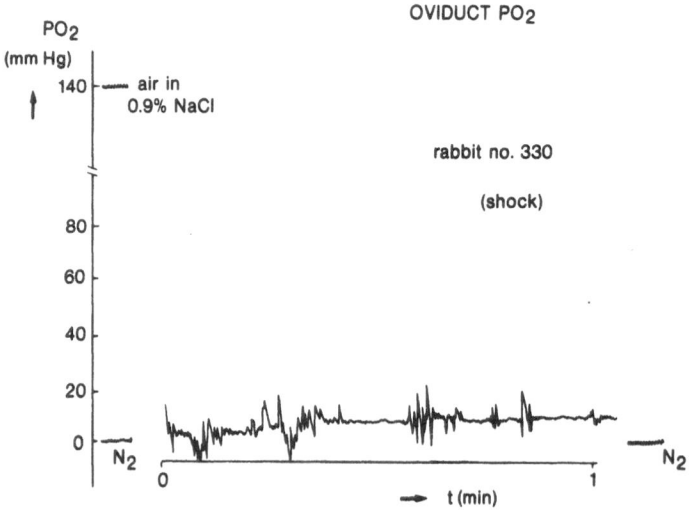

Fig. 2.   Po$_2$ registration obtained from the rabbits' oviduct.
          During laparatomy, circulatory collapse occurred and the
          Po$_2$ values decreased to almost zero level. Registration
          immediately before cardiac arrest.

Type 2. Large slow Po$_2$ oscillations with amplitudes of 20 mm Hg in size and 10-20 sec duration with additional small oscillations of 2-3 mm Hg similar to type 1.

Type 3. Smooth Po$_2$ curves with minimal deviations from the mean Po$_2$. This type was observed under circulatory shock.

Type 4. Slow Po$_2$ waves with small amplitudes of 4-5 mm Hg in the frequency range of 20-30/min (Figs. 1 and 2).

In order to obtain representative mean Po$_2$ values for the rabbit oviduct (Table 1), temporal averaging of the Po$_2$ registrations over 5-10 minutes intervals has been performed.

pH Measurements (Fig. 3)

From pilot studies, the mean pH of the infundibulum tubae with a catheter tip penetration depth of 0.5-1 cm was found to be 7.77. In the ampulla tubae near the ampulla-isthmic junction, a similar pH value of 7.75 was obtained after microsurgical reanastomosis.

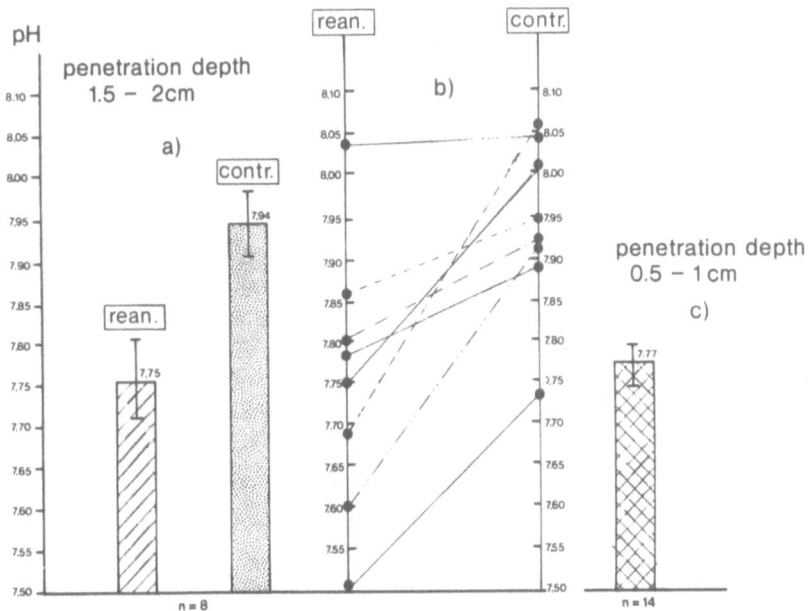

Fig. 3. Summary of the results of the pH measurements within the oviduct of the rabbit. a) Comparison of the mean pH of control and reanastomosed side within a penetration depth of 1.5-2 cm. b) Individual values of column a). c) Mean pH within a penetration depth of 0.5-1 cm.

The results of the control were characterized by a significantly
higher mean pH of 7.94. The difference between the infundibulum
tubae (pH = 7.77) and the pars ampullaris of the undisturbed side
(pH = 7.94) demonstrates an increase of the pH along the distance
of about 1 cm (penetration depth of the first measurements 0.5-1 cm,
of the second 1.5-2 cm). During the shock period in two rabbits a
pH of 7.21 was determined for the control side and 6.8 for the reanastomosed oviduct.

Results from Artificial Fertilization

In three rabbits, which had undergone microsurgery, artificial
insemination was successful and the morulae showed a normal development. The relationship between the number of fertilized ova and
corpora lutae was different in each animal (1:6, 2:9, 3:6) (Table 2).
The control sides however showed balanced quotients between ova and
corpora lutae numbers (4:4, 4:4, 6:6). In two reanastomosed oviducts,
unfertilized ova were observed, the typical outside layer of muco-
proteins were absent while the controls were normal. In one case,
the ovum was unfertilized, the mucine-membrane, however, intact. The
histological results did not correspond in any case with the fertility test. In two rabbits where the serosa of the oviducts showed
extensive adhesions, the fertility was not affected. Two other
rabbits could not be fertilized although the fallopian tube did not
show any deformation. In three rabbits strong adhesions correlated
with low pH values of below 7.68 while the other animals without or
with minor adhesions showed pH values above 7.7.

DISCUSSION

The fallopian tube has often been thought of as essentially a
mechanical connection between the ovary and uterus. It accepts the
egg from the ovary, provides a convenient environment for the interaction between sperm and egg and transfers the fertilized egg as a
morula to the uterus where further development and implantation take
place. The studies of Bishop (1975) and Mastroianni and Jones (1965)
showed that the oviduct is not only a simple physical home but also
a metabolic reaction tube preserving and protecting the various
structures such as sperm, ova, and their combined product during the
period of residence of about three to four days.

Deformations of the oviduct have been found to be a prominent
source of infertility and malfunction of the early embryo development. In order to improve fertility in cases of tubular deformations,
a number of microsurgical procedures have been developed, which have
been accepted as standardized techniques by Intern. Fed. of Fertility Soc. (see Semm, 1979). The aim of this study was to develop a
method which enables us to measure the functional properties of the
oviduct before and after the reanastomosis of short segments of the

Table 2. Comparison of the control and reanastomosed side: Number and developmental stage of the fertilized ova as well as the number of corpora lutea. In addition, the oviduct pH and the severity of the adhesions are listed. +) Oviducts coagulated during another project.

rabbit No.	control side				reanastomosed side				
	cor-pora lutea	n	fertilized ova develop. stage	ovi-duct pH	cor-pora lutea	n	fertilized ova develop. stage	ovi-duct pH	oviduct adhesions
298	4	4	morula 16-32 cells	8.05	6	3	morula	8.04	small
357	4	4	morula	7.95	-	-	-	7.86	tender
342	4	-	- +)	7.93	9	2	morula	7.80	no
363	6	6	morula	8.01	6	5	unfertilized, without mucine-membrane	7.75	no
324	3	-	- +)	8.04	6	1	morula	7.68	pronounced
348	4	2	unfertilized, with mucine-membrane	7.73	3	3	unfertilized, without mucine-membrane	7.51	pronounced

n = number

oviduct. As prominent parameters, the luminal Po_2 and pH were
analyzed by use of small catheter electrodes.

The result from the undisturbed control fallopian tubes of the
rabbits correspond well with those of Bishop (1957) and Mastroianni
and Jones (1965) in their studies on rabbit oviducts. Bishop measured a mean Po_2 of 40 mm Hg in the oestrus period, while in the
luteal phase a lower Po_2 of 30 mm Hg was observed. Mastroianni and
Jones obtained Po_2 values between 40 and 75 mm Hg with no changes
during different stages of the reproductive cycle. In the rhesus
monkey, however, Maas et al. (1976) observed cycle dependent Po_2
values with less than 10 mm Hg in the follicle phase, and a 60-
70 mm Hg Po_2 after ovulation had occurred. According to Daniel
(1968) the optimal Po_2 for in vitro culture of oocytes is 10% O_2
(72 mm Hg), above 95% oxygen, organic defects have been observed in
the examined embryos.

In addition to the observations of other authors the Po_2 values
recorded in our study demonstrate the fluctuations of the mean
values which have been described for other organs as well (Kunze,
1976 for muscle). Changes within the arterial blood vessels might
cause rhythmic flow variations which in turn induce Po_2 oscillations.
It is interesting to note that the Po_2 fluctuations are observed
at low and high Po_2 values. We conclude from these results that Po_2
changes are not instrumented in an oxygen tension feedback control
system. The relatively high oxygen tension values point towards a
more than adequate supply within the oviduct. It is not yet known
how the interaction of the fertilized ovum from the onecell stage
towards the blastocyst stage proceeds, nevertheless, from cell cul-
tures a low blastocyst oxygen consumption seems reasonable. A considerable flow decrease within the tube in response to shock caused
a pronounced Po_2 drop to zero after some minutes. This observation
gives further confidence concerning the catheter electrode response
to changes in the oviduct Po_2.

A considerable alkalotic pH was observed in all rabbits which
was independent of the time of ovulation and far above the pH of the
venous blood. For the rhesus monkey a low pH of 7.1-7.3 was observed
prior to day 14. As soon as ovulation was observed (day 16) the pH
had risen to 7.5-7.7 (Maas et al., 1977). The same authors measured
a high oviduct Pco_2 post ovulation between day 16 and 28. The alkalinity relative to the blood is characteristic of the intact oviduct
because slight acidity occurs not only in the follicular phase of
animals but also in non-ovulatory animals having been ceased to cycle
during the summer and in castrates. It has been hypothetized that
the high pH is necessary to provide the egg with an optimal environment for fertilization and development. At the same time an increase
in fluid secretion, decrease in proteinase inhibitor content and an
increase in pH at both sides of the oviduct (Mastroianni et al.,
1970) occur. However, the interaction of these various processes
has not been explored until now.

SUMMARY

Adhesions of the fallopian tube might be a source of generalized infertility. In order to improve the mechanical and functional
properties of the oviduct, microsurgical replacement of the affected
tubular segments have been applied, however, with contradictory
results. In order to improve the situation and gain further information about this problem, catheter microelectrodes for Po_2 and pH
measurements within the lumen of the fallopian tube have been developed. The validity of the measurements was tested in a series of
rabbits (Dutch belt strain, n = 19, 3.5-4.5 kg bw) which have been
laparatomized unter pentobarbital anesthesia during the oestrus
phase. From one side of the oviduct 1 cm was resected and the free
ends reanastomosed while the other side served as a control. Six to
eight weeks later Po_2 and pH values within the oviduct were measured. After another 4-5 weeks ovulation had been induced and artificial insemination performed. 64 hours later the rabbits were sacrificed. The oviducts were isolated and the corpora lutea as well as
the ova were collected and examined histologically.

The mean Po_2 of the oviduct was 53 ($\pm$ 16) mm Hg (mean $\pm$ SD)
for the control side and 56 ($\pm$ 13) mm Hg (mean $\pm$ SD) within the
anastomosed tubes. The range of values was the same in both groups
(n = 8) 34-75 mm Hg. A T-test comparison of the means of the two
groups gave p-values which were not significant (p > 0.1). Po_2
fluctuations with small-fast amplitudes (1 mm Hg) and additional
slow-large oscillations (about 20 mm Hg) have been observed similar
to the Po_2 waves described for muscle and brain. The mean pH value
of the oviduct a showed slight, statistically significant pH differences (7.75 reanastomosed side, 7.94 control side) whereby the degree
of adhesions was correlated with luminal pH value. In three rabbits
the relationship between corpora lutea and fertilized ova was 1:6,
2:9, 3:6 for the reanastomosed side while the control side showed
balanced quotients of 4:4, 4:4, 6:6. In two rabbits unfertilized ova
have been observed within the reanastomosed tubes where the typical
outside layer of mucoproteids was absent while the control side was
normal. Nevertheless, within both oviducts a continuous layer of
epithelial cells has been observed even after reanastomosis.

REFERENCES

Bishop, D.W., 1957, Metabolic conditions within the oviduct of the
    rabbit, Int. J. Fertil., 2:11.
Daniel, J.C., 1968, Oxygen concentrations for culture of rabbit
    blastocysts, J. Reprod. Fertil., 17:187.
Kunze, K., 1976, Spontaneous oscillations of Pco_2 in muscle tissue,
    Adv. Exp. Med. Biol., 75:631.
Maas, D.H.A., Storey, B.T., and Mastroianni, L., Jr., 1976, Oxygen
    tensions in the oviduct of the rhesus monkey (Macaca mulatta),
    Fertil. Steril., 27:1312.

Maas, D.H.A., Storey, B.T., and Mastroianni, L., Jr., 1977, Hydrogen
    ion and carbon dioxide content of the oviductal fluid of the
    rhesus monkey (Macaca mulatta), Fertil. Steril., 28:981.
Mastroianni, L., Jr., and Jones, R., 1965, Oxygen tension within
    the rabbit fallopian tube, J. Reprod. Fertil., 9:99.
Mastroianni, L., Jr., Urzua, M., and Stambaugh, R., 1970, Protein
    patterns in monkey ovidual fluid before and after ovulation,
    Fertil. Steril., 21:817.
Semm, K., 1979, "Nomenklatur der Eileiterchirurgie (IFFS 1977)",
    Kongreßbericht der deutch-französischen Gesellschaft f. Gynäko-
    logie und Geburtshilfe, München 1979, p. 130.
Vammen, A.N., Gideon, W.P., and Elkins, J.P., 1979, Reanastomosis
    of previously ligated fallopian tube, Fertil. Steril., 32:652.

STUDIES OF HEMORRHAGIC AND TRAUMATIC SHOCK INFLUENCE ON LIVER

OXYGEN TENSION: EFFECTS OF A SINGLE LARGE DOSE OF DEXAMETHASONE

M. Scherf, H.J. Oestern, and H. Metzger

Kinderheilanstalt Hannover, Trauma Dept. and
Medical School Hannover, Physiology Dept.
Karl-Wiechert-Allee 9, 3000 Hannover 61, FRG

INTRODUCTION

Despite stabilized circulatory and respiratory functions, hyperbilirubinemia has been observed in severely traumatized patients which had probably been induced by a lowering of the hepatic microcirculation during the initial period of shock. Improvements in hemodynamic and biochemical parameters by use of an early injection of a single large dose of glucocorticoids are described in the literature, but the reasons for the patients' increased survival chance has as yet not been clarified. The effects of glucocorticoids might be acting via a peripheral vasodilatation, reduction of excess lactate production, protection of the capillary endothelium, lowering of thrombocyte aggregation or conservation of lysosomal integrity (for literature see Glenn, 1975). In order to study the liver oxygen tension as a prominent transport parameter under shock conditions and investigate the influence of glucocorticoids, an experimental shock model was developed by Scherf et al. (1979) which enabled us to study hypovolemic hypotension and trauma as well as drug application under laboratory conditions.

MATERIALS AND METHODS

Male white rats (Wistar strain, 250-300 g b.w.) were anaesthetized by an intraperitoneal injection of ketamine (Ketanest[R] 80 mg/ kg b.w.) and xylazine (Rompun[R] 5 mg/kg b.w.) according to Green et al. (1981). The rats were cannulated by using small teflon catheters placed into the arteria carotis sinistra and vena jugularis dextra for blood pressure monitoring and blood taking. For adequate MAP measurements and bleeding it was necessary to heparinize the animals (1000 IU/kg b.w.). Trauma was induced by fracturing the left tibia and crushing surrounding muscle tissue by means of standardized force of 45 N/2.25 cm^2 tissue for 5 minutes duration. Secondary to the fracture, bleeding was initiated via the vena jugularis. In order to investigate the beneficial effects which a pharmacological dose of glucocorticoids might have, a single large dose of dexamethasone (Decadron[R], Sharp & Dohme) of 8 mg/kg b.w. was applied intravenously.

The rats were divided into three groups: the hemorrhagic group (hs) (n = 16) served as control for the hemorrhagic-traumatic rats (hts) (n = 14); the glucocorticoid group (htsc) (n = 18) was first traumatized and treated with glucocorticoids before slow bleeding was started until a hypotension of 40 mm Hg MAP was achieved. The hemorrhagic-traumatic and the "pure" hemorrhagic groups were infused with a small amount of ringer lactate instead of glucocorticoids in order to compensate for volume effects.

Liver surface oxygen tension (sPo_2) was measured by means of a small surface polyelectrode constructed of six platinum cathodes and a Ag/AgCl reference according to Kessler and Grunewald (1969) and Lübbers (1969). The platinum wires were embedded and isolated in Hysol[R] (Dexter Corp., N.Y.). The surface of the electrodes was covered with a 25 micron teflon membrane. Liver surface temperature was monitored with a small thermistor located immediately at the tip of the electrode. Mean arterial blood pressure (MAP), rectal temperature, heart and respiratory rates were measured at 10 minute intervals. Arterial blood gas and pH values were determined by using small probes with 160 microlitre samples in a blood gas micro system (BMS 3, Mk2, Radiometer, Copenhagen, Denmark).

Time course of the experiments: when anaesthesia became effective, about 10-20 minutes were waited in order to obtain stable MAP and sPo_2 registrations. The animals were traumatized and glucocorticoids were injected; then bleeding was started whereby MAP was lowered at 10 minute intervals to 60 and 40 mm Hg at last. Hypotension of 40 mm Hg was adjusted for a 90 minute duration through reinfusion or withdrawal of small amounts of shed blood. At the end of the hypotensive period, the complete amount of shed blood was retransfused.

RESULTS

Mean Arterial Blood Pressure and Maximal Blood Withdrawal

Initial MAP values of 116 ± 12 mm Hg were recorded in all groups with slightly higher values in the hts rats after trauma induction. A pronounced MAP decrease was registered after the first and second blood withdrawal. At a hypotension level of 40 mm Hg MAP maximum blood withdrawal was 24.4 $\pm$ 2.2 ml/kg b.w. in the hemorrhagic group, 17.9 $\pm$ 3.8 ml/kg b.w. in the hemorrhagic-traumatic group and 20.6 $\pm$ 4.5 ml/kg b.w. in the hemorrhagic-traumatic rats treated with a single large dose of Decadron^R. Reinfusion of the complete volume of shed blood caused a MAP increase to 108 $\pm$ 25 mm Hg in hs rats while hts and htsc rats approximated values of 89 and 85 $\pm$ 29 mm Hg, respectively. Progressive hypotension was registered in hts and htsc animals to 45 $\pm$ 19 mm Hg about 30 minutes later while the hs rats were stable around 91 $\pm$ 39 mm Hg MAP.

Liver Surface Oxygen Tension (sPo$_2$)

A parallel time-course of changes in liver surface oxygen tension in response to progressive hypovolemia as well as during a hypovolemic hypotension of 40 mm Hg and post retransfusion was observed in all groups. Probably, sPo$_2$ reflects the reduced liver blood flow which might be proportional to MAP. sPo$_2$ decreased to anoxic Po$_2$ values (zero level) when MAP was lowered to 40 mm Hg MAP. Only slight improvements of 1 - 2 mm Hg have been observed in the glucocorticoid treated animals at this MAP level. Again, with post reinfusion a parallel MAP and sPo$_2$ increase towards initial levels in hs rats was recorded while in hts and htsc animals much lower values of about 10 mm Hg vs. an initial level of 29 mm Hg were measured (Fig. 1).

Heterogeneities in liver oxygen tension and microcirculation were analyzed on the basis of sPo$_2$ frequency distributions. Similar results were obtained at the beginning of the experiment and with increased hypovolemia. Already moderate hypovolemic hypotension of 60 mm Hg caused a pronounced shift to the left in the sPo$_2$ histogram with more than 40% of the sPo$_2$ values in the 0-5 mm Hg range while at severe hypotension (MAP = 40 mm Hg) sPo$_2$ values above 5 mm Hg were rarely seen. In the htsc group small improvements were registered at 60 mm Hg MAP, however, an overall effect in response to glucocorticoids could not be extracted from the sPo$_2$ data (Fig. 2).

Arterial Blood Gas and pH Values

Blood gas and pH values were measured at the beginning of the experiment, at moderate hypovolemia (MAP = 60 mm Hg) as well as at the beginning and end of hypovolemic hypotension and 10 minutes post retransfusion of the shed blood (Table 1). Hypovolemia, which results in a sympathetic stimulation of the respiratory system, caused hyperventilation and induced a partial compensation of the metabolic acidosis. This effect was especially pronounced in the htsc group and might be one possible explanation for the beneficial effect of glucocorticoids. During the course of the experiments a pH$_a$ drop due to excessive proton production was seen in all groups.

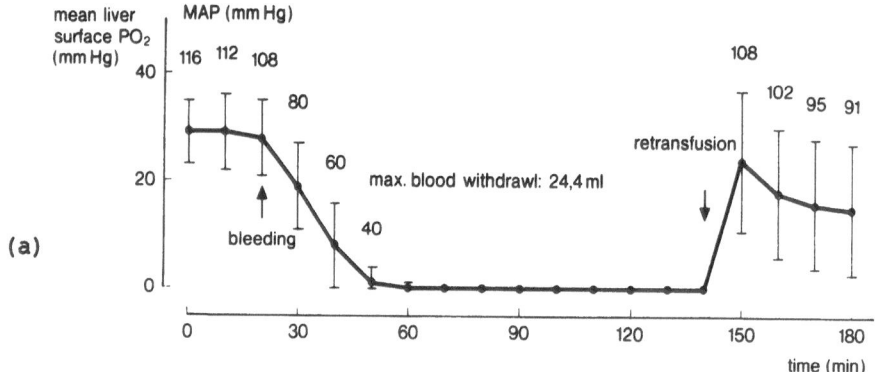

(a)

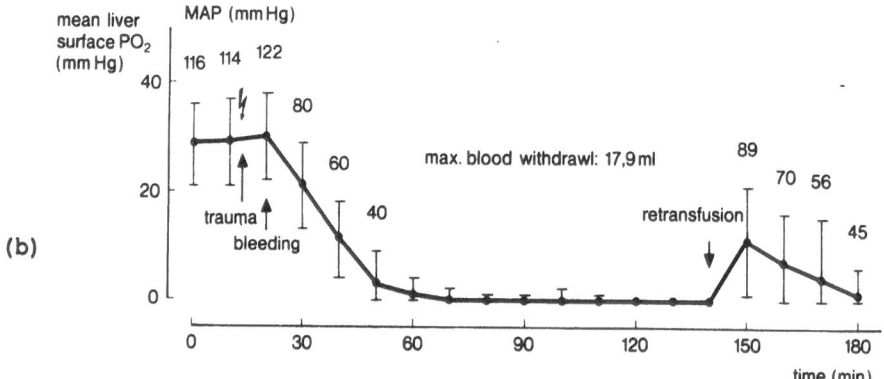

(b)

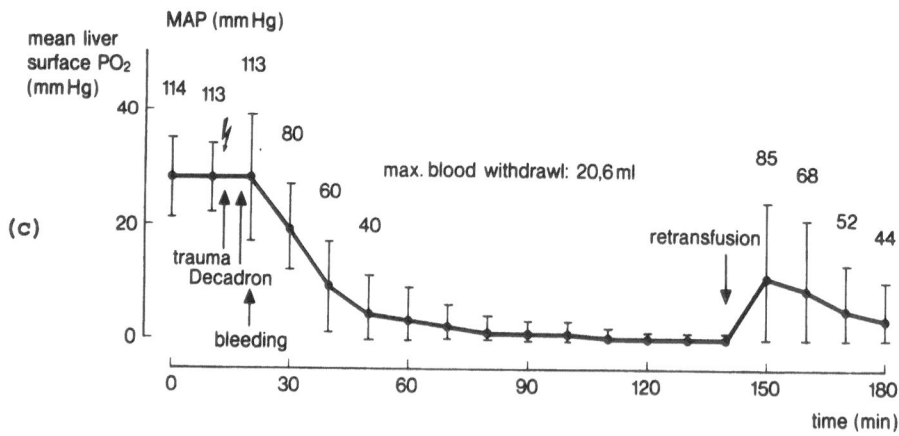

(c)

Fig. 1.   Mean surface Po₂ (sPo₂) and MAP values in three groups of
          rats registered in response to hemorrhagic hypotension (hs
          (a)), hemorrhagic hypotension and trauma influence (hts
          (b)) as well as hemorrhagic hypotension with trauma and
          application of a single large dose of dexamethasone (htsc
          (c)). The numbers at the sPo₂ graphs represent the mean
          arterial blood pressure (MAP). The maximal blood withdrawal
          is related to kg body weight.

Table 1. Arterial blood gas and pH values (P_aO_2, P_aCO_2, pH_a) in response to hemorrhagic shock (hs), hemorrhagic-traumatic shock (hts), and hemorrhagic-traumatic shock with glucocorticoid application (htsc). MAP = mean arterial blood pressure, f_h = heart rate, f_r = respiratory rate. Time is related to the beginning of the experiment when anaesthesia became effective.

	n	initial phase 10 min	moderate shock 40 min	severe shock 90 min	severe shock 130 min	post retransf. 160 min
MAP (mm Hg) hs	16	112±12	60	40	40	102±30
hts	14	114±12	60	40	40	70±23
htsc	18	113±14	60	40	40	68±26
pH_a hs	14	7.30±0.03	7.30±0.06	7.19±0.10	7.10±0.07	7.05±0.11
hts	8	7.29±0.04	7.24±0.06	7.15±0.11	7.08±0.10	7.06±0.05
htsc	8	7.29±0.05	7.36±0.10	7.17±0.05	7.09±0.07	7.13±0.06
P_aO_2 (mm Hg) hs	13	86±10	91±12	92±19	88±19	66±20
hts	8	80±10	86±12	81±15	75±18	49±16
htsc	10	86±17	104±20	89±22	89±15	66±19
P_aCO_2 (mm Hg) hs	13	35±7	32±6	34±7	35±7	43±7
hta	8	38±4	35±8	35±3	36±7	37±13
htsc	10	39±5	27±4	34±7	37±5	41±5
f_h (1/min) hs	16	216±26	179±23	220±32	274±35	300±31
hts	14	210±18	189±21	217±28	270±27	278±30
htsc	18	211±29	176±33	189±28	240±24	270±33
f_r (1/min) hs	16	66±7	67±9	60±7	53±9	71±10
hts	14	64±5	70±13	51±8	49±9	56±15
htsc	18	62±8	64±13	54±8	48±6	53±15

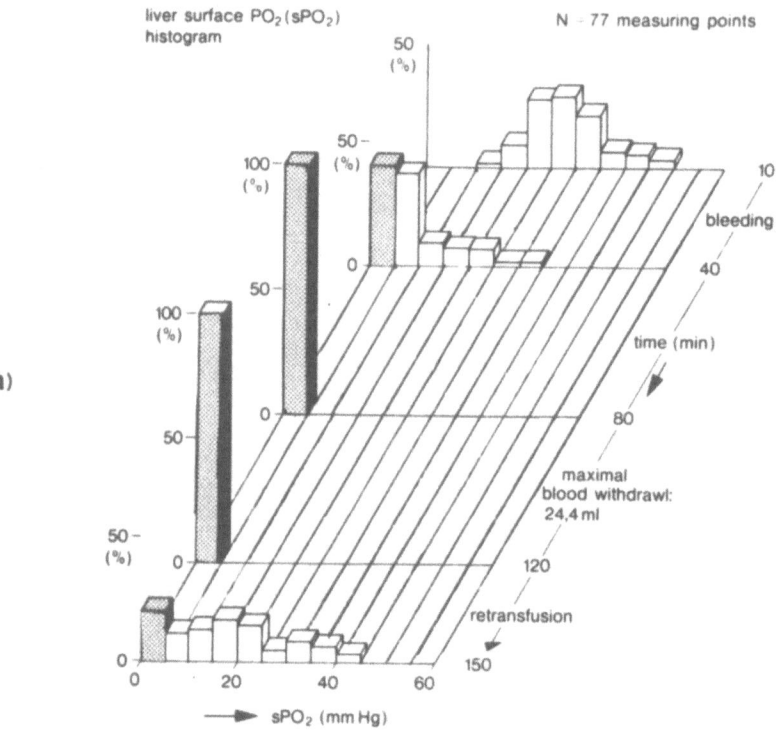

(a)

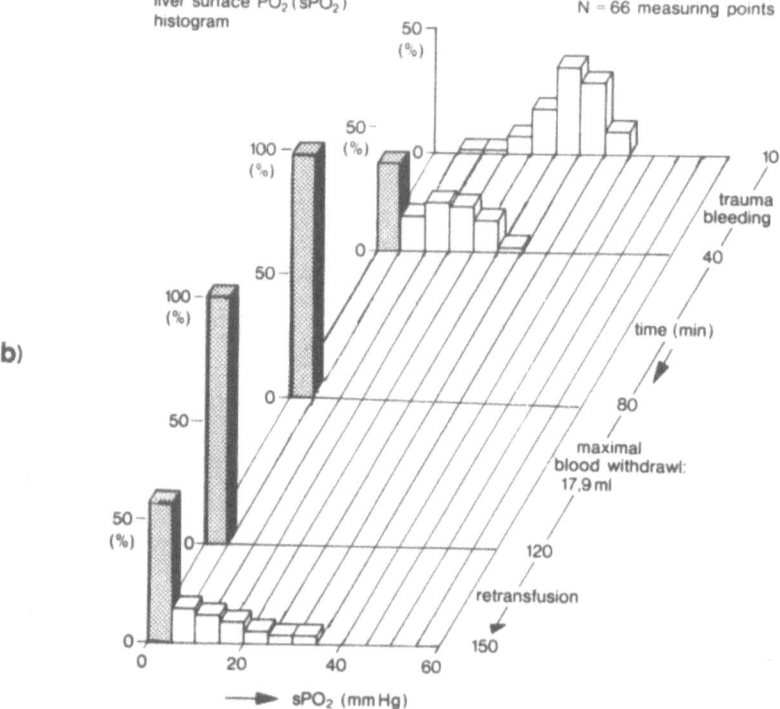

(b)

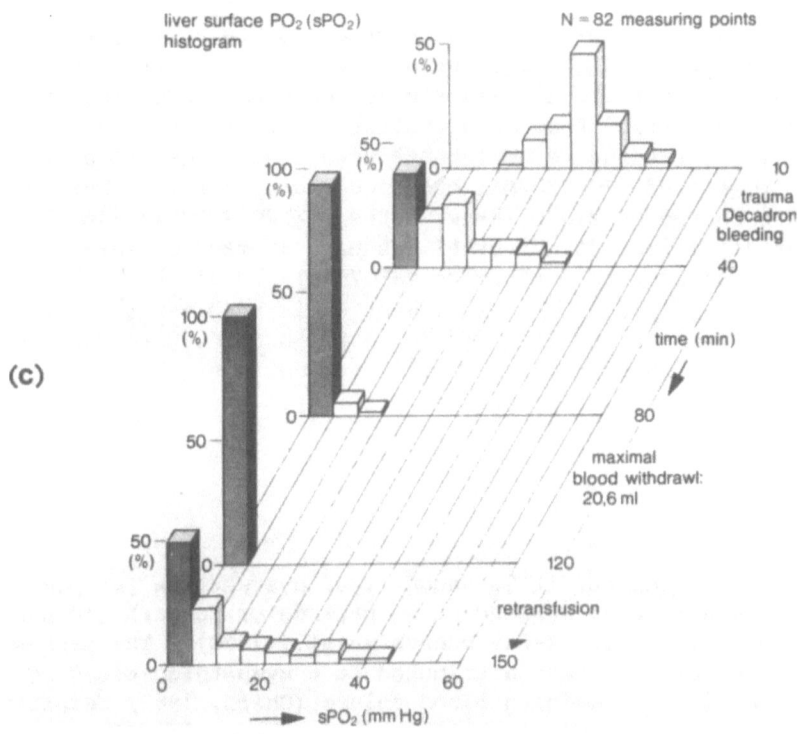

(c)

Fig. 2.   Liver surface Po₂ histograms for three groups of animals
subjected to hemorrhagic shock (hs) (a), hemorrhagic-traumatic shock (hts) (b), and hemorrhagic-traumatic shock with
glucocorticoid application (htsc) (c). The sPo₂ histograms
are plotted for the initial values at the beginning of the
experiment, at the beginning and end of the hypovolemic
period as well as post retransfusion of the shed blood.
Maximal blood withdrawal is related to kg body weight.

This effect was intensified post retransfusion when the peripheral organs were reperfused; nevertheless, the htsc rats improved slightly under these conditions and showed no further pH_a decrease. Furthermore, arterial oxygen tension was stabilized to values above 60 mm Hg while in hts rats P_aO_2 dropped to about 50 mm Hg.

Respiratory and Heart Rates

The ketanest/xylazine anaesthesia caused respiratory frequencies of 65 min which were slightly lowered during the experimental shock and increased post retransfusion to their initial values. The hts and htsc rats showed respiratory rates of as low as 48/min during the shock period and did not recover completely post retransfusion. Initial heart rates of 210-220/min and values of 190-270/min during hypovolemic hypotension of 40 mm Hg MAP were registered in all groups. It is interesting to note that during the hypotensive period an increase to around 240-270/min occurred and post retransfusion the rats were characterized by heart rates of 270-300/min. In another control series pentobarbital was used instead of ketamine/ xylazine. Mean heart rate was higher (350/min) and almost stable during the whole period of the experiment while respiratory rate decreased from an initial value of 85/min to 50/min after trauma and bleeding had been started. Survival rate decreased because the animals died during the shock period. This type of anaesthesia was not used for further shock studies.

DISCUSSION

From clinical reports it is known liver dysfunction and cen-trilobular necrosis occur subsequent to periods of hemorrhagic hypo-tension (Shoemaker et al., 1964; Nunnes et al., 1970)). The patho-logical alterations have been attributed to a sympathico-adrenergic redistribution of the remaining blood volume (Chien, 1967) resulting in tissue hypoxia and anoxia in the liver. From experimental studies in rats the hepatic sinusoidal congestion (Shoemaker et al., 1964), the reduced microcirculation (Koo and Liang, 1977), the decrease in hepatic tissue Po_2 (Miller et al., 1974) and hepato-cyte edema (De Palma et al., 1972) are known.

However, little information is available in the literature about the influences of trauma on liver oxygen supply; furthermore, to our knowledge, the influences of glucocorticoids injected immediately after the trauma induction have not been studied under reproducible conditions.

The results of our study are based upon the comparison of three groups of experiments in which rats were subjected to hemorrhagic shock (hs), hemorrhagic-traumatic shock (hts) and hemorrhagic-trau-matic shock with glucocorticoid treatment (htsc). Identical experimental procedures using the same anaesthesia and animal strain were performed in order to obtain reproducible conditions and fulfill the criteria for a statistical analysis of the results. Mean surface oxygen tension (sPo_2) values and histograms were plotted for the animals of each group.

The shock model: As the basis for comparing the results, the mean arterial blood pressure served as a standard parameter. Instead of the oxygen transport system a more desirable parameter might have been transport capacity and affinity of the arterial blood. However, these values varied among the various groups and complicated the theoretical analysis.

The sPo_2 data of our study are somewhat higher with a shift to the right in the frequency distribution of the sPo_2 values compared with results of other authors (Miller et al., 1974; Lovelace et al., 1979). Probably, the ketamine/xylazine anaesthesia is responsible for these results because respiratory and metabolic depression are much less pronounced than under barbiturates used by other authors. It is interesting to note that the influence of anaesthesia as the most important interacting process which is known to modify almost any physiological investigation has as yet not been studied in other shock models. The main qualities of ketamine are the pronounced analgesia, increase in MAP and heart rate; xylazine applied additionally improves muscle relaxation, but however, acts as an antagonist to most of the circulatory qualities of the ketamine. The combined ketamine/xylazine narcosis showed a sufficient analgesia, hypnosis and relaxation of the animals which tolerated the described manipulations without further complications. Despite careful preparation of the animals a considerable low pH_a was measured which might be caused partially by the surgical procedure, the insertion of the catheters, laparatomy etc.

Control experiments: The high initial sPo_2 values and the distribution of the histogram represent a luxury oxygen supply to the hepatocytes, providing for even the anatomically long sinusoids. The maximum of the histogram was found in the 25 and 35 mm Hg column; sPo_2 values in the 0-5 mm Hg and 5-10 mm Hg range were rarely observed. The correlation coefficient of $r = 0.036$ between MAP and sPo_2 indicates an independency of the MAP. During bleeding and the hypotensive period (MAP = 40 mm Hg) a positive correlation coefficient of $r = 0.776$ was calculated demonstrating the insufficiency of an increase in liver blood flow to counterbalance the reduced perfusion caused by lowering the MAP. Liver sPo_2 reflects mainly the liver perfusion which shows a pressure passive behaviour. A pronounced shift to the left of the sPo_2 histogram is seen already at MAP values around 60 mm Hg. At MAP = 40 mm Hg a generalized anoxia with zero mm Hg sPo_2 was registered. Severe disturbances in metabolism and electrolyte balance were to be expected already at these periods of hypovolemic hypotension.

After retransfusion MAP approximated the initial level; however, the sPo$_2$ histogram showed an altered distribution with more
sPo$_2$ values in the 0-5 mm Hg range which point towards irreversible
damages to the liver parenchyma. These changes might have caused
necrotic alterations with disturbances in the sinusoids perfusion.
Between sPo$_2$ and MAP a correlation coefficient of r = 0.594 has
been found demonstrating a diminution of the pressure, however,
hepatic autoregulation was not approximated according to the initial quality.

Application of glucocorticoids: The early application is
assumed to suppress the circulatory reactions in response to the
trauma. Compared with the controls, heart rates did not increase
significantly. Furthermore, MAP was constant, probably because
glucocorticoids are assumed, to block the alpha-receptors of the
peripheral blood vessels. With progressive hypovolemia a slower decrease in the sPo$_2$ values of the liver surface was observed in the
glucocorticoid group. A significant increase was found at moderate
hypotension (MAP = 40 mm Hg) and at the beginning of severe hypotension (MAP 40 mm Hg). However, during later periods of shock the
influence of dexamethasone disappeared. Nevertheless, 10 minutes
after retransfusion slight improvements in the sPo$_2$ levels were
seen. The main effects of glucocorticoids were the following: 1)
increase of maximal blood volume withdrawn which showed an increase
from about 18 to 24 ml/kg b.w. compared with the hts group; 2)
significant higher pH$_a$ about 25 min post retransfusion of about 0.1
unit; 3) no influences on the respiratory frequency; 4) lower heart
frequency probably due to interaction with catecholamines.

Concerning the flow regulation, a pronounced shift from the
pressure dependent flow control to the autoregulative mechanism was
observed which might be due to protection of the physiological qualities of the blood vessels. In general, the effects of dexamethasone
have been found not as efficient as it would be desirable for the
patients survival. Nevertheless, the increase of the maximal blood
volume withdrawn and the pHa improvement together with a slight trend
towards autoregulation of the liver perfusion might help to overcome
the difficulties of the traumatized patient especially in the first
period of 30-50 minutes.

SUMMARY

Liver surface oxygen tension in response to hemorrhagic shock
(hs) and to combined hemorrhagic-traumatic shock (hts) as well as
to hemorrhagic-traumatic shock with glucocorticoid application
(htsc) has been studied using male rats (Wistar strain, 250-300 g
b.w.). The animals were anaesthetized by i.p. injection of ketamine/
xylazine. All animals were bled through a catheter inserted into
the vena jugularis until a hypotension of 40 mm Hg mean arterial

blood pressure (MAP) was attained. The hts rats were traumatized by fracturing the left tibia and crushing the adjacent muscle tissue with 45 N/2.25 cm^2 for 5 minutes. The htsc animals were traumatized and treated with a single injection of a large pharmacological dose of dexamethasone (6-8 mg/kg b.w.). Initial surface Po_2 of the liver was higher than published by others (29 $\pm$ 7 mm Hg) probably due to the narcotic agents. Maximal blood withdrawal was 24.4 $\pm$ 2.2 ml/kg b.w. for hs, 17.9 $\pm$ 3.8 ml/kg b.w. for hts and 20.6 $\pm$ 4.5 ml/kg b.w. for htsc rats. Liver surface Po_2 decreased to zero mm Hg in response to hypovolemic hypotension in all rats. Retransfusion of the shed blood caused a MAP and surface Po_2 increase to only about half of the initial levels in hts and htsc rats while in hs animals the initial values were approximated. The hts and htsc rats showed signs of progressive hypotension to about 45 mm Hg MAP within 90 minutes post retransfusion. Dexamethasone improved the arterial pH to 7.15 compared with 7.05 of the control. P_aO_2 was elevated to 60 mm Hg vs. 49 mm Hg of the control animals. The beneficial influence of glucocorticoids on liver surface Po_2 has not been substantiated as would have been desirable for the patient. Nevertheless, from a physiological standpoint a positive trend in the liver O_2 supply has been evaluated as a small right shift of the surface Po_2 histogram as well as the blood gas and pH data.

REFERENCES

Chien, S., 1967, Role of sympathetic nervous system in hemorrhage, Physiol. Rev., 47:252.

De Palma, R.G., Holden, W.D., and Robinson, A.V., 1972, Fluid therapy in experimental hemorrhagic shock: Ultrastructural effects in liver and muscle, Ann. Surg., 175:539.

Glenn, T.M., 1975, "Steroids and Shock", Urban & Schwarzenberg, München-Berlin-Wien.

Green, C.J., Knight, J., Precious, S., and Simpkin, S., 1981, Ketamine alone and combined with diazepan or xylazine in laboratory animals: A 10 years experience, Lab. Anim., 15:163.

Kessler, M., and Grunewald, W., 1969, Possibilities of measuring oxygen pressure fields in tissue by multiwire platinum electrodes, Progr. Respir. Res., 3:147.

Koo, A., and Liang, I.Y.S., 1977, Blood flow in hepatic sinusoids in experimental hemorrhagic shock in the rat, Microvasc. Res., 13:315.

Lovelace, D.R., Short, B.L., and Rink, R.D., 1979, Hepatic oxygen supply in reversible and irreversible hemorrhagic shock, J. Surg.Res., 26:130.

Lübbers, D.W., 1969, The measuring of the tissue oxygen distribution curve and its measurement by means of Pt electrodes, in: "Oxygen Pressure Recording in Gases, Fluids and Tissues", F. Kreuzer, ed., Progr. Res., 3:122-123.

Miller, A.T. jr., Shen, A.L., and Bonner, F.B., 1974, Hemorrhagic
    shock in the rat: Metabolic changes in brain and liver, Arch.
    Int. Physiol. Biochem., 82:69.
Numes, G., Blaisdell, F.W., and Margautlen, W., 1970, Mechanisms of
    hepatic dysfunction following shock and trauma, Arch. Surg.,
    100:546.
Scherf, M., Oestern, H.J., and Metzger, H., 1979, Studies of hemor-
    rhagic and traumatic shock influence on liver oxygen tension,
    Ann. Meeting of the ISOTT, La Jolla, USA.
Shoemaker, W.C., Szanto, P.B., Fitch, L.B., and Brill, N.R., 1964,
    Hepatic physiologic and morphologic alterations in hemorrhagic
    shock, Surg. Gynaecol. Obstet., 118:828.

CORRELATION BETWEEN THE REDOX STATE OF NAD(P)H AND TOTAL FLOW IN

THE PERFUSED RAT LIVER

J. Höper

Institut für Physiologie und Kardiologie der Universität
Erlangen, 8520 Erlangen, FRG

INTRODUCTION

It is known that the blood flow in different organs is influenced by the oxygen tension in the blood or in the perfusate. Höper and Kessler (1981) presented evidence that the oxidase involved in this mechanism is MAO-B, an enzyme located in the outer mitochondrial membrane. At the time the question was still open as to whether or not these O_2-dependent flow changes occur after the respiratory chain becomes reduced or not.

Various investigators tried to solve this question by using inhibitors of the respiratory chain (Fay and Jöbsis, 1972; Coburn et al., 1979). Coburn et al. (1979) showed that in the isolated rabbit aorta there is a relation between the inhibition of the respiratory chain and a decrease in the mechanical tension of the smooth muscle. Despite this reaction, an oxygen dependent component still resists.

In this study as well as in others, mechanical tension of smooth muscle or blood flow and inhibition of the respiratory chain were compared under steady-state conditions. The purpose of this study was to investigate the dynamics of changes in blood flow, O_2 uptake and NAD(P)H fluorescence in response to various inhibitors.

METHODS

Experiments were performed in the liver of female rats (Wistar AF, SPF animals, 200 ± 20 g). Prior to the preparation of the organ, the animals were anesthetized with pentobarbital (6 mg/100 g) and heparinized with Liquemin (100 IU/100 g). The liver was prepared according to Miller et al. (1951).

Hemoglobin-free perfusion was performed according to Kessler (Höper and Kessler, in preparation). The perfusion medium (Krebs-Ringer bicarbonate solution containing 35 g bovine albumin/l) was equilibrated with a gas mixture containing 95% O_2 and 5% CO_2. The perfusion temperature was kept constant at 22°C.

Oxygen tension in the perfusate was measured by Clark-type platinum electrodes. The total flow through the organ was measured with an electromagnetic flow meter (Statham). From total flow and the $AVDo_2$ the oxygen uptake rate was calculated.

The NAD(P)H fluorescence was measured with a micro light-guide photometer (Ji et al., 1979).

After preparation, the organs were perfused for 1/2 h with a perfusate equilibrated with a gas mixture containing 95% O_2 and 5% CO_2. Then 50 nmol norepinephrine/l was added to the perfusate. This amount caused a flow decrease of about 25% (flow before the addition of norepinephrine was taken as 100%).

10 minutes after the addition of norepinephrine the different inhibitors were added to the perfusate. At this time the total flow had reached a stable value.

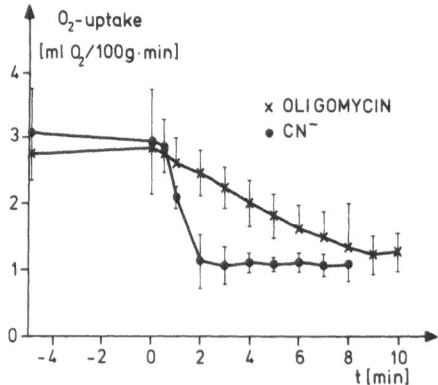

Fig. 1.   Oxygen uptake in the isolated hemoglobin-free perfused rat liver before and after the addition of NaCN (1.5 mmol·l^{-1}) and oligomycin (15 mg/l). The substances were added at t = 0 min (mean values + SD, n = 5).

RESULTS AND DISCUSSION

In Fig. 1 the effect of oligomycin (15 mg/l) and NaCN (1.5 mmol·l^{-1}) on the oxygen uptake rate of the organs is shown. In Fig. 2 the corresponding NAD(P)H fluorescence and in Fig. 3 the flow changes. It is evident that these two substances operate on the O_2 uptake with different time constants. Two minutes after the application, the O_2 uptake rate is significantly lower after NaCN than after oligomycin (p = 0.005).

As shown in Fig. 2 the NAD(P)H fluorescence also behaves significantly different after the addition of NaCN or oligomycin. As one would expect from the behaviour of the O_2 uptake rate, there is a faster increase in NAD(P)H fluorescence with NaCN than with oligomycin.

In contrast to this significantly different effect on the respiratory chain, the total flow through the organs behaves very similarly in response to the different inhibitors (Fig. 3).

Fig. 4 shows the relation between the increase in NAD(P)H fluorescence and the total organ flow. It can be seen that the NaCN-induced increase in flow is slower than the increase in fluorescence, whereas both, flow and fluorescence, seem to change similarly.

This result indicates that there may be a time limiting step between NaCN induced increase in NAD(P)H fluroescence and the flow change.

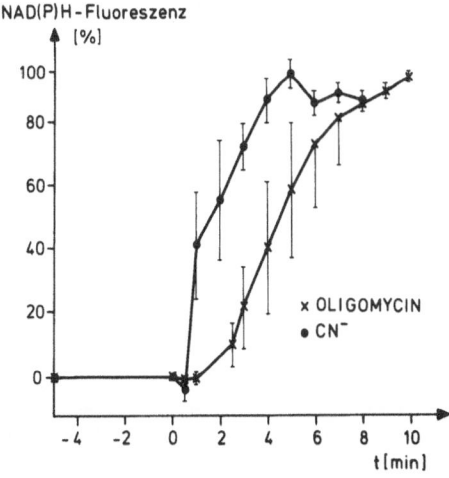

Fig. 2.   Effect of NaCN and oligomycin on NAD(P)H fluorescence, for
           dosage see Fig. 1. (n = 5).

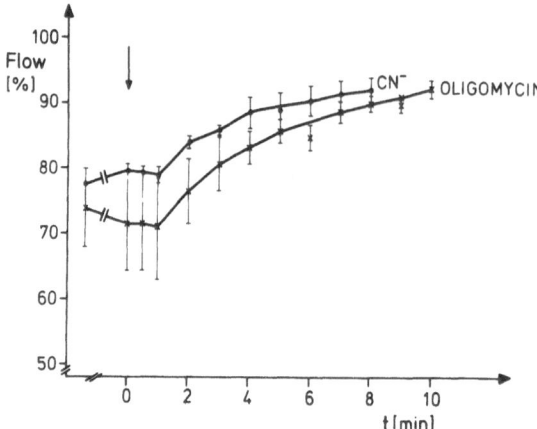

Fig. 3. Changes in total flow through the isolated perfused rat
         liver upon addition of NaCN or oligomycin. The initial flow
         before the addition of norepinephrine was taken as 100%.
         (n = 5).

    In contrast to the inhibitors mentioned above, 2,4-dinitro-
phenol, an uncoupler of the respiratory chain, induced an increase
in O_2-uptake rate and a concomitant decrease in NAD(P)H-fluores-
cence. The O_2-uptake rose by 25%, the fluorescence decreased by
18.4%. Under this condition the total flow does not increase (Fig. 5).

    Another substance causing an oxidation of NAD(P)H is t-buturyl-
hydroperoxide. This oxidation is demonstrated by the decrease in

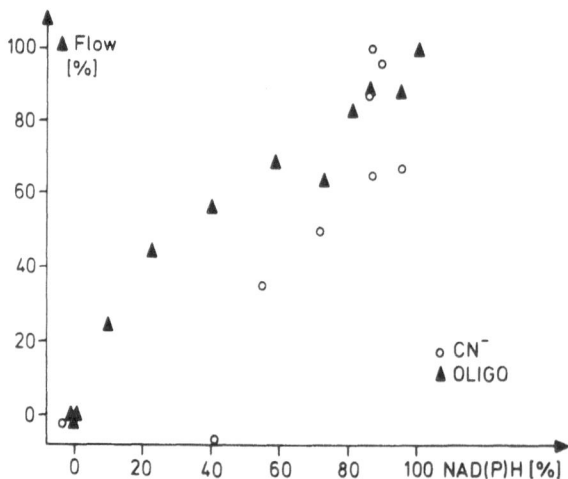

Fig. 4. Relation between the increase in total flow and NAD(P)H
         fluorescence. The figures shows the mean values from the
         preceding figures.

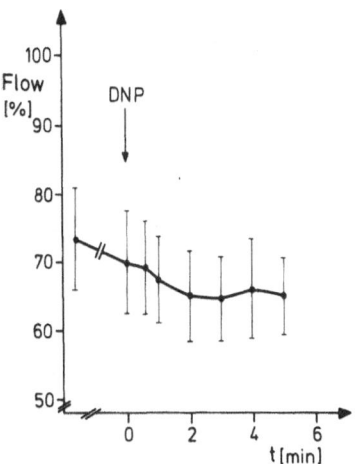

Fig. 5.   Effect of 2,4-dinitrophenol on total flow through the iso-
            lated, hemoglobin-free perfused rat liver (mean values $\pm$
            SD, n = 5).

NAD(P)H fluorescence and is accompanied by a decrease in the total
flow through the organ (Fig. 6).

CONCLUSION

        Summarizing the results, there is evidence that the cellular
redox state of NAD(P)H may modulate the O_2-sensitive regulation of
total flow through the perfused rat liver. This could be independent
of a decrease in ATP, otherwise 2,4-DNP should also cause an in-

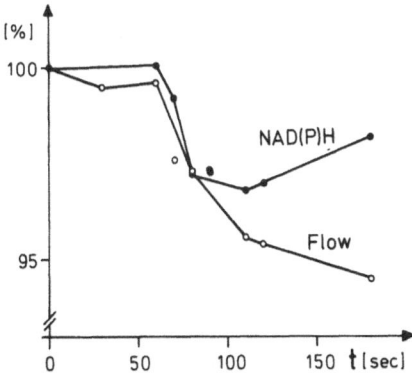

Fig. 6.   Influence of t-buturylhydroperoxide on NAD(P)H-fluorescence
            and total flow through the isolated perfused liver (mean
            values from 3 experiments).

crease in flow. Therefore, another mechanism may connect the redox state of NAD(P)H with the flow regulation.

REFERENCES

Coburn, C.F., Grubb, B., and Aronson, R.D., 1979, Effect of cyanide on oxygen tension-dependent mechanical tension in rabbit aorta, Circ. Res., 44:368-378.

Fay, F.S., and Jöbsis, F.F., 1972, Guinea pig ductus arteriosus. III. Light absorption changes during response to O_2, Am. J. Physiol., 223:588-595.

Höper, J., and Kessler, M., 1981, Po_2 and sodium dependent mechanism regulating liver blood flow, in: "Oxygen Transport to Tissue", Adv. Physiol. Sci., Vol. 25, A.G.B. Kovách, E. Dóra, M. Kessler, I.A. Silver, eds., Pergamon Press, Akadémiai Kiadó, Budapest, pp. 163-164.

Höper, J., and Kessler, M., 1982, Constant pressure perfusion of the isolated rat liver: local oxygen supply and metabolic function, Microcirculation, submitted for publication.

Ji, S., Chance, B., Nishiki, K., Smith, T., and Rich, T., 1979, Micro-light guides: a new method for measuring tissue fluorescence and reflectance, Am. J. Physiol. Cell Physiol., 52:C144-C156.

Miller, L.L., Bly, C.G., Watson, M.L., and Bale, W.F., 1951, The dominant role of the liver in plasma protein synthesis. A direct study of the isolated perfused rat liver with the aid of lysine-E-C[14], J. Exp. Med., 94:431-453.

STUDIES ON THE ABILITY OF KIDNEY CELLS TO RECOVER AFTER PERIODS OF

ANOXIA[*]

G. Gronow[1], F. Meya[1], and Ch. Weiss[2]

[1]Physiologisches Institut der Universität Kiel, FRG
[2]Institut für Physiologie, Medizinische Hochschule
Lübeck, FRG

INTRODUCTION

In shock and during renal transplantation, kidneys may be
exposed to different periods of reduced blood supply. A reversible
or irreversible impairment in renal functions may occur. It is of
clinical and theoretical interest to obtain more detailed information on the ability of kidney cells to recover after different
periods of normothermic anoxia. We compared the recovery of tubular
cell functions such as PAH transport, K^+ accumulation, and gluconeogenesis after different periods of normothermic anoxia in either
a physiological Krebs-Ringer-Bicarbonate (KRB) solution or in a
medium of "intracellular" composition (a modified Collins' solution). Additionally, protein loss and enzyme release from tubular
cells were measured in both incubation media during anoxia.

MATERIALS AND METHODS

Isolated tubular segments of rat renal cortex were prepared by
collagenase treatment as originally described by Burg and Orloff
(1962). Kidney cortex slices of 1 day starved, male Hannover-Wistar
rats (200 - 300 g body weight) were minced (4°C) and subsequently
suspended in an oxygenated buffer medium containing 0.2 g collagen-
ase/100 ml (37°C). After 12 - 15 min the separated tubular segments
were washed twice (4°C) and resuspended in the incubation media
under study (37°C). Details of the preparation have been reported
elsewhere (Gronow et al., 1976). Modifications were as follows:

---

[*] Supported by the Deutsche Forschungsgemeinschaft

1. The Ca^{++} concentration in the collagenase medium was raised to
50 $mmol \cdot l^{-1}$ to increase the collagenase activity. 2. Tubule suspensions in the test media (1.5-2 mg protein/ml) were shaken in 50 ml
Erlenmeyer flasks, and the surfaces of the suspensions were gassed
with either humidified 95% O_2:5% CO_2 (abbreviated as "O_2") or with
humidified 95% N_2:5% CO_2 (abbreviated as "N_2"). 3. Prior to all
experiments a 30 min aerobic pre-incubation was introduced to
stabilize cell functions after the collagenase treatment.

After 30 min pre-incubation the tubular segments were washed
twice (4°C) and equal portions were resuspended at 37°C in three
50 ml Erlenmeyer flasks: a) aerobic control in Krebs-Ringer-Bicarbonate = KRB (with 1 g albumine/100 ml, 10 $mmol \cdot l^{-1}$ lactate, 0.07
$mmol \cdot l^{-1}$ PAH, pH 7.35, 305 m osmol/kg H_2O), b) anoxic incubation in
KRB (as above), c) anoxic incubation in Collins' solution (Collins
et al., 1969), modified as follows: 10 $mmol \cdot l^{-1}$ $NaHCO_3$, 115 $mmol \cdot l^{-1}$
K^+, 57.5 $mmol \cdot l^{-1}$ PO_3^-/PO_4^{3-}, 15 $mmol \cdot l^{-1}$ Cl^-, 206 $mmol \cdot l^{-1}$ mannitol,
with 10 $mmol \cdot l^{-1}$ lactate, 0.07 $mmol \cdot l^{-1}$ PAH, pH 7.35 and 410 m osmol/
kg H_2O. Recovery of cellular functions was tested after different
periods of anoxia in flasks "b" and "c", and after subsequent resuspension of the tubular segments in oxygenated KRB. Additionally,
protein and enzyme losses from anoxic tubules were measured in
flasks "a" and "b".

In all experiments 1 ml samples of tubule suspensions were
analyzed after centrifugation: in the supernatants, enzyme activities and glucose concentrations were measured according to standard
assays (Bergmeyer, 1974). In the sediments, after mechanical cell
disintegration in distilled water, K^+ was measured by flame photometry. ^{14}C-PAH was estimated in the supernatant and the aquaeous
cell extract by liquid scintillation photometry. Total protein
content of the tubular segments was measured according to the method
of Lowry et al. (1951). Cellular K^+, PAH and protein contents were
corrected for the amounts of extracellular K^+, PAH and albumine
which were trapped in extracellular fluid after centrifugation. In
all statistical comparisons a P-value of 0.05 or less was considered
to be an indication of a significant difference (Student's t-test).

RESULTS AND DISCUSSION

The aerobic control incubations (Figs. 1 - 3, open circles)
yielded relatively constant values. In anoxia (Fig. 1, solid
circles) cellular K^+ decreased within 60 min to 28% of the
aerobic control (upper panel). When the isolated tubular segments
were resuspended after different time periods of normothermic
anoxia (solid circles) in oxygenated KRB medium (Fig. 1, upper
panel, open squares), a gradual loss of the ability of tubular
cells to reaccumulate K^+ was observed. After 50 min of anoxia,
for example, only 34% of the aerobic K^+ content was reaccumulated,

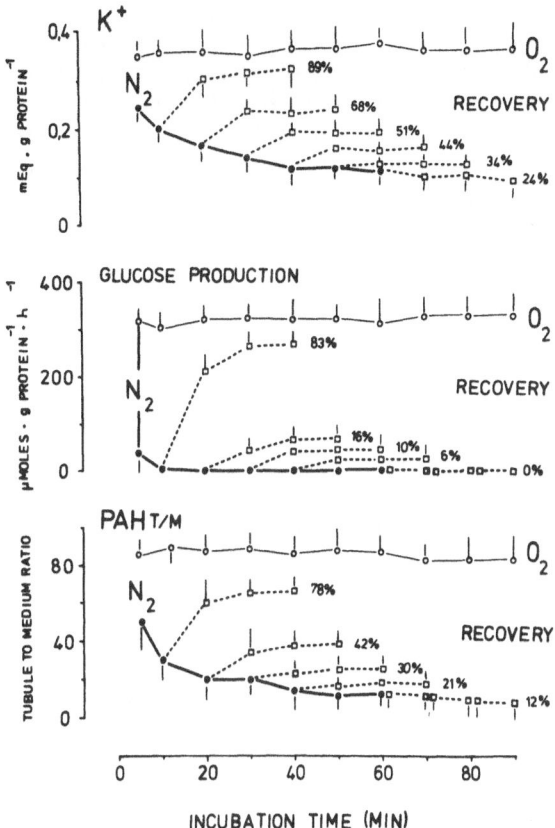

Fig. 1. Recovery of K^+ accumulation, lactate gluconeogenesis,
         and PAH transport in isolated tubular segments of rat
         kidney cortex after different periods of normothermic
         anoxia. All incubations in Krebs-Ringer-Bicarbonate.
         PAH-T/M ratio = ng PAH/mg protein per ng PAH/ml incubation
         medium. Open circles = aerobic control; solid circles =
         anoxic incubation; open squares = cellular functions after
         resuspension of the anoxic tubules in oxygenated medium.
         % recovery refers to the aerobic controls at the same
         point in time. $\bar{x} \pm$ SD (n = 8).

and no statistical significant difference in respect to the anoxic
cells (60 min:28%) could be demonstrated.

     Gluconeogenesis (GNG) fell in anoxic tubular cells within 10
min to zero (Fig. 1, middle panel, solid circles). The recovery of
GNG after different periods of anoxia (open squares), particularly
after 20 min, was more suppressed than the observed recovery of
K^+ accumulation and PAH transport (see below). After 30 min of
anoxia, and following resuspension in oxygenated KRB medium, for

example, GNG was not significantly different from zero. The PAH
tubule to medium ratio (Fig. 1, lower panel) fell in anoxic tubular
segments, in contrast to the immediate anoxic reduction of GNG,
progressively with time and reached after 60 min 14% of the aerobic
control (solid circles). The PAH ratio in tubules after 40 min of
anoxia and resuspension in oxygenated KRB medium (open squares) was
not significantly different from this value. Overall, the reversibility of all tested functions (K^+, GNG, PAH) was markedly impaired
after 20 min of anoxia (recovery = 16 - 68% of aerobic controls),
and no significant reversibility was observed after 30 (GNG), 40
(PAH) and 50 min (K^+) anoxia in the KRB medium.

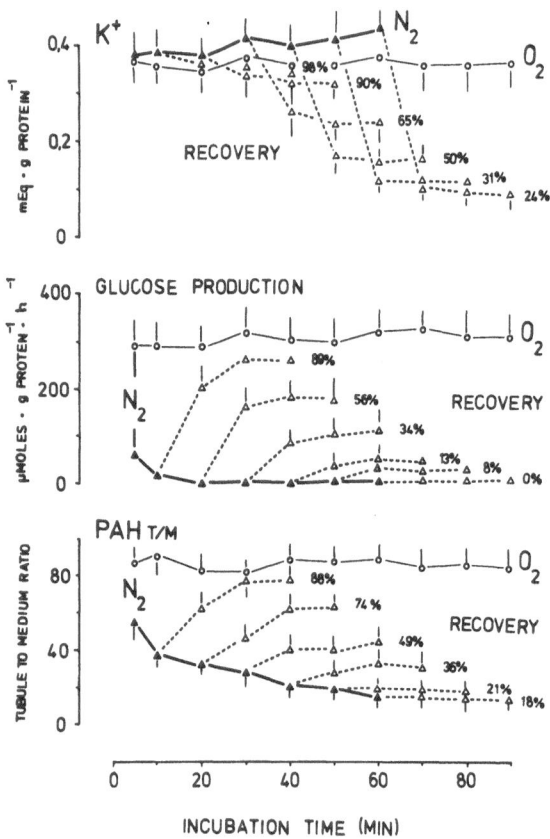

Fig. 2. Recovery of K^+ accumulation, lactate gluconeogenesis,
         and PAH transport in isolated tubular segments of rat
         kidney cortex after different periods of normothermic
         anoxia. PAH-T/M ratio = ng PAH/mg protein per ng PAH/ml
         incubation medium. Open circles = aerobic control in KRB
         medium; solid triangles = anoxic incubation in modified
         Collins' solution; open triangles = cellular functions
         after resuspension of the anoxic tubules in oxygenated KRB
         medium. % recovery refers to the aerobic controls at the
         same point in time. x ± SD (n = 8).

When the tubular segments were incubated during anoxia in an
"intracellular" (IC) solution instead of in "extracellular" (EC)
KRB solution, an improved reversibility of the tested cellular
functions was observed (Fig. 2): due to the high K^+ concentration
in the IC medium (115 mM), the cellular K^+ content of the isolated
tubular segments remained high and, probably due to some exchange of
intracellular Na^+ with extracellular K^+, became slightly elevated
during anoxia (solid triangles, upper panel). After 30 min of
anoxia (in the IC medium) and subsequent resuspension in EC medium,
however, cellular K^+ fell significantly by about 35% below aerobic
control values (open triangles). After 60 min of anoxia and subsequent resuspension, cellular K^+ decreased to a level of 24%
which was identical to the value observed after anoxia in the EC
medium (Fig. 1). Gluconeogenesis (GNG) in the tubular cells fell
within 10 min of anoxia in the IC medium to zero (Fig. 2, middle
panel, closed triangles). Recovery of GNG (open triangles) was
significantly different from zero after 30 min of anoxia. The
PAH-T/M ratio fell markedly in the IC medium during anoxic incubation (Fig. 2, lower panel, solid triangles). The recovery of PAH
transport (open triangles) was 74% after 20 min of anoxia and
became non-significant after 50 min anoxic incubation in the IC
medium.

Thus, in respect to the observed effects of anoxia on cellular
functions in the "extracellular" KRB medium, the incubation of
tubular segments in an "intracellular" medium slowed down, but did
not prevent the anoxic impairment of cellular functions: after 20
min of normothermic anoxia, for example, the recovery of tubules
which had been suspended in the "intracellular" medium was significantly (p < 0.01) improved (K^+ = 68% EC versus 90% IC; GNG = 16%
EC vs. 56% IC; PAH = 42% EC vs. 74% IC, Figs. 1 and 2), and a
measurable reversibility of cellular functions was observed in
tubules from the IC medium even after 40 min (GNG and PAH) of
normothermic anoxia (EC: non-significant recovery).

In order to test whether this improvement in recovery of
cellular functions after up to 40 min of anoxia in an "intracellular" medium could be attributed, at least in part, to a reduction
of anoxic cell swelling and structural damage in the tubular
segments, we compared the losses of total protein and the release
of intracellular enzymes, such as cytoplasmatic lactate dehydrogenase (LDH) (EC 1.1.1.27), mitochondrial glutamate dehydrogenase
(GlDH) (EC 1.4.1.3) and brush border γ-glutamyl transferase (γGT)
(EC 2.3.2.2), into the different incubation media (Fig. 3): the
total protein content of the isolated tubules decreased within 60
min by about 12% in oxygenated EC medium (open circles), and by
about 30% in the desoxygenated EC medium (solid circles). Incubation of tubular segments in the IC medium prevented this marked
anoxic loss of protein (solid triangles). Accordingly, the significant (p < 0.01) anoxic release of about 50% (LDH, γGT) and about

7.5% (GlDH) of initial (aerobic control) enzyme activity into the
EC medium was not observed in the IC medium. No significant differences between enzyme releases from aerobic controls and from anoxic
cells in the IC medium could be calculated. Thus, after anoxic
incubation in the IC medium (Fig. 2), the observed increases in
recovery of cellular functions may have been mediated by a reduced
anoxic swelling of cells and cell organelles.

After 60 min of normothermic anoxia in the IC medium, however,
no significant recovery was observed (Fig. 2), though the anoxic
enzyme release and protein loss was markedly suppressed at this
point in time (Fig. 3). Thus, mechanisms other than a release of
cell constituents into the medium may have reduced functional
recovery after anoxia. Most probably, after 60 min of normothermic
anoxia, an increased activity of lysosomal enzymes (Robinson et
al., 1977), accompanied by mitochondrial membrane alterations and a

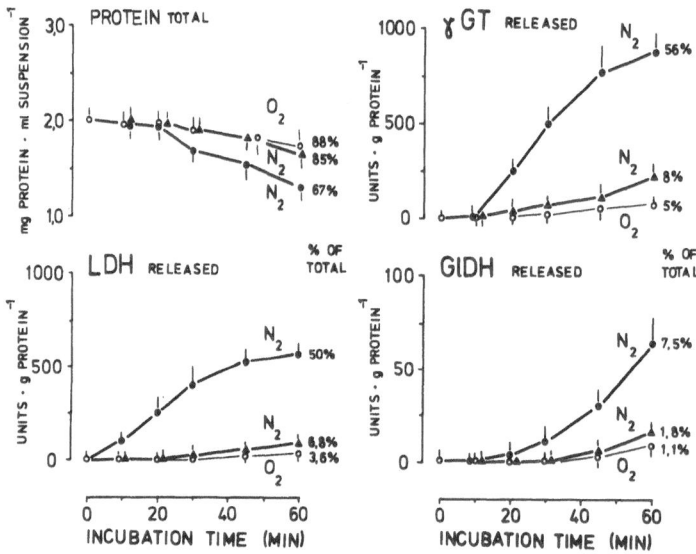

Fig. 3.   Protein loss and enzyme release into the incubation medium
           from isolated tubular segments of rat kidney cortex at
           37°C. LDH = lactate dehydrogenase (EC 1.1.1.27); γGT =
           γ-glutamyltransferase (EC 2.3.2.2.) GlDH = glutamate dehy-
           drogenase (EC 1.4.1.3.). Open circles = aerobic control in
           Krebs-Ringer-Bicarbonate (KRB) medium; solid circles =
           anoxic incubation in KRB medium; solid triangles = anoxic
           incubation in modified Collins' solution. % release refers
           to enzyme activities in the incubation medium in relation
           to the initial activity of LDH (= 1108 units·g protein^{-1}),
           γGT (= 1581 units·g protein^{-1}), and of GlDH (= 733 units·g
           protein^{-1}) in the tubular segments. Note the different
           scale of the GlDH ordinate. x $\pm$ SD (n = 8).

progressive decoupling of respiratory control (Southard et al., 1977) had limited any aerobic recovery of cellular functions in the tubular segments.

REFERENCES

Bergmeyer, H.U., 1974, "Methoden der enzymatischen Analyse", Vol. I and II, Chemie, Weinheim.

Burg, M.B., and Orloff, J., 1962, Oxygen consumption and active transport in separated renal tubules, Am. J. Physiol., 203: 327.

Collins, G.M., Bravo-Shugarman, M., and Terasalsi, P.I., 1969, Kidney preservation for transplantation, Lancet, II:1219.

Gronow, G., Randzio, R., and Weiss, Ch., 1976, Different renal oxidation rates of intramolecular carbon atoms in lactate and pyruvate, in: "Current Problems in Clinical Biochemistry", Vol. 6, U. Schmidt and V.C. Dubach, eds., Hans Huber, Bern, p. 40

Lowry, O.H., Rosenbrough, N.J., Farr, A.L., and Randall, R.J., 1951, Protein measurement with the Folin phenol reagent, J. Biol. Chem., 193:265.

Robinson, J.W.L., Mirkovich, V., and Gomba, S.Z., 1977, Alterations in the dog renal tubular epithelium during normothermic ischemia, Kidney Int., 11:86.

Southard, J.H., Senzig, K.A., Hoffmann, R.M., and Belzer, F.O., 1977, Energy metabolism in kidneys stored by a simple hypothermia, Transpl. Proc., 10:1535.

ERYTHROPOIETIN AND INTRARENAL OXYGENATION IN HYPERCAPNIC VERSUS NORMOCAPNIC HYPOXEMIA[*]

R. Baker, J.R. Zucali[1], B.J. Baker, and J. Strauss

University of Miami, School of Medicine, Miami
FL 33101, U.S.A.
[1]Roswell Park Memorial Institute, New York State
Department of Health, Buffalo, N.Y. 14263, U.S.A.

INTRODUCTION

That the kidney plays the major role in production or activation of erythropoietin (Ep) in adult mammals has been well established through a variety of methods (Krantz and Jacobson, 1970). Although a decrease in intrarenal tissue O_2 availability (Ao_2) has been suggested as the stimulus for Ep production, no measurements of this parameter in Ep studies have been done to date.

Recently the conclusion that hypercapnia may significantly attenuate or even suppress Ep production during hypoxemia has been reported (Miller, 1975; Wolf-Priessnitz et al., 1978). In these studies, however, animals were placed in either hypercapnic or non-hypercapnic hypoxic chambers, and their respiration was not controlled. Since CO_2 stimulates systemic chemoreceptors which, in turn, potentiate hyperpnea, animals placed in hypoxic chambers with CO_2 displayed both higher P_aO_2 and P_aCO_2 levels than those placed in hypoxic chambers without CO_2 (Miller, 1975; Wolf-Priessnitz et al., 1978). Therefore, it is impossible to determine the relative contributions of these two variables (P_aO_2 and P_aCO_2) in attenuating the Ep response.

In the present study, we attempted to control blood gases and compare the effects of normocapnic versus hypercapnic iso-hypoxemia

---

[*] This study was supported by National Institutes of Health Grants SB16 RR 053617 BRSG and AM 18650.

on intrarenal Ao$_2$ levels and net Ep production. An hypoxemic insult
of sufficient strength and duration to acutely produce high levels
of Ep was used in two groups of rabbits. In the first, only hypoxemia was induced; in the second, both hypoxemia and hypercapnia
were induced. Intrarenal Ao$_2$ fluctuation patterns and five-hour
plasma Ep levels from both groups were analyzed and compared.

METHODS

     Twelve experiments were performed in two groups of six male
New Zealand rabbits weighing 3.27 + 0.10 (SEM) kg. In five animals
of each group, the left kidney had been chronically implanted with
a renal arterial occluder (Beran et al., 1968) and O$_2$-sensitive
electrodes (Beran et al., 1968; Strauss et al., 1968; Strauss et
al., 1974) five to ten days prior to the experiment to allow time
for surgical recovery and electrode stabilization.

     Active electrodes consisted of wire made of 80% Pt + 20% Ir and
insulated with Epoxylite[R] except for the exposed sensing tip and
cut to appropriate lengths to measure Ao$_2$ levels simultaneously at
different zones in the kidney: cortex (C, outer medulla (OM), inner
medulla (IM), and papilla (P). All reference electrodes were unin-
sulated Ag-AgCl wires. Electrode pairs were calibrated and selected
as previously described (Strauss et al., 1968; 1974). Oxygen availability was defined as percentage of the O$_2$ current under control
conditions for each region of the kidney. Oxygen-dependent current
was determined by subtracting the current after complete renal
arterial occlusion (RAO) from the total current before RAO since
the former is assumed to be current independent of O$_2$.

     On the day of the experiment, the rabbit was anesthetized with
sodium pentobarbital (25 mg/kg iv) through a peripheral ear vein.
The animal was then placed on a heating pad, and an endotracheal
tube was introduced through a tracheostomy. After bilateral pulmonary ventilation was ascertained by stethoscopic auscultation, the
endotracheal tube was connected to a Harvard respirator (model 607)
which, in turn, was connected to an anesthesia machine, and the
animal was paralyzed with curare. Initially, respirator stroke
volume was set at 25 ml/min and stroke rate at 40/min; the latter
was then adjusted to yield and maintain a proximal airway pressure
of 10-15 cm H$_2$O in order to prevent overinflation of the lungs.
The initial breathing mixture was 75% N$_2$O + 25 O$_2$. The implanted
electrodes were connected to the polarographs and a thermistor
probe, connected to a Yellow Springs Telethermometer (model 42SC),
was placed in the deep subcutaneous tissue for continuous monitoring of body temperature.

     Femoral arterial catheterization was performed, and 10 ml of
heparinized blood were immediately removed for control Ep determi-

nation. After harvesting and freezing the plasma, the packed red cells were reinfused to help maintain control hematocrit (Hct) and hemoglobin (Hb) levels. The catheter was connected to a Statham pressure transducer (model P23AA) which, in turn, was connected to a strain-gauge preamplifier (model 9803) of a Beckman Dynograph (Type R). The catheter was used for continuously monitoring mean arterial pressure (MAP) as well as for sampling. During the experiment, P_aO_2, P_aCO_2, and pHa were measured in a Radiometer Blood Micro System (model BMS-3b) and corrected for body temperature. Hematocrit was measured from blood collected in heparinized microtubes and centrifuged for 10 min at 11.500 rpm, and Hb was measured in an American Optical hemoglobinometer.

After preparation was complete, the respirator and gas mixtures were adjusted to obtain a control temperature-corrected P_aO_2 between 95 and 105 mm Hg and a P_aCO_2 between 25 and 30 mm Hg. The pHa was allowed to fluctuate. These levels were chosen since, in our experience, they represent "normal" levels in unanesthetized rabbits (Strauss et al., 1968). The animal was allowed to stabilize for 30 min within these P_aO_2 and P_aCO_2 ranges. The control Ao_2 levels (taken as 100% values for each area) were determined by performing at least two RAOs no less than 30 minutes apart and averaging the O_2-dependent currents separately for each of the four zones sampled.

In the normocapnic hypoxemic (NH) groups, FIO_2 was then adjusted to achieve a temperature-corrected P_aO_2 between 30 and 35 mm Hg; P_aCO_2 and pHa were allowed to fluctuate. This P_aO_2 was maintained for five hours. Blood samples (0.5 ml) for arterial blood gases, Hct, Hb, and pH determinations were taken every 30 minutes; MAP was noted and RAOs (for intrarenal Ao_2 levels) were done hourly. After five hours of hypoxemia, 10 ml of blood were taken for plasma Ep determination.

In the hypercapnic hypoxemic (HH) group, adjustments in FIO_2 were made as described above. However, in this group, $FICO_2$ was also adjusted to achieve and maintain temperature-corrected P_aCO_2 between 65 and 70 mm Hg for five hours; the pHa was allowed to fluctuate.

Base excess was determined from the Siggaard-Andersen Alignment Nomogram and SaO_2 by using Radiometer Blood Gas Calculator (type BGCI) with the extended cursor for the rabbit. Total arterial O_2 content (CaO_2) was calculated as follows:

$$CaO_2 = SaO_2/100 \ X \ O_2 \ cap + P_aO_2 \ (0.003),$$

where the first term is chemically bound O_2, the second term is physically dissolved O_2, O_2 cap Hb = (in g/dl) X 1.34, and 0.003 is a constant based on the Bunsen solubility coefficient for O_2 in plasma at 37^OC under an atmospheric pressure of 760 mm Hg.

In all 12 experimental animals, Ep levels were determined by
one of us (JRZ) who used the exhypoxic polycythemic mouse assay.
Six-week-old virgin female Swiss mice were subjected to two weeks
of intermittent hypoxia. On days 4 and 5 after the last exposure
to hypoxia, the mice were injected with either 1 ml of saline,
0.05, 0.20, 0.80 units/ml Ep standards (Connaught Lab.) or the
rabbit plasma sample to be tested. The plasma sample was diluted
(1:3) with normal saline so that high levels of Ep could be determined if necessary. On day 6 after the last exposure to hypoxia,
59Fe citrate (0.5 uCi/0.2 ml) was injected i.v. to the assay
animal; 48 hours later, the animal was exsanguinated from the
dorsal aorta and 59Fe uptake was counted from the harvested blood.
Conversion of 59Fe uptake to ESF (in units/ml) was done using the
standard curve from the same group of assay animals. Mean values +
SEM were calculated using four to six assay animals per sample or
standard. Any mouse with a Hct less than 58% was eliminated (Zucali
and Mirand, 1975).

The fact that 59Fe incorporation was caused by Ep and not by
other humoral factors that may have been triggered during the
experiment, was confirmed as follows. Post-hypoxic plasma samples
from two NH rabbits (one implanted and one non-implanted) were
pooled. An attempt was then made to neutralize the Ep activity in
two ways: inhibition by neuroaminidase and neutralization by
anti-Ep antiserum. Plasma samples treated in one of these two ways
were then assayed as described above. Lack of 59Fe incorporation in
the assay animals given the neutralized plasma confirmed that 59Fe
incorporation (in the non-neutralized plasma formerly tested) was
principally related to Ep (Zucali and Mirand, 1975).

RESULTS

The effect of normocapnic and hypercapnic hypoxemia on some
of the systemic factors affecting tissue oxygenation is presented
in Tables 1 and 2. Data in Tables 1 and 2 and Figs. 1 - 3 are only
those from implanted animals. In the NH group, body temperature
fell from a control mean of $37.1^{\circ}C$ to $36.8^{\circ}C$ at the end of the
five-hour exposure period, Hct from 38.7% to 36.6%, and Hb from
11.0 to 10.5 g/dl (representing 5% from control levels) during the
entire experiment. In the HH group, body temperature fell from
$37.7^{\circ}C$ to $36.9^{\circ}C$, Hct from 38.4% to 37.2% and Hb from 11.0 to
10.6 g/dl (representing 3% and 4% falls, respectively). None of
these changes were statistically significant. In both groups there
was a gradual decrease in MAP: from 100 mm Hg during control period
to 86 mm Hg at the end of the five-hour exposure period in the NH
group and from 94 to 86 mm Hg in the HH group.

The attempt to maintain equivalent P_aO_2 levels in both groups
throughout the five-hour exposure period was achieved: at no point

Table 1. Systemic factors affecting tissue oxygenation: normocapnic
         hypoxemic group (n 5)

Sample Time	Body Temp. (°C)	MAP (mm Hg)	Hct (%)	Hb (g/dl)	C_aO_2 (ml/dl)
Control	37.1	100	38.7	11.0	14.7
+ SEM	0.3	5	1.3	0.4	0.5
Hours of Exposure		NORMOCAPNIC HYPOXEMIA INDUCED			
1 hour	37.1	90	37.7	10.8	9.3
+ SEM	0.4	6	1.1	0.3	0.7
2 hours	37.2	89	36.9	10.6	8.9
+ SEM	0.3	5	3.6	0.5	0.9
3 hours	36.9	90	36.9	10.6	9.2
+ SEM	0.2	5	1.9	0.5	1.0
4 hours	36.9	88	36.6	10.5	8.4
+ SEM	0.1	6	1.5	0.4	0.5
5 hours	36.8	86	36.6	10.5	8.0
+ SEM	0.1	7	1.5	0.4	0.8

during the exposure period were the P_aO_2 values significantly different between the two groups (Fig. 1, upper panel). The difference
between S_aO_2 levels of the two groups during the five-hour exposure
on the other hand, was marked (Fig. 1, lower panel). The S_aO_2 of
the NH group fell from a control of 97.4% to 63.5% during the first
hour of exposure and then continued to fall gradually, reaching 56.
4% at the end of the experiment. The S_aO_2 of the HH group fell
from a control of 97.4% to 38.1% during the first hour of exposure
and then rose to 48.9% at the end of the experiment.

The decreased O_2-hemoglobin binding ability of the HH group
must be related to respiratory acidosis partially compensated for
during the five hours of exposure (Fig. 2). The gradual decrease in
S_aO_2 in the NH group may be partially related to a gradual pH
decrease in spite of a relatively stable P_aco_2 (Fig. 2). The lower
S_aO_2 in the HH group than in the NH group (at the same P_aO_2) indicates that the former group had a more right-shifted O_2-dissocia-
tion curve (higher P_{50}) than the latter.

Table 2. Systemic factors affecting tissue oxygenation: hypercapnic
         hypoxemic group (n 5)

Sample Time	Body Temp. ($^{\circ}$C)	MAP (mm Hg)	Hct (%)	Hb (g/dl)	C_aO_2 (ml/dl)
Control	37.7	94	38.4	11.0	14.6
+ SEM	0.4	3	0.9	2.2	0.3
Hours of Exposure		HYPERCAPNIC HYPOXEMIA INDUCED			
1 hour	37.3	88	38.0	10.9	5.6
+ SEM	0.5	8	0.8	0.2	0.6
2 hours	37.0	85	37.8	10.7	6.3
+ SEM	0.3	6	1.0	0.3	0.3
3 hours	37.1	88	37.4	10.7	5.7
+ SEM	0.2	8	1.0	0.3	0.4
4 hours	36.9	87	37.2	10.6	6.0
+ SEM	0.3	10	1.2	0.3	0.6
5 hours	36.9	86	37.2	10.6	7.1
+ SEM	0.2	9	1.2	0.3	0.4

In the NH group, Ao_2 levels in all renal zones fell to, and
remained at, lower levels than in the HH group (Fig. 3). In addition, the Ao_2 levels in the NH group appeared to more closely
parallel P_aO_2 levels than did those of the HH group (Fig. 1). In
the NH group, Ao_2 changes were as follows: C fell to 34% after
one hour of exposure and then fluctuated, ending at 30% after five
hours of exposure; OM fell to 30% ending at 36%; IM fell to 40%
ending at 33%; and P fell to 28% ending at 36%. In the HH group, Ao_2
changes were: C fell to 54% and then gradually rose to 64% after
five hours of exposure; OM fell to 49% then rose to 102%; IM fell to
50% then rose to 99%; and P fell to 54% then rose to 71%. The HH
group consistenly showed higher Ao_2 levels in all regions of the
kidney and at all exposure times than the NH group. However,
statistically significant differences between the two groups could
only be demonstrated in OM at 4 (P<.025) and 5 (P<.010) hours of
exposure and in IM at 3 (P<.025), 4 (P<.050), and 5 (P<.025)
hours of exposure (Fig. 3).

Both groups demonstrated a low plasma Ep concentration during
control period: 0.11 ± 0.02 units/ml in the NH group and 0.09 ±

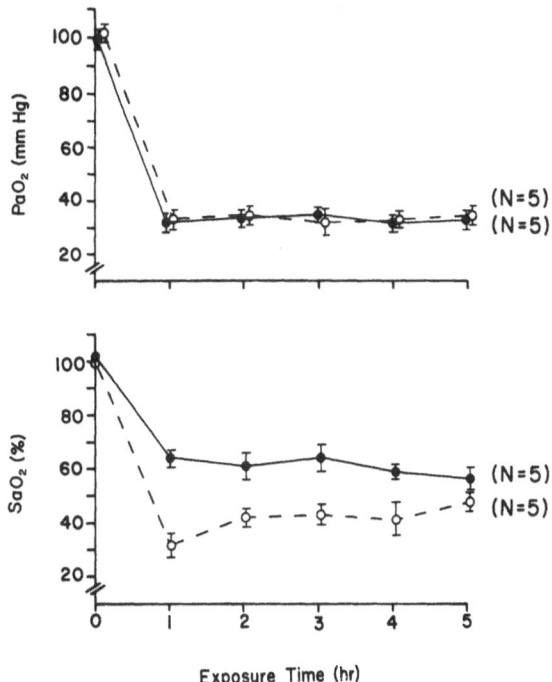

Fig. 1. <u>Upper panel</u>, P_aO_2 levels during normocapnic (close circles) and hypercapnic (open circles) hypoxemia. Note levels are almost identical in both groups. <u>Lower panel</u>, S_aO_2 levels during same exposure period. Values are mean ± SEM.

0.01 units/ml in the HH group; the difference between these values was not significant. Increased Ep production was induced in both groups by hypoxemia: in the NH group, plasma Ep rose to 0.93 ± 0.14 units/ml (an 8.5-fold increase) while that in the HH group rose to only 0.52 ± 0.08 units/ml (a 5.8-fold increase); the difference between these values was significant (P < .01) (Fig. 4).

DISCUSSION

It is now well established that the kidney plays a major role in production or activation of Ep in normal adult mammals. However, extrarenal sources of Ep also exist in most species (Krantz and Jacobson, 1970) and the liver seems to be the principal site during fetal life (Zucali and Mirand, 1975).

The mechanisms by which the kidney increases plasma Ep concentration remain controversial. Kuratowska and co-workers (1964) proposed that the kidney produces a labile inactive precursor of

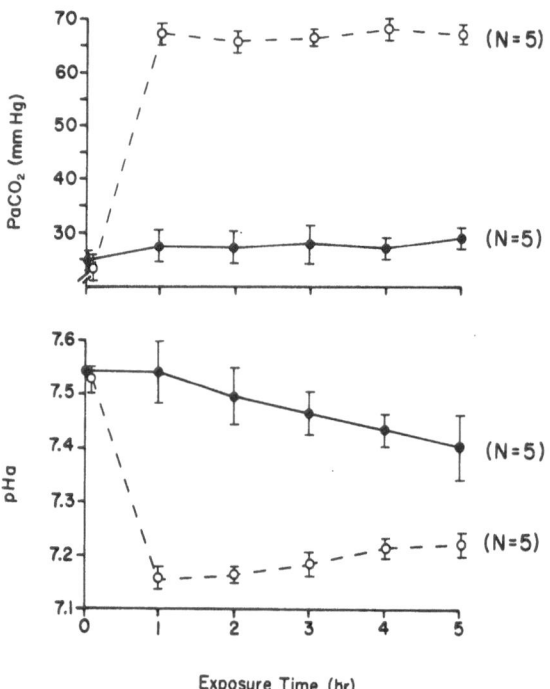

Fig. 2. Upper panel, P_aO_2 levels during normocapnic (closed circles) and hypercapnic (open circles) hypoxemia. Lower
panel, pH_a levels during same exposure period. Values are
mean ± SEM.

Ep which becomes stable and active when exposed to an alpha globulin in normal plasma. Contrera's group (1966) suggested that the
kidney may produce an enzyme which activates an extrarenally produced Ep precursor. Since that time, Erslev (1974) has demonstrated
that true Ep can be synthesized in vitro totally by the kidney
without a serum substrate or precursor.

The triggering event to induce increased Ep production seems
to involve a plethora of stimuli which may either decrease renal
tissue O_2 supply or increase O_2 demand (Krantz and Jacobson, 1970).
This O_2 supply/demand ratio may be reflected by intrarenal Ao_2 as
measured in the experiments reported here.

Our technique measures Ao_2 simultaneously from four different
zones of the kidney which are heterogeneous in their response to
various physiologic or pathophysiologic states. It may therefore be
helpful in establishing if normocapnic hypoxemia and hypercapnic
hypoxemia induce similar or different intrarenal Ao_2 response
patterns. It should be emphasized that the Ao_2 levels reported here
represent only changes in tissue O_2 levels within a given zone.

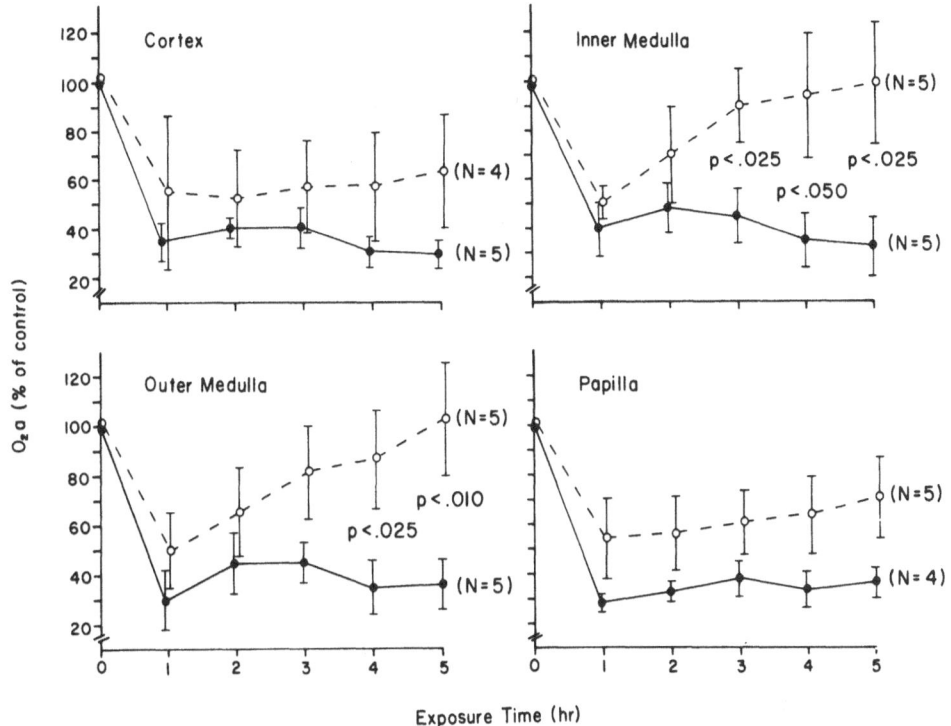

Fig. 3. Comparison of Ao$_2$ levels during normocapnic (closed circles) and hypercapnic (open circles) hypoxemia in four zones of the kidney. Values are mean $\pm$ SEM.

It is also well established that the intrarenal tissue O$_2$ levels vary from zone to zone under normal conditions (Leichtweiss et al., 1969; Baumgärtl et al., 1972).

Two caveats regarding correlations between intrarenal tissue O$_2$ and Ep production should be stressed here. First, even though intrarenal sites governing O$_2$ sensing for Ep production should theoretically demonstrate an inverse correlation between Ao$_2$ and plasma Ep titers, O$_2$ sensing and production sites may or may not be the same since Ep production may be mediated by either locally or remotely produced substances such as prostaglandins (Keighley and Cohen, 1978) or cyclic AMP (Rodgers et al., 1975; Keighly and Cohen, 1978). Second, it is now established that there are two phases involved in Ep production: a programming phase and a synthetic phase (Schooley and Mahlmann, 1972). In the present experiments, the portion of the exposure time related to the programming phase for the postexposure Ep titers is unknown. Therefore, we cannot speculate as to which intrarenal zone(s) is responsible for Ep production.

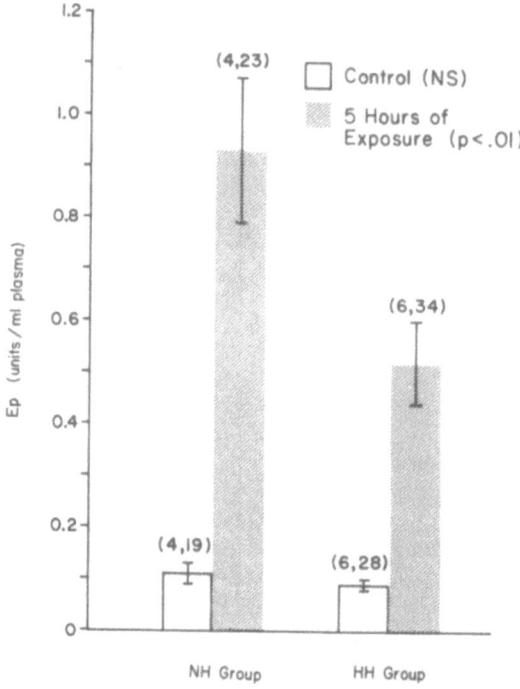

Fig. 4: Comparison of Ep titers before and after hypoxemic exposure
in the two groups studied. Values are mean ± SEM. Numbers
in parentheses above bars indicate the number of experimental animals (rabbits) and assay animals (mice) used for
each mean.

However, it may be stated that the intrarenal O_2 supply/demand
ratio was higher in the HH group than in the NH group since all
zones in the HH group displayed a trend toward higher Ao_2 levels
than those of the NH group throughout the exposure period with significant differences in OM and IM toward the end of the experiment
(Fig. 3). The higher intrarenal Ao_2 in the HH group may have been
responsible for the relatively attenuated Ep production when compared to that of the NH group. The reason Ep production was higher
in the HH group reported here than it was in other studies (Wolf-
Priessnitz et al., 1978) is probably related to the fact that we
kept P_aO_2 at the same level in both groups while others did not.

It has been shown that hypoxic-induced Ep production can be
decreased by hypercapnia or acidosis, or increased by hypocapnia
or alkalosis (Miller et al., 1973; Miller, 1975). One of the
reasons generally given for these observations is that during
respiratory acidosis, the O_2-dissociation curve shifts to the right
thus increasing the P_{50} which facilitates O_2 "unloading" in the
tissues; respiratory alkalosis does the opposite. In' a set of

experiments similar to those reported here, it was demonstrated that in rabbits breathing 8.8% O_2 + 10% CO_2, the P_{50} did increase. However, measurement of intraeryhrocytic organic phosphates revealed that ATP and 2,3-DPG fell; only ADP rose slightly (Strauss et al., 1974). It has been shown that ADP has a much lower hemoglobin binding capacity than 2,3-DPG (Perutz, 1970). Therefore, the P_{50} increase must have been due to decreased pH or increased Pco_2 rather than increased intraerythrocytic organic phosphate levels.

The right-shifted O_2-dissociation curve may be only one of many factors that influence intrarenal Ao_2 levels, and its role in Ep production should not be overemphasized. Miller (1975) has shown that after acute blood loss in cyanate-treated rats, the P_{50} decreased as much as 15 mm Hg, but Ep increases were insignificant if pH_a and P_aco_2 were maintained within normal limits. In addition, the present studies show a greater pH_a (and S_ao_2) disparity between the two groups at the beginning of exposure, indicating a higher P_{50} at that time. The intrarenal Ao_2 differences show the opposite trend.

Carbon dioxide has been shown to increase renal blood flow (Daughtery et al., 1967; Larrieu et al., 1978) and influence tissue metabolism (Craig and Beecher, 1943). Daugherty's group (1967) demonstrated a fall in renal vascular resistance with an elevated Pco_2. Larrieu and co-workers (1978) also showed that hypercapnia increased renal blood flow without changing intrarenal blood flow distribution. In addition, Wolf-Priessnitz and co-workers (1978) showed that whole-blood lactate excess was lower in rabbits breathing 8.8% O_2 + 10% CO_2 than in those breathing 8.8% O_2 alone. This indicated a lower degree of total body tissue hypoxia in hypercapnic than in nonhypercapnic animals. Measurements to determine renal tissue O_2 were not done. However, our data confirm that the kidney has a higher Ao_2 in HH rabbits than in NH rabbits.

In conclusion, plasma Ep titers are significantly lower in rabbits exposed to five hours of hypercapnic hypoxemia than in those exposed to the same duration of normocapnic iso-hypoxemia. This may be related to higher intrarenal Ao_2 levels in the HH group than in the NH group which supports the hypothesis that decreased intrarenal tissue Po_2 is the stimulus for Ep production. The higher intrarenal Ao_2 levels in the HH group reflect a higher O_2 supply/demand ratio. However, further work is needed to determine how specific factors of intrarenal tissue oxygenation are altered during these two conditions.

REFERENCES

Baumgärtl, H., Leichtweiss, H.-P., Lübbers, D.W., Weiss, Ch., and
    Huland, H., 1972, The oxygen supply of the dog kidney: Mea-
    surements of intrarenal Po_2, Microvasc. Res., 4:247-257.
Beran, A.V., Strauss, J., Brown, C.T., and Katurich, N., 1968, A
    simple arterial occluder, J. Appl. Physiol., 24:838-839.
Contrera, J.F., Gordan, A.S., and Weintraub, A.H., 1966, Extraction of
    an erythropoietin-producing factor from a particulate fraction
    of rat kidney, Blood, 28:330-343.
Craig, F.N., and Beecher, H.K., 1943, The effect of carbon dioxide
    tension on tissue metabolism (retina), J. Clin. Invest., 26:
    473-482.
Daugherty, R.M., jr., Scott, J.B., Dabney, J.M., and Haddy, F.J.,
    1967, Local effects of O_2 and CO_2 on limb, renal and coronary
    vascular resistance, Am. J. Physiol., 213:1102-1110.
Erslev, A.J., 1974, In vitro production of erythropoietin by
    kidneys perfused with serum-free solution, Blood, 44:77-85.
Keighley, G., and Cohen, N.S., 1978, Stimulation of erythropoiesis
    in mice by adenosine 3'5'-monophosphate and prostaglandin E_1,
    J. Med., 9:129-138.
Krantz, S.B., and Jacobson, L.O., 1970, "Erythropoietin and Regula-
    tion of Erythropoiesis", University of Chicago Press, Chicago.
Kuratowska, Z., Lewartowski, P., and Lipinski, B., 1964, Chemical
    and biologic properties of an erythropoietic-generating sub-
    stance obtained from perfusates and isolated anoxic kidneys,
    J. Lab. Clin. Med., 64:226-237.
Larrieu, A.J., Newman, G.E., Syracuse, D.C., McClenathan, J.H.,
    Gaudiani, V.A., and Michaelis, L.L., 1978, The effects of
    arterial CO_2 tension on regioanl myocardial and renal blood
    flow: An experimental study, J. Surg. Res., 25:312-318.
Leichtweiss, H.-P., Lübbers, D.W., Weiss, Ch., Baumgärtl, H., and
    Reschke, W., 1969, The oxygen supply of the rat kidney: Mea-
    surements of intrarenal Po_2, Pflügers Arch. Ges. Physiol.,
    309:328-349.
Miller, M.E., 1975, The interaction between regulation of acid-base
    and erythropoietin production, Blood Cells, 1:449-465.
Miller, M.E., Rørth, M., Parving, H.H., Howard, P., Reddington,
    I., and Valeri, C.R., 1973, pH effect on erythropoietin
    response to hypoxia, N. Engl. J. Med., 288:706-710.
Perutz, M.F., 1970, Stereochemistry of cooperative effects in
    haemoglobin, Nature, 228:726-739.
Rodgers, G.M., Fisher, J.W., and George, W.J., 1975, The role of
    renal adenosine 3'5'-monophosphate in the control of erythro-
    poietin production, Am. J. Med., 58:31-38.
Schooley, J.C., and Mahlmann, L.J., 1972, Evidence for the de novo
    synthesis of erythropoietin in hypoxic rats, Blood, 40:662-670.
Strauss, J., Beran, A.V., and Baker, R., 1974, Fabrication of elec-
    trodes for continuous tissue O_2 monitoring, J. Appl. Physiol.,
    37:988-990.

Strauss, J., Beran, A.V., Brown, C.T., and Katurich, N., 1968,
     Renal oxygenation under "normal" conditions, Am. J. Physiol.,
     215:1482-1487.
Wolf-Priessnitz, J., Schooley, J.C., and Mahlmann, L.J., 1978,
     Inhibition of erythropoietin production in unanesthetized
     rabbits exposed to an acute hypoxic-hypercapnic environment,
     Blood, 52:153-162.
Zucali, J.R., and Mirand, E.A., 1975, Biosynthesis of erythropoie-
     tin by mouse fetal liver in culture, Blood Cells, 1:485-494.

TUMOR

EFFECTIVENESS OF RESPIRATORY HYPEROXIA, OF NORMOBARIC AND OF HYPER-BARIC OXYGEN ATMOSPHERES IN IMPROVING TUMOR OXYGENATION

W. Mueller-Klieser[1], P. Vaupel[1], and R. Manz[2]

[1]Dept. of Physiology, University of Mainz
6500 Mainz, FRG
[2]Dept. of Physiology, University of Regensburg
8400 Regensburg, FRG

INTRODUCTION

A restriction and an inhomogeneous distribution of diffusive and convective oxygen supply is considered the main cause for pronounced radioresistance in solid tumors (Thomlinson and Gray, 1955; Tannock, 1972). One way to alleviate this crucial problem in tumor therapy is the direct enhancement of oxygen delivery to the cancer cells. This can be performed either by the tumor host breathing pure O_2 (respiratory hyperoxia, RHO) or by whole body exposure to pure O_2 atmospheres at normobaric or hyperbaric pressures (normobaric oxygenation NMO, and hyperbaric oxygenation, HPO, respectively). Numerous investigations have been undertaken both in animals and in man using RHO, NMO, or HPO as an adjuvant in radiation therapy (Suit and Maeda, 1967; Rubin et al., 1969), yet only equivocal results have been obtained with regard to the benefit of such a combination of treatments (Sause and Plenk, 1979).

Although direct measurements of O_2 tensions in solid tumors during RHO, NMO, or HPO could lead to a quantification of the effectiveness of these treatments, only a few studies using relatively large O_2 electrodes have been reported to date (Cater et al., 1962; Jamieson and van den Brenk, 1963; 1965). Thus, quantitative information about changes in the low range of tissue O_2 partial pressure ($Po_2 < 5$ mm Hg) which could critically influence radiosensitivity has been provided only in a recent study (Mueller-Klieser et al., 1982). Still, no experimental data are available concerning the relative effectiveness of RHO compared to NMO or HPO in terms of tumor oxygenation, even though such data would be of high practical

relevance considering the problems arising during the clinical
application of HPO. These data may also provide evidence for the
relative significance of two mechanisms presumably involved in tumor
tissue oxygenation during exposure to O_2, i.e., (i) the increased
O_2 supply through elevated O_2 tensions in the blood entering the
tumor, and (ii) the enhanced diffusion of O_2 from the surrounding
atmosphere into superficial tumors. To evaluate these questions, the
cryophotometric micromethod according to Grunewald and Lübbers (1976)
has been employed during respiratory normoxia, RHO, NMO, or HPO in
experimental tumors of one cell line implanted at two different
sites.

MATERIALS AND METHODS

       Investigations were carried out on DS-carcinosarcomas either
grown tissue-isolated or implanted subcutaneously in the lateral
femoral region. Details of the implantation techniques used have
been described elsewhere (Vaupel et al., 1971; Mueller-Klieser et
al., 1980). The animals were anesthetized with sodium pentobarbital
(35 mg/kg body weight), put on a heating pad, and blood coagulation
was prevented by injection of Heparin (350 USP-units/kg body weight).
Through cannulating the carotid artery, the mean arterial blood
pressure (MABP) and the relevant parameters of respiratory gas exchange could be monitored throughout the experiments. The animals
were allowed either to breathe air spontaneously or to breathe pure
O_2 from a mask. In addition, animals with subcutaneous tumors were
put into a special pressure chamber where they were exposed to a
pure O_2 atmosphere for 30 minutes either at normal atmospheric
pressure (1 bar) or at 4 bar. A detailed description of the pressure chamber technique has been given recently (Mueller-Klieser and
Vaupel, 1982). The experimental protocol can be summarized by the
following scheme:

Investigations were carried out on animals with:

tissue-isolated tumors
                breathing air spontaneously   (group I)
                breathing pure O_2 from a mask, RHO   (group II)
subcutaneous tumors
                breathing air spontaneously   (group III)
                breathing pure O_2 from a mask, RHO   (group IV)
                in pure O_2 atmospheres at 1 bar, NMO (group V)
                in pure O_2 atmospheres at 4 bar, HPO (group VI)

       The cryophotometric micromethod that has been described in de-
tail previously (Vaupel et al., 1979) was employed to determine the
oxyhemoglobin saturation (HbO_2) of single red blood cells in tumor
microvessels (diameter < 12 um). This parameter has been shown to
be a quantitative measure of tissue oxygenation (Vaupel et al.,

1979; Mueller-Klieser et al., 1981), allowing inferences on the
tumor response to radiation therapy (Mueller-Klieser et al., 1982).

The results are displayed as histograms showing the relative
or cumulative frequencies of occurrence of the HbO_2 values as a
function of HbO_2 saturation. In addition, these curves are characterized by evaluating relevant statistical parameters such as mean,
median or modal class.

RESULTS

The influence of RHO on the oxygenation of tissue-isolated
(group II) and of subcutaneous tumors (group IV) in comparison to
the oxygenation under normoxic conditions (group I + III, respectively) is demonstrated in Fig. 1. It is obvious that there is little
effect of RHO on the O_2 supply to tissue-isolated tumors, whereas
the HbO_2 values in subcutaneous tumors under these conditions are
shifted to considerably higher saturations compared to respiratory
normoxia. This shift is less pronounced in larger tumors (group IVb,
dotted line) than in smaller ones (group IVa) indicated by the
broken curve in Fig. 1b. Differences in HbO_2 saturations between
small and large tumors are evident particularly in the low saturation range.

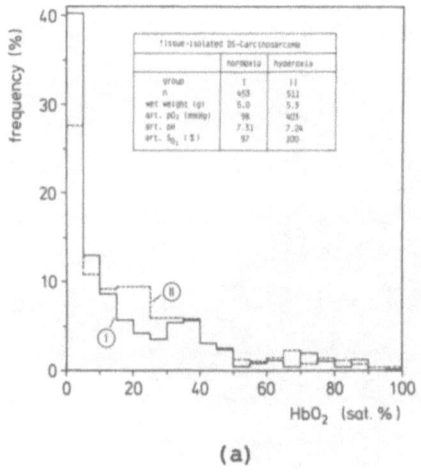

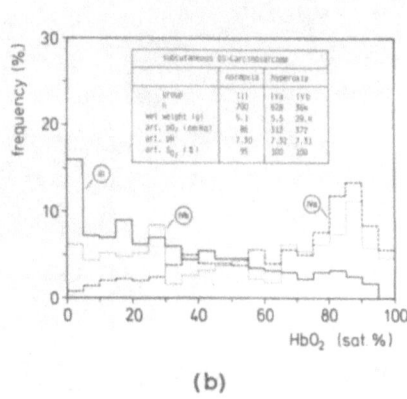

(a)   (b)

Fig. 1.   (a) Frequency distribution of HbO_2 saturations within tumor
               microvessels in tissue-isolated tumors during (I) res-
               piratory normoxia and (II) respiratory hyperoxia.
           (b) Frequency distribution of HbO_2 saturations within tumor
               microvessels in subcutaneous tumors during (III) res-
               piratory normoxia, (IVa) respiratory hyperoxia in small
               tumors, and (IVb) respiratory hyperoxia in large tumors.

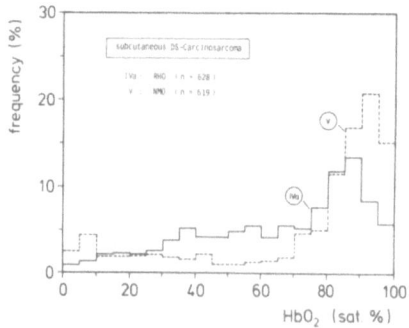

Fig. 2.   Frequency distribution of HbO_2 saturations within tumor
          microvessels in subcutaneous tumors during (IVa) respira-
          tory hyperoxia, and (V) exposure to a normobaric O_2 atmos-
          phere.

        The effect of exposure to pure O_2 atmospheres at 1 bar (NMO,
group V) in comparison to RHO (group IV) on tumor oxygenation is
demonstrated in Fig. 2. During NMO there is a considerable shift of
HbO_2 saturations to higher values in comparison to RHO. This shift
is evident particularly in the medium and higher saturation range,
whereas a considerable part of the values is still to be found in
the low saturation range with 2% of the values below 5% saturation.

        Fig. 3 shows the cumulative HbO_2 frequency distribution curves
from measurements in animals pressurized to 4 bar in a pure O_2 atmos-

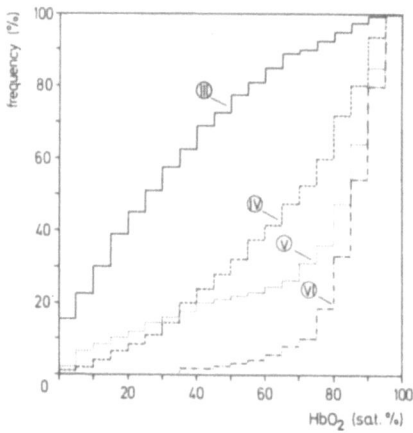

Fig. 3.   Cumulative frequency distibution of HbO_2 saturations within
          tumor microvessels in subcutaneous tumors during (III) res-
          piratory normoxia, (IV) respiratory hyperoxia, (V) exposure
          to an O_2 atmosphere at 1 bar, and (VI) exposure to an O_2
          atmosphere at 4 bar.

Table 1. Relevant statistical parameters of the HbO_2 values
measured in DS-carcinosarcoma under various experimental
conditions of O_2 supply

group	I	II	III	IV	V	VI
mean HbO_2 (sat.%)	17	22	33	65	72	75
median HbO_2 (sat.%)	9	16	22	72	85	86
modal class (sat.%)	0-5	0-5	0-5	85-90	90-95	90-95
values < 5 sat.% (%)	40	28	16	1	2	0

phere (group VI) in comparison to data obtained under normoxic conditions (group III), RHO (group IV) and NMO (group V). In contrast
to all the other experimental conditions, pressurization to 4 bar
in pure O_2 leads to a complete disappearance of values in the low
saturation range with essentially no values present below 35% saturation.

The statistical evaluation of the HbO_2 values measured is summarized in Table 1.

DISCUSSION

The results shown in Fig. 1 clearly demonstrate that the efficacy of RHO is strongly influenced by the site of tumor growth. Even
though the mean arterial O_2 partial pressure is substantially higher
in the animals bearing tissue-isolated tumors during RHO (see the
upper panels in Fig. 1), the latter group exhibits no significant
effect of RHO on tumor oxygenation contrasted by a considerable improvement of tumor oxygenation in subcutaneous tumors during RHO.

The dotted curve in Fig. 1b provides evidence for the tumor
size to be a further critical factor determining the effectiveness
of RHO. These data suggest that RHO may only be effective in a rather
well vascularized tumor rim whereas the enhanced O_2 supply does not
reach the poorly vascularized or necrotic center. Since the volume
fraction of the well supplied rim compared to the total tumor volume
is decreasing with increasing tumor wet weight (Mueller-Klieser et
al., 1980), these changes are paralleled by a decreasing influence
of RHO on the oxygenation of the entire tumor mass.

The data shown in Fig. 2 indicate that there is an improvement of O_2 supply to subcutaneous tumors during NMO compared to RHO, however, this effect is restricted to the high saturation range. Hypoxic areas of comparable extent are to be expected under both experimental conditions. Again, these findings provide evidence for the effectiveness of both RHO and NMO being restricted to the outer better vascularized rim of the tumors. In this tumor region diffusion from the surrounding O_2 atmosphere leads to a considerable enhancement of the O_2 supply to the tumor cells, during NMO compared to RHO, yet the tumor core still remains hypoxic or even anoxic, at least with the exposure time chosen.

During pressurization up to 4 bar for 30 minutes no HbO_2 values were detected in the range of 0 - 35 % saturation (see Fig. 3). It has been shown recently by theoretical considerations (Mueller-Klieser et al., 1982) that radiobiological hypoxia is eliminated under these conditions. Thus, a substantial increase in the radiosensitivity of subcutaneous tumors may be achieved by RHO or NMO. However, the complete eradication of hypoxia can only be provided by HPO at 4 bar, a treatment modality that is definitely connected with many technical and biological problems as discussed in detail elsewhere (Mueller-Klieser et al., 1982).

SUMMARY

Tumor oxygenation during respiratory hyperoxia is dependent on the tumor growth site, on the growth stage, and hence on the vascular pattern. Diffusion of O_2 from the surrounding atmosphere contributes considerably to the oxygenation of subcutaneous tumors during normobaric exposure to a pure O_2 atmosphere. However, hypoxia is still present in the tumor core under these conditions, and pressurization up to 4 bar is required to completely eradicate these hypoxic areas, thus enhancing the radiosensitivity of the tumors.

REFERENCES

Cater, D.B., Schoeninger, E.L., and Watkinson, D.A., 1962, Effect on oxygen tension of tumours breathing oxygen at high pressures, Lancet, 2:381.

Grunewald, W.A., and Lübbers, D.W., 1976, Cryophotometry as a method for analyzing the intracapillary HbO_2 saturation of organs under different O_2 supply conditions, Adv. Exp. Med. Biol., 75:55.

Jamieson, D., and van den Brenk, H.A.S., 1963, Comparison of oxygen tensions in normal tissues and Yoshida sarcoma of the rat breathing air or oxygen at 4 atmospheres, Br. J. Cancer, 17:70.

Jamieson, D., and van den Brenk, H.A.S., 1965, Oxygen tensions in human malignant desease under hyperbaric conditions, Br. J. Cancer, 19:139.

Mueller-Klieser, W., and Vaupel, P., 1982, Tumour oxygenation under
    normobaric and hyperbaric conditions. I. A method for removing
    cryobiopsies from small laboratory animals at liquid nitrogen
    temperature in a pressure chamber, Br. J. Radiol., submitted
    for publication.
Mueller-Klieser, W., Vaupel, P., and Manz, R., 1982, Tumour oxyge-
    nation under normobaric and hyperbaric conditions. II. Oxy-
    haemoglobin saturation of single red blood cells in tumour
    microvessels under hyperbaric conditions, Br. J. Radiol., sub-
    mitted for publication.
Mueller-Klieser, W., Vaupel, P., Manz, R., and Grunewald, W.A., 1980,
    Intracapillary oxyhemoglobin saturation in malignant tumours
    with central or peripheral blood supply, Eur. J. Cancer, 16:195.
Rubin, P., Poulter, C.A., and Quick, R.S., 1969, Changing perspec-
    tives in oxygen breathing and radiation therapy, Am. J. Roent-
    genol. Radium. Ther. Nucl. Med., 105:665.
Sause, W.T., and Plenk, H.P., 1979, Radiation therapy of head and
    neck tumors: A randomized study of treatment in air versus treat-
    ment in hyperbaric oxygen, Int. J. Radiat. Oncol. Biol. Phys.,
    5:1833.
Suit, H., and Maeda, M., 1967, Hyperbaric oxygen and radiobiology
    of a C3H mouse mammary carcinoma, J. Natl. Cancer Inst., 39:639.
Tannock, I.F., 1972, Oxygen diffusion and the distribution of cellu-
    lar radiosensitivity in tumors, Br. J. Radiol., 45:515.
Thomlinson, R.H., and Gray, L.H., 1955, The histological structure
    of some human lung cancers and the possible implications for
    radiotherapy, Br. J. Cancer, 9:539.
Vaupel, P., Guenther, H., Grote, J., and Aumueller, G., 1971, Atem-
    gaswechsel und Glucosestoffwechsel von Tumoren (DS-carcinom-
    sarcom) in vivo. I. Experimentelle Untersuchungen der versor-
    gungsbestimmenden Parameter, Z. Ges. Exp. Med., 156:283.
Vaupel, P., Manz, R., Mueller-Klieser, W., and Grunewald, W.A., 1979,
    Intracapillary HbO_2 saturation in malignant tumors during
    normoxia and hyperoxia, Microvasc. Res., 17:181.

IMPACT OF VARIOUS THERMAL DOSES ON THE OXYGENATION AND BLOOD FLOW

IN MALIGNANT TUMORS UPON LOCALIZED HYPERTHERMIA[*]

P. Vaupel[1], W. Müller-Klieser[1], J. Otte[1], and
R. Manz[2]

[1]Dept. of Physiology, University of Mainz
6500 Mainz, FRG
[2]Dept. of Physiology, University of Regensburg
8400 Regensburg, FRG

Hyperthermia exhibits various direct cytocidal effects (Dickson, 1977; Overgaard, 1977; Suit, 1977). During heat treatment in vivo, several indirect mechanisms enhance the direct cell-killing capacity of hyperthermia. Therefore, the effective use of hyperthermia can overcome some of the well-known problems involved in modern radiation therapy at least in some malignant tumors.

The efficacy of the indirect mechanisms of hyperthermia is largely modified by the microenvironment of the tumor cells. In addition, hyperthermia may influence the pericellular micromilieu and, thus, modulate its own therapeutic capacity.

Amongst the metabolic peculiarities of the interstitial space in tumors a severe acidosis, low oxygen partial pressures and a nutritional deprivation seem to play a paramount role during heat treatment of solid tumors (for reviews see Overgaard and Nielson, 1980; Vaupel et al., 1982a-c). In general, tumor cells living under chronically hypoxic, acidic and nutritionally deprived conditions are more sensitive to heat than are adequately supplied cells.

In solid tumors the metabolic status is largely influenced by the effectiveness of nutritive blood flow. Thus, not only the exploration of the metabolic micromilieu, but also a comprehensive knowledge of hyperthermia-induced flow changes are urgently needed.

[*]Supported by the Deutsche Forschungsgemeinschaft (Va 57/2-2)

From previous experiments there is clear evidence that changes in tumor tissue oxygenation and in tumor blood flow upon hyperthermia can be modified by variations in the tumor tissue temperature level and/or by the heat application time (Vaupel et al., 1982b, c). Based on these results the impression arose that the thermal dose may play a critical role during heat treatment. Therefore, in order to get a better insight into the susceptibility of the relevant parameters to hyperthermia, blood flow as well as oxygenation changes are described as a function of various thermal doses (TD) which are determined by the duration of heating (t) and the tissue temperature increment during heat exposure:

$$TD \ (min \cdot {}^{o}C) = t \cdot (T_H - T_N) \tag{1}$$

where, T_H is the tumor tissue temperature during hyperthermia, and T_N is the "normal" tissue temperature during control conditions (approx. $34{}^{o}C$ in s.c. tumors). This definition for TD is chosen for practical reasons because it consists of quantities that can easily be determined and because it has a practical significance through the application of hyperthermia. The exact definition of TD would be: quantity of heat input per mass.

MATERIAL AND METHODS

1. Animals and Tumors

Inbred Sprague-Dawley rats of both sexes were used. The experimental tumors were grown subcutaneously after injection of ascites tumor cells (0.3 ml) into the dorsum of a hind foot. This tumor location was chosen to separate the site of heat application from the animal's trunk. Tumors were used in experiments when they reached an average volume of approx. 2.5 ml. In the present study both implantation tumors of DS-carcinosarcoma and of Yoshida sarcoma were employed. For tumor tissue heating the tumor-bearing animals were anesthetized by i.p. injection of Na-pentobarbital (35-40 mg/kg body weight). Throughout all experiments, relevant systemic parameters of the host animals, such as mean arterial blood pressure, respiratory gas parameters in the arterial blood as well as the core temperature were monitored.

2. Heating of Tumors and Tissue Temperature Monitoring

For evaluation of the response of tumor blood flow to localized heating, tumor tissue hyperthermia was induced by ultrasound (1.7 MHz) employing a recently developed feed-back control system (Lierke et al., 1982; Müller-Klieser et al., 1982b; Vaupel et al., 1982c). Observations were performed on a total of 134 tumors. Hyperthermia levels used were $40{}^{o}C$, $42{}^{o}C$, $44{}^{o}C$, $45{}^{o}C$ and in a few cases $46{}^{o}C$. Heat application time mostly ranged between 20 and 60 minutes. At

modest hyperthermia ($40^{o}C$ or $42^{o}C$) longer application times were
required to achieve a pronounced shutdown of tumor blood flow.

In order to study the oxygenation changes in tumors upon heat
treatment, local hyperthermia was induced by application of micro-
waves (2.45 GHz) which were produced by a microwave generator and
delivered through a special applicator. The power of the generator
was adjusted as required to hold the measured tumor temperature
constant to within $0.1^{o}C$. Using this system, the tissue temperature
in DS-carcinosarcomas was kept at four levels: approx. $34^{o}C$ (control), $40^{o}C$ (for 30 and 60 minutes), $43^{o}C$ (for 30 and 60 minutes),
and $45^{o}C$ (for 60 minutes). Observations were performed on a total
of 61 tumors.

The ultrasound heating system described above was also used to
study hyperthermic effects on the tissue oxygenation in Yoshida
sarcomas. In this case, the tumor temperature was kept at the
following levels: approx. $34^{o}C$ (control), $40^{o}C$ (for 20 minutes),
$42^{o}C$ (for 20, 40 and 60 minutes), and $44^{o}C$ (for 60 minutes).
Observations were performed on a total of 60 tumors.

The temperature of the tumors was monitored with miniaturized
thermocouples which were connected to an electronically controlled
thermostat supplying a constant reference temperature. In all cases
the thermocouples were inserted into the tissue, so that the tip
was positioned in the center of the tumor. The mean tissue temperature was recorded continuously on a multi-channel recorder (for
details see Vaupel et al., 1982b).

3. Estimation of Tumor Blood Flow Changes during Localized Hyper-
thermia

By means of the 1.7 MHz ultrasound heating system an estimation of changes in tumor blood flow (TBF) during localized hyperthermia is possible when using a microthermocouple for tissue
temperature monitoring, the thermal voltage of which is utilized to
control the electrical power of the sonicator in a way that the
tumor tissue temperature is maintained constant.

As a first approximation and under the experimental conditions
chosen, a linear correlation can be assumed between the tumor perfusion rate and the power required to hold the tumor temperature
constant at a certain level. Thus, the actual tumor blood flow (in
arbitrary units) can be estimated by determining the power input
into the tissue. At the end of the experiment, the residual power
after cessation of the tissue perfusion, i.e., in the dead animal,
is obligatorily recorded. The power input under these conditions
compensates for heat losses from the tumor tissue by conduction and
radiation and is considered the baseline during evaluation of
relative tumor blood flow from the actual ultrasound power input
(Müller-Klieser et al., 1982b).

4. Oxygenation Changes of Tumors upon Hyperthermia

For characterization of the oxygenation status of solid tumors, the oxyhemoglobin saturation (HbO_2) of single red blood cells within tumor microvessels ($\emptyset < 12$ um) was measured using a cryophotometric micromethod (Grunewald and Lübbers, 1976). This method has been proved in previous investigations to be valid for quantifying the oxygenation both in rodent tumors (Vaupel et al., 1979) and in human tumors (Müller-Klieser et al., 1981). The individual steps for the HbO_2 determination upon hyperthermia are described elsewhere (Vaupel et al., 1982b). Tissue sites used for temperature monitoring were not employed for HbO_2 measurements. Thus, possible tissue damage by the thermocouples did not affect the experimental results. Tissue cryobiopsies were taken after heating as soon as tumor temperature had returned to control conditions.

RESULTS

1. Response of Tumor Blood Flow to Localized Hyperthermia

The results concerning the changes in tumor blood flow (TBF) upon hyperthermia are characterized by great interindividual differences. This variability, however, cannot be ascribed to different heating modalities or tumor models.

In about 5% of the tumors investigated no significant changes in TBF upon hyperthermia occur. This TBF pattern can be found quite often especially in those tumors which are treated with moderate hyperthermia (40-42°C) combined with short exposure times, i.e., with low thermal doses.

In approximately 60% of the experiments, TBF exhibits a "biphasic" course upon heating of the tissue. In these tumors the perfusion rate is increased during hyperthermia as long as thermal doses below 250-350 min·°C are applied (see Fig. 1). At higher doses, TBF then decreases steadily. A 50% shutdown of TBF ($TBF_{50\%}$) is achieved at 450-500 min·°C.

The increases in tumor blood flow are subjected to a great interindividual variability. In smaller tumors TBF increases are more pronounced than in larger, partially necrotic ones. Using tumors with wet weights below 2 g, TBF rises up to 100% can be observed. In contrast to this, in larger tumors TBF only increases by 5-30%.

In about 35% of the malignant tumors investigated there is a continuous drop of TBF from the very beginning of hyperthermia. This behaviour can preferentially be observed in those experiments where tissue temperature is adjusted to 44°C or higher.

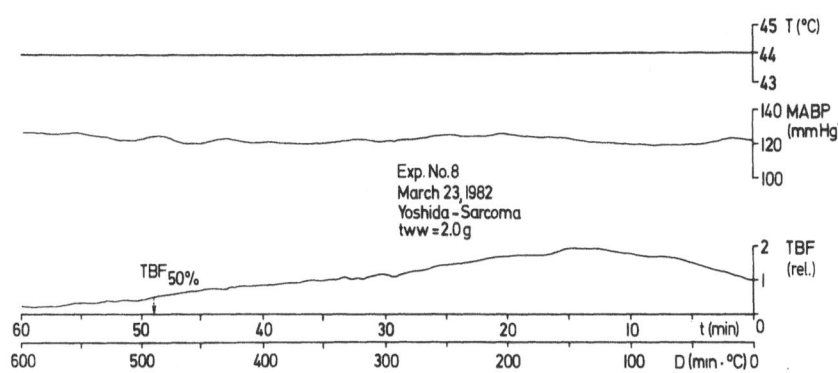

Fig. 1.   Registration of the mean tumor tissue temperature (T; upper
          trace), of the mean arterial blood pressure (MABP; central
          trace), and of tumor blood flow (TBF, in arbitrary units;
          lower trace) upon localized hyperthermia. During heating
          at 44°C for 60 min, the tumor investigated exhibits a
          biphasic change in TBF with a maximum flow at a thermal
          dose of about 150 min·°C and a 50% flow restriction (TBF$_{50\%}$)
          at approximately 480-490 min·°C. Mean tumor temperature
          before heating: 34°C, tumor wet weight (tww): 2 g.

In the latter two groups the thermal doses to achieve a 50%
restriction of TBF, i.e., to obtain a biological isoeffect, decrease
with tissue temperature level increasing from 40 to 46°C. For induction of a 50% shutdown of the circulation the following correlation
between t (the duration of heating) and $\Delta T = T_H - T_N$ is obtained:

$$t = \frac{m}{\Delta T} - b \tag{2}$$

where, m = 1264 min·°C, and b = 84 min.

In the tumor model used it appears that the exposure time to
achieve a biological isoeffect can be halved for every 2.8°C treatment temperature increment, when a linear approximation of correlation (2) in-between 40°C and 46°C is assumed.

2. Tumor Oxygenation upon Localized Hyperthermia

Because tissue oxygenation is known to be one of the critical
determinants of the heat sensitivity of tumor cells, and since there
is only a little information on the interdependence of elevated
tissue temperatures and tissue oxygen supply, one goal of the
present study was to determine changes in tumor oxygenation at
various tissue temperature levels and different heat application
times.

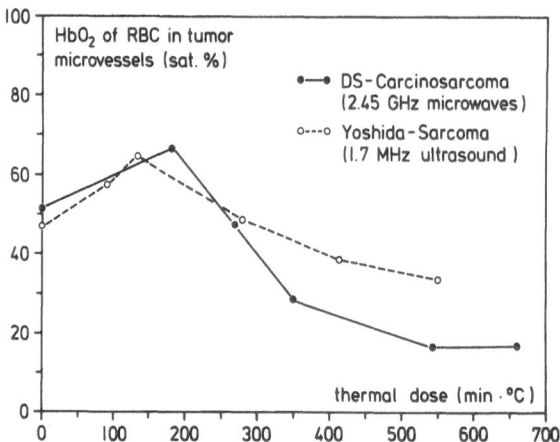

Fig. 2.   Impact of various thermal doses on the oxygenation status
          of s.c. tumors in rats. Solid line: DS-carcinosarcoma,
          broken line: Yoshida sarcoma. For characterization of the
          respective oxygenation status the oxyhemoglobin saturation
          (HbO_2) of single red blood cells within tumor microvessels
          ($\varnothing$ < 12 um) is measured.

     In Fig. 2 the mean HbO_2 saturation in the tumor is plotted as
a function of the thermal dose applied to the tissue. From these
results there is a clear indication that low thermal doses on the
average lead to an improvement in the tumor tissue oxygenation as
compared to physiological control conditions. A maximum oxygenation
is obtained at thermal doses of 100-200 min·°C both for DS-carcinosarcoma and for Yoshida sarcoma. The application of higher doses
leads to a continuous decline of the tissue oxygenation. The starting
level is reached again at thermal doses of 250-300 min·°C.

     In the DS-carcinosarcoma a 50% reduction of the tissue oxygena-
tion is elicited by application of heat doses of about 400 min·°C.
In contrast to this, the deterioration of the oxygenation in the
Yoshida sarcoma is less pronounced. In this case, only a 20% reduction is achieved after application of a comparable thermal dose.

DISCUSSION

     Presumably, several factors have to be considered as causes
for the increased tumor blood flow at lower thermal doses (Vaupel
et al., 1980). Amongst these, a vasodilation of arterial blood
vessels is probably the paramount factor. Tumor vessels arising from
neovascularization are mostly unresponsive to physiological stimuli
since a functioning media in the vessel wall is usually missing.
Therefore, vasodilation during low dose hyperthermia is probably
due to reactions of residual host vessels which still exhibit some

physiological responsiveness. The residual main vascular branches, which have been responsible for the blood supply to the host organ are likely to be situated mainly at the invasion front of the tumor cell mass. These vessels are progressively incorporated into the growing tumor. Perhaps the large variability in the response of TBF to hyperthermia, i.e., the biphasic changes versus the continuous shutdown of TBF, is at least partially induced by the existence or absence of those vessels, by their proportion in the tumor and by their remaining ability to control physiologically local blood flow.

Hyperthermia at higher thermal doses has profound effects on tumor microcirculation, i.e., on the morphology of arterioles, capillaries and venules, on the rheology of blood and on the exchange processes across the vessel walls, thus causing a drastic decrease in tumor circulation. In order to get some data on the pathohisto- logical changes of the microvasculature as a result of hyperthermia, tumors are removed immediately after heating at thermal doses necessary to achieve a 50% flow restriction. The results obtained suggest that under these conditions marked congestion (stasis) and a pronounced vasodilation occur. Tumor capillaries appear to be densely packed with red blood cells forming intravascular aggregates (sludge). There are numerous ruptures of the microvessel endothelium associated with massive hemorrhage into the tumor tissue thus lead- ing to severe tissue necrosis. Microthrombosis can occasionally be observed (for further histopathological studies concerning the effect of hyperthermia on tumor microvasculature see von Ardenne et al., 1979; Eddy, 1980; Emami et al., 1981; Otte et al., 1982).

Up to now there is no clear evidence whether the control or starting temperature before heating plays a critical role for the occurrence or absence of the TBF variations due to hyperthermic treatment. The "normal" tumor temperature may vary over a wide range depending on the tumor location, on tumor blood flow and on its metabolic state, on the heat exchange with the tumor environment as well as on special experimental conditions etc. This ambiguity of "normal" tumor temperature may largely modify the effectiveness of hyperthermic treatment. Therefore, it is necessary that the inter- relationship between "starting temperature" and response to heat is subjected to further investigations.

The changes in the tumor oxygenation during hyperthermic treat- ment seem to be predominantly mediated through changes in TBF. This is substantiated by the fact that the same directional biphasic changes in TBF and in the tissue oxygenation occurred in the majority of the tumors investigated. A similar behaviour can also be found if the mean tissue oxygen partial pressure is measured for charac- terization of the tissue oxygenation during elevated tumor tempera- tures (Müller-Klieser et al., 1982a).

The improvement of the tumor tissue oxygenation which is observed at <u>low thermal doses</u> can be explained by a change in the balance between O_2 availability to the tumor cells and O_2 consumption rate by the cells. An increased O_2 availability which is mostly due to an increased nutritive blood flow obviously prevails over the rise in the O_2 consumption following heat application to the tissue (T_H = 40-42°C). Thus, metabolic effects, i.e., changes in the capacity of the cells to consume O_2, seem to have a minor influence on the HbO_2 saturation distribution. This is substantiated by results from previous investigations which considered the impact of elevated temperatures on the O_2 consumption rate of <u>in vitro</u> tumor cells and of solid tumors in situ (Müller-Klieser et al., 1978; Vaupel et al., 1980).

The restricted oxygenation at <u>higher thermal doses</u> cannot primarily be the result of changes in the O_2 consumption of the tumor cells since a drop in the consumption of the tumor at temperatures higher than 42°C cells would induce an improvement of the tissue oxygenation. On the contrary, the oxygenation status at this hyperthermic level is the result of a restricted nutritive blood flow which predominates over the possible effect of a decrease in the O_2 consumption. This has already been discussed in detail in a preceding paper assuming that the single cell consumption is the same in solid tumors and in cell suspension (Vaupel et al., 1982b).

SUMMARY

Upon localized hyperthermia at modest thermal doses an increase in tumor blood flow can be observed in many tumors which is paralleled by an improvement of the oxygenation status of the tissue. At intermediate or high thermal doses a pronounced restriction of the tumor circulation becomes obvious leading to a deterioration of the tumor oxygenation. As a consequence, a further enhancement of the thermal response of tumors relative to normal tissues has to be expected at intermediate or high thermal doses.

REFERENCES

Ardenne von, M., Lippmann, H.G., Reitnauer, P.G., and Justus, J., 1979, Histological proof for selective stop of microcirculation in tumor tissue at pH 6.1 and 41°C, <u>Naturwissenschaften</u>, 66:59.

Dickson, J.A., 1977, The effects of hyperthermia in animal tumor systems, <u>Recent Results Cancer Res.</u>, 59:43.

Eddy, H.A., 1980, Alterations in tumor microvasculature during hyperthermia, <u>Radiology</u>, 137:515.

Emami, B., Nussbaum, G.H., Hahn, N., Piro, A.J., Dritschilo, A., and Quimby, F., 1981, Histopathological study on the effects of hyperthermia on microvasculature, <u>Int. J. Radiat. Oncol. Biol. Phys.</u>, 7:343.

Grunewald, W.A., and Lübbers, D.W., 1976, Cryophotometry as a method
    for analyzing the intracapillary HbO_2 saturation of organs
    under different O_2 supply conditions, Adv. Exp. Med. Biol.,
    75:55.
Lierke, E.G., Grossbach, R., Sudhof, H., Müller-Klieser, W., and
    Vaupel, P., 1982, A feedback control system for localized
    ultrasonic hyperthermia in tumors. I. Technical and functional
    description of equipment, Strahlentherapie, 158:386.
Müller-Klieser, W., Lierke, E.G., and Vaupel, P., 1982b, A feedback
    control system for localized ultrasonic hyperthermia in tumors.
    II. Application and first experiments in DS-carcinosarcoma,
    Strahlentherapie, 158:387.
Müller-Klieser, W., Manz, R., Otte, J., and Vaupel, P., 1982a,
    Effect of localized hyperthermia on tumor blood flow and
    oxygenation, Proc. 3rd Int. Congr. Thermol., in press.
Müller-Klieser, W., Vaupel, P., Manz, R., and Schmidseder, R., 1981,
    Intracapillary oxyhemoglobin saturation of malignant tumors in
    humans, Int. J. Radiol. Oncol. Biol. Phys., 7:1397.
Müller-Klieser, W., Zander, R., and Vaupel, P., 1978, Oxygen
    consumption of tumor cells suspended in native ascitic fluid
    at 1-42°C, Pflügers Arch., 377:R17.
Otte, J., Vaupel, P., and Gabbert, H., 1982, Changes in tumor
    microvasculature after localized hyperthermia, Proc. Ann.
    Meeting Microcirculat. Soc., Munich.
Overgaard, J., 1977, Effect of hyperthermia on malignant cells in
    vivo, Cancer, 39:2637.
Overgaard, J., and Nielson, O.S., 1980, The role of tissue environ-
    mental factors on the kinetics and morphology of tumor cells
    exposed to hyperthermia, Ann. N.Y. Acad. Sci., 335:254.
Suit, H.D., 1977, Hyperthermic effects on animal tissues, Radiology,
    123:483.
Vaupel, P., Frinak, S., Müller-Klieser, W., and Bicher, H.I., 1982a,
    Impact of localized hyperthermia on the cellular micro-environ-
    ment in solid tumors, Natl. Cancer Inst. Monogr., 60, in press.
Vaupel, P., Manz, R., Müller-Klieser, W., and Grunewald, W.A., 1979,
    Intracapillary HbO_2 saturation in malignant tumors during
    normoxia and hyperoxia, Microvasc. Res., 17:181.
Vaupel, P., Müller-Klieser, W., Otte, J., Manz, R., and Kallinowski,
    F., 1982c, Blood flow, tissue oxygenation, and pH distribution
    in malignant tumors upon localized hyperthermia. Basic patho-
    physiological aspects and the role of various thermal doses,
    Strahlentherapie, in press.
Vaupel, P., Ostheimer, K., and Müller-Klieser, W., 1980, Circula-
    tory and metabolic responses of malignant tumors during
    localized hyperthermia, J. Cancer Res. Clin. Oncol., 98:15.
Vaupel, P., Otte, J., and Manz, R., 1982b, Oxygenation of malignant
    tumors after localized microwave hyperthermia, Radiat. Environ.
    Biophys., in press.

OTHER ORGANS

MEASUREMENT OF LOCAL Po$_2$ AND INTRACAPILLARY HEMOGLOBIN OXYGENATION IN LUNG TISSUE OF RABBITS

H.J. Volkholz[1], J. Höper[2], M. Brunner[2], K.H. Frank[2], D.K. Harrison[2], R. Ellermann[2], and M. Kessler[2]

[1]Institut für Anaesthesiologie der Universität Erlangen-Nürnberg, FRG
[2]Institut für Physiologie und Kardiologie der Universität Erlangen-Nürnberg, FRG

INTRODUCTION

To date investigations of the mechanisms involved in the ventilation and perfusion of the lung have only been carried out by indirect methods. The disadvantage of such methods is that the local distribution of these parameters within the lung, particularly in the alveolar-capillary area, cannot be adequately determined.

In the present study we have used platinum multiwire surface electrodes (Kessler and Lübbers, 1966; Kessler, 1981) and micro light-guide spectrophotometry (Ji and Chance, 1979; Brunner et al., 1981; Frank et al., 1983) to measure local Po$_2$ and intracapillary Hb oxygenation at the lung surface. The investigation examined the influence of altered inspired oxygen concentration on alveolar ventilation. We also studied the influence of perfusion changes in the lung on these same local measurements.

METHODS

The measurements were carried out in thoracotomised rabbits under artificial ventilation (frequency: 35/min, tidal volume: approximately 15 ml). The multiwire oxygen electrode and the micro light-guide were placed on the surface of the lung. The visceral pleura remained intact. In order to keep the lung inflated, PEEP ventilation (+ 2-3 cm H$_2$O) was required.

To ensure proper contact between the measuring devices and the lung surface the multiwire electrode and the micro light-guide were each inserted in a Pertinax[R] (Werner und Hillebrand, Fürth, FRG) ring holder, which was attached to a cellophane membrane. The wetness of the lung provided good adhesion between the cellophane membrane and the lung surface and facilitated placement and fixation of the electrode and the micro light-guide.

Fig. 1 shows that stable absorption spectra could be recorded with micro light-guide spectrophotometry on the surface of the ventilated lung. On the left 80 typical uncorrected HbO_2 spectra measured on the lung surface under controlled ventilation ($FiO_2 = 0.3$) are seen. In the middle one hemoglobin-free spectrum obtained from the rabbit lung tissue after almost total isovolemic hemodilution is displayed. The systemic hemoglobin concentration was < 0.3 Vol % at this point. On the right of Fig. 1 the 80 HbO_2 spectra have been corrected using the hemoglobin-free tissue spectrum from the middle according to the formula (Wodick, 1971)

$$T' = \frac{J}{J'_o}$$

T' = apparent transmission
J = intensity of measured spectra
J'_o = intensity of the spectra after total washout of hemoglobin

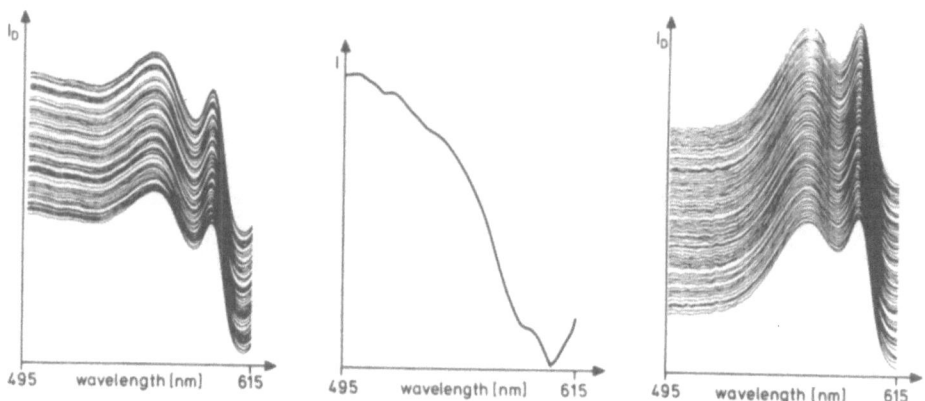

Fig. 1.   Original hemoglobin spectra measured at the lung surface
          (left), spectrum measured after total washout of hemoglobin
          (Hb concentration in the blood < 0.3 Vol.% (middle) and
          corrected spectra (right). Every spectrum is the average
          of 60 single spectra. The time between two spectra is 1 sec.
          $I_D = I_i$ (495) + (i - 1)·C; i = 1,···, n
          $I_i(\lambda)$: intensity
          i   : index of the spectrum
          n   : number of the displayed spectrum
          C   : constant

The results of 3 different procedures are presented, firstly local Po_2 values on the lung surface were measured at inspiratory oxygen concentrations of 30%, 50%, 75%, and 100%. Po_2 histograms were computed for each such inspiratory oxygen concentration.

Secondly local Po_2 was measured under stoppage of the blood flow during a cardiac arrest, while controlled ventilation was continued.

Thirdly, in a three compartment lung model, changes in local Po_2 and Hb spectra in the three lobes of the right lung were measured under different conditions. The lower lobe was used as a control, the middle lobe was blocked by ligature of the vena pulmonalis and the upper lobe was continuously insufflated with pure N_2. The micro light-guide was then placed on the lobe, insufflated with N_2. The reaction was then observed in this lobe when the inspired oxygen concentration in the remaining lobe of the right lung and the other three lobes of the left lung was increased from 30% to 100%.

RESULTS

The Po_2 Histograms

From previous experiments it was noted, that the mean values of the tissue Po_2 histograms, measured during artificial ventilation

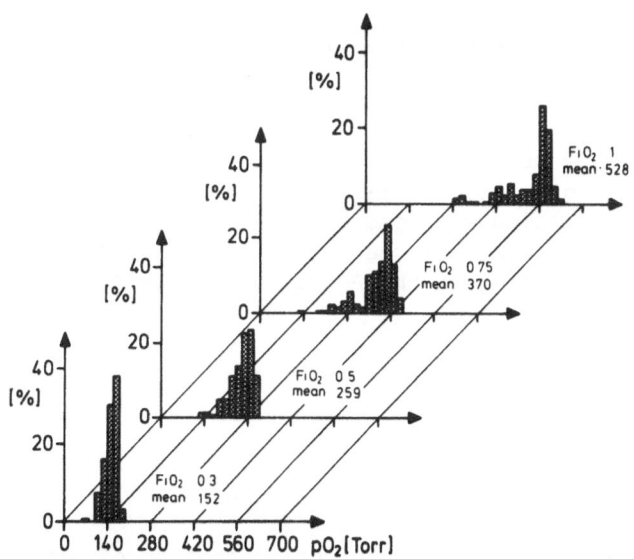

Fig. 2. Po_2 histograms from lung tissue, produced by ventilation with gas mixtures of increasing oxygen content (FiO_2 = 0.3-1.0).

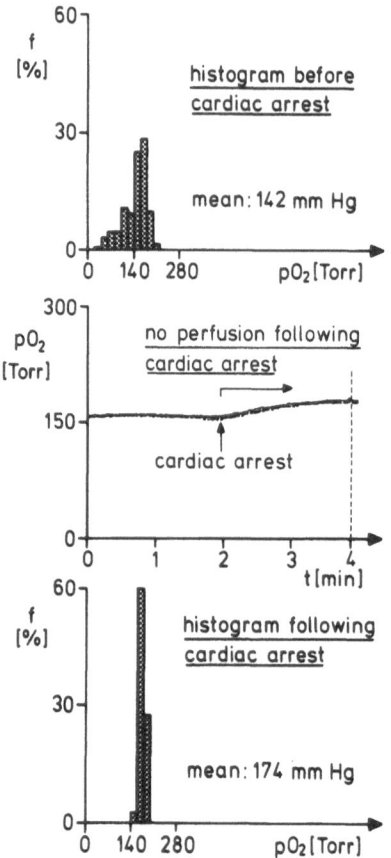

Fig. 3. Local Po_2 measurements of the lung before and during a
cardiac arrest.

($FiO_2 = 0.3$), lie about 20 mm Hg above the arterial Po_2. Fig. 2
(bottom) shows that typical Po_2 values lie in the range of 140 to
200 mm Hg. Ventilation with gas mixtures of high Po_2 content ($FiO_2 =$
1.0) increases the width of the Po_2 histograms. With the control
($FiO_2 = 0.3$) Po_2 histogram the Po_2 lies between 60 mm Hg and
340 mm Hg.

        This figure, therefore, demonstrates the effect of hyperoxic
gas mixtures on pulmonary ventilation, showing an increasing heterogeneity of the alveolar Po_2.

Po_2 Values, before and during a Cardiac Arrest

        Fig. 3 shows the effect of total cessation of blood flow on
local Po_2. Following a cardiac arrest, but under continued ventilation, we found a rise in Po_2 of about 30 mm Hg within a minute.
This could be explained by diminished transportation of oxygen away

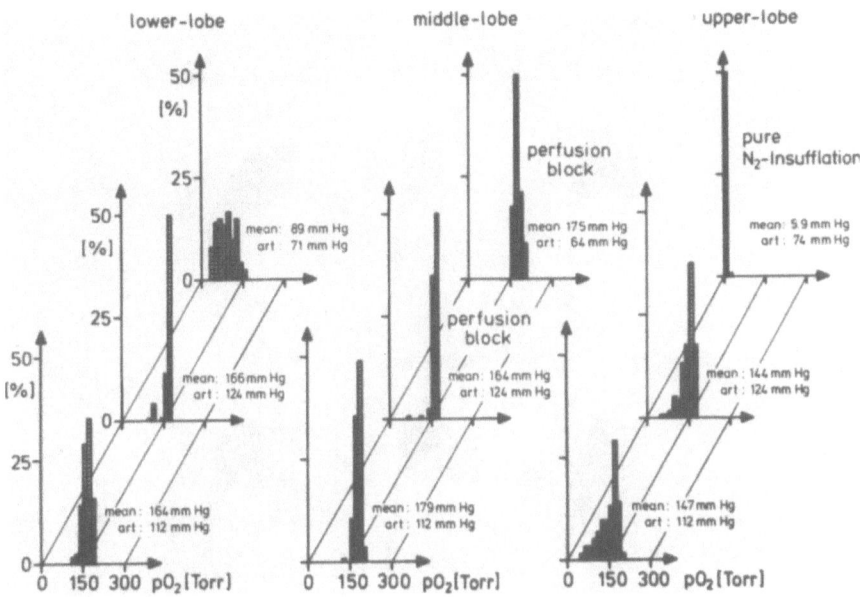

Fig. 4.   Po₂ histograms obtained from the three compartment
          model. Bottom row: control; middle row: perfusion block;
          top row: pure N₂ insufflation of the upper lobe.

from the lung following the stoppage of blood flow. The resultant
histogram still displays a heterogeneity of local Po₂ values.

Three Compartment Lung Model

          a) Po₂ measurements. In a three compartment lung model
we compared three lobes of one of the lungs. Fig. 4 shows the Po₂
histograms, obtained from the three compartment model in one experiment. The bottom row shows control histograms of the three lobes.
The different shapes of the histograms demonstrate, that even with
control ventilation, different patterns of ventilation and perfusion
exist. No obvious changes in the histograms can be observed due to
the perfusion block alone (middle row). However, when the upper lobe
of the lung was insufflated with pure N₂, a total shift to the left
of the upper lobe histogram and a redistribution of the lower lobe
histogram can be clearly seen (top row). Under control conditions
the histograms remain almost unchanged for 6-8 h.

          b) Spectrophotometric measurements. The bottom row of Fig. 5
shows the control spectra from the three lobes. As with the histograms, the different configurations of the spectra shown on the
bottom row can be explained by different levels of ventilation and
perfusion in the three lobes of the lung. During a perfusion block
of the middle lobe it can be seen from the second row of spectra

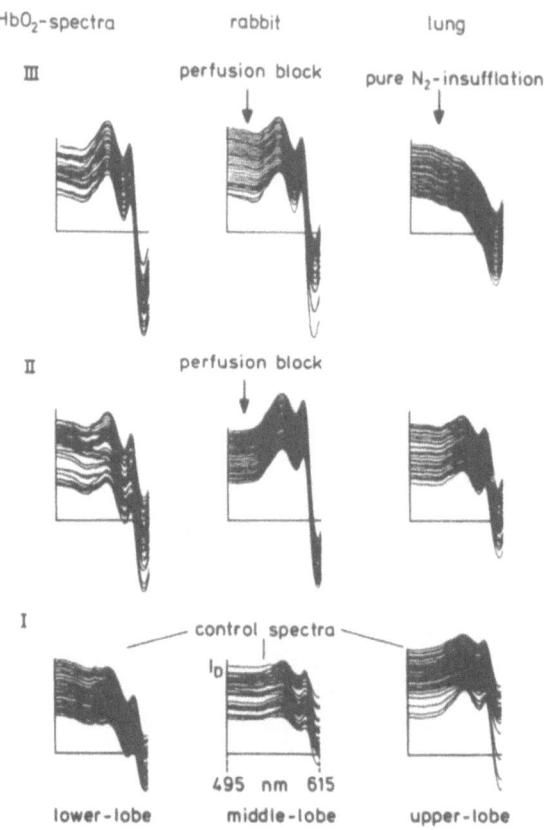

Fig. 5.   Uncorrected HbO$_2$ spectra obtained from the three compart-
          ment model. For explanation of the y-axes see Fig. 1.
          Identical axes for all spectra.

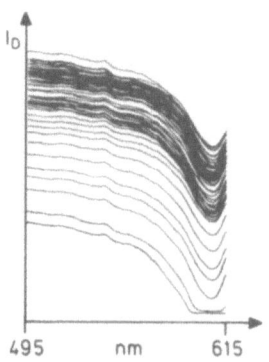

Fig. 6.   60 completely deoxygenated HbO$_2$ spectra (lung) in the
          nitrogen-inflated lobe. For explanation of the y-axes see
          Fig. 1.

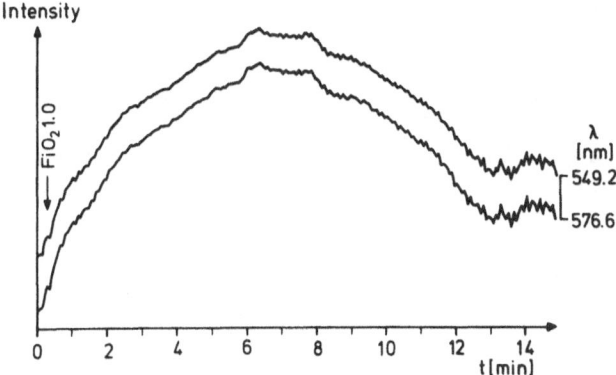

Fig. 7.   The change in intensity of the spectra measured at two
           isosbestic points plotted against time, reveal a change in
           the local Hb concentration over 12 minutes in a N_2-
           insufflated lobe of the right lung.

(II), that a considerable increase in amplitude of the reflected
signal occurs together with an increase in homogeneity of the
spectra in comparison with the control. The increased signal from
the other two lobes indicates an increased Hb content within these
lobes as a result of blocking the middle lobe. In the next stage
(III), pure nitrogen (N_2) insufflation of the upper lobe yielded
totally deoxygenated spectra (top right). No considerable changes
in the spectra from the other two lobes can be seen. As long as the
experimental conditions are unchanged, only minor variations are
obtained in different positions at the same lobe.

Changes of Perfusion in a N_2-Insufflated Lobe

        In a second investigation, mesurements were made on one lobe,
insufflated with pure N_2, in order to examine the effect of removal
of oxygen from the alveolar gas on the perfusion of the isolated
lobe. As shown in Fig. 6 there is no change in the degree of oxygenation seen in the HbO_2 spectra. However, two isosbestic points on
these spectra, plotted against time (Fig. 7) demonstrate an increase
and a subsequent compensatory decrease in local Hb concentration.
This shows that perfusion, in a single lobe being abnormally ventilated, was still under the control of the rest of the lung.

CONCLUSION

        Both, the Po_2 electrode and the micro light-guide are well
suited methods for use on the lung surface in acute animal experiments for observation of changes in ventilation and perfusion.

1. Local Tissue Po$_2$

The single platinum wire of the electrode has a catchment area
of 25-30 um and averages over a volume which is composed of 90%
alveolar air, and 10% cellular tissue and blood. The diameter of
the alveoli ranges from 100-300 um. The measured tissue Po$_2$ values
seem to be well related to alveolar Po$_2$. From this it can be deduced
that the oxygen pressures in lung tissue cells correspond largely
to the alveolar Po$_2$ values due to the very small diffusion gradients. Furthermore, it is well known that the O$_2$ partial pressure
of the blood in the capillaries assumes the alveolar values very
quickly. Consequently the Po$_2$ signal of the electrode on the rabbit
lung surface is influenced by capillary blood Po$_2$ only to an un-
important degree (2-3%) and in fact registers mainly changes in the
alveolar Po$_2$.

2. The Micro Light-guide Spectrophotometry

Local HbO$_2$ spectra within the lung display a characteristic
distorsion due to changes in the base line, which is itself due to
dispersion by the tissue. Correction can be made to the spectra
using a correction factor obtained from spectral measurements on
hemoglobin-free tissue. We obtain pure hemoglobin spectra as demonstrated in Fig. 1. As seen in Fig. 5, changes in perfusion cause
considerable increase in the intensity of the reflected signal together with an increase in homogeneity of the spectra. Fig. 7 demonstrates that changes in the local Hb concentration can be observed
from the intensity of the signal as measured at an isosbestic point.

In summary, we can conclude, that Po$_2$ measurements with the
multiwire electrode on the lung surface can be used to record not
only the degree of tissue oxygenation, but also changes in the
distribution of ventilation.

Hb measurments with the spectrophotometer on the lung surface,
apart from recording the degree of Hb oxygenation, are also particularly applicable to the lung for measuring changes in local Hb
concentration.

REFERENCES

Brunner, M., Kastner, N., Schabert, A., Höper, J., and Kessler,
    M., 1981, On-line Verarbeitung von Hämoglobin-Reflexionsspek-
    tren hoher Repetitionsraten, in: "Medizinische Informatik und
    Statistik 28", S. Koller, P.L. Reichertz, K., Überla, eds.,
    Springer, Berlin-Heidelberg-New York.
Frank, K.H., Friedl, A., Brunner, M., Höper, J., Kerl, G., Schabert,
    A., and Kessler, M., 1983, Correlation between tissue Po$_2$ and
    intracapillary Hb spectra, this volume.

Ji, S., Chance, B., Nishiki, K., Smith, T., and Rich, T., 1979,
    Micro-light guides: A new method for measuring tissue fluores-
    cence and reflectance, Am. J. Physiol. Cell Physiol., 236:
    C144-C156.
Kessler, M., 1981, Grundlegende Prinzipien der Sauerstoffversorgung
    des Gewebes, in: "Mikrozirkulation und arterielle Verschluß-
    krankheiten: Fortschritte in Diagnose und Therapie", K.
    Messmer, B. Fagrell, eds., Karger, Basel-München-Paris-London-
    New York-Sidney.
Kessler, M., and Lübbers, D.W., 1966, Aufbau und Anwendungsmöglich-
    keiten verschiedener Po_2-Elektroden, Pflügers Arch. Ges.
    Physiol., 291:R82.
Wodick, R., 1971, "Neue Auswertverfahren für Reflexionsspektren und
    Spektren inhomogener Farbstoffverteilung, dargestellt am
    Beispiel von Hämoglobinspektren", Inaugural-Dissertation,
    Marburg/Lahn.

DISCUSSION

Question: How many alveoli are covered by a light-guide?

Answer: At the time, experiments are performed to measure the
penetration depth of the light. Therefore, this question cannot be
answered.

CESSATION OF CAPILLARY BLOOD FLOW INDUCED BY LOCALIZED APPLICATION OF CARBON DIOXIDE

T. Koyama, M. Horimoto, and Y. Kikuchi

Research Institute of Applied Electricity, Hokkaido University, 060 Sapporo, Japan

INTRODUCTION

A localized pulmonary hypercapnia can be produced by placing a small plastic ring chamber on an exposed lung surface of anesthetized bullfrogs and by introducing hypercapnic gas mixtures in it (Koyama and Horimoto, 1982). An introduction of 10% CO_2 into the space formed by the ring chamber and the lung surface caused a reduction in the mean flow velocity in both arterioles and capillaries. A decrement in arteriolar diameter was observed at the same time. These results suggested a vasoconstriction of arterioles exposed to the hypercapnia. However, it remained unknown whether the reduction in capillary flow was primarily induced by a direct effect of CO_2 on the capillary bed or secondarily by changes in arterioles. The present study was designed to study this aspect by reducing greatly the test area on the lung surface to such a diameter that the only vessels coming in contact with the gas are lung capillaries.

METHODS

A pore of 1 mm in diameter was bored through a plastic plate. A small gas inlet was made to the pore through the plastic plate. The plastic plate was placed on the surface of an exposed lung. When gas was introduced into the pore, it came into direct contact with the small lung surface area encircled by the pore and exited to the air through the open upper end of the pore. Since the alveolus of the frog lung was larger than 1 mm in diameter, this method of gas application resulted in a highly localized hypercapnic capillary bed. It was possible to exclude arterioles from

643

the test area. The applied gas diffused into the room air without coming into contact with the other portion of the lung surface because of the large area of the plastic plate. When the lower surface of the plastic plate was moistened with a small amount of saline, the gas leakage through the contact between the plastic plate and the lung surface was probably minimal.

Gas analysis of the blood sampled from the ventricle during the localized hypercapnia induced by a 10 minutes application of moistened 100% CO_2 revealed no increase in Pco_2. The Pco_2 remained unaffected at 35.1 $\pm$ 7.5 mm Hg which was similar to the previously reported values (see for example Horimoto et al., 1981). The hypercapnic area was therefore probably too small to evoke measurable changes in blood gases of the whole animal. This observation confirmed the advantage of the present method for the study of the effects of a localized hypercapnia.

The blood flow in the capillary net was observed through the pore by means of a microscope with dark field episcopic illumination (Nikon Metaphot microscope) equipped with a long range working distance objective lens (Nikon BD plan 40, 0.55 LWD x 40) and recorded with video tapes by means of a video camera (Hitachi-Nikon DK 5001) (Koyama et al., 1982). The flow velocity of red blood cells was determined by a slow reproduction of the video tape recordings; the distance which a blood cell under observation traversed during one or two frames on a TV-monitor was measured.

Bullfrogs were anesthetized by immersion in a 0.08% solution of MS 222. The left lung was exteriorized following an incision of the lateral chest wall and was inflated by introducing air through an intra-tracheal balloon catheter placed in the left lung until a transpulmonary pressure (TPP) of 2.5 cm H_2O was attained. TPP was measured with an U-shaped water manometer connected to the outer end of the catheter.

RESULTS

Arterioles distributed on the alveolar surface were folded several times in the basal portion and then stretched out straight on the distal portion. They showed no sequential branchings but had many direct outflow pores to the capillary bed in the straight portion. The outflow pores were arranged at each 30 to 40 um on terminal arterioles. It could be recorded on the video tapes that elongated red blood cells flew out through the pores, just like crowding gold fish.

Examples of the video recording are shown in Fig. 1. The capillary flow was quickly reduced by the local application of wet 100% CO_2. Red blood cells aggregated and finally packed in capil-

laries. Overlapped and vague images of red blood cells can be seen
in Fig. 1A. Five seconds after the replacement of CO_2 with air
the aggregated red blood cells started to disaggregate (Fig. 1B).
Some isolated oval images of red blood cells can be seen. After ten
seconds some red blood cells are slowly moving. Isolated and
elongated oval images are seen, while some blood cells are just
beginning to disaggregate in the area below right (Fig. 1C). High
flow velocity was recovered in 30 seconds. Elongated bamboo-leaf
like images can be seen (Fig. 1D). The pulmonary capillary bed is
not a flat sheet but consists of densely interconnecting pipelines.
Their structure can be seen in Fig. 1C and D, where blood flow was
recovered. The dark areas are alveolar cells concaved from the
swollen capillaries. The diameter of alveolar cells was 39.8 ± 4.9
um (mean $\pm$ SD). Spaces between the alveolar cells (width = $14.6 \pm$
2.8 um) formed an alveolar capillary network, whose geometric
structure could be simulated reasonably well with a hexagonal model
(Weibel, 1964). Red blood cells which were thin and oval in shape
during the stationary state, were elongated to 30 um by the shear
force acting on them in the capillaries. Two red blood cells
strongly deformed in a bamboo-leaf shape could flow together in
parallel in a capillary. Each capillary hexagon, therefore, could
occasionally contain more than four red blood cells at a time.

Flow velocity changes are shown in Fig. 2, where the ordinate
represents the normalized flow velocity and the abscissa the time
elapsed. The thicker and thinner lines represent the time courses
during the application of 100% and 30% CO_2, respectively. The
localized application of 30% CO_2 caused a strong reduction in but
not complete cessation of capillary blood flow, while 100% CO_2
caused a stoppage of the blood flow. In both cases the recovery to
the control velocity was rapid when CO_2 was replaced with air.
But a longer application seemed to slow down the recovery process.

DISCUSSION

Ventilatory hypercapnia is reported to cause pulmonary vasoconstriction in isolated lobes in lungs from the cat (Barer and
McCurrie, 1969), lamb (Hyman and Kadowitz, 1957), and in the
anesthetized dog (Borst et al., 1957), no effects in intact man
(Fishman et al., 1960), in intact dog (Stroud and Rahn, 1953), and
in rapidly frozen preparations of isolated cat lung (Kato and
Staub, 1966), and even some dilator action in cat (Viles and
Shepherd, 1968). The different preparations employed and the
apparent disagreements in the obtained results suggested that the
effect of hypercapnia on the pulmonary microcirculation should be
studied in a localized hypercapnia in experimental animals with
least surgical invasions. Furthermore, it seemed worthwhile to
investigate the effect of CO_2 in different animal species.

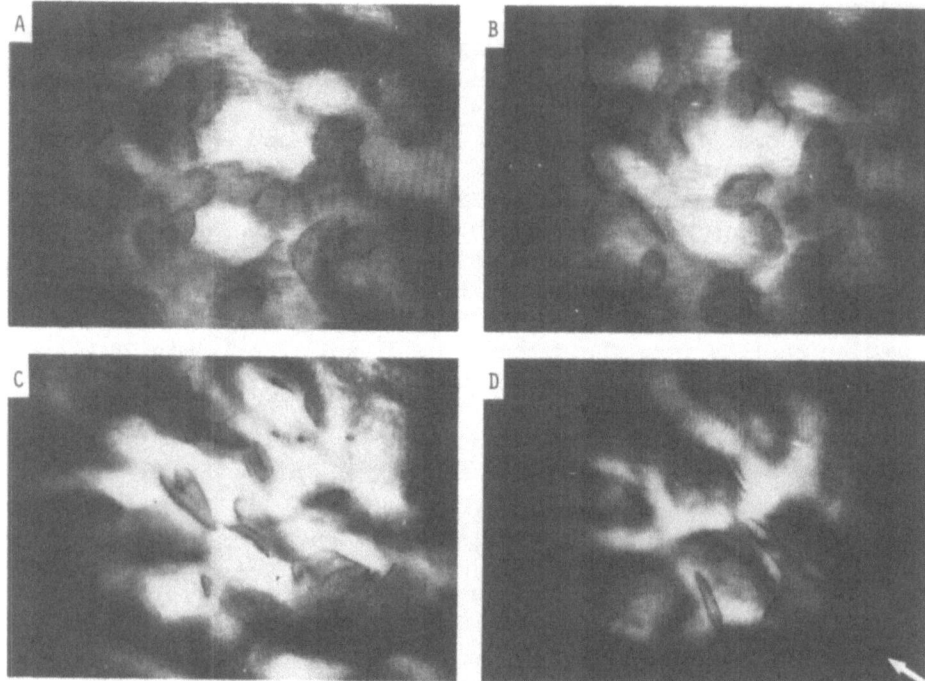

Fig. 1. Photographic reproductions of four frames from a video tape
        recording of a localized capillary bed of the lung of a
        bullfrog.
        A: Capillary blood flow cessation induced by 2 minutes
           localized application of wet 100% CO_2. Red blood cells
           are packed in capillaries.
        B: Five seconds after the replacement of CO_2 with air.
           Red blood cells are slowly disaggregating.
        C: Ten seconds after the reintroduction of air. Some
           separate red blood cells are moving and some others are
           still disaggregating.
        D: Thirty seconds after the reintroduction of air. Capil-
           lary blood flow has recovered. Red blood cells are
           flowing at high velocity. The arrow indicates the
           overall direction of the blood flow.

        In the preceding study a localized hypoxia was induced in a
small area of 6 mm in diameter on the lung surface of anesthetized
bullfrogs. The localized hypercapnic area was smaller than 0.5% of
the total lung surface (Koyama et al., 1981). In the present study
the method was further advanced in which a pore of only 1 mm in
diameter was used. This modification permitted an observation of
flow characteristics in the capillary net, unaffected by changes in
other vessel groups. This modification, however, made the applica-

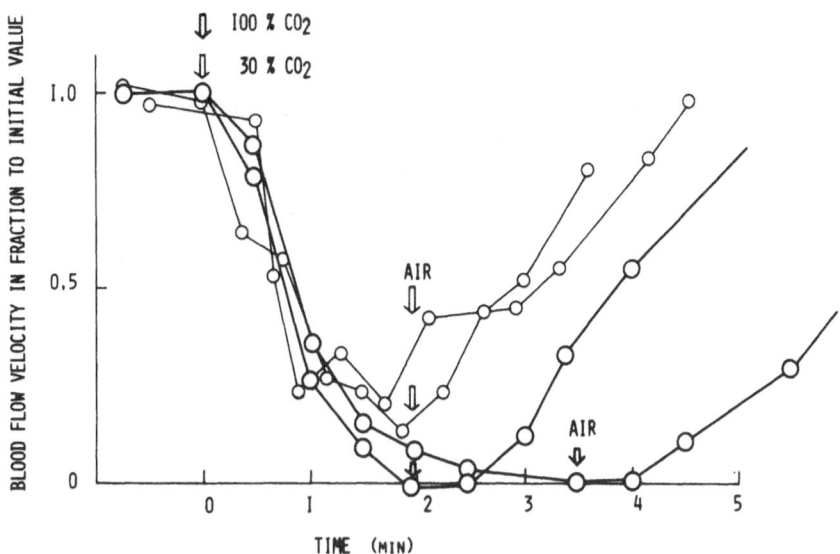

Fig. 2. Time course in mean capillary flow velocity affected by
CO_2.

tion of laser Doppler microscopy geometrically impossible. Therefore, the video camera system was employed in the present study.

Images of red blood cells could be displayed on a TV monitor
with a final magnification of higher than 1500 times. The low flow
velocity was easily determined by measuring the traversed distances
on sequential TV monitor frames. But the high flow velocity during
the control condition was probably underestimated because only the
velocity of red blood cells whose images were clearly reproduced
could be measured and the number of clear images of red blood cells
decreased as flow velocity increased. It seemed probable that most
clear images were attributable to red blood cells whose movement
was somewhat slowed down at the bifurcation of the capillaries. The
underestimation of the flow velocity during the control condition
will therefore cause a relative overestimation of the normalized
flow velocity during hypercapnia. If the flow velocity during the
control condition could have been precisely measured, the percentage reduction in blood flow velocity induced by the localized
application of 30% CO_2 would have appeared larger than shown in
Fig. 2. In some cases of 100% CO_2 application, capillary blood
ceased flowing for a few seconds and then started flowing in the
reverse direction. In any event, it can be said that the high
Pco_2 caused a strong reduction in alveolar capillary blood flow.

The microscope with dark and episcopic illumination revealed
that the lung surface was not flat but it had many concavities

which were probably attributable to alveolar cells. The interspaces among alveolar cells are swollen and form pulmonary capillaries. A capillary bifurcates and each daughter capillary joins in about 20 um with another daughter capillary from a neighbouring mother capillary. The joined capillary bifurcates again in a short distance. Thus, the so-called capillary hexagon is formed. The blood flowing through one mother capillary is divided in two daughter capillaries and then mixed with the blood from the neighbouring daughter capillary. This system provides a high efficiency of blood mixing. Moreover, capillaries are swollen and surround concave alveolar cells just like mountain chains surround a basin. It seems probable that gases diffuse in and out through almost the whole surface of the capillaries. The elongated red blood cells flowing in capillaries can receive oxygen and release carbon dioxide through the elongated whole surface. The same situation can be expected also on alveolar walls in the deeper portion of the lung. Thus, the capillary network can be an effective gas exchanger.

The cause for the hypercapnic reduction in capillary blood flow remains unstudied. But the following possibilities can be mentioned. Narrowings of capillary lumen can be caused by electrical stimulation in the frog mesentery. The nucleus of specialized endothelial cells protrudes, obstructing the capillary lumen (Weigelt et al., 1981; Weigelt, 1982). It may be possible that such specialized endothelial cells are also distributed in the frog pulmonary capillary net and, being stimulated by the hypercapnia obstruct the capillary flow. In addition, the micropore passage of human whole blood was significantly reduced by an exposure to hypercapnia in vitro (Kikuchi et al., 1979). The impaired deformability of red blood cells may reduce capillary blood flow. Thus, the locally applied strong hypercapnia may act to redistribute the pulmonary blood flow to the normocapnic area. This effect is advantageous for the homeostasis in blood P_{CO_2}.

ABSTRACT

A small test area of the pulmonary capillary bed of frog's lung was exposed to carbon dioxide. The capillary bed was observed by means of a dark field episcopal microscope and recorded on video tapes via a TV-camera. Swollen capillaries forming a hexagonal network sourrounded concave alveolar cells. The flow velocity of red blood cells through capillaries was reduced to 17.5% of the flow velocity during the control condition by the application of the moistened gas mixture containing 30% CO_2. The blood flow in capillaries ceased when wet 100% CO_2 was applied to the test area. Red blood cells were packed in capillaries. They disaggregated by the introduction of air in five seconds and were quickly flowing in an elongated leaflet shape in 30 seconds after the introduction of air.

REFERENCES

Barer, G.R., and McCurrie, J.R., 1969, Pulmonary responses in the
    cat; The effects and interrelationship of drugs, hypoxia and
    hypercapnia, Quart. J. Exp. Physiol., 54:156-172.
Borst, H.G., Whittenberger, J.L., Berglund, E., and McGregor, M.,
    1957, Effects of unilateral hypoxia and hypercapnia on pulmo-
    nary blood flow distribution in the dog, Am. J. Physiol., 191:
    446-452.
Fishman, A.P., Fritts, H.W., and Cournand, A., 1960, Effects of
    breathing carbon dioxide upon the pulmonary circulation,
    Circulation, 22:220-225.
Horimoto, M., Koyama, T., Kikuchi, Y., Kakiuchi, Y., and Murao, M.,
    1981, Effect of transpulmonary pressure on blood flow velocity
    in pulmonary microvessels, Respir. Physiol., 43:31-41.
Hyman, A.L., and Kadowitz, P.J., 1957, Effects of alveolar and
    perfusion hypoxia and hypercapnia on pulmonary vascular
    resistance in the lamb, J. Appl. Physiol., 228:397-403.
Kato, M., and Staub, N.C., 1966, Response of small pulmonary
    arteries to unilobular hypoxia and hypercapnia, Circ. Res.,
    19:426-440.
Kikuchi, Y., Horimoto, M., and Koyama, T., 1979, Reduced deform-
    ability of erythrocytes exposed to hypercapnia, Experientia,
    35:343-344.
Koyama, T., Kikuchi, Y., Horimoto, M., Kakiuchi, Y., Tsushima, N.,
    and Nitta, J., 1981, White blood cell adhesion to endothelium
    and rheological behavior in microvessels of overinflated
    frog's lung, Biorheol., 19:221-228.
Koyama, T., and Horimoto, M., 1982, Pulmonary microcirculatory
    response to localized hypercapnia, J. Appl. Physiol., in
    press.
Stroud, R.C., and Rahn, H., 1953, Effect of O_2 and CO_2 tensions
    upon the resistance of pulmonary blood vessels, Am. J. Physiol.,
    172:211-220.
Viles, P.H., and Shepherd, J.T., 1968, Evidence for a dilator
    action of carbon dioxide on the pulmonary vessels of the cat,
    Circ. Res., 22:325-333.
Weibel, E.R., 1964, Morphometrics of the lung, in: "Handbook of
    Physiology", Sect. 3: "Respiration", Vol. 1, pp. 285-307,
    W.O. Fenn, H. Rahn, eds., Am. Physiol. Society, Washington
    D.C.
Weigelt, H., 1982, Die spezialisierte Endothelzelle - erregbare
    Zelle und mechanischer Effektor der Mikrozirkulation, Funkt.
    Biol. Med., 1:53-60.
Weigelt, H., Fujii, T., Lübbers, D.W., and Hauck, G., 1981, Spe-
    cialized endothelial cells in the frog mesentery - attempt of
    an electrophysiological characterization, Biblthca. Anat.,
    20:89-93.

HYPOXIC REDUCTION IN BLOOD FLOW VELOCITY IN PULMONARY ARTERIOLES

AND CAPILLARIES

T. Koyama, M. Horimoto, Y. Shindo, Y. Kikuchi,
Y. Kakiuchi, T. Araiso, and T. Arai

Research Institute of Applied Electricity
Hokkaido University, 060 Sapporo, Japan

INTRODUCTION

The vasoconstrictory effects of hypoxia on pulmonary circulation are well established. The ventilation of a single lung lobe
with nitrogen causes a reduction in the blood flow in the hypoxic
lung lobe. However, the question that remains unsolved is whether a
regional hypoxia to which several alveoli and only small blood
vessels are exposed could reduce the blood flow and affect the flow
velocity contour in the microvessels of that region. We have now
completed a study of the effects of local hypoxia on terminal segments of pulmonary microvessels observable on the lung surface.
The hypoxia to which a small area of the lung surface was exposed
reduced the oxygen tension only in the thin layer just beneath the
lung surface in that region. Therefore, the larger vessels running
in the deeper areas were not influenced by the local hypoxia. In
turn, our technique allowed observations of microcirculatory responses to hypoxia in the absence of general systemic hemodynamic
responses.

METHOD

Bullfrogs, each weighing between 270 and 400 g, were anesthetized by immersion in a 0.08% solution of MS 222. The left lung was
exteriorized following an incision of the lateral chest wall and was
inflated by introducing air through an intra-tracheal balloon catheter until transpulmonary pressure (TPP) reached 2.5 cm H_2O. TPP was
measured with an U-shaped water manometer connected to the outer end
of the catheter. A plastic ring 6 mm in internal diameter covered

651

with a thin glass plate on one side was placed on the exposed lung
surface by means of a manipulator. Hyperoxic gas (90% O_2 and 10% N_2)
and pure nitrogen saturated by bubbling through water, were alternately introduced through a vinyl tube into the space formed by the
small ring chamber and the surface. The plastic ring encircled a
region containing 8 adjacent alveoli in which alveolar, and extra-
alveolar arterioles smaller than 200 um in diameter were observed.
Local hypoxia was produced on the lung surface for 10 to 20 minutes
by introducing pure nitrogen into the ring. The hyperoxic gas mixture was introduced as a control gas, 6 minutes before and after
the introduction of hypoxia. Each hyperoxic period (control condition) continued for from 6 to 10 minutes.

The blood flow velocity and its contour were measured by means
of a laser Doppler microscope (LDM), (Horimoto et al., 1979; 1981;
Koyama et al., 1982) and the time courses of their changes were
examined during the application of the gas mixtures. The experimental arrangement is schematically shown in Fig. 1 and the principle
of the LDM design is as follows.

A 2 mW He-Ne laser beam ($\lambda = 0.63$ um) was split into dual
beams by a beam splitter. The two beams were crossed at an angle of
18.8° at the center of a microvessel in the frog web observed
through a microscope at a magnification of 100 x. The center line

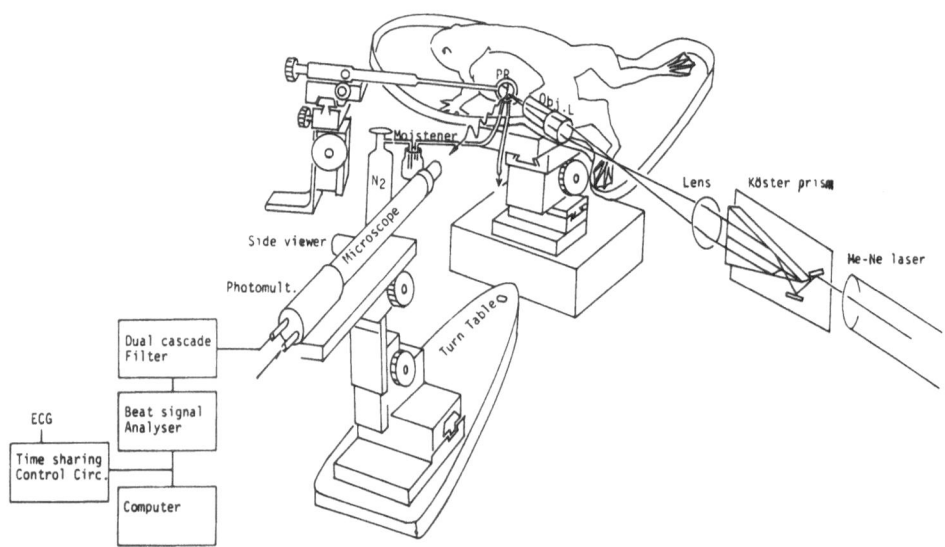

Fig. 1.   Experimental arrangement.
          Lens, Obj. L (= objective lens), Kösterprism (= beam
          splitter) and He-Ne laser stand on a fixed stage, while
          the bullfrog, microscope and PR (= plastic ring chamber)
          are on micro manipulators.

of the intersection of the beams was adjusted to be perpendicular
to the longitudinal axis of the vessel. The dual beams were scattered by the passing red blood cells and each beam underwent a differential Doppler shift, because they were incident in the probing
area with the above intersecting angle. When the scattered light was
collected by a microscope objective lens and detected by a photomultiplier, the differential Doppler shift yielded burst-like beat
signals as blood cells passed one after another.

The relationship between the wave period, Wp, of the signals
and the flow velocity, v, of red blood cells is expressed as follows.

$$1/Wp = fd = v/\lambda \cdot 2 \sin (\theta/2) \tag{1}$$

where fd, θ and λ represent the frequency of a beat signal, the
intersecting angle and the wave length of the incident laser beams,
respectively. Another explanation of the principle of LDM is the
formation of interference fringes which can be confirmed in blood
vessels in vivo (Koyama et al., 1982). They are formed at a definite
distance given by the wave length of laser and intersecting angle
of the dual beams. When blood cells traverse over the interference
fringes, blood cells are alternately bright and dark at the wave
period defined by equation (1).

The output signals from the photomultiplier were transmitted
to a band pass filter to decrease noise. The wave period of the
beat signal was measured by an on-line beat signals analyzer
(Nihon-Kagaku-Kogyo Co., Japan), processed in a time-sharing device
triggered by the R-wave of each ECG, and stored in 16 channels of a
4-k memory as a function of the time elapsed after the R-wave. Each
channel consisted of 256 memories, each of which was programed so
as to store the frequency of the occurrence of an assigned wave
period. These data were then used to determine the pulsatile blood
flow velocity in the vessels. The time-interval covered by each
channel was set at 80 or 100 msec depending on the duration of each
cardiac cycle.

Measurements of the wave period were continued during a number
of cardiac cycles and more than 2 minutes were required to obtain
16 distribution functions showing the frequency of appearance of
each wave period, representing sequential time periods after the
R-wave. The mean value of the wave period for each function was
calculated by the computer system and was substituted into equation
(1). Accuracy and reproducibility of the procedure have been examined previously (Horimoto et al., 1981).

For the preliminary studies the thoracic cavity was opened by
a mediastinotomy in four frogs. An injection needle (gauge #20) was
placed in the ventricle and connected to a pressure transducer
(Statham P23D) with a polyethylene tube (I.D. = 0.5 mm and length

10 cm), so as to measure the ventricular pressure. The systolic ventricular pressure was unaffected by the localized application of nitrogen and remained at 28.1 ± 1.0 Torr. Furthermore, intra-ventricular blood was sampled for gas analysis before and after the local application of pure nitrogen for 20 minutes in order to examine whether the localized hypoxia would induce systemic hypoxia. The blood Po_2 decreased slowly during both the localized hypoxia and hyperoxia. It was not attributable to nitrogen application but to the oxygen uptake from the alveolar gas by the circulating blood. All measurements were made at a temperature of 18 to 20°C.

RESULTS

    The flow velocity contours in pulmonary arterioles (diameter, 31.2 - 101.0 um, 56.9 ± 21.9 um (mean ± SD), and heart rate 30 - 40/min) during local hyperoxia, i.e. the control condition, and hypoxia are shown in Fig. 2A and B, respectively. The mean flow velocity within the test area was 1.98 ± 0.45 mm/sec on an average and ± SD during the control condition. The flow velocity oscillated strongly in accordance with the cardiac events, attaining a maximal value 2.22 ± 0.56 mm/sec (mean ± SD) at 800 msec after the R-wave of ECG during the control condition (Fig. 2A). The localized hypoxia induced by the introduction of nitrogen into the ring caused a reduction in flow velocity in all the frogs studied (Fig. 2B). The mean flow velocity was 1.63 ± 0.32 mm/sec during hypoxia. The replacement of nitrogen with the hyperoxic gas mixture resulted in a high flow velocity which was comparable with the initial control level within a 5% error at most.

    The mean flow velocity (1.52 ± 0.10 mm/sec mean ± SD) and flow pulsation were smaller in capillaries than in arterioles (Fig. 3A). The flow velocity started to rise at 300 msec after the R-wave, which was much later than in arterioles. The localized hypoxia caused a small reduction in all the capillaries studied. The mean flow velocity was 1.33 ± 0.08 mm/sec (mean ± SD) (Fig. 3B).

    The mean flow velocity during hypoxia was 82.3% and 87.5% of the mean flow velocity during the control condition in arterioles and capillaries, respectively. The percentage reduction in the mean flow velocity in arterioles was significantly larger than in capillaries (Student's non-paired t-test, $p < 0.05$). For a better understanding of the characteristics of the flow velocity contour, an attempt was made to try to detect a phase shift in the contour pulsation between control and localized hypoxia. The flow velocity at each time point of the time shared measurements during the localized hypoxia was divided by the flow velocity during the control condition at the corresponding time point. An example of such a normalized curve obtained in an arteriole is shown together with the velocity contours in the control condition and hypoxia in

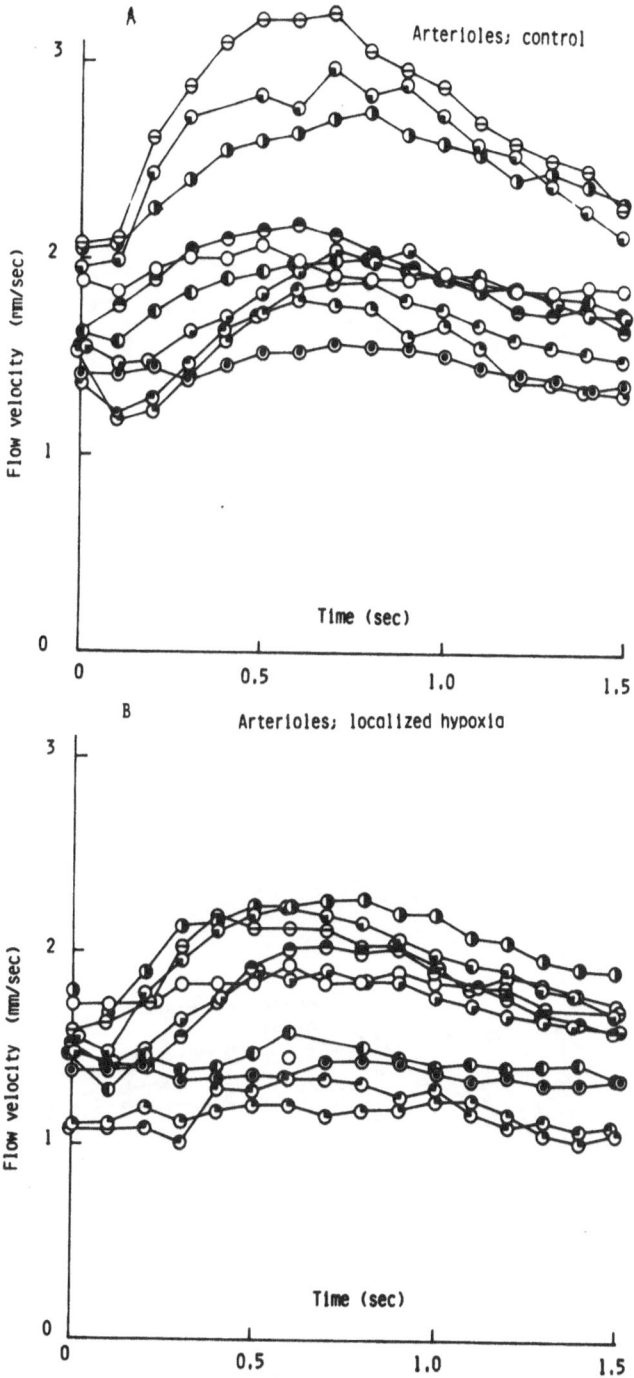

Fig. 2.   Blood flow velocity contours in pulmonary arterioles for
1 cardiac cycle during the control condition (A) and localized hypoxia (B). Each symbol represents data for a
single vessel.

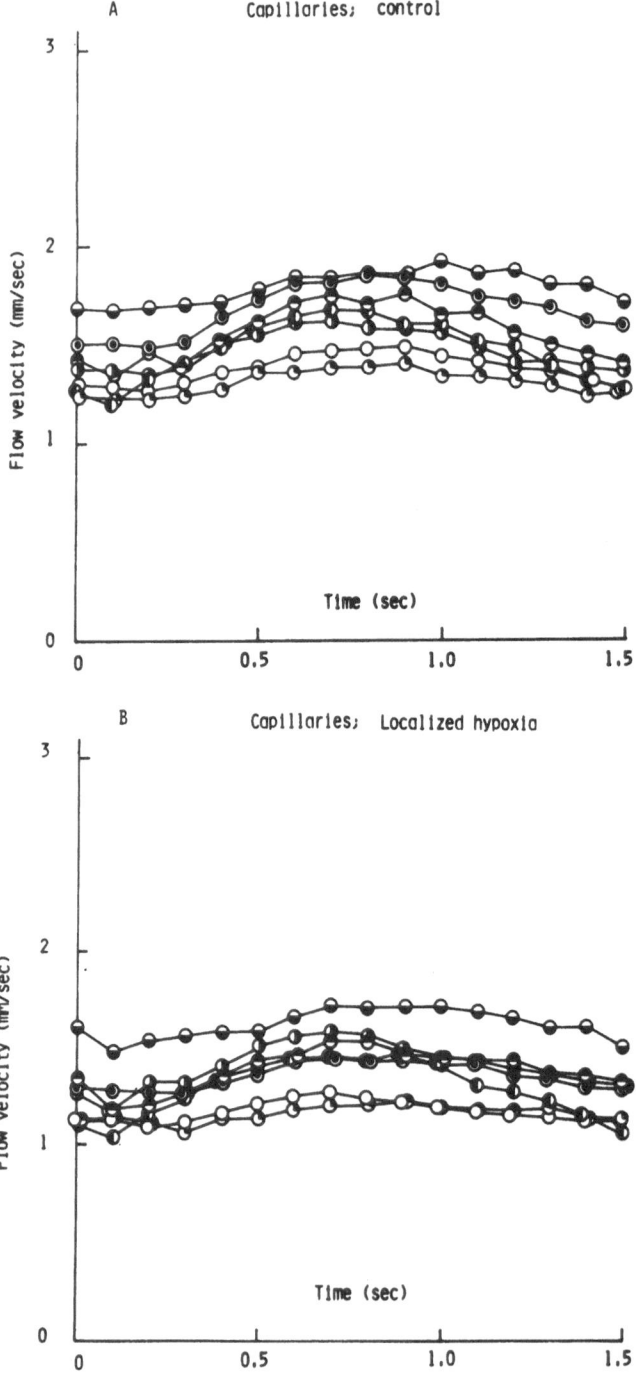

Fig. 3. Blood flow velocity contours in pulmonary capillaries for
1 cardiac cycle during the control condition (A) and localized hypoxia (B). Each symbol represents data for a single
vessel but does not correspond to vessels in Fig. 2.

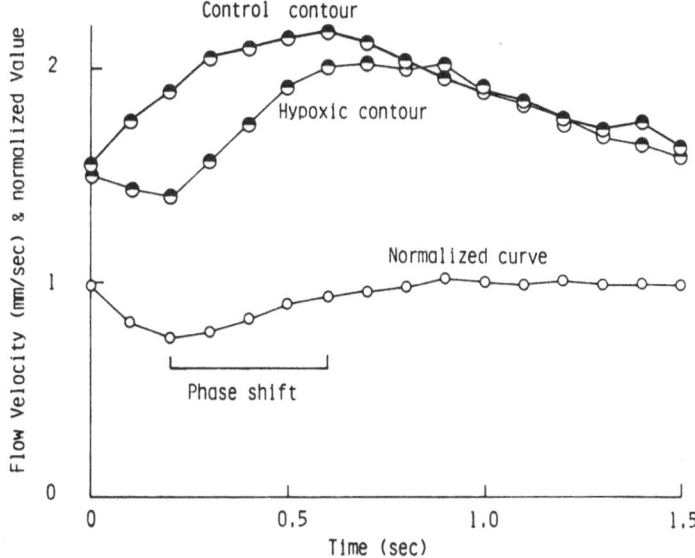

Fig. 4.   An example of a normalized curve obtained in an arteriole,
and the corresponding flow velocity contours during the
control and localized hypoxia. The symbol represents the
corresponding vessel in Fig. 2.

Fig. 4. The maximal flow velocities in both the control and hypoxic
contours appeared at 600 msec and 700 msec, respectively. However,
the minimum of the normalized curve occurred at 300 msec after the
R-wave of ECG, i.e. 300 msec prior to the time of the maximal flow
velocity. This example suggests that the start of the increment in
flow velocity is quicker and sharper during the control condition
than during the hypoxia. The phase shift, i.e. the time interval
between the occurrences of the minimal value on the normalized curve
and the maximal flow velocity on the hyperoxic flow velocity contour
are summarized for arterioles and capillaries in Fig. 5. The plus
on the ordinate indicates that the minimal value appeared on the
normalized curve earlier than the maximal on the flow velocity
contour during the control condition, while the minus indicates the
lag of the appearance of the minimal value on the normalized curve
behind the maximal on the control flow velocity contour. The difference between arterioles and capillaries in the phase shift was
statistically significant by the Student's non-paired t-test.

DISCUSSION

     Many studies have shown that pulmonary vasoconstriction occurs
during ventilatory hypoxia in isolated lungs or in separately perfused lung lobes (Duke, 1954; Sackner et al., 1966; Bergofsky et al.,
1968; Hauge, 1969). In those studies changes in overall pulmonary

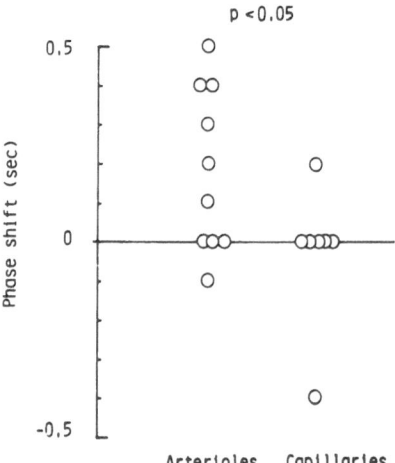

Fig. 5.   Phase shifts in flow velocity contours caused by a loca-
           lized hypoxia.

vascular resistance were measured. But no studies of the effects of
local hypoxia on the blood flow velocity and the flow velocity contour in alveolar microvessels have been reported.

    The pulmonary blood flow velocity strongly depends on the
transpulmonary pressure (TPP). The optimum TPP was 2.8 ± 1.3 cm H_2O.
But there was a large individual variation in TPP in bullfrogs
(Horimoto et al., 1981). The wide variation in arteriolar blood
flow velocity as seen in Fig. 2 was probably attributable to the
individually different dependencies of pulmonary circulation on TPP
in different bullfrogs. In spite of such a wide distribution in
blood flow velocity during the control condition, the localized
hypoxia caused a reduction in blood flow velocity in all the frogs
studied.

    The localized hypoxic area was smaller than 0.2% of the total
lung surface, and only microvessels were exposed to the hypoxia.
The present results show that the exposure of the microvessels to
the localized hypoxia could reduce blood flow velocity in the
microvessels and that the percentage reduction in the amplitude of
the velocity pulsation was larger than the reduction in the mean
flow velocity.

    The detection of the time point at which the maximum flow
velocity was attained was difficult in flow velocity contours whose
higher portion often formed a plateau. For an easier detection of
the phase shift in pulsation, the hypoxic velocity contour was
normalized by the control velocity contour in the present study.
The results shown in Fig. 5 suggest that a delay was produced in
the velocity contour by the localized hypoxia in arterioles. But no

phase shift due to the localized hypoxia was detected on the flow
velocity contour in capillaries. This difference in phase shift
suggests different behaviors of blood flow in these vessel groups.
The capillaries within the small test area form part of a capillary
network which is extensively connected with capillaries outside of
the test area. The interconnection with the normoxic area probably
makes the behavior of blood flow in capillaries different from that
in arterioles. The present result that the reduction in capillary
blood flow velocity was smaller than that in arterioles seems to
support this hypothesis. Pulmonary arterioles constrict when they
are exposed to hypoxia. But the effect of the arteriolar vasoconstriction may be diminished by some compensatory action via the
dense interconnections among pulmonary capillaries.

In conclusion, the hypoxia which was induced in a small
localized area of the lung surface caused a reduction in blood flow
velocity in both pulmonary arterioles and capillaries. It caused a
larger reduction in mean blood flow velocity in arterioles than in
capillaries and produced a phase delay in the flow velocity contour
in arterioles.

ABSTRACT

A small ring chamber (I.D. = 6 mm) was placed on the exposed
lung of anaesthetized bullfrogs. A localized hypoxia was induced in
the ring chamber by introducing nitrogen in it. Blood flow velocity
in pulmonary microvessels was measured by means of a laser Doppler
microscope. The mean blood flow velocity was 1.98 ± 0.45 and $1.52 \pm
0.10$ mm/sec during the control condition in arterioles and capil-
laries, respectively. It was then reduced by the localized hypoxia
to 1.63 ± 0.32 and 1.33 ± 0.08 mm/sec in arterioles and capillaries,
respectively. The reduction, when expressed in the percentage ratio
to the control flow velocity in each blood vessel group, was significantly larger in arterioles than in capillaries. A phase delay in
the pulsation of the flow velocity contour was detected only in
arterioles. These differences between pulmonary arterioles and
capillaries in response to the localized hypoxia may be attributed
to the dense interconnection of capillary network extending beyond
the localized hypoxic area to the normoxic area.

REFERENCES

Bergofsky, E.H., Haas, F., and Porcelli, R., 1968, Determination of
    the sensitive vascular sites from which hypoxia and hypercapnia
    elicit rises in pulmonary arterial pressure, Fed. Proc.,
    27:1420-1425.
Duke, H.N., 1954, Pulmonary vasomotor responses of isolated perfused
    cat lungs to anoxia and hypercapnia, Q. J. Exp. Physiol., 36:
    75-88.

Hauge, A., 1969, Hypoxia and pulmonary vascular resistance. The
relative effects of pulmonary arterial and alveolar Po_2,
Acta Physiol. Scand., 76:121-130.

Horimoto, M., Koyama, T., Kakiuchi, Y., and Murao, M., 1981, Effect
of transpulmonary pressure on blood-flow velocity in pulmonary
microvessels, Respir. Physiol., 43:31-41.

Horimoto, M., Koyama, T., Mishina, H., Asakura, T., and Murao, M.,
1979, Blood flow velocity in pulmonary microvessels of bullfrogs, Respir. Physiol., 37:45-59.

Koyama, T., Horimoto, M., Mishina, H., and Asakura, 1982, Measurements of blood flow velocity by means of a laser Doppler
microscope, Optik, in press.

Sackner, M.A., Will, D.H., and DuBois, A.B., 1966, The site of
pulmonary vasomotor activity during hypoxia or serotonin
administration, J. Clin. Invest., 45:112-121.

RETINAL OXYGEN TENSION IN DIABETIC DOGS FOLLOWING INSULIN INFUSION[*]

T.K. Goldstick[1], J.T. Ernest[2], and R.L. Engerman[3]

[1]Dept. of Chemical Engineering, Northwestern University
Evanston, IL, USA
[2]Dept. of Ophthalmology, University of Illinois
Illinois, Eye and Ear Infirmary, Chicago, IL, USA
[3]Dept. of Ophthalmology, University of Wisconsin
Madison, Wisconsin, USA

SUMMARY

We measured preretinal oxygen tensions using a microelectrode
in dogs made diabetic with alloxan. The intravenous administration
of insulin did not affect preretinal oxygen tension over the two
hours it could be continuously accurately measured. Furthermore, the
oxygen-hemoglobin equilibrium curves measured before and two hours
after insulin administration did not change.

INTRODUCTION

It has been hypothesized that insulin administration may result
in an increased affinity between hemoglobin and oxygen and that this
could cause transient hypoxia (Ditzel et al., 1978). To investigate
this, we studied the acute effect of the infusion of insulin on the
preretinal oxygen tension in diabetic dogs.

---

[*] Presented, in part, at the Spring Meeting of the Association for
Research in Vision and Ophthalmology, Orlando, Florida, May 1980 and
XXVIII International Congress of Physiological Sciences, Budapest,
Hungary, July, 1980.

METHODS

Six female beagles, one to three years of age, weighing approximately 10 kg and having ophthalmoscopically normal eyes, were made
diabetic by an intravenous injection of alloxan monohydrate (55 mg/
kg) following a 24-hour fast. The animals, which were housed in
metabolism cages throughout the study, were allowed food ad lib and
maintained by subcutaneous injections of insulin (up to 7 units NPH
daily) in dosages insufficient to prevent chronic glycosuria and
which caused a gradual increase of glycosylated hemoglobin levels
in their blood. Glycosylated hemoglobin was measured using chromatography columns available commercially (Isolab, Inc., Akron, Ohio),
and was found to rise from 5.6 ± 0.9 ($\pm$ SD) percent before diabetes
was induced to 10.6 ± 0.4 at the time retinal oxygen tensions were
measured, 31 to 36 weeks later.

Of the six dogs used, preretinal oxygen tensions could only be
measured on four. One dog that developed severe cataracts was used
only for the study of its oxygen-hemoglobin equilibrium curve and
one that developed arrhythmias under anesthesia was terminated.
Insulin was withheld for 24 hours preceding the oxygen measurements
and each animal was fasted overnight. The animals were anesthetized
with sodium pentobarbital (25 mg/kg intravenously) and tracheotomized. The animals were then placed on a respirator (Model 607,
Harvard Apparatus Company), and paralyzed with tubocurarine chloride
(0.1 mg/kg) and gallamine triethiodide (0.5 mg/kg). The arterial
blood pressure was monitored from a cannulated femoral artery with
a pressure transducer (Model 1280C, Hewlett Packard). The intraocular pressure was maintained at 15 mm Hg by adjusting the height of
an infusion bottle connected to an anterior chamber cannula and to
a second pressure transducer. The perfusion pressure was defined as
the difference between the mean systemic blood pressure measured in
the femoral artery and the intraocular pressure. The perfusion
pressure was continuously recorded by electronically subtracting
the outputs from the two pressure transducers. All the parameters
were recorded on a polygraph (Model 7758, Hewlett Packard). The
preretinal oxygen tension currents were converted to oxygen tension
using the prior or subsequent calibrations. The animal's rectal
temperature was maintained at $39.0^{\circ}C$ with a thermoblanket. Arterial
blood samples were analyzed throughout each experiment for pH, Pco_2,
and Po_2 with a blood micro system (Models BMS 3 and PHM 72, The
London Company Radiometer A/S) at $39.0^{\circ}C$. The hemoglobin oxygen
saturation was measured with a Co-Oximeter (Model 282, Instrumenta-
tions Laboratory) and averaged 93 ± 2 percent. Although the Co-
Oximeter was not at $39.0^{\circ}C$, there is a negligible anaerobic shift
in saturation with temperature. The animal's pH was maintained at
7.40 ± 0.06 by adjusting the respiratory rate and, with the tidal
volumes used, this resulted in Pco_2 levels of 30 ± 7 mm Hg and Po_2
levels of 86 ± 5 mm Hg. Plasma glucose levels were measured with a
glucose analyzer (Beckman Instruments) and whole blood glucose was

measured with Dextrostix and a reflectance colorimeter (Ames
Division, Miles Laboratories). Plasma osmotic pressure was measured
with an osmometer (Model OS, Fiske).

Simultaneous measurements of blood oxygen tension and percent
saturation were utilized to obtain the P_{50} on four of the diabetic
dogs and on one normal dog. The P_{50} is defined here as the oxygen
tension, corrected to a pH of 7.40 and temperature of 39.0°C, when
the hemoglobin is 50% saturated. The Hill coefficient was also
determined under these same conditions. A modification of the method
of Rossing and Cain (1966), employing a nonlinear regression computer
analysis (NLREG), was used to fit the experimental measurements to
the Hill equation for the oxygen-hemoglobin equilibrium curve
(O'Riordan et al., 1983). While Rossing and Cain pooled all of the
data from 78 dogs (598 blood samples), we measured each dog separately, just before and beginning about two hours after insulin
infusion. Also rather than fitting the data to obtain the best pH
correction factor, we corrected the oxygen tensions to a pH of 7.40
using the Bohr coefficient of -0.48 found by Rossing and Cain (1966)
from the best fit of their data over the pH range of 7.08 to 7.73.
Because Rossing and Cain found that including the blood carbon
dioxide tension in their fitting equation did not significantly
improve the fit, no correction was made for it here. No correction
was needed for temperature because all blood oxygen tensions and pH
were measured at 39°C. To obtain a range of oxygen saturations
comparable to that used by Rossing and Cain, 20 to 98%, the saturations in some blood samples were changed by brief exposure to gases
of different oxygen tensions. The range of pH and Pco_2 of all blood
samples used was 6.86 to 7.66 and 9 to 60 mm Hg, respectively.

The preretinal oxygen tensions were measured using the technique we have previously described (Ernest and Archer, 1979). In
brief, the pupils were dilated with 10% phenylephrine hydrochloride
(Neo-Synephrine) and 1% cyclopentolate hydrochloride (Cyclogyl) and
a cannula containing the oxygen microelectrode passed through the
pars plana into the vitreous cavity. The cannula and the microelectrode were controlled with micromanipulators. The ocular fundus
was observed by axial illumination with an operating microscope
through a contact lens. We used selected, glass insulated, gold
plated, recessed tip microelectrodes (no. 723, Transidyne General
Corp., Ann Arbor, Mich.). These were very slightly tapered with tip
diameters of 2 - 4 um and recess lengths of 5 - 15 um. The voltage-
current relationship (polarogram) of each microelectrode was measured, and a polarization voltage chosen from the middle of the plateau.
The microelectrodes were calibrated in isotonic saline solutions
equilibrated at 39.0°C, with both 100% nitrogen and 5% oxygen in
nitrogen before insertion into the eye and after withdrawal. The
selected microelectrodes had currents which were always less than
10^{-11} amp in nitrogen and sensitivities of the order of 7×10^{-13}
amp/mm Hg (range 1.8 to 12.7×10^{-13} amp/mm Hg). Although the

amplifier (Model 1201, Transidyne General Corp.) and recording
system used made it possible to measure the oxygen tension to an
accuracy of $\pm$ 0.1 mm Hg, the inevitable electrode drift reduced this
accuracy to about $\pm$ 5 mm Hg. The reproducibility of consecutive
readings, however, was usually within $\pm$ 1 mm Hg. Preretinal oxygen
tension was measured in the vitreous humor facing the area centra-
lis. The position of the tip of the microelectrode was determined
by placing it on the internal limiting membrane of the retina
(signaled by a dimpling of the membrane) and then withdrawing it
approximately 100 um. To test the oxygen electrode, and also study
the vascular reactivity as reported elsewhere (Goldstick et al.,
1981), the animals were periodically given 100% oxygen to breathe.

Baseline values were obtained, and then a rapid (30 to 90 seconds)
intravenous infusion of insulin was given. The insulin consisted of
soluble "Regular" insulin (Iletin, Eli Lilly and Co., 40 units per
cc) and was administered either intravenously, or (once) subcutaneously, at a dosage of 1.5 U/kg, 3 U/kg or 9 U/kg given either in
periodic increments over 30 min to 2 hr or as a single bolus of 9
U/kg. Blood glucose concentration was measured periodically before
and after the insulin administration. When measured using Dextrostix,
these whole blood values were converted to plasma values by multiplying by the ratio of plasma glucose to whole blood glucose obtained
from simultaneous measurements of both. The ratio used throughout
was obtained from the average for each dog during the period prior
to the administration of insulin, when the dog's blood glucose
level was relatively constant (approximately 500 mg/dl). For all
five dogs injected with insulin the ratios, in sequence, were 1.37,
1.52, 1.27, 1.32 and 1.34. The average of these ratios, 1.36 $\pm$ 0.09,
agrees well with the value of 1.25 found in rats (Heath and Rose,
1969) with hematocrits (42.5 $\pm$ 1.7%) comparable though lower than
in our dogs. The calculated plasma glucose values following insulin
administration were fitted to a monoexponential by nonlinear
regression analysis which found the best values for the half-time,
the baseline concentration, and the final concentration.

RESULTS

The fitted baseline plasma glucose level $\pm$ SD of the five diabetic dogs which yielded data was 517 $\pm$ 67 mg %, and their plasma
osmolality 313 $\pm$ 14 mosmol/kg (values from comparable normal dogs
are 114 $\pm$ 18 mg % and 196 $\pm$ 8 mosmol/kg). The average of the baseline preretinal oxygen tensions for all four diabetic dogs on which
it was measured was 19 $\pm$ 6 mm Hg on room air and 67 $\pm$ 42 mm Hg on
100% oxygen (values from four comparable normal dogs were 22 $\pm$ 8
mm Hg on room air and 47 $\pm$ 18 mm Hg on 100% oxygen). To demonstrate
the reproducibility and stability of the preretinal oxygen measurements, two of the diabetic dogs were studied continually during the
hour preceding their first insulin injection (Figs. 1 and 2). The

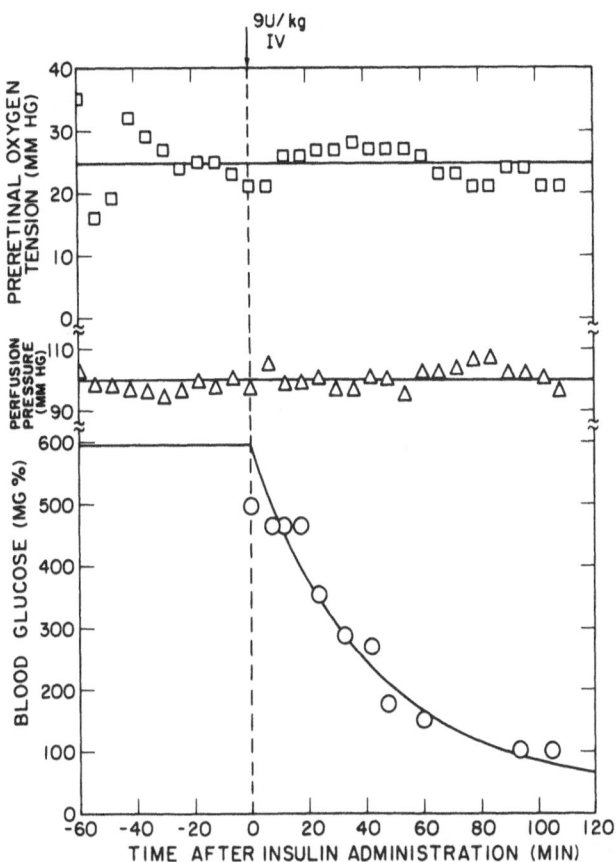

Fig. 1. Effect of bolus injection of insulin on preretinal oxygen
tension and perfusion pressure (mean femoral artery pressure minus intraocular pressure). The horizontal lines
through these data are the means over the entire time
period. The glucose half-time was 28 min and the factor
used to convert the measured whole blood glucose values to
the plasma glucose values shown was 1.32 (see text).

breathing gas was periodically alternated (every 3 or 4 minutes)
from room air (8 values) to 100% oxygen (7 values). The mean
preretinal oxygen tensions ± SD for the dog of Fig. 2 were 11.6 ±
0.8 mm Hg on room air and 49.6 ± 2.9 mm Hg on 100% oxygen.

The preretinal oxygen tension, perfusion pressure and plasma
glucose concentration following the intravenous infusion of a bolus
of insulin are shown in Fig. 1. Similar measurements for a second
dog that was given repeated injections of insulin are shown in
Fig. 2. In all five dogs, the glucose level fell exponentially and
the half-times were always between 28 and 38 minutes (average 32
± 4 min). Approximately 20% of the glucose is inside red blood

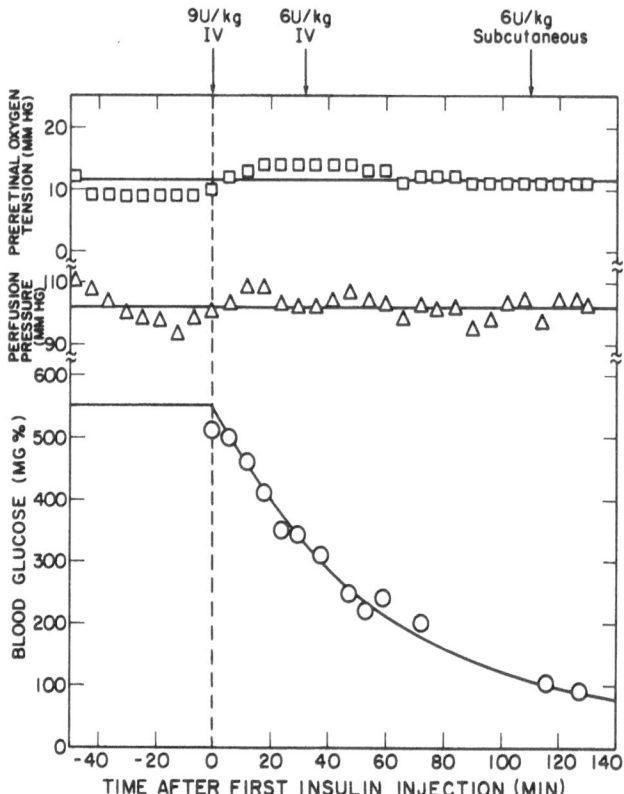

Fig. 2. Effect of repeated insulin injection on preretinal oxygen
        tension and perfusion pressure. The horizontal lines through
        these data are the means over the entire time period. The
        glucose half-time was 38 min and the factor used to convert
        the measured whole blood glucose values to the plasma glucose
        values shown was 1.37 (see text).

cells (Heath and Rose, 1969) and probably not in equilibrium with
the plasma glucose when the latter is rapidly changing. Thus, the
actual plasma glucose concentration probably fell somewhat faster
than indicated by Figs. 1 and 2 in which it was calculated from the
ratio of plasma to whole blood glucose during periods of no change
when the two were in equilibrium. Insulin did not have a significant effect on the perfusion pressure or the preretinal oxygen
tension. One animal was followed for several hours without exhi-
biting any significant changes in these parameters. In other experiments in our laboratory we have found results similar to those
shown in Figs. 1 and 2 with hyperglycemic normal dogs given insulin.

    Fig. 3 is a typical computer-fitted oxygen-hemoglobin equilib-
rium curve from a diabetic dog before and two hours after insulin
administration. The infusion of insulin did not shift the curve.

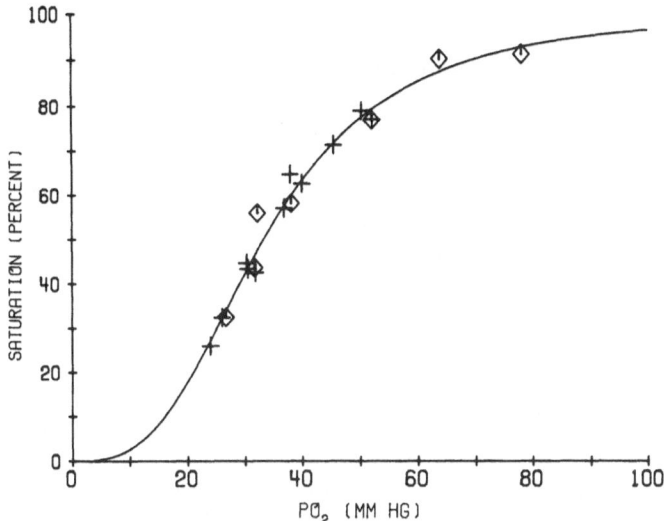

Fig. 3. Typical oxygen-hemoglobin equilibrium data before and after
insulin administration. The curve is the computer generated
fit of all of the data. There was no significant difference
between the data before insulin (crosses) and two hours
after insulin infusion (diamonds). This animal had developed such severe cataracts that preretinal oxygen tension
measurements were not possible.

The mean P_{50} and Hill coefficient both for the normal dog (33.2
mm Hg and 2.8) and for diabetic dogs (32.1 $\pm$ 1.2 mm Hg and 2.6 $\pm$
0.3) agree quite well with the averages one can calculate from
Rossing and Cain (1966). Measurements from their 78 presumably
normal dogs using their best fit equation at a temperature of 39.0°C
and a pH of 7.40, give 31.0 $\pm$ 4.2 mm Hg and 2.52 (SD not given).
The elevated levels of fast hemoglobin may have made the mean P_{50}
in the diabetic dogs slightly lower than that in the normals, as
would be expected from measurements of P_{50} in humans (Ditzel et al.,
1978). Although the data are somewhat scattered, insulin did not
appear to systematically alter P_{50}.

DISCUSSION

     The method is limited by the lack of stability of oxygen micro-
electrodes. Our electrodes were selected, and only about one in five
met our criteria. The polarogram had to have a flat plateau, the
nitrogen currents had to be less than 10^{-11} amp, the sensitivity
less than 1.3×10^{-12} amp/mm Hg, and the current had to be stable
(less than 10% drift/hr). There was always a tendency for the
currents to drift downwards over several hours, and the data were
discarded if the calibration before and after the retinal measure-

ment differed by more than 25 percent. This type of oxygen microelectrode can be calibrated in normal saline and reliably used in
the vitreous body (Alm and Bill, 1972; Schneiderman and Goldstick,
1978). Vitreous body oxygen tension measurements very close to the
retina should reflect the retinal surface oxygen tension both at
steady state and during physiological changes (Linsenmeier et al.,
1981). The validity of our preretinal oxygen tension measurements
is also based on their reproducibility, their agreement with comparable values measured by others (Alm and Bill, 1972; Ernest and
Archer, 1979; Enroth-Cugell et al., 1980; Linsenmeier et al., 1981),
and their independence from both the method of insulin administration (whether bolus, intermittent, or subcutaneous) and rate of
decrease in blood glucose.

Preretinal oxygen tension measurements were made on four diabetic
dogs. They had been chronically hyperglycemic for over half a year,
had twice the normal level of glycosylated hemoglobin and, when the
experimental measurements were started, had fasting plasma glucose
levels approximately five times normal. The diabetic dog is an
appropriate if not unique model of the human disease, since it has
been shown to develop diabetic retinopathy similar to that seen in
man (Engerman et al., 1977). In the present study, however, vascular
lesions in the retina were neither expected nor observed since the
pathology does not appear for three to five years. The baseline
preretinal oxygen tensions for all dogs studied were quite similar
to those found in other animals. The effects of the insulin infusions were also very similar in all dogs whether given intermittent-
ly or in a bolus.

The intravenous infusion of insulin over a wide range of dosages
had no significant effect on either perfusion pressure or preretinal
oxygen tensions (Figs. 1 and 2). In fact, the electrode current was
so stable following insulin that we tested it periodically (every 6
to 8 minutes) by the inhalation of 100% oxygen to be sure that the
electrode was responding. The results when breathing 100% oxygen
were reported elsewhere (Goldstick et al., 1981). Moreover, there
was no change in the P_{50} and therefore no shift in the oxygen-hemoglobin equilibrium curve before or after insulin administration
(Fig. 3). It is thus not so surprising that insulin infusion did
not change retinal oxygen tension over the time we measured it.

Ditzel et al. (1978), studying human diabetics, have shown
that the P_{50} is routinely lowered a few mm Hg following insulin
administration. They propose that insulin administration causes a
phosphate shift resulting in a decrease in red blood cell 2,3-DPG
and in oxygen-releasing capacity. Their measurements in nonketotic
diabetics were made before and approximately one day after insulin
infusion, and it may be that the measurements reported here were
not carried out over a long enough time to demonstrate changes in
the P_{50} or retinal oxygen tension. In their juvenile diabetics,

however, their measurements were made before and three-four hours after the patients had received their usual insulin dose. It might also be that dog blood is less sensitive to the changes in blood chemistry elicited by insulin administration or that our methods were not sensitive enough to detect a small change. Nonetheless, in our studies the diabetes was relatively severe, the percentage of glycosylated hemoglobin high, and the insulin infusions massive, yet there was no significant change in the P_{50} or the preretinal oxygen tension.

ACKNOWLEDGEMENTS

This research was supported by National Eye Institute Research Grants EY 04085 (J.T.E.) and EY 00300 (R.L.E.) and a RPB Research Professorship (J.T.E.). We are also grateful for the technical assistance of Debbie Wing, M. Larson, and J.F. O'Riordan, and for the discussions in early phases of planning with R.A. Linsenmeier.

REFERENCES

Alm, A., and Bill, A., 1972, The oxygen supply to the retina. I. Effects of changes in intraocular and arterial blood pressure, and in arterial Po_2 and Pco_2 on the oxygen tension in the vitreous body of the cat, Acta Physiol. Scand., 84:261-274.

Ditzel, J., Jaeger, P., and Standl, E., 1978, An adverse effect of insulin on the oxygen-release capacity of red blood cells in non-acidotic diabetes, Metabolism, 27:927-934.

Engerman, R., Bloodworth, J.M.B., and Nelson, S., 1977, Relationship of microvascular disease in diabetes to metabolic control, Diabetes, 26:760-769.

Enroth-Cugell, C., Goldstick, T.K., and Linsenmeier, R.A., 1980, The contrast sensitivity of cat retinal ganglion cells at reduced oxygen tensions, J. Physiol. (Lond.), 304:59-81.

Ernest, J.T., and Archer, D.B., 1979, Vitreous body oxygen tension following experimental branch retinal vein obstruction, Invest. Ophthalmol. Vis. Sci., 18:1025-1029.

Goldstick, T.K., Ernest, J.T., and Engerman, R.L., 1981, Impaired retinal vascular reactivity in diabetic dogs, Invest. Ophthalmol. Vis. Sci. (ARVO Suppl.), 20:92.

Heath, D.F., and Rose, J.G., 1969, The distribution of glucose and $[^{14}C]$ glucose between erythrocytes and plasma in the rat, Biochem. J., 112:373-377.

Linsenmeier, R.A., Goldstick, T.K., Blum, R.S, and Enroth-Cugell, C., 1981, Estimation of retinal oxygen transients from measurements made in the vitreous humor, Exp. Eye Res., 32:369-379.

O'Riordan, J.F., Goldstick, T.K., Ditzel, J., and Ernest, J.T., 1983, Diabetic oxygen-hemoglobin equilibrium curves evaluated by non-linear regression of the Hill equation, Adv. Exp. Med. Biol., this volume.

Rossing, R.G., and Cain, S.M., 1966, A nomogram relating Po_2, pH, temperature and hemoglobin saturation in the dog, J. Appl. Physiol., 21:195-201.

Schneiderman, G., and Goldstick, T.K., 1978, Oxygen electrode design criteria and performance characteristics: recessed cathode, J. Appl. Physiol., 45:145-154.

A MECHANISM FOR OXYGEN DAMAGE TO THE IMMATURE RETINAL VASCULATURE

R.W. Flower

The Applied Physics Laboratory and The Wilmer
Ophthalmological Institute of the John Hopkins University and Hospital Baltimore, Maryland, USA

Although it is generally believed that direct cytotoxic effects of oxygen on capillary endothelial cells may be involved in the vascular degenerative process of retrolental fibroplasia (RLF), the precise mechanism of the oxygen vaso-obliteration which preceeds it has not been firmly established. It has also long been recognized that risk factors other than sustained exposure to high oxygen concentration environments may contribute to this retinopathy, but no solid evidence of other risk factors has emerged.

For nearly forty years, ophthalmologists - and in recent years, neonatologists also - have sought to elucidate the mechanism of RLF pathogenesis. Unfortunately, the difficulties of studying RLF in the clinical environment remain overwhelming today. Moreover, given that RLF is a disease of the immature eye, the paucity of our know-ledge concerning normal physiology of the immature eye makes it un-likely that abnormal developmental processes will be understood even when they are observed.

In our laboratory, we have improved an animal model of RLF and are using it to determine how physiological responses of the immature eye differ from those of the adult and to study how changes in various physiological parameters may affect development of the im-mature retinal vasculature. The animal model we use was introduced in the 1950's by Norman Ashton and his co-workers (1953) who were aware that full-term, newborn kittens and puppies have retinas whose vascularization approximates that of the seven-month-gestation human eye. They observed that the retinal vessels of kittens exposed to high oxygen concentrations undergo the same vasoconstriction and vaso-obliteration observed clinically. Newborn kittens were exposed

to various oxygen concentrations for various lengths of time, after
which they were returned to room air until they matured. At maturity,
the animals were euthanized, their blood vessels filled with india
ink, and their retinas removed and placed flat on glass slides for
examination.

One improvement we made to this model was to replace the ink
injection method for visualizing microscopic retinal vessels by one
in which 500 mg/kg horseradish peroxidase (HRP) is injected i.v.
into the anesthetized animal and permitted to circulate for 15 minutes. Following euthanasia, the retinas are prepared for flat-mount
examination according to a modified version of a method suggested
by Raviola and Freddo (1980). Only the functionally open retinal
vascular bed is stained, and erythrocytes, within vessels or extravascular, can also be seen. This technique is superior to the india
ink injection one formerly used in that the stain is circulated by
natural heart action at normal physiological pressures.

More recently, we developed a lead method for ATPase whereby
the vasculature of whole retinas from decapitated animals can be
stained. With this newest method, death is quicker than that produced by drug injection, thereby permitting retinas to be obtained
from animals whose physiological parameters are minimally changed
at the time of death. Although this stain is not as dense as that
obtained from HRP, cells of the blood vessel walls also can be
easily seen with this preparation, and a more accurate "picture" of
the retinal vascular state under normal as well as abnormal conditions can be routinely obtained.

The chief criticism of the RLF animal model, in general, has
been its failure to produce the cicatricial, or end-stage forms of
the retinopathy. Therefore, our recently reported production of cicatricial RLF in the beagle puppy retina (Flower et al., 1981) is
significant in that it increases confidence in the animal model,
and the method by which it was produced suggested a specific mechanism for damaging immature retinal vessels. We determined a regimen
of oral aspirin administration which apparently inhibits the normal
vasoconstriction of the retinal vasculature in response to sustained
elevation of arterial Po_2. Compared to their unmedicated, oxygenated
littermates, puppies given oral aspirin doses during oxygenation
developed significantly more severe retinopathy, including cicatricial RLF. We postulated from these results that retinal vasoconstriction is a protective response to hyperoxia.

A WORKING HYPOTHESIS: RETINAL VASOCONSTRICTION AS A NORMAL PHYSIO-
LOGICAL EVENT

The vasoconstriction observed in oxygenated premature infants
may in fact be only an extreme of a normal physiological response

by which retinal blood flow is modulated during the period of in
utero development. During that period, retinal vascular development
is such that retinal tissue gradually ceases to be totally dependent
for maintenance upon the adjacent choroidal and hyaloid blood flows.
This same mechanism may also be active throughout the perinatal
period.

The possibility then arises that susceptibility of an eye to
oxygen-associated retinopathy may depend upon the extent to which
a protective vasoconstriction response is functional at birth as
well as upon the degree of retinal maturity attained. From this point
of view, it may be argued that cases of "spontaneous" RLF reported
in premature infants never administered oxygen (Foos, 1975), or in
full-term infants (Brockhurst et al., 1975), are simply examples of
inadequate retinal vasotonia at birth for protection of structurally
immature vessels. We also speculated that the apparently strong vasoconstriction response normally present in both the puppy and kitten
may account for the fact that cicatricial RLF was never produced in
these animals until vasoconstriction was apparently inhibited by
aspirin administration.

An implicit assumption of the foregoing is that vasoconstriction
is in fact a physiological property of the perinatal retinal vasculature. Justification for making such an assumption is found in the
recent fetal and neonatal lamb retinal blood flow studies of Peeters
et al. (1980). Their results indicate that autoregulation of blood
flow takes place in the perinatal eye. They reported that in the
lamb eye, when arterial blood pressure increases at birth, choroidal
blood flow increases but retinal blood flow does not. The simplest
explanation for this observation is that some degree of retinal vasoconstriction occurs at birth concomitantly with the rise in blood
pressure, and thereby the retinal vasotonia recognized as clinically
normal is established.

Compared to its blood supply after birth, the immature retina
receives venous-like blood in utero. Approximate values for those
parameters that characterize arterial blood in utero are: Po_2 =
25 Torr, Pco_2 - 45 Torr, and blood pressure = 25 to 30 Torr. But at
birth, blood oxygen content rises (Po_2 = 70 Torr) as does blood
pressure (35 to 40 Torr), and Pco_2 drops to about 35 Torr. It therefore seems naive to ignore that breathing room air causes blood gas
levels to assume values that are quite abnormal for a premature
infant. The effects of arterial hyperoxia (especially during oxygen
breathing) on immature retinal vessels have long been the subject of
a great deal of interest, but the possible consequences of changes
in arterial Pco_2 and blood pressure that occur at birth should also
be considered.

On the assumption that retinal vasotonia is a predisposing
factor in development of RLF, two mechanisms associated with peri-

natal changes in those factors that characterize the arterial circulation can be invoked to explain immature-retina vascular damage.
The first concerns elevated arterial Po_2 and attributes vascular
damage to increased contact between developing peripheral capillary
walls and blood-borne oxygen known to be capable of damaging cell
membranes. This could explain the more severe retinopathy in the
oxygenated puppies whose vasoconstriction was aspirin-inhibited. On
this basis, the animal model data might support the suggestion of
Ashton and Peddler (1962) that retinal vessel damage may result from
direct cytotoxic effects of oxygen on endothelial cells, although
extensive electron microscopic evaluation of our animal retinas has
not been completed. Compared to those of littermates in which normal
vasoconstriction occurred during oxygen breathing, the dilated
retinal blood vessels of the aspirin-treated puppies would have
allowed a greater blood flow through the immature vessels, those
farthest from the optic nerve. Walls of the immature capillaries in
the vasoconstriction-inhibited animals would therefore have been in
contact with greater concentrations of oxygen; endothelial cells of
the vessel wall can be damaged by oxygen-induced peroxidation of
lipids contained in them.

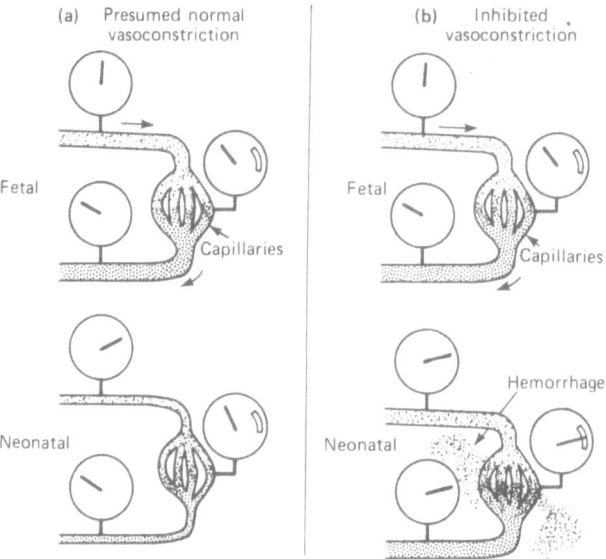

Fig. 1. Schematic representation of the changes in transmural pressure that might occur in the peripheral retinal vasculatures
of fetal and neonatal eyes. (a) Pressure relationships resulting from a presumed normal degree of vasoconstriction.
(b) Pressure relationships resulting from inhibited vasoconstriction; under such circumstances it is possible for
hemorrhage to occur.

The second mechanism takes into account occurrence of a presumed normal degree of vasoconstriction at birth concomitant with
rise in arterial blood pressure as shown in Fig. 1a. However, vasoconstriction would not be uniform throughout the vasculature as
shown. (This schematic simplification was used only in order to make
a point about the peripheral capillary transmural pressure.) Vasoconstriction, nevertheless, could add resistance to blood flow
throughout the retinal vasculature in such a way that the most peripheral, and hence the most structurally immature, capillaries would
experience the smallest increase in transmural pressure. However,
failure of such vasoconstriction to occur at birth (Fig. 1b) could
result in excessively high transmural pressures, possibly producing
retinal capillary hemorrhage. This second mechanism might conceivably
work independently of or concomitantly with the first.

TESTING THE WORKING HYPOTHESIS

Again we returned to the experimental RLF animal model in order
to examine the effects of changes in those parameters that characterize arterial blood during the perinatal period.

Fig. 2 summarizes results of experiments performed to show how
the immature retinal vasculature responds to manipulation of arterial
Pco_2 levels while holding blood pressure constant. The middle fundus
photograph shows the condition of the retinal vessels of an anesthetized puppy while it breathed room air. After the oxygen content
breathed was increased, causing an elevation in arterial Po_2 without
changing either Pco_2 or blood pressure, the retinal vessels constricted and some became obliterated as shown in the bottom photograph. Then the breathing gas mixture was manipulated so that arterial
Pco_2 was also elevated while arterial Po_2 remained high and blood
pressure remained unchanged. The top fundus photograph in the figure
shows that, with elevated arterial Pco_2, those vessels that constricted or became obliterated reopened to calibers exceeding those
observed during room-air breathing. The vasodilation persisted in
the presence of elevated arterial Po_2 even when the applied arterial
Pco_2 dropped as low as 70 Torr. Results of these experiments led to
the impression that the tendency for immature retinal vessels to
dilate in response to elevated arterial Pco_2 is greater than their
tendency to constrict in response to elevated arterial Po_2.

Thus, in keeping with the aforementioned hypothesis that vasoconstriction plays an important role in the developing retinal
vasculature, the response of the RLF animal model to changes in
arterial Pco_2 can be considered analogous to changes in retinal
vascular status throughout the perinatal period. The top fundus
photograph of Fig. 2 can be thought of as representing the retinal
vasculature in utero. The middle fundus photograph then represents
retinal vasculature status at birth (i.e., clinically normal retinal

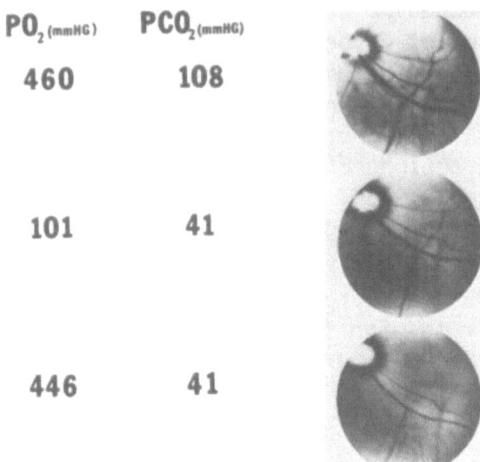

PO$_2$ (mmHG)   PCO$_2$ (mmHG)

460   108

101   41

446   41

Fig. 2.   Response of the immature retinal vasculature to manipulation of arterial Pco$_2$ level while holding systemic blood
pressure constant. The middle fundus photograph shows the
condition of the retinal vessels of an anesthetized puppy
that was breathing room air. The bottom fundus photograph
shows the constricted state of the same retinal vessels
after the puppy breathed oxygen raising the arterial Po$_2$
but not affecting arterial Pco$_2$. The top fundus photograph shows retinal vessels, both arteries and veins, more
dilated then during either air-breathing or oxygen-breathing
states as a result of simultaneously breathing a high oxygen
and high CO$_2$ concentration gas mixture.

vasotonia), and the bottom fundus photograph is consistent with the
known retinal vasculature response in the newborn breathing a high
concentration of oxygen.

These observations led to the performance of experiments aimed
at producing retinopathy via the latter of the two hypothetical
mechanisms proposed above. Specifically, as an alternative to
aspirin administration, hypercapnia produced by chronic inspiration
of 10% CO$_2$ was used as a method of inducing vasodilation. Litters of
newborn puppies were randomly divided into two groups. The first
group was reared for a period of three days in an environment of
humidified 10% CO$_2$/90% O$_2$, and the other littermates were reared in
an environment of humidified 10% N$_2$/90% O$_2$. The mother nursed each
group alternately during the three day period, and then both groups
were returned to her and allowed to mature to 20 days of age. At
that time the puppies were euthanized and their retinas were
examined.

An evaluation of the preliminary data suggested that the vascular patterns in both groups of littermates were abnormal, but the CO_2/O_2-reared groups had vascular anomalies distinctly different from those in the N_2/O_2-reared groups. The major retinal vessels of the CO_2/O_2 puppies were significantly dilated in comparison to those of the N_2/O_2 littermates, even though the animals had been removed from the high CO_2 environment for more than two weeks before euthanasia. Moreover, whereas the capillary densities in the N_2/O_2 puppies tended to be uniform throughout the retina, capillary densities in the CO_2/O_2 littermates were significantly higher in the peripheral retina than elsewhere. At the microscopic level, the retinas of the N_2/O_2 puppies showed characteristic oxygen-induced vascular changes. Note the randomness of the capillary net pattern and the presence of an abnormal shunt-like vessel (arrow) permitting virtually direct communication between artery and vein in Fig. 3 (N_2).

In Fig. 3 (CO_2), the peripheral retina of a CO_2/O_2 littermate is shown. In this case the peripheral capillaries have such large diameters that they resemble the sinusoid-like capillaries of the choroid more than they do the retinal capillaries. Their appearance is consistent with the pressure within them having risen to the point where they are dilated beyond the elastic limits of their walls, resulting in their ballooned-out appearance.

Although this interpretation of the data supports the working hypothesis, it represents a concept sufficiently contrary to tradi-

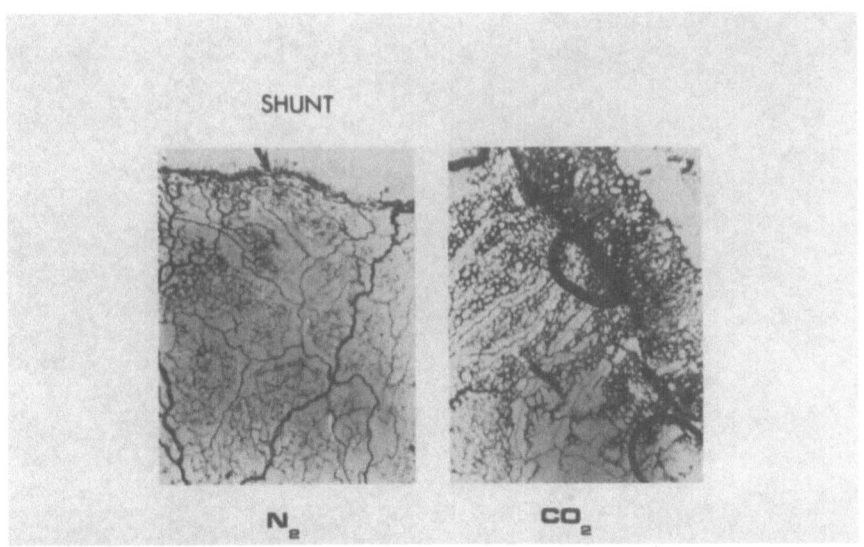

Fig. 3. Micrographs comparing the peripheral retinal capillaries of two puppy littermates: (left) maintained for 3 days in a 10% N_2/90% O_2 environment, and (right) maintained for 3 days in 10% CO_2/90% O_2 environment.

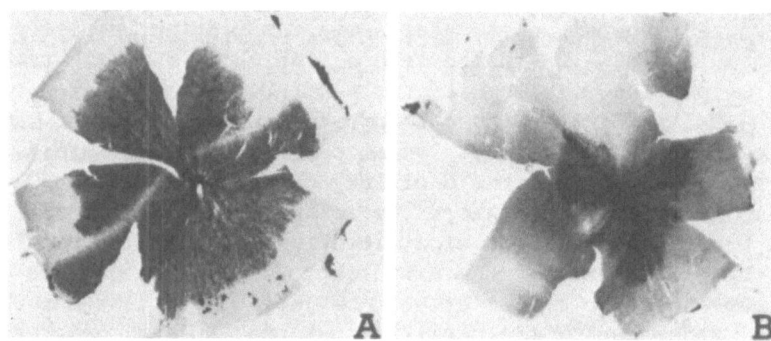

Fig. 4.   Flatmounted whole retinas from a 2-day old beagle puppy (A)
          and a 2-day old kitten (B).

tional thinking about RLF that still further investigation of the
transmural pressure damage hypothesis was warranted.

NORMAL PATTERN OF RETINAL VASCULAR DEVELOPMENT

     In order to provide baseline data on the normal pattern of
beagle puppy retinal vascular development, retinas from litters of

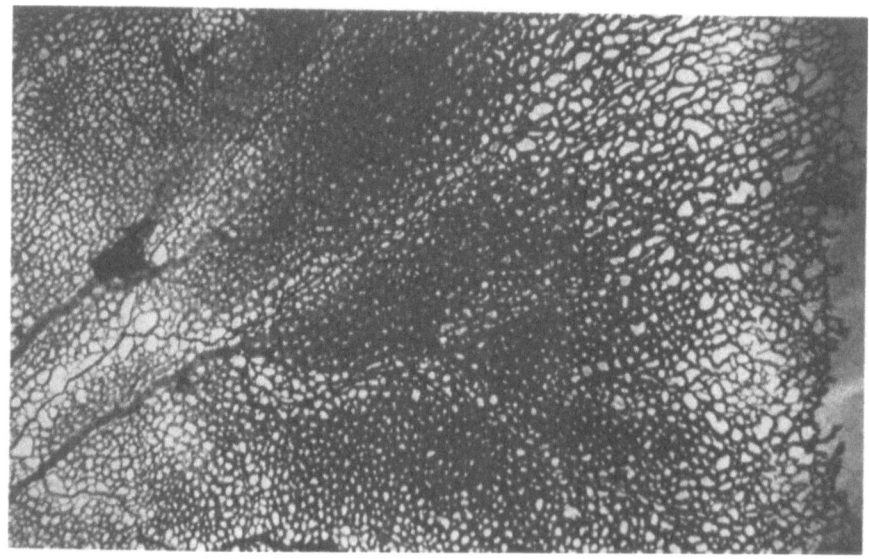

Fig. 5.   Micrograph of a beagle puppy retina showing the various
          stages of retinal vascular development. The most mature
          vessels are near the disc toward the left, and the most
          immature vessels are at the far periphery toward the right.

puppies reared in room air only were examined at various ages from one- to 22-days old. (Initially retinas prepared by the HRP method were evaluated, but when the lead method was eventually perfected, it was used to prepare retinas from other air-reared litters. When comparisons were made between retinas prepared by the two methods but from identically exposed animals, no significant differences were observed in the gross patterns of anatomical structures which characterize retinal vascular development.)

At full-term birth vascularization of the beagle puppy retina (A) is significantly more advanced than that of the newborn kitten (B), as demonstrated in Fig. 4. In fact, the extent of puppy retinal vascularization more closely resembles that of the seven-month-gestation human infant than does that of the kitten, suggesting that the puppy retina may be the better model to study. By day two in the puppy, the sides of three major lobes of vascularization which are still quite evident in the kitten have coalesced to form a retinal

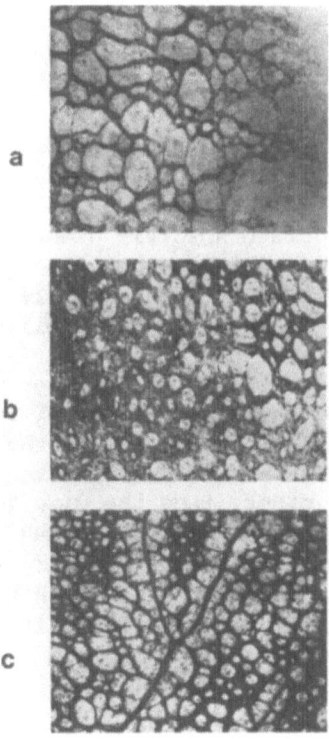

Fig. 6.   Higher magnifications of representative areas of the retina in Fig. 5. (A) Mature retinal vasculature. (B) The middle zone of large diameter immature retinal vessels. (C) Most immature peripheral retinal vessels.

vascular pattern nearly radially symmetrical with respect to the disk.

In general, vascularization of the retina tends to acquire radial symmetry at all angles with respect to the disk as vessel growth progresses from the disk towards the periphery. The developing puppy retina can be typified in terms of three approximately concentric annuli, or zones, centered about the disk. All three zones are represented in the segment of retina shown in Fig. 5, and each of the three are shown at higher magnification in Fig. 6.

The annulus nearest the disk consists of fairly mature vessels, as demonstrated in Fig. 6A, resembling those of the adult. Narrow capillary segments, relatively long with respect to their diameters, assume fairly random orientations, and the spaces between adjacent capillaries are large. The impression gained is that most of the intercapillary spaces have five or more sides. In this figure, a developing narrow capillary-free zone is identifiable on both sides of an arteriole. Right-angle branchings from arterioles into the capillary beds may be seen also.

Distal to this zone of fairly mature vessels is a zone consisting of complexes of large diameter vessels resembling the "immature retinal vessels" identified by Michaelson (1954) in india-ink-injected kitten retinas (see Fig. 6B). As Michaelson observed, these large vessel complexes tend to be more closely associated with the venous rather than the arterial circulation, but for the most part it was difficult to associate specific elements within these complexes in the puppy retinas with one or the other circulation. The intervascular spaces defined by these vessels tended to be mostly round or oval and small relative to the capillary vessel diameters.

At the distal edge of the large diameter vessel complexes lies a narrow zone containing the most peripheral vessels. These formed a network of smaller diameter vessels like the ones in Fig. 6C. It appears that as age increases, the various zones described above, expanded one behind the other, from the disc toward the retinal periphery. As this process proceeded, various stages of remodelling apparently take place within the zone of larger diameter "immature" vessels. The impression gained was that within those complexes of vessels, arterioles became differentiated first, then capillaries, and finally venules, ultimately producing a vascular bed like that shown in Fig. 5.

EFFECTS OF EXCESSIVE TRANSMURAL PRESSURE ON SUBSEQUENT RETINAL VASCULAR DEVELOPMENT

Once the normal pattern of retinal vascular development was established, it became possible to attempt to separate the presumed

oxygen toxic effect from that presumably associated with excessive
transmural capillary pressure. Randomly selected puppy littermates
were reared for 3 days in humidified 10% CO_2/90% air, and the
remaining littermates served as air controls. At approximately 3
weeks of age, retinas of the exposed and control puppies were examined and evaluated.

Resistance to blood flow through mature vascular beds like
those of Fig. 6A would be considerably higher than through the vascular beds like those of Fig. 6B. Consequently, were it not for the
resistance to blood flow normally imposed by the retinal arteries
and arterioles which distribute blood to various segments of the
retinal vasculature, blood flow would simply be through beds of the
least resistive vessels. In the case of the immature retina, the
most compliant and least resistive vessels would likely be those
with the least structurally mature walls; such vessels largely make
up the peripheral retinal vasculature. If resistance to blood flow
through most major retinal arteries and arterioles were significantly lowered, then blood flow through the immature peripheral vessels
would increase, and higher transmural pressures placed across their
walls could cause them to dilate.

Such a description seems compatible with conditions which may
have existed in the retinal vasculatures of puppies reared in 10%
CO_2/air for three days and then allowed to mature prior to euthanasia.

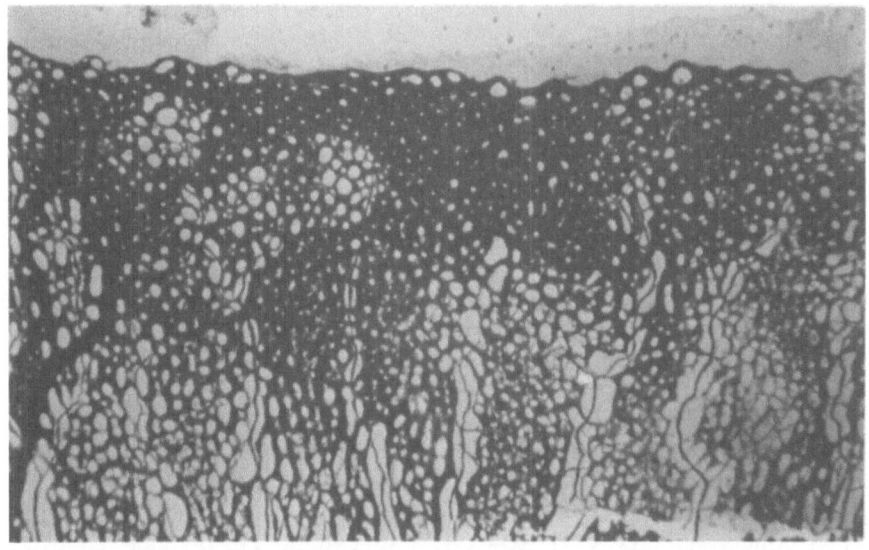

Fig. 7.   The peripheral retinal vessels of a beagle puppy (approxi-
          mately 3 weeks of age) reared initially for 3 days in 10%
          CO_2/90% air.

The main difference between their retinas, represented by the one
in Fig. 7, and those of air-reared littermates appeared to be the
absence of the narrow zone of peripheral small diameter vessels
(c.f. Fig. 5). Disappearance of that zone could be simply explained
by the diameters of the peripheral vessels having increased to equal
those of the adjacent, larger diameter vessel complexes.

In two of the CO_2/air-reared littermates whose ocular media
were sufficiently clear to permit ophthalmic examination prior to
euthanasia, abnormal vessels resembling those found in clinical RLF
were observed and photographed. These particular vessels were also
found to leak intravenously injected sodium fluorescein dye. Conceivably, break down of the blood/retina barrier also might be linked
to high transmural pressures across retinal vessels, causing them to
dilate and producing excessively high tensions in their walls.

HYALOID ARTERY HEMORRHAGE

Additional evidence to support the concept that excessive
transmural pressure may be a mechanism for damaging immature retinal vessels came from a retrospective review of beagle puppy fundus photographs.

At two days of age, when exposure to the various breathing gas
mixtures began in our experiments, both the hyaloid artery and
tunica vasculosa lentis are still patent in the normal puppy. By
five to seven days of age, atrophy of the hyaloid artery advances to
the stage that blood in the tunica no longer significantly inhibits
fundus ophthalmoscopy, but pulsating blood in the hyaloid artery can
often be seen extending more than 1 mm into the vitreous. Beyond
this stage, atrophy of the residual stalk of the artery appears to
be highly variable, but by 25-30 days of age it usually is not easily
identifiable.

A review of the color fundus photographs made prior to euthanasia of 62 puppies reared in either 10% N_2/90% O_2 or 100% O_2 indicated
that in 31 (50%), hemorrhage had occurred at the base of the hyaloid
artery. Considering the N_2/O_2 puppies only, 8 of 13 had hemorrhaged
(62%); and considering the O_2 puppies only, 23 of 49 had hemorrhaged
(47%).

However, hemorrhage occurred in only one of 24 puppies (4%)
reared in 10% CO_2/90% O_2. These data were interpreted as supporting
the idea that administration of 10% CO_2/90% O_2, even when administered concomitantly with 90% O_2, can reduce resistance to blood
flow through the retinal vasculature. In a puppy breathing 10% N_2/
90% O_2 or 100% O_2 continuously for three days, severe constriction
of the retinal vasculature must reduce blood flow and result in an
elevation of the pressure head in retinal arteries near the disc.

Of these, the atrophied hyaloid artery is most likely to leak if
the intra-arterial pressure head rises sufficiently. The same puppy
breathing 10% CO_2/90% O_2 would be spared such a blood pressure rise
in vessels near the disc if elevated Pco_2 inhibited downstream vasoconstriction.

The interpretations made of the foregoing data are consistent
with the idea that redistribution of pressure gradients across
various regiments of the immature retinal vasculature and subsequent
elevation of transmural pressures is a mechanism by which normal
vascular growth may be altered. The effects of elevated arterial
Pco_2 may be viewed as destructive due to the fact that they ultimately led to production of abnormal peripheral vessels, presumably
by either of the two mechanisms hypothetized in the beginning of
this paper. That is, CO_2-inhibited vasoconstriction could have increased the amount of blood-borne O_2 in contact with developing
vessel walls excessively beyond that which would have occurred with
O_2 alone (mechanism 1), or excessive transmural pressure alone
could have mechanically produced the damage (mechanism 2, as in
Fig. 1). The fact that similar damage was observed in puppies exposed to elevated CO_2 but not O_2, however tends to favor the latter
mechanism.

Previous investigators concluded that 5% CO_2 inhalation had
no effect on development of RLF in the animal model (Patz, 1955;
Ashton et al., 1954). Results of this preliminary study indicate
that 10% CO_2 inhalation does. Conclusions based upon data obtained
from retinas nearly three weeks after exposure regarding the specific mechanisms by which inspired CO_2 affects vascular growth are
compelling but tentative. Nevertheless, elevated arterial Pco_2 clearly must be considered a potential factor which in addition to elevated Po_2 may affect RLF pathogenesis.

RETINAL VASCULAR STATUS DURING EXPOSURE

Improved methods for visualization of the immature animal
retinal vasculature (especially the lead method) made it possible for
the first time to investigate the status of the retinal vasculature
during exposure to various breathing gas mixtures. This was recently
done in a series of experiments in which divided litters of puppies
were euthanized without interruption of their ambient environments
at the end of three day's exposure to 10% CO_2/90% O_2 and 10% N_2/90%
O_2. The retinas shown in Fig. 8 typify the results. Clearly, the
peripheral retinal vessels from animals A, B, and C which were in
10% N_2/90% O_2 became constricted and signs of vessels obliteration
are also evident. By comparison, peripheral vessels from littermates
D, E, and F which were in 10% CO_2/90% O_2 remained dilated; in fact,
the peripheral capillaries - especially those immediately adjacent
to the vascularized retina - appeared dilated even in comparison

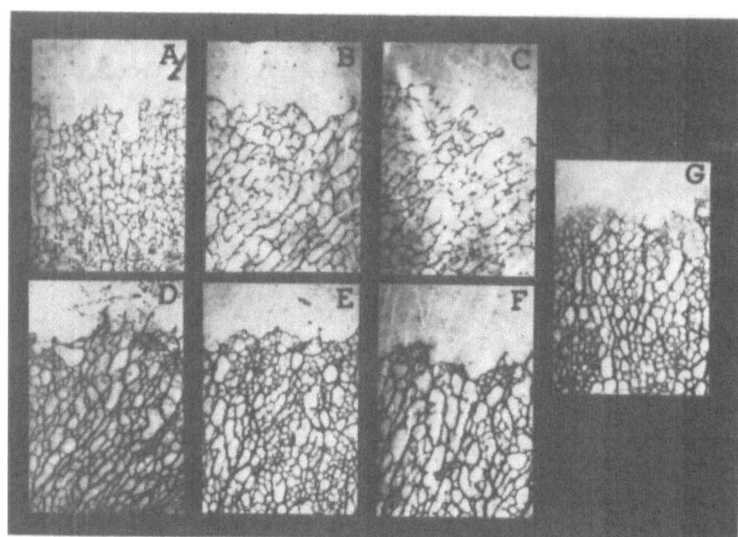

Fig. 8.   The peripheral retinal vasculatures of three puppy litter-
          mates following 3 days exposure to 10% N_2/90% O_2 (A, B,
          and C) compared to those of three littermates following 3
          days exposure to 10% CO_2/90% O_2 (D, E, and F). The retina
          of an air control littermate of the same age (G).

to an air control littermate G. This behaviour is in agreement with
that previously described in animals at a significant time following
exposure.

     The most interesting data, however, came from retinas taken
from puppies after only 4 hours of exposure. Retinas from these
puppies exposed to mixtures containing CO_2 had features compatible
with those from puppies immediately following 3 days exposure or even
with those which matured to three weeks of age following exposure.
However, retinas from puppies placed in 100% O_2 showed signs of
vessel constriction, including shutdown of capillary beds, within
the zone of most mature vessels near the disk and dilation of the
entire surrounding peripheral vasculature. In Fig. 9 the peripheral
vasculature of an O_2 exposed puppy (A) is compared to that of an
air-control littermate (B). Note the absence of capillary filling
around the large vein in the lower right-hand corner of photograph A.
Fig. 10 shows the peripheral capillaries of those same retinas at
much higher magnification; note the "leakage" of erythrocytes in
the O_2-exposed puppy retina (A).

     Apparently, those major retinal vessels in the O_2 exposed
puppies mature enough to respond to changes in blood gas levels
began to constrict. Then initially, when capillary beds lying in the

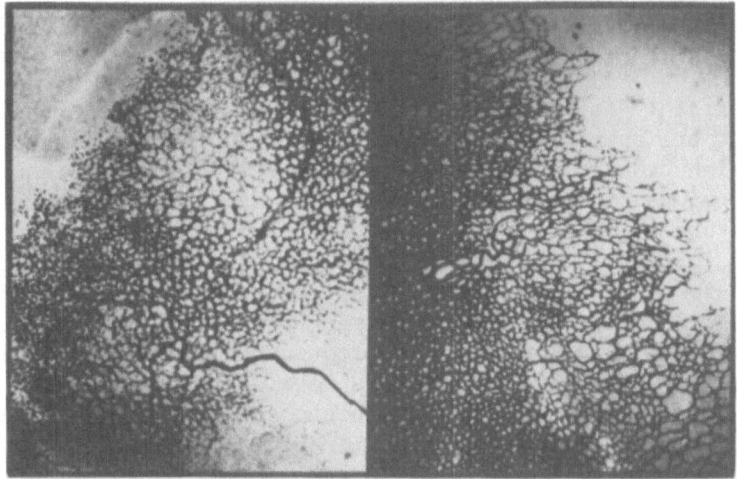

Fig. 9.   The peripheral retinal vasculature of an O_2 exposed puppy
           (A) compared to that of an air control littermate (B).

zone nearest the disk shut down, blood flow was shunted through the
peripheral vasculature; this condition could conceivably persist
until the mature major vessels constricted to the extent that
resistance to flow through them became sufficiently high to reduce
blood flow to the entire vasculature. Thus, with respect to the
transmural pressure experienced by the most peripheral immature

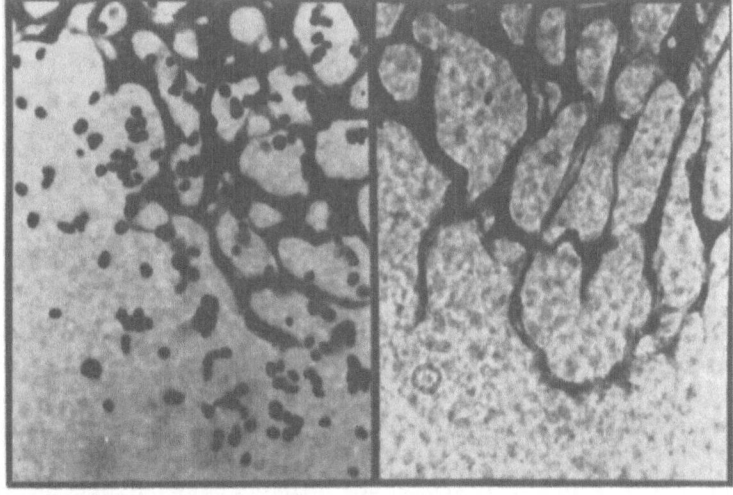

Fig. 10.   Higher magnification micrographs of the peripheral areas
           of the retinas shown in Fig. 9.

retinal vessels, the effect of the <u>initial</u> vasoconstrictive response of the immature retinal vasculature to elevated P_{O_2} could be identical to that produced by elevation of P_{CO_2}.

Certainly at this time data on retinal vascular status during exposure are preliminary and not conclusive, but in light of the data which precede them, they too support the concept that excessive transmural pressure may be an initial insult to the immature retinal vasculature which triggers subsequent development of retinopathy. The possibility that factors other than oxygen alone may contribute to genesis of RLF is compelling; so is the possibility that clinically observed RLF may be only a narrow range of a broad spectrum of retinopathies that could be produced in the immature retina if appropriate physiological parameters were varied outside the current clinically acceptable range. In this context, the significance of these preliminary investigations is that investigators - and clinicians especially - become altered to look for correlations between incidence of RLF and factors other than oxygen alone.

ACKNOWLEDGEMENT

This work was supported in part by NIH grants EY-02482 and EY-00205.

REFERENCES

Ashton, N., and Pedler, C., 1962, Studies on developing retinal vessels: IX. Reaction of endothelial cells to oxygen, <u>Br. J. Ophthalmol.</u>, 46:257.

Ashton, N., Ward, B., and Serpell, G., 1953, Role of oxygen in the genesis of retrolental fibroplasia: a preliminary report, <u>Br. J. Ophthalmol.</u>, 38:397-432.

Ashton, N., Ward, B., and Serpell, G., 1954, Effect of oxygen on developing retinal vessels with particular reference to the problem of retrolental fibroplasia, <u>Br. J. Ophthalmol.</u>, 38:397.

Brockhurst, R.J., and Christi, M.I., 1975, Cicatricial retrolental fibroplasia: Its occurrence without oxygen administration and in full term infants, <u>Albrecht von Graefes Arch. Klin. Exp., Ophthalmol.</u>, 195:113.

Flower, R.W., and Blake, D.A., with Wajer, S.D., Egner, P.G., McLeod, D.S., and Pitts, S.M., 1981, Retrolental fibroplasia: Evidence for a role of the prostaglandin cascade in the pathogenesis of oxygen-induced retinopathy in the newborn beagle, <u>Pediatr. Res.</u>, 15:1293.

Foos, R.Y., 1975, Acute retrolental fibroplasia, <u>Albrecht von Graefes Arch. Klin. Exp. Ophthalmol.</u>, 195:87.

Michaelson, I.C., 1954 "Retinal Circulation in Man and Animals", Charles C. Thomas, Springfield.

Patz, A., 1955, Experimental studies, Am. J. Ophthalmol., 40:174.

Peeters, L.L.H., Sheldon, R.E., Jones, M.P. jr., and Battaglia, F.C., 1980, Retinal and choroidal blood flow in unstressed fetal and newborn lambs, Pediatr. Res., 14:1047.

Raviola, F., and Freddo, T.F., 1980, A simple staining method for blood vessels in flat preparations of ocular tissues, Inv. Ophthalmol. Vis. Sci., 19:1518.

REACTIONS OF CAROTID BARORECEPTORS TO ANOXIA AND ISCHAEMIA

U. Müller, and W. Wiemer

Institut für Physiologie, Universitätsklinikum
Hufelandstr. 55, 4300 Essen, FRG

Already in 1935, Bogue and Stella recorded carotid sinus nerve activity during ischaemia, and observed a long lasting activation which they attributed to carotid chemoreceptors. However, according to experiments by Wiemer and Ott (1963) this activity originated mainly in baroreceptor fibres whereas chemoreceptor activity decreased after a short initial excitation. Having studied the response of chemoreceptors to anoxia and ischaemia in more detail (Wiemer et al., 1981; Müller et al., 1982), we now extended these investigations to carotid baroreceptors.

METHODS

The carotid sinus region of anaesthetized rabbits was exposed and functionally isolated from its arterial connections. Following the cannulation of the common carotid artery the carotid bifurcation could be perfused with (a) normoxic blood from the systemic circulation, (b) anoxic blood (Po_2 < 1 Torr, Pco_2 34.5 - 40.7 Torr) provided by a blood gas exchanger, or (c) was rendered ischaemic by occlusion of the normal blood supply. Perfusion pressure was controlled by means of a pressure generator which permitted the application of defined pressure steps. In addition, during normoxia the carotid sinus could be subjected to the systemic pulsatile pressure via a silicon rubber tubing connecting the sinus area to the common carotid artery. The effluent blood drained out through a catheter inserted in the external carotid artery and was passed to the external jugular vein via an arterio-venous shunt. Blood flow was kept constant (0.3 ml/min) during normoxia and anoxia as well. When ischaemia was produced the shunt was fully opened in order to drop carotid sinus pressure to 0 mm Hg.

Activity was recorded from baroreceptor fibres of the cut carotid sinus nerve by means of bipolar platinum electrodes. Most experiments were done on single fibre preparations, but in a few cases recordings were also performed on filaments containing two or three fibres, if the amplitudes of the action potentials were different enough to permit a clear distinction between the discharges of individual fibres.

The experiments regularly began with control recordings of the baroreceptor activity under normoxic conditions; these included recordings during both normal pulsatile pressure and steady pressure levels varied in steps of 20 mm Hg between 0 and 160 mm Hg. The adapted discharge rate was quantified 30 s after the step variation and was taken as a measure of the "static response" characteristics. In some experiments using ramp-shaped increasing and decreasing pressure steps the "dynamic responses" were also measured. Subsequently, the perfusion pressure was adjusted to a constant level between 50 and 90 mm Hg.

In 20 experiments anoxia was performed by switching the preparation to anoxic blood without changing the perfusion pressure. While in part of the preparations the pressure was kept permanently on this level, in 7 preparations static responses, and in 5 also dynamic responses were tested as under control conditions. After the fibres had stopped firing during steady pressure, the excitability was again tested by a series of pressure steps, and - if no responses could be evoked any more - the preparation was reconnected to normoxic blood at the same level of steady pressure. After the activity had returned, in 9 of the preparations static and dynamic responses were tested again.

Ischaemia was produced by decreasing the carotid sinus pressure from the level of the initial normoxic perfusion to 0 mm Hg. The experiments were complicated by the fact that part of the baroreceptors discontinue their activity below a certain pressure level ("threshold" fibres), while others only decrease their firing rates and, after a transient undershoot or short silent period during the pressure fall, continue to fire at reduced rates even at 0 mm Hg ("non-threshold" fibres). For better assessment of the reactions at this low level, 22 "non-threshold" fibres in comparison to only 4 "threshold" fibres were used in these experiments. In most cases the pressure was kept at zero level until the activity ceased (in "non-threshold" fibres) or returned temporarily and ceased again (in "threshold" fibres). In some experiments the ischaemia was interrupted by single pressure steps testing the dynamic response. After cessation of the activity, most preparations were reconnected directly to normoxic blood at control pressure; in other experiments, the ischaemia was terminated by a series of increases and decreases of the sinus pressure between 0 and 80 mm Hg, or by switching the preparation back to normal pulsatile, instead of standard steady pressure.

RESULTS

Lack of Significant Reactions in the Beginning of Anoxia, or Ischaemia.

In most experiments with anoxia, the activity of baroreceptor fibres did not change markedly upon switching the preparation from normoxic to anoxic blood, as long as the sinus pressure was kept constant. In 12 of these preparations, the initial firing rate was maintained rather steadily within the first minutes of anoxic perfusion. In 2 preparations the activity decreased more or less continuously; 6 preparations showed irregular fluctuations of the firing rate which became especially distinct after 12 - 18 min of anoxia.

Similarly in the experiments with ischaemia, most of the "non-threshold" fibres maintained a remarkably constant "basic" firing rate, whereas 7 fibres showed a gradual decrease, and 2 fibres irregular fluctuations of activity.

Response Changes and Final Paralyzation during Prolonged Anoxia, or Ischaemia.

In all preparations subjected to anoxia or ischaemia, the initial phase of the relative stable firing described above, finally ended in cessation of activity.

During anoxia, the loss of activity of most baroreceptor fibres occurred rather suddenly, following one of two distinct patterns: In part of the preparations, the activity stopped simply by a rapid

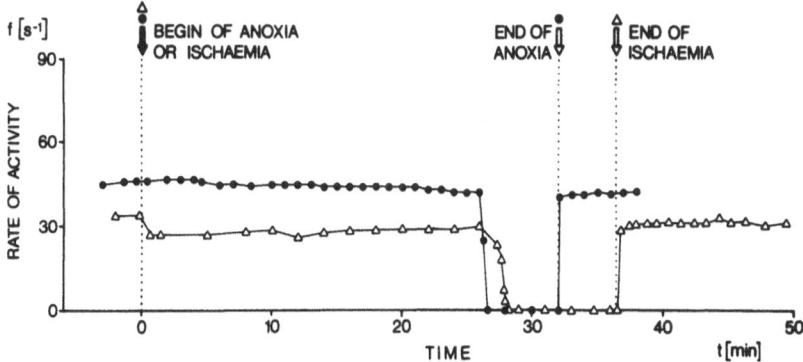

Fig. 1.   Time course of the discharge rate of two baroreceptor fibres showing abrupt cessation of activity during anoxia (closed circles), and ischaemia (open triangles; "non-threshold" fibre). Normoxic blood supply was interrupted in both experiments at time 0, and restored at respective arrows.

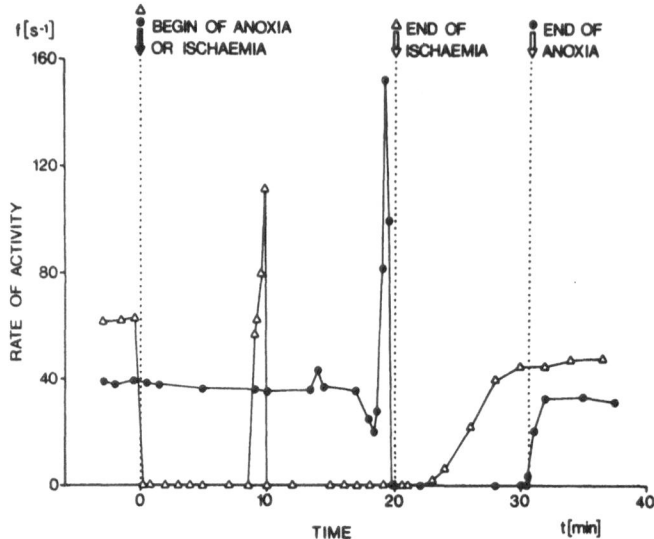

Fig. 2.   Time course of the discharge rate of two baroreceptor
          fibres showing pre-final increase followed by abrupt
          cessation of activity during anoxia (closed circles), and
          ischaemia (open triangles; "threshold" fibre).

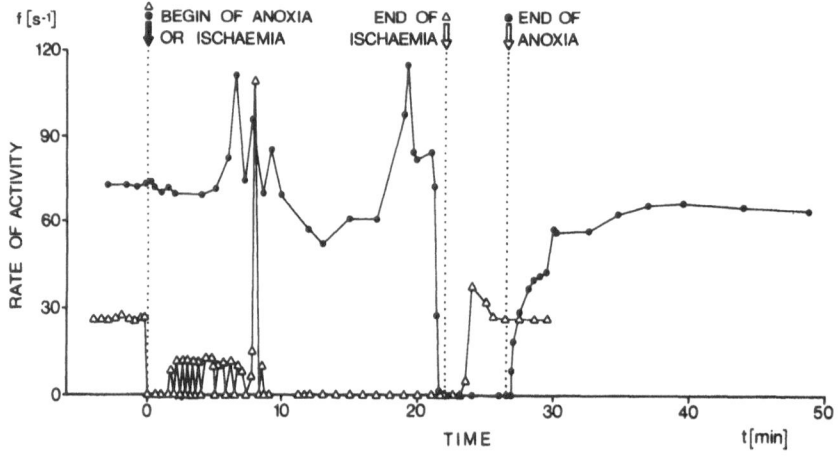

Fig. 3.   Time course of the discharge rate of two baroreceptor
          fibres showing irregular firing and pre-final increase
          followed by abrupt cessation of activity during anoxia
          (closed circles), and ischaemia (open triangles; "threshold"
          fibre).

decline of the firing rate (Fig. 1); in others this breakdown was
preceded by a sudden steep rise of the firing rate (Fig. 2). Only
in a few preparations, the cessation developed from a gradual decrease or from irregularities of the discharge frequency (Fig. 3).

Prolonged anoxia also affected the static and dynamic sensitivity of the receptors: Out of the 7 preparations in which static
responses were tested before cessation of activity, 5 showed an
increase of the absolute threshold and a shift of, at least, the
upper part of the response curve to lower frequency values. Frequently, the fibres also became more and more unable to maintain high
firing rates. An example is presented in Fig. 4: After the disappearance of electrical activity under a steady pressure of 50 mm Hg
(evidently due to anoxic threshold elevation), an increase of the
pressure level to 75 mm Hg was able to restore the firing (A). A
further increase of the pressure to 100 mm Hg (B) still caused a
corresponding initial increase of the firing rate, but was followed
by a complete breakdown of the activity.

Similar to the effects of anoxia on static discharge, the
dynamic response to pressure changes became definitely smaller with
continuing anoxia. Due to the inability of the fibre to maintain
high firing rates, the activity then ceased abruptly during the
increase (Fig. 4, C), and returned during the decrease of pressure
(appearance of an off-response, Figs. 4, C and 5, A_2). Even after
the activity under steady pressure had ceased completely, in some
preparations pressure variations still evoked transient responses.
Subsequently, the preparations became completely insensitive to any
form of pressure stimulus (Fig. 4, D).

The effects of prolonged ischaemia on baroreceptor discharge
proved to be similar to those observed during anoxia: All preparations but one "non-threshold" fibre lost the activity during the
exposure to ischaemia. This one baroreceptor fibre fired rather
steadily for 180 minutes, and was then abandoned. The cessation of
activity observed in other fibres developed - on the average - from
lower firing rates than during the anoxic perfusion. However, it
followed the same main patterns: A rather rapid decline of the
activity was observed in 11 preparations (Fig. 1); 9 preparations
showed a steep rise preceding the abrupt breakdown of the discharge
rate. The recordings of the remaining 6 preparations revealed
gradual decreases and/or fluctuations before the activity was
finally lost. The latter two patterns could even be recognized in
the 4 "threshold" fibres, as these interrupted the prevailing
silence by transient bursts or a steep rise of activity which afterwards ended in cessation (Figs. 2, 3). The dynamic responses to
pressure stimuli during ischaemia (Fig. 6) were affected in a similar way as in anoxia. Following the stop of the static discharge,
most fibres became totally insensitive to further pressure stimuli.

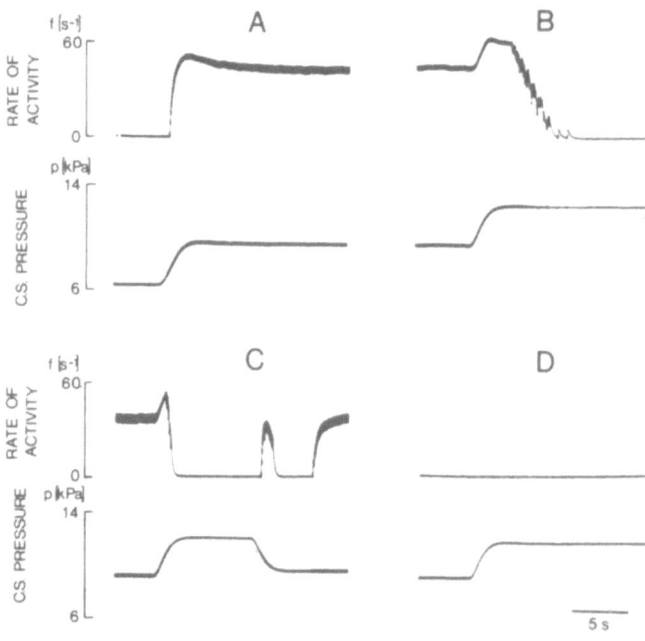

Fig. 4.   Response of a baroreceptor fibre to step variations of
          carotid sinus pressure during continuing anoxia after 50
          (A), 55 (B), 60 (C), and 66 min of exposure (D). Prior to
          the first recording, the activity during constant perfusion
          at 50 mm Hg (6.7 kPa) had ceased after 45 min of anoxia.

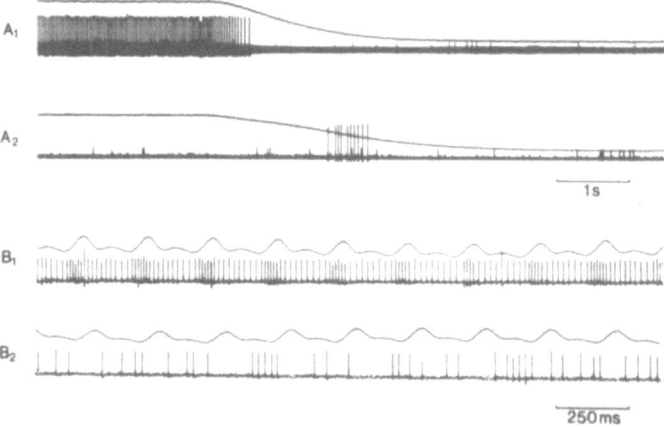

Fig. 5.   Responses of baroreceptor fibres A_1, A_2: to a ramp decrease
          of carotid sinus pressure during normoxia (A_1) and after
          80 min of anoxia (A_2). B_1, B_2: to normal pulsatile pressure
          before (B_1) and after 118 min of ischaemia, normoxic blood
          supply having been restored for 40 sec (B_2).

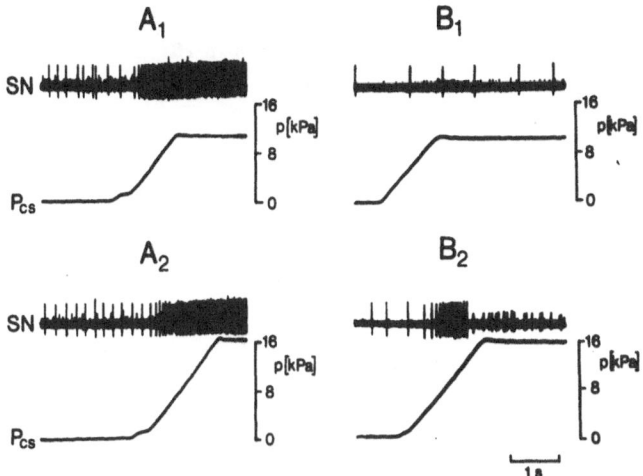

Fig. 6.   Responses of a baroreceptor fibre to ramp increases of
          carotid sinus pressure during continuing ischaemia; A_1,
          A_2: about 1 min after reducing sinus pressure to O.; B_1,
          B_2: after 31 min, and 31.5 min of exposure, respectively.
          A_1, B_1: pressure increase to 80 mm Hg (10.7 kPa); A_2, B_2:
          pressure increase to 150 mm Hg (16.0 kPa).

        The survival time of the static activity during anoxia varied
widely between 3.2 and 85 minutes (mean $\pm$ S.D.: 29.7 $\pm$ 23.5 min).
However, the majority of the fibres lost their function after anoxic
periods ranging from 10 to 30 minutes. During ischaemia the survival
times of the baroreceptor discharge were observed to vary between
7.7 and 111.5 minutes (mean 37.0 $\pm$ 35.8 min). Again, most fibres
ceased to function after 10 to 30 minutes of ischaemia. One preparation which at first was subjected to anoxia and subsequently -
after a period of recovery - underwent ischaemia revealed only
slightly different survival times amounting to 19.8 and 21.5 minutes,
respectively. In addition, the distribution of the survival times
in ischaemia proved to be very similar to that observed in fibres
subjected to anoxia (Fig. 7). Moreover, the survival times were
found to be correlated neither to the initial firing rate (varying
between 28.0 and 77.5 imp/s) nor to the total number of action
potentials (9 500 - 251 000) produced during the anoxic perfusion.

Recovery after Return to Normoxia

        After reconnecting the preparations to normoxic blood (3 to 84
min after cessation of activity, average 15.7 min), all but 3 preparations regained their activity (the latter preparations were abandoned after 30, 50, and 60 min of observation). Latencies until the
beginning of firing varied from 5 sec to 11.3 min (average 2.0 $\pm$
3.3 min) in the anoxic, and 5 sec to 10 min (average 1.3 $\pm$ 2.3 min)

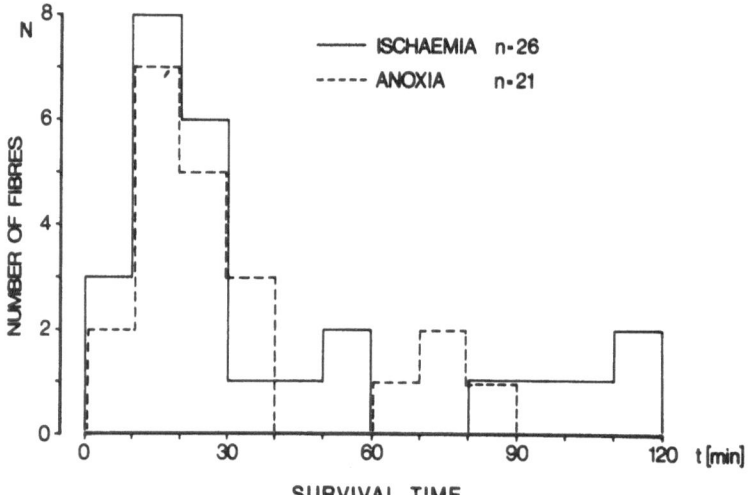

Fig. 7. Distribution of survival times (latencies between stop of
         normoxic blood supply and cessation of activity) in baro-
         receptor preparations submitted to carotid sinus anoxia,
         or ischaemia.

in the ischaemic experiments (Fig. 8). No relation could be detected
between the total duration of anoxia or ischaemia and the time
delay for the occurrence of the first impulse discharged after the
restitution of the normoxic blood supply. In most experiments the
resumption of activity was followed by a rather rapid rise of the
discharge rate which further increased to values equal to or slightly

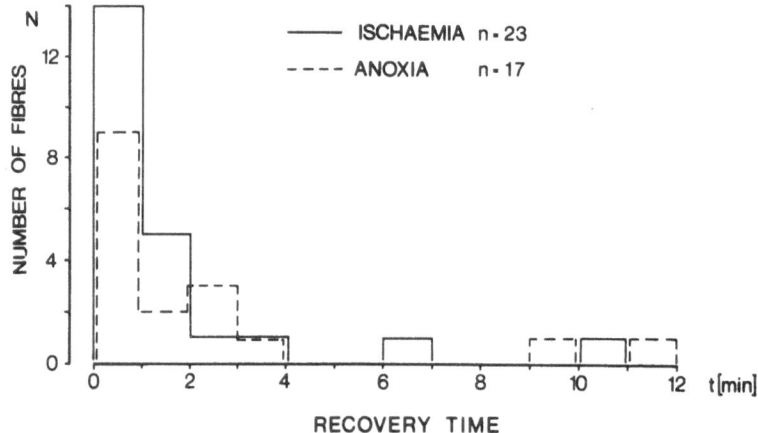

Fig. 8. Distribution of recovery times (latencies between restora-
         tion of normoxic blood supply and return of activity) in
         baroreceptor preparations after exposure to carotid sinus
         anoxia, or ischaemia.

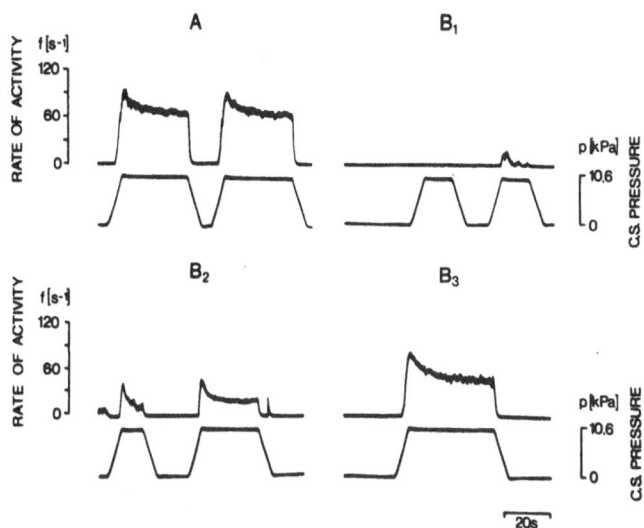

Fig. 9. Recovery of a baroreceptor fibre from prolonged ischaemia.
A: Response to ramp pressure steps immediately after
dropping carotid sinus pressure to 0; B_1 - B_3: response
to a sequence of 5 similar pressure steps after 20 min of
ischaemia.

below the pre-anoxic control levels. Occasionally, a transient
overshoot of the firing rate above the control level could be observed immediately after return to normoxia. In the course of this
recovery, the stages of paralyzation of receptor sensitivity were
passed in reverse sequence: Thus in one experiment, in which normoxic blood was returned to the ischaemic preparation by successive
single pressure pulses instead of steady pressure, the first pulse
proved ineffective, the second evoked only a transient "on"-response,
whereas the fourth already caused a (however reduced) reaction of
the normal type (Fig. 9). Similarly, ischaemic preparations reconnected to normal pulsatile pressure at first proved unable to
respond to the systolic pressure increases, producing only diastolic
bursts in reverse to the normal pattern of baroreceptor activity
(Fig. 5, B). The time until a new steady state of activity was
reached varied from 10 sec to 17 min (average 6.9 ± 6.1 min) in the
post-anoxic, and 20 sec to 16.0 min (average 4.4 ± 4.3 min) in the
post-ischaemic preparations. At this stage the response curve to
static pressure was found to be shifted only slightly to lower frequency values as compared to the pre-anoxic state.

Consequently, in experiments with anoxia as well as ischaemia,
the sequence of paralyzation and successive recovery could be repeated several times in the same preparation. The resulting survival
times, although varying greatly from one preparation to the other,
proved to be remarkably constant in the same preparation. The

recovery times, however, varied similarly as in experiments done on
different preparations and showed no clear tendency to longer time
values with respect to the progressive number of anoxic or ischaemic
periods.

DISCUSSION

    The experiments demonstrate that carotid baroreceptors can
sustain activity for a considerable period of oxygen lack, although
oxidative metabolism is principally indispensable also for the
function of these receptors. Our results have shown that there are
two different patterns of cessation of activity during prolonged
anoxia, or ischaemia: with and without pre-final increase of activity. Similar patterns were previously found in other mechano-
receptors, as in muscle spindles (Mathhews, 1933), Lorenzini receptors (Hensel, 1957), stretch receptors (Giacobini and Przybylski,
1971), or in central neurons (Kolmodin and Skoglund, 1959; Speckmann
et al., 1970). In view of the present results, the question arises
whether different processes are involved in both alternatives. We
suppose that at least two different processes participate in the
action of oxygen lack. A first process may affect the receptor
itself and produce a gradual depolarization of the membrane as known
for spinal and cortical neurons (Collewijn and Harreveld, 1966 a,
b). As a consequence the fibre discharge will be enhanced (pre-final
increase) until the spike generating mechanism is inactivated. A
second process may directly depress the spike-initiating zone or
the terminal portion of the sensory axon, hence elevating the threshold for impulse generation and conduction (Paintal, 1959). As a
result the fibre discharge would be abolished abruptly.

    This concept is also compatible with the findings that the
responses of baroreceptor fibres to static pressure and pressure
variations are markedly altered during anoxia or ischaemia. The
elevation of the threshold pressure as well as the abrupt cessation
of activity in response to high pressure steps may be due to the
action of oxygen deficiency on the terminal portion of the fibre.
Thus the failure of the fibres to maintain high firing rates may be
interpreted as a limitation of the maximum transmissable frequency
of discharge as it has been described by Paintal (1967) for the
effect of cold on the conduction of action potentials. Alternatively,
the effect may be attributed to a gradual anoxic depolarization of
the receptive ending. Additional depolarization elicited by the
stimulus may then depress the spike generation. Accordingly, the
paradoxical off-response to decreasing pressure can be explained by
a hyperpolarization, which may transiently compensate for the per-
sistent anoxic depolarization and thus enables the spike-initiating
zone to produce a burst of action potentials.

The survival times in anoxia or ischaemia were shown to be
quite constant in an individual fibre, but they were also shown to
differ very markedly from one fibre to another. This variability
can be interpreted in several ways: (1) Structural and functional
differences among the receptors; (2) different receptor locations,
and (3) different environmental parameters.

Regarding the first point, the great intervals between short
and long survival times suggest that there are discrete types of
baroreceptors with high and low susceptibilities to oxygen deficiency. Such types, however, could not be defined on the basis of the
observed frequency distributions of the survival times. In contrast,
the shapes of the histograms were rather homogeneous (Fig. 7).
Nevertheless, functional and morphological differences between individual receptors might not always be discrete but rather continuous.
A corresponding continuous variability was found in some aspects of
the receptors responses to adequate stimuli (Wiemer et al., 1974;
Müller and Wiemer, 1982). In addition, histological observations
revealed a diversity of forms of baroreceptors rather than distinct
types (Rees, 1967). However, we failed to correlate the survival
times to parameters, such as threshold pressure, static or dynamic
response, and the total number of action potentials produced in
anoxia. Thus the variability of survival times should be attributed
to differences in the metabolism of the receptors, or to other
factors than the receptors themselves.

With respect to the second point, it is known that carotid
baroreceptor terminals are confined to the deep adventitia and to
the medio-adventitial border of the sinus wall. Thus the distances
of individual receptors to the vessel lumen should not vary by more
than 20 um. These differences seem too small to explain the great
variability of survival times by a gradient of oxygen tension.

Concerning the third point, influences of environmental parameters cannot be excluded. The receptive structures are embedded
in a tissue which is presumably not metabolically inert. In addition, it is conceivable that baroreceptors are shielded by terminal
Schwann cells in a more or less effective manner. This mechanism,
however, is expected to affect both survival time and recovery time
uniformly. Yet such an effect was not found. On the other hand, the
preparation of the sinus might cause an artificial difference
between the ventral and dorsal portion of the sinus. The latter is
attached to the underlying tissue which is well supplied with
normoxic blood, thus receiving oxygen by means of diffusion. But we
do not think that these differences are great enough to account for
the marked variability of survival times.

The similarity between the distributions of the survival times
in anoxia and ischaemia can theoretically be due to random coinci-
dences. The observations, however, indicate that both hypoxic

conditions have nearly the same effects on the activity of the
receptors. Thus pre-final increases and irregularities of activity
correspond in both conditions, and in an experiment including anoxia
and subsequent ischaemia the survival times proved to be very
similar. This suggests that the most important factor influencing
the activity of baroreceptor fibres in both anoxia and ischaemia is
the lack of oxygen.

ACKNOWLEDGEMENT

The investigation was supported by the Deutsche Forschungsgemeinschaft (Wi 165 and SFB 114).

REFERENCES

Bogue, J.Y., and Stella, G., 1935, Afferent impulses in the carotid
        sinus nerve (nerve of Hering) during asphyxia and anoxaemia,
        J. Physiol., 83:459.
Collewijn, H., and Harreveld, A. van, 1966 a, Membrane potential of
        cerebral cortical cells during spreading depression and
        asphyxia, Exp. Neurol., 15:425.
Collewijn, H., and Harreveld, A. van, 1966 b, Intracellular recording
        from cat spinal motoneurons during acute asphyxia, J. Physiol.,
        185:1.
Giacobini, E., and Przybylski, A., 1971, Studies on hypoxia in
        isolated neurons: II. The effect of hypoxia on the impulse
        activity and intracellular K/Na ratio of the slowly adapting
        stretch receptor neuron (SRN), Int. J. Neurosci., 1:163.
Hensel, H., 1957, Die Wirkung verschiedener Kohlensäure- und
        Sauerstoffspannungen auf isolierte Lorenzinische Ampullen von
        Selachiern, Pflügers Arch. ges. Physiol., 264:228.
Kolmodin, G.M., and Skoglund, C.R., 1959, Influence of asphyxia on
        membrane potential level and action potentials of spinal moto-
        and interneurons, Acta Physiol. Scand., 45:1.
Mathhews, B.H.C., 1933, Nerve endings in mammalian muscle, J.
        Physiol., 78:1.
Müller, U., Prühs, D., Oehlke-Unterholzner, C., and Wiemer, W.,
        1982, Comparison of carotid chemoreceptor and baroreceptor
        reactions to anoxia and ischaemia (in press).
Müller, U., and Wiemer, W., 1982, Stimulus-response characteristics
        of carotid baroreceptors in the rabbit, Pflügers Arch., 394:R53.
Paintal, A.S., 1959, Facilitation and depression of muscle stretch
        receptors by repetitive antidromic stimulation, adrenaline and
        asphyxia, J. Physiol., 148:252.
Paintal, A.S., 1967, Block of conduction in mammalian myelinated
        nerve fibres by low temperature, J. Physiol., 180:1.
Rees, P.M., 1967, Observations on the fine structure and distribu-
        tion of presumptive baroreceptor nerves at the carotid sinus,
        J. Comp. Neurol., 131:517.

Speckmann, E.-J., Caspers, H., and Sokolov, W., 1970, Aktivitäts-
    änderungen spinaler Neurone während und nach einer Asphyxie,
    Pflügers Arch., 319:122.
Wiemer, W., Kaack, D., Kezdi, P., Brügge, C., and Zmijewski, M.,
    1974, Response characteristics of carotid baroreceptors to
    steady pressure, in: "Symposium Mechanoreception", J. Schwartz-
    kopff, ed., Abh. Nordrhein. Westf. Akad. Wiss., 53:77.
Wiemer, W., and Ott, N., 1963, Erregung und nachfolgende Lähmung
    der Chemorezeptoren durch Ischämie des Sinus Caroticus, Z.
    Biol., 113:395.
Wiemer, W., Prühs, D., and Oehlke-Unterholzner, C., 1981, Carotid
    chemoreceptor reactions to ischaemia and anoxia, in: "Arterial
    Chemoreceptors", C. Belmonte, D.J. Pallot, H. Acker, S. Fidone,
    eds., Leicester University Press, Leicester, p. 344.

O_2 CHEMORECEPTORS

O_2 CHEMORECEPTION OF THE CAT CAROTID BODY IN VITRO

M.A. Delpiano, and H. Acker

Max-Planck-Institut für Systemphysiologie
Rheinlanddamm 201, 4600 Dortmund 1, FRG

Since the fundamental research of de Castro (1926) and of
Heymans and Bouckaert (1930), the carotid body has been considered
to be a peripheral chemoreceptor which transduces changes in arterial Po_2 and arterial Pco_2/pH into nerve signals. These signals
predominantly regulate ventilation via the respiratory center and
thus help to control the arterial blood gas level. However other
organs such as the heart and the kidneys can also be influenced by
nerve signals from the carotid body (Daly and Scott, 1958; Korner,
1963). Recently it has been found that the carotid body seems to be
involved in very common diseases such as hypertension (Honig et al.,
1981; Trzebski et al., 1982) as well as in sudden infant death syndrome (Naeye et al., 1976). The transducing process enabling the
chemoreceptor to respond to changes in blood gases is unknown. If
one regards the different cell types in the carotid body tissue as
a complex of cells which collaborate in the chemoreceptive process,
it can be postulated that there exists a Po_2 or Pco_2/pH dependent
transmitter release from these cells which excites nerve fibres connected to these cells (Hayashida et al., 1981). It is our intention,
to support this idea with our measurements of the tissue Po_2, extracellular Ca^{2+} activity and cyclic AMP content during hypoxia and
hypercapnia in the cat carotid body in vitro.

The first figure shows an electron micrograph of the cat carotid
body cell elements. The type-I cell is characterized by many dense-
cored vesicles whereas the glia-cell-like type-II cell is free of
dense-cored vesicles and surrounds both the type-I cell and its adjacent nerve endings (McDonald and Mitchell, 1975; Verna, 1979).

The isolated superfused carotid body in vitro according to
Eyzaguirre and Lewin (1961) is a good model for investigating the

705

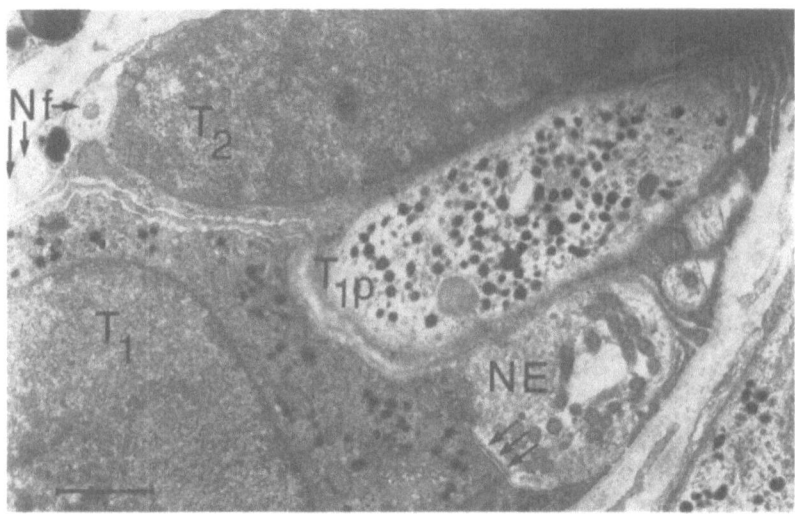

Fig. 1.   Electron micrograph of a cat carotid body showing a type-I
          cell (T_1) with nucleus and dense-cored vesicles in the
          cytoplasm, a type-I cell process (T_{1p}), a type-II cell
          (T_2), unmyelinated nerve fibres (Nf) and a nerve ending
          (NE) showing a synaptic connection with a type-I cell
          (arrows). Bar = 1.0 um.

direct interaction between stimuli such as Po_2 or Pco_2/pH and the
chemosensory response. This <u>in vitro</u> model is no longer influenced
by changes in either total or local blood flow or by efferent nerve
discharge which could make the interpretation of the data more
complex.

      Hypoxia was produced in the superfused preparation either by
altering the Po_2 in the superfusion medium which was equilibrated
with different gas mixtures by two gas mixing pumps (Wösthoff,
Bochum, FRG) or by stopping superfusion. The Po_2 of the medium was
continuously monitored by a Po_2 catheter electrode (Lübbers et al.,
1969) and the Po_2 in the carotid body tissue was measured polarographically with platinum electrodes according to Baumgärtl and
Lübbers (1973). The chemoreceptor discharge was recorded from multi-
fibre filaments from the cut end of the sinus nerve as reported previously (Delpiano and Acker, 1980). The extracellular ion activity
in the carotid body tissue was measured with two-channel glass ion-
selective microelectrodes (Dufau et al., 1980). The cyclic AMP
content of the whole cat carotid body was determined with a very
sensitive radioimmunoassay kit from New England Nuclear (NEN).

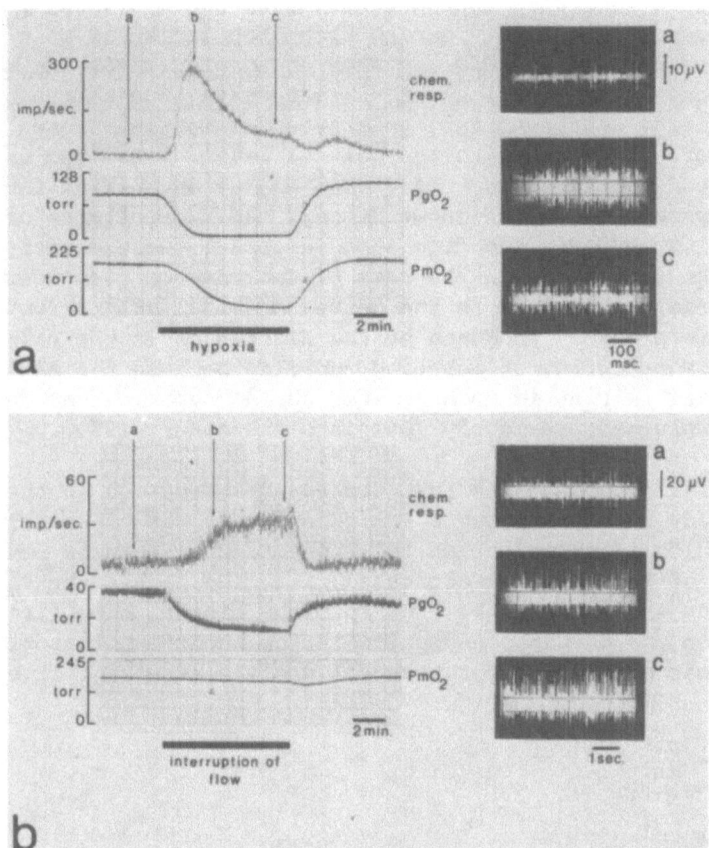

Fig. 2. Carotid body responses to hypoxia and interruption of flow. Fig. 2a shows the relationship between chemoreceptor discharge (summated activity, imp/s), tissue Po_2 in the carotid body (P_go_2/Torr) and Po_2 in the superfusion medium (P_mo_2/Torr) under hypoxia. Fig. 2b shows the same parameters as in Fig. 2a but now during interruption of flow. Bar indicates 2 min. At a, b, and c original registration of the nerve discharge.

Fig. 2a shows a typical example of a hypoxic response. When the medium Po_2 (P_mo_2) was decreased, tissue Po_2 (P_go_2) decreased as well. The chemoreceptor nerve activity increased about half a minute after the initial P_go_2 decrease. After a maximum had been reached, the chemoreceptor nerve activity declined in spite of the low P_go_2. However, the nerve activity remained higher than in the control. On returning to the normal solution, P_go_2 increased and the nerve activity in the first instance decreased and then slightly increased before returning to the control level. Fig. 2b shows that when the superfusion flow was stopped, the chemoreceptor nerve discharge,

after a delay increased concomitantly with the declining P_gO_2.
The chemoreceptor activity during flow stop increased more slowly
than during hypoxia and did not show an adaptation during the
stimulation. After the flow was started again, the chemosensory
nerve activity decreased very quickly and sometimes showed an undershoot before it returned to the control level. These two examples
demonstrated the importance of tissue Po_2 as an initial factor in
Po_2 chemoreception in the carotid body. The two patterns of responses, i.e. hypoxic and flow stop responses, were clearly different. The difference in the rate of increase of chemosensory nerve
activity and the decline in the nerve activity (adaptation) during
hypoxia are probably produced by the difference in the rate of decrease in P_gO_2 during stimulation and also because the magnitude of
the decrease in P_gO_2 is much greater during hypoxia than during flow
stop as explained previously (Delpiano and Acker, 1980).

In Fig. 3 one can see that the Po_2 distribution in the cat
carotid body in vitro shows a Po_2 gradient in the tissue, the steepness of which is dependent on the P_mO_2. As the P_mO_2 was lowered the
steepness of the gradient decreased, i.e. the higher the P_mO_2, the
steeper the gradient of the P_gO_2 into the tissue. As a first approximation the oxygen consumption in the superfused carotid body
in vitro was calculated for each Po_2 profile (see Fig. 3) by taking

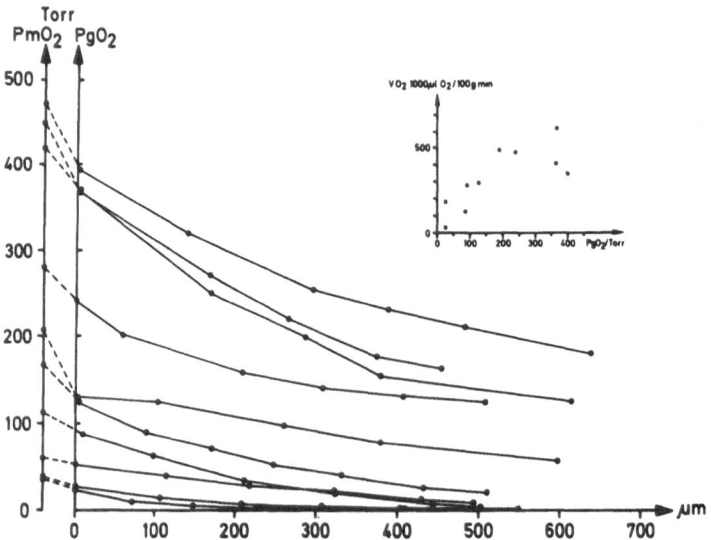

Fig. 3.   Tissue Po_2 (P_gO_2) gradients in the cat carotid body in re-
          lationship to the medium Po_2 (P_mO_2). The P_gO_2 and P_mO_2 are
          given in Torr on the y-axis. The depth of puncture of the
          Po_2 needle electrode (um) is shown on the x-axis. The
          inset shows carotid body oxygen consumption versus the
          tissue Po_2 at the surface.

the P_{O_2} difference between the surface and a depth of 400 um, according to Krogh (1919a, b). The calculated mean oxygen consumptions are given in the inset of Fig. 3. It shows that the calculated value of the mean oxygen consumption is dependent on the P_{O_2} at the surface of the carotid body. That means, in a first estimation for the superfused carotid body in vitro, that the steepness of the P_{O_2} profiles in the tissue is not only determined by oxygen diffusion into the tissue at different $P_{m}O_2$, but also by an oxygen consumption which is dependent on the oxygen pressure in the medium. The fact that the low values of the $P_{g}O_2$ in the center of the carotid body (Fig. 3) are in the range of about 3-2 Torr but did not reach zero, supports this conclusion. It must be investigated in this respect as to whether each P_{O_2} profile itself results from a P_{O_2} dependent oxygen consumption in the tissue as found in vivo and in the perfused carotid body in vitro (Acker and Lübbers, 1977). The absolute value of oxygen consumption in vitro is about three times lower than that found in vivo. Although there is no complete explanation of why the carotid body in vitro shows a lower metabolic rate, one can propose several factors for this. These include a possible substrate deficiency (Starlinger and Lübbers, 1976) and that it lacks both vascular (Purves, 1970) and efferent nerve inputs (Mitchell and McCloskey, 1974). In spite of this lower metabolic rate, the carotid body in vitro shows a relationship between oxygen partial pressure and chemosensory nerve discharge similar to that found in vivo (Hornbein, 1968).

By changing the P_{O_2} field in the carotid body evidence exists that the type-I cells release transmitters which probably then generate action potentials in the neighbouring nerve endings (see for review Eyzaguirre and Fidone, 1980). Fidone et al. (1981) could show that dopamine is liberated during hypoxic stimulation of the rabbit carotid body in vitro. Since it is known that such transmitters like dopamine and noradrenaline transfer their information into the cell with the aid of the second messenger cyclic AMP, we investigated the c-AMP content of the whole carotid body in connection with P_{O_2} changes.

Cyclic AMP is increased during hypoxia in the cat carotid body in vitro as shown in Fig. 4. After two hours of superfusion under normoxic conditions, two minutes of hypoxic stimulation induce a small but significant increase ($p < 0.05$, Lord-test, unilateral) in the cyclic AMP content of the cat carotid body (Delpiano et al., 1982, in press). These findings are consistent with a neurohumoral release of catecholamines during hypoxia in the carotid body. This process seems to be regulated by Ca^{2+} like in other organs (Harvey and MacIntosh, 1940). It is known from synapses and secretory cells that the level of cytosolic calcium can determine the amount of transmitters which are released. This was also shown by Grönblad et al. (1980b) for the rat carotid body. They demonstrated by electron microscopy that the application of the Ca^{2+}-transporting ionophore

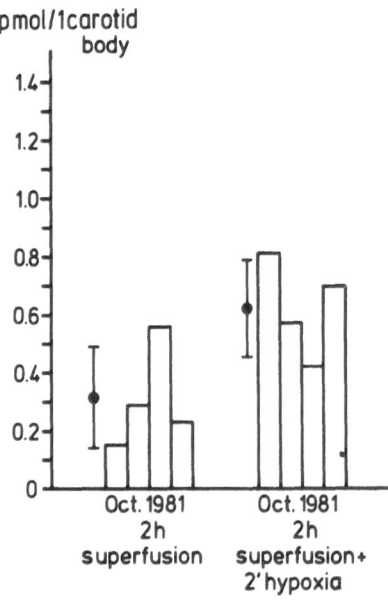

Fig. 4. Cyclic AMP of the cat carotid body <u>in vitro</u> under normoxia
(P_mO_2: 180 Torr) and hypoxia (P_mO_2: 20 Torr), after 2 hours
of superfusion.
Each bar represents the mean value of two carotid bodies
taken from the same cat. Point represents means $\pm$ SD.

A23187 in the presence of calcium in the medium, induces an increased exocytosis of dense-cored vesicles which contain catecholamines. The level of cytosolic calcium is known to be regulated by
different cellular mechanisms such as calcium exchange by the mitochondria, active and passive ion exchange across the cell membrane
and calcium binding proteins in or near the cell membrane. Although
the extracellular calcium concentration is normally stable, each
functional alteration of the cell associated with activation of such
cellular processes can be reflected in variation of the extracellular ionic Ca^{2+} (Nicholson, 1980). Following this idea, we measured
the extracellular Ca^{2+} activity in the tissue of the carotid body
<u>in vitro</u> under hypoxia and hypercapnia using the double-barrelled
ion-selective micro-electrodes. With a calcium concentration in the
medium of 2.0 mmol·l^{-1}, hypoxia induces a decrease in the extracellular Ca^{2+} activity which starts always in all experiments some
seconds later than the typical chemosensory nerve response, as shown
in the lower part of Fig. 5. During hypercapnia, the calcium activity is generally increased. Since these ionic changes could be
also observed in the chronically denervated carotid body (Delpiano
and Acker, 1982, in press), one can assume that these calcium changes
are produced by chemosensory cells. This assumption is supported by
our findings that $CoCl_2$ totally blocks the hypoxic and partially

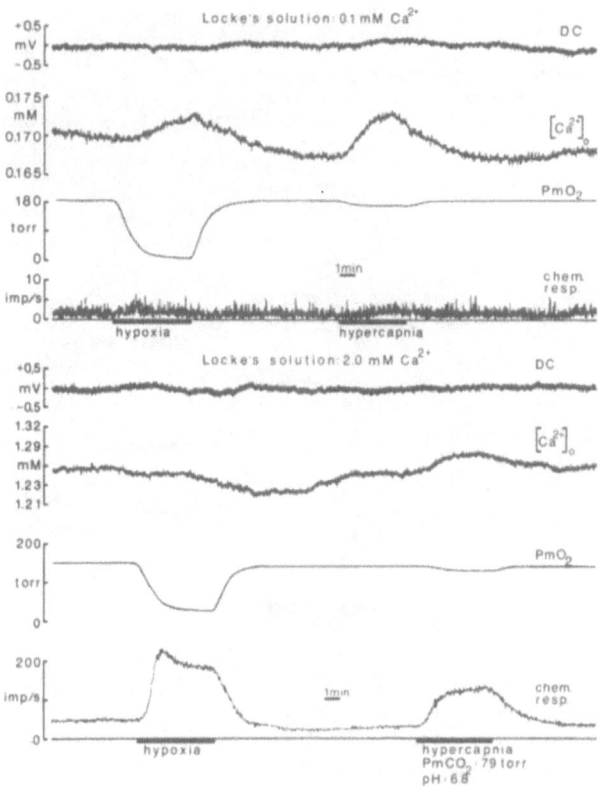

Fig. 5.   Extracellular Ca^{2+} activity response of the intact (un-
denervated) cat carotid body during hypoxia and hypercapnia under normal (2.0 mmol·l^{-1}, lower part) and low calcium (0.1 mmol·l^{-1}, upper part) conditions. Traces from
the top are: DC potential, extracellular calcium activity
$[Ca^{2+}]_o$, P_mO_2 and chemosensory nerve response.

blocks the hypercapnic extracellular Ca^{2+} changes (Delpiano and
Acker, 1981). With respect to the evoked afferent discharge produced
by hypoxia and hypercapnia, it is evident that they are highly dependent on the calcium concentration in the medium as shown in the
upper part of Fig. 5. Here, under a low critical calcium concentration in the medium of 0.1 mmol·l^{-1}, neither a Po_2 - nor a Pco_2/pH
- chemosensory nerve response could be elicited. These experiments
are in accordance with the results of Fidone et al. (1981) who found
that the dopamine release in the carotid body induced by hypoxia was
nearly completely blocked when the calcium in the medium was nearly
zero. In the upper part of Fig. 5 it can also be seen that in spite
of the missing chemosensory nerve response during hypoxia and hypercapnia, the extracellular changes in Ca^{2+} activity are still present

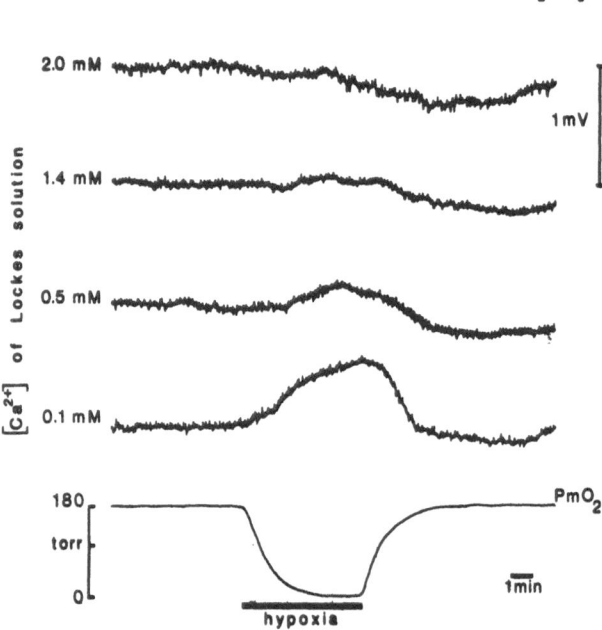

Fig. 6.   Changes in the carotid body Ca^{2+} activity during hypoxia
          versus the calcium concentration in the medium. Calcium
          concentration in the medium is shown at the left. P_mO_2 is
          shown at the lower part of the figure.

but reversed during hypoxia. The dependence of this change of the
extracellular Ca^{2+} activity during hypoxia is shown in Fig. 6 more
detailed at different Ca^{2+} concentration in the medium. As the Ca^{2+}
concentration in the medium was lowered, the extracellular Ca^{2+} activity during hypoxia is reversed gradually. With a normal calcium
concentration of 2.0 mmol·l^{-1} in the medium, the Ca^{2+} activity in
the carotid body decreases during hypoxia. This decrease however
seems to be accompanied by an increase which is only more evident
when the calcium concentration in the medium is lowered (1.4 - 0.5
mmol·l^{-1}), until a distinct increase in the extracellular Ca^{2+} activity can be observed with the low calcium concentration of 0.1
mmol·l^{-1} (Fig. 6).

These response patterns were always observed in each experiment. These extracellular responses of the Ca^{2+} activity during
hypoxia can be inhibited by calcium antagonists such as $CoCl_2$ and
D-600 in a dose-dependent fashion (Delpiano and Acker, 1981; 1982,
in press). Therefore and because of the delayed onset and recovery
of these Ca^{2+} responses, we assume that these extracellular Ca^{2+}
changes are similar to the basically biphasic ion fluxes found in
secretory glands after stimulation (Poggioli et al., 1981). Kondo

and Schulz (1976), using the radioactive Ca^{2+} method for measuring
ion fluxes on isolated pancreas cells of the rat, showed that the
Ca^{2+} influx but not the efflux is decreased by lowering the calcium
in the medium. The increase in Ca^2 activity during hypoxia seen
with low calcium concentration in the medium (0.1 $mmol \cdot l^{-1}$) could
be indicative of an unmasking of a Ca^{2+} efflux which is visible now
due to a decreased Ca^{2+} influx (Fig. 6). The inhibition of this Ca^{2+}
response (increase of Ca^{2+} activity during hypoxia with 0.1 $mmol \cdot l^{-1}$
calcium in the medium) with $CoCl_2$ in a low concentration (2 $mmol \cdot l^{-1}$)
and the fact that $CoCl_2$ in the same concentration also reduced the
activity of a Ca^{2+} dependent ATPase in the cat carotid body
(Starlinger, 1982) supported our assumption. The increase in the
extracellular Ca^{2+} activity observed here during hypercapnia probably represents calcium liberation from some proteins produced by acidification during hypercapnia. The Ca^{2+} ion carrier itself used in
these experiments is unaffected by changes in pH in the range produced with hypercapnia (pH: 7.4 - pH: 6.8) (Dufau, Delpiano, unpublished observation). Our observation that $CoCl_2$, which blocks
the Ca^{2+} reponse during hypoxia, only weakly blocks the Ca^{2+} activity increase during hypercapnia supported this possibility of
the delivery of calcium by acidity. However, the origin of this Ca^{2+}
increase is not yet clear. One possibility could be, that these
changes are analogous to those seen by Petersen et al. (1981) on
isolated mouse pancreatic fragments. They observed that hypercapnia
causes release of $^{45}Ca^{2+}$ from prelabelled pancreatic fragments and
showed evidence indicating that this release is probably due to an
intracellular acidification brought about by the hypercapnia.

We could show here how different factors such as tissue Po_2,
extracellular Ca^{2+} activity and cyclic AMP content of the carotid
body are interconnected with the mechanisms of chemoreception. The
tissue Po_2 in the carotid body in vitro is determined by the Po_2 of
the superfusion medium and the oxygen consumption of the tissue. A
lower oxygen pressure induced a lower oxygen consumption and vice
versa. It would be of interest to investigate how this possible
Po_2-dependent oxygen consumption could be involved in the transduction process of chemoreception. The findings of Mulligan et al.
(1981) suggested that O_2 chemoreception is directly connected to
the energy production and oxygen consumption of the mitochondria of
the chemoreceptor cells. These authors found that oligomycin which
is a relatively specific inhibitor of mitochondrial oxidative
phosphorylation of ADP and ATP stimulated vigorously the afferent
nerve discharge of the sinus nerve but reduced or abolished the
chemosensory nerve response to hypoxia. Although it does not seem
to be, that this effect of oligomycin is produced by the disturbance of the mitochondrial Ca^{2+} exchange as proposed by Roumy and
Leitner (1977), little is known about the possible connection
between the depletion of the energy phosphate pool produced by
oligomycin and the disturbance of Ca^{2+} fluxes at the plasma membrane. Our results with the lowering of the calcium concentration

in the medium and the effect of calcium antagonists support the
findings of Eyzaguirre and Zapata (1968) and of Fidone et al. (1981)
that calcium plays an important role in the Po_2- or Pco_2/pH-depen-
dent transmitter release of the carotid body. At the moment it is
not certainly known which transmitter is involved in this process
of the chemoreception but the changes in the cyclic AMP content of
the cat carotid body in vitro during hypoxia is an additional evidence for the participation of catecholamine in the chemoreception
as well as other transmitters (Eyzaguirre and Fidone, 1980).

When the afferent chemosensory response of the carotid body
was abolished during the superfusion with low calcium concentration
of 0.1 $mmol \cdot l^{-1}$ in the medium (Fig. 6), the extracellular changes
of the Ca^{2+} activity were still present during the hypoxic and
hypercapnic stimulation under these conditions. This fact and the
long time (minutes) taken by these extracellular Ca^{2+} activity
changes to return to control levels, contradict the possibility
that they represent fast ionic changes connected with the neuro-
transmission in chemoreception. More probably, as discussed here
previously, they represent slow ionic changes connected to secretory
or restorativ processes induced by the chemoreception itself.

Finally CO_2 chemoreception seems to have a partially separate
pathway than the O_2 chemoreception. Although the chemoreceptor
fibres were stimulated during hypercapnia, the tissue Po_2 did not
change in this case as under hypoxia (Delpiano and Acker, 1980).
This different pathway was also suggested by Mulligan and Lahiri
(1982) when they demonstrated that CO_2 chemoreception was not
abolished after inhibition of O_2 chemoreception by oligomycin.

ACKNOWLEDGEMENT

The authors are grateful to Dr. H. Starlinger for discussion
and determining the cyclic AMP content in the carotid body, to Dr.
E. Mulligan for helpful criticism and translation and to Dr. D.
Schäfer for electron microscopy. The generous supply of Gallopamil-
Hydrochloride (D-600) from Prof. Kretzschmar of the Knoll AG
Laboratories (Ludwigshafen, FRG) is also gratefully acknowledged.

REFERENCES

Acker, H., and Lübbers, D.W., 1977, The kinetics of local tissue
    Po_2 decrease after perfusion stop within the carotid body of
    the cat in vivo and in vitro, Pflügers Arch., 369:135.
Baumgärtl, H., and Lübbers, D.W., 1973, Platinum needle electrodes
    for polarographic measurement of oxygen and hydrogen, in:
    "Oxygen Supply", M. Kessler, D.F. Bruley, L.C. Clark Jr., D.W.
    Lübbers, I.A. Silver, eds., Urban & Schwarzenberg, München,
    p. 130.

Daly, M., and Scott, M.J., 1958, The effects of stimulation of the carotid body chemoreceptors on the heart rate of the dog, J. Physiol. (Lond.), 144:148.

De Castro, F., 1926, Sur la structure et l'innervation de la glande carotidienne (glomus caroticum) de l'homme et des mammifères, et sur un nouveau système d'innervation autonome du nerf glossopharyngien, Trab. Lab. Invest. Biol. Univ. Madrid, 24:365.

Delpiano, M.A., and Acker, H., 1980, Relationship between tissue Po₂ and chemoreceptor activity of the carotid body in vitro, Brain Res., 195:85.

Delpiano, M.A., and Acker, H., 1981, Extracellular Ca^{2+} and K^+ activity changes in vitro in intact and denervated cat carotid body under hypoxic and hypercapnic stimulation, Pflügers Arch. Suppl., 391:R48.

Delpiano, M.A., and Acker, H., 1982, Extracellular Ca^{2+} and K^+ activities in the cat carotid body in vitro and their relationship to chemoreception, in: "Arterial Chemoreceptors", D.J. Pallot, ed., Leicester University Press, Leicester (in press).

Delpiano, M.A., Starlinger, H., Fischer, M., and Acker, H., 1982, The c-AMP content of the cat carotid body in vivo and in vitro under normoxia and after stimulation by hypoxia, in: "Arterial Chemoreceptors", D.J. Pallot, ed., Leicester University Press, Leicester (in press).

Dufau, E., Acker, H., and Sylvester, D., 1980, Double-barrel ion-sensitive microelectrodes with extra thin tip diameters for intracellular measurements, Med. Progr. Technol., 7:35.

Eyzaguirre, C., and Fidone, S.J., 1980, Transduction mechanisms in carotid body: glomus cells, putative neurotransmitters, and nerve endings, Am. J. Physiol., 239:C135.

Eyzaguirre, C., and Lewin, J., 1961, Effect of different oxygen tension on the carotid body in vitro, J. Physiol. (Lond.), 159:238.

Eyzaguirre, C., and Zapata, P., 1968, A discussion of possible transmitter or generator substances in carotid body chemoreceptors, in: "Arterial Chemoreceptors", R.W. Torrance, ed., Blackwell Scientific Publications, Oxford, p. 213.

Fidone, S., Gonzalez, C., and Yoshizaki, K., 1981, A study of the relationship between dopamine release and chemosensory discharge from the rabbit carotid body in vitro: preliminary findings, in: "Arterial Chemoreceptors", C. Belmonte, D.J. Pallot, H. Acker, S. Fidone, eds., Leicester University Press, Leicester, p. 218.

Grönblad, M., Åkerman, K.-E., and Eränkö, O., 1980b, Exocytosis of amine-storing granules from glomus cells of the rat carotid body induced by incubation in potassium media or media containing calcium and ionophore A23187, in: "Histochemistry and Cell Biology of Autonomic Neurons, SIF-Cells and Paraneurons", O. Eränkö, S. Soinila, H. Päivärinta, eds., Raven Press, New York, p. 227.

Harvey, A.M., and MacIntosh, F.C., 1940, Calcium and synaptic transmission in a sympathetic ganglion, J. Physiol. (Lond.), 97:408.

Hayashida, Y., Koyano, H., and Eyzaguirre, C., 1981, Intracellular recording from chemoreceptor afferent fibres and terminals, in: "Arterial Chemoreceptors", C. Belmonte, D.J. Pallot, H. Acker, S. Fidone, eds., Leicester University Press, Leicester, p. 362.

Heymans, C., and Bouckaert, J.J., 1930, Sinus caroticus and respiratory reflexes, J. Physiol. (Lond.), 69:254.

Honig, A., Habeck, J.O., Pfeiffer, C., Schmidt, M., Huckstorf, Ch., Rotter, H., and Eckermann, P., 1981, The carotid bodies of spontaneous hypertensive rats (SHR). A functional and morphological study, Acta Biol. Med. Germ., 40:1021.

Hornbein, T.F., 1968, The relation between stimulus to chemoreceptors and their response, in: "Arterial Chemoreceptors", R.W. Torrance, ed., Blackwell Scientific Publications, Oxford, p. 65.

Kondo, S., and Schulz, I., 1976, Ca^{++} fluxes in isolated cells of rat pancreas. Effect of secretagogues and different Ca^{++} concentrations, J. Membr. Biol., 29:185.

Korner, P.J., 1963, Effects of low oxygen and carbon monoxide on the renal circulation in unanesthetized rabbits, Circ. Res., 12:361.

Krogh, A., 1919a, The rate of diffusion of gases through animal tissue with some remarks on the coefficient of invasion, J. Physiol. (Lond.), 52:391.

Krogh, A., 1919b, The number and distribution of capillaries in muscles with calculations of the oxygen pressure head necessary for supplying the tissue, J. Physiol. (Lond.), 52:409.

Lübbers, D.W., Baumgärtl, H., Fabel, H., Huch, A., Kessler, M., Kunze, K., Riemann, H., Seiler, D., and Schuchardt, S., 1969, Principles of construction and application of various platinum electrodes, Progr. Respir. Res., 3:136.

McDonald, D.M., and Mitchell, R.A., 1975, The innervation of glomus cells, ganglion cells and blood vessels in the rat carotid body: a quantitative ultrastructural analysis, J. Neurocytol., 4:177.

Mitchell, J.H., and McCloskey, D.I., 1974, Chemoreceptor response to sympathetic stimulation and changes in blood pressure, Respir. Physiol., 20:297.

Mulligan, E., Lahiri, S., and Storey, B.T., 1981, Carotid body O_2 chemoreception and mitochondrial oxidative phosphorylation, J. Appl. Physiol., 51:438.

Mulligan, E., and Lahiri, S., 1982, Separation of carotid body chemoreceptor responses to O_2 and CO_2 by oligomycin and by antimycin A, Am. J. Physiol., 242:C200.

Naeye, R.L., Fisher, R., Rysler, M., and Whalen, P., 1976, Carotid body in the sudden infant death syndrome, Science, 191:567.

Nicholson, Ch., 1980, Modulation of extracellular calcium and its functional implications, Fed. Proc., 39:1519.

Petersen, O.H., Collins, R.C., and Findlay, I., 1981, Effects of CO_2, acetylcholine and caerulein on ^{45}Ca efflux from isolated mouse pancreatic fragments, Pflügers Arch., 392:163.

Poggioli, J., and Putney Jr., J.W., 1982, Net calcium fluxes in rat parotid acinar cells. Evidence for a hormone-sensitive calcium pool in or near the plasma membrane, J. Physiol. (Lond.), 281:383.

Purves, M.J., 1970, The effect of hypoxia, hypercapnia and hypotension upon carotid body blood flow and oxygen consumption in the cat, J. Physiol. (Lond.), 209:395.

Roumy, M., and Leitner, L.M., 1977, Role of Ca^{++} ions in the mechanisms of arterial chemoreceptor excitation, in: "Chemoreception in the Carotid Body", H. Acker, S. Fidone, D.J. Pallot, C. Eyzaguirre, D.W. Lübbers, R.W. Torrance, eds., Springer, Berlin, p. 257.

Starlinger, H., 1982, ATPases of the cat carotid body and of the neighbouring ganglia, Z. Naturforsch., 37c:532.

Starlinger, H., and Lübbers, D.W., 1976, Oxygen consumption of the isolated carotid body tissue (cat), Pflügers Arch., 366:61.

Trzebski, A., Malgorzata, T., Zoltowski, M., and Przybylski, J., 1982, Increased sensitivitiy of the arterial chemoreceptor drive in young men with mild hypertension, Cardiol. Res., 16:163.

Verna, A., 1979, Ultrastructure of the carotid body in the mammals, Int. Rev. Cytol., 60:271.

LOCAL VARIATIONS OF OXYGEN CONSUMPTION WITHIN MULTICELLULAR

SPHEROIDS CALCULATED FROM MEASURED Po_2 PROFILES

U. Großmann, P. Winkler, J. Carlsson[1], and H. Acker

Max-Planck-Institut für Systemphysiologie
Rheinlanddamm 201, 4600 Dortmund, FRG
[1]Dept. Rad. Biol., National Defence Res. Inst.
S-90182 Umeå, Sweden

INTRODUCTION

The influence of oxygen supply on the development of necrotic zones in tumors is still unknown. It is therefore of considerable interest to know whether oxygen consumption in tumors decreases with increasing distance to the site of oxygen supply. If so, O_2 consumption variations must be considered when the size of oxygen depleted zones is calculated for tumor tissue (Tannock, 1968; Thomlinson and Gray, 1959). In order to study these questions Carlsson et al. (1979) measured Po_2 profiles inside multicellular spheroids. In a previous paper (Großmann et al., 1981) trends of oxygen consumption variation within different types of spheroids were calculated from these Po_2 profiles. In the present paper we show an improved method to calculate oxygen consumption profiles. Moreover, we studied the influence of varying O_2 diffusion coefficient and O_2 solubility coefficient on the Po_2 profile and the calculated O_2 consumption curve.

In the first section the mathematical model which describes the O_2 supply of a spheroid is defined. In the next section, in part (a), some theoretical results concerning the influence of varying diffusion and solubility coefficient on the Po_2 profile and the calculated O_2 consumption curve are presented. In part (b), O_2 consumption profiles for spheroids of type U118 MG (malignant glioma) and HTh 7 (human thyroidea carcinoma) are shown.

THE MODEL

In this section the transport and consumption of oxygen in a multicellular spheroid is described mathematically. A multicellular spheroid is considered to be a sphere of radius R. Within this sphere oxygen is assumed to be transported by diffusion. The transport of oxygen by extracellular fluid streams (i.e. convection) is neglected. Oxygen is consumed within the viable cells inside the sphere. Under the assumption of spherical symmetry the steady-state oxygen concentration profile satisfies the following boundary value problem (Crank, 1955):

$$\frac{1}{r^2} \cdot \frac{d}{dr}\left(r^2 \cdot D(r) \cdot \frac{dC}{dr}(r)\right) = \dot{V}_{O_2}(r) \quad ; \quad 0 < r < R, \tag{1}$$

$$\lim_{r \to 0} \left(r^2 \cdot D(r) \cdot \frac{dC}{dr}(r)\right) = 0, \tag{2}$$

$$C(R) = C_O, \tag{3}$$

where $C(r)$ indicates the oxygen concentration (ml O_2/ml) at distance r (um) from the center of the sphere. C_O is the oxygen concentration (ml O_2/ml) at the surface of the sphere, R (um) the radius, $D(r)$ (um^2/sec) the diffusion coefficient depending on r, $\dot{V}_{O_2}$ (ml O_2/(g·min)) the oxygen consumption at distance r from the center. Equation (2) means that the total oxygen diffusion flux at the center of the spheroid vanishes.

Using Henry's law we obtain

$$P(r) = \frac{C(r)}{\alpha(r)} \quad ; \quad 0 \leq r \leq R ; \tag{4}$$

the oxygen partial pressure $P(r)$ (Torr). The oxygen solubility coefficient (ml O_2/(g·atm)) is indicated by $\alpha(r)$.

If $D(r)$, $\alpha(r)$, $\dot{V}_{O_2}$, R, C_O are known, the boundary value problem (1) - (3) has an unique solution. Equation (4) yields the oxygen partial pressure $P(r)$. Thus, boundary value problem (1) - (3) together with (4) may be used to study the influence of local variations of the diffusion coefficient D, the solubility coefficient α, and the oxygen consumption $\dot{V}_{O_2}$ on the Po_2 profile $P(r)$.

a) Let us assume a constant oxygen consumption within the spheroid

$$\dot{V}_{O_2}(r) = \dot{V}_{O_2} \quad ; \quad 0 \leq r \leq R \tag{5}$$

and diffusion coefficient D and solubility coefficient α depending linearly on r,

$$D(r) = D_O + (D_R - D_O) \cdot \frac{r}{R} \qquad ; \qquad O \leq r \leq R \qquad (6)$$

$$\alpha(r) = \alpha_O + (\alpha_R - \alpha_O) \cdot \frac{r}{R} \qquad ; \qquad O \leq r \leq R, \qquad (7)$$

where D_O, D_R indicate the diffusion coefficients at the center and at the boundary of the spheroid. α_O and α_R are similarly defined.

The solution C(r) of (1) - (3) together with (5) - (7) is

$$C(r) = C_O - \frac{\dot{V}_{O_2} \cdot R^2}{3 \cdot (D_R - D_O)} \cdot \left[1 - \frac{r}{R} + \frac{D_O}{D_R - D_O} \cdot \ln \left(\frac{D_O + \frac{(D_R - D_O) \cdot r}{R}}{D_R} \right) \right] \qquad (8)$$

$$O \leq r \leq R$$

Constant diffusion coefficient D_R yields

$$\bar{C}(r) = C_O - \frac{\dot{V}_{O_2}}{6 \cdot D_R} \cdot (R^2 - r^2) \qquad ; \qquad O \leq r \leq R \qquad (9)$$

Using Henry's law (equation (4)) we obtain the corresponding Po_2 profile

$$P(r) = \frac{C(r)}{\alpha(r)} \qquad ; \qquad O \leq r \leq R \qquad (10)$$

and with constant oxygen solubility α_R

$$\bar{P}(r) = \frac{C_O}{\alpha_R} - \frac{\dot{V}_{O_2}}{6 \cdot D_R \cdot \alpha_R} \cdot (R^2 - r^2) \qquad ; \qquad O \leq r \leq R \qquad (11)$$

b) If measured Po_2 values along a central line within the spheroid are available, and if the courses of α and D are known, we are able to calculate an oxygen consumption profile by using equation (1). For this purpose we used two approaches:

- The measured O_2 concentration profile was approximated by a fourth order orthogonal polynomial series using least square methods.

- The measured O_2 concentration profile was approximated by piecewise cubic spline functions.

Both techniques are described in detail by Späth (1973; 1974), and with both techniques we obtain a sufficiently differentiable function, to which the differential operator on the left side of equation (1) can be applied.

The second approach has the advantage of showing local variations of oxygen consumption in more detail, but on the other hand has the disadvantage of increased scatter, since the third derivative of the cubic spline functions may show large jumps.

RESULTS

In the first part (a) of this section we study the influence of varying α and D on the Po_2 profile and on the oxygen consumption curve. The second part (b) of this section includes oxygen consumption profiles calculated from measured Po_2 profiles.

(a) Influence of Varying α and D

Carlsson (1982) measured in two types of spheroids (U118 MG, HTh 7) the percentage of extracellular space at different distances from the center, $g(r)$. In both types he found a linear correlation between the percentage of extracellular space and the distance. In fact, in both cases the percentage of extracellular space increases towards the center of the spheroid. They found the correlation

$$g(r) = 0.51 + 0.0005058 \cdot (R-r) \tag{12}$$

in the case of HTh 7 and

$$g(r) = 0.184 + 0.001362 \cdot (R-r) \tag{13}$$

in the case of U118 MG. With the assumption that α and D are linear combinations of the values for cells and water we have:

$$D(r) = D_{water} \cdot g(r) + D_{cell} \cdot (1-g(r)) \tag{14}$$

$$\alpha(r) = \alpha_{water} \cdot g(r) + \alpha_{cell} \cdot (1-g(r)). \tag{15}$$

Oxygen diffusion coefficient and solubility coefficient in water at a temperature of 20^oC was given by Grote (1967):

$$D_{water} = 2.3 \cdot 10^{-5} \ cm^2/sec \tag{16}$$

$$\alpha_{water} = 3.1 \cdot 10^{-2} \ ml \ O_2/(g \cdot atm) \tag{17}$$

and in tumor tissue (DS-carcinosarcoma) at 20^oC by Grote et al. (1977):

$$D_{cell} = 1.25 \cdot 10^{-5} \ cm^2/sec \tag{18}$$

$$\alpha_{cell} = 2.24 \cdot 10^{-2} \ ml \ O_2/(g \cdot atm) \tag{19}$$

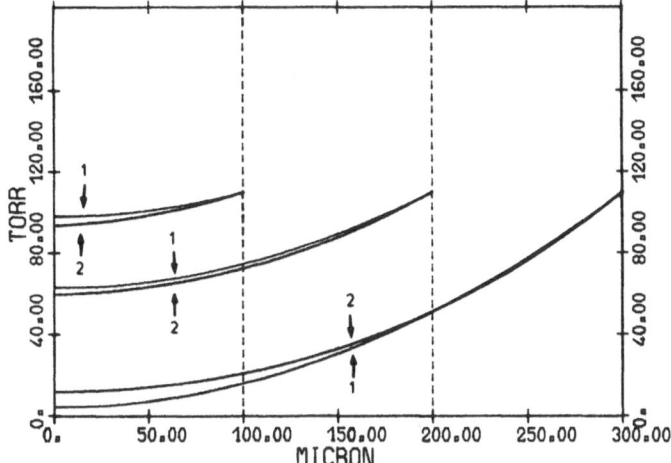

Fig. 1.   Simulated Po_2 profiles within multicellular spheroids (1:
constant α and D, 2: linearly increasing α and D towards
the center)

For the oxygen consumption we used a value of

$$\dot{V}_{O_2} = 0.0185 \text{ ml } O_2/(g \cdot min), \tag{20}$$

also given by Grote et al. (1977).

     Fig. 1 shows Po_2 profiles for a constant oxygen consumption
and for three sizes of spheroids (100, 200 and 300 um radius).
Profile 1 is calculated for constant oxygen diffusion and solubility
coefficient, profile 2 for varying α and D according to eqs. (14,
15). We used the data of U118 MG since they show the largest variation. The figure shows that for small radii profile 1 lies above
profile 2 and that the order changes with increasing radius. This
is due to the fact that an increase of α at the same oxygen concentration results in a decrease of Po_2, but an increase of D results
in an increase of Po_2. These two effects counteract each other. For
small radii the α-effect prevails, for larger radii the D-effect.
The figure shows that for small radii the difference between the two
curves is negligible or lies in the range of the measured Po_2. In
the case of larger radii the difference in the neighbourhood of the
center is remarkable, but in the region from 150 um to the surface
the difference may also be neglected.

     Profile 2 of Fig. 1, which was computed with the assumption
of constant oxygen consumption and varying oxygen diffusion and
solubility coefficient, was used to recalculate the oxygen consumption profile if instead it were assumed that α and D were constant.
The calculations were again done for three radii using the data of

U118 MG. The largest deviations of the recalculated oxygen consumption profile we observe in the neighbourhood of the center. The
curves show at first a slight decrease towards the center and a
strong increase in the neighbourhood of the center. Over a wide
range the deviation from the original constant consumption curve is
not appreciable. Thus, in the following we assume constant α and D.

(b) <u>Oxygen Consumption Profiles Calculated from Measured Po$_2$ Values</u>

Using the methods described in part (b) of section 1, oxygen
consumption profiles are calculated from measured Po$_2$ values inside
of multicellular spheroids. The Po$_2$ values were measured in spheroids of type U118 MG and HTh 7 (Carlsson et al., 1979). Fig. 2 shows
a representative result for the case of HTh 7. At the left vertical
axis of the diagram a scale is drawn for Po$_2$ and at the right vertical axis a scale for O$_2$ consumption in units of ml O$_2$/(100 g·min).
At the horizontal axis the distance to the center of the spheroid
is drawn in units of micron. The upper profile of Fig. 2 represents
the approximated Po$_2$ curve. The measured Po$_2$ values are signed by
circles. Curve 1 shows the oxygen consumption profile obtained by
the method of orthogonal polynomials, curve 2 the O$_2$ consumption
profiles obtained by using cubic splines. Both curves are drawn in
units of ml O$_2$/(100 g$_{tissue}$·min). They both include the possible
variation of density of cells. Using equation (12), and later on in
the case of U118 MG using equation (13), we corrected these values
to ml O$_2$/(100 g$_{cell}$·min) by dividing through the factor 1 - g(r).

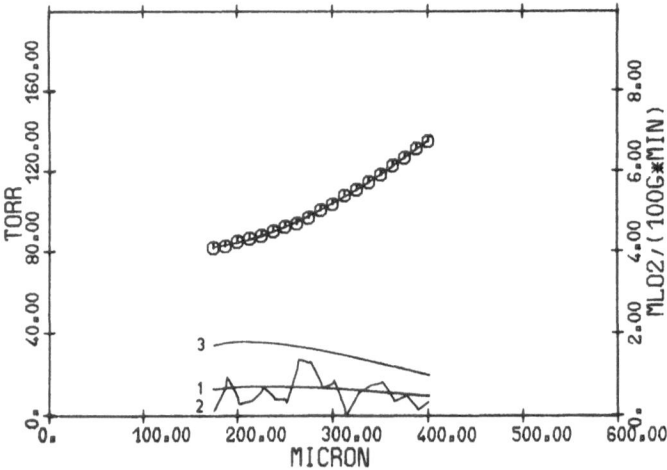

Fig. 2.   Measured Po$_2$ values and calculated O$_2$ consumption profile
          within a spheroid of type HTh 7 (circles: measured Po$_2$
          values; line 1: O$_2$ consumption profile (orthogonal poly-
          nomials); line 2: O$_2$ consumption profile (cubic splines);
          line 3: corrected O$_2$ consumption profile).

The result is shown by curve 3. Fig. 2 shows a nearly constant oxygen consumption profile, which is in good agreement with our previous results (Großmann et al., 1981). This behaviour is characteristic of all spheroids of this type. Profile 2 shows a larger scattering but the same trends. The corrected profile 3 increases slightly towards the center, but this is not significant.

In the case of U118 MG three different types of oxygen consumption profiles are observed. Fig. 3 shows the first type, which is similar to the results of HTh 7, namely a more or less constant oxygen consumption profile. It is of interest that the absolute value in both cases lies in the range of the value given by Grote et al. (1977). Fig. 4 shows the second type of O_2 consumption profile in the case of U118 MG. Here, a decrease of oxygen consumption towards the center is observed.

Fig. 5 shows the third type observed in the case of U118 MG. The oxygen consumption decreases considerably from the surface in direction of the center. Moreover, this decrease of oxygen consumption in units of ml $O_2/(100\ g_{tissue}\cdot min)$ is not caused by a decrease of density of cells, since profile 3 shows the same behaviour. Also the oxygen consumption in units of ml $O_2/(100\ g_{cell}\cdot min)$ decreases dramatically.

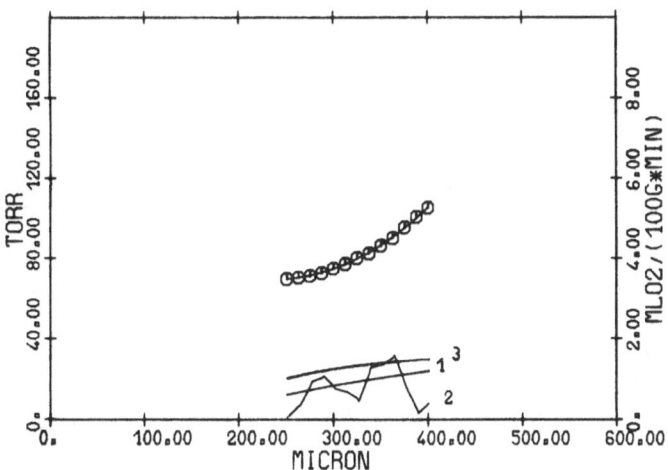

Fig. 3.   Measured Po_2 values and calculated O_2 consumption profile within a spheroid of type U118 MG (circles: measured Po_2 values; line 1: O_2 consumption profile (orthogonal polynomials); line 2: O_2 consumption profile (cubic splines); line 3: corrected O_2 consumption profile).

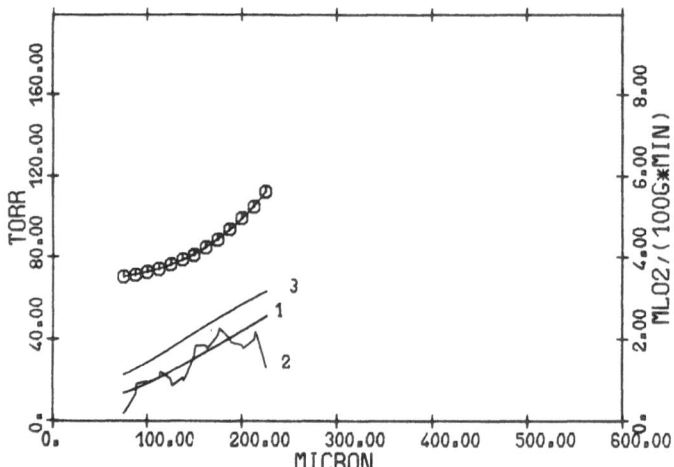

Fig. 4. Measured Po$_2$ values and calculated O$_2$ consumption profile
within a spheroid of type U118 MG (circles: measured Po$_2$
values; line 1: O$_2$ consumption profile (orthogonal polynomials); line 2: O$_2$ consumption profile (cubic splines);
line 3: corrected O$_2$ consumption profile).

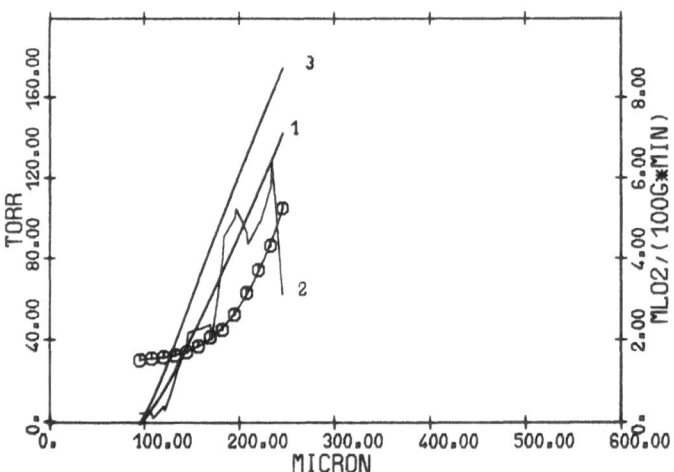

Fig. 5. Measured Po$_2$ values and calculated O$_2$ consumption profile
within a spheroid of type U118 MG (circles: measured Po$_2$
values; line 1: O$_2$ consumption profile (orthogonal polynomials); line 2: O$_2$ consumption profile (cubic splines);
line 3: corrected O$_2$ consumption profile).

DISCUSSION

The calculations using the model described in section 1 yield
results showing that in the case of spheroids of type U118 MG and
HTh 7 it is not necessary to consider varying oxygen diffusion coefficients and solubility coefficients. The influence on the calculated O_2 consumption profile is small, since the variations in α
and D are small in both cases. Only in the neighbourhood of the
center may these be of considerable influence. Furthermore, the
density of living cells, i.e., the effect of changing from a tissue
to a cell weight basis does not change the calculated O_2 consumption
profile appreciably. In both cases we observe only a slight increase
of the absolute value.

Both methods used to calculate oxygen consumption profiles
(orthogonal polynomials, cubic splines) yield similar results.

All spheroids of type HTh 7 show a more or less homogenous
oxygen consumption in the range of 2 ml $O_2/(100$ g·min). This value
is in good agreement with that given by Grote et al. (1977). In the
case of spheroids of type U118 MG we observed three different types
of oxygen consumption profiles:
1) a similar one as in the case of HTh 7, i.e. homogenous
oxygen consumption of similar size,
2) an O_2 consumption decreasing towards the center with similar surface O_2 consumption in comparison to 1), and
3) an O_2 consumption with a steep decrease towards the center
and a considerably higher surface O_2 consumption in comparison to
1) and 2).

These results are in accordance with those of our previous
paper (Großmann et al. 1981). At least in the case of U118 MG one
cannot assume, a priori, a homogenous O_2 consumption. A forthcoming
paper will show results concerning the O_2 consumption of spheroids
of other types (V79, U178 MG, OS393, CHEL).

The authors are indebted to Prof. Lübbers for valuable suggestions
and helpful criticism of the paper.

REFERENCES

Carlsson, J., 1982   (personal communication)
Carlsson, J., Stalnacke, C.-G., Acker, H., Haji-Karim, M., Nils-
    son, S., and Larsson, B., 1979, The influence of oxygen
    variability and proliferation in cellular spheroids, Int. J.
    Radiat. Oncol. Biol. Phys., 5:2011-2020.
Crank, J., 1955, "The Mathematics of Diffusion", At the Clarendon
    Press, Oxford.

728   U. GROSSMANN ET AL.

Großmann, U., Carlsson, J., and Acker, H., 1981, Oxygen consumption
    profiles inside cellular spheroids calculated from Po_2 profiles,
    in: "Oxygen Transport to Tissue IV", H.I. Bicher and D.F. Bruley,
    eds., Plenum Press, New York (in print).
Grote, J., 1967, Die Sauerstoffdiffusionskonstanten im Lungengewebe
    und Wasser und ihre Temperaturabhängigkeit, Pflügers Arch.
    Ges. Physiol., 295:245-254.
Grote, J., Süsskind, R., and Vaupel, P., 1977, Oxygen diffusivity
    in tumor tissue (DS-carcinosarcoma) under temperature condi-
    tions within the range of 20 - 40°C, Pflügers Arch., 372:
    37-42.
Späth, H., 1973, "Algorithmen für elementare Ausgleichsmodelle",
    Oldenbourg, München-Wien.
Späth, H., 1974, "Algorithmen für multivariable Ausgleichsmodelle",
    Oldenbourg, München-Wien.
Tannock, I., 1968, The relation between proliferation and the
    vascular system in a transplanted mouse mammary tumor, Br.
    J. Cancer, 22:258-273.
Thomlinson, R.H., and Gray, K.H., 1959, The histologic structure
    of some human lung cancers and the possible implications for
    radiotherapy, Br. J. Cancer, 9:539-549.

DISCUSSION

T. Goldstick: Why occurs the factor r^2 in the limit equation (2).
The gradient in c, as r approaches zero, should approach zero. Making
this change would require recomputation of the entire simulation or
has it already been taken into account. In any event, making this
change would probably not affect the analysis of any of the experimental results because they do not extend close to r = 0.

U. Großmann: Equation (1) means that the divergence of the oxygen
diffusion flux is equal to the amount of O_2 consumed. This is
derived from the continuity equation under steady state conditions.
Under spherical symmetry it is physically obvious that the total
diffusion flux at the center vanishes (equation (2)). Equation (2)
is equivalent to the condition that the gradient of c vanishes at
the center except the case of D going to infinity at the center. We
excluded this case. All calculations remain valid if we omit r^2 in
equation (2).

MICROCIRCULATION

THE FINE ADJUSTMENT OF CAPILLARY BLOOD FLOW THROUGH EXCITATION OF

THE CAPILLARY WALL

H. Weigelt, and D.W. Lübbers

Max-Planck-Institut für Systemphysiologie
Rheinlanddamm 201, 4600 Dortmund 1, FRG

Within the system of blood vessels controllers are inbuilt which provide for an economic distribution of blood. These controllers are mechano-effectors and are morphologically vascular smooth muscle cells.

In the region of microcirculation of blood, i.e. the region in an organ in which the blood volume is controlled by the arterioles, vascular smooth muscle cells are responsible for how much blood reaches the exchange area. Distally from the arteriole we find the precapillary sphincter which in a classical sense is the last smooth-muscular controller of microflow (Rhodin, 1967).

But from the knowledge of Po_2 measurements with needle micro-electrodes it is assumed that a local final regulation of oxygen supply exists in adaptation to the needs of the tissue. These fine regulations are independent of blood flow through bigger vessels which lie proximally to the supplied region (Lübbers, 1974).

Because there are no known smooth muscular mechano-effectors in the capillary region, the question arose whether the cells from which the true capillaries are derived, i.e. the endothelial cells and the pericytes could be regarded as mechano-effectors in microcirculation. The basis for such a speculation would be the comparability of the reactivity of the capillary wall with the reactivity of larger blood vessels. Therefore the capillary should be excitable.

The aim of the present study was to show the excitability of the capillary wall with direct electrical stimulation and by stimulation with serotonin as a mediator of inflammation (Majno et

731

al., 1969). Additionally indirect stimulation was applied via the splanchnic nerve.

In a second study the reactivity of capillaries to adrenaline was tested (Dietrich and Weigelt, 1982, this volume).

MATERIAL AND METHODS

General Methods

1. Animals: For the studies frogs (Rana esculenta, L.) were used. The male-female ratio was about 1:1. Body weight varied from about 40 g to 90 g, head-crotch length varied from 6.5 to 9.5 cm. The animals were supplied by Fa. Stein (Lauingen/Donau, FRG) in summer and winter. They were captured wild.

2. Anaesthesia was brought about by an injection of 0.3 to 0.5 ml of 2% Tricaine-methane-sulfonate (Fa. Sigma, Munich, FRG), dissolved in 0.9% NaCl solution in the dorsal lymphsack.

3. Preparation of the mesentery: The surgical steps were identical for all studies up to the exteriorization of the mesentery:
    a) fixing the frog in supine position
    b) medial-ventral incision through the skin
    c) medial-ventral incision parallel to the linea alba and laparatomy
    d) exteriorization of the mesentery
    e) keeping the mesentery moist either with Frog-Ringer solution or by covering with liquid paraffin.

4. Intravital microscopy of the mesentery was performed according to Weigelt and Schwarzmann (1981) with the combined microscopes allowing simultaneous observation using incident light and trans-illumination of the capillaries. This setup allows the simultaneous observation of the capillary net in an overview and the stimulated capillary segment at high magnifications. Therefore the amount of influence from neighbouring arterioles spreading to the capillary can be estimated.

Special Methods

1. Direct electrical stimulation: Mesenteric capillaries were stimulated with Ag/AgCl covered needle micro-electrodes (Weigelt et al., 1979; Weigelt et al., 1981). Stimulation current was supplied by a constant current source (Iontophoresis Programmer, WP Instruments, Connecticut, USA) which was pulsed via an impulse generator (Stimulator II, Hugo Sachs Elektronik, Hugstetten, FRG). The reference electrode was an Ag/AgCl wire which touched the intestinal wall.

2. Nervous stimulation (indirect stimulation): In order to stimulate the splanchnic nervous plexus surrounding the Arteria mesenterica cranialis (A.m.c.) the squamous mesothelium was incised and the nervous plexus was marked with a thin surgical thread. This procedure produced sufficient mobility of the nerves to allow satisfactory positioning of the bipolar stimulation electrode beyond the A.m.c. The procedure also increases the mechanical stability and allows the stimulation of transmural nerves (Campbell and Jackson, 1979).

3. Application of serotonin was performed with microliter pipettes in concentrations ranging from 10^{-3} mol$\cdot$l^{-1} to 10^{-2} mol$\cdot$l^{-1}. The volume of each topically applied dose on the mesentery was about 50 ul.

RESULTS

1. Capillary constrictions result from direct electrical stimulation. Fig. 1 shows characteristic examples of time courses of constrictions. The initial diameter of the capillaries was normalized to 100%. The capillaries had an average diameter of 16 um $\pm$7 um (SEM). The stimulation parameters are listed in the legend to Fig. 1.

2. With serotonin mesenteric capillaries also constrict. Characteristic examples of time courses of capillary constriction are shown in Fig. 2. The concentration of serotonin amounted to 10^{-2} mol$\cdot$l^{-1}.

3. Behaviour of capillaries to indirect, nervous stimulation is as follows: While stimulating the nervous plexus of the splanchnic nerves around the cranial mesenteric artery, capillaries react independently from arterioles or small veins. In a more or less uniform manner capillaries constrict at the beginning of stimulation. In a few cases constriction is so strong that the passage of red cells is considerably hindered.

At prolonged stimulation over 20 sec. capillaries dilate and in some cases the diameter exceeds 50% of the original diameter. Fig. 3 shows three characteristic examples of responses of frog mesenteric capillaries to nervous stimulation. The stimulation parameters are shown in Fig. 3.

4. The previous three examples of capillary stimulation showed that the capillary is able to constrict. This shows that the cells of capillaries can change their shape as a reaction to excitation and therefore should not be considered to make up the walls of inelastic tubes.

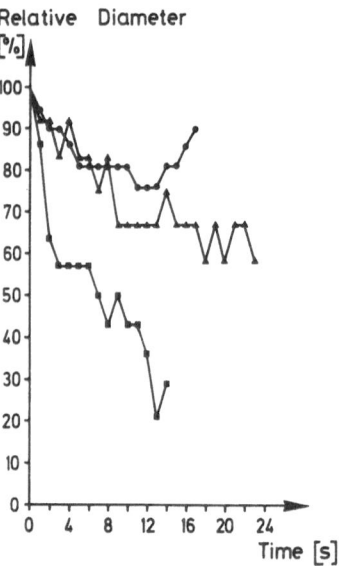

Fig. 1.   Normalized time courses of capillary constriction. The
           capillaries were stimulated directly electrically.
           Stimulation conditions:
           ● : stimulation current intensity +400 nA; stimulation
               duration 17 sec.
           ▲: stimulation current intensity +300 nA. Stimulation
               duration 23 sec.
           ■: stimulation current intensity +300 nA. Stimulation
               duration 15 sec.

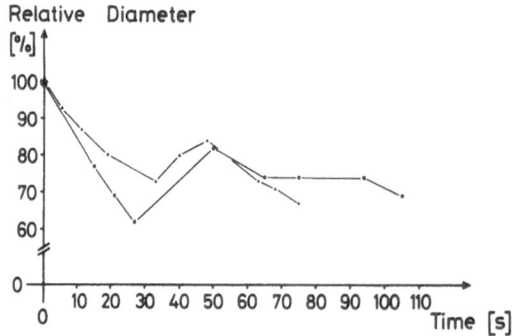

Fig. 2.   Characteristic examples of the time course of constriction
           of frog mesenteric capillaries after topical application
           of 10^{-2} mol·1^{-1} serotonin solution.

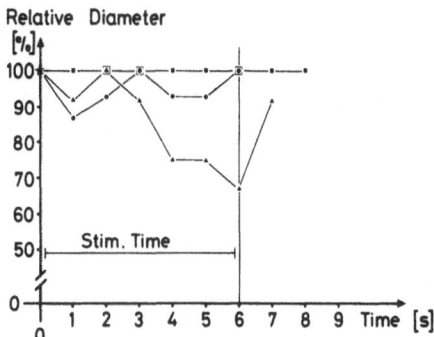

Fig. 3.   Time courses of diameter changes in mesenteric capilla-
          ries of the frog during stimulation of the splanchnic ner-
          vous plexus of the A.m.c. After the end of stimulation the
          original capillary diameter is restored.

        In a further series of experiments the degree of capillary
constriction with stimulation strength was examined to test whether
a relationship similar to that for excitable cells (Aidley, 1978)
is obtainable. If so changes in stimulation duration should have a
similar effect to changes in stimulation strength. Fig. 4 shows
with the example of three capillaries that with increasing strength
of direct stimulation current the degree of capillary constriction
also increases.

        From the responses to stimulation it is apparent that the de-
gree of constriction also depends on the duration of stimulation,
since the degree of constriction slowly saturates to a maximum over
a period of several seconds.

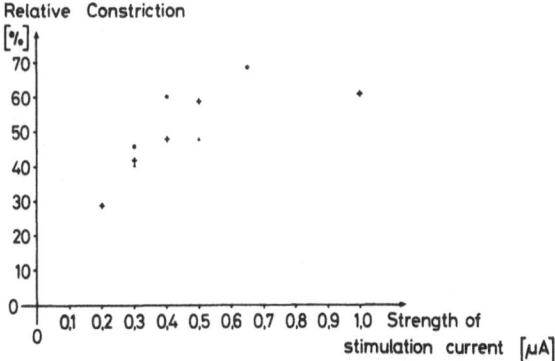

Fig. 4.   Characteristic example of the dependency of the degree of
          constriction in capillaries from the strength of direct
          stimulation current. The single values represent the
          highest degree of constriction at a certain stimulation
          current strength for 3 different capillaries.

DISCUSSION

The previous findings show, that within the capillary bed it
is possible to directly influence capillary diameter and capillary
blood flow, distally from the precapillary sphincter.

With the direct electrical stimulation it has to be taken into
account that the electrical resistance of the circuit apart from
other factors also depends on the fine structure of the tissue and
that therefore there are big differences from site to site. It
cannot be assumed that with comparable stimulation conditions one
can obtain comparable degrees of constriction in frog mesenteric
capillaries, or that the effective part of the stimulation current
strength which acts at the excitable cell will always be the same.
But the fact that frog mesenteric capillaries constrict under indirect nervous stimulation is of greater physiological importance
than are the different degrees of constriction at direct electrical
stimulation.

Additionally one may assume that in the inflammatory process
if the inflamed tissue area has to be blocked from a surrounding
tissue to prevent a spreading of the cause of imflammation (Letterer, 1959) serotonin plays an essential role. During the action of
serotonin it could be observed that there is an outflow of plasma
through endothelial gaps in frog mesenteric capillaries (Weigelt
et al., 1979b), this causes edema and blocking of the microcirculation in the inflamed tissue area. In mammalian capillaries it
could be shown additionally that mediators of inflammation such as
histamine and serotonin cause gap formation between the endothelial
cells through which blood plasma flows in the interstitial spaces.

Combining the findings with those made by Riedel (1980) the
results may be put together to form a unifying hypothesis: Riedel
was able to show that within the nervous system of the splanchnic
nerves, e.g. in the mesentery, there exist together with adrenergic nerve fibres also histaminergic and serotonergic nerve fibres.
These three types therefore would allow the control of capillary
blood flow via changes in capillary diameter and capillary permeability, all three having one common mechano-effector namely the
endothelial cell. With a local reduction of capillary diameter the
hydrostatic pressure would increase and therefore the pressure gradient across the endothelial cell would be the driving force for
an increased plasma outflow through the endothelial gaps (Majno et
al., 1969) into the interstitial tissue.

REFERENCES

Aidley, D.J., 1978, "The Physiology of Excitable Cells", 2nd edition,
    Cambridge University Press, Cambridge.

Campbell, W.B., and Jackson, E.K., 1979, Modulation of adrenergic
    transmission by angiotensins in the perfused rat mesentery,
    Am. J. Physiol., 236 (N2):H211-H217.
Dietrich, H.H., and Weigelt, H., 1982, Effect of adrenaline on
    capillary diameters in the frog mesentery, this issue.
Letterer, E., 1959, "Allgemeine Pathologie. Grundlagen und Proble-
    me", Thieme, Stuttgart.
Lübbers, D.W., 1974, Das O$_2$-Versorgungssystem der Warmblüterorgane,
    in: Jahrbuch der Max-Planck-Gesellschaft zur Förderung der
    Wissenschaften e.V., München, pp. 87-112.
Majno, G., Shea, M., and Leventhal, M., 1969, Endothelial contrac-
    tion induced by histamin-type mediators. An electron micro-
    scopical study, J. Cell Biol., 41:647-672.
Rhodin, J.A.G., 1967, The ultrastructure of mammalian arterioles
    and precapillary sphincters, J. Ultrastruct. Res. 18:181-223.
Riedel, W., 1980, Zwei efferente Systeme des Sympathikus: Die dif-
    ferenzierte Steuerung der lokalen Kreislauffunktion, Habilita-
    tionsschrift, Universität Gießen.
Weigelt, H., Addicks, K., Hauck, G., and Lübbers, D.W., 1979b,
    Vitalmicroscopic studies in regard to the role of intraendo-
    thelial reactive structures in the inflammatory process, Bibl.
    Anat., 17:11-20.
Weigelt, H., Baumgärtl, H., Hauck, G., and Lübbers, D.W., 1979a, A
    vitalmicroscopic set-up to study vasomotion induced by elec-
    trical stimulation with microelectrodes, Bibl. Anat., 18:
    81-84.
Weigelt, H., Fujii, T., Lübbers, D.W., and Hauck, G., 1981b,
    Specialized endothelial cells in the frog mesentery - attempt
    of an electrophysiological characterization, Bibl. Anat., 20:
    89-93.
Weigelt, H., and Schwarzmann, V., 1981a, A new method for the
    simultaneous presentation of low and high magnifications of
    microscopic specimens. Application to in vivo studies of
    mesenterial capillaries, Microsc. Acta, 85/2:161-173.

INTERMEDIATE SIZE FILAMENTS AND MICROFILAMENTS IN FROG MESENTERIC

CAPILLARIES

H. Weigelt

Max-Planck-Institut für Systemphysiologie
Rheinlanddamm 201, 4600 Dortmund 1, FRG

INTRODUCTION

The majority of mesenteric capillaries in frogs contract if
they are directly electrically stimulated (Weigelt et al., 1981).
They also show changes in diameter under conditions of nervous
stimulation. This contractile set of the total capillary population
is termed reactive (Weigelt et al., 1982). Because of the similarity of the time course of capillary luminal constriction compared
with the time course of arteriolar contraction it was speculated
that both mechanisms may have the same molecular basis. In the case
of arteriolar smooth muscle cells, contraction is based on the
action of intracellular actin-myosin filaments. The restitution of
the cell shape may be due to the action of elastic elements, e.g.,
intermediate size filaments (Franke, 1978a/b). The aim of this
study was to explain the observed contractile properties by
examining the content of endothelial cells in electrically stimulated capillaries. Therefore, we applied immunofluorescence methods
(Franke et al., 1976, 1978a/b) and transmission electronmicroscopi-
cal methods to capillary endothelial cells in the mesentery of the
frog (Rana esculenta, L.).

METHODS

Frogs (Rana esculenta, L.) of both sexes were used with an
average head-rump length of 8 cm and an average weight of approximately 60 g. The mesentery was exteriorized and its microcirculation was observed in a microscope chamber (Weigelt et al., 1981).
Electrical stimulation of capillaries was performed with needle
micro-electrodes (Weigelt et al., 1979). Immunological methods were

performed according to Franke et al. (1978a) to show the existence
of intermediate size filaments and tonofilaments (Osborn et al.,
1980). The f-actin was shown with immunological methods according
to Franke et al. (1976) and with cytochemical decoration of f-actin
with heavy-meromyosin. In addition, cross-striated arrays in
microfilament bundles were shown in electrically stimulated endothelial cells with electron microscopical methods.

RESULTS

In frog mesenteric capillaries which reacted with luminal
constriction to direct electrical stimulation we were able to
identify by transmission electronmicroscopy intermediate size
filaments and microfilaments. With immunofluorescence methods it
was possible to show that the majority of the intermediate size
filaments are vimentin-type filaments (Franke et al., 1978b) and
tonofilaments (Franke et al., 1978c).

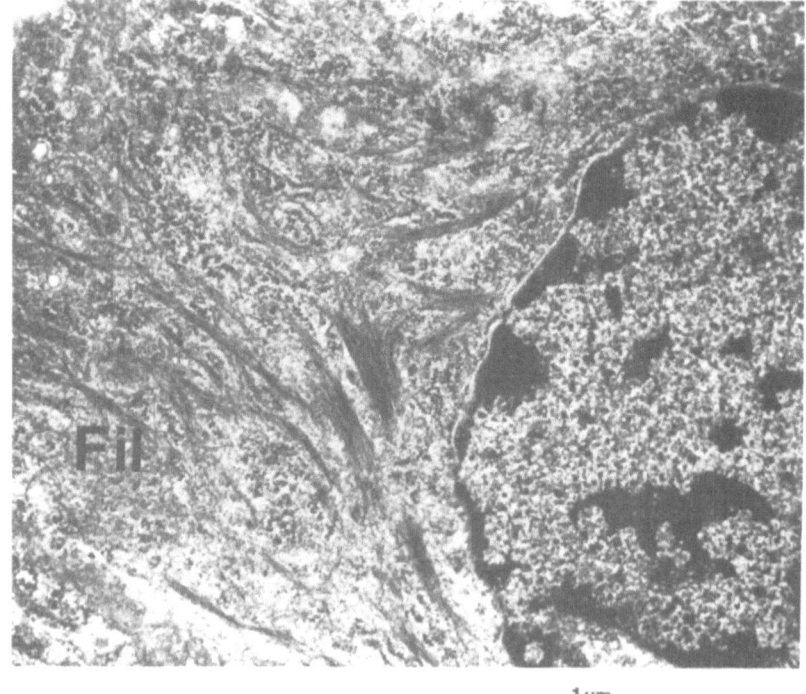

1μm

Fig. 1.   Characteristic example of a reactive capillary site. The
          endothelial cell contains filaments of different size (Fil).
          (Photograph taken by courtesy of Prof. Dr. K. Addicks,
          Anatomical Institute, University Cologne.)

Fig. 2 shows the vimentin-type intermediate size filament of
mesenchymal origin. In Fig. 3 it could be shown by immunofluorescence that tonofilaments exist which react with antibodies to
prekeratin (Franke et al., 1978c).

The existence of f-actin is necessary for cellular motility
or contraction. Therefore, antibodies to filamentous actin were
used to show the existence of actin in endothelial cells. Fig. 4
shows that in reactive capillary sites there are bundles of fluorescent arrays oriented parallel to the main capillary axis.

As in smooth muscle cells microfilament bundles could be found
which show cross striations in endothelial cells. This type of filament arrangement could only be found in reactive capillaries
(Weigelt et al., 1981). The dense bodies presumably represent
Weibel-Palade-bodies which are characteristic of endothelial cells
(e.g. Krstić, 1978).

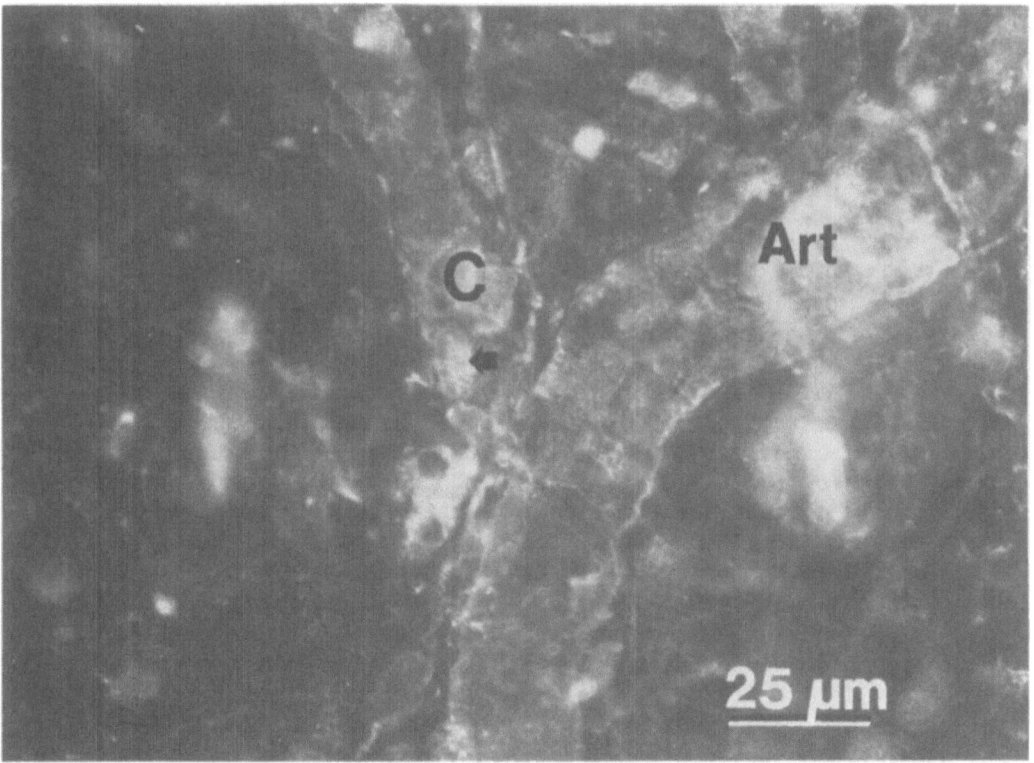

Fig. 2.   Immunofluorescence microscopical presentation of inter-
           mediate size filaments of mesenchymal origin (arrow)
           vimentin-type (Osborn et al., 1980) in frog mesenteric
           capillaries (C). Art = arteriole.

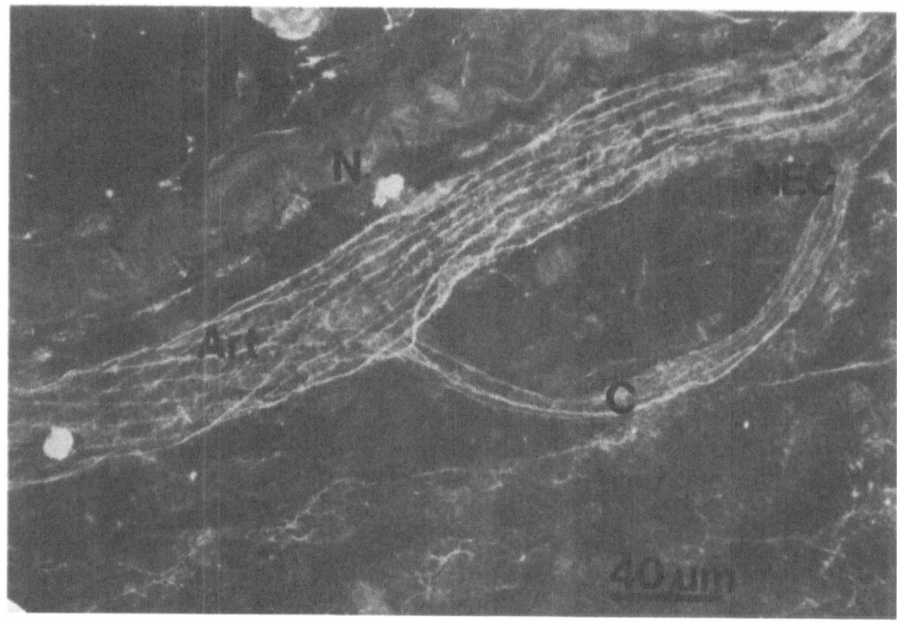

Fig. 3.   Immunofluorescence microscopical presentation of tonofila-
          ments (prekeratin-type) in frog mesenterial capillaries
          (C), NEC = region of the endothelial nucleus. In the ad-
          jacent arteriole (Art) tonofilaments form a meshwork. N =
          nerve bundle.

DISCUSSION

     Frog mesenteric capillaries show contractile properties if
they are stimulated either electrically or by mediators of inflammation (Weigelt et al., 1979b). If this reaction is a rather
common feature of capillaries, their endothelial cells and pericytes should contain contractile and elastic proteins similar to
those in vascular smooth muscle cells. Unlike the contraction of
arteriolar smooth muscle cells where a considerable proportion of
the contractile effort is exerted circumferentially around the
vessel, the endothelial cell contraction has a major component of
the force it develops exerted longitudinally along the main capillary axis. The contractile proteins are generally located at the
abluminal part of the endothelial cell, therefore while shortening
the nucleus protrudes into the capillary lumen.

     The time courses of capillary constriction lay in the same
range as the spontaneous flow rhythmicity in capillaries found by
Tyml and Groom (1980). From the similarity of the time course of
capillary constriction and luminal restitution and from the filament

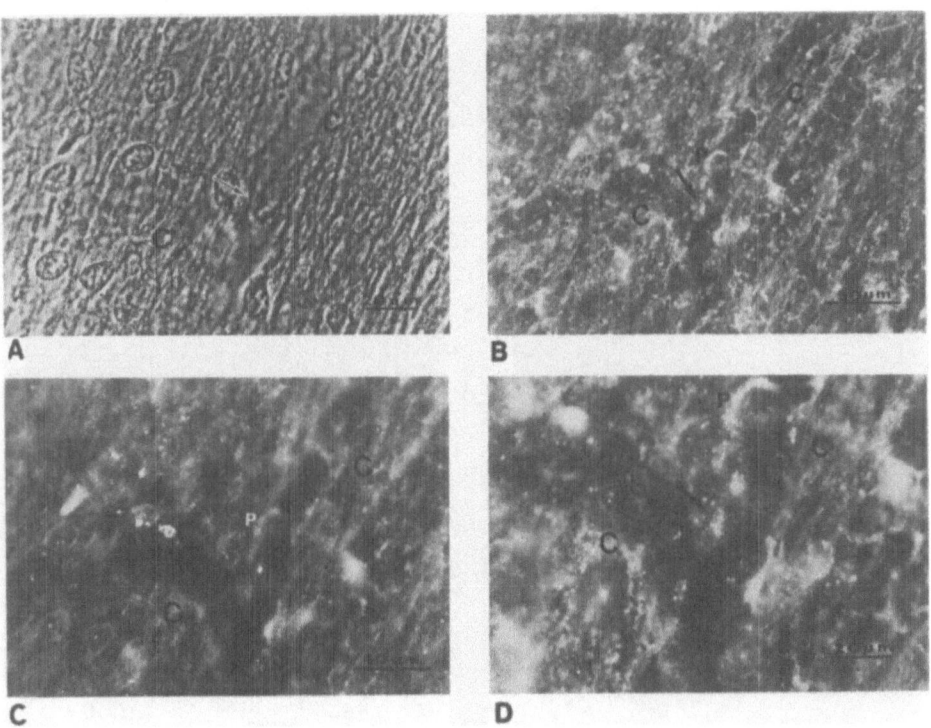

Fig. 4.   Immunofluorescence microscopical presentation of f-actin
          in frog mesenteric capillaries (C).
          Arrow indicates a dense array of fluorescence in an endo-
          thelial cell showing a strong reaction of f-actin antibo-
          dies to f-actin. The pericyte (P) also shows an intensive
          fluorescence of f-actin antibodies bound to f-actin. Part A
          of the figure shows the capillary branching in a phase con-
          trast microscopical presentation. Part B shows the upper
          surface layer of the mesentery, part C shows the lower
          surface layer. In both parts B and C the main fluorescence
          arises from epithelial cells. In part D the capillary
          branching is shown in a higher magnification showing more
          clearly the fluorescent patches of f-actin in an endothe-
          lial cell (= arrow) and in a pericyte (P).

content of pericytes and endothelial cells it is assumed that there
exists a more refined control of capillary perfusion distally from
the precapillary sphincter and that, therefore, the endothelial cell
may contribute to the control of the homeostasis of the interstitial
fluid.

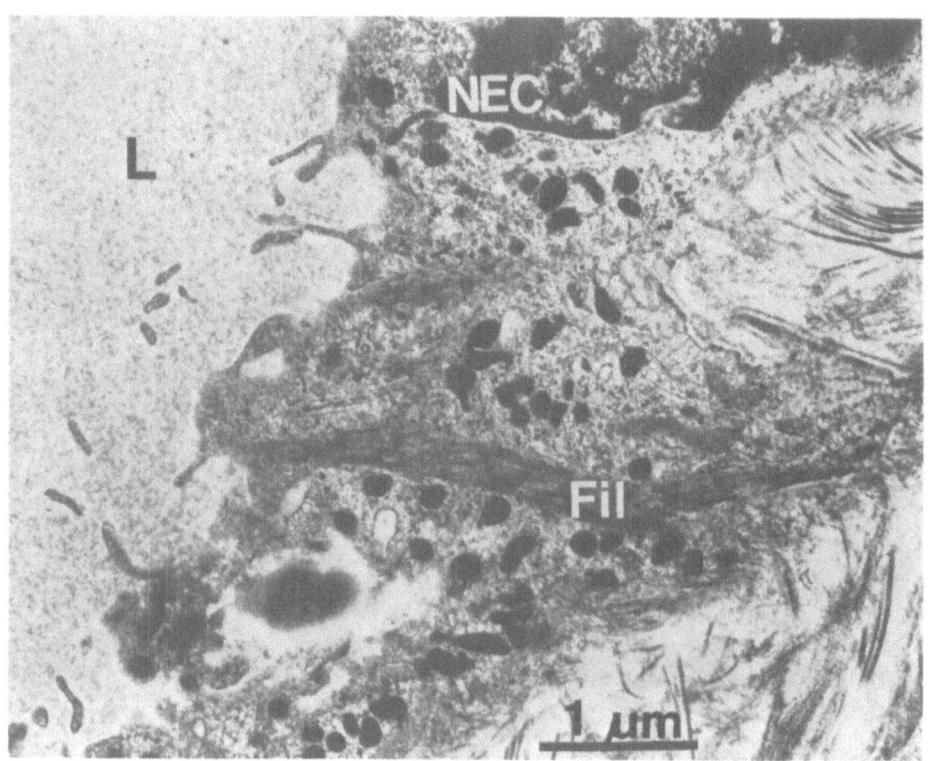

Fig. 5. Electronmicroscopic presentation of array of cross-striat-
         ed microfilament bundles in electrically stimulated capil-
         laries in the mesentery of the frog (photograph taken by
         courtesy of Prof. Dr. med. K. Addicks, Anatomical Insti-
         tute, University of Cologne, FRG). L = capillary lumen,
         Fil = micro filament bundles, NEC = endothelial nucleus.

SUMMARY

     Intermediate size filaments of mesenchymal origin in endothe-
lial cells as well as tonofilaments could be identified by immunofluorescence. Both types of filaments form bundles which run in
parallel to the main capillary axis. Filamentous actin could also
be detected by this technique. The immunofluorescence of f-actin
forms patch-like arrays irregularly distributed along the capillary
and striated arrays which are oriented in some cases in parallel to
the main capillary axis.

ACKNOWLEDGEMENT

The author is greatly indebted to the scientific help of
Prof. Dr. W.W. Franke who enabled the application of the immunological methods in his laboratory (German Cancer Research Center,
Heidelberg), and Prof. Dr. K. Addicks who provided the electronmi-
crographs.

REFERENCES

Franke, W.W., Grund, C., Osborn, M., and Weber, K., 1978a, The
    intermediate-sized filaments in the rat Kangarro PtK$_2$ cells.
    II. Structure and composition of isolated filaments, Cyto-
    biologie, 17:365.
Franke, W.W., Rathke, P.C., Seib, E., Trendelenburg, M.F., Osborn,
    M., and Weber, K., 1976, Distribution and mode of arrangement
    of microfilamentous structures and actin in the cortex of
    amphibian oocyte, Cytobiologie, 14:111.
Franke, W.W., Schmid, E., Osborn, M., and Weber, K., 1978b, Diffe-
    rent intermediate-sized filaments distinguished by immuno-
    fluorescence microscopy, Proc. Nat. Acad. Sci. USA, 75/10:
    5034-5038.
Franke, W.W., Weber, K., Osborn, M., Schmid, E., and Freudenstein,
    C., 1978c, Antibody to prekeratin decoration of tonofilament-
    like arrays in various cells of epithelial character, Exp.
    Cell Res., 116:429-445.
Krstić, R.V., 1978, "Ultrastruktur der Säugetierzelle", Springer,
    Berlin-Heidelberg-New York.
Osborn, M., Franke, W.W., and Weber, K., 1980, Direct demonstration
    of the presence of two immunobiologically distinct intermedi-
    ate-sized filament systems in the same cell by double-immuno-
    fluorescence microscopy, Exp. Cell Res., 125:37-46.
Tyml, K., and Groom, A.C., 1980, Regulation of blood flow in
    individual capillaries of resting skeletal muscle in frogs,
    Microvasc. Res., 20:346-357.
Weigelt, H., Addicks, K., Hauck, G., and Lübbers, D.W., 1979b,
    Vitalmicroscopic studies in regard to the role of intraendo-
    thelian reactive structures in the inflammatory process, Bibl.
    Anat., 17:11-20.
Weigelt, H., Baumgärtl, H., Hauck, G., and Lübbers D.W., 1979a, A
    vitalmicroscopic set-up to study vasomotion induced by elec-
    trical stimulation with microelectrodes, Bibl. Anat., 18:
    81-84.
Weigelt, H., Fujii, T., Lübbers, D.W., and Hauck, G., 1981,
    Specialized endothelial cells in the frog mesentery - attempt
    of an electrophysiological characterization, Bibl. Anat., 20:
    89-93.

H. WEIGELT

Weigelt, H., Preuß-Nowotny, A.; and D.W. Lübbers, 1982, Excitation
of splanchnic nervous plexus (SNP) causes diameter changes in
frog mesenterial capillaries (C), Pflügers Arch., 392:R9/35.
Weigelt, H., and Schwarzmann, V., 1981, A new method for the
simultaneous presentation of low and high magnifications of
microscopic specimens: Application to in vivo studies of
mesenterial capillaries, Microsc. Acta, 85/2:161-173.

EFFECT OF ADRENALINE ON CAPILLARY DIAMETER IN THE FROG MESENTERY

H.H. Dietrich, and H. Weigelt

Max-Planck-Institut für Systemphysiologie
Rheinlanddamm 201, 4600 Dortmund 1, FRG

INTRODUCTION

Investigations by Stricker (1876), Field (1935), Lübbers et al. (1979), and recently Weigelt and Schwarzmann (1981), have shown that capillaries constrict under electrical stimulation. These studies suggest regulation of capillary lumen by humoral, metabolic and/or neuronal signals. The evidence for humoral and/or neuronal regulation has been described for arterial vessels, including pre-capillary sphincters, in warm-blooded as well as in cold-blooded animals (Zweifach, 1937; Furness and Marshall, 1974; Altura, 1978; Riedel, 1980). A direct effect on true capillaries of warm-blooded animals, however, has not been demonstrated (Furness and Marshall, 1974; Altura, 1978). In capillaries of cold-blooded animals the effect has been in question because of contradictory observations of changes in capillary diameter (Krogh, 1929; Field, 1935; Zwei-fach, 1937; Chambers and Zweifach, 1944). The present study was undertaken to contribute to the solution of this problem. In our experiments the vasoconstrictor effect of adrenaline was investi-gated because of its role as the main transmitter in the frog (Taxi, 1976).

METHODS

The experiments were carried out on the mesenteries of 16 frogs (Rana esculenta). They were "summer frogs" of both sexes, head to crotch length 8.5 $\pm$ 1.1 cm, body weight 64.7 $\pm$ 18.6 g and were obtained from Stein (Lauingen, FRG). The animals were anaes-thetized by injection of a 0.4 to 0.6 ml 2% solution of 3-amino-benzoic acid-ethylester-methanesulfonate (Sigma, München, FRG) into

the dorsal lymph sack. The abdominal skin was cut ventero-laterally
and the peritoneum was opened with an incision made laterally to
the Linea alba. The duodenal loop was exteriorized and the mesentery was placed on the transparent window of a microscope chamber.
The studies were carried out with an inverted camera microscope
ICM 405, (C. Zeiss, Oberkochen, FRG) which was combined with an
incident light microscope (C. Zeiss). This arrangement allowed one
to observe the capillary bed simultaneously at low and high magnification (Weigelt and Schwarzmann, 1981). The images of the two
microscopes were monitored with videocameras TC 1005, (RCA,
Lancaster, PA, USA), connected via a video-mixer (Pieper, Schwerte,
FRG) and recorded on a 1/2"-videorecorder (Grundig, Nürnberg, FRG).
A black and white monitor was used to observe the video images.
Objectives with magnification power of 16:1 to 32:1 were used in
the transmitted light microscope which allowed final magnifications
of 500 to 1000-fold on the video screen.

Adrenaline (Suprarenin 1:1000, Hoechst, Frankfurt, FRG) was
diluted with phosphate buffered saline (PBS, pH 7.36) in proportions
of 1:10,000 to 1:50,000, to give concentrations of 1.1 to $5.5 \cdot 10^{-7}$
mol/ml adrenaline. The solutions were stained with 1-3% Evan's blue
(Merck, Darmstadt, FRG) and filtered through filters with pore
diameter 0.45 um (Millipore, Bedford, MA, USA). The adrenaline was
delivered to the vessels under study by means of a nanoliter pump
(Hampel, Neu-Isenburg, FRG), the pumping system of which was movable
in all three planes by a mechanical micromanipulator (Prior, H.
Albrecht, München, FRG) and in one plane by a hydraulic microdrive
(D. Kopf, Tujunga, CA, USA). To apply adrenaline locally, micro-
pipettes were produced from microliterpipettes (10 ul volume, Brand,
Wertheim, FRG). The tips were extracted to tip diameters of 2-5 um
with an electrode puller (Max-Planck-Institut für Systemphysiologie,
Dortmund, FRG). The pumping rate was 10-50 nl/min. The pumping time
varied between 2-33 seconds and generally ended at the beginning of
the adrenaline effect. The start of the adrenaline effect could be
determined by effusion of the Evan's blue cloud out of the micro-
pipette because adrenaline and Evan's blue were present in the same
solution. Furthermore, the spread of adrenaline could be estimated
by the spread of the color cloud. Between each application of adrenaline the mesentery was rinsed with aerated frog Ringer solution
(22°C). The period between the single applications ranged between
5-30 minutes. The dose of adrenaline was calculated from the pumping
rate, the pumping time and the adrenaline concentration. While
playing back the video recordings the vessel diameters were measured
by a transparent ruler from the videoscreen. Diameters of the precapillary sphincters and true capillaries (Chambers and Zweifach,
1944; Rhodin, 1967) were measured every ten seconds and the maximum
change in diameter was determined within the first minute after the
adrenaline effect.

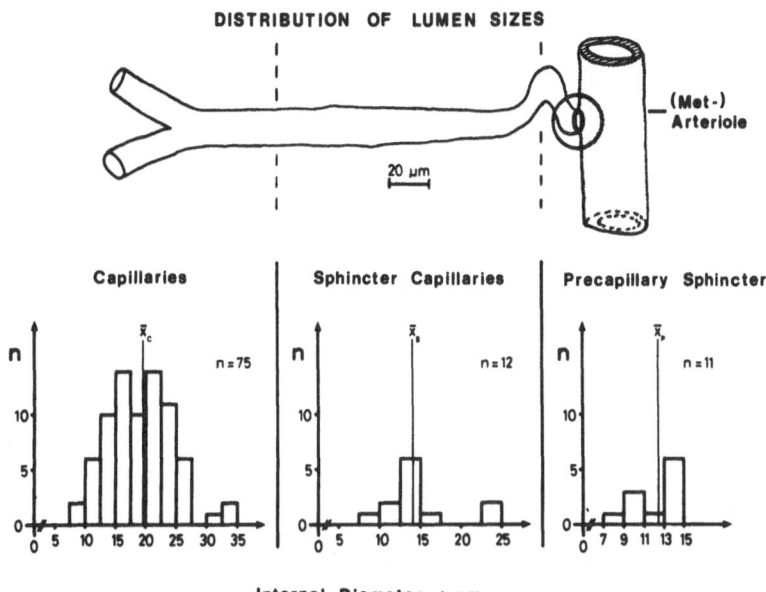

Fig. 1.   Histograms showing initial diameters of the vessel types
          studied. The histograms are associated with their schemat-
          ic vessel segment. $\bar{X}_C$ = average capillary diameter, $\bar{X}_S$ =
          average sphincter capillary diameter and $\bar{X}_p$ = average
          precapillary sphincter lumen.

RESULTS

        Initial diameters of the precapillary sphincters studied varied
from 7.5 to 14.9 um with an average diameter of $\bar{X}_p$ = 12.4 ± 2.6 um
(n = 11). Diameters of sphincter capillaries varied from 8.5 to
22.4 um with an average of $\bar{X}_S$ = 14.3 ± 4.1 um (n = 12) and for true
capillaries from 7.5 to 33.2 um with an average of $\bar{X}_C$ = 19.3 ± 5.7
um (n = 75) (Fig. 1). The applied adrenaline doses varied between
0.2 pmol and 5.6 pmol. The spread of the Evan's blue mixture depended on the pumping time, the pumping rate and the secretion of fluid
through the squamous mesothelium. In general, the color cloud spread
out over an area of approximately 0.1 mm by 0.1 mm within the first
minute of the experiment.

        Within the applied range of adrenaline doses the precapillary
sphincters and sphincter capillaries always constricted. Reactions
of precapillary sphincters to locally applied adrenaline are shown
in Table 1, maximally constricted down to 30 percent of the initial
diameter in the examined dose range of 0.8 pmol to 1.4 pmol adrenaline. Reactions of sphincter capillaries are shown in Table 2. The

Table 1. Precapillary sphincter constriction to adrenaline.

order of reactivity	1	2	3	4	5	6	7	8	9	10	11
initial diameter [/um]	14.9	14.1	12.5	14.0	9.1	10.0	10.8	7.5	14.9	13.3	14.9
minimum diameter [/um]	6.6	4.2	4.2	5.0	4.2	5.0	5.8	3.3	9.1	11.6	13.3
dose of adrenaline [p mol]	1.3	5.5	0.8	1.4	1.7	1.0	1.1	1.3	0.8	0.6	0.8
decrease of diameter [%]	77.9	70.2	66.4	63.8	53.8	49.4	46.2	44.0	38.9	12.8	10.7

maximum constriction was down to 35% of the initial diameter at doses of about 0.5 pmol to 2 pmol adrenaline.

The reaction of the true capillaries is shown in Fig. 2. In total, there were 34 non-reactive capillaries, so designated because their maximum diameter change was less than 5% of the initial diameter, and 41 reactive capillaries. Diameters of two of the reactive capillaries apparently widened at doses of 0.2 pmol to 0.6 pmol. Thirty-nine capillaries narrowed by varying amounts, the most extreme case being to 60% of the initial diameter.

The constrictions could be divided into two forms as seen by Weigelt and Schwarzmann (1981). The first type of constriction is a local one where the vessel narrowing is restricted to an area of 30 um to either side of the location of maximum constriction. The second is a decremental one where the narrowed zone extended more than 60 um.

Table 2. Sphincter capillary constriction to adrenaline.

order of reactivity	1	2	3	4	5	6	7	8	9	10	11	12
initial diameter [/um]	16.6	13.3	22.4	14.9	12.5	20.8	8.3	10.0	13.3	13.3	14.9	11.6
minimum diameter [/um]	4.2	4.2	9.1	8.3	4.2	12.5	5.0	6.6	10.0	10.0	13.3	10.8
dose of adrenaline [p mol]	0.6	0.8	2.2	0.4	5.5	1.1	1.1	0.6	0.8	1.0	0.6	1.1
decrease of diameter [%]	74.7	68.4	59.4	44.3	40.0	39.9	39.8	34.0	24.8	24.8	10.7	6.9

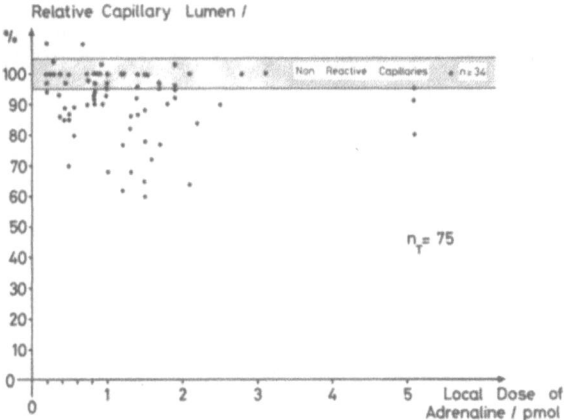

Fig. 2.   Relative capillary diameter after local adrenaline appli-
          cation (initial diameter = 100%). The maximum diameter
          change within 60 seconds after adrenaline effect is marked
          with o. Hatched area = non-reactive capillaries, blank
          area = reactive capillaries.

DISCUSSION

     The results demonstrate that locally applied adrenaline can
affect the lumen of vessel segments of the frog mesentery distal
to the (met)-arterioles. Of these vessel segments, only the sphincter capillaries possess smooth muscle cells, namely the precapillary sphincters (Chambers and Zweifach, 1944; Rhodin, 1967), while
succeeding capillaries do not possess smooth muscle cells. The
smooth muscle cells, like the precapillary sphincters are the
mechano-effectors which show vasoactivity under adrenaline effect.
The studies demonstrate that the precapillary sphincter reacts
quickly and very sensitively enabling fast adjustment of capillary
blood flow in the adjoining capillary segments. The unexpected
reaction of the adjoining capillary segments, the sphincter capillaries and the true capillaries, to adrenaline, when no effect on
precapillary sphincters was observed shows the independence of
these segments. The independence from the precapillary sphincters,
suggests additional effector mechanisms on the capillary level.

     In comparison to the precapillary sphincters, the reaction of
the sphincter capillaries and true capillaries to adrenaline is
weaker. Possibly the reason for this is that these vessel segments
do not possess smooth muscle cells and consequently the ability to
constrict has to be explained by the activity of the smaller actin-
and myosin content of pericytes and endothelial cells (Hama, 1961;
Le Beux and Willemont, 1978a, 1978b;). True capillaries of the frog
mesentery reacted less uniformly to local adrenaline application.

While the precapillary sphincter and the sphincter capillaries
constricted at all adrenaline doses used, about 50% of the 75 true
capillaries did not react at all. Two reactive capillaries widened,
the other thirty-nine narrowed. The results also show that at low
adrenaline doses capillaries can dilate which could not be observed
at higher doses.

REFERENCES

Altura, B.M., 1978, Humoral, hormonal and myogenic mechanism in
    microcirculatory regulation, in: "Microcirculation", Vol. II,
    G. Kaley and B.M. Altura, eds., University Park Press, Balti-
    more.
Chambers, R., and Zweifach, B.W., 1944, Topography and function of
    the mesenteric capillary circulation, Am. J. Anat., 75:173-205.
Field, M.E., 1935, The reaction of the blood capillaries of the
    frog and rat to mechanical and electrical stimulation, Scand.
    Arch. Physiol., 72:175-191.
Furness, J.B., and Marshall, J.M., 1974, Correlation of the direct-
    ly observed responses of mesenteric vessels of the rat to
    nerve stimulation and noradrenaline with the distribution of
    adrenergic nerves, J. Physiol.,239:75-88.
Hama, K., 1961, On the existence of filamentous structures in endo-
    thelial cells of the amphibian capillary, Anat. Rec., 139:
    437-441.
Krogh, A., 1929, "Anatomie und Physiologie der Kapillaren", Springer,
    Berlin.
Le Beux, Y.J., and Willemont, J., 1978a, Actin- and myosin-like
    filaments in rat brain pericytes, Anat. Rec., 190:811-826.
Le Beux, Y.J., and Willemont, J., 1978b, Actin-like filaments in the
    endothelial cells of adult rat brain capillaries, Exp. Neurol.,
    58:446-454.
Lübbers, D.W., Hauck, G., Weigelt, H., and Addicks, K., 1979,
    Contractile properties of frog capillaries tested by electri-
    cal stimulation, Bibl. Anat., 17:3-10.
Riedel, W., 1980, Zwei efferente Systeme des Sympathikus: Die dif-
    ferenzierte Steuerung der lokalen Kreislauffunktion, Habilita-
    tionsschrift, Gießen.
Rhodin, J.A.G., 1967, The ultrastructure of mammalian arterioles
    and precapillary sphincters, J. Ultrastruct. Res., 18:181-223.
Stricker, S., 1876, Untersuchungen über die Contractilität der
    Capillaren, Sitzungsberichte der Wiener Akademie der Wissen-
    schaften, math.-naturwiss. Klasse, 74, Abt. 3, 313-332.
Taxi, J., 1976, Morphology of the autonomic nervous system, in:
    "Frog Neurobiology", R. Llinás and W. Precht, eds., Springer,
    Berlin, pp: 93-150.
Weigelt, H., and Schwarzmann, V., 1981, A new method for the
    simultaneous presentation of low and high magnification of
    microscopic specimens: Application to in vivo studies of
    mesenterial capillaries, Microsc. Acta, 85/2:161-173.

Zweifach, B.W., 1937, The structure and reactions of the small blood
    vessels in amphibia, <u>Am. J. Anat.</u>,60:473.

RHYTHMICITY AND TENSION DEVELOPMENT OF SPONTANEOUS CONTRACTING

CAPILLARIES IN THE MESENTERY OF FROGS

H. Weigelt, and K.E. Wohlfarth-Bottermann[1]

Max-Planck-Institut für Systemphysiologie
Rheinlanddamm 201, 4600 Dortmund 1, FRG
[1]Institute of Cytology, University of Bonn
Ulrich Haberlandstr. 61 a, 5600 Bonn 1, FRG

INTRODUCTION

The existence of capillary contractility is still in debate
(Hammersen, 1980). Although evidence exists that the endothelial
cell and the pericyte contain contractile filaments (Le Beux, 1978a,
1978b), direct evidence that capillary contractility exists in
vivo, is lacking.

The problem has been re-examined using the recently developed
microtensiometer method (Wohlfarth-Bottermann, 1977; 1979). Microtensiometry is a method commonly used for the measurement of non-
muscular movement (Wohlfarth-Bottermann, 1977). It is mainly applied
to directly measure the tension development by single protoplasmic
strands (so-called veins) of slime moulds, e.g. Physarum polycepha-
lum. Because measurements made using microtensiometry are sensitive
the method was introduced to measure the tension development of
single blood capillaries in the mesentery of the frog. The aim of
this study was to provide quantitative data of the tension developed
by single capillaries and to show that capillary contractility is
more than simply a phenomenon occurring upon electrical stimulation
(Stricker, 1865; Field, 1935; Weigelt, 1981).

METHODS

A microtensiometry device was used as described by Wohlfarth-
Bottermann (1977). The tension development of frog mesenteric capillaries was studied. In order to adapt blood capillaries to the
microtensiometer system, they had to be isolated from the surround-

755

ing interstitial tissue. After loosening the capillaries from the
interstitial tissue of the mesentery they still remained in connection with other blood vessels so that the blood flow through the
capillaries was still intact, though flow may be reduced.

   Frogs (Rana esculenta, L.) of both sexes were used, average
head-crotch length 7.5 cm, average weight $\sim$ 45 g. Anaesthesia was
brought about with Tricaine-methane-sulfonate (Sigma) in doses of
0.2 mg/g body weight. The procedure of capillary isolation was as
follows:
   1) Dislocation of the frog's mesentery through a lateral
abdominal incision.
   2) The mesentery is bathed in a solution containing 0.84%
mannose and 0.06% inulin. By this procedure the squamous mesothelium
which covers the mesenteric interstitial tissue loosens and can be
taken away with fine foreceps.
   3) Bathing of the mesentery with frog-Ringer solution for ca.
15 min.
   4) Treatment of the mesentery with collagenase-dispase solution for 10 to 30 minutes in order to loosen the connective tissue
fibres. The microcirculation still remains intact.
   5) After the enzymatic treatment the mesentery is bathed in
so-called "recovery medium" which essentially is frog-Ringer
solution containing 1% bovine serum albumin and 0.5% glucose.
   6) Using a Zeiss dissection microscope (150 x magnification),
a hook consisting of a bent piece of platinum wire (101 um diameter),
attached to the microtensiometer device, is hooked onto a selected
capillary as illustrated in Fig. 1.

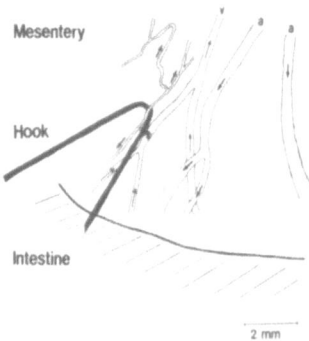

Fig. 1.   Experimental situation of the microtensiometer measurement.
          It shows the main vessels which are close to the measured
          capillary under which the microtensiometer hook is lying.
          v = vein; a = small artery; arrows indicate blood flow
          directions. The capillary diameter was about 20 um.

For the microtensiometer measurements it was found to be necessary for tension development in the capillaries that their microcirculation remains intact.

The measurements were only performed on capillaries with remaining blood flow.

In Fig. 1 an experimental situation is demonstrated using a semischematic drawing from a frog's mesentery just before the onset of the microtensiometer measurement. The drawing is a projection of the original situation.

RESULTS

The tension development of isolated frog mesenteric capillaries was measured under a traction force of the microtensiometer of 200 - 350 uN. After a silent period of about 2 - 5 minutes the capillary spontaneously developed rhythmical tensions with an average strength of 6.8 uN $\pm$ 2.4 uN. Fig. 2 shows a tension record which is characteristic of spontaneous tension development in a frog mesenteric capillary. The graph was taken from an original registration.

By measuring the times between the maxima of spontaneous tension developments and plotting these on a frequency histogram, the variation in time between contractions can clearly be seen. In Fig. 3 the time between successive tension developments in three capillaries are compiled, the times range between 0.2 min to 2.7 min. The average interval amounts to 1.09 min $\pm$ 0.19 min (SEM).

If capillaries which show the spontaneous rhythmicity in tension development are stimulated directly electrically while they are hanging on the microtensiometer hook they show a clear relationship of the tension development with intensity of stimulation current. The strength of tension increased with increasing stimulation current. Fig. 4 shows two characteristic examples of this relationship.

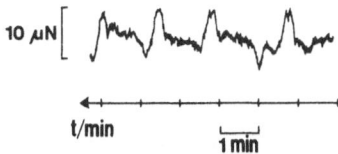

Fig. 2.   Characteristic example of a recording from a frog mesenteric capillary. The drawing was made from the original registration. The microtensiometer traction force in this case was 280 uN.

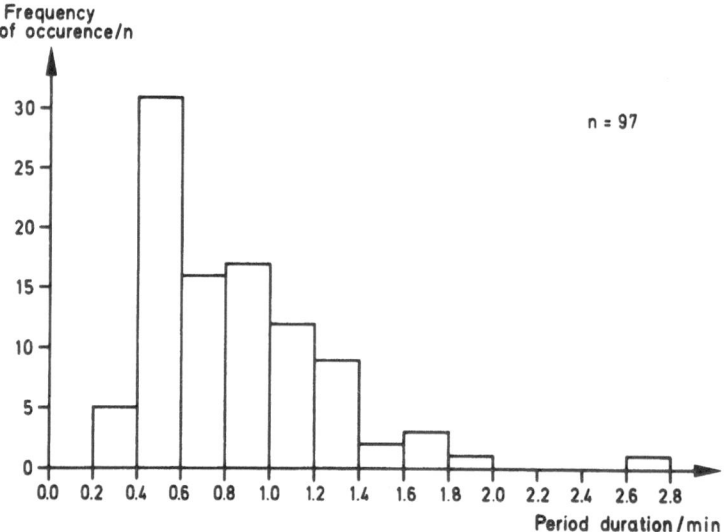

Fig. 3.   Frequency distribution of the time between the maxima of
           spontaneous tension fluctuations in frog mesenteric capil-
           laries. Values were obtained from three individual capil-
           laries.

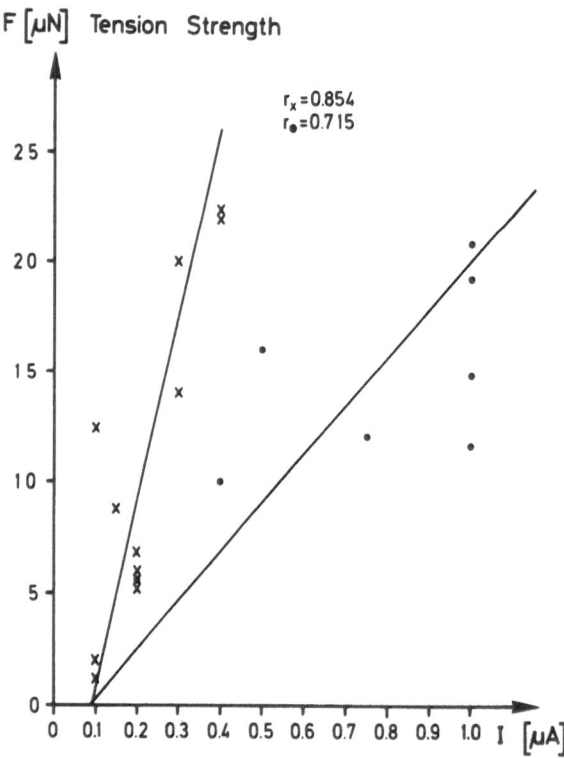

Fig. 4.   Dependency of tension strength from stimulation current
           intensity (DC) during direct electrical stimulation.

DISCUSSION

The measured spontaneous tension development in frog mesenteric capillaries was independent of arteriolar or venular constriction. It was also independent of changes in blood pressure due to the heart beat and from respiratory movements. Peristalsis of the gut showed a periodicity of seven minutes and was therefore clearly distinguishable from the periodicity of capillary tension development.

The findings provide evidence that capillaries possess contractile features which resemble those of vascular smooth muscle cells (Altura, 1971). They also support the view that capillaries should not be considered as rigid tubes (Fung, 1966). From our observations there exists a contractile process with the possible potential for control of the microcirculation distal from the precapillary sphincter.

REFERENCES

Altura, B.M., 1971, Chemical and humoral regulation of blood flow through the precapillary sphincter, Microvasc. Res. 3:361-384.
Field, M.E., 1935, The reaction of the blood capillaries of the frog and rat to mechanical and electrical stimulation, Skand. Arch. f. Physiol. 72:175-191.
Fung, Y.C., Zweifach, B.W., and Intaglietta, M., 1966, Elastic environment of the capillary bed, Circ. Res., 19:441-461.
Hammersen, F., 1980, Endothelial contractility - does it exist? in: "Advances in Microcirculation", B.M. Altura, ed., Karger, Basel, pp. 95-134.
LeBeux, Y.J., and Willemot, J., 1978a, Actin- and myosin-like filaments in the rat brain pericytes, Anat. Res., 190, No. 4: 811-826.
LeBeux, Y.J., and Willemot, J., 1978b, Actin-like filaments in the endothelial cells of adult rat brain capillaries, Exp. Neurol., 58:446-454.
Stricker, S., 1865, Studien über den Bau und das Leben der capillaren Blutgefäße, Sitzungsber. Wiener Akad. Wiss., math.-naturwiss. Klasse, 52, Abt. 2:379-394.
Weigelt, H., Fujii, T., Lübbers, D.W., and Hauck, G., 1981, Specialized endothelial cells in the frog mesentery - attempt of an electrophysiological characterization, Biblthca. anat., 20: 89-93.
Wohlfarth-Bottermann, K.E., 1977, Oscillating contractions in protoplasmatic strands of Physarum: Simultaneous tensiometry of longitudinal and radial rhythmus, periodicity analysis and temperature dependence, J. Exp. Biol., 67:49-59.
Wohlfarth-Bottermann, K.E., 1979, Oscillating contraction activity in Physarum, J. Exp. Biol., 81:15-32.

PRACTICAL ASPECTS

TRANSCUTANEOUS Po_2 DURING EXERCISE

J.M. Steinacker, and R.E. Wodick

Division of Sports Medicine, Department of Applied
Physiology, University of Ulm, 7900 Ulm/Do., FRG

INTRODUCTION

During the performance of work, the energy needed for muscle
concentration is mainly provided by the metabolism of glucose or
fats with oxygen. For that reason the oxygen uptake Vo_2 is closely
related to physical working capacity. During long-lasting exercise
$\dot{V}o_2$ can increase 15-20 times compared to the value at rest. The
maximal $\dot{V}o_2$ obtained, is defined as the aerobic capacity of an
athlete. During strenuous and short duration exercise there are increasing lactic acid concentrations observed in the blood (Margaria
and Edwards, 1933). These lactic acid concentrations indicate a high
energy demand, which is not immediately provided for by oxygen
uptake, so that the percentage of glycolysis in total energy production must increase. The balance between this anaerobic and aerobic
energy production is influenced by muscle tissue, cardiopulmonary
function and neuromuscular coordination (Margaria and Edwards, 1933;
Åstrand, 1952; Keul and Doll, 1973). Oxygen transport in the blood
is another factor, therefore, the Po_2 was observed in arterial blood
samples. This method only allows discontinuous measurements to be
made. Arterial puncture is necessary, because samples from capillary
blood are of limited value (Siggard-Andersen, 1968). Continuous measurement has been attempted, using intravascular measuring probes or
by continuously analyzing blood sucked out from the indwelling canula
(Schwarz and Fabel, 1976). All these methods require well-trained
personnel and/or expensive equipment, they are invasive and are,
therefore, not recommended for routine use in exercise testing.

A new approach to such measurements is the method of continuous, transcutaneous determination of arterial Po_2 (P_ao_2) on hyperemic
skin using a heated Clark-electrode, according to the method pro-

763

posed by R. Huch, A. Huch and D.W. Lübbers (Huch et al., 1973; 1981; Lübbers, 1979). The usage of this method has become common in perinatal medicine (Huch et al., 1973). There are some problems in the examination of P_aO_2 with transcutaneous PO_2 methods ($tcPO_2$) in adults (Löllgen et al., 1979; Lübbers, 1979; Tremper et al., 1980; Steinacker and Wodick, in press). We have demonstrated in previous studies that $tcPO_2$ methods are usable in adults during exercise tests. PO_2 values from arterial samples were compared with those from transcutaneous determinations and good individual correlations from 0.77 to 0.99 were found in the electrodes used (Steinacker and Wodick, in press; Wodick and Steinacker, in press). We observed changes in $tcPO_2$ during exercise, which were classified into different types. In this paper we will discuss the possibilities of using $tcPO_2$ measurements in exercise testing and will present some of our findings in athletes, nontrained persons and patients.

MATERIAL AND METHODS

The principles of transcutaneous oxygen monitoring are based on the well known finding, that oxygen diffuses through intact skin, and on observations that the PO_2 level of the skin rises with a local hyperemia of the measuring area (Keul and Doll, 1973; Löllgen et al., 1979; Lübbers, 1979; Schonfeld et al., 1980).

In heat-induced hyperemia, the flow in the capillary loops of the upper epidermis increases to high values. There is a lower level of oxygen consumption in tissue than of the oxygen transport to tissue by the blood, hence an oxygen pressure can be measured on the surface of the skin. Under these conditions, $tcPO_2$ is dependent on the local blood flow. Small changes in perfusion pressure during normal macrocirculation cause no alterations in the $tcPO_2$ because the local blood flow rises during vasodilatation in heat hyperemia from normally 5.0 ml/(100 g·min) to 150 ml/(100 g·min) and more (Lübbers, 1979). Severe disturbances of blood flow during shock or in arterial occlusive disease, will reduce the possibility of measuring $tcPO_2$ values related to arterial PO_2 (Lübbers, 1979; Tremper et al., 1980; Huch et al., 1981). It is possible to measure in these cases only the local oxygenation of tissue (Tremper et al., 1980). This $tcPO_2$ method is very well explained in the literature (Huch et al., 1973, 1981; Lübbers, 1979).

In our experiments commercially available $tcPO_2$ electrodes with a heated area of 55 mm^2 were used during exercise tests. We fixed the electrode, after a two-point calibration, with an adhesive ring on the upper thorax of the subject. After an equilibration time of about 6 minutes it was possible to begin with the test. The electrode was used at a temperature of 45°C, which seems necessary in short-time measurements (Lübbers, 1979; Schonfeld et al., 1980). We examined $tcPO_2$ in three groups of people:

1) Patients with clinical signs of impaired cardiac output
   (n = 11, 37-62 years old)
2) Healthy, untrained persons (n = 15, 28-56 years old)
3) Well-trained athletes (n = 11, 20-31 years old)

All participants on this study had no disorders of pulmonary function. The experiments were carried out on a bicycle ergometer, work was increased in steps of 3 minutes duration, the increase in work for each step being defined, according to the classification of the subject from 25 Watt for patients, to 1 W/kg body weight for well-trained athletes. All tests were carried out to exhaustion. During all tests, heart rate, ECG, blood pressure and $tcPo_2$ were recorded. The oxygen consumption, ventilation, exspiratory gas concentrations and other respiratory parameters were measured with an open spirometric system (Jaeger, mod. Fleisch-type) and recorded together with heart rate and $tcPo_2$ by means of a microcomputer. ΔPo_2, as the difference between the mean $tcPo_2$ in a control period before exercise and the $tcPo_2$ value at the end of exercise was determined. Tan α as a relative index for ΔPo_2 is calculated as ΔPo_2 divided by the maximum work load P (see Fig. 5). These values were compared statistically using a t-test.

RESULTS

In the Figs. 1-4, some results are shown from our measurements in sportsmen and patients. We have previously found (Steinacker and Wodick, in press) a good correlation of transcutaneous and arterial values in well-trained persons during hard exercise (Fig. 2). Increases in $tcPo_2$ were seen in members of a rehabilitation group,

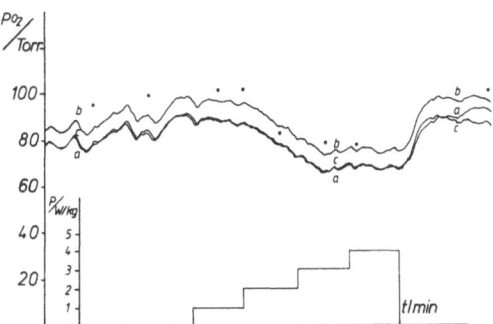

Fig. 1. $TcPo_2$ during rest and exercise in a trained athlete, the points indicate values of P_aO_2; a, b, and c are different types of electrode. The values of the ordinate as power (measured in Watt) have been divided by Watt per body weight (in kg).

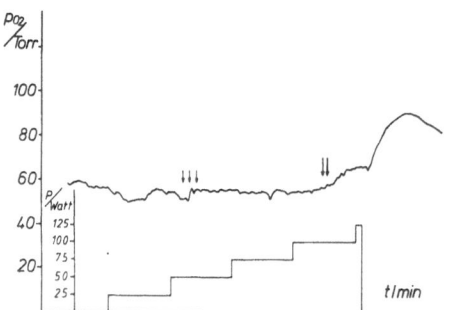

Fig. 2.   TcPo$_2$ during exercise in a 53 years old patient (myocardial
          infarction two years ago, medication: Acebutol 400 mg/day),
          max. power load 125 Watt, ↓↓↓ mild angina pectoris, ↓↓ in-
          creasing chest pain and dyspnea.

which had suffered from a myocardial infarction. In one member of
this group (Fig. 2) the normal initial drop in Po$_2$ was not very
clearly observed. The patient showed mild angina pectoris the first
time after 1 min of 50 W exercise (↓↓↓). In the last minutes of exercise (at 100 W) chest pain and dyspnea increased (↓↓) and a rise in
tcPo$_2$ was recorded, until exercise was terminated by exhaustion.
Typical records from a well-trained athlete are shown in Fig. 3. In
the upper axes tcPo$_2$ and work load are displayed. In the central axes
the almost linear increase in heart rate and a more steeply rising
alveolar ventilation can be seen. In the lower axes tcPo$_2$ and the
index Vo$_2$ per heart beat, as an indicator of the cardiopulmonary
efficiency are graphed (Åstrand, 1952; Doll et al., 1966). Changes
in Vo$_2$ and tcPo$_2$ can be seen in intermittent, stepwise increases in
exercise (Huch et al., 1973). In the groups examined we found the
following results, which are shown graphically in Figs. 5 and 6.

The following values are the means:
    1. Patients with lowered cardiac output:
       Max. work load: 62.5 Watt
       ΔPo$_2$: 8.4 Torr
       tan α: 0.13
    2. Untrained:
       Max. work load: 155 Watt
       ΔPo$_2$: -0.2 Torr
       tan α: 0.001
    3. Well-trained:
       Max. work load: 330 Watt
       ΔPo$_2$: -4.8 Torr
       tan α: 0.0146

The differences in ΔPo$_2$ between each group are statistically significant (p = 0.001).

DISCUSSION

The aim of this work was to examine the changes in P_aO_2 during exercise, which are discussed by other authors (Doll et al., 1966; Reichel, 1975; Schwarz and Fabel, 1976; Young and Woolcock, 1978) and which can be observed with transcutaneous methods (Borgia and Horvath, 1978; Löllgen et al., 1979; Schonfeld et al., 1980; Wodick and Steinacker, in press). Using several different $tcPo_2$ electrodes simultaneously, we found parallel changes in transcutaneous measurements as shown in Fig. 1. The divergence between arterial and transcutaneous values varies for each type of electrode and each skin area, but will remain constant during the whole measuring time. A sufficient hyperemia in the measuring area is important, as confirmed in other studies (Lübbers, 1979; Huch et al., 1981; Steinacker and Wodick, in press). Our experiments demonstrate the dependency of $tcPo_2$ on the heated area, which influences both the values of measurable $tcPo_2$ and the duration of the arterialization times, a very important point in routine measurements. We can demonstrate good correlations and little differences between P_aO_2 and $tcPo_2$, short arterialization times and good practicability in routine use of the technique. This is confirmed by some other investigators, who measured $tcPo_2$ during exercise (Borgia and Horvath, 1978; Löllgen et al., 1979; Schonfeld et al., 1980). The $tcPo_2$ method enables the usage of computer-aided $tcPo_2$ recording along with other respiratory values (Figs. 3 and 4).

In the three groups studied differences were found which were related to the state of physical fitness of the tested persons (Figs. 5 and 6), the decrease in tan α being particularly noticeable in the trained subjects. I like to mention that we found in another very well-trained group a large change in $tcPo_2$ of about minus 10-30 Torr.

Similar results were found by other investigators using invasive methods (Doll et al., 1966; Keul and Doll, 1973; Young and Woolcock, 1978). In the control group (healthy untrained) no significant systematic change in ΔPo_2 can be seen. In patients with a clinical diagnosis of reduced cardiac output, only an increasing Po_2 is found (Fig. 5). Fig. 2 demonstrates that such an increase seems to be related to heavy cardial dyspnea. The observed differences are very significant. It has been said that changes in Po_2 during exertion are not remarkable (Borgia and Horvath, 1966; Reichel, 1975; Schonfeld et al., 1980) in healthy persons. But this cannot be generalized, since it holds only for the group of untrained healthy persons, but not for the group of well-trained athletes and of persons with lowered cardiac output. We suggest for the individual characterization of observed $tcPo_2$ changes the tan α value, where ΔPo_2 is divided by the level of work P. With this value a real comparison of the change in Po_2 is possible. For example, a ΔPo_2 of 10 Torr at 50 Watt is not the same as such a change at 200 Watt. It

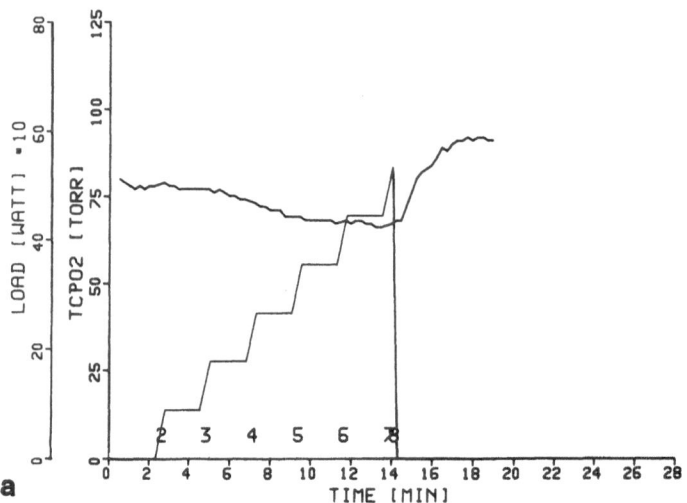

a

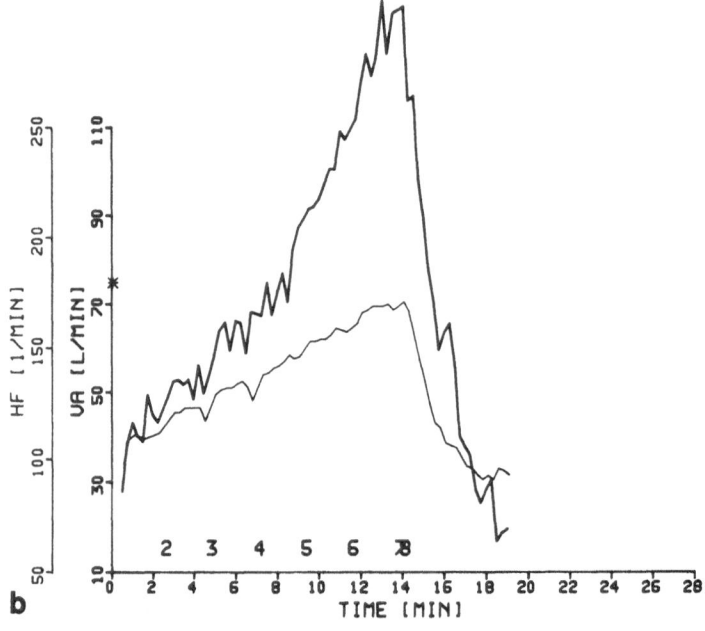

b

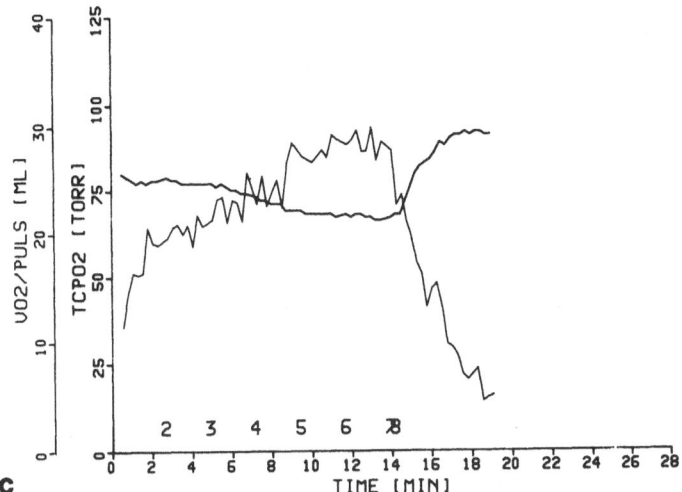

Fig. 3. Examination in a well-trained athlete: a) tcPo₂ and work load, b) $\dot{V}_A$ and heart rate HR, c) tcPo₂ and Vo₂/HR

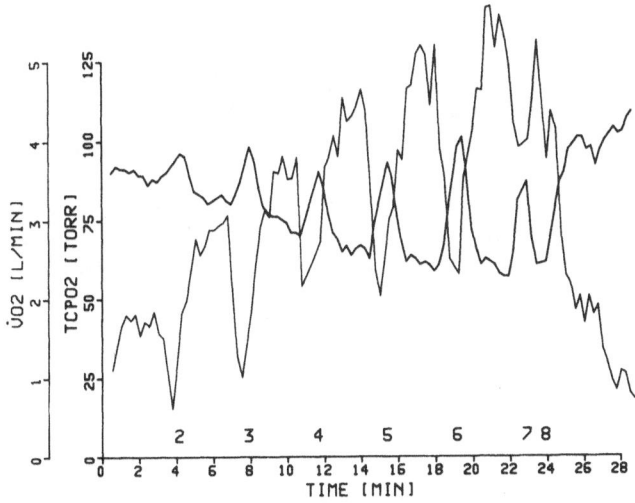

Fig. 4. TcPo₂ and $\dot{V}o_2$ during treadmill exercise in a well-trained runner. The treadmill speed is increased beginning with 6 km/h in 2 km/h-steps of 3 min 30 sec duration. Between the steps there are breaks of 30 seconds for sampling capillary blood. During the breaks Vo₂ decreases whereas tcPo₂ markedly increases. The tcPo₂ starts with a value of 90 Torr and amounts to 109 Torr at the end.

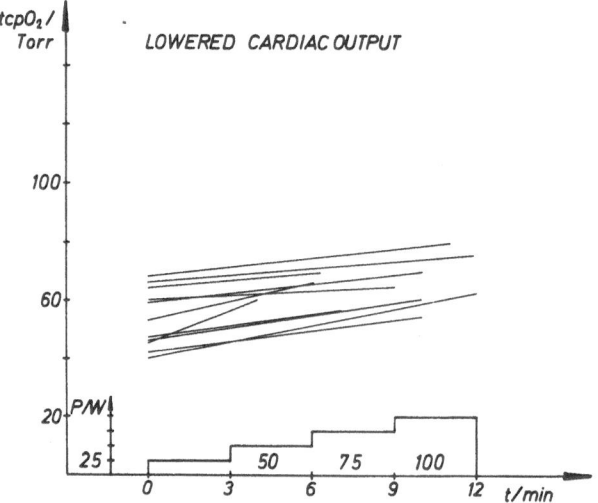

Fig. 5.   TcPO$_2$ changes in three groups of tested persons (see text).
           The initial and end values of the PO$_2$ are directly connected
           to demonstrate the trend of the change. Tan α is the steep-
           ness of these lines.
           (a) n = 11; (b) n = 15; (c) n = 11.

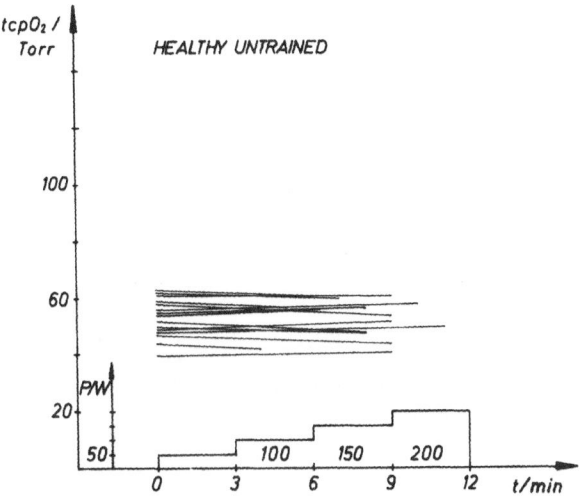

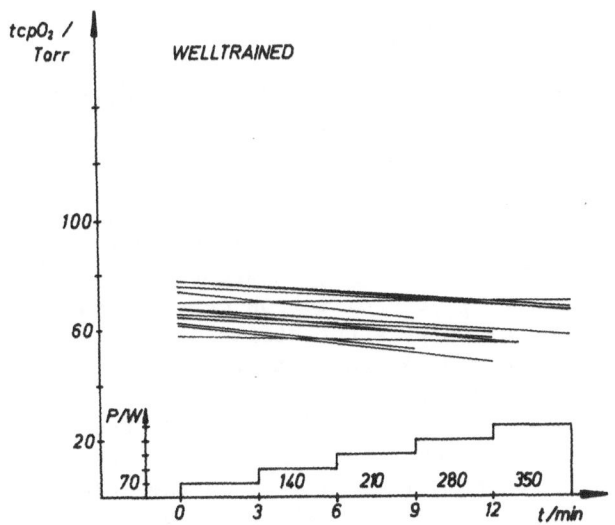

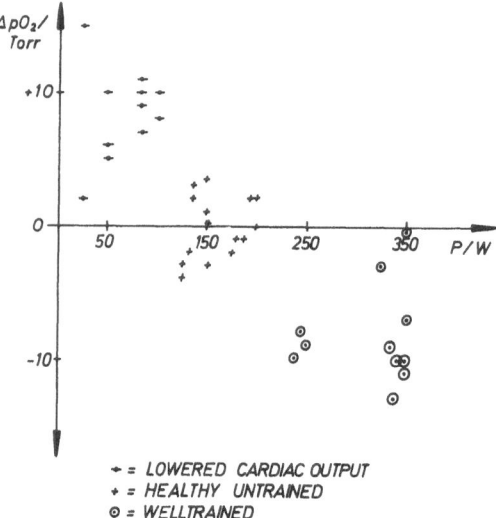

Fig. 6.   Change in tcPo₂ related to the max. sustained work load P.
          (-o- n = 10; + n = 16; o n = 10).

seems evident that the discontinuous measurement methods used, involving arterial puncture and blood gas analysis, are not sufficient
for exact determination of rapid changes in P_ao_2, as shown in the
present results. Such rapid changes occur usually very soon after
the termination of exercise, Po_2 increases to exceed preexercise
levels (Figs. 1, 2, 3, 4).

    During intermittent exercise (Fig. 4) new levels for $\dot{V}o_2$ and
tcPo₂ are reached at each step. During the breaks, tcPo₂ rises for
this short time. This observation seems to indicate that P_ao_2 is
regulated during exercise by cardiopulmonary processes. The changes
in tcPo₂ are reproducible in re-testing. Mechanisms for adjusting
oxygen transport and Po₂ in the arterial blood may be dependent on
some of the following parameters, to greater or lesser extents: the
alveolar ventilation $\dot{V}_A$, the diffusion capacity D, the pulmonary
perfusion $\dot{Q}_1$, the pulmonary shunt diffusion $\dot{Q}_sh$, the arterial-
venous O_2 difference avD and the hemoglobin concentration Hb .
Each of these (and also other) factors influence P_ao_2. In subjects
with healthy lungs and normal diffusion membranes the relationship
$\dot{V}_A/\dot{Q}_1$ seems to be very important well allover the lung as for a
local mismatch in these values. In Fig. 3 the changes in the index
$\dot{V}o_2$/heart rate will reflect reciprocal changes in tcPo₂, thereby
the close relationship between cardiopulmonary capacity and exercise
Po₂ is stressed, this has been shown for well-trained persons by
other investigators (Åstrand, 1952; Doll et al., 1966). In trained
persons $\dot{V}o_2$/HR during muscular exercise is higher than in untrained
ones or in patients with lowered cardiac performance. The relation
between $\dot{V}o_2$ and the oxygen transport value Po₂ is demonstrated in

Fig. 4. Remarkable seems to be the increase in $tcPo_2$ ($\downarrow\downarrow$) in a patient, who suffered from a myocardial infarction, during exercise (Fig. 2), where this change appeared together with cardiac dyspnea. This can indicate a change in Q_1, which can alter both local V_A/Q_1 or the transit time in the lung capillaries.

Further investigations will be necessary to examine and explain the changes in the oxygen transport systems during exercise. Transcutaneous measurements of Po_2 may be a useful tool in these experiments.

REFERENCES

Åstrand, P.O., 1952, "Experimental Studies of Physical Working Capacity in Relation to Sex and Age", Mungsgaard, Kopenhagen.

Borgia, J.F., and Horvath, S.M., 1978, Transcutaneous, noninvasive Po_2 monitoring in adults during exercise and hypoxemia, Pflügers Arch., 377:143-145.

Doll, E., Keul, J., Maiwald, C., and Reindell, H., 1966, Das Verhalten von Sauerstoffdruck, Kohlensäuredruck, pH, Standardbicarbonat und base excess im arteriellen Blut bei verschiedenen Belastungsformen, Int. Z. Angew. Physiol., 22:327-355.

Huch, R., Huch, A., and Lübbers, D.W., 1973, Transcutaneous measurement of blood Po_2 ($tcPo_2$) - Method and application in perinatal medicine, J. Perinat. Med., 1:183-191.

Huch, R., Huch, A., and Lübbers, D.W., 1981, "Transcutaneous Po_2", Thieme-Stratton, Stuttgart-New York.

Keul, J., and Doll, E., 1973, Intermittent exercise: Metabolites, Po_2 and acid-base equilibrium in the blood, J. Appl. Physiol., 34:220-225.

Löllgen, H., Nieding, G.V., Kersting, F., and Just, H., 1979, Transcutaneous measurement of Po_2 in adults: Exercise testing and monitoring in acute myocardial infarction, Med. Prog. Technol., 6:43-52.

Lübbers, D.W., 1979, Cutaneous and transcutaneous Po_2 and their measuring conditions, Birth Defects, 15/4:13-31.

Margaria, R., Edwards, H.T., and Dill, D.B., 1933, The possible mechanism of contracting and paying the oxygen debt and the role of lactic acid in muscular contraction, Am. J. Physiol., 106:689-715.

Reichel, G., 1975, Die Bedeutung der arteriellen Blutgasanalyse für die Lungenfunktionsdiagnostik und Begutachtung, Wien. Med. Wochenschr. (Suppl.), 28:4-10.

Schwarz, W., and Fabel, H., 1976, Das arterielle Sauerstoffdruckprofil unter Belastung und in der Erholungsphase - fortlaufende Sauerstoffpartialdruckmessung bei Lungengesunden und Bronchitikern, Pneumonol. (Suppl.), 5:216-227.

Schonfeld, T., Sargent, C.W., Bautista, D., Walters, M.A., O'Neal, M.H., Platzker, A.C.G., and Keens, T.G., 1980, Transcutaneous oxygen monitoring during exercise stress testing, Am. Rev. Respir. Dis., 121:457-462.

Siggard-Andersen, O., 1968, Acid-base and blood gas parameters - arterial or capillary blood?, Scand. J. Clin. Invest., 21:289-292.

Steinacker, J.M., and Wodick, R.E., Transcutaneous measurement of Po_2 in adults: Design of an improved electrode, in: "Continuous Transcutaneous Blood Gas Monitoring", R. Huch, A. Huch, J.F. Lucey, eds., Dekker, New York (in press).

Wodick, R.E., and Steinacker, J.M., Better clinical diagnostics of the physical condition of adults using transcutaneous Po_2-measurement during exercise, in: "Continuous Transcutaneous Blood Gas Monitoring", R. Huch, A. Huch, J.F. Lucey, eds., Dekker, New York (in press).

Tremper, K.K., Waxmann, K., Bowman, R., and Shoemaker, W.C., 1980, Continuous transcutaneous oxygen monitoring during respiratory failure, cardiac decompensation, cardiac arrest and CPR, Crit. Care Med., 8:377-381.

Young, I.H., and Woodcock, A.J., 1978, Changes in arterial blood gas tensions during unsteady exercise, J. Appl. Physiol., 44:93-96.

QUANTIFICATION OF THE UTILIZATION OF RESERVE OXYGEN TRANSPORT

CAPACITY: INITIAL STUDIES IN CRITICALLY ILL PATIENTS

K. Farrell, R. Bowen, and J. Beatty

Departments of Surgery and Anaesthesiology
West Virginia University, Morgantown, West Virginia
USA

Reserve oxygen transport is defined as oxygen transport that is available for utilization but is in excess of actual tissue requirements (Shoemaker et al., 1976). A decrease in oxygen delivery (Do_2) below 'normal' or an increase in oxygen consumption ($\dot{V}o_2$) above 'normal' are factors in causing the utilization of reserve oxygen transport. It is postulated that the ability to quantify the utilization of reserve oxygen transport capacity may lead to:

1) better clinical management of some critically ill patients;
2) an improved understanding of the interrelationships of a variety of oxygenation and hemodynamic parameters.

Utilization of reserve oxygen transport capacity can occur in the following settings:
   a) 'normal' $\dot{V}o_2$ and decreased Do_2;
   b) 'normal' Do_2 and increased $\dot{V}o_2$;
   c) decreased Do_2 and increased $\dot{V}o_2$;
   d) increased Do_2 and increased $\dot{V}o_2$ where the percent increase in $\dot{V}o_2$ is greater than the percent increase in Do_2;
   e) decreased Do_2 and decreased $\dot{V}o_2$ where the percent decrease in Do_2 is greater than the percent decrease in $\dot{V}o_2$.

For these clinical settings the equations for the calculation of the percent utilization of reserve oxygen transport capacity ($\%URO_2TC$) are defined as follows:

a) for 'normal' $\dot{V}o_2$ and decreased Do_2
   $\%URO_2TC = [(Do_2)_1/(Do_2)_2 - 1] \times 100$
b) for 'normal' Do_2 and increased $\dot{V}o_2$
   $\%URO_2TC = [(\dot{V}o_2)_2/(\dot{V}o_2)_1 - 1] \times 100$
c) for decreased Do_2 and increased $\dot{V}o_2$
   $\%URO_2TC = \{[(Do_2)_1/(Do_2)_2 - 1] + [(\dot{V}o_2)_2/(\dot{V}o_2)_1 - 1]\}$
   $\times 100$
d) for increased Do_2 and increased $\dot{V}o_2$
   $\%URO_2TC = \{[(Do_2)_2/(Do_2)_1 - 1] - [(\dot{V}o_2)_2/(\dot{V}o_2)_1 - 1]\}$
   $\times 100 \times (-1)$
e) for decreased Do_2 and decreased $\dot{V}o_2$
   $\%URO_2TC = \{[(\dot{V}o_2)_1/(\dot{V}o_2)_2 - 1] - [(Do_2)_1/(Do_2)_2 - 1]\}$
   $\times 100 \times (-1)$

If a negative number is obtained for equations (d) and (e) then there is no utilization of reserve oxygen transport capacity.

The subset (1) values are 'normal' values for Do_2 and $\dot{V}o_2$. These 'normal' values are based on known metabolic rates for age and sex (Wilmore, 1977), the assumption or measurement of respiratory quotient (R.Q.), and a normal oxygen utilization coefficient. $\dot{V}o_2$ ($ml \cdot min^{-1} \cdot m^{-2}$) can be calculated as follows:

$$\dot{V}o_2 (ml \cdot min^{-1} \cdot m^{-2}) = \frac{\text{Basal Metabolic Rate } (kCal/m^2hr \times 100\ ml/Lit)}{60\ (min/hr) \times k\ Cal\ Equivalent\ (kCal/Lit)}$$

kCal Equivalent = $3.91 + (1.1 \times R.Q.)$

The R.Q. can be assumed or measured. If an R.Q. of 0.8 is used then the kCal Equivalent is 4.78 kCal/Lit. Using a normal oxygen utilization coefficient of 0.25 (Bryan-Brown, 1980) then 'normal' Do_2 can be calculated as follows:

$$Do_2 = \dot{V}o_2 \times 4$$

The subset (2) values are those observed in patients.

142 sets of hemodynamic and oxygenation parameters from 40 patients in a multidisciplinary intensive care unit are the basis of this report. These 142 sets of data are those (from a larger set) in which the utilization of reserve oxygen transport capacity was found to occur.

A pulmonary artery catheter was used to determine cardiac output (thermodilution method) and to obtain mixed venous blood. Mixed venous Po_2 (P_vo_2) was obtained by direct measurements. Arterial and mixed venous oxygen contents were calculated from measured hemoglobin concentrations, measured arterial and mixed venous Po_2 values, and calculated hemoglobin-oxygen saturations. Do_2 and $\dot{V}o_2$ were calculated from the cardiac index and arterial and mixed venous oxygen contents.

$$Do_2 (ml \cdot min^{-1} \cdot m^{-2}) = C.I. \times Ca \times 10$$

$$\dot{V}o_2 (ml \cdot min^{-1} \cdot m^{-2}) = C.I. \times (Ca - Cv) \times 10$$

```
C.I. = cardiac index
Ca   = arterial oxygen content
Cv   = mixed venous oxygen content
```

RESULTS

When P_vO_2 is plotted vs %URO$_2$TC the following equation from a linear regression analysis was obtained:

$$P_vO_2 = (-.121)(\%URO_2TC) + 39.2 \quad (n = 142, r = 0.81).$$

When the subgroups of determinations with $Do_2 > 500$ ml·min^{-1}·m^{-2} and with $Do_2 < 350$ ml·min^{-1}·m^{-2} the plot of P_vO_2 vs. %URO$_2$TC is shown in Fig. 1 (see Table 1 for data). From linear regression analyses the following equations were obtained;

for $Do_2 > 500$ ml·min^{-1}·m^{-2} $\quad P_vO_2 = (.094)(\%URO_2TC) = 40.4$
$(n = 36, r = 0.72)$

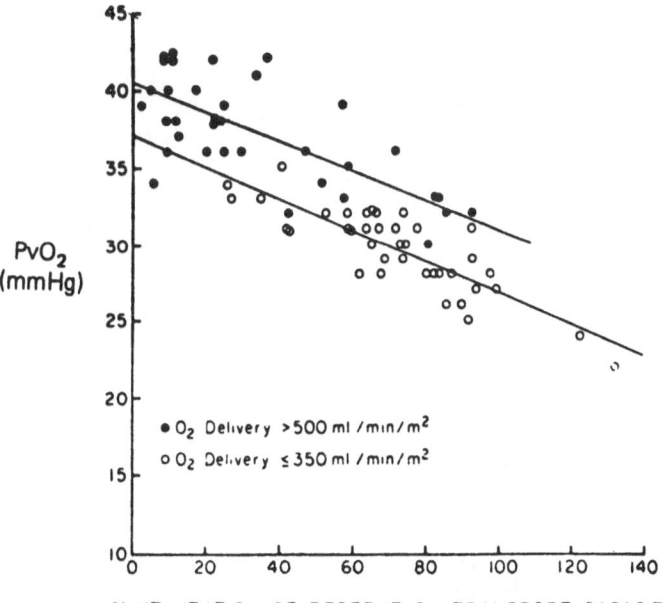

Fig. 1. P_vO_2 vs. the % utilization of reserve oxygen transport capacity in two groups of patients with different oxygen deliveries.

Table 1.   %URO$_2$TC and corresponding P$_v$O$_2$ in two subgroups of
           patients with different oxygen deliveries.

| For Oxygen Delivery ≤ 350 ml·min^{-1}·m^{-2} | | For Oxygen Delivery > 500 ml·min^{-1}·m^{-2} | |
Utilization of Reserve Oxygen Transport Capacity	P$_v$O$_2$ mm Hg (kPa)	% Utilization of Reserve Oxygen Transport Capacity	P$_v$O$_2$ mm Hg (kPa)
59	31 (4.1)	59	35 (4.6)
66.4	32 (4.2)	47	36 (4.8)
87.3	28 (3.7)	23.8	38 (5.0)
58.8	31 (4.1)	9.5	40 (5.3)
132	22 (2.9)	36.7	42 (5.6)
77.9	31 (4.1)	24.7	36 (4.8)
35.2	33 (4.4)	80.6	30 (4.0)
42.8	31 (4.1)	22	42 (5.6)
27	33 (4.4)	83	33 (4.4)
74.4	32 (4.2)	34	41 (5.4)
83.6	23 (3.7)	83	33 (4.4)
40.6	35 (4.6)	8.8	42 (5.6)
26	34 (4.5)	25	39 (5.2)
67.6	31 (4.1)	11	42 (5.6)
80.5	28 (3.7)	8.4	42 (5.6)
58.5	32 (4.2)	52	36 (4.8)
68.9	29 (3.8)	86	32 (4.2)
62	28 (3.7)	57.5	39 (5.2)
74	30 (4.0)	9.55	36 (4.8)
122.5	24 (3.2)	58	33 (4.4)
42.6	31 (4.1)	51.7	34 (4.5)
52.7	32 (4.2)	11.9	38 (5.0)
63.9	31 (4.1)	17.7	50 (5.3)
65.5	30 (4.0)	29.8	36 (4.8)
74	30 (4.0)	12.4	37 (4.9)
92	25 (3.3)	5.9	34 (4.5)
66	32 (4.2)	22.6	38 (5.0)
73.9	29 (3.8)	93	32 (4.2)
93	31 (4.1)	42.6	32 (4.2)
64	32 (4.2)	20.6	36 (4.8)
85.8	26 (3.5)	21.5	38 (5.0)
93.3	29 (3.8)		
98	28 (3.7)		
94.5	27 (3.6)		
72	31 (4.1)		
67.8	38 (3.7)		
99.5	27 (3.6)		
83.6	28 (3.7)		

for $Do_2 < 350$ ml$\cdot$min$^{-1}\cdot$m^{-2} $P_v O_2 = (-.1026)(\%URO_2TC) + 37.05$
(n = 39, r = 0.85)

For $Do_2 > 500$ ml$\cdot$min$^{-1}\cdot$m^{-2} the mean $\%URO_2TC$ is 34.38 + 26.85
(S.E. = 4.47)
   For $Do_2 < 350$ ml$\cdot$min$^{-1}\cdot$m^{-2} the mean $\%URO_2TC$ is 72.23 + 22.94
(S.E. = 3.67)
   P is < .01 for the difference of the means. (t-test)

DISCUSSIONS AND CONCLUSIONS

The percent utilization of reserve oxygen transport capacity
is a new term that potentially interrelates a number of hemodynamic
and oxygenation parameters, (i.e., cardiac output, hemoglobin concentration $P_v O_2$, Do_2 and $\dot{V}o_2$, arterial-venous oxygen content difference). As such it may prove to be of value in the care of critically
ill patients especially in determining the value of various therapeutic interventions.

There was a significant correlation between $P_v O_2$ and $\%URO_2TC$.
There was also a significant difference between the mean values for
the $\%URO_2TC$ in the two subgroups of patients (34.38 for those with
$Do_2 > 500$ ml$\cdot$min$^{-1}\cdot$m^{-2} and 72.23 for those with $Do_2 < 350$ ml$\cdot$min$^{-1}\cdot$
m^{-2}. For the given $\%URO_2TC$ those patients with higer Do_2's
(> 500 ml$\cdot$min$^{-1}\cdot$m^{-2}) tended to have a higher $P_v O_2$ than those with
lower Do_2's (< 350 ml$\cdot$min$^{-1}\cdot$m^{-2}). Similarly for a given $P_v O_2$ those
patients with lower Do_2's tended to have a lower $\%URO_2TC$.

REFERENCES

Bryan-Brown, C.W., 1980, Chapter on gas transport and delivery, in:
      "Critical Care State of the Art", Vol. I, W.C. Shoemaker,
      W.L. Thompson, eds., Society of Critical Care Medicine,
      Fullerton, California.
Shoemaker, W.C., Launder, M.D., Castagna, J., and State, D., 1976,
      Method for estimation of the perfusion defect in shock,
      J. Surg. Res., 20:77.
Wilmore, D., 1977, "The Metabolic Care of the Critically Ill",
      Plenum Medical Book Company, New York.

ON A GENERALIZED BIOENERGETIC CONTROL MECHANISM OF MICROCIRCULATION

WITH DIFFERENT EFFECTS IN THE LUNG AND THE OTHER PARTS OF THE BODY

M. von Ardenne

Forschungsinstitut Manfred von Ardenne, Dresden, GDR

Particular studies (von Ardenne and Klemm, 1982) have shown that under appropriate circumstances (no congestion, no asymmetric physical exercise, resting state) the relatively simple determination of the venous oxygen partial pressure P_vO_2[1] in blood taken from the cubital vein permits a sufficiently exact assessment (standard deviation $\pm$ 2.5 mm of mercury) of the mixed central P_vO_2 and its changes. This observation led to the discovery that states of physical weakness after stress, diseases, operations etc. are correlated with both diminished values of arterial oxygen partial pressure P_aO_2[1] (von Ardenne, 1980; 1981a, b) and, primarily, elevated values of P_vO_2. Moreover, it was found that the critical reduction of η, i.e., the utilization factor of the oxygen-binding capacity of blood, can be obviated within less than one hour by means of the 15 min-oxygen multistep[2] quick procedure according to Table 1 (von Ardenne, 1983). Characteristic findings demonstrating both the elevation of P_aO_2 persisting for month and the decline of P_vO_2 are shown in Fig. 1. Results obtained from a group of randomized individuals are statistically evaluated in Table 2. Two examples indicating Po_2 measurements in states of weakness, during the OM quick procedure, and after recovery are given in Figs. 2 and 3. In such cases, when a patient is bedridden due to diseases, crises, weakness etc., that means when the energy consumption is minimal, a decrease of η below 15% has always been observed. The drop of this factor being mainly indicative of oxygen transport to tissues is caused primarily by elevated values of P_vO_2. As shown in Figs. 3 and 4, the value of η can be re-elevated within 1 hour from, e.g., 13 to 30% (and beyond) for weeks and months by means of the 15 min-procedure

[1] All P_vO_2 or P_aO_2 values given here represent resting values if not otherwise stated
[2] Abbreviated as OM

Table 1. Protocol of the 15 min-oxygen multistep quick procedure for sufficiently mobile individuals producing a rather long lasting effect.

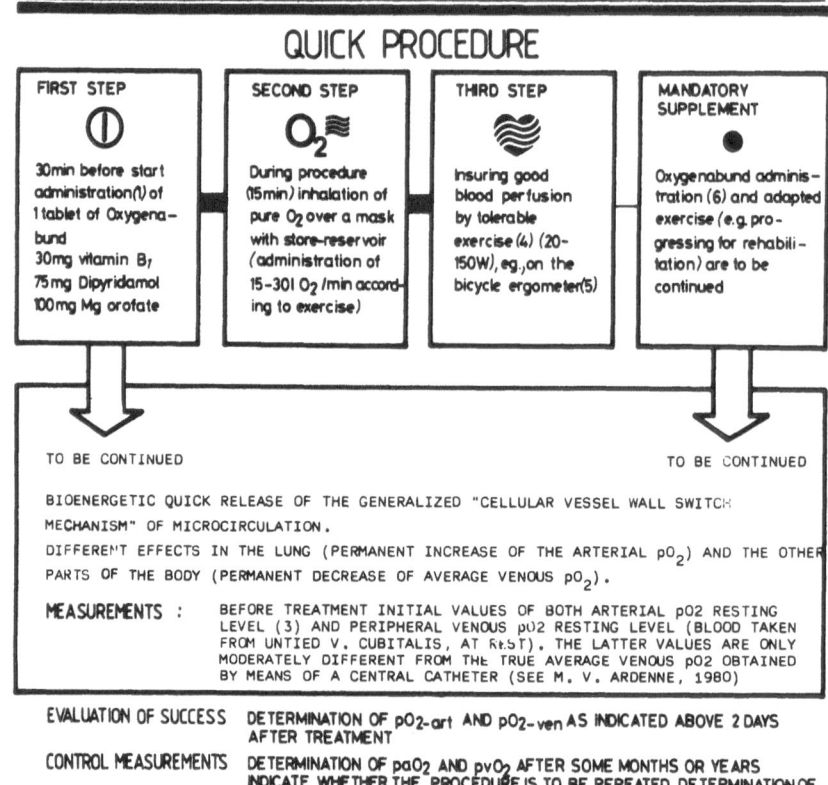

QUICK PROCEDURE

FIRST STEP	SECOND STEP	THIRD STEP	MANDATORY SUPPLEMENT
30min before start administration(1) of 1 tablet of Oxygenabund 30mg vitamin B₁ 75mg Dipyridamol 100mg Mg orotate	During procedure (15min) inhalation of pure O₂ over a mask with store-reservoir (administration of 15-30l O₂ /min according to exercise)	Insuring good blood perfusion by tolerable exercise (4) (20-150W), eg.,on the bicycle ergometer(5)	Oxygenabund administration (6) and adapted exercise (e.g. progressing for rehabilitation) are to be continued

TO BE CONTINUED   TO BE CONTINUED

BIOENERGETIC QUICK RELEASE OF THE GENERALIZED "CELLULAR VESSEL WALL SWITCH MECHANISM" OF MICROCIRCULATION.

DIFFERENT EFFECTS IN THE LUNG (PERMANENT INCREASE OF THE ARTERIAL pO_2) AND THE OTHER PARTS OF THE BODY (PERMANENT DECREASE OF AVERAGE VENOUS pO_2).

MEASUREMENTS :   BEFORE TREATMENT INITIAL VALUES OF BOTH ARTERIAL pO2 RESTING LEVEL (3) AND PERIPHERAL VENOUS pO2 RESTING LEVEL (BLOOD TAKEN FROM UNTIED V. CUBITALIS, AT REST). THE LATTER VALUES ARE ONLY MODERATELY DIFFERENT FROM THE TRUE AVERAGE VENOUS pO2 OBTAINED BY MEANS OF A CENTRAL CATHETER (SEE M. V. ARDENNE, 1980)

EVALUATION OF SUCCESS   DETERMINATION OF pO_{2-art} AND pO_{2-ven} AS INDICATED ABOVE 2 DAYS AFTER TREATMENT

CONTROL MEASUREMENTS   DETERMINATION OF paO_2 AND pvO_2 AFTER SOME MONTHS OR YEARS INDICATE WHETHER THE PROCEDURE IS TO BE REPEATED. DETERMINATION OF THE OXYGEN STATE BY MEANS OF THE 17 NOMOGRAM

(1) Add 1 g vitamin C, if necessary

(2) Always after severe distress (physical inactivity, diseases etc.) One treatment or repetition are often successful

(3) Measurement of paO2 capillary blood taken from the ear-lobe after arterialization and 10 min rest; about at the same time of day (no meals, coffee, tea etc. before). Instrument: MO 10 Universal pO2 Meter, made by VEB Pracitronic, Dresden/DDR, or others

(4) Exercise adapted to pulse rate ≈ 180 minus age

(5) 5 min before start of treatment exercise is slowly increased (initial phase); If exercise by using a bicycle ergometer (e.g., Type Golf, with Watt indication, made by Kettler, D-4763 Ense Parsit/W.-Germany) is not (yet) possible(acceleration of rehabilitation after dieseases, operations etc.), a manual fitness device based on the air pump principle (e.g., Type Maxi Power, with adjustable power from 1 - 70 kg, made by Maxi Power HB, S-50002 Boras/Sweden) is a good alternative

(6) Daily administration of Oxygenabund 30 min before start of exercise effects decrease of pO2-ven

including physical exercise for expanding both the cardiac minute volume and the ventilation rate (Table 1). The oxygen supply up to 30 l/min during treatment corresponded almost exactly to the increased ventilation rate necessary for the exercise actually performed.

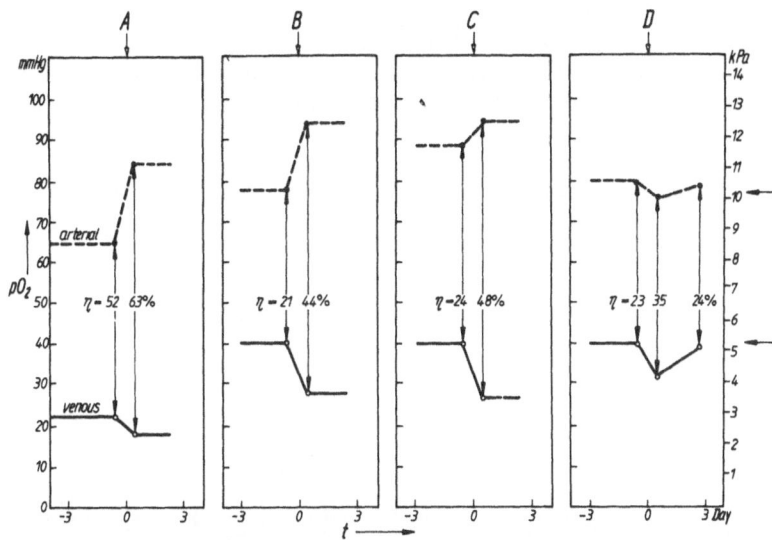

Fig. 1.   Increase of arterial and decrease of peripheral venous P_{O_2}
resting values by the 15 min-Oxygen Multistep Quick Procedure including exercise on a bicycle ergometer at 100 W.
A - C: Typical responses of three individuals (53, 52 and
31 yr. old) after treatment at day 0 (open arrows). -
Compare with D: Mean values of an age-matched control
group (n = 3) that accomplished the same exercise as above
but received no oxygen. - The vertical double arrows indicate the values of η before and after treatment. For A - C
a persisting "switch" effect is supposed, whereas in D the
training-dependent effect disappears after few days. - The
arrows at the right ordinate indicate the expected values
for arterial and venous P_{O_2}.

The reactivity of P_aO_2 and P_vO_2 is attributed to a generalized
control or "switch" mechanism of microcirculation by which the narrowest capillary cross sections are altered in dependency on the
energetic (oxygenation) state of vessel wall cells (endothelial
cells, pericytes). Results indicating the oxygen state during the
OM quick procedure and the 36 hr-variant of the OM therapy (von
Ardenne, 1980), respectively, are summarized in Fig. 4. We suggest
that the marked elevation of P_{O_2} has a direct effect on the venous
ends of capillaries. The electronmicrograph (Fig. 5) made by Löwe,
Blasig and Modersohn (1980) shows the targets on the cellular level.
Stress effects provoke tumescence and, inversely, the OM therapy
generates detumescence of endothelial cells. Swelling of hypoxic
endothelial cells is an effect in the capillary region, which corresponds probably to the rapid unspecific mesenchyme reaction in
the arterial region. This effect discovered by Hauss (1970) is
characterized by swelling of arterial vessels under hypoxic condi-

Table 2. Po$_2$ measurement showing the (permanent[1]) elevation of arterial Po$_2$ resting level and the statistically highly
significant drop of venous Po$_2$ resting level on randomly
selected individuals after single (first) performance of
the 15 min-oxygen multistep quick procedure. Values of
column 3 determined earliest one day after treatment using
blood from the same, mostly right arm. - 4/28/82

     1) Effect disappears only after severe distress but not
       under healthy conditions of life. This is the decisive
       difference toward the 'training effect' that is known
       to be declining after about two days (cp. Fig. 6)

     2) t $\triangleq$ significant at a level of 1%

1 Individuals			2 before 15min-O₂M quick proc.			3 after 15min-O₂M quick proc.			4 Difference		
N°	Age	Sex	pO$_2$-art	pO$_2$-ven	η	pO$_2$-art	pO$_2$-ven	η	pO$_2$-art	pO$_2$-ven	η
	Years	♂ ♀	mmHg	mmHg	%	mmHg	mmHg	%	mmHg	mmHg	%
1	75	♂	75	48	13 (Flu)	84	34	32	+ 9	- 14	+ 19
2	32	♂	84	50	12 (Flu)	88	35	30	+ 4	- 15	+ 18
3	65	♀	82	40	22	86	29	40	+ 4	- 11	+ 18
4	63	♂	64	20	57	77	17	66	+ 13	- 3	+ 9
5	58	♀	64	22	52	84	18	64	+ 20	- 4	+ 12
6	70	♂	66	30	34	72	16	67	+ 6	- 14	+ 33
7	45	♂	80	42	19	82	32	33	+ 2	- 10	+ 14
8	46	♂	70	44	15	76	22	53	+ 6	- 22	+ 38
9	71	♂	78	35	27	70	18	62	- 8	- 17	+ 35
10	52	♀	80	30	37	75	25	47	- 5	- 5	+ 10
11	39	♂	79	32	33	84	26	46	+ 5	- 6	+ 13
12	58	♂	72	36	25	74	29	37	+ 2	- 7	+ 12
13	66	♂	71	34	27	74	26	45	+ 3	- 8	+ 18
14	58	♀	78	32	33	80	25	48	+ 2	- 7	+ 15
15	72	♂	64	40	17	64	28	37	0	- 12	+ 20
$\overline{x}$ 15	58,0	N♂ =11 N♀ =4	73,8	35,7	28,7	78,0	25,3	47,1	+ 4,2	- 10,3 [2]	+ 18,9
S_x -	12,9		7,0	8,6	13,3	6,7	6,1	12,8	± 6,7	± 5,4	± 9,2

tions and after the action of 'sclerogenous' noxae which are mostly
accompanied by hypoxia.

The discovered bioenergetic control and controllability of
microcirculation in the capillaries of the organism has different
effects on the lung and the other parts of the body. In the lung,
P_aO_2 is changed and, consequently, its arterialization function. In
the other parts of the body, the mixed P_vO_2 is changed inversely
and, hence, the oxygen utilization of tissues is influenced.

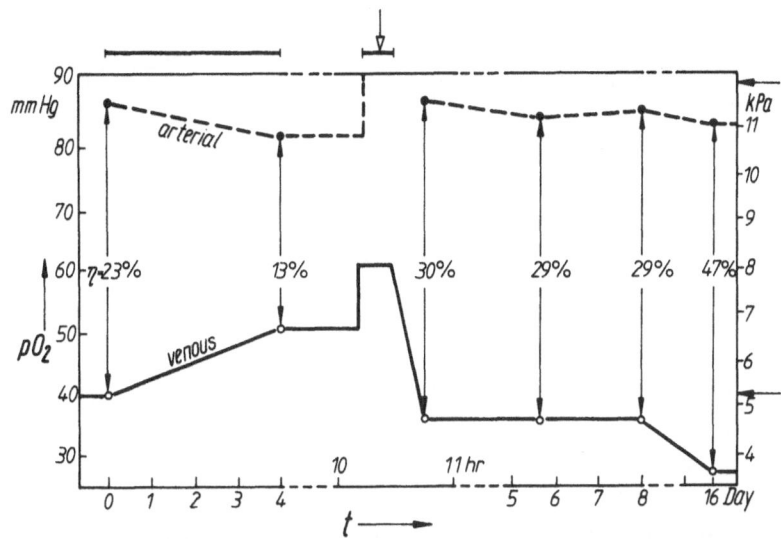

Fig. 2.   Response of arterial and venous Po$_2$ of a 32 year old male
patient on bed rest due to influenza with febrile episodes
up to 39.5°C. Critical decrease of η between day 0 and day
4 (marked by the horizontal bar) and rapid and persisting
restoration of the normal values of η after the 15 min-
Oxygen Multistep Quick Procedure including exercise on a
bicycle ergometer at 100 W (open arrow). - The arrows at
the right ordinate indicate the expected (normal) Po$_2$
values of the respective individual.

It is surprising and of great practical importance for public
health care as well that the effect of a relatively short bioenergetic control process is permanent for weeks and months, and under
appropriate circumstances even for years (von Ardenne, 1980). Therefore, we refer to it as a "switch mechanism". The reason why a
mechanism like this was expected here is that a feedback mechanism,
which is always a kind of switch operation, is involved in changes
of microcirculation. This statement becomes immediately obvious when
one considers that oxygen deficiency provokes a marked decrease of
blood flow by diminishing the a priori narrowest capillary cross
sections. By this effect, the blood flow decelerates additionally
due to the increase of apparent blood viscosity. This potentiated
decline of blood microcirculation results in a further and, hence,
critical increase of the primary oxygen deficiency. Exactly the
reverse is true for temporarily improved oxygen supply by the
several variants of the OM therapy. The mechanism outlined briefly
here and its effects concern a system having feedback or switch
characteristics.

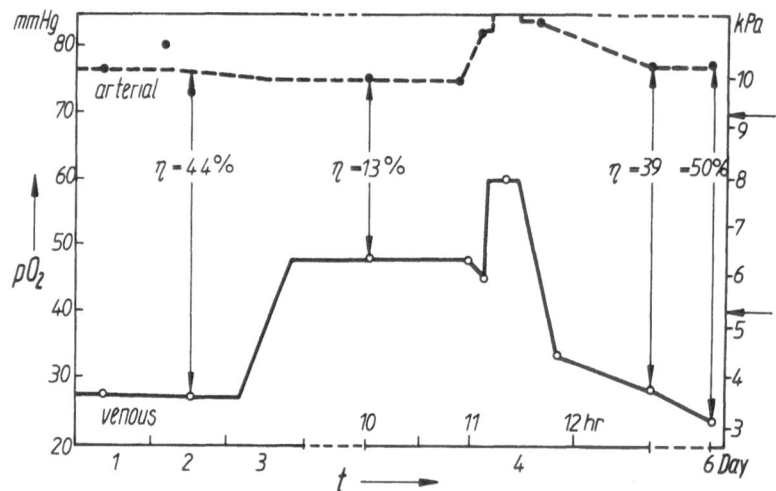

Fig. 3.   Course of the Po$_2$ values of a 75 yr. old male individual,
           their disturbation due to physical and mental overstrain
           and the rapid recovery by means of the 15 min-Oxygen Multi-
           step Quick Procedure. Day 1-2: normal life, good physical
           conditions (η = 44%!) after preceding oxygen therapy per-
           formed weeks ago. - Day 3: straining influences, during the
           following night pectanginous complaints, circulatory dis-
           orders, diarrhea, nausea. - Day 4: application of the
           procedure with 15 min pure oxygen inhalation (25 l/min)
           and physical exercise at 50 W (hometrainer) resulting in
           the immediate recovery of the pre-strain conditions (η =
           39%). - Day 6: persisting effect; η remained approx. con-
           stant as was checked 7 weeks later. - The arrows at the
           right ordinate indicate the expected (normal) values for
           75 yr. of age.

        A direct proof of the permanently improved oxygen transport
into tissues by means of the OM quick procedure or - more exactly -
of an elevated value of η (von Ardenne and Klemm, 1982) correlating
to the former is given by the results shown in Fig. 6 which demonstrate the enhanced physical efficiency one day or two weeks after
treatment.

        Most recently further treatment variants of the OM therapy have
been developed for improving blood microcirculation. These short-time
variants manage without physical exercise, but certain drugs are
used for temporary enlargement of cardiac minute volume (von Ardenne,
1983). The control and "switch" mechanism of microcirculation discussed here merits due consideration because it might be crucially
involved in the reversible phases of myocardial infarction and the
shock syndrome. Moreover, it seems to represent a basic physiological process that plays an important part, e.g., in genesis and

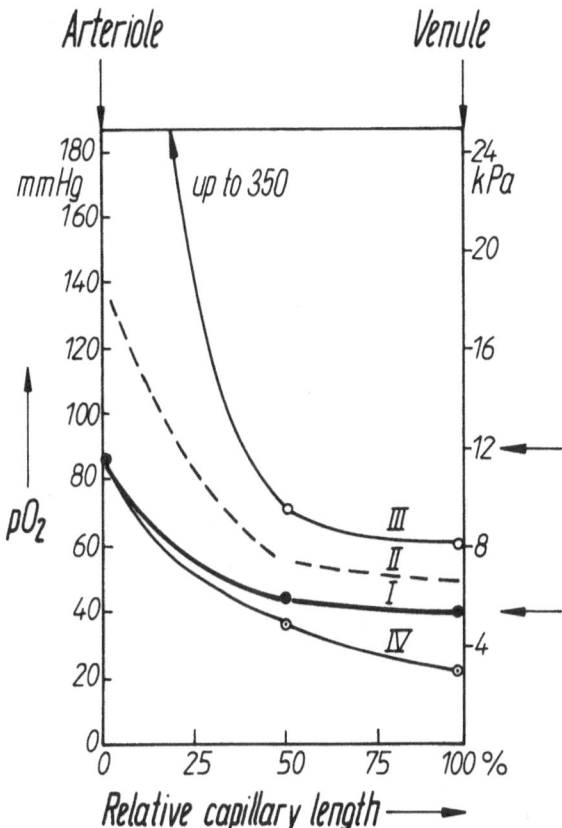

Fig. 4. Assessment of the Po_2 profiles in a capillary between its
arterial and venous ends under different conditions. I:
standard curve for healthy individuals; II: during the 36
hr-Oxygen Multistep (O_2M) Procedure; III: during the 15
min-O_2M Quick Procedure; IV: after O_2M therapy. – The
arrows at the right ordinate indicate the mean values of
arterial and venous Po_2 for normal individuals.

abolition of all kinds of weakness (problem of rehabilitation), in
onset and removal of distress sequelae, in generation and elimina-
tion of peripheral circulatory disorders as well as in the development and temporal repression of many maladies, complaints and
diseases in aging.

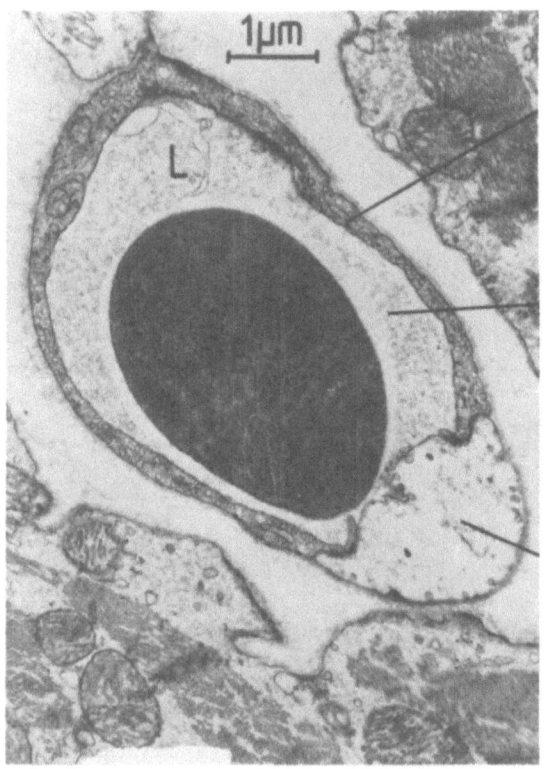

Fig. 5. The target of the oxygen multistep therapy shown in the
         electronmicroscopic picture (from: Löwe, Blasig and Moder-
         sohn, 1980). - It was found that the not too far advanced
         tumescence of endothelial cells can be persistently reduced
         by means of the OM procedure. Improved blood microcircula-
         tion in all perfused capillaries of the organism.
         L = capillary lumen, E = erythrocyte

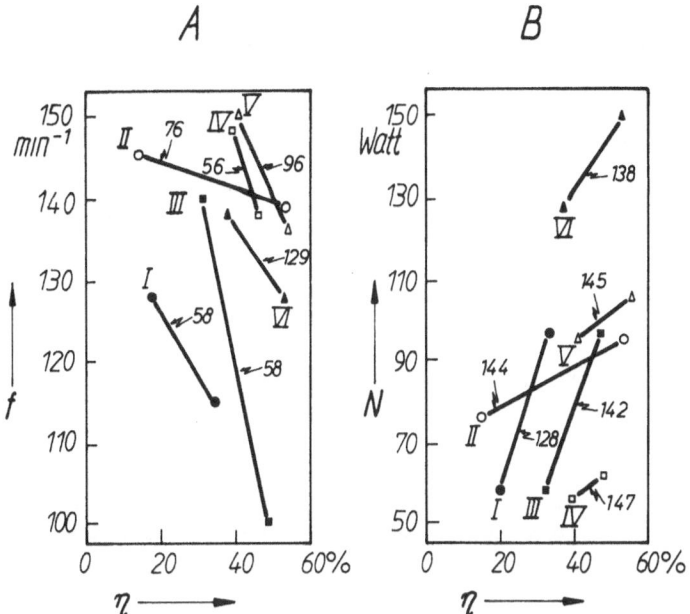

Fig. 6. Results of ergometry in six individual cases after the
Oxygen Multistep Quick Procedure. - Diagram A shows the decrease of pulse rate f as function of η at constant individual exercise measured in Watt as is indicated for each
case. - In diagram B, the increase of physical capacity N
as function of η is given at constant individual pulse
rates.- Measurements are made one day after oxygen treatment; almost the same values were found again 14 days
thereafter. The findings demonstrate the "permanant effect"
of this treatment in contrast to the common training effect
which is known to disappear one or two days later provided
the training is not continued regularly.

REFERENCES

von Ardenne, M., 1980, "Sauerstoff-Mehrschritt-Therapie", 2. Aufl.,
     Thieme, Stuttgart.
von Ardenne, M., 1981a, Measurement and removal of certain stress
     effects. Synergism, physical exercise and oxygen multistep
     therapy, Stress, 2:25-35.
von Ardenne, M., 1981b, Zellulärer Gefäßwand-Schaltmechanismus der
     Mikrozirkulation, Biomed. Tech., 27:111-118.
von Ardenne, M., 1983, "Sauerstoff-Mehrschritt-Therapie", 3. Aufl.,
     Thieme, Stuttgart, in press.
von Ardenne, M., and Klemm, W., 1982, Grundlagen der O_2-Mehr-
     schritt-Prozesse mit lang anhaltender Vergrößerung der arterio-
     venösen Po_2-Differenz, Klin. Wochenschr., 60, in press.

Hauss, W.H., 1970, Rolle der Mesenchymzellen in der Pathogenese der
    Arteriosklerose, Festvortrag Bad Nauheim 7.10.1970, Doc.
    Angiol., 2.
Löwe, H., Blasig, I., and Modersohn, D., 1980, Zur Bedeutung der
    Mikrozirkulation bei myokardialen Durchblutungsstörungen, Acta
    Biol. Med. Ger., 39:419-431.

DISCUSSION

Question to the author:
    Since, even under normal conditions, the venous Po_2 of e.g.
heart and brain are quite different, the central venous Po_2 is a
mixed value: thus, it is not always allowed to draw conclusions from
the mixed venous Po_2 about the O_2 supply of single organs. This
holds even more for the peripheral venous Po_2. The venous Po_2 in
the cubital vein mirrors the O_2 supply situation of the arm: is it
allowed to use this value as a representative mean for all the other
organs, especially in sick patients, to obtain a meaningful value
of η?

Answer of the author:
    The determination of Po_2 in central venous mixed blood allows
for conclusions to be drawn about the O_2 extraction from all
tissues. But its measurement needs a heart catheterization which is
too risky for a routine application. We looked for a routine method
which may be ambulatorily applied and found that, under certain
conditions, the venous Po_2 at rest in the cubital vein can be taken
as a sufficiently representative measuring value. In general, the
difference between peripheral and central venous Po_2 in the mean
only amounts to 2 mm Hg (v. Ardenne and Klemm, 1982; see Fig. 5;
v. Ardenne, 1983 (in print)), while the variations of the peripheral
venous Po_2 found in the same patients was higher by about a factor
of ten (weakness, variation in the 24-hour-cycle, distress, decrease
through training and O_2-multistep-processes). Values of the venous
Po_2 at rest taken from the cubital vein (freely flowing blood!)
may - in general - therefore be accepted as sufficiently representative for deriving a meaningful η.

    In the blood of the cubital vein there are - apart from local
specific factors - also general factors measurable which concern
blood flow and O_2 supply of the whole organism. Influences of
muscle work and heat regulation are excluded because measurements
are done at rest and under equal conditions of the surrounding.
Blood samples are taken after declining of any vasoconstrictory
reactions which are sometimes produced by puncture of the vein.

Question to the author:
    How can you draw conclusions from the value of η (arterio-venous
oxygen difference) without knowing anything about the behaviour of

blood flow? Does the mean blood flow under your test conditions remain so constant that its effect can be neglected?

Answer of the author:

The quantity η is derived from the arterio-venous Po_2 difference corrected according to the hemoglobin-oxygen binding curve and, therefore, corresponds to the absolute O_2 difference between arterial and venous blood. Apart from the cardiac output and hemoglobin content of blood, the quantity η provides for the main contribution to calculate O_2 transport in the whole organism and is, therefore, of great diagnostic values when estimating the actual O_2 supply status or energetic status, respectively. Different measurement of flow in single organs would, of course, be appreciable but the latter is not easily possible in ambulatory practice. A number of external non-invasive methods only measure the macrocirculation which does not give any information about the actual nutritive blood flow, th microcirculation of an organ.

Question to the author:

To understand the described permanent effects it would be helpful to know the abosulte O_2 consumption during and after your treatment:

1. Does the behaviour of the O_2 consumption of the treated person correspond to the theory presented in this paper?

2. Is an improvement of η accompanied by an increased O_2 consumption?

3. If not, why should a person then benefit from increased arterial Po_2 and η?

Answer of the author:

The important increase of the η value by processes of the oxygen multistep-therapy leads to a significant increase of O_2 transport in the organism which can be equalized to oxygen uptake, the reason being – according to our measurements – the change of the other factors of O_2 transport (cardiac output and hemoglobin content) remaining small in comparison to changes of η. In our institute, as well as in different sport-medical partner institutes, we are, at present, directly measuring the O_2 uptake and physical efficiency (ergometry) after different variations of the oxygen-multistep-therapy.

OXYGEN PROFILES AND STRUCTURE OF PENICILLIUM CHRYSOGENUM PELLETS

H. Baumgärtl[1], R. Wittler[2], D.W. Lübbers[1], and
K. Schügerl[2]

[1]Max-Planck-Institut für Systemphysiologie
Rheinlanddamm 201, 4600 Dortmund, FRG
[2]Institut für Technische Chemie der Universität
300 Hannover, FRG

The efficiency of large Penicillium chrysogenum pellets in biosynthesis of penicillin depends very much on a sufficient oxygen supply. It is known that at sites where the O_2 pressure decreases below 6 kPa the penicillin production stops. Since the mechanisms of O_2 delivery inside the pellet are almost unknown (Phillips, 1966; Huang and Bungay, 1973; Miura et al., 1975), we investigated this problem by measuring the O_2 pressure field within the pellet with Po_2 microneedle electrodes. It is found that under our experimental conditions the oxygen supply within the pellet occurs not only by molecular diffusion, but also by turbulence and convection.

MATERIALS AND METHODS

The mycelial pellets were prepared by a controlled fermentation in an 80 l tower-loop fermenter at the "Institut für technische Chemie" in Hannover. The measurements were carried out in a simple air lift loop glass reactor (Fig. 1) which allowed to vary flow direction, turbulence, temperature and O_2 content of the medium. The pellets were fixed in a pellet holder. A turbulence probe (DISA) was fixed near the pellet. The aerators allowed to equilibrate the medium with different gas mixtures. By the alternative use of one of the aerators, flow direction could be reversed. The Po_2 profiles were measured polarographically by a microcoaxial membranized needle electrode (tip diameter about 0.7 um) (Baumgärtl and Lübbers, 1983). The electrode was mounted in a micromanipulator. The puncture was made under microscopical control and performed as perpendicular to the surface as possible.

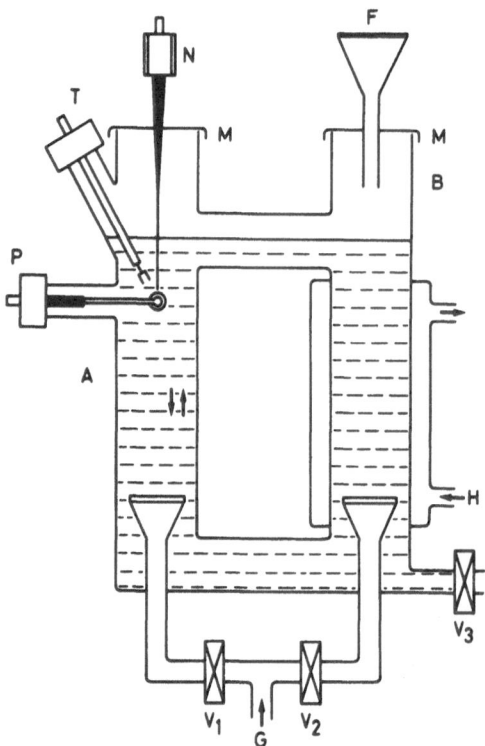

Fig. 1. Experimental setup to measure Po_2 profiles in pellets. The
          setup consists of two glass cylinders (A, B) connected in
          such a way that by proper opening and closing of the valves
          V_1, V_2, the flow direction of the medium can be changed.
          F: funnel for filling; G: gas inlet; H: heat exchanger for
          thermostating the system; M: covering film; N: Po_2 elec-
          trode; P: pellet holder; T: turbulence probe; V_3: outlet of
          medium.

        For histological investigation the pellets were embedded in
Tissueteck[R] at -20°C. 10 um cryocuts were prepared and stained with
cresylviolet.

RESULTS AND DISCUSSION

Structure of the pellet

        The pellets of Penicillium chrysogenum consist of a relatively
loose wickerwork of mycelial hyphae. From the pellet surface free
hyphae protrude into the surrounding so that there is not a very
sharp boderline (Fig. 2). Pellets up to a size of 0.4 mm have a more
or less homogeneous structure. The larger pellets, as they were used

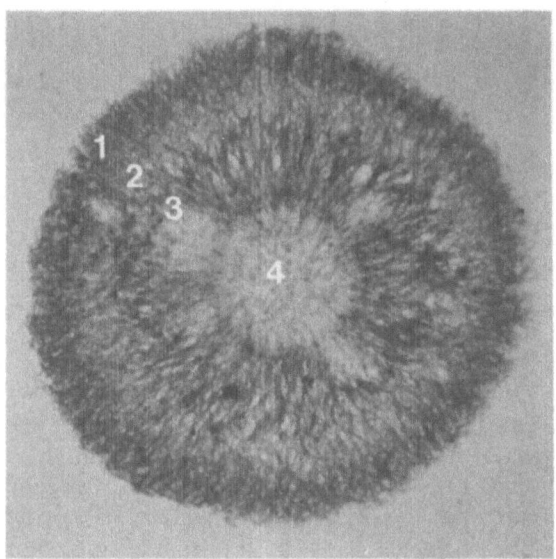

Fig. 2.   Light microscopic picture of a central slice of a pellet
          of Penicillium chrysogenum. The pellet consists of a loose
          wickerwork of hyphae. There are 4 zones (1, 2, 3, 4). Some
          hyphae protrude from the surface into the surroundings
          (for further explanation see text).

in our experiments, show four different zones. (The pellet shown in
Fig. 2 is from a 68 hours culture).

Outer Zone: 1 Thickness: 160-280 um; volume fraction: 50%. In
this zone due to decreasing Po_2 (see Figs. 3 and 4) conditions
change from cell growth to maintenance.

Dead mycelia zone: 2 Thickness: 200-400 um; volume fraction:
25-50%. In this zone the hyphae have lost their cytoplasm and the
remaining cell walls have been slowly destroyed by exoenzymes of
the cells.

Inner maintenance zone: 3 Thickness: 0-200 um. This zone does
not always exist. The maintenance conditions may be possible due to
a slight supply with oxygen through channels in zones 2 and 3.

Mycelia-free centre: 4 Thickness: 250-500 um; volume fraction
2-18%. In this zone the hyphae are completely degenerated. Some un-
metabolized residues of nutrients and some salt crystals may remain.
In our experiments we focussed on zone 1 in which penicillin production may occur.

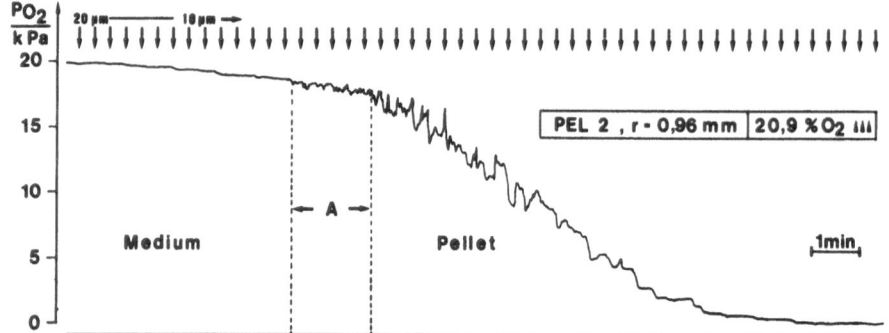

Fig. 3.   Original record of a Po₂ profile of a pellet of Penicillium
chrysogenum (fixed in the holder, see Fig. 1). r: radius of
the pellet; perpendicular arrows mark the steps of penetration (10 um); PEL 2: pellet number 2. Zone A: in this zone
free hyphae protrude into the medium. Ordinate: Po₂ in kPa.
The medium was equilibrated with 20.9% O₂ in N₂.

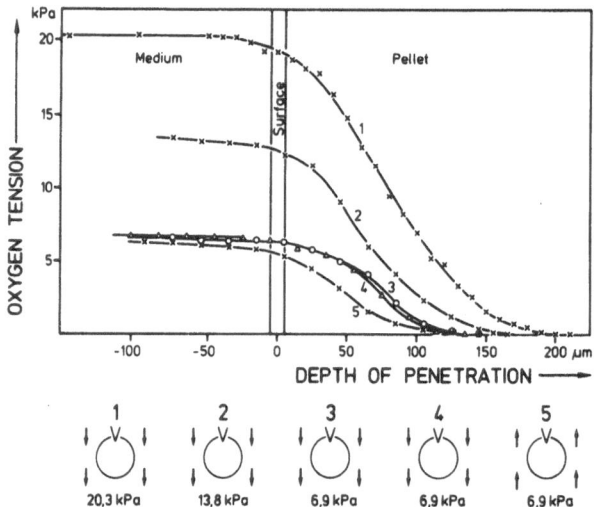

Fig. 4.   Oxygen pressure profiles within the pellet of the Penicillium chrysogenum at different Po₂ values of the outer
medium with intermediate turbulence. Curve 1, 2, 3, 5: measurements during inserting the electrode, curve 4 measurement during withdrawing of the electrode (same puncture as
in curve 3). In curve 1, 2, 3, 4, the flow was downwards
(in the direction of the puncture, curve 5 downwards (as
shown at the bottom by the direction of the arrows).

Po_2 profiles in the pellets

A typical Po_2 profile of a pellet fixed in a holder in a medium with low turbulence is shown in Fig. 3. The Po_2 of the medium is 20.0 kPa (150 mm Hg). In the zone "Medium" the Po_2 decreases only by a small amount and without distinct fluctuations. In zone A the electrode tip approaches the border zone of the pellet in which free hyphae protrude into the medium. In zone A the Po_2 decrease becomes steeper and the signal shows small fluctuations. After penetration into the pellet (zone "Pellet"), the steepness of the Po_2 profile increases as well as the amplitudes of the fluctuations. In a distance of 260 um from the pellet surface the Po_2 reaches zero. In approaching zero, the fluctuations becomes smaller and at zero they disappear.

In measurements without turbulence and convection (not shown here) fluctuations were not observed. They occur with increasing turbulence, but particularly at low turbulence values. Since the used electrodes are membranized and have only a small "stirring" effect, the fluctuations correspond to changes in local Po_2. The measurements show that the intensity of the fluctuations depends on the local Po_2 gradient and on the distance from the pellet surface. This demonstrates that the turbulence of the outer medium penetrates into the interior meshwork of the pellet. By this mechanism the O_2 supply is improved as compared to the molecular diffusion.

In Fig. 4 the dependence of the Po_2 profile within the pellet on the Po_2 of the outer medium is shown. The experiments were carried out at intermediate turbulence. The medium was equilibrated with 21% O_2 (20.2 kPa = 151.5 mm Hg; curve 1), 14% O_2 (13.8 kPa = 103.5 mm Hg; curve 2) and 7% O_2 (6.9 kPa = 51.8 mm Hg; curve 3, 4, 5) in nitrogen. The given Po_2 values are averaged values. Under these turbulence conditions curve 1 begins to decrease in a distance of - 50 um from the surface zone (marked by two lines, 10 um apart). The middle between the two lines is taken as begin of the pellet (marked by zero). Curve 1, 2, 3, 5 are measured when inserting the electrode. Curve 3 and 4 are Po_2 measurements of the same puncture: in curve 3 the electrode was inserted, in curve 4 withdrawn. The data show that the Po_2 profiles are practically identical. Fig. 4 shows that in a pellet of this size all Po_2 profiles reach zero. By changing the Po_2 of the outer medium the depth of the penetration of O_2 as well as the steepness of the O_2 profiles are changed. Knowing from the Po_2 profiles the site at which the local Po_2 falls below 6 kPa, it is possible to estimate the width of the zone in which penicillin may be produced. Table 1 shows the corresponding data. The values clearly demonstrate that the Po_2 of the outer medium has a strong effect on the oxygen supply.

Curve 3 and 5 show the effect of flow direction on the Po_2 profile. In curve 3 the medium moves from above to the bottom, in the

Table 1. Depth of O_2 penetration into the pellet of Penicillium
          chrysogenum at different Po_2 values of the outer medium.

Po_2 of the medium (in kPa)	Depth of O_2 penetration (in um)	Difference (in um)	Width of the penicillin producing zone (in um)	Difference (in um)
20.2	200		110	
		35		45
13.8	165		65	
		15		50
6.9	150		15	

same direction as the electrode puncture is performed, whereas in
curve 5 the medium moves upwards in opposite direction. By the
different effect of convection in curve 5 the Po_2 profile is shifted
by about 30 um to the left.

With molecular diffusion or with equal turbulence a smaller Po_2
decrease inside the pellet would mean that the amount of transported
oxygen at a Po_2 of 13.8 kPa (103.5 mm Hg) is smaller than at 20.2 kPa
(151.5 mm Hg) Po_2. This would point to a Po_2-dependent oxygen consumption of the pellet. However, further experiments are necessary
to prove this hypothesis.

SUMMARY AND CONCLUSIONS

Oxygen profiles were measured polarographically with microneedle electrodes in single fixed pellets of Penicillium chrysogenum.
The oxygen concentration, the grade of turbulence and the flow conditions of the outer medium were varied. Pellets were then histologically investigated. Under our experimental conditions the pellets
were found to contain a 200 um thick active outer zone with intact
cells. In this zone, the oxygen partial pressure decreased from the
initial values to zero. The results show that the Po_2 profile within the pellet is strongly influenced by the Po_2 of the outer medium.
Apart from molecular diffusion, surface turbulence as well as a convective flow through the pellets are of importance for the oxygen
transport.

ACKNOWLEDGEMENT

We like to thank P.D. Dr. Th. Bär and his coworker Dr. A. Budi Santoso for the skillful histological preparation of the pellets.

REFERENCES

Baumgärtl, H., and Lübbers, D.W., 1983, Microcoaxial needle sensor for polarographic measurement of local O_2 pressure in the cellular range of living tissue. Its construction and properties, in: "Polarographic Oxygen Sensors", E. Gnaiger, H. Forstner, eds., Springer, Berlin-Heidelberg-New York, pp. 36-65.

Huang, M.J., and Bungay III, H.R., 1973, Microprobe measurements of oxygen concentrations in mycelial pellets, Biotechnol. Bioeng., 15:1143-1197.

Miura, Y., Miyamoto, K., Kanomori, T., Teramoto, M., and Ohira, N., 1975, Oxygen transfer within fungal pellet, J. Chem. Eng. (Japan), 8:300-304.

Phillips, D.H., 1966, Oxygen Transfer into mycelial pellets, Biotechnol. Bioeng., 8:456-460.

INSTRUMENTATION
AND
METHODS

SOME ASPECTS OF SIGNAL ANALYSIS APPLIED TO INTRACAPILLARY HEMOGLOBIN

SPECTRA

M. Brunner, R. Ellermann, and M. Kessler

Institut für Physiologie und Kardiologie der Universität
Erlangen-Nürnberg, Waldstr. 6, 8520 Erlangen, FRG

INTRODUCTION

The light absorption of hemoglobin depends on its degree of
oxygenation. Different levels of oxygenation with their different
absorption bands produce obvious changes in the shape of the reflected spectra between 500 nm and 600 nm. Therefore, it is possible
to make an assessment concerning the degree of oxygenation by means
of the reflection photometry.

Reflection Photometer

The micro light-guide photometer (Brunner et al., 1981) which
we used consists of an illuminating part and a unit for measuring
the reflected light. A rotating disc interference filter selects
the wavelengths; according to the angle of rotation light of different wavelengths can pass through. From 0° to 180° the wavelength
range between 495 nm and 615 nm is scanned in increasing order, and
from 180° to 360° in decreasing order. The intensity of the transmitted light is measured by a photomultiplier and, in the form of a
continuous voltage signal via an anti-aliasing filter and an analog-
digital converter, it is sent to a computer for future processing
(Brunner, 1980).

Time Domain

Processing with a digital computer requires the converting of
the continuous voltage signal to discrete values. The sampling
interval must be short enough to avoid loss of information (Jerry,
1977). In our investigations we satisfied this condition by using
sampling frequencies up to 8192 Hz.

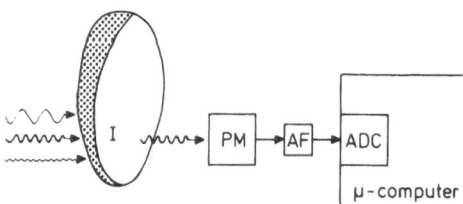

Fig. 1. Micro light-guide photometer: receiver part
         I: interference filter, PM: photomultiplier, AF: anti-
         aliasing filter, ADC: analog-digital converter.

     Fig. 2 shows on the one hand that the periodicity of the input
signal depends on the rotation of the interference filter and on
the other hand that it is superimposed by considerable noise.

     In order to analyse the data of the digitized signal, separa-
tion into periods is necessary.

     Two problems arise from this: 1) the period length is unknown;
and 2) the lengths of different periods are not identical.

     In order to solve these problems we developed a computer pro-
gram which first detects a "reference period" and then determines
the most similar sections by means of a correlation technique
(Fig. 3).

     Our experiments have shown that noise reduction can be achieved
by averaging several periods. The result of averaging 50 periods is
presented in Fig. 4.

     The function in Fig. 5 demonstrates the dependence of noise on
the number of periods averaged. This curve results from a series of
measurements (n > 100).

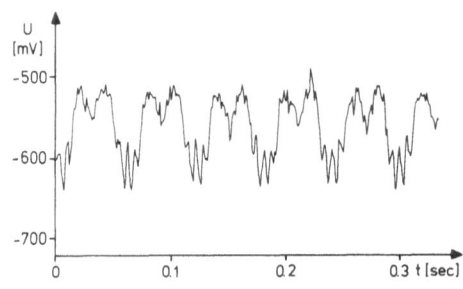

Fig. 2. Original signal from the reflection photometer.

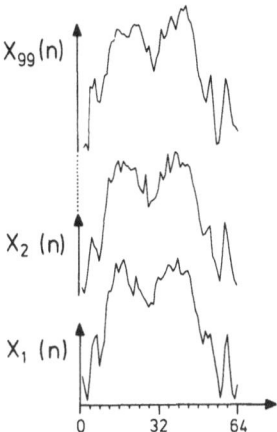

Fig. 3.   Sampled values in periods.

Frequency Domain

The aims of the analysis in the frequency domain were the
determination of both the optimum sampling frequency as a function
of the angular velocity of the interference filter, and the relation
between the frequency spectrum and degree of oxygenation.

The input signals are periodic and frequency band-limited;
therefore it is possible to perform a conversion from the time
domain by means of Fourier Transform into the frequency domain. The
transformation produces a representation of the function as a frequency amplitude diagram (Brigham, 1974).

In order to determine the optimum sampling frequency, angular
velocities and sampling intervals were varied at constant hemoglobin
oxygenation. In every case the frequency amplitude diagrams, when
normalized, showed nearly identical results (Fig. 6). The aliasing
effect was avoided by using very short sampling intervals. It is
evident from Fig. 6 that only the first 16 normalized frequency
components make a significant contribution to the signal.

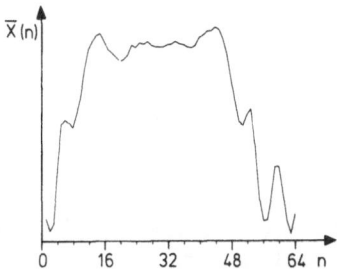

Fig. 4.   Average of 50 periods.

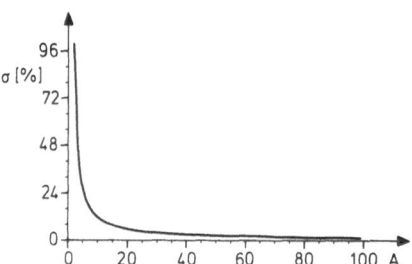

Fig. 5.   Standard deviation of the noise signal as a function of
          the number of periods averaged.
          A: number of periods averaged;
          σ: standard deviation.

        Changes in hemoglobin oxygenation cause significant changes in
the corresponding frequency amplitude diagrams. Again only the
first 16 normalized frequency components are important (Figs. 7, 8,
9).

        The optimum sampling rate sr can be computed as

        sr = w ·   16 ·   2   (Hz)   (1)
             │   │   │
             │   │   └─── requirement of sampling theorem
             │   └─── highest normalized frequency component
             └─── angular velocity

        The use of a RC-low-pass filter (Fig. 1) causes an additional
restriction for the angular velocity of the disc interference filter.

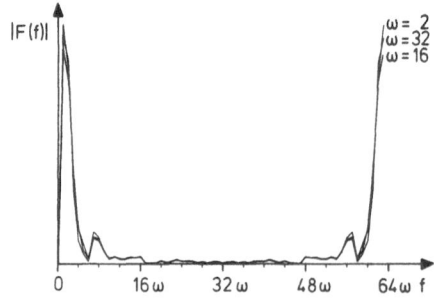

Fig. 6.   Frequency amplitude diagram of input signals with three
          different period lengths.

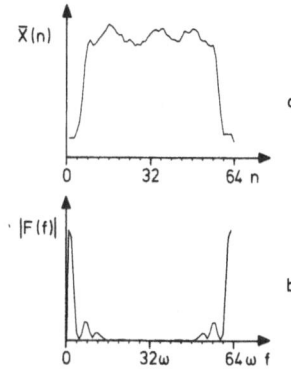

Fig. 7. Oxygenation hemoglobin
          a) input signal; b) frequency amplitude diagram.

A method for quantifying hemoglobin spectra with unknown degree of
oxygenation

    Minor changes in oxygenation also cause only small changes in
the input signal. Differences which are hardly perceptible can be
precisely determined by quantifying the spectra. The following procedure is proposed for spectra recorded with constant hemoglobin
concentration where, therefore, the only variable is the degree of
oxygenation.

    a) noise reduction by means of averaging $\longrightarrow$ m(t)
    b) definition of the fully oxygenated spectrum h(t)
    c) definition of the fully deoxygenated spectrum l(t)
    d) computation of the amount of the component h(t) from m(t)

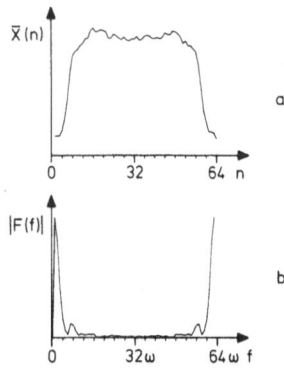

Fig. 8. Partly oxygenated hemoglobin
          a) input signal; b) frequency amplitude diagram

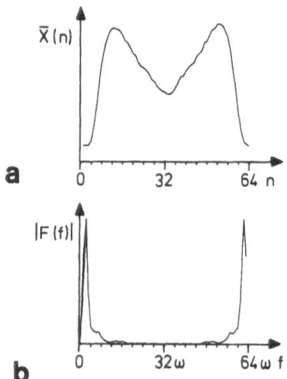

Fig. 9.   Deoxygenated hemoglobin
          a) input signal; b) frequency amplitude diagram

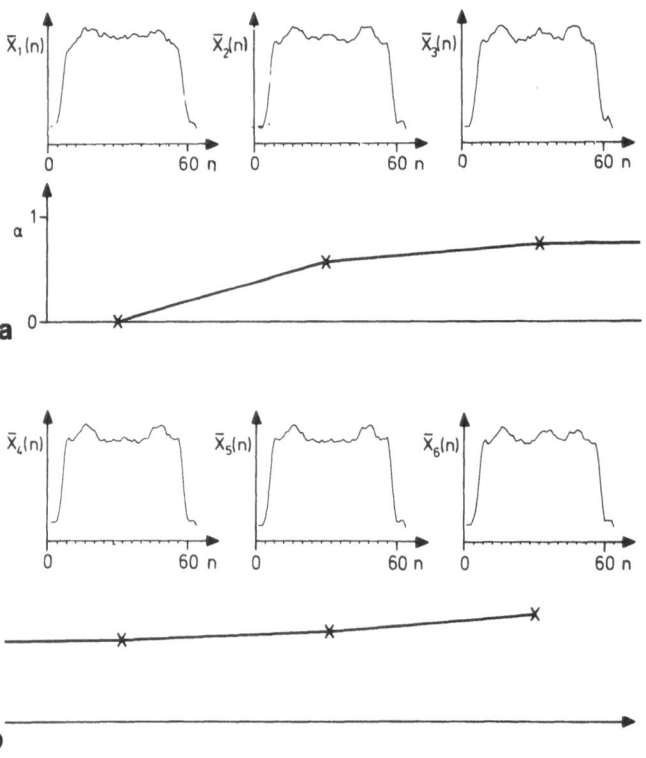

Fig. 10. Series of hemoglobin spectra with different degrees of
          oxygenation.
          a) input signal
          b) quantitative value of the fraction oxygenated, α.

where

$$m(t) = \alpha \cdot h(t) + (1-\alpha) \cdot l(t) \qquad\qquad (2)$$

Here, α is a measure of the oxygenated, and $(1-\alpha)$ of the deoxygenated portion of the spectrum $m(t)$. α can be computed by correlating equation (2) with $h(t)$ as well as with $l(t)$.

$$\alpha = \frac{\int l(t)m(t)dt \cdot \int h(t)l(t)dt - \int h(t)m(t)dt \cdot \int l(t)l(t)dt}{\int l(t)h(t)dt \cdot \int h(t)l(t)dt - \int h(t)h(t)dt \cdot \int l(t)l(t)dt} \qquad (3)$$

where: $\int x(t)y(t)dt$

is the correlation function between $x(t)$ and $y(t)$ without time-shift.

Fig. 10 shows a series of hemoglobin spectra and the corresponding α values. The first and last spectra represent the spectra defined in c) and b) above.

The procedure above produces a relative measure of the degree of oxygenation within $h(t)$ and $l(t)$. A method for measurement of absolute values could subsequently be obtained by calibrating the spectra $h(t)$ and $l(t)$.

ACKNOWLEDGEMENT

We wish to thank Dr. M. Berg for his advice and helpfulness.

REFERENCES

Brigham, E.O., 1974, "The Fast Fourier Transform", Prentice-Hall, Englewood Cliffs, N.J.

Brunner, M., 1980, On-line Verarbeitung von Hämoglobin-Reflexions-spektren mit einem Mikro-Lichtleiter-Spektrometer hoher Repititionsrate, Diplomarbeit, Universität Erlangen-Nürnberg, Erlangen.

Brunner, M., Kastner, N., Schabert, A., Höper, J., and Kessler, M., 1981, On-line Verarbeitung von Hämoglobin-Reflexionsspektren hoher Repititionsraten, in: "Medizinische Informatik und Statistik", 28, S. Koller, P.L. Reichertz, K, Überla, eds., Springer, Berlin-Heidelberg-New York.

Jerry, A.J., 1977, The shannon sampling theorem - Its various extensions and application: A tutorial review, Proc. IEEE 65: 1565-1596.

DISCUSSION

Question: Why was the frequency domain analysis not used to assess degree of oxygenation as stated on page 3.

Answer: The frequency domain analysis was used to show the relation between degree of oxygenation and the corresponding frequency amplitude diagram (see Figs. 7, 8, 9).

Question: The last paragraph is unclear. What are the authors trying to say about the difference between relative and absolute values? Is the oxygenated fraction, α, relative or absolute?

Answer: The application of this method requires the satisfaction of the following conditions:

$$DO\ (l(t)) \leq DM\ (m(t)) \leq DO\ (h(t))$$

with DO: degree of oxygenation.
α values are absolute for DO $(l(t)) = 0\%$ and DO $(h(t)) = 100\%$ and are relative for DO $(l(t)) \neq 0\%$ and/or DO $(h(t)) \neq 100\%$.

Question: Is the evaluation method invariant against changes of hemoglobin concentration?

Answer: Changes of hemoglobin concentration within the range for which equation (1) is true (see answer to next question) causes changes of the amplitudes in the form $c \cdot |F(f)|$, where c is a concentration-dependent constant. The evaluation method is invariant against changes of hemoglobin concentration, when c is taken into account.

Question: How does inhomogeneous distribution of Hb and light scattering effect the evaluation?

Answer: Inhomogeneous distribution of Hb effects the evaluation when equation (1) is not satisfied, i.e. when m(t) consists of additional terms. When these parts are very small in comparison with h(t) respectively l(t), they can be neglected (Brunner, 1980). Otherwise the proposed method is not applicable.

Question: What is the smallest HbO_2 saturation difference you can distinguish with this method?

Answer: The smallest HbO_2 saturation difference, which can be distinguished with this method, depends on the accuracy of the computer used for the investigation. The evaluation method is more exact than the measuring system; therefore the limitation for distinguishing different HbO_2 saturations is given by the accuracy of the measuring system.

CORRELATION BETWEEN TISSUE Po$_2$ AND INTRACAPILLARY Hb SPECTRA

K.H. Frank, A. Schabert, A. Friedl, M. Brunner,
J. Höper, G. Kerl, and M. Kessler

Institut für Physiologie und Kardiologie der Universität
Erlangen-Nürnberg, Waldstr. 6, 8520 Erlangen, FRG

INTRODUCTION

The development of highly flexible micro light guides by Ji et
al. (1979) made possible the construction of a spectrophotometer for
measuring intracapillary hemoglobin oxygenation. Due to the flexi-
bility, good contact between the light guides and the tissue is
ensured. The photometer has already been used successfully for measurements in the beating heart, the lung and the skeletal muscle.

The present study proposes a simple method for estimating the
intracapillary hemoglobin oxygenation, using a micro-light guide
spectrophotometer in the isolated perfused rat liver.

METHODS

Measurements of Intracapillary Hemoglobin Oxygenation

Intracapillary hemoglobin oxygenation was measured using reflection spectrophotometry. The tissue was illuminated with light,
limited from 480-630 nm by filters. This being the part of the
spectrum where oxygenated and deoxygenated hemoglobin show distinct
characteristic absorption bands. Tissue components, erythrocytes
and hemoglobin cause absorption, scattering and reflection of the
irradiated light. The spectra of the modulated reflected light are
then measured.

For the spectrophotometric measurements, a micro light guide
spectrophotometer, developed in our laboratory by Kessler and

811

Schabert (unpublished), was used. It consists of three parts, a source of light, a system of light guides and a detection and recording system.

A xenon high pressure arc lamp with a high light intensity serves as a source of light. Light is band pass filtered and focused on a micro light guide of 70 um diameter. This light guide is the central fiber of an arrangement of seven light guides, which transmit the illuminating light from the lamp to the tissue as well as the reflected light back from the tissue to the spectrophotometer (see the schematic drawing in Fig. 1). The six detection light guides are placed circularly around the illuminating light guide. Fig. 2 shows a photomicrograph of the sensor tip. The whole arrangement of the seven light guides is highly flexible and can be of variable length. The tip of the sensor is placed perpendicularly on the tissue surface. The illumination characteristics of the illuminating light guide are approximately conic. The circular arrangement of the detection light guides assures, that a maximum of the reflected light is picked up.

The detection and measuring system consists of a rotating band pass interference filter (scanning range 120 nm, 495-615 nm) and a photomultiplier tube. At each position of the filter disk the transmission is maximal for one distinct wavelength. Thus, with rotation of the disk, the spectrum between 495 and 615 nm can be scanned continuously. The reflected light has to pass through the filter disk and the transmitted monochromatic light is measured by a photomultiplier tube. The resulting periodic wavelength dependent signal and a trigger signal marking the beginning of each spectrum are processed by an on-line computer (DEC LSI 11/2) (Brunner, 1981).

Po_2 measurements: Local tissue Po_2 was measured with Pt-multi-wire surface electrodes (Kessler and Lübbers, 1966; Kessler and Grunewald, 1969). Alternating measurements of Po_2 and intracapillary hemoglobin spectra were performed on the surface of the same liver segment.

Preparation and perfusion technique: The measurements were carried out in the isolated perfused liver of femal Wistar rats, using the technique, described by Höper and Kessler (1982).

The perfusate was equilibrated with different gas mixtures containing O_2, N_2 and 5% CO_2. The Po_2 of the perfusate was varied between 5 Torr and 500 Torr. After the control spectra, (apparatus + tissue spectra I_o, in the following called tissue spectra) were recorded, washed bovine erythrocytes were added to the perfusate. The hemoglobin concentration amounted to 4-7g%.

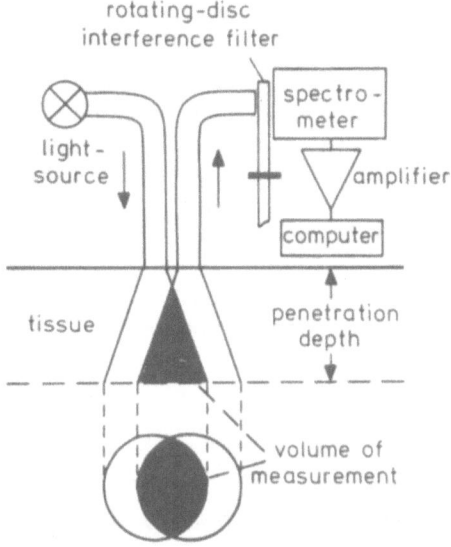

Fig. 1.   Schematic drawing of the micro light guide spectrophotometer.

Fig. 2.   Photomicrograph of the tip of the micro light guide
arrangement.

RESULTS

        After half an hour of normoxic perfusion control Po_2 histograms
and tissue spectra were measured on the hemoglobin-free perfused
liver surface. Spectra were obtained from three different sites or
more. The lower part of Fig. 3 shows four tissue spectra obtained
under these conditions. It can be seen that the intensity of the
reflected light is different for each site under investigation. This
is assumed to be due to different optical properties of the tissue.

        These spectra can be transformed almost completely one into
the other by multiplication with a constant factor. We therefore
assume a linear relation:

$$\frac{I'_o(\lambda)}{I''_o(\lambda)} = \text{const} \tag{1}$$

where $I'_o(\lambda)$ and $I''_o(\lambda)$ are the light intensities of two different
homoglobin-free tissue spectra. It follows that, for two wavelengths
λ_1 and λ_2

$$\frac{I'_o(\lambda_1)}{I''_o(\lambda_1)} = \frac{I'_o(\lambda_2)}{I''_o(\lambda_2)} = \text{const} \tag{2}$$

and that

$$\frac{I'_o(\lambda_1)}{I'_o(\lambda_2)} = \frac{I''_o(\lambda_1)}{I''_o(\lambda_2)} = \text{const}' \tag{3}$$

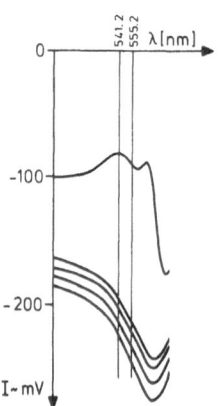

Fig. 3.   Absorption spectra measured by the micro light guide
           spectrophotometer. In the lower part of the figure, spectra
           from the hemoglobin-free liver at four different sites are
           displayed, the upper spectrum was measured after the addi-
           tion of hemoglobin.

In the following these assumptions are applied to hemoglobin spectra called $I(\lambda)$. A typical intracapillary hemoglobin spectra is shown in the upper part of Fig. 3.

We studied the relation $I(\lambda_1)/I(\lambda_2)$ for two wavelengths which show characteristic changes with different degrees of oxygenation. $\lambda_1 = 555.2$ nm is the wavelength where deoxygenated hemoglobin has its maximal absorption and $\lambda_2 = 541.2$ nm is one of the two wavelengths where oxygenated hemoglobin has its maximal absorption.

We assumed a correlation between the ratio of $I(\lambda_1)/I(\lambda_2)$ and the degree of hemoglobin oxygenation and found the logarithm of the ratio proportional to hemoglobin oxygenation (SO_2).

$$\ln \frac{I(\lambda_1)}{I(\lambda_2)} \sim SO_2. \tag{4}$$

This linear relation was obtained from the data measured in our experiments in the isolated perfused liver. Fig. 4 shows a typical example of our findings. On the left three intracapillary hemoglobin spectra are displayed. On the right tissue Po₂ histograms measured

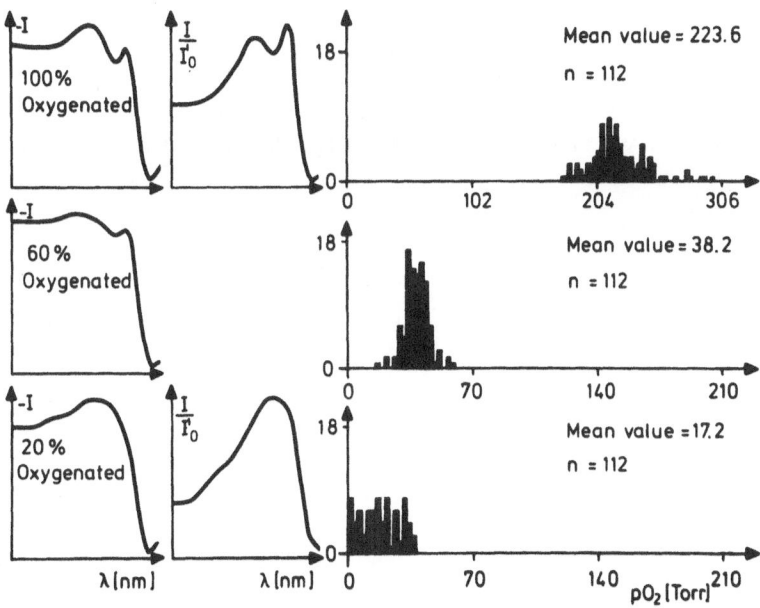

Fig. 4. Original hemoglobin spectra (left part) and corresponding Po₂ histograms (right part) measured at different Po₂ values of the perfusate. In the middle two spectra obtained by dividing $I(\lambda)/I'_o(\lambda)$ are shown.

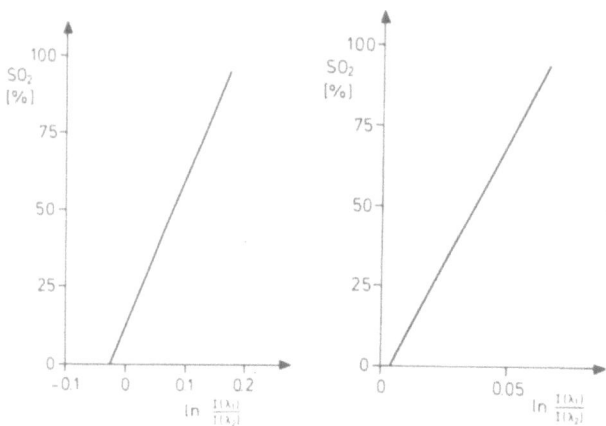

Fig. 5.   Calibration curves for two separate experiments (for
           details see text).

at the same site and under the same experimental conditions are
shown. From the Po_2 histogram a mean Po_2 was calculated and using
this Po_2 the degree of hemoglobin oxygenation (SO_2), marked on
Fig. 4 for each spectra, was derived from hemoglobin dissociation
curves (Bartels and Harms, 1959). These SO_2 values and the logarithm
of $I(\lambda_1)/I(\lambda_2)$ of the corresponding spectra were underlying data for
the calibration curves in Fig. 5. From two SO_2 values (0% and 95%)
and the logarithm of $I(\lambda_1)/I(\lambda_2)$ of the two corresponding spectra,
a calibration curve was constructed. Fig. 5 shows two such curves.

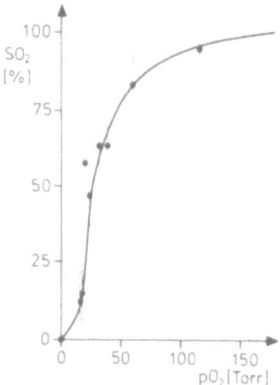

Fig. 6.   Correlation between SO_2, derived from the left calibration
           curve shown in Fig. 5, and mean local tissue Po_2, calcu-
           lated from the histograms. The bars indicate the standard
           deviation from at least 3 measurements.

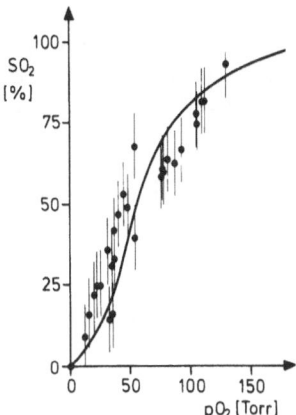

Fig. 7. Correlation between SO_2, derived from the right calibration curve shown in Fig. 5. For further explanation see Fig. 6.

In our experiments the Po_2 of the perfusate and thereby the tissue Po_2 was varied. Thus, a set of mean tissue Po_2 values between zero and 300 Torr were measured. The mean local Po_2 values and SO_2 values measured under the same conditions were correlated. The SO_2 values were determined from the measured change of $\ln\left[(I(\lambda_1)/I(\lambda_2))\right]$ using the individual calibration curve (see Fig. 5). The resulting "dissociation curves" are shown in the Figs. 6 and 7.

The differences between the two displayed sigmoid curves could be due to the difference in the pH of the perfusate and heterogeneities of microcirculation in the perfused organs.

CONCLUSION

The proposed method allows the estimation of the intracapillary hemoglobin saturation from spectra measured with micro light guide spectrophotometry.

REFERENCES

Bartels, H., and Harms, H., 1959, Sauerstoffdissoziationskurven des Blutes von Säugetieren, Pflügers Arch. Ges. Physiol., 268:334.
Brunner, M., Kastner, N., Schabert, A., Höper, J., and Kessler, M., 1981, On-line Verarbeitung von Hämoglobin-Reflexionsspektren hoher Repetitionsrate, in: "Medizinische Informatik und Statistik", S. Koller, P.L. Reichartz and K. Überla, eds., Springer, Berlin, pp. 384-389.

Höper, J., and Kessler, M., 1982, Constant pressure perfusion of
    the isolated rat liver. Local oxygen supply and metabolic
    function, (to be submitted to Microcirculation).

Ji, S., Chance, B., Katsuyuki, N., Smith, T., and Rich, T., 1979,
    Micro-light guides: a new method for measuring tissue fluores-
    cence and reflectance, Am. J. Physiol., 5(2):C144–C156.

Kessler, M., and Grunewald, W., 1969, Possibilities of measuring
    oxygen pressure field in tissue by multiwire platinum elec-
    trodes, Prog. Respir. Res., 3:147–153.

Kessler, M., and Lübbers, D.W., 1966, Aufbau und Anwendungsmöglich-
    keiten verschiedener Po_2-Elektroden, Pflügers Arch. Ges.
    Physiol., 291:R82

APPENDIX

$I_o'(\lambda)$: apparatus + tissue spectra

$I(\lambda)$: hemoglobin spectra

λ_1 = 555.2 nm, the wavelength of the absorption maximum for the deoxygenated hemoglobin spectrum

λ_2 = 541.2 nm, the wavelength of one absorption maximum for oxygenated hemoglobin spectrum

SO_2 : hemoglobin oxygen saturation in %

$\sim$: proportional

EVALUATION OF REFLECTION SPECTRA OF THE ISOLATED HEART BY MULTI-

COMPONENT SPECTRA ANALYSIS IN COMPARISON TO OTHER EVALUATING METHODS

H.R. Figulla, J. Hoffmann[1], and D.W. Lübbers[1]

Dept. of Internal Medicine, Cardiology Division
University of Göttingen, 3400 Göttingen, FRG
[1]Max-Planck-Institut für Systemphysiologie
Rheinlanddamm 201, 4600 Dortmund 1, FRG

INTRODUCTION

Satisfactory quantification of reflection spectra has been a challenging problem for several decades (Kortüm, 1969). Reflected light from organ surfaces gives an indication of the biochemical properties of the underlying living tissue without destruction or even affecting sensitive structures.

In recent years, efforts were made to find a better method for the quantification of multicomponent reflection spectra. Technical facilities advanced from a dual wavelength method to a multiwavelength method which allowed more sophisticated data assembling and evaluation. It is the purpose of this presentation to compare the multicomponent analysis with the dual wavelength method and the tangential method.

METHODS

Experiments were performed with isolated blood-free perfused Langendorff guinea pig hearts. A light guide of the rapidspectrometer T13/3 (Lübbers and Niesel, 1957) was attached to the left ventricle surface and reflection spectra from the heart were recorded. The spectra were digitized and stored in a computer (Fig. 1). The method is described in detail by Figulla et al. (1981) and Lübbers and Hoffmann (1981). The data were evaluated with 3 different analyzing methods:

PERFUSION (CONSTANT PRESSURE, CONSTANT FLOW)

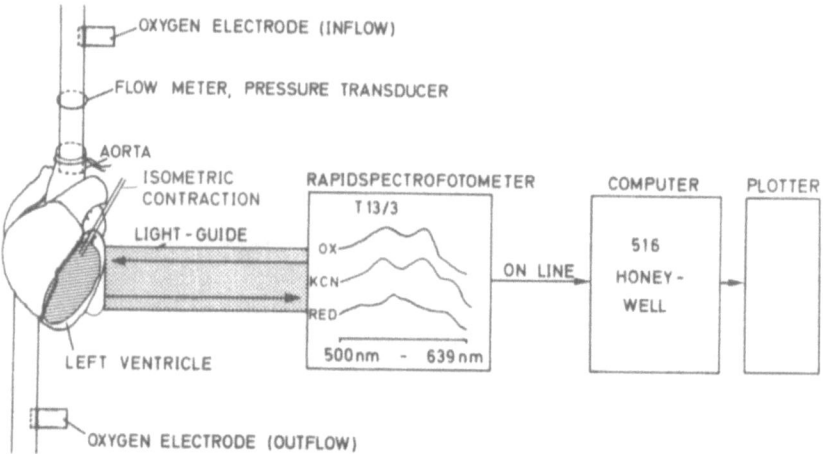

Fig. 1.   Setup for measurement and evaluation of the reflection
           spectra of the guinea pig heart. The blood-free perfused
           Langendorff guinea pig heart is suspended in nitrogen. A
           non-flexible light guide, outer diameter 1 cm, is attached
           to the left ventricular epicardium. The spectra of the
           reflected light are recorded and then stored in a computer.
           Evaluation is performed off line (OX: fully oxygenated and
           oxidized; KCN: poisoned with KNC; RED: desoxygenated and
           reduced).

        1. dual wavelength analysis (DWLA),
        2. tangential analysis (TANA),
        3. linear multicomponent analysis, using reference spectra
(LMCA(R)).

1. Dual wavelength analysis (DWLA)

        The dual wavelength method has been introduced in biology in
the thirties by Kramer (1934) and Millikan (1942); it was very much
improved by Chance (see 1954). The method uses two wavelengths: one
of them, called measuring wavelength, is selected so that a distinct
signal change is obtained when the component to be measured changes,
whereas the other wavelength, called isosbestic wavelength, remains
unchanged. By the signal difference between these two wavelengths,
the concentration of the component is determined. The advantage of
using the difference is that disturbances which affect both wavelengths in the same way, are cancelled. For the heart muscle, Tamura
et al. (1978) discussed the interrelationship of the spectral components in the multicomponent spectra of the cytochromes and used
the wavelength pair 605 and 620 nm to determine the cytochrome aa_3
redox state and the wavelength pair 587 and 620 nm to determine
myoglobin-O_2 saturation. These pairs of wavelengths were chosen to

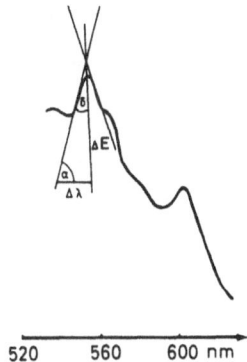

Fig. 2. The tangential analysis (for details see text).

minimize the posible error which might result from mutual influences
of the components.

2. Tangential analysis (TANA)

The tangential method to evaluate reflection spectra which was
proposed by Wodick and Lübbers (see Schwickardi, 1968) can be
applied if spectra have distinct peaks or shoulders. Fig. 2 shows,
as an example, the reflection spectra of hemoglobin-free brain
tissue with the cytochrome c and cytochrome aa_3 peak and the cytochrome b shoulder. To evaluate the cytochrome c peak, tangents are
drawn so that a triangle with the sides $\Delta\lambda$ and ΔE and the angles α
and δ is formed. If cytochrome c is oxidized the peak becomes
smaller and also the angle α. According to the Lambert-Beer law,
the tangent is proportional to the extinction

$$\frac{dE(\lambda)}{d\lambda} = cz\,\frac{d\,(\lambda)}{d\lambda} = tg\alpha = ctg\delta \tag{1}$$

with

$E(\lambda)$ extinction,
λ wavelength,
c   concentration,
z   thickness of the sample,
$\epsilon(\lambda)$ molar extinction coefficient,
α, δ angles (see Fig. 2).

Is δ_m the angle for maximal reduction and δ its actual value,
we find for cz

$$cz = \frac{ctg\,\delta}{ctg\,\delta_m} \tag{2}$$

(for further detail, see Schwickardi, 1968).

3. Linear multicomponent analysis using reference spectra

The analysis is based on the assumption that the actually measured reflection spectra of the heart can be composed by a linear combination of the reflection spectra of the reduced cytochromes as well as of the desoxygenated myoglobin and of the oxidized respective oxygenated spectra of the same tissue. These spectra are obtained from the reference spectra (see Figs. 1 and 5): 1. oxidized cytochromes and oxygenated myoglobin (OX) 2. oxygenated myoglobin and reduced cytochromes (produced by KCN poisoning, KCN) 3. reduced cytochromes and desoxygenated myoglobin. The recalculation of intermediate states of oxidation showed that the error between measured and recalculated spectra was within the noise level (for further details see Figulla et al., 1981).

RESULTS

For comparison, all three evaluation methods were applied to the same spectra recorded by the rapid spectrometer and stored in the computer.

In Figs. 3 and 4 examples of the results of the different evaluation methods are shown: in A the results of evaluation according to the double wavelength analysis (DWLA), in B the results of evaluation according to the tangential analysis (TANA) and in C the results of evaluation according to the linear multicomponent analysis using reference spectra (LMCA(R)). The recording in Fig. 3 starts with a 15 s control period. In this period, all evaluation methods (A, B, C) result in about the same steady state value of 80% oxidation or oxygenation, respectively. The difference is less than 10%. After stopping the flow (15th to 150th second) with DWLA, MbO_2 and $cytaa_3(ox)$ became simultaneously desoxygenated and reduced, reaching zero at the same time. With LMCA(R) only during the first seconds the decrease is similar, then the reduction of $cytaa_3(ox)$ becomes slower than the change of MbO_2 saturation. After reperfusion with DWLA, $cytaa_3(ox)$ shortly overshoots MbO_2 saturation, whereas with LMCA(R) the opposite occurs. When the heart is arrested by injecting potassium chloride to a final concentration of 24 $mmol \cdot l^{-1}$, with DWLA MbO_2 becomes 100% saturated, whereas $cytaa_3(ox)$ remains at 80% oxidation, however, with LMCA(R) both components reach 100%. With TANA at first the recording is similar to that of LMCA(R) but after reperfusion the results differ from both the other methods.

In Fig. 4 an experiment with KCN poisoning of the respiratory chain followed by a flow stop is shown. Now the steady state values differ much from each other as it is often observed in longer lasting

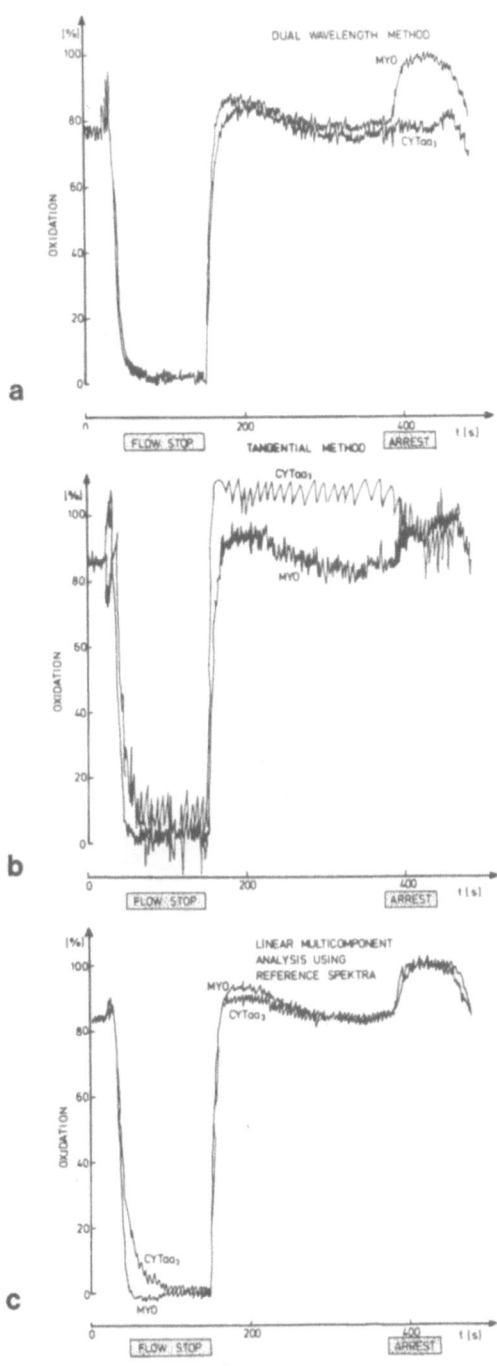

Fig. 3.   Reflection spectra of the isolated heart evaluated with
          three different methods, with: A dual wavelength method,
          B tangential method, C linear multicomponent analysis. A
          flow stop and cardiac arrest experiment is shown.

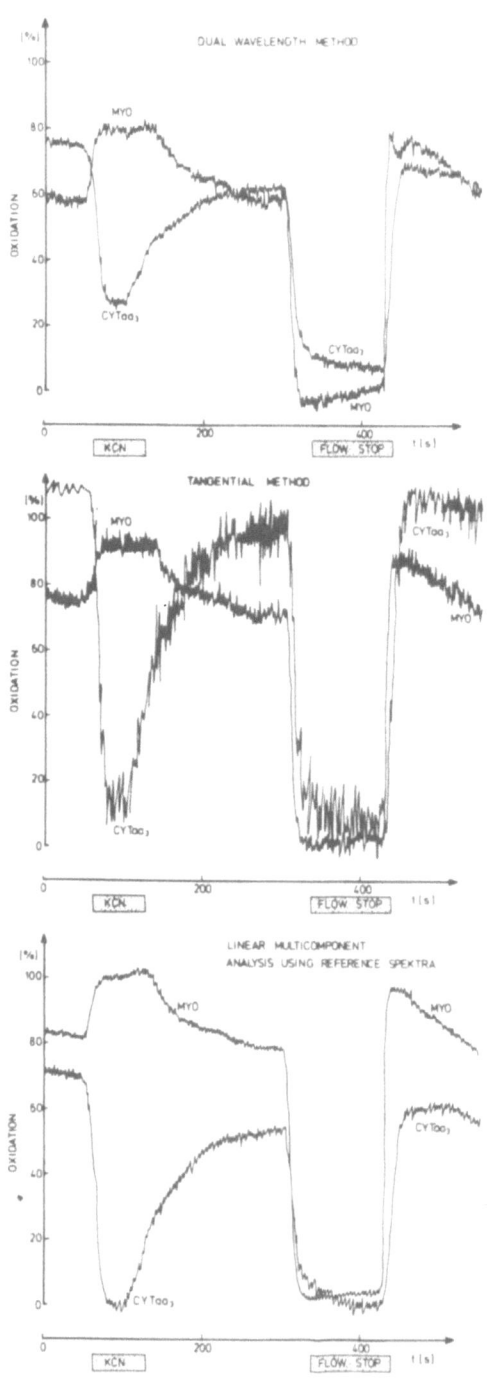

Fig. 4. Reflection spectra of the isolated heart evaluated with three different methods, with: A dual wavelength method, B tangential method, C linear multicomponent analysis. A cyanide poisoning and flow stop experiment is shown.

experiments. With LMCA(R) MbO_2 saturation is similar, cytaa$_3$(ox)
decreased about 10-15%. With DWLA MbO_2 saturation decreased by about
20%, whereas cytaa$_3$(ox) remained the same. With TANA MbO_2 saturation decreased but cytaa$_3$(ox) increased over 100%. At the end of
the records in Fig. 4 with LMCA(R) the decreasing MbO_2 trace reaches
just about the same value as at the beginning, cytochrome aa$_3$ is
about 10-15% less oxidized. With DWLA MbO_2 saturation and cytochrome
aa$_3$ oxidation have the same value although, at the beginning, they
were about 20% apart. With TANA values at the end are about the same
as in the beginning. KCN (1 mmol·l^{-1}) is infused from the 30th to
90th second. With LMCA(R) Mb becomes 100% saturated with oxygen,
whereas with DWLA only 80%. With LMCA(R) cytaa$_3$(ox) reaches 0%, with
DWLA only about 25%. During infusion and washout of KCN the dynamic
of MbO_2 and cytaa$_3$(ox) is similar with LMCA(R) and DLWA. The same
holds for the dynamic behaviour during the following flow stop. However, with DWLA cytochrome aa$_3$ did not reach zero as in the flow
stop in Fig. 3. At reperfusion MbO_2 overshoots cytaa$_3$(ox) in recording A and C, but the reached percentage values differ. As compared
to the flow stop in Fig. 3, the recordings C are similar, whereas
in the recordings A cytaa$_3$(ox) shows a different dynamic behaviour.
With TANA the dynamic is correctly described, but the absolute values
differ from the values obtained by the other evaluation methods.

DISCUSSION

    For the evaluation the linear multicomponent analysis needs
reference spectra to extract the spectra of Mb and MbO_2 as well as
of cytaa$_3$(ox) and cytaa$_3$(red). The state of oxygenation of the heart
muscle is achieved by KCN arrest, maintaining high flow perfusion
with oxygenated medium (95% O_2 and 5% CO_2). The state of anoxia is
reached during the flow stop. A similar spectra is obtained by perfusion with sodium dithionite. Considering the KCN reference spectra
one has to be careful to avoid changes of spectral properties of
cytochrome aa$_3$. To obtain the reference spectra, cyanide was added
to the fully oxidized respiratory chain. Cyanide bound to the oxidized cytochrome aa$_3$ does very little effect the spectral properties
(Erecińska and Wilson, 1980). We tested our cyanide dosis and application method with isolated mitochondria, and could not detect differences between the reduced spectra of cytochrome aa$_3$ by withdrawing oxygen or by KCN poisoning. DWLA and TANA need only the OX-
and RED-state as reference.

    The results show that the different evaluation methods dis-
tinctly gave different results. Considering the differences the
question arises which results are the most reliable ones. The
advantage of the TANA is that it can be evaluated by graphical
analysis, however, the error which is made by drawing the tangents
can be considerable. Differentiation by the computer yields a large
error, larger than that with the other two methods.

A judgement should be possible as to which evaluation method
comes close to the truth, if by suitable experiments predictable
changes of the spectra are produced.

It is known that under our experimental conditions, during flow
stop, tissue oxygen pressure drops to zero, consequently myoglobin
must be fully deoxygenated and cytaa3(ox) fully reduced. At cardiac
arrest, the O_2 consumption of the heart is very much decreased at
maintained flow; thus, myoglobin must be fully oxygenated and cytochrome aa3 oxidized (Fig. 3). The O_2 consumption can by similarly
reduced by a proper dosis of KCN; in this case myoglobin will be
fully oxygenated but cytaa3(ox) fully reduced. The LMCA(R) describes
all these experiments sufficiently well, whereas DWLA is only partly
satisfactory. This is understandable if the spectral changes under
different conditions are taken into account (Fig. 5). With DLWA
620 nm is used as isosbestic wavelength. This wavelength is isosbestic if cytochromes are reduced simultaneously with a deoxygenation

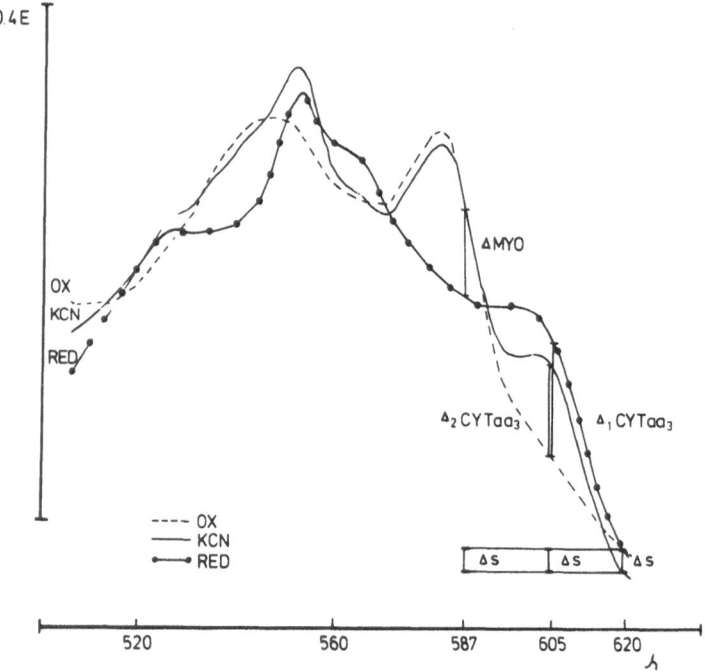

Fig. 5. Reflection spectra of oxidized (oxygenated), reduced
         (deoxygenated) and cyanide poisoned heart muscle. The
         wavelength pair 587 and 620 nm was used for evaluation of
         myoglobin-O_2 saturation. The wavelength pair 605 and
         620 nm was used for evaluation of cytochrome aa3 redox
         state. Note the change of extinction at the so-called
         isosbestic point at 620 nm, Δs, and the different extinc-
         tion changes at 605 nm: Δ_1 cytaa3 = Δ_2 cytaa3 + Δs.

of myoglobin. However, if MbO_2 remains oxygenated and the cytochromes are reduced, the extinction at λ = 620 nm remains not unchanged but decreases by - Δs (Fig. 5). Under similar conditions the
extinction at 605 nm changes by about Δ_2 cytaa$_3$, however, the extinction change is greater by + Δs if simultaneously MbO_2 is deoxygenated: Δ_1 cytaa$_3$ = Δ_2 cytaa$_3$ + Δs. The extinction change for myoglobin at 587 nm is almost independent of the redox state of cyto-
chromes. The mutual interdependences of the extinctions are the
reason for the observed discrepancies between LMCA(R) and DWLA. The
DWLA can only be applied under very special conditions in which the
essential presupposition of this method, the existance of an isosbestic wavelength, is fulfilled. The spectra in Fig. 5 show how
difficult it is to find suitable isosbestic wavelengths in systems
with myoglobin, particularly when the relative concentrations can
change. The same holds for systems with hemoglobin.

REFERENCES

Chance, B., 1954, Spectrophotometry of intracellular respiratory
    pigments, Science, 120:767.

'Erecińska, M., and Wilson, D.F., 1980, Inhibitors of cytochrome c
    oxidase, Pharmac. Ther., 8:1-20.

Figulla, H.R., Hoffmann, J., and Lübbers, D.W., 1981, Coronary
    conductivity and tissue oxygenation as measured by the myoglo-
    bin O_2 saturation and cytochrome aa$_3$ redox state in the Langen-
    dorff guinea pig heart preparation, in: "Oxygen Transport to
    Tissue IV", D.F. Bruley, H.I. Bicher, eds., Plenum Press, New
    York, in print.

Kortüm, G., 1969, "Reflexionsfotometrie", Springer, Berlin-Heidel-
    berg-New York.

Kramer, K., 1934, Fortlaufende Registrierung der Sauerstoffsättigung
    im Blut an uneröffneten Blutgefäßen, Klin. Wschr., 13:379.

Lübbers, D.W., Figulla, H.-R., Hoffmann, J., and Wodick, R., 1981,
    Histotoxic hyperoxia and its effect on blood flow contractility
    in the Langendorff heart, in: "Oxygen Transport to Tissue IV",
    D.F. Bruley, H.I. Bicher, eds., Plenum Press, New York, in
    print.

Lübbers, D.W., and Hoffmann, J., 1981, Absolute reflection photo-
    metry at organ surfaces, in: Adv. Physiol. Sci., Vol. 8,
    "Cardiovascular Physiology of Heart, Peripheral Circulation
    and Methodology", A.G.B. Kovách, F. Monos, G. Rubányi, eds.,
    Pergamon Press, Akadémiai Kiadó, Budapest, pp. 353-361.

Lübbers, D.W., and Niesel, N., 1957, Ein Kurzzeitspektralanalysator
    zur Registrierung schnell verlaufender Änderungen der Absorp-
    tion, Naturwissenschaften, 44:60.

Millikan, G.A., 1942, The oximeter, an instrument for measuring
    continuously the oxygen saturation of arterial blood in man,
    Rev. Sci. Instrum., 13:434.

Schwickardi, D., 1968, Konzentration und Kinetik der Atmungsfermente
      am isoliert perfundierten Meerschweinchengehirn in vivo und
      Hyperthermie von 18°C, Dissertation, Marburg.
Tamura, M., Oshino, N., Chance, B., and Silver, I.A., 1978, Optical
      measurements of intracellular oxygen concentration of rat
      heart in vitro, Arch. Biochem. Biophys., 191:8-22.

QUANTITATIVE ANALYSIS OF REFLECTION SPECTRA OF THE SURFACE OF THE

GUINEA PIG BRAIN

J. Hoffmann[1], R. Wodick[2], F. Hannebauer[3], and
D.W. Lübbers[1]
[1]Max-Planck-Institut für Systemphysiologie
Rheinlanddamm 201, 4600 Dortmund 1, FRG
[2]Physiologie II der Universität, Oberer Eselberg
7900 Ulm, FRG
[3]AEG, 6000 Frankfurt, FRG

SUMMARY

For a nearly constant light scattering coefficient a method is
presented which allows the quantitative analysis of reflection
spectra of the brain using a nonlinear multicomponent analysis
(NLMCA).

INTRODUCTION

Light reflected from tissue and light transmitted through
tissue is related to the absorption coefficient and to the scattering coefficient of radiation transfer. In earlier experiments with
mitochondrial suspensions, tissue homogenates and tissue slices we
have shown (Heinrich, 1981; Heinrich et al., 1981; Lübbers and
Hoffmann, 1981) that reflection spectra of these samples can be
quantitatively evaluated with sufficient accuracy by applying the
two flux theory of Kubelka and Munk (1931). Based on the results of
these experiments a new evaluation method will be presented which
allows to quantitatively evaluate tissue reflection spectra of the
surface of the blood-free perfused guinea pig brain.

THEORY

According to the Kubelka-Munk theory reflectance (R), which is
the ratio of reflected light intensity to the incident beam intensity, and transmittance (T), which is the ratio of the transmitted

831

light intensity to the incident beam intensity, of a layer of thickness d is

$$R = \sinh(kd)/\sinh(kd + y) \tag{1}$$

$$T = \sinh(y)/\sinh(kd + y) \tag{2}$$

with

$k = \sqrt{a(a + 2s)}$
$y = -\ln((a + s - k)/s)$
a = absorption coefficient
s = scattering coefficient

In the case of infinite thickness of the layer reflectance is

$$R_\infty = (a + s - k)/s \qquad \text{or} \qquad -\ln R_\infty = y \tag{3}$$

From characteristic absorption coefficients or the absorption spectra the (chemical) composition of the sample can be determined. Considering randomly distributed non-interacting substances, the total absorption is

$$a(\lambda) = \sum_i c_i \, a_i(\lambda) \tag{4}$$

where $a_i(\lambda)$ denotes the absorption of the i-th component ("book spectra"), dependent on wavelength (λ), and c_i its concentration.

The values c_i in eq. (4) can be related to physiological parameters (such as the oxidation states of the cytochromes, which is the ratio of the oxidized concentration to the total concentration). The aim of the following quantitative analysis was to derive the values c_i from the measured tissue spectra.

In the case of finite thickness and non-vanishing transmittance and reflectance eqs. (1) and (2) can be solved to find the absorption and scattering coefficients. The quantitative analysis of the spectra can be performed in two steps: 1) calculation of the true absorption coefficients from transmittance and reflectance, and 2) the analysis of the true absorption spectra according to book components. In the following, the equations for infinite thickness, according to the Kubelka-Munk theory, will be discussed and as a result, an approximation method will be described which, under certain conditions, allows the evaluation of reflection spectra of samples of infinite thickness.

With constant scattering coefficient, s, there is a one-to-one relationship (transformation) between absorption and reflectance:

$$-\ln R_\infty = y = H(a(\lambda)) \tag{5}$$

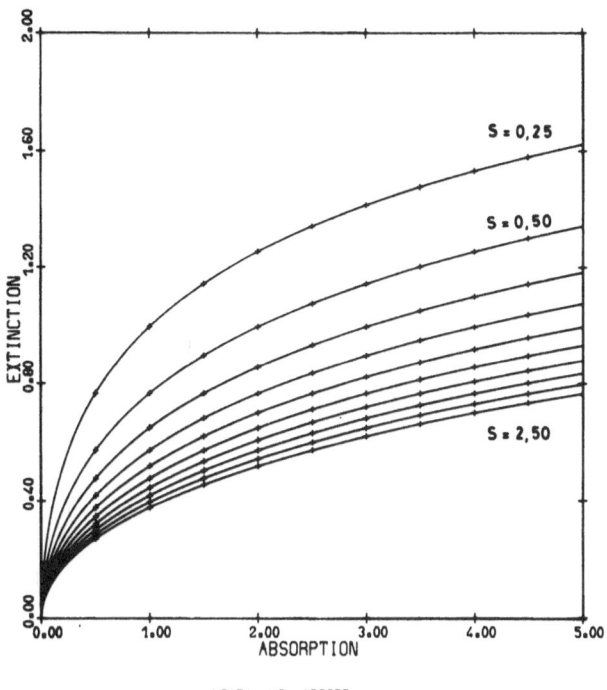

-LG(R) .VS. ABSORP

Fig. 1. Relationship between absorption coefficient (abscissa) and
         reflectance -ln R_∞ (ordinate), light path infinite,
         according to Kubelka and Munk (1931) (eq. (1)), for
         various values of scattering coefficient, s. This nonlinear
         relationship can only be approximated in a limited range
         by a straight line; in all other cases a nonlinear analysis
         is necessary.

        The transformation H is a function of the scattering coeffi-
cient. If the scattering coefficient is wavelength-independent, the
transformation will also be wavelength-independent. Figs. 1 and 2
give plots of -ln R_∞ versus absorption coefficient, and of -ln T
versus absorption coefficient for different constant scattering
coefficients, s.

        The one-to-one relationship eq. (5) is not only due to the
Kubelka-Munk theory but holds also for the radiation transfer
equation:

$$dI = - K\rho \, I \, dz \qquad\qquad (6)$$

(see Kortüm (1969); K = mass absorption coefficient describing
absorption and scattering, ρ = density of the material, I = beam
intensity, z = path length). If the absorption is homogeneous and
the scattering coefficient does not depend on the wavelength, then

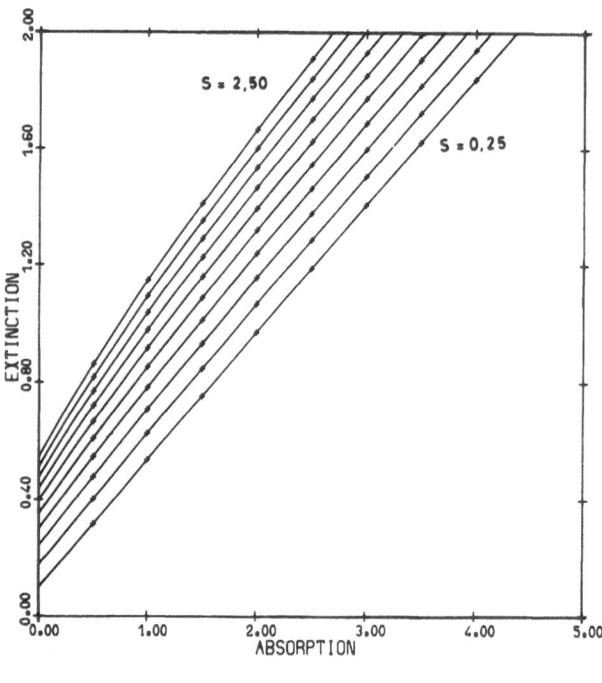

Fig. 2.   Relationship between absorption coefficient (abscissa) and
          transmittance -ln T (ordinate) for various values of
          scattering coefficient according to Kubelka and Munk
          (1931) (eq. (2)). Almost linear relationship for small
          values of the scattering coefficient shows that in this
          range it is allowed to apply for evaluation of absorption
          changes the Lambert-Beer-law.

a formal solution of eq. (6), according to Wodick and Lübbers
(1973) is given by

$$I_q = I_0 \int_0^\infty \Psi_q(s, z) \exp(-az) \, dz \qquad (7)$$

where $\Psi_q(s, z)$ denotes a (nonnegative) light path distribution.
The subscript q refers to the measuring conditions, reflectance or
transmittance. From this equation we see that a monotonic relationship between I_q/I_0 and the absorption coefficient (a) exists. The
Kubelka-Munk theory can be considered as a special case of the
light path distribution theory. With eqs. (3) and (7) we obtain an
analytic expression for the reflectance light path distribution of
a sample of infinite thickness:

$$\Psi_{R_\infty}(s, z) = (1/z)\ I_1(sz)\ \exp(-sz)$$

(I_1 is the modified Bessel function) which yields

$$R_\infty = \int_0^\infty \Psi_{R_\infty}(s, z)\ \exp(-az)\ dz$$

A nonconstant s, for example a wavelength-dependent scattering coefficient disturbs this one-to-one transformation. However, if scattering can be described as

$$s(\lambda) = s_0 + s_1 \lambda \quad ; \quad |s_1 \Delta\lambda / s_0| \ll 1 \tag{8}$$

then the one-to-one relationship can be sufficiently well preserved by adding a further component to the total absorption (eq. 4):

$$-\ln R_\infty = H(a(\lambda) + C\,\lambda)$$

where C   represents the additional effects of variable wavelength-dependent scattering.

Using the inverse transformation H^{-1} and substituting eq. (4) we get

$$H^{-1}(y(\lambda)) = \sum_i\ (c_i\ a_i(\lambda) + C\,\lambda) \tag{9}$$

where H^{-1} denotes a nonlinear scaling of the spectra $y(\lambda)$.

Eq. (8) describes a nonlinear multicomponent analysis (NLMCA). When $y(\lambda)$ denotes the measured spectra, with no further assumptions, the NLMCA can be applied to samples of finite as well as infinite thickness (or even to samples with light path distribution).

PRACTICAL REALIZATION OF EVALUATION

H^{-1} is a monotonic function (eq. (9)). In practice it has been useful to approximate $H^{-1}(y)$ with a rational function (hyperbola):

$$H^{-1}(y) = (\alpha y + \beta)/(\gamma y + 1) \tag{10}$$

The NLMCA computes the coefficients of the hyperbola (10) and the concentrations c_i in eq. (9) iteratively. For the first step the linear multicomponent analysis is done, i.e. setting $\alpha = 1$ and $\beta = \gamma = 0$. The concentrations c_i can only be estimated besides an arbitrary factor, the evaluation of the concentrations are found using a nonnegative least-squares method. Because the approximation eq. (10) is invariant to linear scaling we chose all spectra (i.e. $y(\lambda)$, $a_i(\lambda)$, and $C\lambda$) to have the mean zero.

RESULTS

To test the described nonlinear multicomponent analysis (NLMCA) tissue reflection spectra have been simulated according to eqs. (3) and (4), based on the two flux theory of Kubelka and Munk (1931). As spectra ($c_i \cdot a_i$, $C\lambda$) we used the spectra of the pure components ("book spectra") of the basic components of oxidized and reduced cytochrome aa_3, c, b and combined with them the various values of the scattering coefficient, s, so that the simulated spectra were similar to the spectra recorded from the surface of the guinea pig brain. The simulated spectra were computed in the wavelength range of 450-650 nm (step size: 0.5 nm). Using these simulated "noiseless" spectra the NLMCA found the given degrees of oxidation with an error less than 0.1% degree of oxidation. Thus, from the formal point of view, the NLMCA gives very satisfactory results for the investigated range of absorption and scattering coefficient.

In the next step, reflection spectra of blood-free perfused guinea pig brain were evaluated. The spectra were measured with the rapid spectrometer in the wavelength range from 500-620 nm (Lübbers and Niesel, 1957), digitized and stored for computer analysis. The basic components for the analysis were the book spectra of cyt aa_3, c and b (oxidized and reduced) as well as the spectra of oxygenated and deoxygenated hemoglobin. The latter was included since even in the blood-free perfused brain mostly traces of hemoglobin were found.

In Fig. 3, as an example, a measured brain spectra ($-\ln R_\infty$) is shown and compared with recalculated spectra (H(a)). Recalculation is carried out in two different ways: 1. by the linear multicomponent analysis (see eq. (10); $\alpha = 1$; $\beta = \gamma = 0$; Fig. 3, top) and 2. by the NLMCA (Fig. 3, bottom).

The applicability and exactitude of the analyzing method can be judged from the error, that is the deviation between measured and recalculated spectra: $-\ln R_\infty - H(a)$ (right hand panels). With the first analyzing method (top) the amplitude of the error is distinctly wavelength-dependent, i.e. systematical deviations remain between measured and recalculated spectra. This is a sign for the fact that by this method it is not satisfactorily possible to simulate tissue spectra. In contrast to the linear analysis by applying NLMCA one obtains an excellent fit (right hand pannel, bottom). It yields a nonlinear transformation (middle panel, bottom) abolishes the wavelength dependency of the error and reduces the amplitude to noise level ($\pm$ 0.0015 O.D.). For this

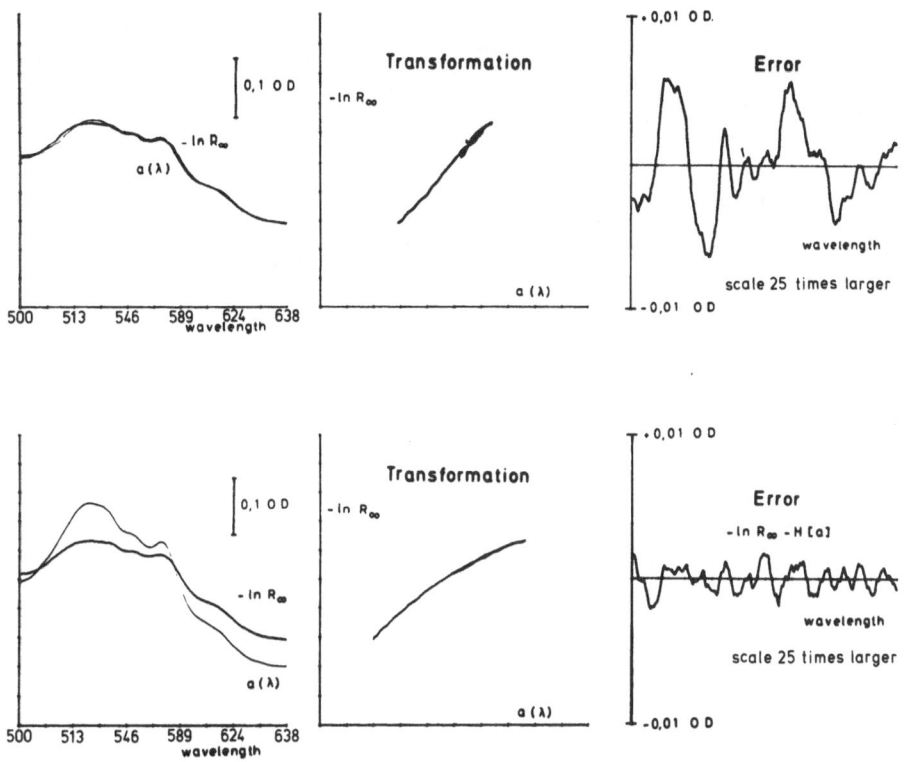

Fig. 3. Linear and nonlinear evaluation spectra of the blood-free
         perfused guinea pig brain. Notations: -ln R_∞ measured
         reflection spectra, blood free perfused guinea pig brain,
         a(λ) recalcualted absorption spectra according to book
         components. Top: linear evaluation. Bottom: nonlinear
         evaluation. For explanation see text. In the right hand
         panels the sensitivity is 25 times larger than in the left
         hand panel.

spectra of a well oxygenated brain the following redox states were
calculated by NLMCA: cyt aa_3 96%, cyt c   85% and cyt b   67%;
the rest hemoglobin was 84% oxygenated. The linear multicomponent
analysis yielded clearly erroneous results: cyt aa_3 3%, c
100%, b   0% and HbO_2 47%. With the NLMCA the evaluation of
spectra of fully oxidized brain (oxidized by H_2O_2, Schwickardi,
1968) and of fully reduced brain (reduced by dithionite) yielded
the expected redox states.

DISCUSSION

     By mixing a known substance to turbid solutions it was demon-
strated that within certain limits the evaluation of the transmitted

light applying the Lambert-Beer Law yielded correct results,
particularly if extinction differences are evaluated as it is done
for example with the two-wavelength method which uses the difference
between a measuring and an isosbestic wavelength. Being so, it is
often assumed that the same holds for reflected light.

Application of the two flux theory shows indeed that for
transmitted light and constant light scattering the relationship
between absorption coefficient and transmittance can be well
approximated by a straight line (see Fig. 2). Thus, for such a
system a linear multicomponent analysis will give satisfactory
results. However, this does not hold for reflected light: the
assumption of (constant) scattering forces a nonlinear evaluation
method. In Fig. 1 the relationship between absorption and reflectance is depicted. For constant scattering there is a nonlinear
one-to-one transformation between $a(\lambda)$ and $-\ln R_{\infty}$.

As it is shown in this paper such an one-to-one relationship
follows also from the more general radiation transfer equation
considering the light path distribution in the sample (Wodick and
Lübbers, 1974).

The different behaviour of transmitted and reflected light is
understandable since in transmission most of the scattered light
does not reach the photosensor, whereas in reflection only scattered
light is measured.

If the scattering coefficient changes with the wavelength then
the one-to-one transformation can be strongly disturbed. For a
practical realization of the evaluation of reflection spectra of the
brain we made the following assumptions:

1) the scattering coefficient can be approximated by a large
part, which is constant and a small part, which is wavelength-dependent (see eq. (8)) and
2) the transformation, H^{-1} can be approximated by a hyperbola
(see eq. (10)).

If the first assumption is fulfilled, the one-to-one relationship is maintained, which then allows the approximation with a
hyperbola. The hyperbolic approximation leads to a stable algorithm
for performing the nonlinear multicomponent analysis.

The NLMCA is performed with a set of linearly independent
spectra of the components (book spectra). By the analysis the
"best" concentrations (c_i in eq. (9)) are determined which give
the best last square fit. For getting satisfactory results it is
important

1) that the components have well distinguished spectra and
2) that the number of components is not too large.

This can be achieved

1) by a proper selection of the wavelength range in order to
obtain the most different spectra and
2) by an effective reduction of the number of components. This
for example is the case when the total concentration of the components remains constant.

The good results which we obtained with the brain (see Fig. 3,
bottom) show that the assumptions are justified and that the above
mentioned conditions for the application of the NLMCA are fulfilled.

REFERENCES

Heinrich, U., 1981, Untersuchungen zur quantitativen photometrischen
      Analyse der Redox-Zustände der Atmungskette in vitro und in
      vivo am Beispiel des Gehirn, Dissertation, Ruhr-Universität
      Bochum.
Heinrich, U., Hoffmann, J., Lübbers, D.W., and Hannebauer, F.,
      1981, Quantitative analysis of reflection spectra by simulating
      experiments on tissue, Adv. Physiol. Sci., 25:35-36.
Kortüm, G., 1969, "Reflexionsspektroskopie", Berlin, Heidelberg,
      New York, Springer.
Kubelka, K., and Munk, F., 1931, Ein Beitrag zur Optik der Farb-
      anstriche, Z., Tech. Phys., 11a:593-603.
Lübbers, D.W., and Niesel, W., 1957, Ein Kurzzeit-Spektralanalysator
      zur Registrierung rasch verlaufender Änderungen der Absorption,
      Naturwissenschaften, 4:59-60.
Lübbers, D.W., and Hoffmann, J., 1981, Absolute reflection photo-
      metry at organ surfaces, Adv. Physiol. Sci., 8:353-361.
Schwickardi, D., 1968, Konzentration und Kinetik der Atmungsfermente
      am isolert perfundierten Meerschweinchengehirn in vivo und in
      Hypothermie von 18°C, Dissertation, Marburg.
Wodick, R., and Lübbers, D.W., 1973, Methoden zur Bestimmung des
      Lichtweges bei der Photometrie trüber Lösungen oder Gewebe mit
      durchfallendem oder reflektiertem Licht, Pflügers Arch.,
      342:29-40.
Wodick, R., and Lübbers, D.W., 1974, Quantitative evaluation of
      reflexion spectra of living tissues, Hoppe-Seyler's Z. Physiol.
      Chem., 355:583-594.

ASSESSMENT OF BRAIN OXYGENATION: A COMPARISON BETWEEN AN OXYGEN ELECTRODE AND NEAR-INFRARED SPECTROPHOTOMETRY

P.H. Mook[1], H.J. Proctor[2], F. Jöbsis[3], and
Ch.R.H. Wildevuur[1]

[1]Department of Experimental Surgery, University of
Groningen, Groningen, The Netherlands
[2]Trauma Section, Department of Surgery, University of
North Carolina, Chapel Hill, N.C. 27514, USA
[3]Department of Physiology, Duke University, Durham
N.C. 27710, USA

INTRODUCTION

The inability of the brain to withstand more than a brief
period of hypoxia is often the limiting factor determining survival
for a variety of clinical conditions, most notably the adult respiratory distress syndrome and the hypoxic-hypotension related to
major trauma. The capability for continuously and non-invasively
monitoring cerebral oxygen availability within the mitochondria
would be extremely desirable. Prior to the introduction of near-
infrared spectrophotometry (niroscopy) by Jöbsis (1977) such a
capability did not exist. Subsequent work (Proctor et al., 1982)
demonstrated the feasibility of monitoring cerebral cytochrome a,a_3
redox states using in vivo spectrophotometry in a variety of experimental models, designed to mimic frequently encountered clinical
conditions. An immediate problem became the interpretation and
validation of the data in terms of a more conventional and commonly
accepted method of assessing tissue oxygenation. As one of a series
of investigations designed to clarify the role of niroscopy with
regard to future applications, the present study correlates cytochrome a,a_3 redox state and the intracerebral hemoglobin (HbO_2) as
measured by niroscopy with simultaneously performed measurements
of cerebral cortex surface Po_2 obtained using a polarographic oxygen
electrode.

METHODS

A. Equipment

    The Po_2 measurements were performed with the eight cathode sur-
face electrode as described by Kessler and Lübbers (1966). Rotation
of the electrode on the surface of the tissue 13 times allows the
construction of Po_2 histograms based upon 104 data points (Kessler
et al., 1976).

    The near-infrared spectrophotometer employed was an Omni 4 (In-
ternational Instrument Co., Durham, N.C. 27710, U.S.A.). Since 813 nm
represents an absorption band of oxidized cytochrome a,a_3, detection
of the amount of transmitted and/or reflected light at this frequency
using a photomultiplier tube represents the quantity of oxidized
cytochrome a,a_3 in the illuminated tissue. The spectra of Hb and
HbO_2 overlap that of cytochrome a,a_3 necessitating two additional
simultaneously presented wavelengths (770, 905 nm) which, through
appropriate algorithms, allow for the correction of the hemoglobin
contribution to the cytochrome signal, while at the same time providing separate signals denoting the quantity of Hb and HbO_2 in the
illuminated tissue.

B. Experimental procedure

    After anesthetizing cats (n = 6) with an intraperitoneal injec-
tion of sodium pentobarbital (30 mg kg^{-1} b.w.) catheters were placed
in a femoral vein and artery for taking blood samples, measuring
arterial pressure, and the administration of drugs. The cats were
then connected to a Harvard animal respirator by means of a tracheostomy, and paralysed with tubocurarine chloride (0.2 mg kg^{-1} b.w.).
The heads of the cats were placed in a stereotaxic device and after
a midline incision the skin and muscle were retracted exposing the
parietal skull. Two 1.5 cm burrholes were drilled on one side to
place the Po_2 electrode on the parieto-occipital cortex (after
removing the dura) and the photomultiplier tube. The fiber optic
bundle through which light was presented, was placed on the opposite
side of the skull at a 90° angle to the photomultiplier tube. The
cats were then ventilated with a mixture of 95% O_2 and 5% CO_2 and
the resulting HbO_2 and cytochrome a,a_3 signals were arbitrarily
defined as 100% oxidation. The cats were subsequently made hypoxic
for short periods of approximately 15 minutes by ventilating with
gas with O_2 concentrations between 5 and 15%, alternating with room
air. When the niroscopic and electrode tracings indicated a stable
state, a Po_2 histogram was recorded and an arterial blood sample
was taken to measure the Po_2, Pco_2 and pH. Each animal was subjected
to ten episodes of graded hypoxia. At the end of the experiment the
cats were ventilated with 100% nitrogen and the resulting cytochrome
and hemoglobin signals arbitrarily defined as 100% reduction (i.e.,
0% oxidation). All values for the cytochrome a,a_3 redox state and

hemoglobin saturation are expressed as a percent of this scale. The degree of correlation between the data obtained niroscopically and those obtained by the electrode was calculated using linear regression with computation of r values. Significance was tested for using a t-test.

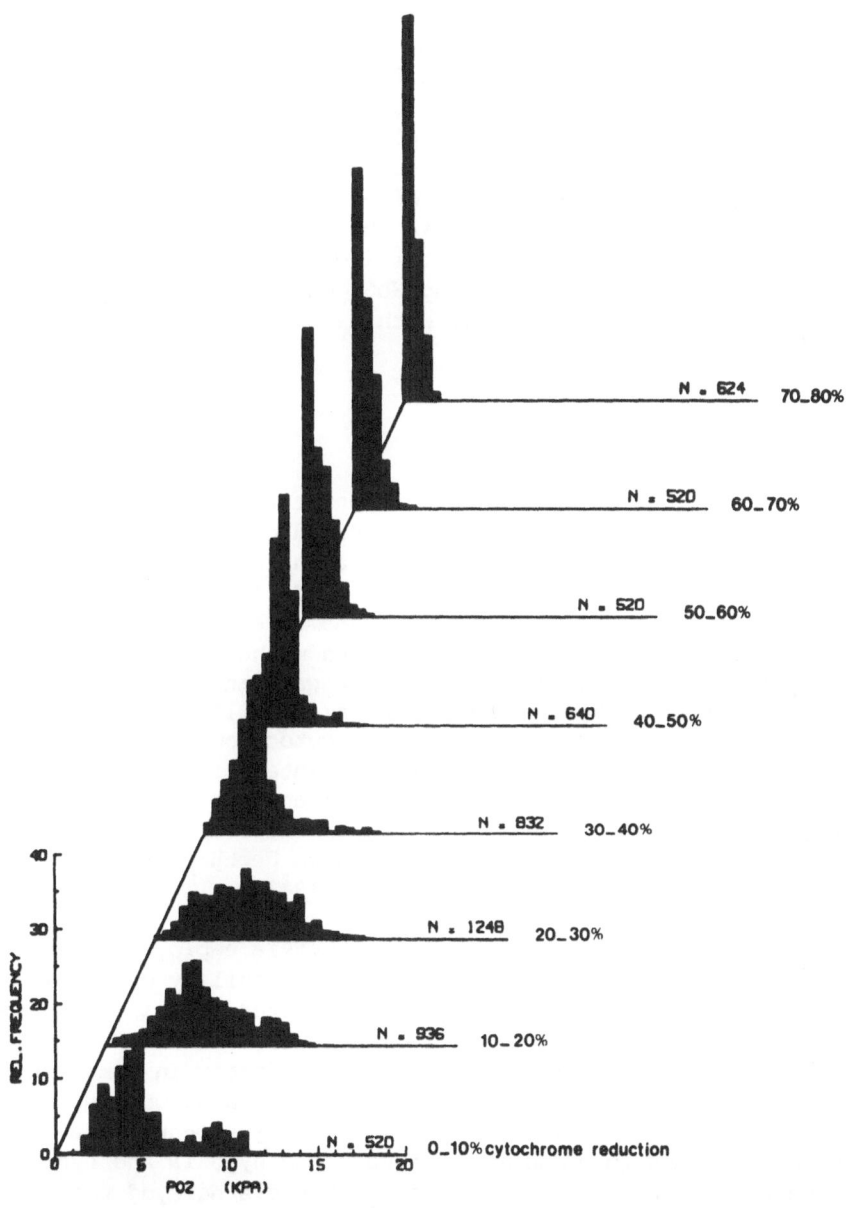

Fig. 1. Accumulated Po$_2$ histograms for percent cytochrome a,a$_3$ reduction values from between 0 - 10% up to between 70 - 80%. 100% = initial value.

RESULTS

With increasing hypoxia the histograms shifted to the left with an associated increased reduction of the cytochrome a,a$_3$ and a decreasing hemoglobin saturation.

In Fig. 1 the histograms from all cats corresponding to cyto-chrome reduction data points grouped from 0 - 10%, 10 - 20%, etc., are accumulated to one histogram and plotted for each cytochrome data group. The correlation between the left shift of the histograms and the increasing reduction of the cytochrome a,a$_3$ is also ex-pressed in the correlation diagram plotting the percentage of cyto-chrome a,a$_3$ reduction against the mean tissue Po$_2$ (Fig. 2a) demon-strating a correlation coefficient of -0.742 (r^2 = 0.551). The in-creasing percentage anoxic values in the Po$_2$-histograms (between 0 and 0.5 kPa) are plotted versus the increasing cytochrome a,a$_3$ reduction (Fig. 2b). The decreasing HbO$_2$ correlated well with a de-creasing mean tissue Po$_2$ (Fig. 3) with a correlation coefficient of -0.777 (r^2 = 0.604).

DISCUSSION

The high degree of correlation demonstrated in this study be-tween tissue Po$_2$, the niroscopically derived cytochrome a,a$_3$ redox state, and intracerebral HbO$_2$ is very reassuring in as much as it conforms to accepted theory and as such lends support to the valid-ity of the optical method. It must be remembered, however, that a direct correlation need not exist, as both methods assess oxygen availability in tissue in quite different ways and at different sites; the optical method assessing the availability of oxygen in the mitochondria indirectly by measuring cytochrome a,a$_3$ redox state whereas the polarographic method measures interstitial oxygen ten-sion on the surface of the cortex with perhaps some artifact intro-duced by pial vessels. A surface oxygen profile derived in this manner has been shown to be similar to an in depth profile as measured with needle electrodes (Lübbers, 1981).

During the short periods of graded hypoxia employed in this study no permanent alterations in oxygen availability or utilization were incurred as illustrated by the prompt return of all parameters to baseline upon reinstitution of room air ventilation. However, after longer periods of hypoxia or anoxia, possibly in combination with hypotension, the correlation between results of both methods may be less clear or not present at all. In experiments in which rats were submitted to 30 minutes of combined hypoxia and hypoten-sion, the reduced brain cytochrome a,a$_3$ showed a delayed recovery towards oxidation compared to intracerebral HbO$_2$ upon restoration of the blood pressure and the arterial Po$_2$ to normal values (Proctor et al., 1982). While in those experiments no tissue Po$_2$ measurements

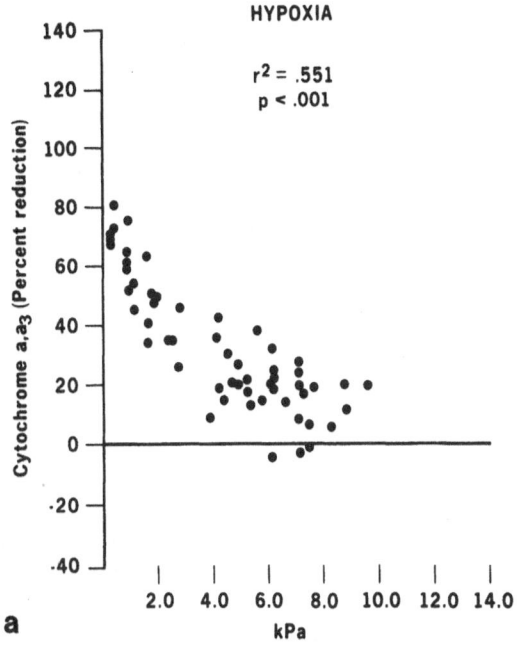

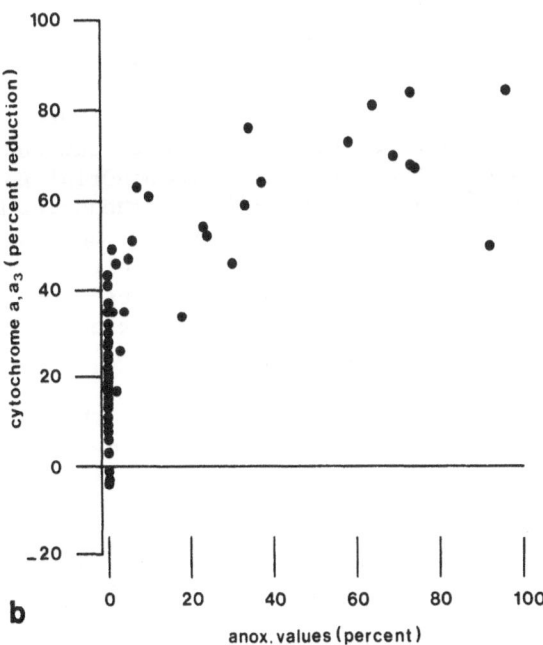

Fig. 2.   The percent cytochrome a,a$_3$ reduction values plotted against
          the mean Po$_2$ values of the histograms (a) and against the
          percentage anoxic Po$_2$ values (0–0.5 kPa) in the histograms
          (b) (100% = initial value).

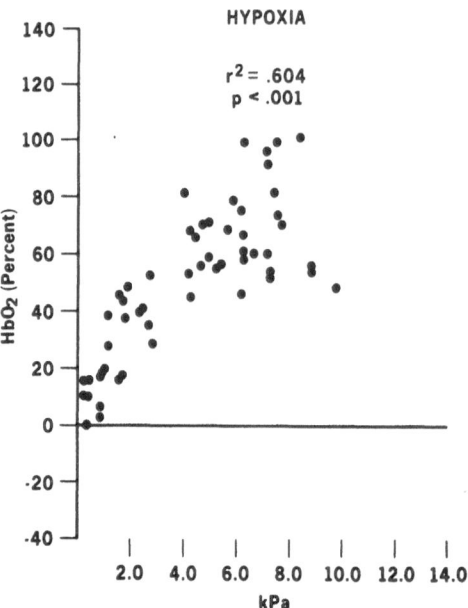

Fig. 3. The hemoglobin saturation percentage plotted against the
mean P_{O_2} values of the histogram (100% = initial value).

were performed they did illustrate the ability of optical monitoring
to provide valuable insight into mitochondrial function. One of the
ultimate goals of assessing tissue oxygen is, after all, to insure
adequate supply of oxygen to the cell to carry out cell function and
in this regard a method that assesses mitochondrial function is pref-
erable to monitoring oxygen outside the mitochondria. Moreover, the
optical method has a most important feature of special clinical importance since it may be utilized non-invasively, although it was
not applied in this manner in this study, whereas the polarographic
method requires craniotomy and tedious precautions to preserve ste-
rility.

     Further differences include the fact that the polarographic
method measures the tissue P_{O_2} over a clearly defined area of the
organ, the assumption being that this area is in some way representative of the organ as a whole whereas the optical method assesses a
less clearly defined volume. Finally, the optical method has, as an
additional feature, the capability of providing data concerning
oxygen in tissue by also measuring the oxygen bound to hemoglobin.

     In summary, the close correlations obtained between polaro-
graphically and niroscopically derived data strongly support the
validity of the principles of niroscopy as setforth above, and this
coupled with the non-invasive nature of niroscopy along with its
ability to assess mitochondrial function make it an attractive
method for future clinical application.

REFERENCES

Jöbsis, F.F., 1977, Noninvasive, infrared monitoring of cerebral
    and myocardial oxygen sufficiency and circulatory parameters,
    Science, 198:1264.
Kessler, M., and Lübbers, D.W., 1966, Aufbau und Anwendungsmöglich-
    keit verschiedener Po_2-Elektroden, Pflügers Arch. Ges.
    Physiol., 289:R98.
Kessler, M., Höper, J., and Krumme, B.A., 1976, Monitoring of tissue
    perfusion and cellular function, Anesthesiology, 45:184.
Lübbers, D.W., 1981, Quantitative measurement of tissue oxygen
    supply by Po_2 histogram, in: "Monitoring of Vital Parameters
    during Extracorporeal Circulation", H.P. Kimmich, ed., Karger,
    Basel.
Proctor, H.J., Sylvia, A.L., and Jöbsis, F.F., 1982, Failure of
    brain cytochrome a,a_3 redox recovery after hypoxic hypoten-
    sion as determined by in vivo reflectance spectrophotometry,
    Stroke, 13:89.

CRYOGENIC MICROSPECTROPHOTOMETRY

T.E.J. Gayeski

University of Rochester, Department of Anesthesiology
Rochester, NY, USA

An optical method and the theoretical basis for measuring hemoglobin (Hb) and myoglobin (Mb) saturation in frozen tissue via reflected light has been applied to Mb saturation in subcellular volumes of dog gracilis muscle, a pure red muscle. This cryogenic microspectrophotometric method utilizes light intensity measurements at four quasi-monochromatic wavelengths and a modified Lambert-Beer law to calculate saturation values. Transformation from Mb saturation to tissue Po_2 is based upon a Mb P_{50} value of 5.3 Torr. Both macrocirculation and microcirculation oxygen delivery parameters are monitored as covariates. The physiologic status of Mb saturation and microcirculatory covariates at a given instant of time are obtained by rapid freezing. A 5 cm copper cube, pre-cooled to liquid nitrogen temperature, is applied to the surface of the gracilis muscle with a pressure of 0.1 kg/cm^2. Cooling to 0^oC occurs at a rate of approximately 5-10 um/ms. This rate slows as the freezing front advances into the tissue due to the decrease in temperature gradient. At 500 um deep to the surface of the muscle, measurements indicate that -40^oC is reached in < 500 ms. All measurements are made within this distance from the surface. Specimen and copper cube are maintained in contact and submerged in liquid nitrogen. The specimen remains under liquid nitrogen from this submersion until final preparation for microspectrophotometry. The final preparation involves transfer of the specimen from liquid nitrogen to 95% ethanol at -75^oC so that its surface can be prepared for reflected measurements. Time spent at -75^oC is typically 1 min. Experiments indicate that more than 3 minutes at -75^oC are necessary for an increase in measured Mb saturation. After surface preparation, the specimen is transferred to a cold stage, regulated at $-120^o \pm 5^oC$. No change in saturation of heme pigments was observed over a 2-hour period at this temperature. Thus, the calculated Mb saturation represents the Mb saturation present in the tissue at the moment of freezing.

The accuracy of myoglobin saturation measurements has been estimated by studying the effects of myoglobin concentration in vitro on the calculated saturation and comparing this in vitro result with the range of saturation observed in a cyanide poisoning muscle allowed to equilibrate with perfusate. The range of myoglobin saturation attributable to a 0.1 $mmol \cdot l^{-1}$ to 1.0 $mmol \cdot l^{-1}$ change was +10% while that observed in the cyanide muscle was +3%. In addition, hemoglobin saturation measured by this method agreed to within +3% of values measured on an I.L. Co-Oximeter standardized to +0.5% of Hb saturation. Thus, Mb saturation differences of < +5% can be measured within a muscle. In comparing saturation values between muscles, qualitative data suggests that +5% change in saturation is also measurable.

Measurements on chicken breast, a Mb-free muscle, demonstrate that a 5 um measuring spot could be brought to within 5 um of small pre- or post-capillary vessels before an influence of oxygenated or deoxygenated Hb could be detected. Moreover, Mb saturation in potassium cyanide treated muscles was the same for cells surrounded by 0-8 capillaries. Saturation at a fixed location is identical for measurement apertures from 1x1 um to 15x15 um. Thus, spatial resolution in the plane of observation depends on the size of the measuring aperture, but not on the size of the illumination spot. Depth of penetration depends on the scattering properties of the tissue, the Mb concentration, and size of ice crystals. Ice crystals measure 0.76 um + 0.25 um in electron photomicrographs. Theoretical calculations indicate that light is attenuated by more than 90% after being internally reflected in 3 crystals. The depth of penetration is therefore thought to be approximately 2 um.

The conclusion based on theoretical considerations and experimental results are 1) the physiologic state of Mb is halted in the time at the moment of freezing, 2) the accuracy of measured Mb saturation is < +5%. As a result, the lower range of calculable O_2 tension is 0.25 Torr, 3) the volume of tissue included in a light intensity measurement can be as small as 3 um x 3 um x 3 um. The combination of 2 and 3 makes mapping of intracellular Po_2 gradients feasible.

A MEASURING DEVICE TO DETERMINE A UNIVERSAL PARAMETER FOR THE FLOW CHARACTERISTICS OF BLOOD: MEASUREMENT OF THE YIELD SHEAR STRESS IN A BRANCHED CAPILLARY

H. Radtke, R. Schneider, R. Witt, H. Kiesewetter, and
H. Schmid-Schönbein

Medizinische Fakultät der RWTH Aachen, Abteilung
Physiologie, Melatenerstr. 211, 5100 Aachen, FRG

INTRODUCTION

There has been an increasing attempt in recent years to search
for alterations in hemorheological parameters in a variety of disease
states, characterized by an increase in the viscosity of whole blood
(Volger, 1980). Factors which influence this gross parameter include:
hematocrit, plasma viscosity, extent of erythrocyte aggregation, and
red cell deformability. A number of controversial methods exist to
quantify these factors, but until recently it has been impossible to
evaluate the total extent of a rheological alteration or to quantify
the influence of a change in several parameters on the whole blood
viscosity. The most commonly used measurement of whole blood viscosity – "rotational viscosimetry" – seems to be inappropriate because
certain pathological states are concealed by phase separation in the
measuring chamber due to sedimentation of large erythrocyte aggregates. This results in an apparent whole blood viscosity, as measured
by rotational viscosimetry, much lower than the true viscosity in
samples with greatly altered rheological parameters.

We chose a new experimental method which permits the universal
estimation of flow characteristics: measuring the yield shear stress
in a y-shaped branched capillary viscosimeter (Radtke, 1982).

The appearance of a yield point implies that the fluid exhibits
solid body behavior at pressure differences other than zero (greater
than or less than zero); that is, the fluid does not flow in spite
of a finite pressure gradient. The yield shear stress is the marginal

851

shear stress at the transition point between solid body and fluid behavior (Kiesewetter et al., 1982c).

The yield shear stress can be directly measured and does not have to be estimated by a disputed method as is the whole blood viscosity. Experiments on suspension models have demonstrated that the yield shear stress is dependent on the rheological factors described above, especially the erythrocyte aggregation and the red cell deformability (Radtke, 1982). It can be inferred from this, that the yield shear stress represents a universally applicable measurement of the flow characteristics of whole blood. Below we describe the results of our measurements of blood samples from various patients.

MATERIALS AND METHODS

We tested our method in a blind study with sodium heparinate anticoagulated venous blood from patients with a variety of neurological diseases. The hematocrit was always adjusted to 45% using an impedance method (Kiesewetter et al., 1982b).

In addition to the yield shear stress we measured the plasma viscosity (according to the method of Harkness (1963)), the erythrocyte aggregation (using an aggregometer (Kiesewetter et al., 1982d)) and the erythrocyte deformability (in a single erythrocyte rigidometer (Kiesewetter et al., 1982a)).

The heart of our measuring system is a measuring chamber with a y-shaped branched tube (Fig. 1) (Radtke, 1982). The measuring chamber is constructed of polyester resin; unstretched polyamide fibers are cast in the resin, after hardening the fibers are stretched at their ends and pulled from the block. The diameter of the completed channels is approximately 84 um.

The chamber must be completely filled before beginning the measurements. This is accomplished by forcing the blood sample under high pressure through a teflon tubing (which also serves as the blood reservoir) into the chamber until it exits at the outflow.

The measurement begins with establishing a zero-point: the driving force is slowly increased from a negative force value until a flow stop occurs in the shorter branch of the measuring chamber - this is the zero point. The driving force is then increased in steps of 2.5 Pa and the movement of the erythrocytes in the long branch is measured, using a calibrated ocular and a stopwatch. If there is a movement of the erythrocytes in the longer channel we conclude that the measured blood has no yield shear stress. If there is no visible movement of the red blood cells - meaning that the linear displacement of single cells is less than 0.5 um per minute - the pressure difference is elevated in steps of 2.5 Pa until a visible cell move-

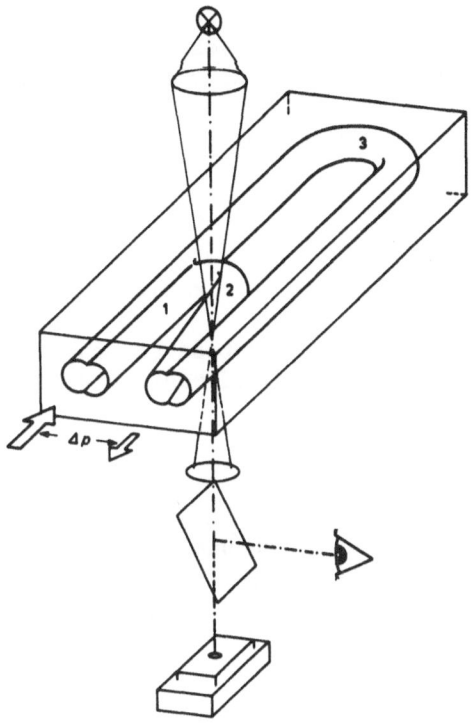

Fig. 1. Schematical drawing of the measuring chamber.

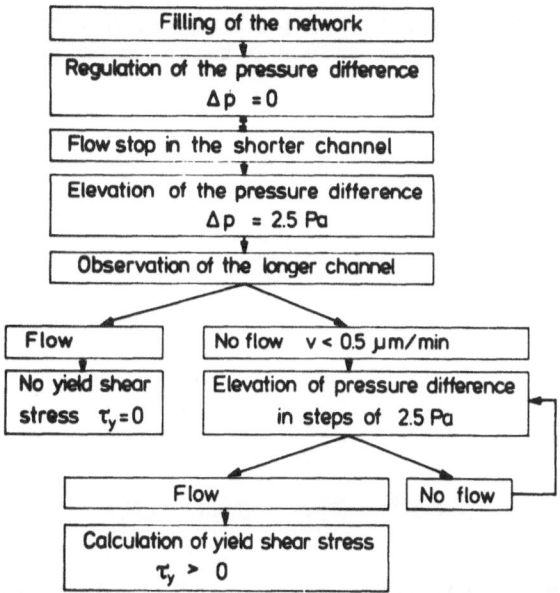

Fig. 2. Flow diagram of the measurement procedure.

ment is noted. The yield shear stress is calculated from the last pressure difference at which no movement was detectable. Our measurement procedure is illustrated as a flow diagram in Fig. 2.

The marginal shear stress is calculated as follows:

$$\tau = \frac{\Delta p \cdot r}{2l}$$

where Δp = pressure difference, r = radius, l = length.

The difference in pressure across the parallel channels is calculated by means of the equation of continuity:

$$\Delta p = \frac{2l_s}{l_i + 2l_s} \Delta p_o$$

where l_s = length of the short channnel, l_i = length of the initial channel, Δp_o = pressure difference across the whole measuring chamber.

RESULTS

The following diagrams depict the yield shear stress τ_y (Fig. 3), the standardized plasma viscosity η_{pl} (mean plasma viscosity SEVI) (Harkness, 1963) (Fig. 4), the standardized erythrocyte aggregation index SEA (Kiesewetter et al., 1982d) (Fig. 5), and the standardized erythrocyte deformability index SEVI (Kiesewetter et al., 1982a) (Fig. 6) for samples from patients with various neurological diseases compared to normal values obtained from clinically normal volunteers.

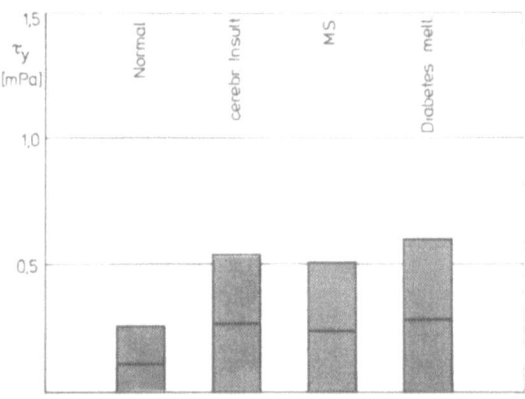

Fig. 3. Mean yield shear stress (τ_y) and standard deviation of pathological and normal samples.

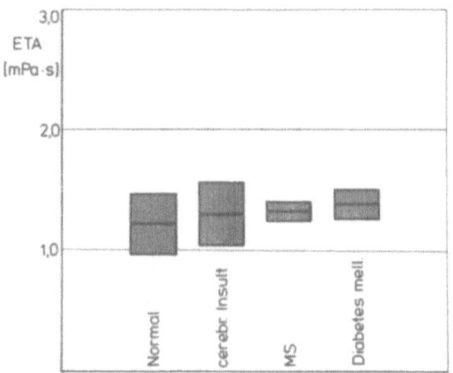

Fig. 4. Mean plasma viscosity (ETA) and standard deviation of patho-
         logical and normal samples.

DISCUSSION

    The mean values for the yield shear stress for all the diseases
sampled here are clearly elevated in comparison to the normal values.
The large standard deviations are also seemingly conspicuous; these
can be explained by the heterogeneity of the patient groups (this
also applies to the standard deviation of the other parameters
examined) as well as by the measurement procedure followed, which
progresses in discrete steps (see Methods) and which basically
yields a yes/no result in differentiating between rheologically
conspicuous and rheologically "healthy" samples. This differentiation is seemingly accurate (a significance level of $p < 0.05$ for
the pathological blood sample compared to the "normal" values);
blood samples with a yield shear stress greater than zero were

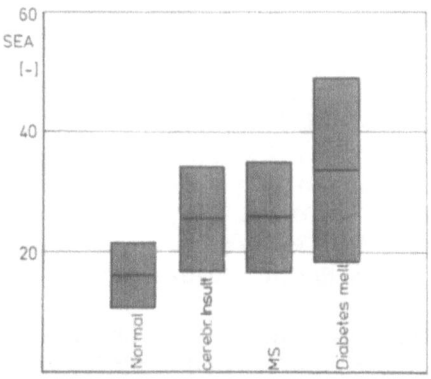

Fig. 5. Mean standardized erythrocyte aggregation (SEA) index and
         standard deviation of pathological and normal samples.

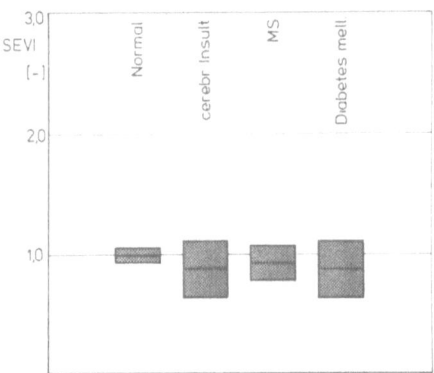

Fig. 6. Mean standardized erythrocyte deformability index (SEVI)
         and standard deviation of pathological and normal samples.

always associated with at least one rheological parameter in the
pathological range, and no blood sample with conspicuous rheological
values had a yield shear stress equal to zero.

        These results would seem to indicate that the yield shear stress,
directly measured in a branched capillary viscosimeter, can be used
as an universally applicable description of the flow characteristics
of blood, as it allows one to draw conclusions as to the end effect
of all the hemorheological parameters. For a general discussion of
the flow point phenomenon please refer to Kiesewetter et al. (1982c;
see also Radtke, 1982).

REFERENCES

Harkness, J., 1963, A new instrument for the measurement of plasma
        viscosity, Lancet, 2:280.
Kiesewetter, H., Dauer, U., Teitel, P., Schmid-Schönbein, H., and
        Trapp, R., 1982a, The single erythrocyte rigidometer (SER) as
        a reference for RBC deformability, Biorheology, 19:737-753.
Kiesewetter, H., Lazar, H., Radtke, H., and Thielen, W., 1982b,
        Hämatokritbestimmung durch Impedanzmessung, Biomed. Tech., 27:
        171-175.
Kiesewetter, H., Radtke, H., Jung, F., Schmid-Schönbein, H., and
        Wortberg, G., 1982c, Determination of yield point: Methods and
        review, Biorheology, 19:363-374.

Kiesewetter, H., Radtke, H., Schneider, R., Mussler, K., Scheffler,
    A., and Schmid-Schönbein, H., 1982d, Das Mini-Erythrozyten-
    Aggregometer: Ein neues Gerät zur schnellen Quantifizierung·des
    Ausmaßes der Erythrozytenaggregation, Biomed. Tech., 27:209-213.
Radtke, H., 1982, Bestimmung der Fließschubspannung von Blut und
    Erythrozytensuspensionen in einem Kapillarviskosimeter mit
    y-förmiger Verzweigung, Dissertation, RWTH Aachen.
Volger, E., 1980, Experimentelle und klinische Untersuchungen über
    die Rheologie des Blutes bei kardiovasculären Erkrankungen und
    deren Risikofaktoren, Habilitationsschrift, TU München.

TELEVISION FLUORESCEIN-ANGIOGRAPHY OF THE RETINA WITH ON-LINE

MEASUREMENT OF THE DILUTION CURVES

H. Kiesewetter, F. Jung, N. Körber, and M. Reim

Abt. Physiologie der RWTH Aachen, 5100 Aachen, FRG

For a long time television techniques have been employed for the imaging of the human retina. Lidley (1950) reported as early as 1950 about the application of television systems in ophthalmology. During the last years the low light level technique has rendered possible examinations with a low light stress; and computers allow an objective evaluation of the data. Van Heuven et al. (1972), Yuhasz et al. (1973) and Wessing et al. (1974) have made recordings

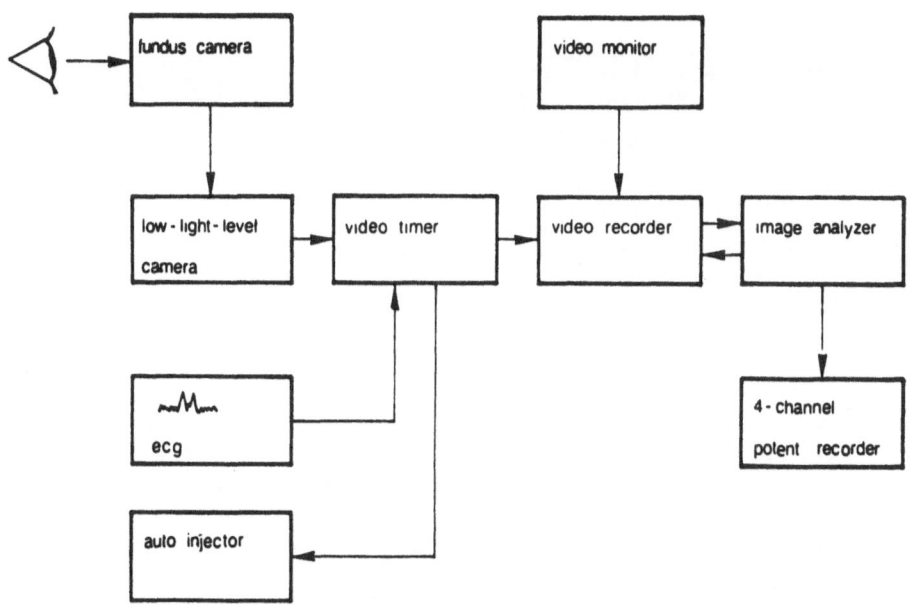

Fig. 1. Block diagram of the whole system.

of retinal fluorescein angiograms with a good resolution and a tol-
erable light stress. Some procedures are able to measure blood flow
parameters of the human retina on-line as well as off-line. The
essential disadvantages of these methods - especially the on-line
methods - are the very high light stress and the missing reproducibility in the evaluation of the data. This leads to the fact that
none of these methods are used in clinical routine as yet. The image
analyzer Mikro-Videomat 3 (Carl Zeiss AG) renders the measurement
of objective data from the recorded television fluorescein angiograms
possible (Jung et al., 1980; Körber and Gesch, 1982). The recording
method is shown in Fig. 1.

   The fundus camera (Carl Zeiss AG) is directly attached to a low
light level TV camera (SIT). The retinal area, which shall be investigated, is focused under monitor control. The images are recorded on a 3/4 U-matic video tape. Figs. 2 - 5 show photographs of the
retina in different phases of the filling process of the vessels.

   For reproducible evaluations of dye dilution curves it is essential to inject the dye at defined points in time always with the same
injection speed (Vyska, 1976). Therefore, an injector is started by

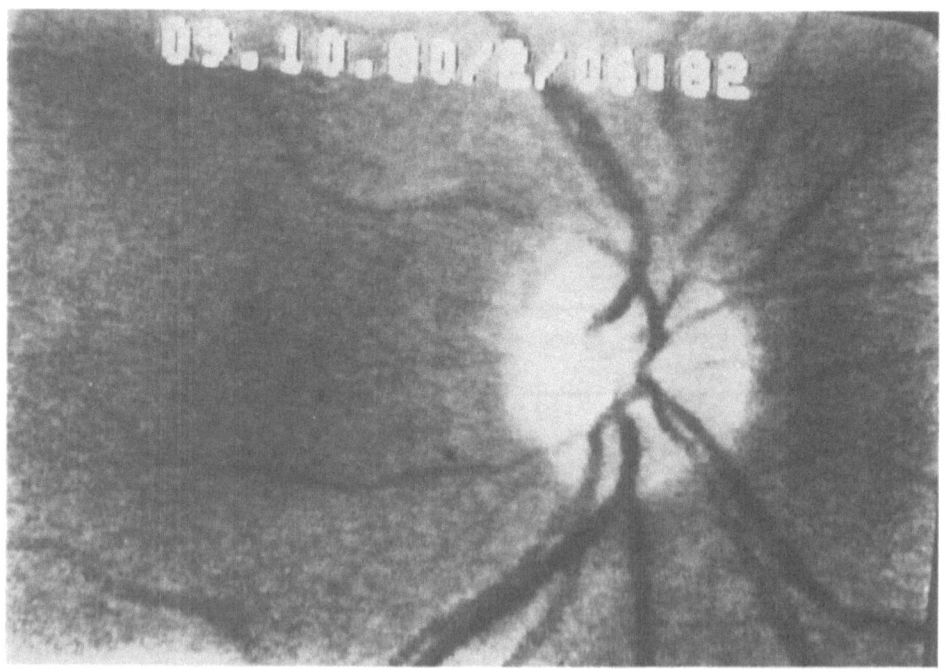

Fig. 2.   Part of the retina of a healthy person before dye injec-
          tion (initial phase).

the videotimer, triggered by the ECG. For the registration of the
dye dilution curves, areas of measurement of arbitrary size are positioned in areas of the recorded part of the fundus by the image
analyzer. Thus, the dilution curve is registered as an intensity
curve over the time axis. During the recording of the dye dilution
curves the area of measurement must remain the position elected at
the beginning of the measurement.

    The measuring area must follow continuously the eye movements
of the patient and the microsaccades especially. In order to realize
this, a coordinate system is defined on the image in a way that its
origin lies in the optic nerve head, the point of strongest bright-
ness. This point, and thereby the coordinate system, is always positioned in real time on every new image. Thus, the areas of measurement, positioned over arterioles and corresponding venoles at the
beginning of the measurement, are always located over the same
points of the vessels because the distances of all arterioles and
venoles in relation to the optic nerve head do not change. The image
analyzer is built up as a modular hardware system and is controlled
by two-stage software. A block diagram of the Mikro-Videomat 3 is
shown in Fig. 6.

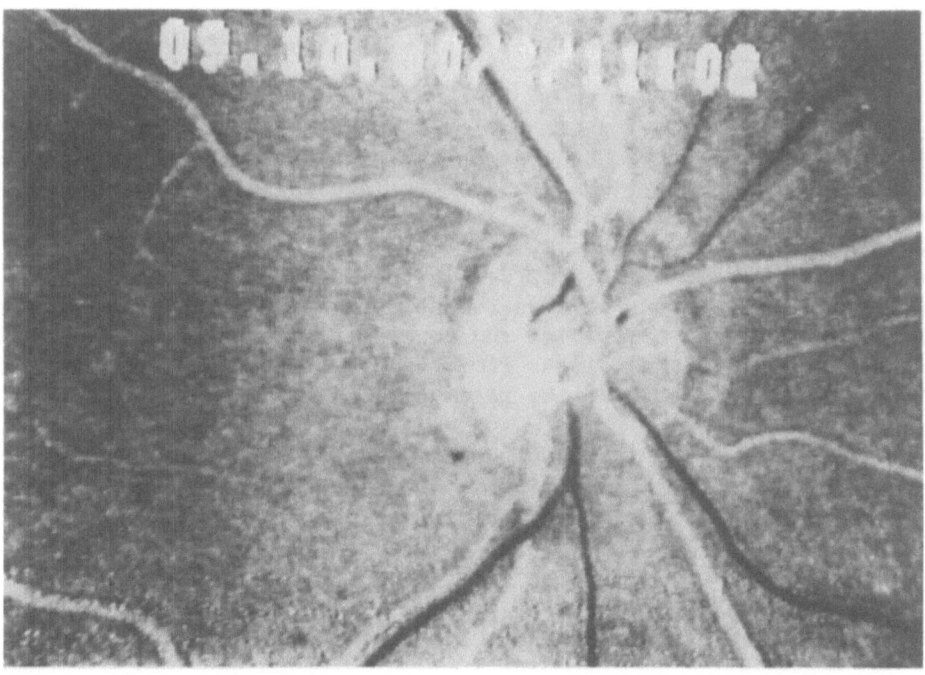

Fig. 3. Same part of the retina after the dye is in the arterioles
        (arterial phase).

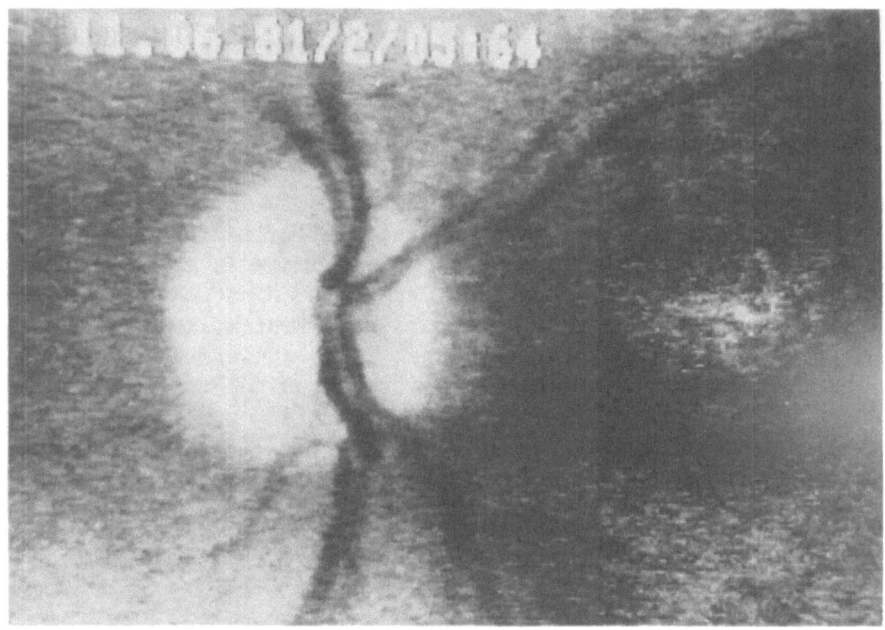

Fig. 4. Part of the retina of a patient with diabetes mellitus
        (type II) before dye injection (initial phase).

     The complete device was employed for the measurement of fluo-
rescein dilution curves from human retinal vessels. Two groups were
investigated: healthy young volunteers (age 23 - 29 years, n = 30)
and patients with diabetes mellitus (d.m.) type II, not insulin
dependent (age 60 - 72 years, n = 10). The elected television fluorescein angiograms were recorded in the Dept. of Ophthalmology,
RWTH Aachen. The areas of measurement were near the optic nerve head
over corresponding arterioles and venoles in the upper or lower
nasal quadrant of the retina. From the dye dilution curves provided
by the analyzer the following parameters are taken:

1) The time required for the dye to pass the areas of microcirculation between arteriole and corresponding venole: the
   arterio-venous passage time (AVP).

2) The mean slope of the dilution curves in arterioles and
   venoles (MS). The mean gradient is calculated by the intensity values at 90% and 10% of the dilution curve.

3) The mean plasma velocity ($\bar{v}_{pI}$) in the venole of the inverted
   quadrant. The mean plasma velocity is calculated from the
   displacement of the peak maximum on the time axis, which is

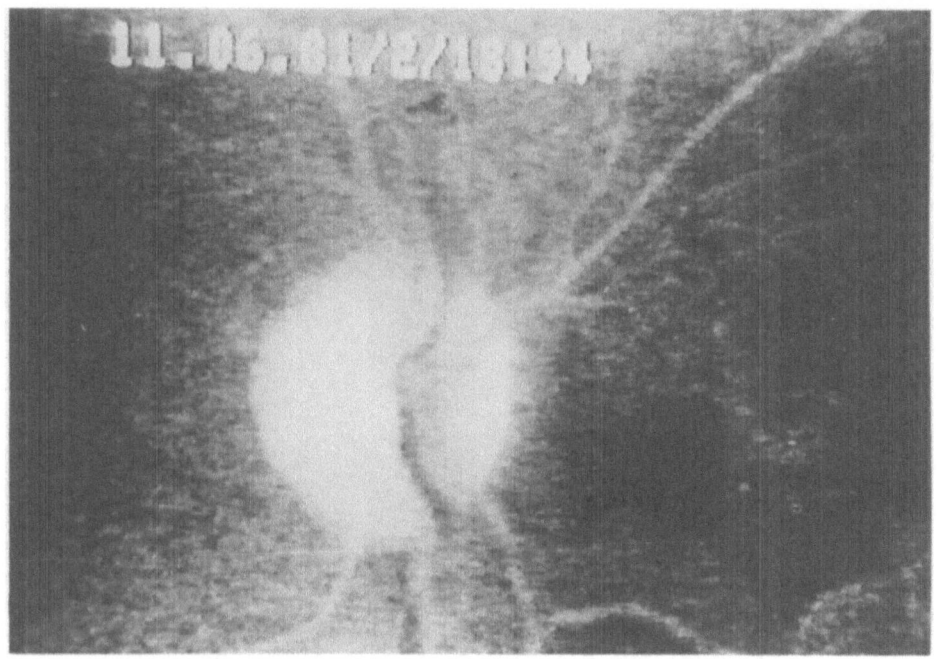

Fig. 5.   Same part of the retina after the dye is in the arterioles
          (arterial phase).

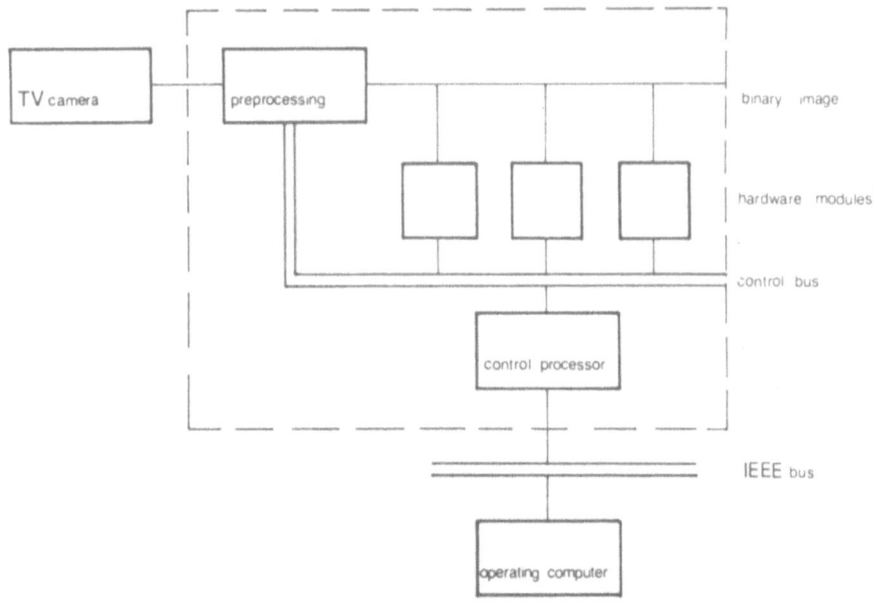

Fig. 6.   Block diagram of the image analyzer.

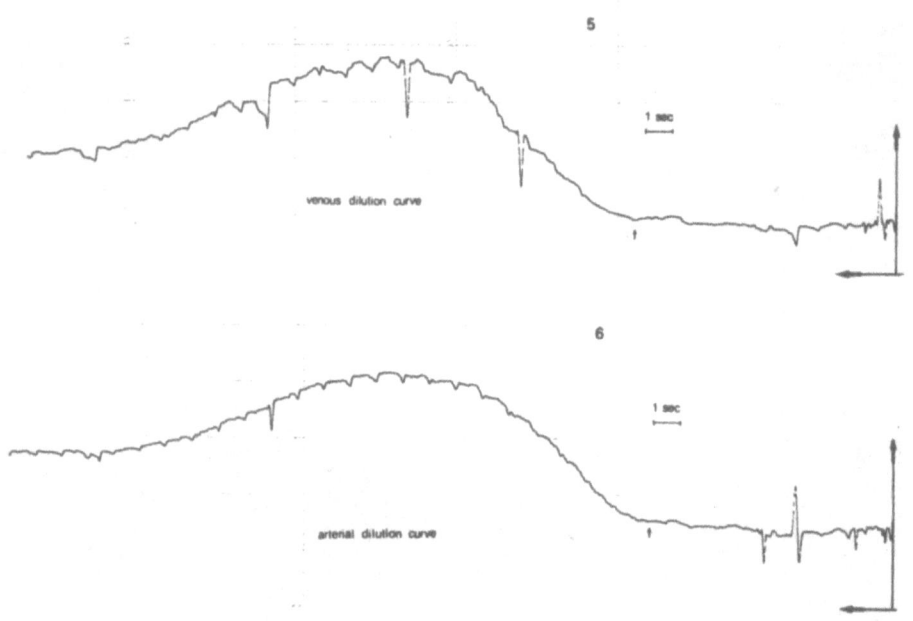

Fig. 7.   Typical dye dilution curves of a healthy person.

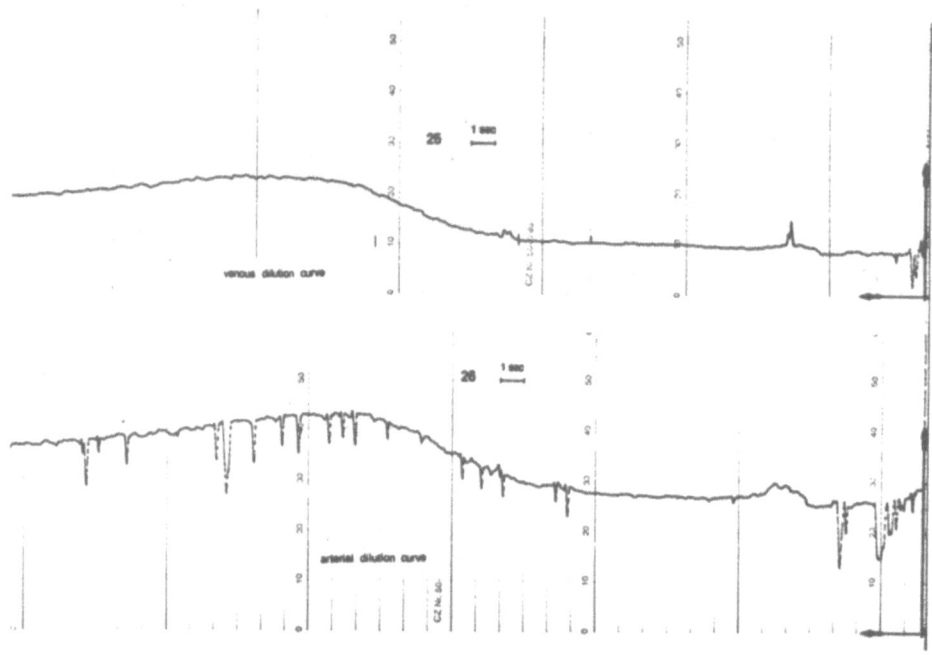

Fig. 8.   Typical dye dilution curves of a patient with diabetes
          mellitus.

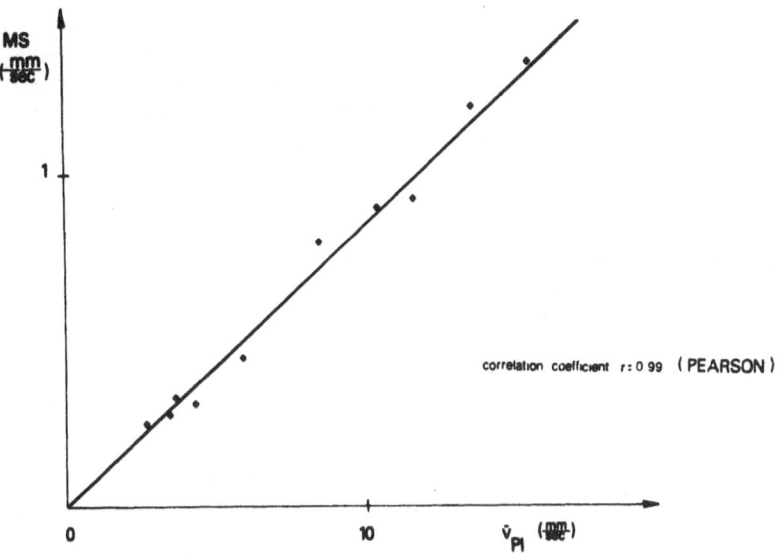

Fig. 9.   Correlation of the mean slope of the dye dilution curves
          and the mean plasma velocity.

          apparent when dye dilution curves are recorded on a vessel
          at two different points separated by a known distance.

       Figs. 7 and 8 show typical dye dilution curves in an arteriole
and a corresponding venole for a normal person and a patient with
diabetes mellitus, respectively.

	n	AVP	MS arterial	MS venous	$\check{v}_{Pl}$ venous
healthy volunteers	30	1.5 ± 0.35	1.2 ± 0.5	1.0 ± 0.3	12.3 ± 3.4
patients with diab. mell.	10	3.5 ± 1.9	0.6 ± 0.5	0.5 ± 0.3	5.5 ± 3.7

•   p < 0.002   (Wilcoxon; Mann; Wnitney)
••   p < 0.05
•••   p < 0.1
•••• p < 0.01

Fig. 10. Mean values and standard deviations of the arterio-venous
          passage time (AVP), of the mean slope of the dilution
          curves in arterioles and venoles (MS) and of the mean
          plasma velocity for both groups.

The mean arm retina time (ART) (the time the dye needs from injection into the cubital vein to the appearance in the vessels of the retina) of the normal group is significantly smaller than the mean ART of the d.m. group. The mean slope of the normal dilution curve is significantly higher than the slope of the curve of the patients with d.m.

The mean plasma velocity and the mean slope of the dye dilution curves are highly correlated ($r = 0.99$, Fig. 9). Thereby it is possible to calculate approximately the mean plasma velocity in the arterioles. The mean values and standard deviations for AVP, MS and $\bar{V}_{PI}$ are shown in Fig. 10. The arterio-venous passage time (AVP) as well as the mean plasma velocity is significantly changed in diabetic patients compared to healthy volunteers.

It can be expected that only a sufficient blood flow through the retinal vessels, especially in the macular area guarantees normal visual acuity. A reduction of blood flow under a yield flow leads to a permanently insufficient supply of O_2 to axons and ganglions of the nerve fibers. One of the pathophysiologic changes in diabetic microangiopathy is - caused by the changed metabolic situation - a circulatory disease, which is compensated in the first stages and decompensates in later stages.

The flow measurements in the retinas of patients could lead to a detection of the early stage of decompensation in order to cause a therapeutic prolongation of this event. Our data provided from major retinal vessels are not representative of the situation of blood perfusion in the real microcirculation, the place of the metabolic exchange. For this reason the image resolution should be enhanced so that the vessels of the microcirculation can be resolved in the video image. Then the image analyzer could allow the exact quantification of the flow in human capillaries.

With this device it is possible to measure retinal dye dilution curves with a very small light stress. In addition, we are able to monitor the circulation before and during a therapy objectively.

REFERENCES

Heuven, van, W.A.J., and Schaffer, C.A., 1972, Advances in televised fluorescein angiography, in: "Proc. Int. Symp. Fluorescein Angiography", K. Shimizu, ed., Igaku, Shoin, Tokyo, pp. 10-14.
Jung, F., Körber, N., Kiesewetter, H., and Reim, M., 1982, Television fluorescein-angiography with on-line evaluation of the dilution curves, Int. J. Microcirc., 1, 3:284.
Körber, N., and Gesch, M., 1982, Möglichkeiten und Grenzen der Fernsehfluoreszenzangiographie, Klin. Monatsbl. Augenheilkd., 180: 100-102.

Lidley, H., 1950, Television in ophthalmology, Acta Consil. Oph-
    thalmol., 16:1397.
Vyska, K., 1976, Theorie der Indikatorverdünnungsmethode, Habilschr.
    KFA Jülich.
Wessing, A., 1974, Fluorescein-Fernseh-Angiographie der Netzhaut
    und der vorderen Augenabschnitte mit hochempfindlichen TV-Auf-
    nahmeröhren, Klin. Monatsbl. Augenheilkd., 165:817.
Yuhasz, Z., Akashi, R.H., Urban, J.C., and Müller, M.M.H., 1973, A
    new apparatur for video tape recording of fluorescein angio-
    grams, Arch. Ophthalmol., 90:481.

MICRO, SURFACE, AND NEEDLE OXYGEN ELECTRODES: COMPARISON OF PHYSIO-
LOGICAL RELEVANCE AND CLINICAL ACCEPTANCE

A.J. van der Kleij[1], H.P. Kimmich[2], R.J.A. Goris[1],
F. Kreuzer[2], J. de Koning[2], and G. Beerthuizen[1]

[1]Department of Surgery, Faculty of Medicine, University
of Nijmegen, The Netherlands
[2]Department of Physiology, Faculty of Medicine
University of Nijmegen, The Netherlands

The aim of a clinical determination of tissue Po_2 is to discriminate between levels of tissue oxygenation found in normal
subjects and those levels found in patients with tissue hypoxia or
anoxia where other diagnostic tests and clinical symptoms provide
insufficient information. Micro, surface, and needle electrodes are
available to determine Po_2 in tissue polarographically. A polarographic electrode may be called a micro electrode (Kunze, 1966)
when the outside diameter of the glass-insulated cathode is several
um's. This type of Po_2 electrode may be ideal for tissue Po_2 measurements (Silver, 1966), but unfortunately it is too fragile for
routine clinical application. Therefore, for clinical application
two different types of polarographic electrodes have been developed.
The construction of a Po_2 micro electrode in a silver needle
(Schuchhardt, 1971), or a stainless-steel needle (Whalen and Spande,
1980; de Koning and van der Kleij, 1980) provides a Po_2 electrode
with an outside diameter which corresponds to the outside diameter
of the needle. This type of needle electrode is rugged enough to be
used in clinical situations. The outer type is the multi-wire-surface
Po_2 electrode (Kessler and Lübbers, 1966), which requires direct
contact with the organ. In other words, a fasciotomy is necessary
to measure the Po_2 of muscle tissue.

The best way to express the oxygenation of an organ is by the
construction of a Po_2 histogram (Lübbers, 1981). The Po_2 measuring
device must be able to collect a number (approximately 100) of
tissue Po_2 values from different locations in order to construct
a corresponding Po_2 histogram. Using the multi-wire-surface Po_2

869

Fig. 1. The percentage change compared to the control level,
         (= mean of t0, t1, t2, t3) of: mean arterial blood pres-
         sure, cardiac output, lactate/pyruvate ratio, number of
         leucocytes, art.-ven. Po_2 difference, and the median of
         the cumulative Po_2 histograms of skeletal muscle. Septic
         shock was induced by the infusion of live Escherichia coli
         for 5 hours from t3 to t13. Note that in the early stage
         (t4) the change in the median of skeletal muscle Po_2 is
         greatest (-58.5%).

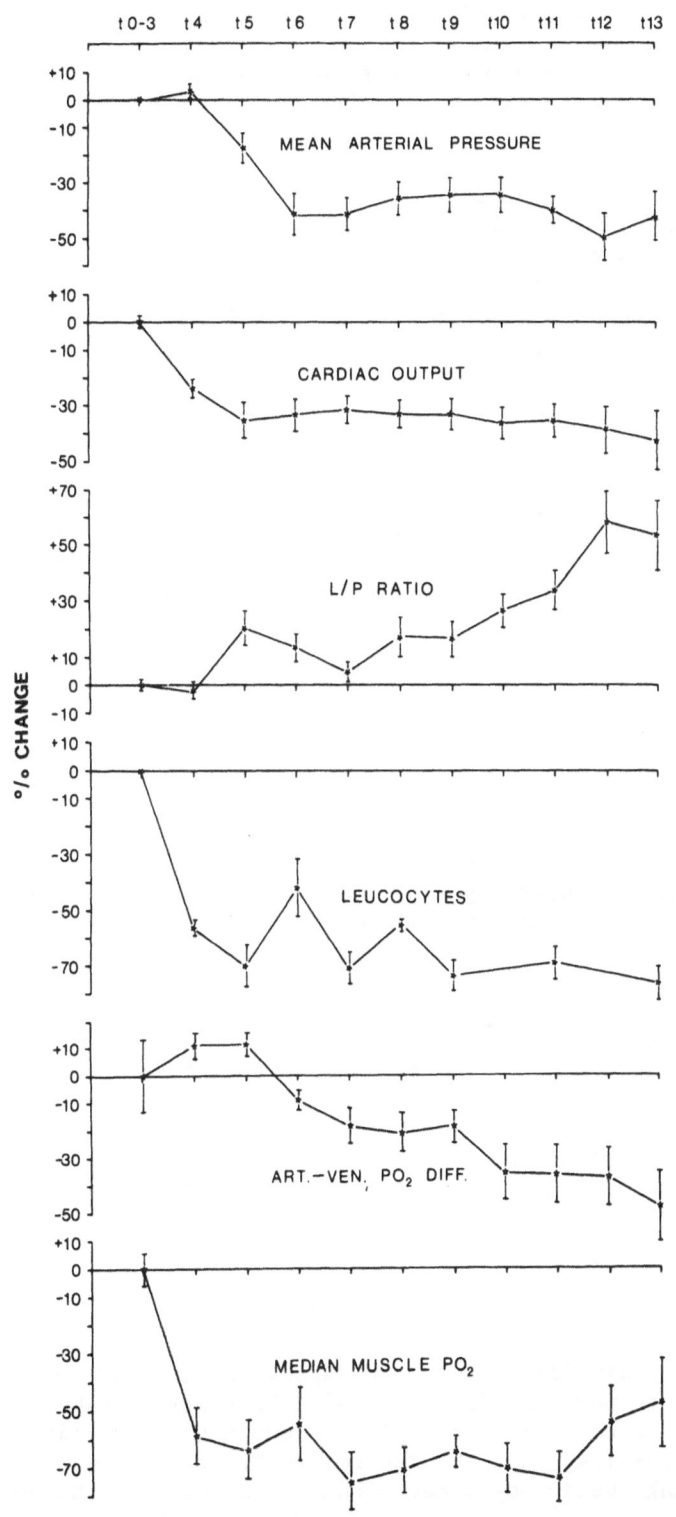

electrode, this number of Po_2 values is acquired by rotating the
electrode, whereas using a needle Po_2 electrode it is obtained by
stepwise withdrawal from the tissue. Both types of electrodes measure
the averages of the oxygen pressures in the vicinity of the measuring cathode.

From a clinical point of view any measuring device must fulfil
several conditions to be accepted by the physician and the patient.
Minimal invasiveness for the patient, easy handling of the device,
reliability, instantaneous results after the measurement and short
measuring time are all important clinical criteria. A polarographic
needle electrode meets each of these criteria. It has been shown by
van der Kleij et al. (1982) that a left shift in a cumulative Po_2
histogram of skeletal muscle, obtained with a polarographic needle
electrode, is a sensitive indicator to detect an early hemorrhagic
hypovolemia.

The purpose of this paper is to show that the determination of
skeletal muscle Po_2 with a polarographic needle electrode enables
the detection of an early septic shock and a compartmental syndrome.

MATERIALS AND METHOD OF SEPTIC SHOCK EXPERIMENT

Splenectomized dogs (n = 9, weight 21.8-24.2 kg) were subjected to a 5-hour infusion of live Escherichia coli (10^{10} m.o./kg body
weight). After induction with thiopentone (15 mg/kg) the trachea
was intubated and artificial ventilation was carried out with a
Harvard respirator. Cannulation of the femoral artery, superior
caval vein, and pulmonary artery allowed continuous monitoring and
recording of the mean arterial blood pressure, the central venous
blood pressure, the mean pulmonary arterial blood pressure, and the
pulmonary wedge pressure. Cardiac output was obtained by thermodilution using a Swan-Ganz catheter. The heart rate was derived from
the E.C.G. Hematological and biochemical variables as well as arterial and mixed venous blood gas values were determined at 30-minute
intervals (t0-t13). Skeletal muscle Po_2 measurements were done in
the quadriceps muscle as previously described (van der Kleij and de
Koning, 1981). The processing of the one hundred Po_2 values was
performed on line by a microcomputer. The total time from the onset
of the measurement until the final result was 8 minutes.

RESULTS

Two animals died during the experiment (at t12 and t13), whereas the others were sacrificed after t13. Early septic shock was
noticeable in the first hour after the onset of the live Escherichia
coli infusion (t4, t5). The control level of the medians (p 50) of
the skeletal muscle Po_2 measurements, derived from the cumulative

Po_2 histograms, showed a large inter-individual variation between dogs. However, as soon as the infusion with live Escherichia coli started, we observed comparable changes in these medians. The decrease of the medians was directly related to the extent of the circulatory deterioration. The percentage change in all variables compared to the control level (= mean of t0, t1, t2, t3) was calculated. In Fig. 1 the percentage change in several important variables is shown. During early septic shock the highest change (-58.5%) appeared in the median derived from the cumulative Po_2 histograms of skeletal muscle. Simultaneously we observed a widening of the arterial-venous Po_2 difference. This became smaller after the pulmonary function deteriorated as indicated by a decrease in the arterial Po_2 (Fig. 1: t6, t7).

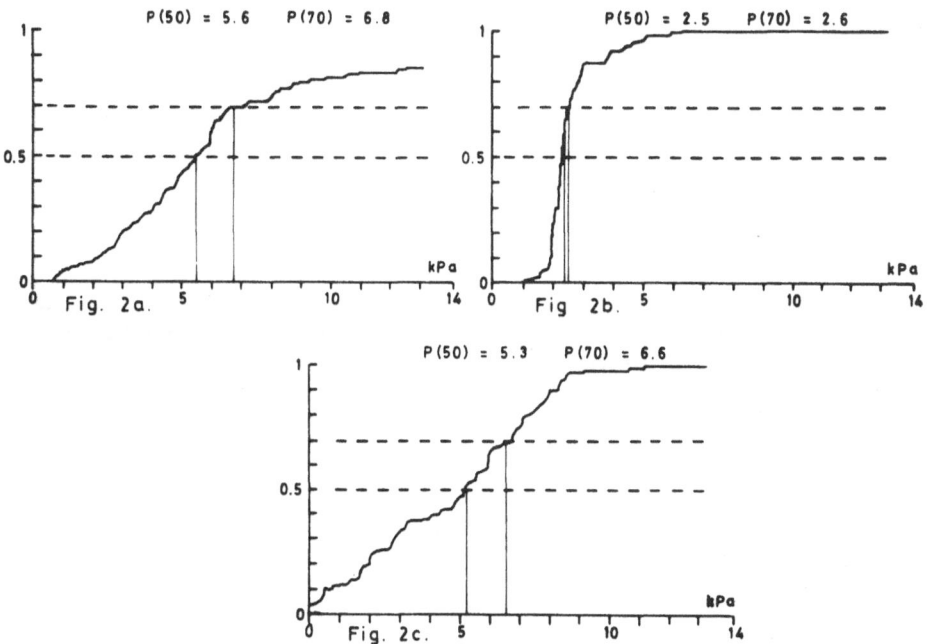

Fig. 2.   Fig. 2a cumulative Po_2 histogram of the left anterior tibial muscle. This is a normal distribution. Fig. 2b is a cumulative Po_2 histogram of the right tibial muscle 12 hours after an embolectomy. Note the left shift with the correspondingly decreased P_{50} (2.5 kPa). Fig. 2c is a cumulative Po_2 histogram of the right side 36 hours after a fasciotomy. Note that the curve has nearly the same shape as the left side, indicating the therapheutic effect of the fasciotomy.

COMPARTMENTAL SYNDROME

A 45-year-old woman became septic postoperatively after an appendectomy. Pulmonary insufficiency required artificial ventilation. The sepsis was accompanied by atrial fibrillation (300 beats/min). On the third postoperative day clinical examination showed signs and symptoms of arterial obstruction in the right leg. An embolectomy was done through a femoral arteriotomy and several fresh clots were removed. The next day a skeletal muscle Po_2 measurement of the left and the right anterior tibial muscle was performed in order to detect a possible compartmental syndrome. A cumulative Po_2 histogram of the left side showed a normal Po_2 distribution (Fig. 2a). The affected side showed a left shift in the cumulative Po_2 histogram (Fig. 2b) reflecting hypoxia in the muscle tissue. A fasciotomy was done to decompress the anterior tibial compartment. A skeletal muscle Po_2 measurement done 36 hours later showed a normal Po_2 distribution (Fig. 2c).

CONCLUSION

Skeletal muscle Po_2 measurements, obtained with a polarographic needle electrode (de Koning and van der Kleij, 1980), was studied in a septic shock experiment. This study showed clearly that a polarographic needle electrode (outside diameter 0.5 mm) allows the collection of a number of Po_2 values in order to construct a cumulative Po_2 histogram. Also it has been shown that the shift in a cumulative Po_2 histogram of skeletal muscle enables the detection of an early septic shock and compartmental syndrome. The major advantage of the needle Po_2 electrode as compared to the multi-wire surface electrode is in the minor invasiveness for the patient. Therefore, the use of a polarographic needle electrode appeared to meet the clinical criteria to be accepted by the clinician as well as by the patient.

REFERENCES

Kessler, M., and Lübbers, D.W., 1966, Aufbau und Anwendungsmöglich-
    keiten verschiedener Po_2-Elektroden, Pflügers Arch. ges.
    Physiol., 291:R82.
Kleij, van der, A.J., and Koning, de, J., 1981, Tissue oxygen elec-
    trode for routine clinical application, in: "Monitoring of
    Vital Parameters during Extracorporeal Circulation", H.P.
    Kimmich, ed., Karger, Basel, pp. 95-100.
Kleij, van der, A.J., Koning, de, J., Beerthuizen, G., Goris, R.J.A.,
    Kreuzer, F., and Kimmich, H.P., 1982, Early detection of hemor-
    rhagic hypovolemia by muscle Po_2 assessment, Surgery, in press.
Koning, de, J., and Kleij, van der, A.J., 1980, An electrode for
    clinical Po_2 monitoring, Arzneim.-Forsch. (Drug Res.), 30(11),
    12:14.

Kunze, K., 1966, Die lokale, kontinuierliche Sauerstoffdruckmessung
     in der menschlichen Muskulatur, Pflügers Arch., 292:151.
Lübbers, D.W., 1981, Quantitative measurement of tissue oxygen
     supply by Po_2 histogram, in: "Monitoring of Vital Parameters
     during Extracorporeal Circulation", H.P. Kimmich, ed., Karger,
     Basel, p. 67.
Schuchhardt, S, 1971, Po_2-Messung im Myocard des schlagenden
     Herzens, Pflügers Arch. ges. Physiol., 322:83.
Silver, I.A., 1966, The measurement of oxygen tension in tissues,
     in: "Oxygen Measurements in Blood and Tissues", J.P. Payne and
     D.W. Hill, eds., Churchill, London, pp. 135-145.
Whalen, W.J., and Spande, J.I., 1980, A hypodermic needle Po_2
     electrode, J. Appl. Physiol.: Respirat. Environ. Exercise
     Physiol., 48(1):186.

AN IMPROVED POLAROGRAPHIC MULTIWIRE SURFACE Po$_2$ ELECTRODE,

PARTICULARLY FOR MEASUREMENT OF HIGH Po$_2$ VALUES

Bi Yu[1], H. Baumgärtl[2], and D.W. Lübbers[2]

[1] Wuhan Medical College, Wuhan, PRC
[2] Max-Planck-Institut für Systemphysiologie
Rheinlanddamm 201, 4600 Dortmund 1, FRG

INTRODUCTION

In 1956 Clark described a polarographic electrode which
allowed to measure the Po$_2$, particularly in biological materials
with an exactitude and stability which hitherto was not possible.
This was achieved by a membrane which separated electrode and biological material from each other so that the polarographic reaction
at the Pt surface and the current transfer at the reference electrode took place in a stable environment. However, our own experiments had shown (Lübbers et al., 1969) that at high Po$_2$ values
even Clark type electrodes with an Ag/AgCl reference electrode and
a buffered 0.1 mol KCl solution may drift so much that longer
lasting experiments are impossible. Since we wanted to measure the
Po$_2$ histogram of the guinea pig brain perfused with a blood-free
perfused medium (Heinrich et al., this volume) which was equilibrated with carbogen (95% O$_2$ and 5% CO$_2$), we investigated possibilities
to improve the stability of the electrode, particularly at high Po$_2$
values.

It is known that several factors influence the stability of
the polarographic Po$_2$ electrode. The most important ones are 1. the
material of the polarographic electrode and its surface structure,
2. the reference electrode, 3. the composition of the electrolyte.
The O$_2$ reduction at the electrode surface can be improved by polishing the Pt wire or by plating the Pt wire with gold (Baumgärtl and
Lübbers, 1983). In a relatively large electrolyte vessel with a
separate stable reference electrode, such electrodes have an excellent polarogram and show a minimal drift. Since the same electrode
has a large drift if the electrolyte space is made small and thin
and is covered by a teflon membrane, we focussed our experiments

877

on the improvement of the reference electrode and on the selection
of a proper electrolyte.

MATERIAL AND METHODS

The improved Po_2 electrode consists of the following parts:

1. Platinum Electode: The multiwire platinum electrode consists
of 8 platinum wires with a diameter of 15 um (Kessler and Lübbers,
1966; Lübbers, 1977). The wires are fused together in glass and the
glass rod with the infused Pt wires included in a tightly fitting
silver tube. Then the front surface is carefully polished first with
emerald paper and water and then finished with a polish rouge on a
piece of lime wood. In addition, if necessary, the wires are electrically polished (Baumgärtl and Lübbers, 1983).

2. Reference Electrode: As reference electrode, silver/silver
oxide is used. Silver/silver oxide is formed in a H_2O_2 solution
(30% H_2O_2, perhydrol p.a., Fa. Merck, Darmstadt) using an electrical circuit consisting of the Ag surface of the electrode (positive
pole) and a Pt wire (0.5 mm Ø, negative pole) (Fig. 1). To achieve
the proper current density (10 uA/mm^2 Ag) a variable electrical
resistance of 100 k with a 9 V battery is used. During the formation of the silver oxide layer the metallic lustre of silver disappears and a pure milky surface is formed. The formation of this
layer takes about 10-20 minutes. Then the electrode is washed with
distilled water, dried at room temperature. After the formation of
the silver oxide layer the surfaces of the Pt wires have to be
cleaned. For example by gently polishing on a piece of lime wood.
Then the electrode is washed and dried again.

3. Electrolyte Solution: As electrolyte solution borax-biphosphate buffer (pH 8) in the following composition was used:

    55 ml 0.05 mol $Na_2B_4O_7$
    45 ml 0.1 mol KH_2PO_4

After mixing, the solution has to be shaken thoroughly.

Perfusion Experiments

Experiments were performed in 5 guinea pigs of body mass
between 200 and 300 g. Before beginning the experiments the animals
were given an injection of Liquemin i.m. (0.1 ml/100 g), anaesthetized with Nembutal (60 mg/kg b.w.) and 0.1 ml atropine (1 mg/ml)
i.p.. The preparation of the animals was performed as described by
Heinrich et al. (this volume). The compostion of the perfusion
medium was the following: Glucose (5.5 mmol·1^{-1}) + KCl (4.7 mmol·1^{-1}),
CaCl (1.25 mmol·1^{-1}), KH_2PO_4 (1.2 mmol·1^{-1}), $NaHCO_3$ (24.9 mmol·1^{-1}),

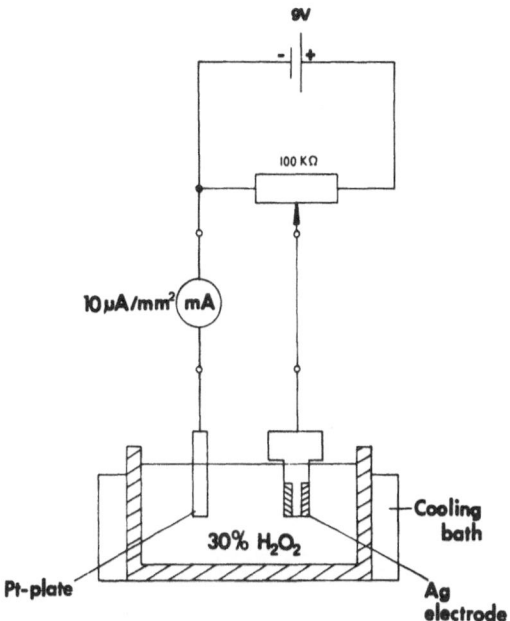

Fig. 1.   Setup used to produce the silver/silver oxide reference
           electrode. The Ag electrode is connected with the positive
           pole of a 9 v battery, the platinum plate serves as nega-
           tive pole. The electrical current is adjusted by a variable
           resistance (100 kΩ) to a current density of about 10 uA/mm²
           Ag. mA: milliampere meter.

$MgSO_4$ (0.3 mmol·l^{-1}), pyruvate (2.0 mmol·l^{-1}) + papaverine (4.0 mg/
100 ml perfusion solution). The medium was equilibrated with 95%
O_2 + 5% CO_2. The perfusion was carried out at temperatures of 18°C,
24°C and 37°C. The perfusion rate was 30 ml/min.

RESULTS AND DISCUSSION

        The improvement of the electrode is achieved first by a stable
reference electrode and secondly by a suitable electrolyte.

1. The Reference Electrode

        After mechanically polishing, the silver surface remains rather
inhomogeneous as it is shown in an electronmicroscopic photograph
(Fig. 2). But in spite of these inhomogeneities a homogeneous oxide
layer is formed if a proper current density is chosen. Such a layer
consists of many fine grains of about 1 um (Fig. 3). There are
several silver oxides described in the literature (Ag_2O, AgO
(Fleischmann et al., 1968), Ag_2O_3 (Feller-Kiepmeier et al., 1967)
and AgO_x (Gossner and Polle, 1968)). We are not able to give the

Fig. 2.   Electron micrograph of a mechanically polished Ag surface
          with scratches. The scratches and holes on the Ag surface
          may disturb the current transfer layer which connects the
          electrolyte and the metal.

formula for the oxide layer on the basis of its production. It may
be that the layer is a mixture of different oxides.

        With regard to the H_2O_2 solution, it has to be mentioned that
there are two different kinds of H_2O_2 solutions: one is stable at
room temperature (30% H_2O_2 stabilized for increased storing temperature, E. Merck, Darmstadt) whereas the other one must be kept in the
refrigerator. Trying to form a silver/silver oxide electrode with
the first solution, one obtains an oxide layer with a dark black
colour. This kind of reference electrode is unstable and cannot be
used. The milky white colour mentioned above, appears only with the
second kind of H_2O_2. The stability of the reference electrode can
be indirectly tested by measuring its electrical resistance. This
is achieved by applying a voltage of for example -700 mV and measuring the current.

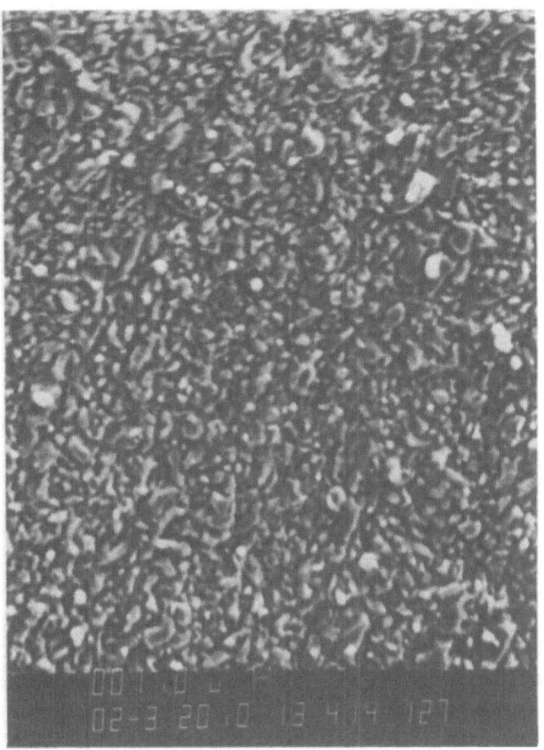

Fig. 3.   Electron micrograph of the surface of the silver/silver
          oxide reference electrode. The silver/silver oxide layer
          is formed in 30% H_2O_2 as described in the text. The surface
          consists of fine grains (diameter of about 1 um).

2. The Electrolyte

     To obtain stable Po₂ electrodes a choice of suitable electro-
lytes is similarly important as the stability of the reference
electrodes. We have tried many electrolytes and buffers such as
KOH, KCl, NaOH, D-mannite + H_2BO_3, glucose + H_3BO_3, H_3BO_3 + NaOH,
pyrocatechol + boric acid, ethyleneglycol buffer, veronal buffer,
borax-biphosphate buffer, etc. A part of these systematical investigations are shown in Table 1 which gives the width of the polarographic plateau (in air and O_2), the current in N_2 and O_2, the sensitivity in nA/mm Hg Po₂ and the drift in air per hour. We found
that borax-biphosphate buffer (Table 1, 5) is better than the other
ones. An electrode built with borax-biphosphate buffer 1) has a
higher nA output than electrodes built with other electrolytes, 2)
can work steadily over a long period of time, and 3) seems helpful
in decreasing the velocity of Ag deposit on the surface of the Pt
electrode during the polarization process.

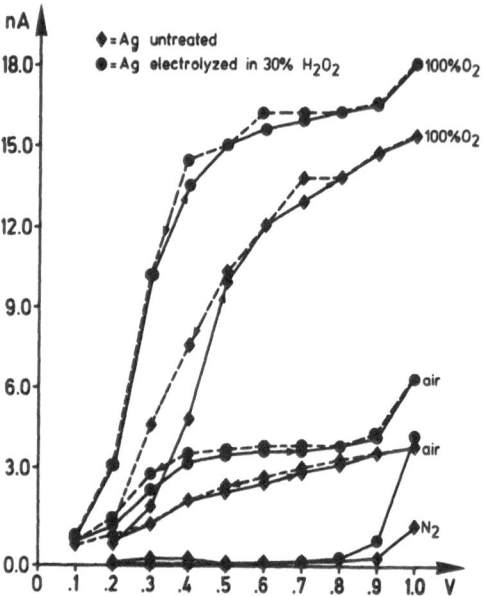

Fig. 4. Polarograms of two different oxygen electrodes. One
         electrode had the described silver/silver oxide reference
         electrode (Ag electrolysis in 30% H_2O_2). The other one was
         only mechanically polished (Ag untreated). Both sensors
         had the same electrolyte. ($\longrightarrow$ 0 - 1.0 V (increasing vol-
         tage); $- - \rightarrow$ 1.0 - 0 V (decreasing voltage)).

The performance of the electrode

     For Po_2 measurements the electrode prepared as above described
is covered by a 12 um cellophane membrane to form a stable electrolyte space and by a Teflon membrane (Clark principle). Our experiments showed that the electrode must be polarized in a saline solution equilibrated with 100% O_2 at least for about 10 hours. Then
the electrical resistance of the silver/silver oxide layer remains
constant and the full stability of the electrode is reached. Why
this long period is necessary is unknown.

     Fig. 4 shows the polarograms of this electrolyte (Ag electro-
lyte in 30% H_2O_2) compared with that of a simple Ag reference electrode. The polarogram of the electrode with the particularly prepared silver/silver oxide reference electrode is broader and
flatter, the current per mm Hg Po_2 is larger.

     In Fig. 5 original recordings in saline equilibrated with 100%
oxygen are shown over a period of 86 hours. After the polarization
period of 16 hours the drift only amounts to 0.1-0.27%/h (see
legend of Fig. 5).

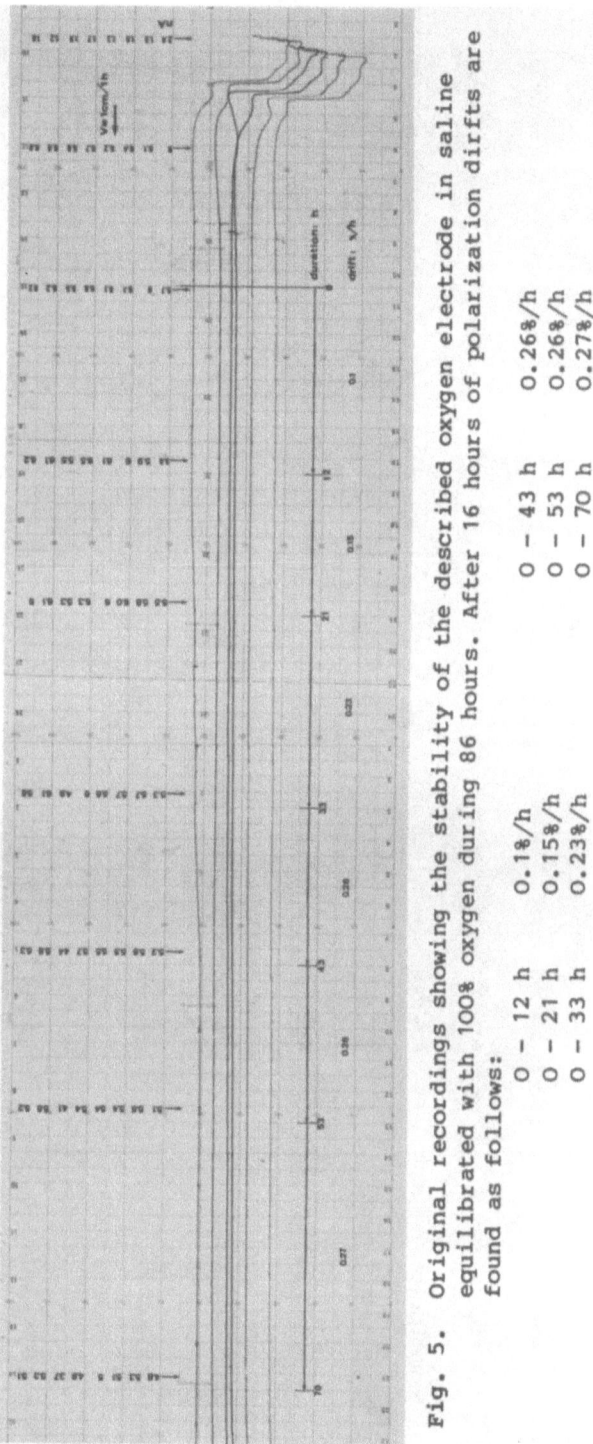

Fig. 5. Original recordings showing the stability of the described oxygen electrode in saline equilibrated with 100% oxygen during 86 hours. After 16 hours of polarization dirfts are found as follows:

0 – 12 h	0.1%/h
0 – 21 h	0.15%/h
0 – 33 h	0.23%/h

0 – 43 h	0.26%/h
0 – 53 h	0.26%/h
0 – 70 h	0.27%/h

Table 1. The plateau, sensitivity and drift of the Po$_2$ electrode (15 um Pt wire and silver/silver oxide reference electrode) with different electrolyes.

electrolyte composition	pH	air mV	air slope nA/100 mV	plateau 100% O$_2$ mV	plateau 100% O$_2$ slope nA/100 mV	E 1/2 mV air	E 1/2 mV 100% O$_2$	current nA 100% N$_2$	current nA 100% O$_2$	sensitivity (air) nA/mmHg	drift (air) %/h	remarks
0.1 m KOH	12.5	1100-1300	0.125	800-900	0.32	800	610	/	/	/	+3.8	overshoot in 100% O$_2$
0.2 m KCl	6	900-1000	0.32	/	/	600	/	0.16	5.75	0.0076	-2.1	no plateau in 100% O$_2$
Na-barbituratbuffer (6)	7	900-1000	0.14	700-800	0.79	420	440	0.025	18	0.024	-1.47	long response time
borax-ethylenglycol (4)	6	600-800	0.084	900-1000	0.165	420	560	0.01	2.49	0.003	+1.2	low sensitivity
boric acid-glycerine (2)	9-5	800-1000		700-1000	0.24	480	470	0.14	9.28	0.0125	+0.89	the average of two experiments high zero current
boric acid-mannite (3)		900-1100	0.05	900-1100	0.1	500	500	0.043	4.12	0.005	+0.44	good plateau with shift to higher potantials
boratbuffer (1)	8	800-1000	0.225	900-1000	0.99	430	515	0.053	21.63	0.028	+0.255	the average of two experiments bad polarogram with high slope in 100 % O$_2$
borax-biphosphate (5)	8	450-800	0.06	600-850	0.04	280	260	0.09	17.8	0.024	0	good plateau and high stability

1 = boratbuffer according to Clark and LUBS
0.1 m H$_3$BO$_3$ + 0.1 m NaOH

2 = boric acid-glycerine
80 ml 0.1 m H$_3$BO$_3$
20 ml 87% glycerine + 1 m NaOH

3 = boric acid-mannite
1 mol/liter D-mannite
0.5 mol/liter H$_3$BO$_3$

4 = borax-ethylenglycol
0.2 m/liter Na$_2$B$_4$O$_7$ in 95% ethylenglycol

5 = borax-biphosphate after KOLTHOFF
55 ml 0.05 m Na$_2$B$_4$O$_7$ + 45 ml 0.1 m KH$_2$PO$_4$

6 = Na-barbital buffer
1.85 g barbital acid
10.3 g Na-barbital ad 1000 ml

Table 2. The mean oxygen pressure $\overline{Po_2}$, median and the modul of the
          Po_2 histogram of the blood-free perfused brain cortex at
          different temperatures.

Po_2 Temp.	$\overline{Po_2}$ mm Hg	kPa	Median mm Hg	kPa	Modul mm Hg	kPa
18°C	416	55.5	346	46.1	526–560	70.1–74.7
24°C	219	29.2	130	17.3	351–385	46.8–51.3
37°C	66	8.8	42	5.6	0–35	0–4.7

Measurements of Po_2 histograms of the blood-free perfused guinea pig brain

Since measurements were performed at 18°C, 24°C and 37°C, the
electrodes were calibrated at three temperatures; the calibration
curves before and after the experiment are practically identical.
Table 2 shows the mean Po_2 values of the Po_2 histograms, the
median and modul.

Since by the temperature increase from 18°C to 37°C the oxygen
consumption increases, the O_2 supply of the brain cortex becomes
worse i.e. the Po_2 values decrease.

CONCLUSION

This paper describes a polarographic Pt electrode to measure
high Po_2 values. The improved stability was achieved by a silver/
silver oxide electrode (produced in 30% H_2O_2) and by borax-biphosphate buffer as electrolyte. The electrode needs a polarizing period
of at least 10 hrs in saline equilibrated with 100% oxygen. The
usefulness of a Po_2 electrode, stable at high Po_2 values, was proved
by measurements of Po_2 histograms of blood-free perfused guinea pig
brain.

REFERENCES

Baumgärtl, H., and Lübbers, D.W., 1983, Microcoaxial needle sensor
    for polarographic measurement of local O_2 pressure in the
    cellular range of living tissue. Its construction and proper-
    ties, in: "Polarographic Oxygen Sensors", E. Gnaiger, H.
    Forstner, eds., Springer, Berlin-Heidelberg-New York, pp.
    36-65.
Clark, L.C., jr., 1956, Monitor and control of blood and tissue
    oxygen tension, Trans. Am. Soc. Art. Int. Organs, 2:41.
Feller-Kniepmeier, M., Feller, H.G., and Titzenthaler, E., 1967,
    Eletronenoptische und röntgenographische Untersuchungen an
    Silberkatalysatoren für Oxydation des Äthylens zu Äthylenoxid,
    Ber. Bunsenges Phys.-Chem., 71:606.
Fleischmann, M., Lax, D.J., and Thirsk, H.R.,1968, Electrochemical
    studies of the Ag_2O/AgO phase change in alkaline solutions,
    Trans Faraday Soc., 64:3137.
Gossner, K., and Polle, H., 1968, Über ein neues, bei Oberflächen-
    reaktionen entstehendes Silberoxid, Z. Phys. Chem., 54:93-100.
Heinrich, U., Yu, B., Hoffmann, J., and Lübbers, D.W., 1982,
    The effect of glucose on the oxygen supply of the blood-free
    perfused guinea pig brain as measured by reflection spectra
    and Po_2 histograms, in: "Oxygen Transport to Tissue V", D.W.
    Lübbers, H. Acker, T.K. Goldstick, E. Leniger-Follert, eds.,
    this volume.
Lübbers, D.W., 1977, Quantitative measurements and descriptions of
    oxygen supply to the tissue, in: "Oxygen and Physiological
    Function", F.F. Jöbsis, ed., Professional Information Library,
    Dallas, pp. 254-276.
Lübbers, D.W., Baumgärtl, H., Fabel, H., Huch, A., Kessler, M.,
    Kunze, K., Riemann, H., Seiler, D., and Schuchhardt, S., 1969,
    Principle of construction and application of various platinum
    electrodes, in: "Oxygen Pressure Recording in Gases, Fluids,
    and Tissues", F. Kreuzer, ed., Progr. Resp. Res., Vol. 3, pp.
    136-146, Karger, Basel-New York.

STEADY STATE CONDITION - WHAT DOES IT MEAN DURING INVESTIGATION OF

OXYGEN SUPPLY?

J. Hauss, H.-U. Spiegel, and K. Schönleben

Chirurgische Universitätsklinik Münster, FRG

Measuring local tissue Po_2 with platinum multiwire electrodes according to Kessler and Lübbers (1966) has become a well-established method in experimental research. At the Surgical Clinic of Münster University it was introduced first into clinical practice in 1975 by investigating intensive care patients (Hauss et al., 1978, 1978a, 1978b, 1979, 1980, 1980a, 1982).

Since that time by recording about 400 Po_2 histograms in 92 patients a lot was learned about the changes in the microcirculatory bed, for instance during shock situations or hypovolemia, during application of vasoactive drugs, during operations or during mechanical ventilation. But always - also in animal experiments - there are two difficulties: The first one is to decide precisely whether a Po_2 histogram is still "normal" reflecting physiologic conditions or is already "pathologically disturbed", when the borderline to a pathologic status is passed? And the second is: Everybody speaks about "steady state conditions" during an experiment, but when is this state really reached and how long can it be maintained? We did lots of experiments - in fact 12 different ways of anesthesia or analgesia were tested - to be able to produce steady state conditions and to maintain them for a longer period. Some results concerning animal investigations are presented.

MATERIAL AND METHODS

3 groups of 7 mongrel dogs (22-28 kg) were premedicated with 0.2 ml/kg Thalamonal[R]. 30 minutes after premedication, anesthesia was initiated with 0.7 mg/kg piritramide and 0.25 mg atropine sulfate given intravenously. After relaxation with 0.15 mg/kg

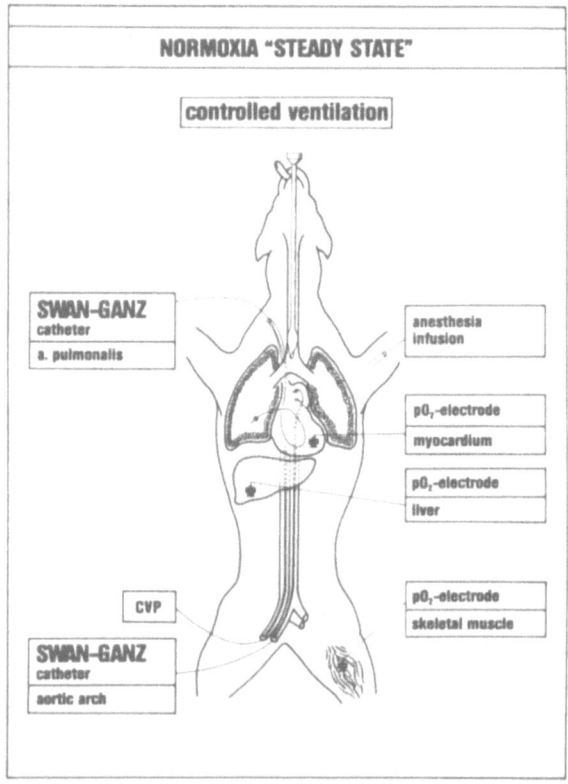

Fig. 1.   Position of catheters and electrodes during the investiga-
          tion.

Pancuronium bromide i.v. orotracheal intubation and controlled
ventilation by means of a Servo 900 C with 20.9% O_2 was performed.
During the entire investigation arterial Po_2 was stabilized at
70-90 mm Hg, arterial Pco_2 at 36-40 mm Hg, both were controlled by
blood gas analysis. Basic anesthesia was maintained by continuous
infusion of 1 mg/(kg·h) piritramide and 0.08 mg/(kg·h) Pancuronium
bromide according to Zimmermann et al. (1977).

     Hemodynamic and respiratory parameters were measured by cathe-
ters as demonstrated in Fig. 1. Catheters were inserted into the
pulmonary artery, left heart, and vena cava. Continuous measurements
recorded the systemic arterial pressure, pulmonary arterial pressure,
central venous pressure, heart rate, ECG, central and rectal temperature.

     Discontinuous measurements registered the cardiac output, wedge
pressure, arterial and mixed venous blood gas analysis, O_2 content,
hemoglobin concentration, hematocrit, electrolytes, lactate, and

Fig. 2.   The electrode on the surface of skeletal muscle.

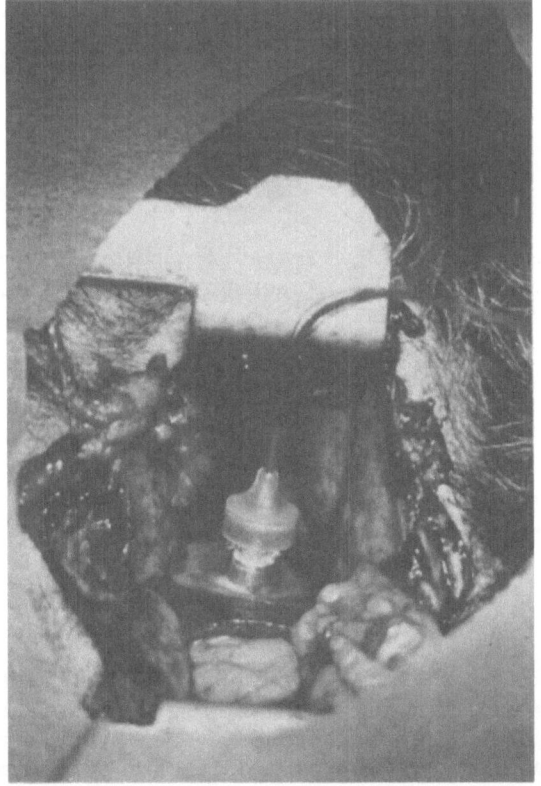

Fig. 3.   The electrode on the liver surface.

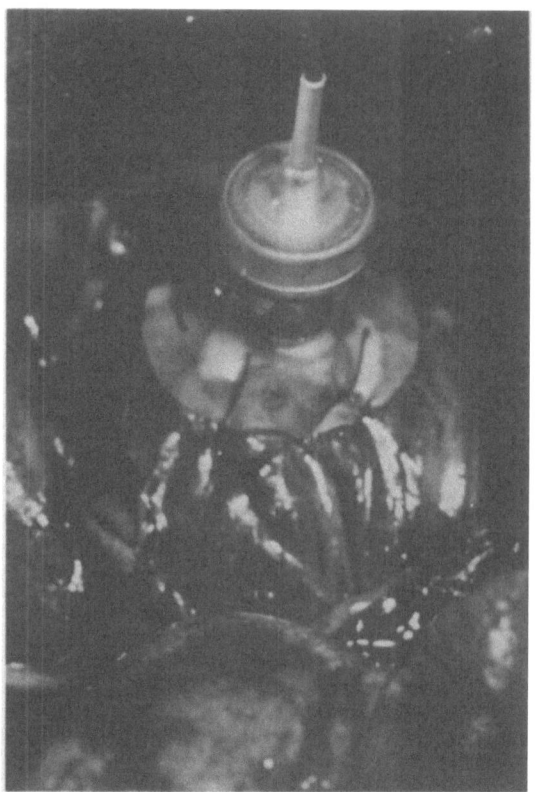

Fig. 4.   The electrode on the myocardium (left ventricle).

pyruvate. As an essential parameter, measurements with multiwire surface electrodes recorded tissue Po_2 on the gracilis muscle, on the liver, and on the left ventricle (Figs. 2, 3, 4).

In the diagram (Fig. 5) the chronologic course of the investigation is demonstrated:

Po_2 histograms were made at the beginning, and 1, 2, and 3 hrs later; continuous Po_2 measurements were taken over the entire period.

RESULTS

Mean arterial pressure, total peripheral resistance, mean pulmonary arterial pressure, and pulmonary resistance are shown in Fig. 6. The dashed lines represent the mean values calculated from seven dogs measured on skeletal muscle, the full-drawn lines represent the condition of dogs subjected to upper abdominal laparotomy

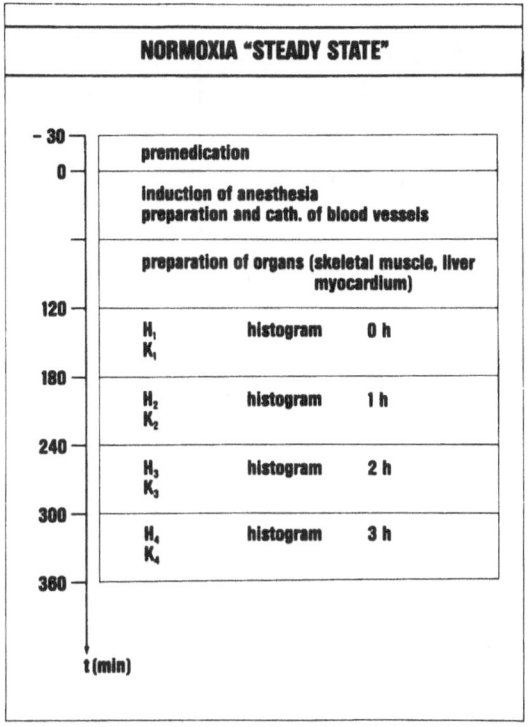

Fig. 5.   Experimental protocol and chronologic course of the
          investigation.

for exposure of the liver surface. The dotted lines show the mean
values of 7 dogs which, for measurements of myocardial Po_2, were
subjected to lateral thoracotomy in the fifth left intercostal space
and opening of the pericardium for access to the left ventricle.
Arterial and pulmonary pressure patterns and the assessed wedge
pressure values revealed remarkable stability and congruence in the
different groups of dogs. The rise of pulmonary wedge pressure in
the thoracotomized dogs was certainly provoked by the opening of
the thorax and reduced ventilation of the right lung.

      Behavior of heart rate and heart index, together with oxygen
consumption and oxygen transport are summarized in Fig. 7. Once
more it can be noted that heart rates in the thoracotomy group are
again on a higher level which reflects the severity of the surgical
intervention.

      Respiratory parameters are presented on the next diagram (Fig.
8). Arterial Po_2 in the thoracotomy group is relatively low due
to the above-mentioned reasons, but all other results show a striking stability over the whole period.

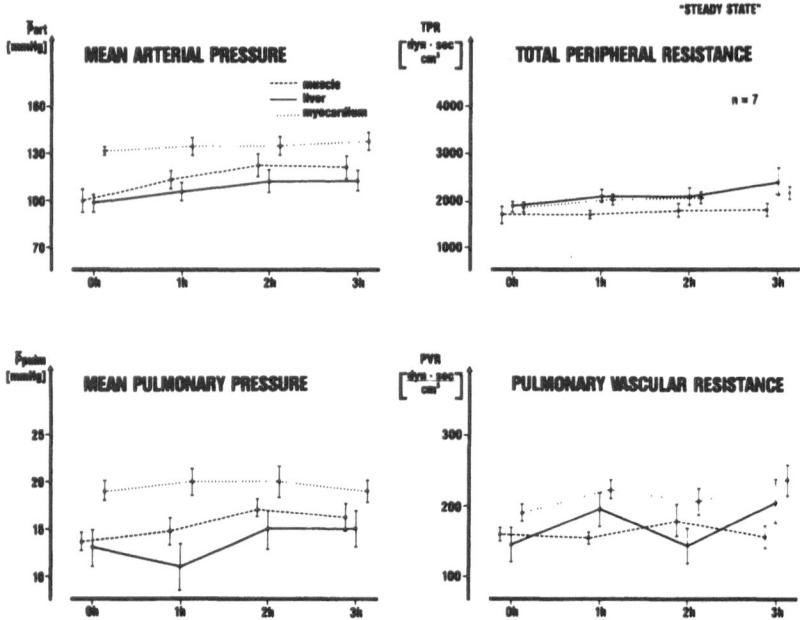

Fig. 6. Behavior of arterial pressure, resistance and pulmonary
         pressure, (n = 7 per group; X̄ + SEM).

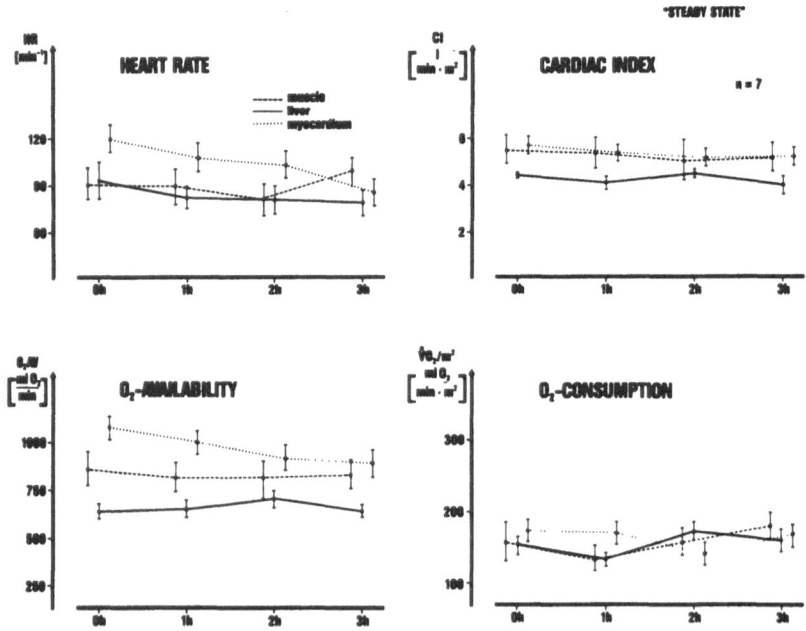

Fig. 7. Behavior of heart rate, cardiac index, oxygen availability
         and oxygen consumption (n = 7 per group, X̄ + SEM).

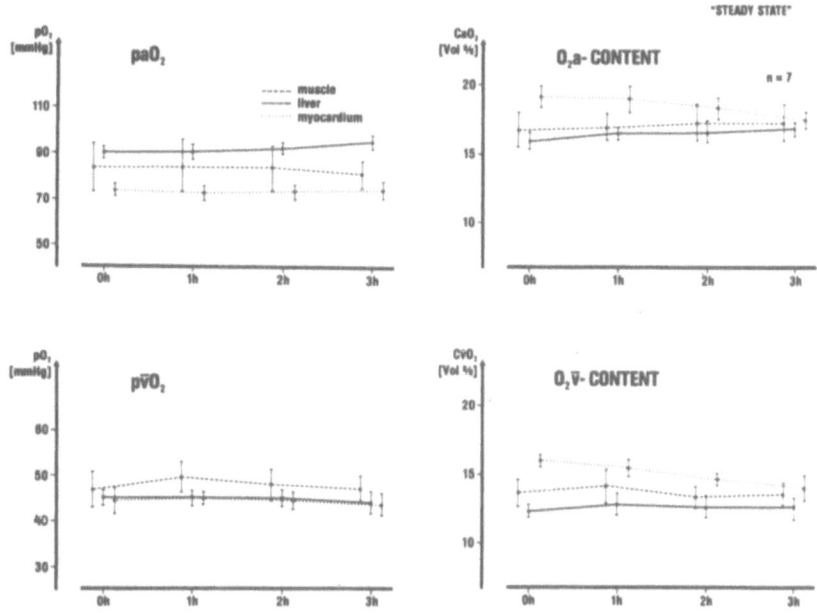

Fig. 8.   Respiratory parameters of the muscle, liver and myocardium
          group (n = 7 per group; X̄ + SEM).

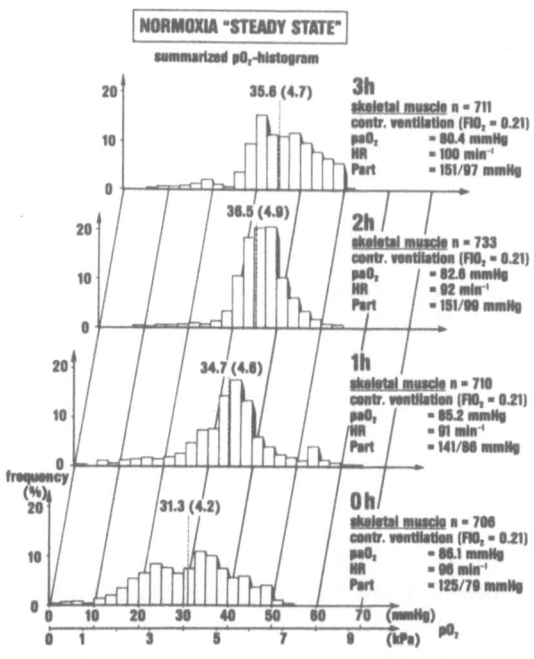

Fig. 9.   Summarized Po₂ histograms of the "skeletal muscle group".

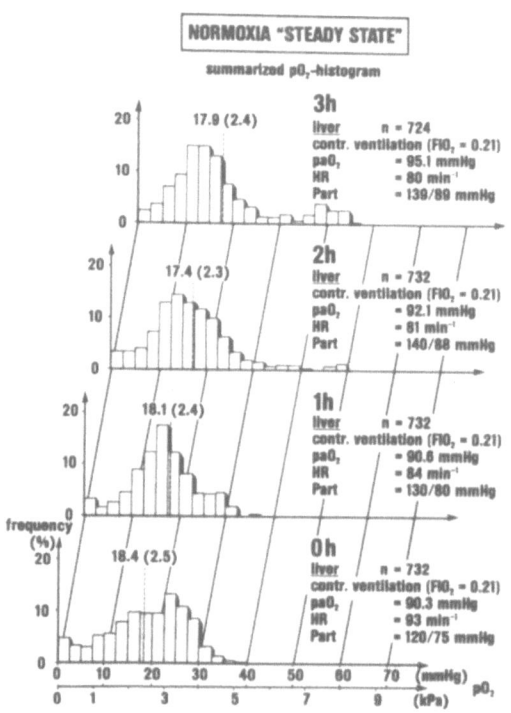

Fig. 10.   Summarized Po$_2$ histograms of the "liver group".

The summarized Po$_2$ histograms of each group, first of the 7 dogs whose skeletal muscle Po$_2$ was recorded, are demonstrated in Fig. 9. At the bottom is the basic Po$_2$ histogram, above the subsequent histograms taken 1, 2 or 3 hrs later. The mean values of some other relevant parameters recorded simultaneously, are listed on the right side. The basic Po$_2$ histogram has a mean value of 31.3 mm Hg. No significant alterations were recorded over the whole period of the experiment.

The results of the liver group were similar (Fig. 10): Nevertheless, we noted that all Po$_2$ histograms had some classes with anoxic or near-anoxic values. Such findings were confirmed by other authors, for instance by the Messmer group (1973). In the myocardium animals too, the mean values of summarized Po$_2$ histograms were practically unaltered (Fig. 11).

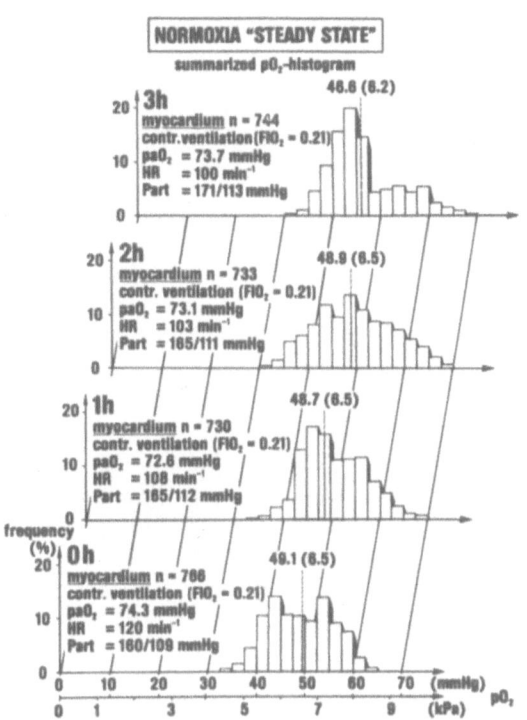

Fig. 11. Summarized Po$_2$ histograms of the "myocardium group".

SUMMARY

Judging from the presented data it was possible to maintain steady state conditions during high dose piritramide anesthesia for a period of three and a half hours. Under these carefully defined experimental conditions it will be possible to actually correlate any changes observed in the microcirculatory bed with medication or other interventions. So-called "normal Po$_2$ histograms" taken from different organs of dogs under the above-defined experimental conditions (basic anesthesia with high-dose piritramide, controlled artificial ventilation with 21% O$_2$) are presented in Fig. 12. We feel that any comparison of results from different working groups must be based on exactly defined and standardized experimental conditions, because otherwise the methodological differences would prevent adequate comparability.

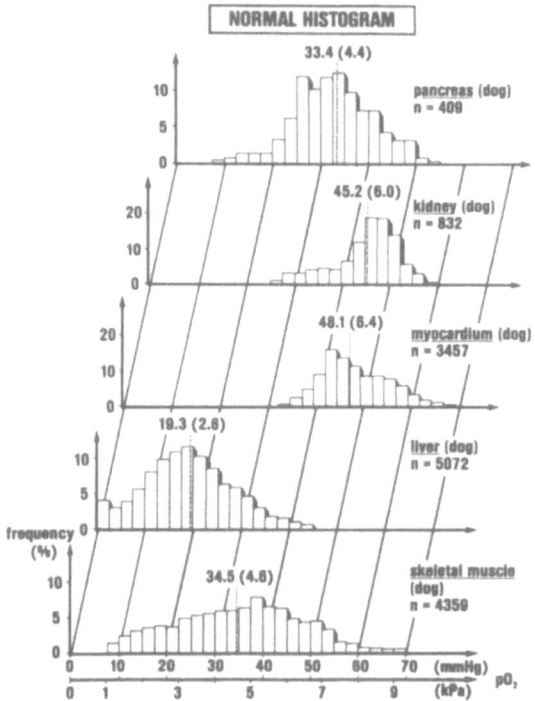

Fig. 12.   "Normal Po_2 histograms" taken from different organs of
           dogs.

REFERENCES

Hauss, J., Schönleben, K., Bünte, H. Spiegel, H.-U., Wendt, M., and
    Themann, H., 1980, Nitroglycerin and nitroprusside induced
    hypotension, Drug Res., 30 (II) 12: 2204-2221.
Hauss, J., Schönleben, K., and Spiegel, H.-U., 1982, Therapiekon-
    trolle durch Überwachung des Gewebe-Po_2, Hauss, ed., Huber,
    Bern-Stuttgart-Wien.
Hauss, J., Schönleben, K., Spiegel, H.-U., and Kessler, M., 1978,
    Measurement of local oxygen pressure in skeletal muscle of
    patients suffering from disturbances of the arterial circula-
    tion, in: "Oxygen Transport to Tissue", I.A. Silver, M.
    Erecińska, H.J. Bicher, eds., Plenum Press, New York-London,
    pp. 419-422.
Hauss, J., Schönleben, K., Spiegel, H.-U., Wendt, M., and Hartenauer,
    U., 1978a, Die kontrollierte Hypotension mit Natriumnitro-
    prussid, Herz/Kreisl., 8:379-387.

Hauss, J., Schönleben, K., Spiegel, H.-U., Wendt, M., and Hartenauer, U., 1978b, Änderung der Hämodynamik und der lokalen Sauerstoffversorgung der Gewebe während der kontollierten Hypotension mit Natriumnitroprussid, in: "Verhandlungsband der Jahrestagung der Deutschen Gesellschaft für Angiologie", Band 3, Hild, ed., Witzstrock, Wiesbaden.

Hauss, J., Schönleben, K., Spiegel, H.-U., Wendt, M., and Hartenauer, U., 1979, Die Beeinflussung der Hämodynamik und der Mikrozirkulation durch kontollierte Hypotension mit Natriumnitroprussid, in: "Nitrate, Wirkung auf Herz und Kreislauf", W. Rudolph, ed., Urban und Schwarzenberg, München-Berlin-Wien, pp. 74-78.

Hauss, J., Schönleben, K., Wendt, M., Hartenauer, U., and Spiegel, H.-U., 1980a, Die Wirkung der kontrollierten Hypotension mit Natriumnitroprussid auf Hämodynamik und Mikrozirkulation, in: "Intensivmedizin-Notfallmedizin-Anästhesiologie, Band V, P. Lawin, V. von Loewenich, H. Stoeckel, eds., Thieme, Stuttgart, pp. 115-125.

Kessler, M., and Lübbers, D.W., 1966, Aufbau und Anwendungsmöglichkeit verschiedener Po_2-Elektroden, Pflügers Arch Ges. Physiol., 291:R82.

Messmer, K., Sunder-Plassmann, L., Jesch, F., Görnand, L., Sinagowitz, E., and Kessler, M., 1973, Oxygen supply to the tissue during limited normovolemic hemodilution, Res. exp. Med., 159:152.

Zimmermann, G., Hess, W., Johannsen, H., and Patschke, D., 1977, Der Einfluß der inspiratorischen N_2O-Konzentration auf das kardiovaskuläre System. Tierexperimentelle Untersuchungen in hochdosierter Piritramid-Basisnarkose, Anaesthesist, 26:257-263.

EVIDENCE FOR BOUNDARY LAYER EFFECTS INFLUENCING THE SENSITIVITY OF

MICROENCAPSULATED O_2 FLUORESCENCE INDICATOR MOLECULES

N. Opitz, and D.W. Lübbers

Max-Planck-Institut für Systemphysiologie
Rheinlanddamm 201, D-4600 Dortmund, FRG

As Kautsky and Hirsch (1935) and later also Vaughan and Weber (1970) have shown fluorescence quenching by molecular oxygen is often a diffusion-controlled collisional process. Thus, it is not surprising that the oxygen solubility coefficient , and the oxygen diffusion coefficient D of various solvents are of great importance with regard to the O_2 sensitivity of the indicator molecules. Fig. 1 shows this effect on different calibration curves for the indicator pyrenebutyric acid dissolved in alkaline water and di-octylphthalate and embedded in a silicone membrane. Calibration has been performed by means of the so-called "Po$_2$ optode", a schematic drawing of which is shown in Fig. 2 (Lübbers and Opitz, 1975a, b; Opitz and Lübbers, 1976). This optical device incorporates a thin indicator film (1 - 5 um) between a radiation permeable quartz window and a membrane which separates the fluorescence probes from the medium to be measured (i.e., different partial oxygen pressures in N_2-O_2 gas mixtures).

We have also used microencapsulated droplets (Fig. 3) (Lübbers et al., 1977) containing PBA dissolved in dioctylphthalate, encapsulated by polyurethane, and suspended or embedded into the above mentioned solvents or matrix. Contrary to our expectations, we obtained calibration curves whose sensitivities deviated significantly from that of the free indicator dissolved in dioctylphtha-late (Fig. 4). This has also been found for nanocapsules prepared according to an encapsulation process described by Podgorski et al. (1981). These latter capsules contained pyrene dissolved in paraffin oil, which had been encapsulated in polyacrylamide. Contrary to our observations, those authors reported that the sensitivity of the capsules did not change when altering the O_2 solubility coefficient of the medium in which they were suspended. Because we also

899

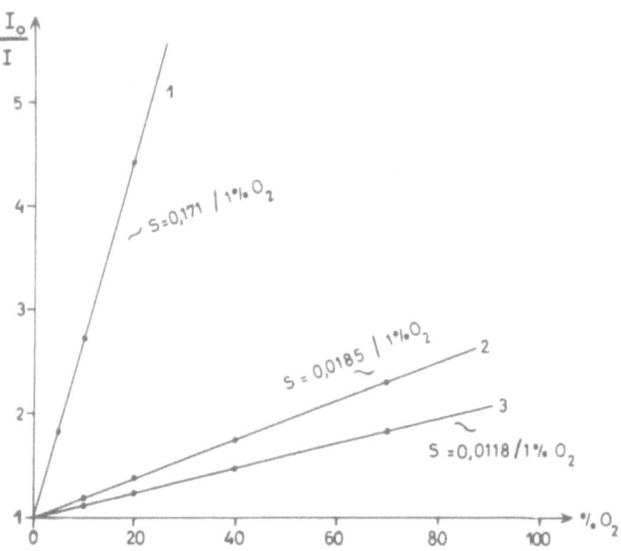

Fig. 1. Calibration curves for PBA embedded in 1) silicone mem-
         brane and dissolved in 2) dioctylphthalate and 3) alkaline
         H_2O. S = Steepness of the calibration curve.

observed varying oxygen sensitivity with different membrane materials, i.e., teflon, polycarbonate, polyurethane and copolyetherester,
using an optode with an indicator film of PBA in dioctylphthalate,
we have assumed that boundary layer effects are responsible for
these effects on capsule oxygen sensitivity.

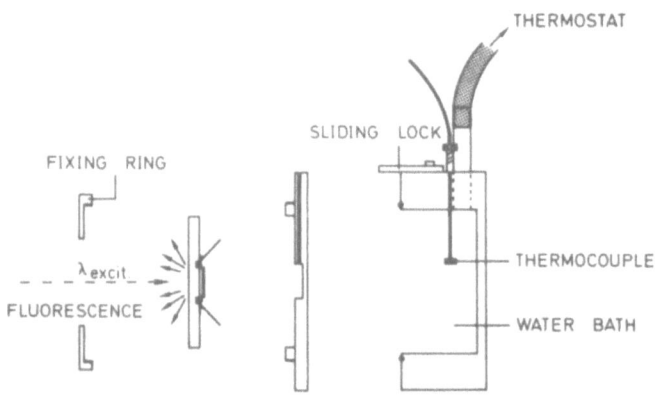

Fig. 2. Schematic representation of the Po_2 optode

Fig. 2. Schematic representation of the Po_2 optode

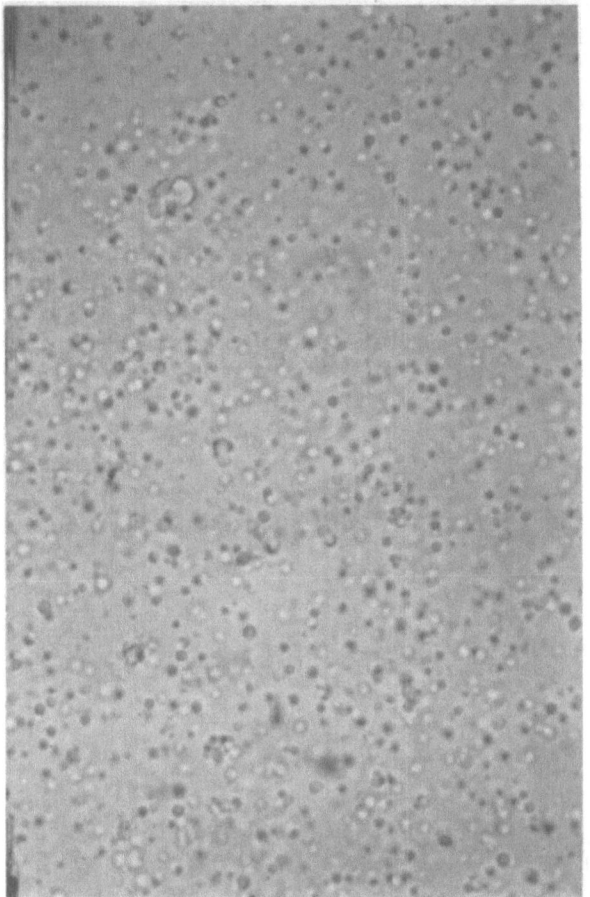

Fig. 3. Light microscopic photograph of a microcapsule preparation
         containing PBA dissolved in dioctylphthalate.

     To obtain a better understanding of the observed phenomenon,
we used the following model for a theoretical analysis. Fig. 5 shows
an enlarged view of a section of the optode with the indicator layer
adjacent on one side to glass and on the other to a membrane or
medium to be measured.

     Within the indicator layer there are statistically distribut-
ed indicator molecules. The mean diffusion paths of the O_2 molecules
are spherical with a mean radius R, given by $R = \sqrt{6 \cdot D_I \cdot \tau_0}$, where
D_I is the oxygen diffusion coefficient of the indicator layer and
τ_0 is the mean lifetime of the excited state of the indicator
molecule in the absence of oxygen. Thus, oxygen molecules within
this sphere are principally capable of quenching the excited state
of the fluorescence indicator molecule via diffusion by a collisional

process during its mean lifetime of excitation. Indicator molecules
which are in the boundary zones within the indicator layer show a
partial overlap of spheres with the adjacent materials which have
different α's and D's. In the case of glass, a spherical segment is
lost, since the α and D of glass can be assumed to be zero. In the
case of a membrane, the spherical segment is diminished for $D_m < D_I$
or enlarged for $D_m > D_I$. Compared to the indicator molecules in the
deeper layers of the indicator film, the number of oxygen molecules
which could disturb the excited state of the fluorescence molecule
is therefore changed for indicator molecules within the boundary
zone, however, the theory shows that there is an exception in the case
that $\alpha_m^2 \cdot D_m \approx \alpha_I^2 \cdot D_I$. Thus the effective oxygen concentration becomes
dependent on the position of the indicator molecule within the bound-
ary zone and therefore, the Stern–Volmer equation, $I_0/I = 1 + K \cdot Co_2$
(Stern and Volmer, 1919) which is valid only for deeper layers of
the indicator film outside of the boundary zones, cannot be applied.
With the above mentioned assumptions and for a homogeneous distribution of indicator molecules within the boundary zone, one finds
instead:

$$\frac{I_0}{I} = 1 + K \cdot Po_2 \cdot (\alpha_I \cdot \frac{13}{16} + \alpha_m \cdot \frac{1}{32} \cdot (\sqrt{(\frac{D_m}{D_I})^3} + 5 \cdot \sqrt{\frac{D_m}{D_I}})) \quad (1)$$

with: K   = "overall" quenching constant;

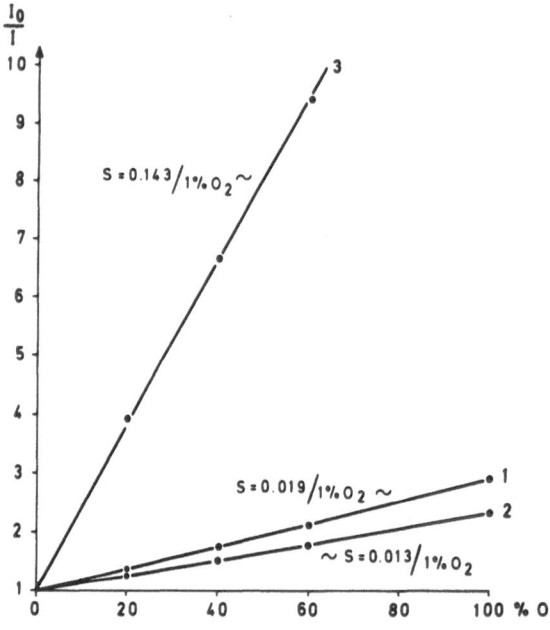

Fig. 4.   Calibration curves for microcapsules suspended in 1)
          dioctylphthalate and 2) H_2O and embedded into a 3) sili-
          cone membrane. S = Steepness of the calibration curve.

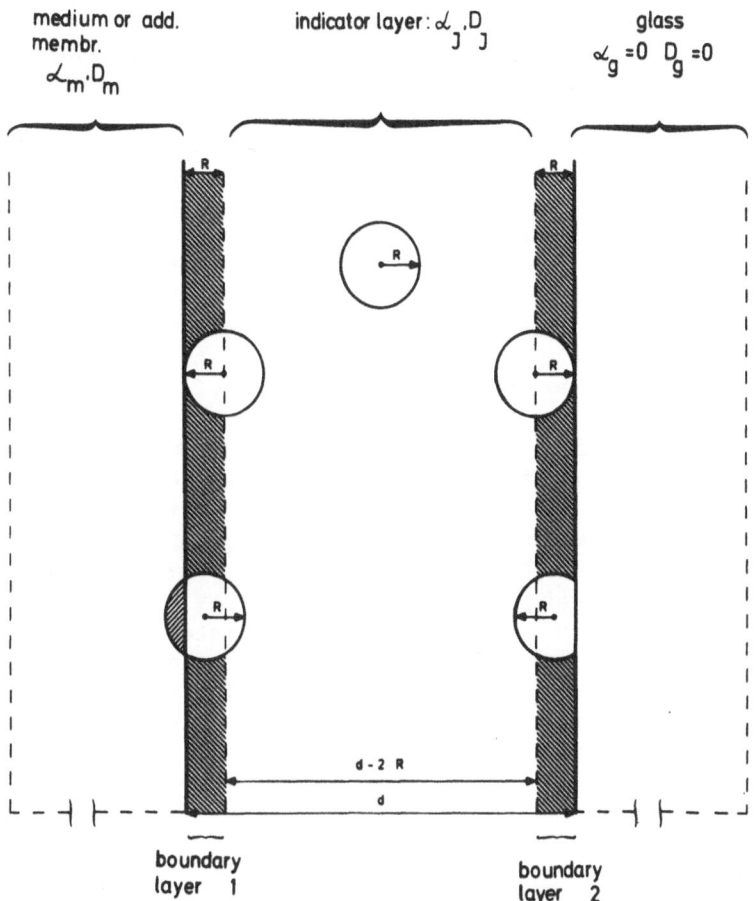

Fig. 5.   Schematic enlargement of the indicator layer between glass
          and membrane (see text).

From equation (1) one recognizes that, in the trivial case
when $\alpha_I = \alpha_m$ and $D_I = D_m$, the boundary layer influence vanishes.
Furthermore, equation (1) states that, within the boundary layer,
the oxygen sensitivity of the fluorescence probes is determined by
the coefficients α_I and D_I of the matrix in which they are dissolved as well as by the coefficients α_m and D_m of the adjacent
material.

With respect to boundary layer effects one now obtains for the
whole indicator layer of thickness d:

$$\frac{I_0}{I} = 1 + K \cdot Po_2 \cdot \frac{R}{d} \cdot (\alpha_I \cdot (\frac{d}{R} - \frac{3}{8}) + \alpha_m \cdot \frac{1}{32} \cdot (\sqrt{(\frac{D_m}{D_I})^3} + 5 \cdot \sqrt{\frac{D_m}{D_I}}))   \quad (2)$$

From equation (2) it follows that boundary layer effects are negligible for d   R. This could be experimentally proved with rather thick indicator layers of d   20 um.

According to our theoretical considerations and experimental results, we want to point out that, for micro- and especially for nano-encapsulated O_2 sensitive fluorescence molecules, boundary layer effects which influence the sensitivity of the capsules increase with decreasing capsule diameter and/or shell thickness. According to our theoretical model, the effect will be greater for capsules, since the overlapping volume of the spheres with the adjacent material, and moreover with the medium to be measured, is greater due to the geometrical configuration of the capsules.

The effect of boundary layers could be reduced by encapsulating solvents with low O_2 diffusion coefficients which, unfortunately, would simultaneously reduce the sensitivity, so that a higher O_2 solubility coefficient would be necessary to compensate for this loss. In order to eliminate the influence of body fluids and tissues on capsule oxygen measurements, it is necessary to produce shell thickness greater than $6 \cdot D_m \cdot 0$ (i.e., greater than about 0.1 um) so that the spheres around the indicator molecules will not overlap with the measuring medium. Since capsule shells have thicknesses of about 10 nm, the above requirement could be achieved by an additional coating of the capsules with a proper polymer material that would preserve the oxygen sensitivity of the capsule.

ACKNOWLEDGEMENT

We are indebted to the Research Institute Obernburg of ENKA AG, 5600 Wuppertal, FRG for providing polymer materials and for advice and discussion.

REFERENCES

Kautsky, H., and Hirsch, A., 1935, Original observation of effect-dyes absorbed on silica gel, Z. Anorg. Allg. Chem., 222: 126-134.

Lübbers, D.W., and Opitz, N., 1975a, Die Pco$_2$-/Po$_2$-Optode: Eine neue Pco$_2$- bzw. Po$_2$-Meßsonde zur Messung des Pco$_2$ oder Po$_2$ von Gasen und Flüssigkeiten, Z. Naturforsch., 30c:532.

Lübbers, D.W., and Opitz, N., 1975b, The "Po$_2$-optode", a new tool to measure Po$_2$ of biological gases and fluids by quantitative fluorescence photometry, Pfügers Arch., 359:R145.

Lübbers, D.W., Opitz, N., Speiser, P.P., and Bisson, H.J., 1977, Nanoencapsulated fluorescence indicator molecules measuring pH and Po$_2$ down to submicroscopical regions on the basis of the optode-principle, Z. Naturforsch., 32c:133.

Opitz, N., and Lübbers, D.W., 1976, Simultaneous measurement of blood gases by means of fluorescence indicators, Pflügers Arch., 362:R52.

Podgorski, G.T., Longmiur, I.S., Knopp, J.A., and Benson, D.M., 1981, Use of an encapsulated fluorescent probe to measure intracellular Po_2, J. Cell Physiol., 107:329.

Stern, O., and Volmer, M., 1919, Über die Abklingungszeit der Fluoreszenz, Z. Physik, 20:183.

Vaughan, W.M., and Weber, G., 1970, Oxygen quenching of pyrenebutyric acid fluorescence in water. A dynamic probe of the microenvironment, Biochemistry, 9:464.

A CORRECTION METHOD FOR IONIC STRENGTH-INDEPENDENT FLUORESCENCE PHOTOMETRIC pH MEASUREMENT

N. Opitz, and D.W. Lübbers

Max-Planck-Institut für Systemphysiologie
Rheinlanddamm 201, D-4600 Dortmund 1, FRG

The pH fluorescence indicator molecules could serve as sensitive sensors for an optical pH measuring technique in various body fluids and tissues (Tageeva et al., 1971; Lübbers and Opitz, 1976; Opitz, 1976; Lübbers et al., 1977; Peterson et al., 1980), if one could eliminate ionic strength interference which causes spectral changes indistinguishable from those caused by pH. To obtain the true optically measured pH value, a correction method for ionic strength changes has been developed based on a double pH indicator system, i.e., hydroxypyrenetrisulfonic acid (HPTS, introduced by Wolfbeis et al., 1983) and ß-methylumbelliferone (ß-M), (Bülow and Dieck, 1928; Chen, 1968; Nakashima et al., 1972).

Figs. 1 and 2 show several corrected excitation spectra of the indicators at varying pH and constant ionic strength and temperature, found with the spectrofluorometer developed by Boldt (1971) and Lübbers. Evaluation of the spectra is done by taking the quotient of the signal at two wavelengths, λ-evaluation and λ-isosbestic, which, after normalization against the maximal quotient, is plotted versus pH (Fig. 3). As the plots show, HPTS and ß-M behave like weak electrolytes, which can be described by pH dependent dissociation curves, $\alpha(pH) = (1 + 10^{pK-pH})^{-1}$. Thus, a linearization of the pH calibration curves can be achieved by plotting $\log((1-\alpha)/\alpha)$, or by plotting the fluorescence signals, $\log((S_d-S_x)/S_x)$, versus pH (Fig. 4).

At constant pH and temperature, the spectral changes with ionic strength depend on the valence of the indicator. Compared to ß-M (valence, -1), the ionic strength influence on HPTS (valence, -4) is rather strong. Fig. 5 shows the effect of ionic strength on HPTS. However, the influence is nevertheless signifi-

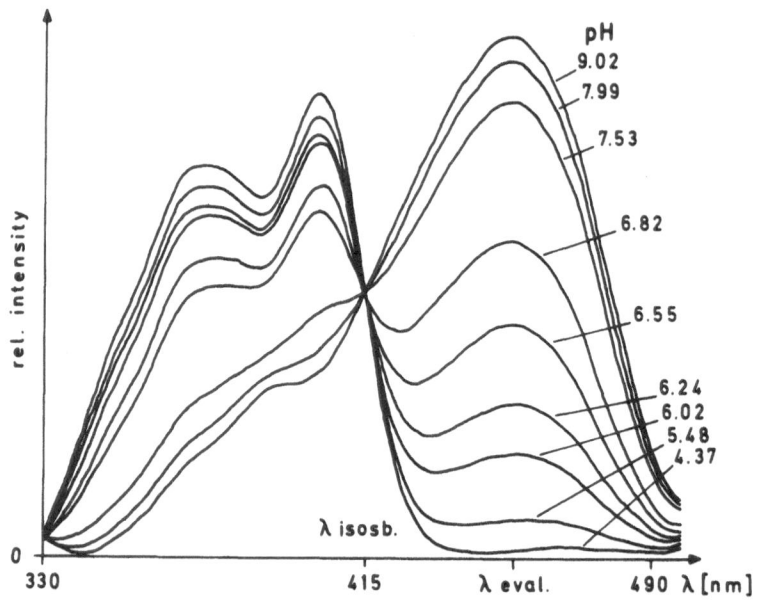

Fig. 1. Corrected excitation spectra of the indicator HPTS at
         varying pH and constant ionic strength and temperature,
         (λ_{em} = 510 nm) (Wolfbeis et al., 1983).

cant for ß-M. Since one cannot discriminate between ionic strength
and pH induced spectral changes, a variation in ionic strength at
constant pH and temperature yields an apparent pH value, pH'. The
apparent pH change, $\Delta pH = pH'-pH$, is caused by a pK shift ΔpK of
the pH indicator. The dependence of pK on ionic strength follows
from Debye-Hückel-Onsager's theory (Rauen, 1964):

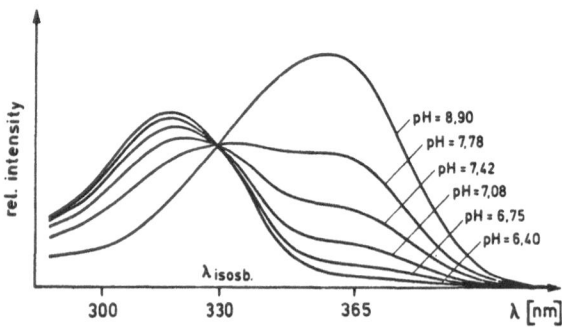

Fig. 2. Corrected excitation spectra of the indicator ß-M at
         varying pH and constant ionic strength and temperature,
         (λ_{em} = 445 nm).

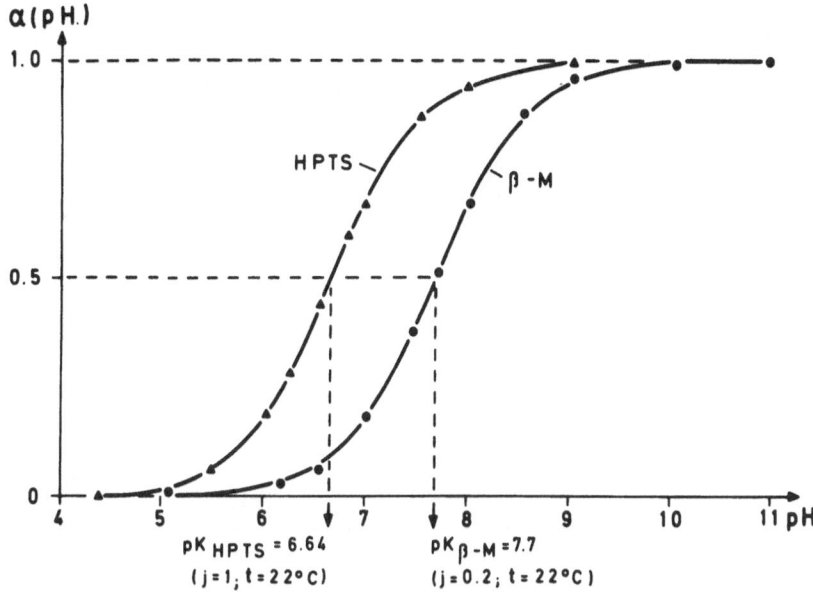

Fig. 3. Dissociation curves of the indicators HPTS and ß-M. (For explanation see text).

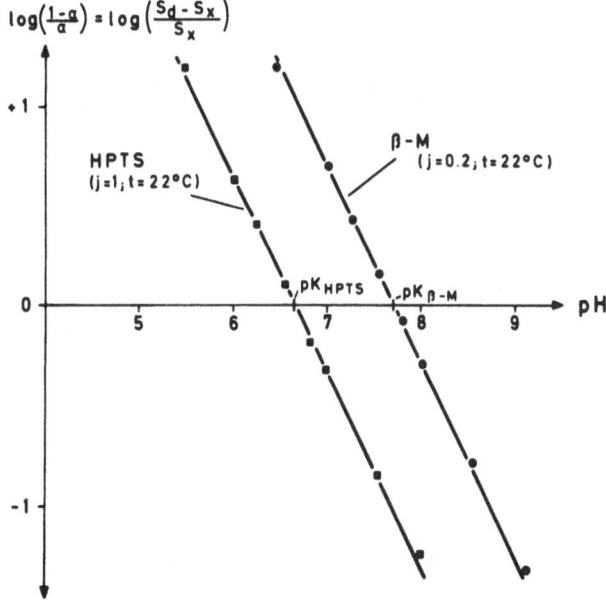

Fig. 4. Linearized pH calibration curves for HPTS and ß-M. (For explanation see text).

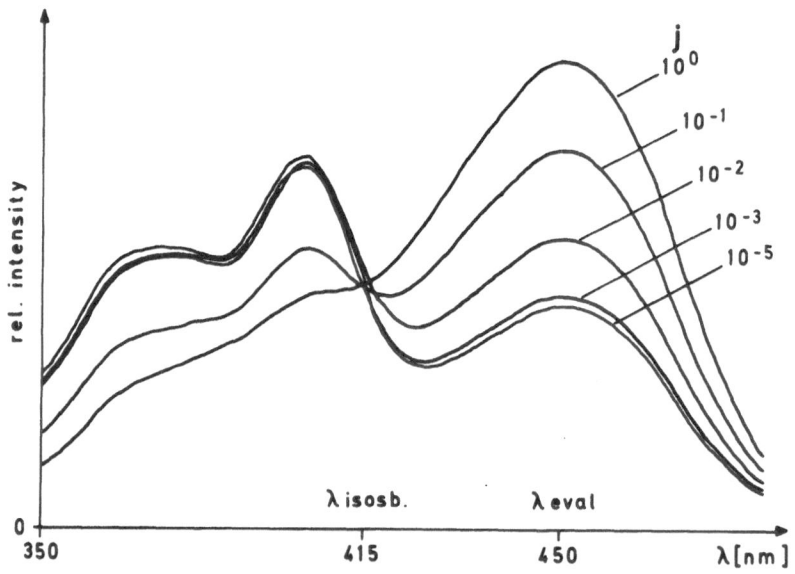

Fig. 5.   Spectral changes of HPTS with ionic strength, j, for both
          constant pH and temperature (λ_{em} = 510 nm).

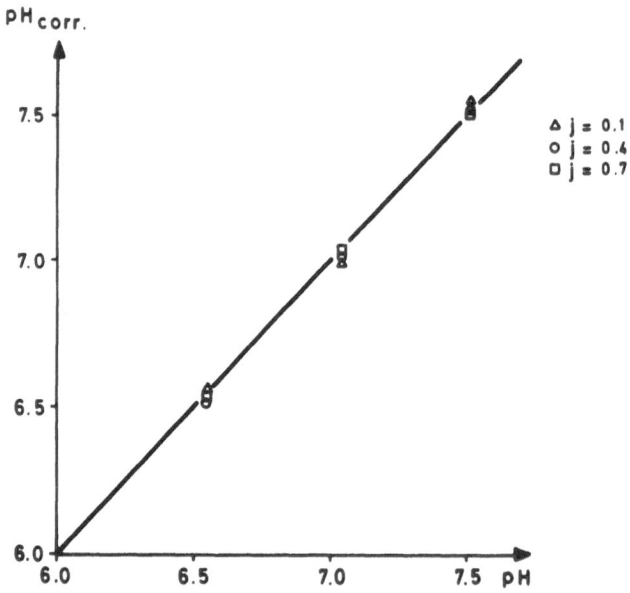

Fig. 6.   Corrected fluorescence measured pH values versus pH values
          measured with a glass electrode for different ionic
          strengths.

$$\Delta pK = -(2 \cdot v + 1) \cdot f(j) \tag{1}$$

with: $f(j) = A \cdot \sqrt{j}/(1 + B \cdot d \cdot \sqrt{j})$ and $j = (1/2) \cdot \sum_i c_i \cdot v_i^2$

where: j = ionic strength; v = valence; A = temperature dependent constant; B, d = molecule parameters.

Thus, taking into account that $f_{HPTS}(j) \approx f_{\beta-M}(j)$, one finds for the optically measured corrected pH value with the indicator pair HPTS and ß-M:

$$pH = pH'_{\beta-M} - 0.17 \cdot (pH'_{HPTS} - pH'_{\beta-M}) \tag{2}$$

Equation (2) states that a double pH fluorescence indicator system, e.g., HPTS and ß-M allows an ionic strength-independent fluorometrical pH measurement. Fig. 6 shows preliminary results of fluorescence measured pH values corrected according to equation (2) versus pH values measured with a glass electrode for three different ionic strength values. As can be seen, the corrected measured points are well within the theoretically expected curve. Table (1) represents an example of uncorrected pH' values of HPTS and ß-M.

The pH measuring range of this method is determined:
1. by the overlapping range of sensitivities of the indicators, which can be approximately found from Fig. 3 between pH 6 and pH 9,
2. by the required resolution range in pH determination and
3. by the signal-to-noise ratio of the fluorescence photometrical detection system.

On the basis of this correction method, optical fluorescence pH measurements by means of optode devices (Lübbers and Opitz, 1975) should be reliable. Moreover, the correction principle may be applicable to similar problems in fluorescence photometry, e.g., with fluorescing chelate complexes measuring Na^+, K^+ or Cl^- ion activities, but these unfortunately suffer from low specifity.

Table 1. Uncorrected pH' values of HPTS and ß-M.

j	pH electrode	pH'_{HPTS}	$pH'_{\beta-M}$	
0.1	7.51	7.02	7.45	
0.4	7.51	7.32	7.48	
0.7	7.51	7.46	7.50	
1	7.51	7.51	7.51	calibration

REFERENCES

Boldt, M., 1971, Ein korrigiert messendes Spektrofluorometer zur
    Messung an trüben Medien, Dissertation, Marburg/Lahn.
Bülow, C., and Dieck, W., 1928, ß-Methylumbelliferon als fluores-
    zierender Indikator, Fresenius Z. Anal. Chem., 75:81-86.
Chen, R.F., 1968, Fluorescent pH indicators: Spectral changes of
    4-methylumbelliferone, Anal. Letters, 1(7):423-428.
Lübbers, D.W., and Opitz, N., 1975, Die Pco₂-/Po₂-Optode: Eine neue
    Pco₂- bzw. Po₂-Meßsonde zur Messung des Pco₂ oder Po₂ von
    Gasen und Flüssigkeiten, Z. Naturforsch., 30c:532-533.
Lübbers, D.W., and Opitz, N., 1976, Quantitative fluorescence
    photometry with biological fluids and gases, in: "Oxygen
    Transport to Tissue II", Adv. Exp. Med. Biol., 75, J. Grote,
    D. Reneau, G. Thews, eds., Adv. Exp. Med. Biol., 75, Plenum
    Press, New York and London, pp. 65-68.
Lübbers, D.W., Opitz, N., Speiser, P.P., and Bisson, H.J., 1977,
    Nanoencapsulated fluorescence indicator molecules measuring pH
    and Po₂ down to submicroscopical regions on the basis of the
    optode-principle, Z. Naturforsch., 32c:133-134.
Nakashima, M., Sousa, J.A., and Clapp, R.C., 1972, Spectroscopic
    species of 4-methylumbelliferone in water and ethanol, Nature
    (London), 235:16.
Opitz, N., 1976, Messung biologischer Gase und Flüssigkeiten mit
    Hilfe der quantitativen Fluoreszenzfotometrie mit Indikatoren,
    Dissertation, Marburg/Lahn.
Peterson, J.I., Goldstein, S.R., and Fitzgerald, R.V., 1980, Fiber
    optic pH probe for physiological use, Anal. Chemistry, 52:
    864-869.
Rauen, H.M., 1964, "Biochemisches Taschenbuch" (2. Teil), Springer,
    Berlin, Göttingen, Heidelberg, New York.
Tageeva, S.V., Kosheleva, G.N., and Dubrov, A.P., 1971, Determina-
    tion of hydrogen ions concentration in some biological objects
    with fluorescent indicators (russ), Tsitologiia, 13(1):
    122-125.
Wolfbeis, O.S., Fürlinger, E., Kroneis, H., and Marsoner, H., 1983,
    Fluorimetric analysis: 1. A study on fluorescent indicators
    for measuring near neutral ("physiological") pH-values, Fres.
    Z. Anal. Chemie, 314:119-124.

INFLUENCE OF ENZYME CONCENTRATION AND THICKNESS OF THE ENZYME LAYER

ON THE CALIBRATION CURVE OF THE CONTINUOUSLY MEASURING GLUCOSE

OPTODE

N. Uwira, N. Opitz, and D.W. Lübbers

Max-Planck-Institut für Systemphysiologie
Rheinlanddamm 201, 4600 Dortmund 1, FRG

INTRODUCTION

Since the first construction of enzyme electrodes by Updike and Hicks (1967) and Clark and Sachs (1968) bioelectrodes have been used in physiology, for example for the determination of glucose in blood and serum. One of the main advantages is that measurements can be made continuously for longer times. In these applications an enzyme reaction is coupled to the electrode process. A glucose electrode consists of a membrane of immobilized glucose oxidase, E.C. 1.1.3.4.. The enzyme catalyzes the oxidation of glucose to gluconic acid by dissolved oxygen from blood or serum. The difference in Po_2 generated by the enzyme layer is a measure for glucose concentrations. The Clark electrode uses the reaction product H_2O_2.

$$\text{glucose} + O_2 \xrightarrow{\text{glucose oxidase}} \text{gluconic acid} + H_2O_2 \qquad (1)$$

To avoid enzyme inactivation by hydrogen peroxide catalase E.C. 1.11.1.6. can be immobilized in the same enzyme layer, but sensitivity gets lower as O_2 is generated.

$$\text{glucose} + O_2 \xrightarrow[\text{catalase}]{\text{glucose oxidase}} \text{gluconic acid} + 1/2\ O_2 + H_2O \qquad (2)$$

The electrical device of electrodes can be substituted by an optical device, which is also sensitive to oxygen. Lübbers and Opitz (1975) used the fluorescent dye pyrene butyric acid in bis-2-ethyl-hexylphthalate, which is quenched by oxygen molecules

913

as described by Vaughan and Weber (1970). Lübbers and Opitz (1975) called the fluorescent oxygen sensor "optode". This sensor was used by Völkl et al. (1980; Völkl, 1980) for substrate determinations like an electrode by coupling it to an oxygen consuming enzyme reaction.

The indicator reaction itself is an equilibrium process and does not consume oxygen. The oxygen uptake from the sample is determined only by the diffusional flux through the membranes and the enzyme reaction.

The generation of a signal change is caused by a complex process; diffusional and solubility characteristics of the substrates in the different membranes are involved as well as the kinetic characteristics of the immobilized enzyme. Since understanding of the membrane processes is one of the most important prerequisites for an optimal optode construction, special optodes with glucose oxidase were tested to see the effects of the concentrations of both substrates, i.e. oxygen and glucose, and of the thickness of the enzyme layer and of the enzyme concentration in the layer in detail.

MATERIALS AND METHODS

Glucose oxidase (grade II), catalase and NADH were obtained from Boehringer, Mannheim, Ethocel from FLUKA, Neu-Ulm, bovine serum albumin (BSA) from SERVA, Heidelberg, and glutaraldehyde from Sigma, München, D+glucose and Bis(2-ethylhexyl)phthalate from Merck, Darmstadt. All chemicals were of analytical grade.

Enzyme immobilization by glutaraldehyde in BSA matrix was done according to Thomas et al. (1972).

Glucose oxidase 2-25 mg/ml, catalase 0.4 mg/ml, and BSA 47-70 mg/ml were dissolved in 0.1 mol$\cdot$l^{-1} phosphate buffer pH 7.2. Immediately after mixing with 20 ul/ml glutaraldehyde a defined volume of the viscous immobilization mixture was spread on a cuprophane membrane (12 um) (see below). The mixture was left overnight at room temperature in a humid atmosphere. The enzyme layer was rinsed by immersing the cuprophan support into phosphate buffer which was changed three to four times.

Different enzyme concentrations in a layer of approximately the same thickness were obtained by spreading equal volumes of the immobilization mixture to the same area. The concentrations of glucose oxidase and BSA were changed in a way that the total protein content was kept constant.

Enyzme layers of different thickness were obtained by spreading different volumes of the immobilization mixture to the same area. The area was limited by a square of adhesive tape on the cuprophane membrane. In a typical experiment 0.5 ml of the immobilization mixture were spread to 36 cm^2. The actual thickness after immobilization was not determined.

Enyzme optodes were constructed according to Völkl et al. (1980) but using the immobilization method described above. One drop of a highly viscous solution of 10^{-2} mol pyrene butyric acid in Bis-(2-ethylhexyl)phthalate containing 2% ethocel was spread on the optode front plate and covered with a 12 um Teflon membrane.

The cuprophane membrane was placed directly onto the Teflon membrane with the immobilized enzyme layer facing the indicator. The optode front plate bearing the membranes was pressed against a sealing ring around the optode chamber and fixed by screws.

Fluorescence measurements were done in an Aminco Bowman SPF by mounting the optode into the sample compartment at an angle of about 45 degree with its indicator layer facing to the excitation light source in a way that reflection light did not fall on the emission monochromator slit. The walls of the perfusion chamber were painted black to eliminate influences of reflected or scattered light.

Fluorescence was also measured on a Chance light guide fluorometer with rotating filters (MB2/TSFI Air Turbine Fluorometer, Johnson Foundation, University of Pennsylvania). The light guide was mounted vertically to a miniaturized optode construction. At this side the glass fibers for excitation and emission light were homogeneously mixed over a cross-section of 0.4 cm.

The optode chamber was perfused by solutions equilibrated at $37^\circ C$ to a defined Po_2 with gas mixtures of O_2 and N_2 using a gas mixing pump (Fa. Wösthoff, Bochum, FRG). Calibration was done by setting the signal intensity obtained with pure N_2 (I_0) equal to 100% and recording the relative fluorescence intensities (I) at a series of O_2 mixtures up to 100% O_2. Corrections were made for the drift of I_0 which were in the range of 2 to 6% per hour.

RESULTS

1. Po$_2$ calibration curves without and with addition of glucose

An optode built with immobilized glucose oxidase was perfused with 0.1 $mol \cdot l^{-1}$ phosphate buffer pH 7.1 containing no glucose. The perfusion buffer was kept at constant Po_2 by equilibrating it with gas mixtures of O_2 and N_2 at a temperature of $37^\circ C$.

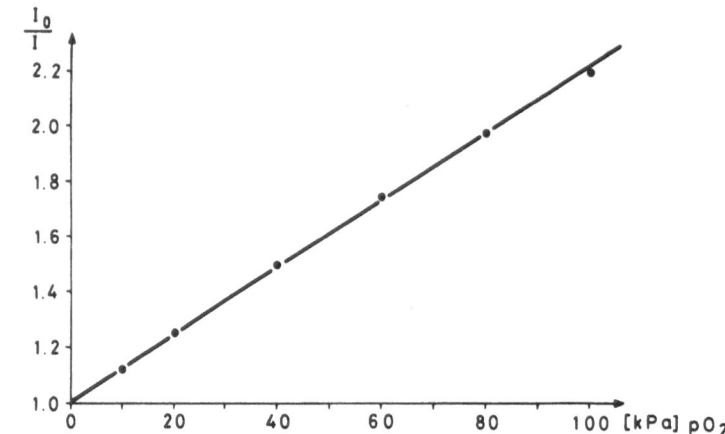

Fig. 1.   Oxygen calibration curve of the glucose optode: Relative
fluorescence intensity (I_O/I) vs. Po_2 in kPa.

The highest relative fluorescence intensity, I_O, is reached
in the absence of oxygen, if the optode is perfused with buffer with
100 kPa N_2 partial pressure. When Po_2 in the perfusion solution is
increased, diffusion leads to a new equilibrium of oxygen, resulting
in a lower fluorescence intensity, I. Plotting of I_O/I vs. Po_2 shows
a linear oxygen calibration curve (Fig. 1), following the Stern-
Volmer quenching formula (1919):

$$I_O/I = 1 + K \cdot Co_2 = 1 + K \cdot \alpha \cdot Po_2 \qquad (3)$$

with K   = quenching constant
     Co_2 = oxygen concentration
     α = solubility coefficient of oxygen in the indicator layer.

If glucose is added to the buffer the relative fluorescence
intensity increases at a given Po_2 as oxygen is consumed by the
enzyme reaction. Optodes constructed for analytical application in
the physiological range of glucose concentration show a set of
nearly parallel lines in the plot of I_O/I vs. Po_2 for different
glucose concentrations, as compared to the oxygen calibration curve
(Fig. 2). There is a roughly linear response to glucose at sufficient oxygen supply. But in some optode measurements deviations
from linearity could be seen especially with high glucose concentrations and at low values of I_O/I, when there is not enough oxygen
left for the enzyme reaction.

2. Effect of substrate saturation

a) Glucose: From the results with special optodes an insight
into the kinetic of the glucose oxidase reaction as part of the
optode measurements can be derived.

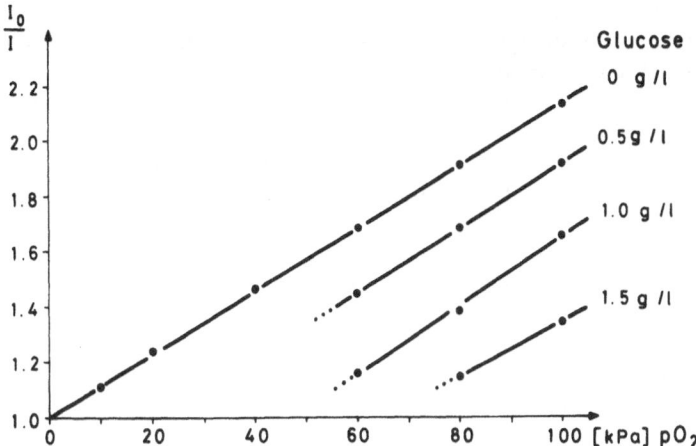

Fig. 2.   Oxygen calibration curves in the presence of different
          glucose concentrations. Relative fluorescence intensity
          (I_0/I) vs. Po_2 in kPa.

     At low influx rates of substrates limitation by diffusional
properties within the optode layers should not develop. These
conditions can be realized by thin enzyme layers and low enzyme
concentration.

     For this purpose instead of 0.5 ml only 0.25 ml of immobiliza-
tion mixture containing 2 mg/ml of glucose oxidase were spread on
an area of 36 cm^2. Glucose concentration in the perfusion buffer
varied up to 10 g/l at two different Po_2 values, i.e. 60 and 100 kPa
(Fig. 3).

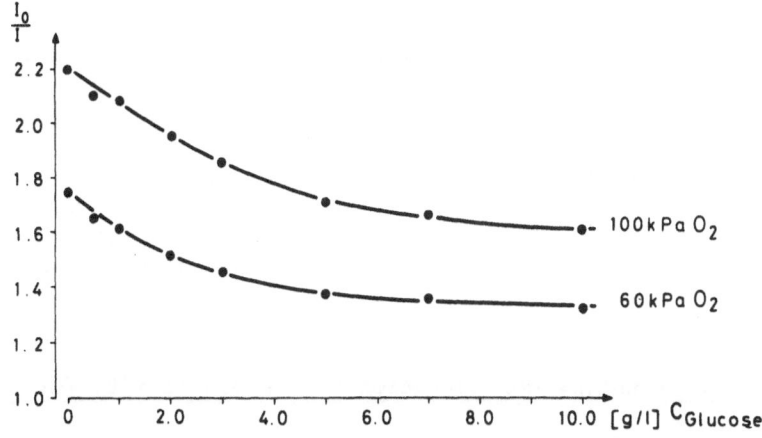

Fig. 3.   Relative fluorescence intensity (I_0/I) vs. glucose concen-
          tration at 60 and 100 kPa Po_2. The enzyme layer was made
          by spreading 0.25 ml of immobilization mixture with 2 mg/ml
          of glucose oxidase on an area of 36 cm^2 ("thin enzyme
          layer").

The curves clearly demonstrate a substrate saturation of the
enzyme layer: at high glucose levels oxygen consumption becomes
independent of glucose concentration. Only the very first part of
the curves shows a linear response to glucose. The curves resemble
the hyperbolic Michaelis-Menten substrate saturation curves of
enzyme reactions with the change from a first order rate law to a
zero order rate law at high substrate concentrations.

Double reciprocal treatment of these values in plots of $1/(I_o/I)$
vs. $1/c_{gluc.}$ according to Lineweaver and Burk (1934) results in
straight lines.

b) Oxygen: With regard to oxygen saturation with higher oxygen
pressures of the perfusion buffer the difference in I_o/I produced
by addition of glucose becomes larger (see Fig. 3). From the curves
shown in Fig. 3 and from further measurements can be concluded, that
the kinetic state of the immobilized glucose oxidase will only then
be close to oxygen saturation, if the perfusion buffer is equilibrated with oxygen at 1 atmosphere and if the glucose concentration
is less than 2 g/l.

3. Influence of thickness of the enzyme layer on the sensitivity for glucose

The influence of the thickness of the enzyme layer was tested
by spreading 0.5 ml of an enzyme immobilization mixture with 2 mg/ml
glucose oxidase to an area of 36 cm^2. This enzyme layer should
have approximately twice the thickness compared to the layer described above. The measurements wer performed under similar conditions.

Fig. 4 shows the plot of I_o/I vs. $c_{gluc.}$. The upper curve is
taken from Fig. 3. Comparison with the lower curve shows that
doubling of the thickness of the enzyme layer leads to higher differences in I_o/I values pressures. Saturation with respect to glucose
can be clearly seen. The doubling of layer thickness does not result
in doubling of O_2 consumption.

Approximately twice the oxygen consumption is reached only by
a combination of high oxygen supply and low glucose concentration.
Under these conditions the sensitivity of glucose determinations by
optode measurements can be increased by using thicker enzyme
layers. But the sensitivity increase is not linear in the range of
physiological glucose concentrations.

4. Influence of enzyme concentration on the sensitivity for glucose

A changed sensitivity of optodes for glucose can also be
obtained by increasing the enzyme concentration at constant thickness of the enzyme layer. In Fig. 5 I_o/I vs. $c_{gluc.}$ at 100 kPa Po_2

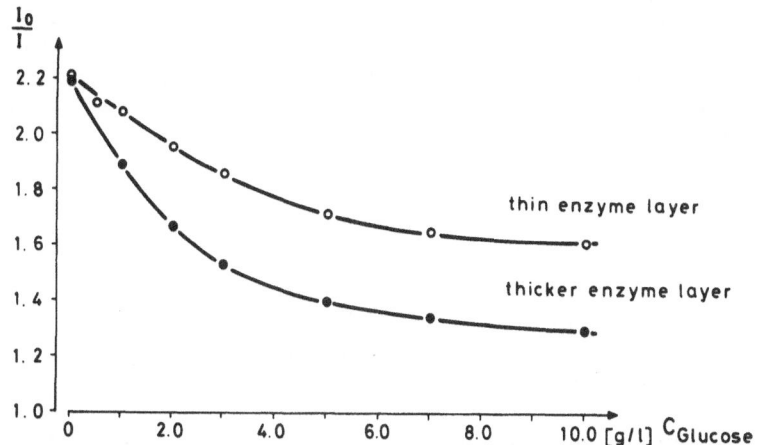

Fig. 4.   Relative fluorescence intensity (I_0/I) vs. glucose concen-
          tration at 100 kPa Po_2.
          The thin enzyme layer was made by spreading 0.25 ml of
          immobilization mixture (2 mg/ml glucose oxidase) on an area
          of 36 cm^2. The thicker enzyme layer was made by spreading
          0.5 ml of the same immobilization mixture on an area of
          36 cm^2.

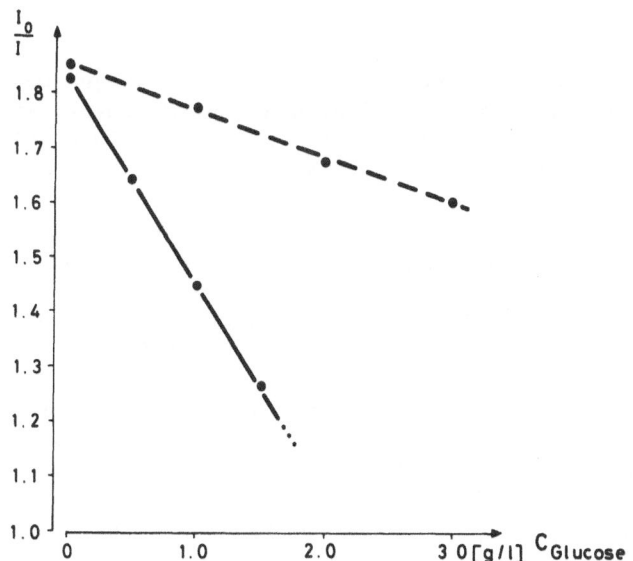

Fig. 5.   Relative fluorescence intensity (I_0/I) vs. glucose concen-
          tration at 100 kPa Po_2.
          Dotted line: Thicker enzyme layer with 2 mg/ml glucose
          oxidase. Straight line: Thicker enzyme layer with 25 mg/ml
          glucose oxidase.

is plotted with concentrations of 2 mg/ml and 25 mg/ml glucose
oxidase in the immobilization mixture so that a "thicker" enzyme
layer is produced. It shows that by increasing the enzyme concentration the sensitivity for glucose is distinctly increased, but that
at the same time the measuring range becomes smaller.

DISCUSSION

   The analytical system of the enzyme optode is based on diffusion of substrates and products of the enzyme reaction perpendicular
to the plane of the membranes and thus differs from most flow
reactors with or without enzymes.

   The advantage of the optode is that the fluorescent indicator
layer shows no oxygen consumption. The binding of oxygen is fully
reversible. At any diffusional steady state the indicator measures
the Po_2 at the Teflon layer which covers the highly viscous indicator layer. Diffusion or flux of O_2 in the enzyme layer is not
influenced during steady state measurements.

   The products of the enzyme reaction build up a gradient in the
opposite direction, i.e. from the indicator layer to the perfusion
solution. During steady state conditions there are no changes of
the fluxes of substrates and products with time and the fluorescence
quenching of the indicator reaches a stable value, measured as a
quotient of relative fluorescence intensities I_0/I.

   Our experiments show that without glucose oxygen calibration
curves are identical without and with an enzyme layer. When the
perfusion buffer contains glucose the quenching signal of the
fluorescent indicator (I) is reduced and the value of I_0/I increases. This increase could be correlated to the substrate concentration in the perfusion buffer, and to a couple of combinations of
enzyme concentration and thickness of the enzyme layer.

   There is a concentration range of glucose in which I_0/I
showed an approximately linear response to glucose.

   However, the signal change is complex and is determined by the
diffusional characteristics of all the layers of the optode (i.e.
the cuprophane membrane, the immobilized enzyme layer, and the
Teflon membrane) as well as by the velocity of glucose oxidation
within the enzyme layer, i.e. the kinetic characteristics of the
immobilized enzyme layer.

   The amount of the difference in I_0/I shows the extent of
substrate consumption under steady state diffusional conditions. As
this difference changes with the supply of either substrate it can
be taken to characterize the kinetic state of the immobilized

enzyme layer. Substrate saturation effects can be clearly seen for glucose (see Fig. 3). The supply with oxygen is sufficient for glucose up to 2.0 g/l at 60 kPa to 100 kPa Po_2, but not for higher glucose concentration or at lower oxygen saturation. Only in this range of substrate concentrations a linear response to glucose can be expected, as the signal change is no longer dependent on O_2 as the second substrate. This would represent the "ideal" range for analytical application. In Fig. 2 "forbidden" areas for analytical application can be seen, where the lines are nearly parallel to the axis of glucose concentration. Then the optode is insensitive to glucose. Up to 5.0 g/l of glucose there is a (nonlinear) response to glucose at oxygen saturations of 100 kPa or 60 kPa Po_2.

Since oxygen and glucose are strongly coupled by the stoichiometric reaction the difference in Po_2 which mirrors the actual glucose consumption is mainly influenced by the kinetics of the immobilized enzyme layer. Limitations by diffusional resistances can be eliminated by building optodes with low fluxes of substrates.

Elaborating the kinetic data from such measurements will hopefully allow using the known diffusion coefficients for the substrates, defined thickness of the enzyme layer and its concentration to calculate the exact shape of the gradients of Po_2, glucose and gluconic acid, and to establish theoretical models for optode reactions including both, diffusion and enzyme kinetics.

REFERENCES

Clark, L.C., and Sachs, G., 1968, Bioelectrodes for tissue metabolism, Am. N.Y. Acad. Sci., 148:133.

Lineweaver, H., and Burk, D., 1934, J. Am. Che. Soc., 56:658-666.

Lübbers, D.W., and Opitz, N., 1975, Die Pco_2-/Po_2-Optode: Eine neue Pco_2- bzw. Po_2-Meßsonde zur Messung des Pco_2 oder Po_2 von Gasen in Flüssigkeiten, Z. Naturforsch., 30c:532.

Stern, O., and Volmer, M., 1919, Über die Abklingungszeit der Fluoreszenz, Physik Z., 20:183.

Thomas, D., Broun, G., and Selegny, E., 1972, Monoenzymatic model membranes: Diffusional reaction, kinetics and phenomena, Biochimie, 55:229-244.

Updike, S.J., and Hicks, G.P., 1967, The enzyme electrode, Nature, 214:986.

Vaughan, W.M., and Weber, G., 1970, Oxygen quenching of pyrenbutyric acid fluorescence in water. A dynamic probe of the microenvironment, Biochemistry, 9:464.

Völkl, H.K., 1980, Die Enzymoptode, Dissertation, Bochum.

Völkl, H.K., Opitz, N., and Lübbers, D.W., 1980, Continuous measurement of concentration of alcohol using a fluorescence-photometric enzymatic method, Z. Anal. Chem., 301:162.

PARTICIPANTS

Prof. Dr. H. Acker
Max-Planck-Institut für Systemphysiologie,
Dortmund, FRG

Dr. H.R. Ahmad
Max-Planck-Institut für Systemphysiologie,
Dortmund, FRG

Prof. Dr. M. von Ardenne
Forschungsinstitut Manfred von Ardenne
Dresden, GDR

Dr. A. Baker
Deptartment of Physiology, University of Bristol,
Bristol, Great Britain

Dr. H.I. Bicher
Hypothermia Clinic, Western Tumor Medical Groups
Van Nuys, California, USA

Prof. Dr. D. Bingmann
Physiologisches Institut,
Münster, FRG

Dr. R. Bourgain
Faculteit Geneesk. en Farm., Lab. Fysiologie en Fysiopathol.,
Vrije Universiteit Brussels,
Brussels, Belgium

Dr. S. Th. Bouwer
Department of Physiology, University of Nijmegen,
Nijmegen, Netherlands

Dr. W. Breull
Department of Physiology, University of Bonn,
Bonn, FRG

Dr. D.F. Bruley
Biomedical Engineering Dept., Louisiana Tech. University,
Ruston, Lousiana, USA

M. Brunner,
Institut für Physiologie und Kardiologie,
University of Erlangen-Nürnberg,
Erlangen, FRG

Dr. A.-C. Bylund-Fellenius
Department of Surgery I, University of Göteborg,
Göteborg, Sweden

Prof. Dr. S.M. Cain
Department of Physiology, University of Alabama,
Birmingham, Alabama, USA

Prof. Dr. B. Chance
Johnson Research Foundation, University of Pennsylvania
Philadelphia, Pennsylvania, USA

Prof. Dr. F. Colin
Faculté de Medicine, Université Libre de Bruxelles,
Service de Physiologie Générale,
Brussels, Belgium

Dr. M.A. Delpiano
Max-Planck-Institut für Systemphysiologie,
Dortmund, FRG

Dr. E. Dóra
Experimental Research Department, Semmelweis Medical University,
Budapest, Hungary

Dr. A. Elander
Department of Surgery I, University of Göteborg,
Göteborg, Sweden

R. Ellermann
Institut für Physiologie und Kardiologie
University of Erlangen-Nürnberg
Erlangen, FRG

F. Fallenstein
Perinatal Physiology Unit, Dept. of Obstetrics and Gynaecology,
University Hospital,
Zürich, Switzerland

Dr. K. J. Farrell
Department of Surgery, West Virginia University,
Morgantown, West Virginia, USA

Dr. H.R. Figulla
Universitätsklinik, Abt. Kardiologie,
Göttingen, FRG

W. Fleckenstein
Physiologisches Institut, University of Kiel,
Kiel, FRG

Prof. Dr. R.W. Flower
John Hopkins University, Applied Physics Laboratory,
Laurel, Maryland, USA

K.-H. Frank,
Institut für Physiologie und Kardiologie,
University of Erlangen-Nürnberg,
Erlangen, FRG

Dr. Y.A. Gauduel,
Laboratoire de Biophysique, Hopital Fernand Widal,
Paris, France

Dr. T.E.J. Gayeski
Department of Anesthesiology, University of Rochester,
Rochester, New York, USA

Prof. Dr. T.K. Goldstick,
Department of Chemical Engineering, Technological Institute,
Northwestern University,
Evanston, Illinois, USA

Dr. H. Grewe,
Chirurg. Universitätsklinik,
Münster, FRG

Dr. J. Gronczewski
Max-Planck-Institut für Systemphysiologie,
Dortmund, FRG

Dr. G. Gronow
Physiologisches Institut, University of Kiel,
Kiel, FRG

Dr. U. Großmann
Max-Planck-Institut für Systemphysiologie,
Dortmund, FRG

Prof. Dr. Dr. J. Grote
Physiologisches Institut, University of Bonn,
Bonn, FRG

H. Günther
Institut für Physiologie und Kardiologie,
University of Erlangen-Nürnberg,
Erlangen, FRG

Dr. J. Hamar
Experimental Research Dept., Semmelweis Medical University,
Budapest, Hungary

Dr. D. Harrison,
Institut für Physiologie und Kardiologie,
University of Erlangen-Nürnberg,
Erlangen, FRG

P.D. Dr. J.P. Hauss
Chirurg. Universitätsklinik,
Münster, FRG

Dr. U. Heinrich
Max-Planck-Institut für Systemphysiologie,
Dortmund, FRG

J. Hoffmann
Max-Planck-Institut für Systemphysiologie,
Dortmund, FRG

Dr. C.R. Honig
University of Rochester, School of Medicine and Dentistry,
Rochester, New York, USA

Dr. L.J.C. Hoofd
Department of Physiology, University of Nijmegen,
Nijmegen, Netherlands

J. Höper
Institut für Physiologie und Kardiologie,
University of Erlangen-Nürnberg,
Erlangen, FRG

L. Hubl
Max-Planck-Institut für Systemphysiologie,
Dortmund, FRG

Dr. J.P. Idström
Surgical Metabolic Research Laboratory, Sahlgrenska Sjukhuset,
Göteborg, Sweden

Dr. U. Jensen
Institut für Anaesthesiologie, Klinikum München Großhadern,
München, FRG

Dr. K.-H. Jeroschewski
Universitätsklinik,
Münster, FRG

Prof. Dr. M. Kessler
Institut für Physiologie und Kardiologie,
University of Erlangen-Nürnberg,
Erlangen, FRG

Dr. H. Kiesewetter,
Abt. Physiologie der RWTH,
Aachen, FRG

Prof. Dr. P. Kiwull
Institut für Physiologie, Ruhr-Universität,
Bochum, FRG

Prof. Dr. H. Kiwull-Schöne
Institut für Physiologie, Ruhr-Universität,
Bochum, FRG

Dr. A.J. van der Kleij
Afd. Algemene Chirurgie, St. Radbondziekenhuis,
Nijmegen, Netherlands

Dr. T. Koyama
Research Institute for Applied Electronics, Hokkaido University,
Sapporo, Japan

Prof. Dr. F. Kreuzer
Department of Physiology, University of Nijmegen,
Nijmegen, Netherlands

Prof. Dr. K. Kunze
Neurologische Universitätsklinik Hamburg-Eppendorf,
Hamburg, FRG

Prof. Dr. W. Kuschinsky
Physiologisches Institut, University of München,
München, FRG

Dr. J.C. La Manna
Department of Neurology, Case Western Reserve University,
University Hospitals,
Cleveland, Ohio, USA

Prof. Dr. E. Leniger-Follert
Max-Planck-Institut für Systemphysiologie,
Dortmund, FRG

Prof. Dr. I.S. Longmuir
Department of Biochemistry, NC State University,
Raleigh, North Carolina, USA

Prof. Dr. D.W. Lübbers
Max-Planck-Institut für Systemphysiologie,
Dortmund, FRG

Dr. J. Lutz
Physiologisches Institut der Universität,
Würzburg, FRG

Prof. Dr. J. Manil
Faculteit Geneeskunde en Farmacie, Vrije Universiteit Brussels,
Brussels, Belgium

Dr. C. Marconi,
Centro per lo studio del lavoro muscolare,
C.N.R. Università di Milano,
Milano, Italy

Prof. Dr. D. Mayer,
Bayer AG, Institut für Pharmakologie,
Wuppertal, FRG

W. Menke
Physiologisches Institut, Free University,
Berlin, FRG

Prof. Dr. K. Meßmer
Abt. für Experimentelle Chirurgie, Klinikum der Univ. Heidelberg,
Heidelberg, FRG

Prof. Dr. H. Metzger
Department of Physiology, Hannover Medical School,
Hannover, FRG

Dipl.-Ing. H.J. Meuer
Department of Physiology, Hannover Medical School,
Hannover, FRG

Dr. E.M.H. Mitnick
Department of Physiology, Duke University,
Durham, North Carolina, USA

P.H. Mook
Exp. Surgery, University of Groningen,
Groningen, Netherlands

Dr. J. Moravec
Pathol. Cardiovasculaire, INSERM U2, Hopital Léon Bernard,
Limeil-Brévannes, France

Dr. K. Mottaghy
Abt. Physiologie, RWTH Aachen,
Aachen, FRG

Dr. W. Müller-Klieser
Institute of Physiology, University of Mainz,
Mainz, FRG

Dr. E. Mulligan
Max-Planck-Institut für Systemphysiologie,
Dortmund, FRG

S. Ohkawa
Sumitomo, Electr. Ind. Ltd.,
Osaka, Japan

J. O'Riordan
Chemical Engineering Deptartment, The Technological Institute,
Northwestern University,
Evanston, Illinois, USA

Prof. Dr. J. Piiper
Abt. Physiologie, Max-Planck-Institut für exptl. Medizin,
Göttingen, FRG

Prof. Dr. H.J. Proctor
Department of Surgery, University of North Carolina,
at Chapel Hill Burnett, Chapel Hill, North Carolina, USA
and, University of Groningen, Netherlands

Prof. Dr. K. Rakusan
Department of Physiology, University of Ottawa,
Ottawa, Ontario, Canada

Dr. R. Reichl,
Abt. Pharmakologie, Boehringer Ingelheim KG,
Ingelheim, FRG

Prof. Dr. H.M. Rhee
Department of Pharmacology, Oral Roberts University,
Tulsa, Oklahoma, USA

Prof. Dr. P. Scheid
Abt. Physiologie, Max-Planck-Institut für experimentelle Medizin,
Göttingen, FRG

Dr. K. Schlossmann
Institut für Pharmakologie, Bayer AG,
Wuppertal, FRG

Dr. H.-J. Schmidt
Physiologisches Institut I, University of Bonn,
Bonn, FRG

Prof. S. Schuchhardt
Physiologisches Institut, Free University,
Berlin, FRG

Dr. P. Scotto
Ist Fisiologia Umana,
Napoli, Italy

Dr. E. Seidl
Max-Planck-Institut für Systemphysiologie,
Dortmund, FRG

Prof. Dr. G. Siegel
Physiologisches Institut, Free University,
Berlin, FRG

Dr. K.H. Sinagowitz
Friedrichshafen, FRG

Dr. K. Skolasinska
Institut für Physiologie und Kardiologie,
University of Erlangen,
Erlangen, FRG

Dr. J. Sossinka
Eppendorf Gerätebau,
Hamburg, FRG

Dipl.-Ing. H.U. Spiegel
Chirurg. Universitätsklinik,
Münster, FRG

Prof. Dr. B.T. Storey
Department of Physiology, University of Pennsylvania,
Philadelphia, Pennsylvania, USA

P.D. Dr. Dr. K. Strein
Boehringer Mannheim GmbH,
Mannheim, FRG

Prof. Dr. K. Sugioka
Department of Anesthesiology, University of North Carolina,
Chapel Hill, North Carolina, USA

Prof. Dr. Dr. G. Thews
Physiologisches Institut, University of Mainz,
Mainz, FRG

Dr. J. Tozer
Richard Dimbleby, St. Thomas Hospital,
London, Great Britain

Dr. Z. Turek
Department of Physiology, Fac. Medicine, University of Nijmegen,
Nijmegen, Netherlands

Prof. Dr. P.W. Vaupel
Physiologisches Institut, University of Mainz,
Mainz, FRG

Ind. Eng. F. Vereecke,
Lab. Fysiologie en Fysiopathol., Vrije Universiteit Brussels,
Brussels, Belgium

Dr. H. Vermarien
Lab. Fysiologie en Fysiopathol., Vrije Universiteit Brussels,
Brussels, Belgium

P.D. Dr. F. Vetterlein
Institut für Pharmakologie und Toxikologie,
Göttingen, FRG

Dr. H. Vogel
Institut für Physiologie und Kardiologie,
University of Erlangen-Nürnberg,
Erlangen, FRG

Dr. H.-J. Volkholz
Institut für Physiologie und Kardiologie,
University of Erlangen-Nürnberg,
Erlangen, FRG

Prof. Dr. H. Wayland,
I. Physiologisches Institut, University of Heidelberg,
Heidelberg, FRG

Dr. H. Weigelt,
Max-Planck-Institut für Systemphysiologie,
Dortmund, FRG

Prof. Dr. Ch. Weiss
Physiologisches Institut der Med. Hochschule Lübeck
Lübeck, FRG

Prof. Dr. W. Wiemer
Institut für Physiologie,
Universitätsklinikum der Gesamthochschule Essen,
Essen, FRG

Dr. N. Wiernsperger
Preclin. Research, Sandoz Ltd.,
Basel, Switzerland

Prof. Dr. D.F. Wilson
Department of Biochemistry and Biophysics,
University of Pennsylvania,
Philadelphia, Pennsylvania, USA

Dipl-Chem. R. Wittler
Institut für technische Chemie,
Hannover, FRG

Prof. R. Wodick
Physiologie II der Universität,
Ulm, FRG

Prof. Dr. R.D. Woodson,
Department of Med/Hematology, University of Wisconsin,
Madison, Wisconsin, USA

Dr. B. Yu
Max-Planck-Institut für Systemphysiologie,
Dortmund, FRG

Prof. Dr. R. Zander
Physiologisches Institut, University of Mainz,
Mainz, FRG

CONTRIBUTORS

Acker, H., 705, 719
Alvarez, J.G., 433
Anderer, W., 331
Arai, T., 651
Araiso, T., 651
Ardenne von, M., 781

Baker, B.J., 597
Baker, R., 597
Bär, Th., 281
Baumgärtl, H., 793, 877
Beatty, J., 775
Beerthuizen, G., 869
Beier, I., 69
Bicher, H.I., 327
Bingmann, D., 215
Blockeel, E., 231
Bourgain, R.H., 231
Bowe, C., 359
Bowen, B., 775
Breepoel, P., 133
Breull, W., 369
Bruley, D.F., 103
Brunner, M., 331, 477, 633, 803, 811
Budi Santoso, A.W., 281
Busch, N.H., 103
Bylund-Fellenius, A.-C., 489, 507

Cain, S.M., 381
Carlsson, J., 719
Cerretelli, P., 487
Chance, B., 489
Chapler, C.K., 381
Clark, H., 23
Clark, P., 23

Coakham, H.B., 241
Codas, C., 327
Colin, F., 231
Cooper, J., 389
Coremans, J., 231
Cremer, J., 175

Danz, C., 297
Delpiano, M.A., 705
Derissen, W., 175
Dézsi, L., 551
Dietrich, H.H., 747
Ditzel, J., 187
Dóra, E., 81, 305, 315
Dujovny, H., 327
Duvelleroy, M.A., 421

Ebeling, B.J., 515
Elander, A., 507
Ellermann, R., 331, 477, 633, 803
Engerman, R.L., 661
Erecińska, M., 351
Ernest J.T. 187, 661

Farrell, K., 775
Federspiel, W., 23
Feńvray, D., 359
Figulla, H.R., 821
Fleckenstein, W., 447
Fletcher, J.E., 145
Flower, R.W., 671
Frank, K.H., 331, 477, 633, 811
Friedl, A., 811
Fritz, H., 341
Funk, R., 271

933

INDEX